DOLOMITES

A VOLUME IN HONOUR OF

DOLOMIEU

DOLOMITES

A VOLUME IN HONOUR OF DOLOMIEU

Edited by
Bruce Purser, Maurice Tucker
and Donald Zenger

SPECIAL PUBLICATION NUMBER 21 OF THE
INTERNATIONAL ASSOCIATION OF SEDIMENTOLOGISTS
PUBLISHED BY BLACKWELL SCIENTIFIC PUBLICATIONS
OXFORD LONDON EDINBURGH BOSTON
MELBOURNE PARIS BERLIN VIENNA

and published for them by
Blackwell Scientific Publications
Editorial Offices:
Osney Mead, Oxford OX2 0EL
25 John Street, London WC1N 2BL
23 Ainslie Place, Edinburgh EH3 6AJ
238 Main Street, Cambridge
Massachusetts 02142, USA
54 University Street, Carlton
Victoria 3053, Australia

Other Editorial Offices:
Librairie Arnette SA
1, rue de Lille
75007 Paris
France

Blackwell Wissenschafts-Verlag GmbH
Düsseldorfer Str. 38
D-10707 Berlin
Germany

Blackwell MZV
Feldgasse 13
A-1238 Wien
Austria

First published 1994

Set by Excel Typesetters Company, Hong Kong
Printed and bound in Great Britain
at the University Press, Cambridge

DISTRIBUTORS

Marston Book Services Ltd
PO Box 87
Oxford OX2 0DT
(*Orders*: Tel: 0865 791155
Fax: 0865 791927
Telex: 837515)

USA
Blackwell Scientific Publications, Inc.
238 Main Street
Cambridge, MA 02142
(*Orders*: Tel: 800 759-6102
617 876-7000)

Canada
Oxford University Press
70 Wynford Drive
Don Mills
Ontario M3C 1J9
(*Orders*: Tel: 416 441-2941)

Australia
Blackwell Scientific Publications Pty Ltd
54 University Street
Carlton, Victoria 3053
(*Orders*: Tel: 03 347-5552)

A catalogue record for this title
is available from the British Library

ISBN 0-632-03787-3

Library of Congress
Cataloging-in-Publication Data

Dolomites: a volume in honour of Dolomieu/
edited by Bruce Purser, Maurice Tucker,
and Donald Zenger.
p. cm.
(Special publication no. 21 of the
International Association of Sedimentologists)
Includes bibliographical references
and index.
ISBN 0-632-03787-3
1. Dolomite.
I. Dolomieu, Déodat de, 1750–1801.
II. Purser, B. H. III. Tucker, Maurice E.
IV. Zenger, Donald H.
V. Series: Special publication . . . of the
International Association of Sedimentologists; no. 21.
QE471.15.D6D63 1994
552′.58 – dc20

Contents

Introduction

3 Problems, progress and future research concerning dolomites and dolomitization
B.H. Purser, M.E. Tucker and D.H. Zenger

21 Dolomieu and the first description of dolomite
D.H. Zenger, F.G. Bourrouilh-Le Jan and A.V. Carozzi

29 Summary
B.H. Purser, M.E. Tucker and D.H. Zenger

Sabkha, Evaporitic and Reflux Dolomitization Models

37 Salina sedimentation and diagenesis: West Caicos Island, British West Indies
R.D. Perkins, G.S. Dwyer, D.B. Rosoff, J. Fuller, P.A. Baker and R.M. Lloyd

55 Mechanisms of complete dolomitization in a carbonate shelf: comparison between the Norian Dolomia Principale (Italy) and the Holocene of Abu Dhabi Sabkha
S. Frisia

75 Changing dolomitization styles from Norian to Rhaetian in the southern Tethys realm
A. Iannace and S. Frisia

91 Distribution, petrography and geochemistry of early dolomite in cyclic shelf facies, Yates Formation (Guadalupian), Capitan Reef Complex, USA
M. Mutti and J.A. Simo

Mixing-Zone and Seawater Dolomitization Models

111 Dolomitization by near-normal seawater? Field evidence from the Bahamas
F.F. Whitaker, P.L. Smart, V.C. Vahrenkamp, H. Nicholson and R.A. Wogelius

133 Late Cenozoic dolomites of the Bahamas: metastable analogues for the genesis of ancient platform dolomites
V.C. Vahrenkamp and P.K. Swart

155 Dolomitization caused by water circulation near the mixing zone: an example from the Lower Viséan of the Campine Basin (northern Belgium)
P. Muchez and W. Viaene

Burial Dolomitization Models

169 Burial dolomitization of the Middle Ordovician Glenwood Formation by evaporitic brines, Michigan Basin
J.A. Simo, C.M. Johnson, M.R. Vandrey, P.E. Brown, E. Castrogiovanni, P.E. Drzewiecki, J.W. Valley and J. Boyer

187 Petrographic, geochemical and structural constraints on the timing and distribution of postlithification dolomite in the Rhaetian Portoro ('Calcare Nero') of the Portovenere Area, La Spezia, Italy
J.K. Miller and R.L. Folk

203 Has burial dolomitization come of age? Some answers from the Western Canada Sedimentary Basin
E.W. Mountjoy and J.E. Amthor

231 Burial and hydrothermal diagenesis of Ordovician carbonates from the Michigan Basin, Ontario, Canada
M. Coniglio, R. Sherlock, A.E. Williams-Jones, K. Middleton and S.K. Frape

255 Progressive recrystallization and stabilization of early-stage dolomite: Lower Ordovician Ellenburger Group, west Texas
J.A. Kupecz and L.S. Land

Dolomite Reservoirs

283 Nature, origins and evolution of porosity in dolomites
B.H. Purser, A. Brown and D.M. Aissaoui

309 Permeability and porosity evolution in dolomitized Upper Cretaceous pelagic limestones of Central Tunisia
M.H. Negra, B.H. Purser and A. M'Rabet

325 Porosity evolution through hypersaline reflux dolomitization
F.J. Lucia and R.P. Major

Petrology and Geochemistry of Dolomites

345 Synthesis of dolomite and geochemical implications
E. Usdowski

361 Discontinuous solid solution in Ca-rich dolomites: the evidence and implications for the interpretation of dolomite petrographic and geochemical data
A. Searl

377 Rates of dolomitization: the influence of dissolved sulphate
D.W. Morrow and H.J. Abercrombie

387 Pervasive dolomitization of a subtidal carbonate ramp, Silurian and Devonian, Illinois Basin, USA
J.M. Kruger and J.A. Simo

Dolomitization and Organic Matter

409 Organic matter distribution, water circulation and dolomitization beneath the Abu Dhabi Sabkha (United Arab Emirates)
F. Baltzer, F. Kenig, R. Boichard, J.-C. Plaziat and B.H. Purser

429 Burial dolomitization of organic-rich and organic-poor carbonates, Jurassic of Central Tunisia
M. Soussi and A. M'Rabet

447 Index

Introduction

Spec. Publs Int. Ass. Sediment. (1994) **21**, 3–20

Problems, progress and future research concerning dolomites and dolomitization

B.H. PURSER,* M.E. TUCKER† *and* D.H. ZENGER‡

* *Laboratoire de Pétrologie Sédimentaire, Université de Paris Sud, 91405 Orsay, France;*
† *Department of Geological Sciences, University of Durham, DH1 3LE, UK; and*
‡ *Department of Geology, Pomona College, 91711 Claremont, California, USA*

INTRODUCTION

At the 8th Bathurst Meeting of Carbonate Sedimentologists in Liverpool, July 1987, the editors-to-be decided to hold a conference on dolomitization to honour Déodat de Dolomieu on the 200th anniversary of his 1791 classic paper describing dolomite in detail for the first time. Independently, another group of carbonate workers (A. Bosellini, R. Brandner, E. Flügel and W. Schlager) also decided to pay tribute to Dolomieu by holding a conference on carbonate platforms. Collaboration between these two groups led to the successful Dolomieu Conference on Carbonate Platforms and Dolomitization held in September 1991 in Ortisei, Italy, in the magnificent setting of the Dolomite Mountains. This conference was sponsored by the International Association of Sedimentologists (IAS) and the Society for Sedimentary Geology (SEPM).

Much had been published on dolomites and dolomitization. The proceedings of three SEPM symposia were published as SEPM Special Publications: No. 13 (Pray & Murray, 1965); No. 28 (Zenger, Dunham & Ethington, 1980); and No. 43 (Shukla & Baker, 1988). In addition to many published research papers, numerous reviews have appeared over the years, including those of Steidtmann (1911), Van Tuyl (1916), Fairbridge (1957), Ingerson (1962, pp. 830–837), Sonnenfeld (1964), Friedman and Sanders (1967), Bathurst (1971, pp. 517–543), Zenger (1972b), Chilingar *et al.* (1979), Morrow (1982a,b), Zenger and Mazzullo (1982), Land (1985), Machel and Mountjoy (1986), Hardie (1987), Tucker and Wright (1990, pp. 365–400), Braithwaite (1991) Fowles (1991) and Mazzullo (1992). As a result of the many excellent presentations at Ortisei, and the continued interest in dolomite, it was decided to publish a volume honouring Dolomieu.

Many contributions to the knowledge of ancient dolomites are the product of North American research and our present understanding is probably influenced significantly by examples of Palaeozoic age. This bias reflects, at least in part, the preferential dolomitization of Palaeozoic rocks in North America. Whatever the reason, the preponderance of Palaeozoic examples may mean that some of the mineralogical and petrographic properties of dolomite may be less typical of younger dolomites. Mesozoic and Tertiary dolomites have been less studied, and it is of considerable interest that certain geochemical, petrographic and petrophysical properties may differ from those of the older, and sometimes more deeply buried, Palaeozoic dolomites. Quaternary and Recent dolomites have been studied on a worldwide scale. These studies have demonstrated that dolomite forms under varied chemical and physical conditions, thus leading to the formulation of a series of dolomite models. Other models are based on the interpretation of ancient dolomites (seepage–reflux, *dorag* and burial). One of the main conclusions of Chilingar *et al.* (1979, pp. 485–486) and of SEPM Special Publication No. 28 (Zenger & Dunham, 1980, p. 7) was that there are 'dolomites and dolomites'. The Dolomieu Conference helped to give a more balanced picture, although clearly much remains to be done, especially outside North America. Of the 90 oral and poster presentations, about 50 considered post-Palaeozoic examples, including the Triassic 'Dolomites'.

The relatively limited number of publications concerning the nature and origins of porosity in dolomite reservoirs may be partly due to 'confidentiality' and, possibly, to the relatively homogeneous nature and relatively low porosities of many Palaeozoic reservoirs. Studies of Mesozoic and Tertiary dolomites, some aspects of which are presented in this volume, show that porosities, even in pure dolomites, are variable, both in percentage and type (Plate 1, opposite). The key to understanding porosity development may lie in the study of relatively young dolomites and the characteristics of the predolomite sediments.

In spite of numerous publications, there remain a considerable number of themes and specific problems that merit discussion. Our choice of important problems is obviously very subjective, and the editors are unable to define precisely 'the state of the dolomite art'. In presenting certain problems, most of which are already well known, we will emphasize post-Palaeozoic rocks, for reasons already noted. Rather than a thorough review, we present our material in two parts. The first concerns important advances and gaps in our knowledge of dolomitization; the second treats certain specific problems discussed at Ortisei. We hope that our admittedly biased approach involving some undoubtedly controversial subjects will spark thought and debate.

SOME SPECIFIC ASPECTS OF DOLOMITE AND ITS ORIGINS: PROGRESS, PROBLEMS AND SPECULATIONS

Concepts and problems not specifically discussed during the Dolomieu Conference

The 'dolomite question'

The 'dolomite question', as initially envisaged by Fairbridge (1957), essentially concerned its origins. He stressed two basic problems:

1 That, in spite of the existence of massive dolomites in the geological record, modern dolomite seemed to be limited to traces forming in deep marine environments; peritidal dolomites were unknown (Fairbridge, p. 126). This contradiction has since been resolved, although there remains the apparent discrepancy between the relatively limited amount of modern surface dolomite and the great quantities of dolomite formed during particular geological epochs (Zenger, 1972b). However, these differences may be somewhat exaggerated, as discussed below.

2 The difficulty with which dolomite is synthesized under laboratory conditions (Fairbridge, p. 128). This 'problem' has also been discussed by McKenzie (1991). The relatively high (100°C) temperatures and pressures (20 atmospheres) required for experimental synthesis (Graf & Goldsmith, 1956) indeed remain a problem when one considers the natural conditions under which surface dolomite is forming today.

In addition to these two basic questions there are many other well known problems, not the least of which concerns the relative importance of replacement dolomite versus dolomite cements. Some of these are examined in the following pages.

Dolomite forms from various kinds of water under many different environmental conditions. This situation tends to perpetuate the 'problem', which may be largely an artificial one, for at least two basic reasons:

1 Dolomite (like feldspar; Land, 1985, p. 33) is highly variable in composition, not only in terms of Ca:Mg ratios and degree of order, but also in terms of other elements, notably iron. These mineralogical variations, which affect the solubility of the mineral (Carpenter, 1980; Land, 1980; Lumsden & Chimahusky, 1980) indicate that dolomite forms under a variety of conditions. A unique 'magic' dolomitizing fluid is obviously an illusion; it is not surprising that different kinds of dolomite form in quite different sedimentary and diagenetic settings.

2 In spite of the highly variable chemical and physical conditions under which dolomite is known to form, there must exist geochemical and thermodynamic/kinetic 'rules' which are common to all dolomites. Although these are partly understood in theory, there is considerable disagreement among specialists concerning which factors are important in particular situations.

One of the more flagrant contradictions between basic theory and geological 'fact' concerns the rates and temperatures of formation of dolomite. Experimental conditions (Graf & Goldsmith, 1956; Katz & Mathews, 1977; Gaines, 1980; Bullen & Sibley, 1984; Sibley *et al.*, 1987; Sibley, 1990; and Usdowski, this volume) indicate that the time required for dolomitization at near-surface temperatures is long, notably in its 'induction stage'. However, modern dolomite, i.e. less than 4000 years old, occurs within many peritidal environments. In Qatar and Abu Dhabi, it contributes to a sheet of sediment which may exceed 10 km in width. The initial formation of modern dolomite must require considerably less than 1000 years, since it occurs in actively accreting seaward margins of sabkhas within living microbial mats. It is also interesting to note that 'instant'

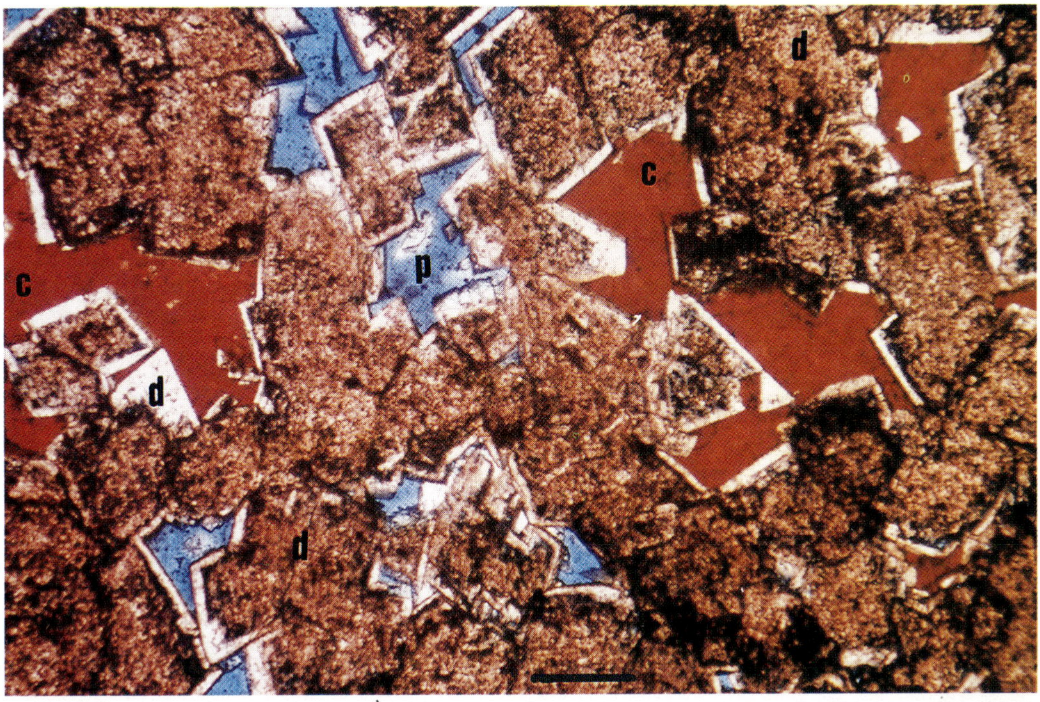

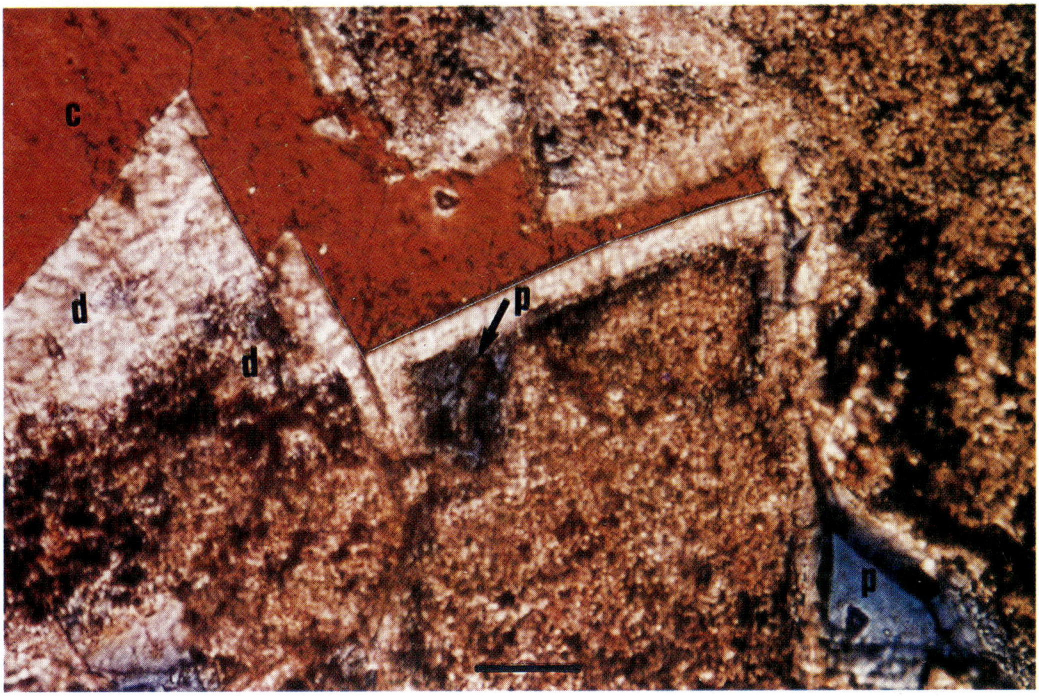

Plate 1 Cretaceous oil reservoir from Syria demonstrating the diagenetic and petrophysical complexity of dolomites; brownish, fabric-destructive dolomite (d) has replaced sediments. Rhombohedra have subsequently grown into pore space as clear dolomite cement (d). Most postdolomite voids are filled with late sparitic calcite (red, c). Porosity (blue) is both intercrystalline and intracrystalline (p, in lower photo), the latter having formed by selective dissolution of the less stable 'replacement' dolomite. (Scale bars: upper photo = 300 μm; lower photo = 100 μm).

[*facing page 4*]

dolomite appears to precipitate, both within the Coorong of South Australia (Von der Borch, 1965) and within pits dug on Umm Said sabkha, Qatar (Shinn, personal communication).

The apparent contradictions between the conditions of formation of dolomite in the laboratory and in nature may be due to the fact that the composition of dolomite formed under relatively high temperatures may not be identical to that formed under natural conditions, and that the seed from which a dolomite crystal develops is more likely to precipitate under various natural conditions, even though its exact nature remains to be established.

Although the evolution of dolomite crystals in modern sediments has been studied (McKenzie, 1981; Gregg *et al.*, 1992), the nature of the initial crystal phases is not well understood. Discerning the role of the substrate and other factors associated with the early stages of crystal formation may be one of the keys to understanding the factors controlling dolomite nucleation. It is possible that, once the conditions of initial growth have been satisfied, subsequent growth of the dolomite lattice requires less stringent conditions and dolomitization may proceed under varying hydrochemical conditions and rates (Sibley *et al.*, 1987).

In view of the highly variable composition of the mineral, it is clear that its origins will never be explained in terms of one unique dolomitizing model.

Modern dolomites and their role in understanding ancient dolomites

There are well-documented geological, mineralogical and petrographic differences between modern and ancient dolomites. An actualistic approach is frequently criticized, both with respect to dolomite and to sedimentation in general, from at least two points of view. First, no modern equivalents of the vast dolomitized platforms of the past are known, and thus modern dolomites apparently cannot provide scaled analogues. Logically, one should compare the comparable! Modern dolomitization, by definition, has been active over a very short period of time, in spite of which there are many cubic kilometres of sediment already partially dolomitized, notably on the sabkhas of the Arabian Gulf. Given a period of several million years, it is not illogical to imagine the creation of dolomite bodies formed under conditions analogous to those operating in modern environments, whose dimensions could be somewhat more comparable with those of some ancient dolomite formations. Furthermore, the thick carbonate platforms of the past may not, in many cases, have been dolomitized during a single sustained hydrological system; the individual rock bodies of ancient platforms and ramps may well have been comparable in size to those existing today.

Secondly, the petrographic textures and mineralogical composition of modern dolomites differ significantly from those of many ancient ones. As stressed by Vahrenkamp and Swart (this volume) and others, modern metastable dolomites will evolve into more stoichiometric and ordered crystals. This being the case, the statement that 'one cannot compare the chemistry of modern and ancient dolomites' is basically true. To interpret the nature of parental waters of many fossil dolomites as a function of 'modern analogues' is risky. However, in spite of this incongruity, the study of Recent dolomites is essential, if only to define reference points from which certain fossil dolomites have evolved.

Although the understanding of modern dolomites is a prerequisite to the understanding of ancient dolomites, notably in terms of processes, the global scale of dolomitization may have varied through time. Given the immense nature of epeiric seas of the past, dolomitizing environments were probably more extensive at certain periods. Dolomite formation, although fairly common today, was probably even more widespread during the Pleistocene and Late Neogene (Vahrenkamp *et al.*, 1991). Processes were probably similar, but rates and regions affected seem to have been different.

It is worth noting that dolomite is also being precipitated at the present time in locations other than sabkhas and tidal flats. In Kuwait, and elsewhere in the Middle East, it occurs in soil profiles as dolocretes and is formed from the evaporation of mixed meteoric–marine groundwaters (e.g. El-Sayed *et al.*, 1991). Dolomite is also being extensively precipitated in some modern lakes (Last, 1990).

The significance of dolomite fabrics and mineralogy

During the Dolomieu Conference this topic did not receive the attention we think it merited, possibly because these fundamental aspects have been the subject of several recent publications (e.g. Gregg & Sibley, 1984; Sibley & Gregg, 1987; and others). However, certain aspects deserve comment. The volumetric importance of certain dolomite fabrics and their geological significance may have been somewhat exaggerated by studies emphasizing Palaeozoic rocks, with insufficient consideration of Tertiary, Quaternary and Recent dolomites. The dense, non-planar fabrics of many Palaeozoic dolomites seem to be less abundant in Mesozoic and Tertiary dolomites, in which planar fabrics, often forming highly porous reservoirs, are common.

It is important to understand what factors determine the preservation or destruction of primary sedimentary fabrics during dolomitization. Two aspects are considered:

1 In any given rhombohedron the inner parts of the crystal are often cloudy, owing to the presence of numerous inclusions which may partially preserve the primary sedimentary fabric (Fig. 1). The peripheral part of the same crystal may be limpid (because it is probably a cement) and therefore does not preserve primary fabric.

2 In certain samples the nature of the sediment or the predolomite diagenetic fabric (e.g. submarine cement) may be well preserved in spite of total diagenetic replacement by dolomite. In other samples of crystalline dolomite, the primary fabric is totally destroyed.

Do these basic petrographic differences reflect fundamental differences in the processes of dolomitization? There are at least three possible explanations of fabric-preserving and fabric-destroying dolomitization. The first, invoked by Sibley (1980), involves the saturation state of parental fluids. It is suggested that fabric-preserving dolomitization is associated with waters that are saturated with respect to calcite (which is incorporated as inclusions; Fig. 1), whereas limpid dolomite is formed from solutions undersaturated with respect to calcite. Curiously, within any given crystal the situation is essentially invariable, crystals nearly always exhibiting cloudy centres and clear rims, but rarely the contrary.

The second explanation, also discussed by Cullis (1904), Sibley (1990) and Tucker & Wright (1990, p. 373), concerns the original mineralogy of the sediment. If dolomitization is early and thus affects primary carbonate minerals, HMC (high Mg-calcite) tends to be dolomitized with retention of primary fabric (Fig. 1), whereas aragonite and, to a lesser degree LMC (low Mg-calcite), tend to be dolomitized with fabric destruction. Thus, both timing of dolomitization and primary mineralogy are important.

The third explanation concerns the dissolution of

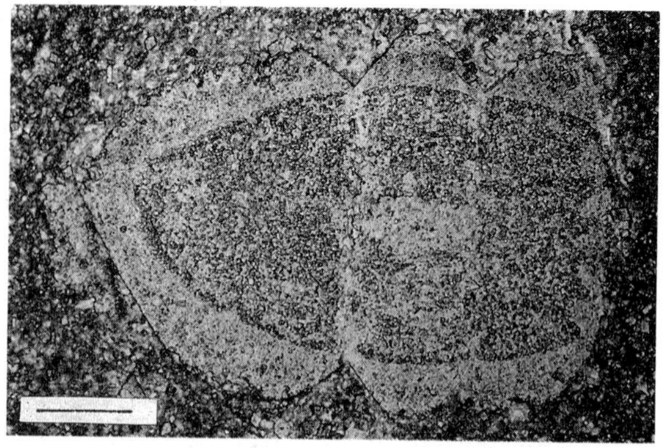

Fig. 1. An example of fabric-preserving dolomite: fragment of dolomitized pelmatozoan stem with overgrowth; inclusions, most of which are calcite, define original structure of stereom. Burrowed member, 'C' zone, U. Ordovician, Red River Fm., Williston Basin, E-Central Montana. Scale bar = 200 μm (with permission of Unocal Oil Co.).

predolomite components. In many cases dolomitization involves dissolution of the precursor carbonate followed by the precipitation of dolomite. One may presume that the relative importance of these two interrelated processes varies. Where dissolution is 'balanced' by precipitation of dolomite, the intervening void is small and inclusions are incorporated into the resultant dolomite, preserving the fabric (Fig. 1). However, if dissolution occurs more rapidly the intervening void will be larger and inclusions will not be incorporated into the dolomite, which will be a cement.

This brings us to the problem of dolomite cements. If we agree that a 'cement' is a mineral phase growing into a void (whatever its size), then dolomite cements are volumetrically very important. Where dolospar lines vugs or has an equant drusy fabric, its cement origin is beyond doubt (Fig. 2). Furthermore, when a porous dolomite becomes occluded with calcite or anhydrite we also term this a cement. However, this later cement may also be *dolomite*, in which case 'replacement' dolomite is followed by cement dolomite. This transition may involve a single crystal, the peripheral parts of which are a limpid cement. Because replacement dolomitization involves the dissolution of a precursor carbonate and the precipitation of dolomite (via a solution film), it is normally associated with an intervening void, whatever its size. Dolomite then grows into this void (Figs 3 and 4). Thus, it could be

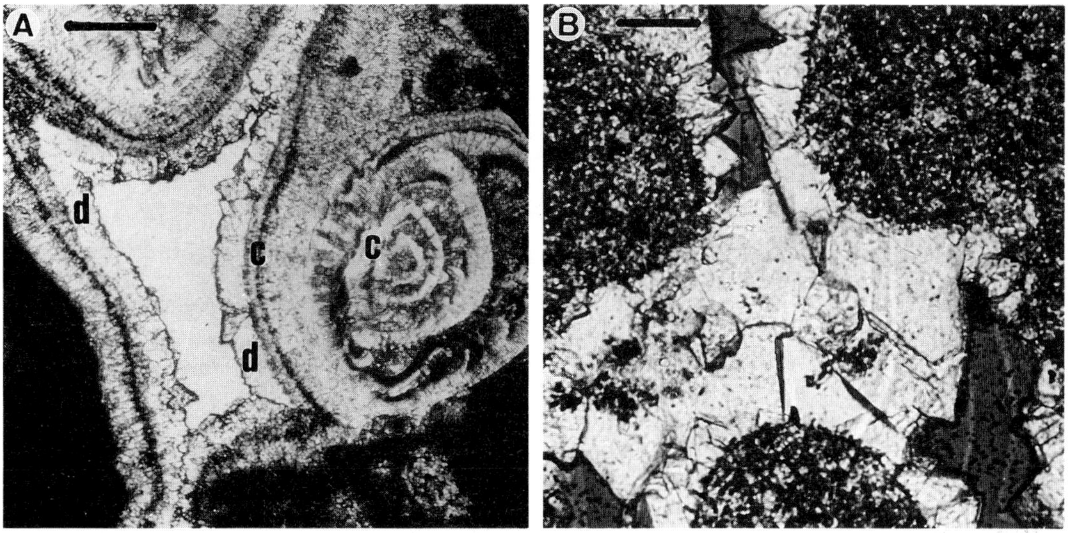

Fig. 2. Dolomite cement, Plio-Quaternary, Mururoa Atoll. (A) Foraminifera cemented with fibrous calcite (c) not affected by dolomitization, followed by an isopachous layer of dolomite cement (d). Scale = 500 μm. (B) Coral debris replaced by inclusion-rich dolomite (dark). Residual voids are cemented with clear dolospar following the total dolomitization of the coral.

Fig. 3. Dolomite 'replacing' fine lagoonal sediments, Mururoa Atoll. Note that rhombohedra grow into microvoids as a 'nanocement'. Ultimately, all precursor sediment will be 'replaced' by microcrystalline dolomite as the pre-existing carbonate is dissolved. Scanning electron micrograph. Scale bar = 10 μm (photo, Aissaoui).

argued that virtually all replacement dolomite is a cement. Where do we draw the boundary? This problem is discussed further in connection with reservoirs.

Dolomite distribution and basin morphology

Since the development of thick dolomite formations requires the passage of large volumes of fluids through

Fig. 4. Scanning electron micrograph of 'tight', completely dolomitized burrow fill (*Thalassinoides?*), lower right half, and porous dolomite matrix (upper left half). Porosity in matrix interpreted as due to postdolomitization dissolution of calcite matrix. U.Ordovician, Steamboat Point Member, Bighorn Dolomite, Wind River Gorge, W Central Wyoming. Scale bar = 100 μm.

a permeable substrate, the system may depend in part on morphological relief. With the exception of late fracture-controlled dolomites, the morphotectonic control of dolomitization appears to have received little attention. Because hydrology depends in part on geomorphology, one may wonder whether the stratigraphic and/or geographic distribution of many dolomite formations is not, at least partially, dependent on tectonics and/or regional patterns of uplift and subsidence. For example, the hydrology of continental or mixed waters will be more dynamic when the basin is bordered by positive relief. Although in theory this relief may favour the predominance of terrigenous sedimentation, this need not necessarily be the case, the Red Sea being a classic example. The predominance of dolomitized carbonates in the Neogene synrift formations of the Northern Red Sea and Gulf of Suez region may in part be due to this tectonically stimulated relief (Aissaoui *et al.*, 1986; Purser *et al.*, 1990).

One can speculate that, on a larger scale, the relatively high geothermal gradients favoured by particular tectonic settings may stimulate thermal convection systems to which certain dolomite bodies may be related. Clearly, many (possibly most) dolomite formations are created by processes which are independent from, or only remotely related to, tectonics. However, the influence of structural and geothermal gradients as stimulants for interstitial water flow and related dolomitization merits consideration.

Global factors that may influence dolomite distribution

A larger-scale approach to the understanding of dolomite distribution in time and space may be useful. The most obvious question concerns whether or not dolomite was more abundant during certain geological periods. This question has been evaluated by Given and Wilkinson (1987), who concluded that dolomite abundance has fluctuated through geological time and that its relative abundance in older rocks is not the result of burial or accumulated time. However, as noted by Zenger (1989), the data upon which Given and Wilkinson (1987) based their conclusions are incomplete. The secular variation of dolomite abundance and its possible causes require further investigation. There is a general feeling that dolomites are particularly well represented during the Proterozoic, Cambro-Ordovician, Middle and Upper Devonian and, possibly, Miocene, carbonates of this last epoch being dolomitized in the Mediterranean, Middle East (Iran) and Pacific atolls. However, the factors favouring the abundance of dolomite or, conversely, its rarity during particular stratigraphic intervals, are probably multiple, and may depend on the following factors:

1 Climate. This is probably the factor most frequently invoked to explain the regional distribution of dolomite. Both temperature and humidity determine chemical reactions, including dolomitization. Today, most dolomites, and indeed carbonates in general, are forming in subtropical or tropical latitudes, often under arid climates. As has been suggested by Sibley (1980) and discussed in detail by Tucker and Wright (1990, p. 364), warm, possibly arid, climates may be a key factor in the development of many major dolomite bodies.

2 Global sea-level fluctuations. Dolomitization may depend on sea-level stability. Sibley (1991) has suggested that a stable sea level may favour the lateral accretion of carbonate platforms and thus the development of wide, flat surfaces, which may in turn favour the formation of brines or mixed dolomitizing waters. This subject is discussed in the section on sequence stratigraphy.

3 Evolution of the world ocean and atmosphere. Changes in $p\text{CO}_2$ have probably influenced changes in carbonate mineral saturation in the oceans (e.g.

Tucker, 1992). These global changes may also have influenced carbonate mineralogy, as has been suggested by Sandberg (1983). Metastable HMC inverts rapidly to LMC, aragonite being somewhat more stable. In that dolomite generally replaces aragonite before it replaces LMC, it is possible that the 'aragonite seas' were somewhat more favourable for dolomitization.

4 Geological time. This may be a factor determining the abundance of dolomite in older rocks. As has been suggested by Fairbridge (1957) and others, the older the rock the greater the chance that it will have been affected by dolomitizing fluids, especially during burial.

The above factors are the more classic ones thought to influence global dolomitization. However, the fact that a particular stratigraphic unit, e.g. the Early Ordovician, is composed of dolomite does not necessarily mean that this diagenetic mineral is Early Ordovician in age, as noted by Zenger and Dunham (1980). Indeed, many contributions to this volume stress the polyphased nature of individual dolomite formations.

Some specific concepts and problems discussed during the Dolomieu Conference

Many aspects of dolomite and dolomitization were discussed, of which the official abstract volume records only a part. In addition to oral presentations and posters there were informal discussion sessions, and the editors of this volume distributed a questionnaire comprising a dozen points. This brief review is limited to a number of topics considered to mark progress and controversy.

An evaluation of specific hydrodynamic models

A special session was devoted to hydrodynamic models, during which diverse dolomite bodies were explained in terms of evaporative reflux, mixing zones, seawater pumping, basinal compaction, etc. One of the models not discussed was the thermal (Kohout) convection model, perhaps suggesting that this fluid circulation mechanism does not represent a substantiated explanation for many dolomites. Of the hydrological concepts presented, two provoked enthusiastic discussion: dolomitization from normal seawater, and dolomitization from mixed waters.

Dolomitization from normal seawater. Although this is not a new concept (see Van Tuyl, 1916; Atwood & Bubb, 1970; Zenger, 1972a; Saller, 1984; Smart & Whitaker, 1990; Tucker & Wright, 1990; etc.), the possibility that normal seawater is an important dolomitizing agent has recently received considerable acclaim, thanks partly to the presentations of Carballo *et al.* (1987) and Land (1991). Based on petrography and C, O and Sr isotopes, dolomitization by normal seawater has the obvious advantage of appearing to explain the thick, areally extensive dolomitized platforms of the past, notably those lacking evidence for evaporative reflux. Further support is derived from the fact that seawater must be the only major source of Mg.

As has been suggested by Alton Brown (questionnaire), massive replacement dolomitization involves the dissolution of a precursor carbonate and the precipitation of dolomite. It is difficult to envisage *normal* near-surface seawater dissolving carbonate on a large scale. However, as Land (1991) has noted, the precipitation of dolomite from seawater will lead to undersaturation with respect to $CaCO_3$, so that dissolution of the host rock could take place.

There are several major problems concerning dolomitization by normal seawater. If seawater is an important reactant, why are most carbonate platforms not dolomitized? The possible answer (Land, 1991; and others) is that dolomitization also requires an efficient hydrodynamic drive ('pump'), which may not necessarily be associated with all platforms. One newly considered mechanism for driving marine groundwaters through a platform relates to an overlying mixing zone; circulation in the latter results in active movement in the former. This has been documented by Whitaker *et al.* (this volume) in the Bahamas, and invoked by Hein *et al.* (1992) to account for dolomitization of Quaternary reefs in the Cook Islands, S. Pacific. Dolomite does form under oceanic conditions (Lumsden, 1985) but generally in modest quantities, probably because the only 'pump' operating is molecular diffusion. In the Pacific atolls, Neogene and Quaternary carbonates are often dolomitized, notably around their oceanic peripheries (Aissaoui *et al.*, 1986a). However, at Mururoa (French Polynesia; Aissaoui *et al.*, 1986b), Maré (Loyalty Group; Carrière, 1987) and Enewetak (Marshall Group; Saller, 1984), dolomite does not extend to the surface. In spite of these obvious problems, many participants at the Dolomieu Conference were convinced that normal seawater has been a major dolomitizing medium.

Dolomitization from mixed waters (Hanshaw et al., *1971).* There were both convinced critics and advocates for 'mixed-water' dolomitization. The critics question the thermodynamics (e.g. Miriam Kastner, questionnaire). Hans Machel suggested that the mixing zone is thermodynamically efficient but kinetically slow, so that dissolution of calcite is favoured but only slow precipitation of dolomite can take place. Others point out that the mixing zone requires a freshwater lens, the associated mixing zone being in large part oblique. These factors are difficult to reconcile with the great thickness of seemingly homogeneous dolomitized platforms. Most ancient dolomites exhibit clear traces of dissolution, leached fossils being typical (Fairchild *et al.*, 1991). However, dissolution may precede, postdate or be contemporaneous with dolomitization. Both meteoric and diluted seawater, unlike normal seawater, are generally undersaturated with respect to aragonite and calcite, and thus have the potential to dissolve. All three editors of this volume consider that mixed waters are *potentially* capable of dolomitizing. However, as noted above, the significance of the mixing zone may be more in its role of inducing fluid movement in the marine groundwater below.

Perhaps some of us have missed the point. As Duncan Sibley (questionnaire) has pointed out, the word 'origin' is not very precise. The fact that dolomite may indeed form from normal or mixed seawaters is not proof that the chemical properties we normally associate with these waters (temperature, salinity etc.) are prerequisites for dolomite formation. Clearly, dolomites may form from many different types of waters and, as already noted, there is no unique fluid or model. The current popularity of one or other system is ephemeral.

The importance of organic matter for dolomite formation

The potential importance of organic matter in the precipitation of dolomite, although a relatively recent concept (Garrison *et al.*, 1984), has been invoked by a number of workers, often concerning limited quantities of Neogene dolomite forming discrete beds or nodules within hemipelagic muds otherwise poor in carbonates. Baker and Burns (1985) described a positive correlation between dolomite and organic matter in DSDP cores, and Burns *et al.* (1988) and Compton (1988) have demonstrated the role of bacterial reduction of sulphates and the precipitation of dolomite within Miocene clays of coastal California. However, it is not entirely clear whether the process involves only bacterial reduction of sulphate within the interstitial waters (leading to increased alkalinity and supersaturation with respect to dolomite), or whether, in addition, the oxidation of organic matter, relatively abundant in regions of coastal upwelling, is a major factor. Slaughter and Hill (1991) suggested that decomposition of organic matter by sulphate-reducing bacteria and, specifically, by the production of ammonia by the enzymatic degradation of protein, is vital to organogenic dolomitization. This process increases both the alkalinity and pH of porewaters, which in turn provide the necessary dissolution and surface chemistries for dolomitization to occur.

Dolomite is commonly associated with phosphorites (BHP, personal observation). Certain black shales rich in organic matter are similarly associated with modest amounts of dolomite (Soussi & M'Rabet, this volume). However, all these associations concern open-marine systems, often with pelagic affinities. Furthermore, the dolomite generally exists as thin discrete layers, often within non-calcareous shales. These deeper marine organogenic dolomites are very different from the massive dolomites that generally replace ancient shallow-marine carbonates.

The decomposition of organic matter within sabkhas and non-evaporitic tidal flats may also be important for the precipitation of modern dolomites, as suggested by McKenzie (1981). The bacterial reduction of sulphates leading to increased alkalinity, and oxidation of microbial mats and mangrove soils (Baltzer *et al.*, this volume) are processes intimately associated with the distribution of dolomite in the sabkhas of Abu Dhabi, where depletion of ^{13}C within the dolomite may reflect a small contribution of organic carbon to the dolomite lattice.

In spite of this documentation, it is not entirely clear whether the organic matter contributes only to the nucleation of crystals, or whether its presence is important for sustained dolomitization. The general feeling of participants at the Dolomieu Conference, almost without exception, was that the role of organic matter was probably important, although most confessed that its precise function remains to be determined.

The modification and evolution of dolomite

Perhaps one of the more important advances in our understanding of dolomite is the demonstration that the petrographic, mineralogical and geochemical properties of this mineral have all evolved during burial. Dolomite diagenesis has been demonstrated in the past, and is well documented in two SEPM Special Publications (Zenger *et al.*, 1980; Shukla & Baker, 1988). The concept was confirmed by a number of presentations at Ortisei. The implications relating to this evolution are numerous and important.

The properties of dolomite have evolved in two basically different ways. First, dolomite tends to change or 'mature' with time (Vahrenkamp & Swart, this volume). Initially metastable and generally calcium-rich, poorly ordered dolomite is relatively soluble and thus susceptible to partial dissolution of the Ca-rich parts of the crystal (Ward & Halley, 1985). Metastable dolomite may recrystallize relatively early in its history (Gregg *et al.*, 1992) as well as later, resulting in the re-equilibration of its trace elements and isotopic ratios (Mazzullo, 1992; Banner *et al.*, 1988). Thus, oxygen isotopes may be reset, although carbon tends to be more stable, this evolution often coinciding with a depletion of Sr (Land, 1980). As shown by Vahrenkamp and Swart (this volume), Pliocene dolomites have already been modified in this manner. With time, a given dolomite crystal will grow, as evidenced by zoned crystals. During burial, dolomites may recrystallize further, producing the non-planar fabrics typical of many Palaeozoic dolomites. Thus, the end-product resulting from multiphased diagenesis involving both recrystallization, dissolution and successive phases of crystal growth in changing environmental settings, may be a dolomite whose properties are quite different from those of the initial product.

Secondly, dolomites have evolved, probably to a relatively modest degree, because of slight changes in composition of the world's atmosphere and ocean. Although this point may be disputed, there is evidence that both the C and Sr isotopic composition and Sr concentrations in oceanic waters have evolved with ever-changing world climate which, together with the global tectonic evolution of oceans, has modified water composition and circulation patterns. Since most dolomite precipitates from some form of seawater, these global changes affect the original isotopic and, possibly, the mineralogical composition of dolomites.

This picture may be further complicated by the relationships between the mineralogical and isotopic compositions of a given dolomite and the chemical composition of its parental fluids, these relationships depending upon the stoichiometry of the dolomite. The surface structure of the crystal will vary according to its composition, as has been suggested by Reeder (1991) and by Searl (this volume), influencing subsequent growth rates and composition of the successive crystal layers. This implies that a fluid of constant composition may produce dolomites of varying composition.

The above considerations have obvious implications concerning the interpretation of ancient dolomites, notably those of Proterozoic and Palaeozoic age. They clearly imply that comparison between modern and ancient dolomites, notably in terms of isotopic signals and trace elements, and to a lesser degree, petrography, has its limitations.

The importance of burial dolomitization

Dolomite cements with light oxygen isotopic signatures and undulatory extinction, generally filling fractures, are typical of relatively deep burial conditions. They are commonly associated with Mississippi Valley-type mineralization. However, current thinking is that late (burial) dolomitization is extensive, although there is considerable uncertainty concerning its nature. Perhaps the most important question is whether massive dolomitization of limestones occurs at depths exceeding 1000 m. Mattes and Mountjoy (1980), Zenger (1983) and Mountjoy and Amthor (this volume) have shown widespread replacement of limestones at burial depths estimated to be in the order of at least 1000 m, but it is not entirely clear whether or not the parental fluids and formation water flow are totally independent of near-surface conditions.

A second problem naturally concerns the definition of 'burial dolomitization': is dolomite which is formed at, say, 500 m the product of burial processes? There probably exist several definitions of burial, but the most acceptable may be to limit 'deep burial' to dolomites formed within the mesogenetic zone of Choquette and Pray (1970). 'Shallow burial' should be applied to dolomites which, although recording somewhat higher than surface temperatures, nevertheless can be related to artesian lenses. Obviously, the distinction between shallow and deep is not easy; some relatively shallow dolomites may form independently of freshwater lenses and solutions generated at the contemporaneous surface.

Another debate concerns whether burial dolomitization involves mainly the diagenesis of pre-existing dolomite. As discussed in the preceding section, the recrystallization of a pre-existing dolomite under burial conditions is important, notably in Palaeozoic dolomites where non-planar fabrics dominate. However, many current workers consider that the diagenesis of pre-existing dolomite, although important, is not the principal expression of deep burial. Indeed, several studies based essentially on cathodoluminescence petrography have shown that the dolomite filling fractures may also precipitate within the rock matrix, regardless of whether it is calcite or dolomite. Highly luminescent dolomite often occupies the fracture, and also forms the final zone of matrix dolomite crystals within Mississippian dolomites of the Wyoming Overthrust belt (Bureau, 1988; Choquette *et al.*, 1992).

In addition to the recrystallization fabrics, much deep-burial dolomite postdates earlier phases of dolomitization, some of which occurred under near-surface conditions. The possible replacement of a precursor limestone under deep-burial conditions poses obvious problems concerning both the source of magnesium and its hydrodynamic supply. The subject has been reviewed by Machel and Mountjoy (1986), by Mazzullo (1992) and by Mountjoy and Amthor (this volume). Interestingly, Zenger and Dunham (1988), in their study of a deep core of Silurian–Devonian carbonates in New Mexico, concluded that neither geochemically nor petrographically could they distinguish between a dolomite formed by replacement in the mesogenetic zone and one that was formed early but was subsequently neomorphosed during burial, nor, for that matter, some combination of these two end-member models. Defining more conclusive ways to make this distinction is an important challenge in studies of dolomitization.

Sequence stratigraphy and dolomitization

There were few papers at the Dolomieu meeting discussing dolomitization within a sequence-stratigraphic framework. However, since several of the popular models for dolomitization, namely sabkha, reflux, mixing-zone and seawater circulation, are penecontemporaneous or very early diagenetic near-surface processes, they can be integrated into the sequence-stratigraphic succession. For the development of pervasively dolomitized carbonate platforms by any of the early diagenetic models, one of the main factors, in addition to the efficient pumping of the dolomitizing pore fluids through the carbonate sediments, is the lateral migration of the dolomitizing zone. Such fluid movements may take place during periods of relative sea-level change (rising or falling), or during periods of sea-level stability and platform progradation (Tucker, 1993).

Three principal scenarios can be envisaged:

1 During stillstands and relative sea-level falls, and under humid climates, dolomitization may take place in association with the mixing zone.

2 During stillstands or relative sea-level falls, and under arid climates, supratidal evaporative and reflux dolomitization by marine water may take place.

3 During relative sea-level rises, dolomitization may take place through circulating seawater (Fig. 5).

With the first model, the meteoric groundwater zone migrates basinwards during the late highstand as the platform progrades, and dolomitization takes place in the mixing zone or, more likely, within the zone of circulating marine pore fluids ahead of the mixing zone (Fig. 6). There are many examples of pervasively dolomitized carbonate platforms lacking evaporites where dolomitization appears to relate to major exposure horizons and unconformities (e.g. Ordovician of Nevada; Dunham & Olson, 1980). One feature of this type of dolomitization is that it may be followed by meteoric diagenesis as the groundwater zones continue to migrate basinwards. Thus, any porosity in the dolomites may be occluded by meteoric calcite cements and there may be some dedolomitization.

With the second model, evaporative-sabkha dolomitization occurs in the high intertidal and supratidal zones, and under conditions of stable sea-level/slight fall and platform progradation, during which very extensive dolomites can be generated. There are many ancient examples of massive dolomites associated with evaporites interpreted to have formed during a relative sea-level fall (e.g. the Zechstein of western Europe, the Permian of the Delaware Basin and the Silurian of the Michigan Basin).

In the third model, marine porewaters move landwards within the platform as the sea level rises, pushing the mixing and meteoric zones ahead through the transgressive systems tract sediments and underlying sequence. The active circulation in the marine porewater zone, and in the vicinity of the mixing zone, could lead to pervasive dolomitization. The dolomitization in this scenario will take place after the sediments of the earlier sequence have been affected by meteoric diagenesis, through exposure at the sequence boundary.

Many carbonate sequences consist of shallowing-

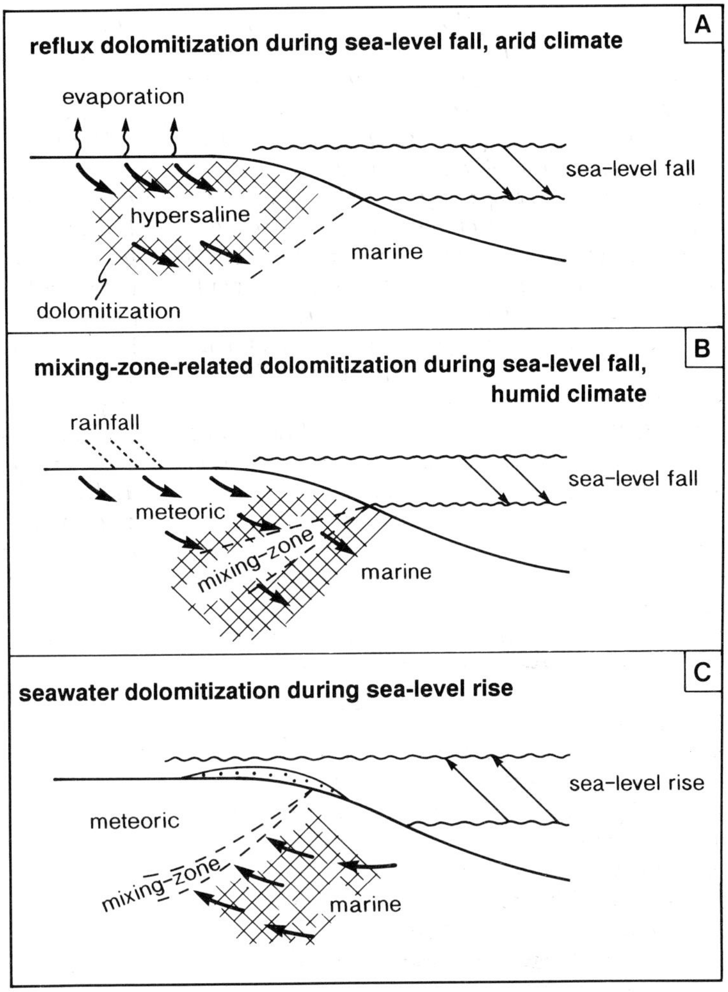

Fig. 5. Models for dolomitization induced by relative changes in sea level. Mixing-zone-related dolomitization refers to dolomitization taking place within the mixing zone and to dolomitization within the circulating marine groundwaters ahead of the mixing zone.

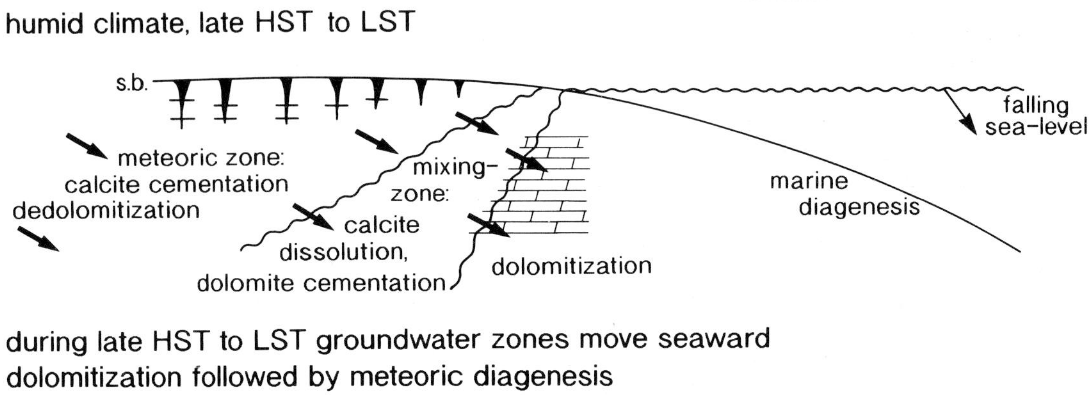

Fig. 6. Model for dolomitization, showing migration of waters during lowering of sea level.

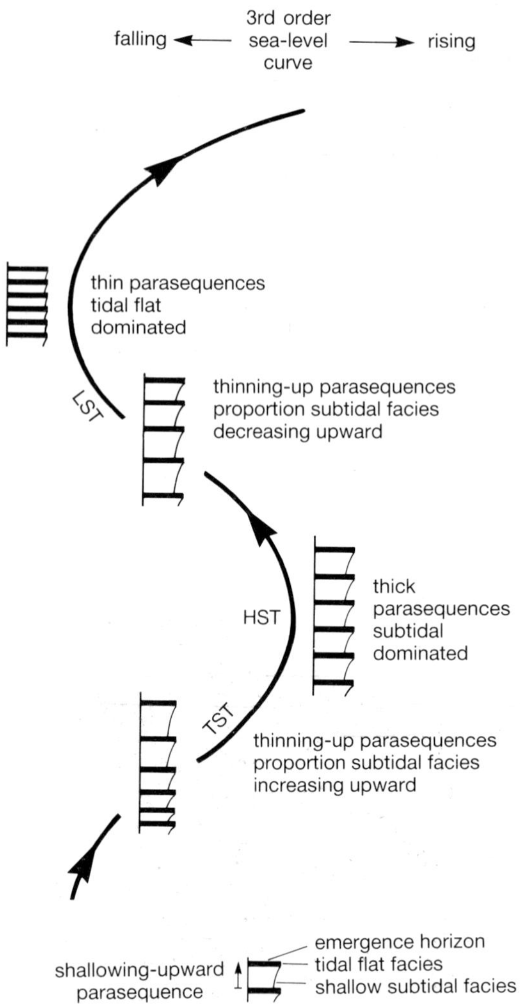

Fig. 7. Schematic stacking patterns for parasequences deposited on a carbonate platform subjected to high-frequency low-amplitude sea-level changes against a background of lower-frequency higher-amplitude sea-level change.

upward cycles or parasequences produced by relative sea-level changes on a shorter timescale. Parasequences commonly display systematic vertical changes in thickness and facies through a sequence (Fig. 7) and, in association with these, there may be systematic changes in the type and degree of early surface-related diagenesis. Thus, in terms of dolomitization, peritidal dolomitization should be more extensive in the parasequences of the third-order sea-level fall, compared to those of the third-order sea-level rise. Where carbonate platforms were very extensive there should be lateral variations in the degree of peritidal dolomitization, with the more landward parts of the platform showing more supratidal dolomitization if the climate were arid (Fig. 8), as for example in the Ordovician Knox Group, Appalachians (Montañez & Read, 1992).

Porosity evolution and reservoirs in dolomite

There appear to have been few important advances concerning the nature and origin of porosity in dolomites since the classic papers of Murray (1960) and Weyl (1960). It is a difficult subject because so many factors control porosity development in dolomites. The subject was discussed actively during the Dolomieu Conference, the principal results being included in this volume. The main question concerns whether dolomitization rearranges or improves initial porosity, or whether the process is entirely destructive in terms of reservoir potential. Although a number of participants felt that dolomitization tends to destroy porosity, many indicated that dolomite essentially 'rearranges' pre-existing porosity. Others suggested that early dolomitization, at best, preserves porosity during burial compaction, and most participants considered that dolomitization may generate, preserve or destroy porosity, depending on the fabrics and textures being replaced, rate, fluid composition, water–rock system and duration of dolomitization.

There are at least two important factors that determine whether dolomitization has a positive or a negative effect on porosity. The first concerns the various dissolution phenomena, as most dolomites generally exhibit moulds formed during selective dissolution of carbonate particles. Dissolution of aragonite during dolomitization does appear to be a major porosity-generating process (Sun, 1992). This phenomenon may predate dolomitization, especially when it results from the selective dissolution of aragonite skeletons and has little bearing on porosity evolution during dolomitization. Dissolution may also postdate dolomitization (Figs 4 and 9) and, again, have little to do with it, but may form important reservoirs. However, when dissolution occurs during dolomitization it is clearly of prime importance. Determining the timing of dissolution is indeed difficult. Where dissolution is limited to those parts of the carbonate sequence that have been dolomitized, the contemporaneity of the two processes (dissolution and dolomitization) is probably, but not invariably, the case. In that the replacement of precursor carbonate involves both

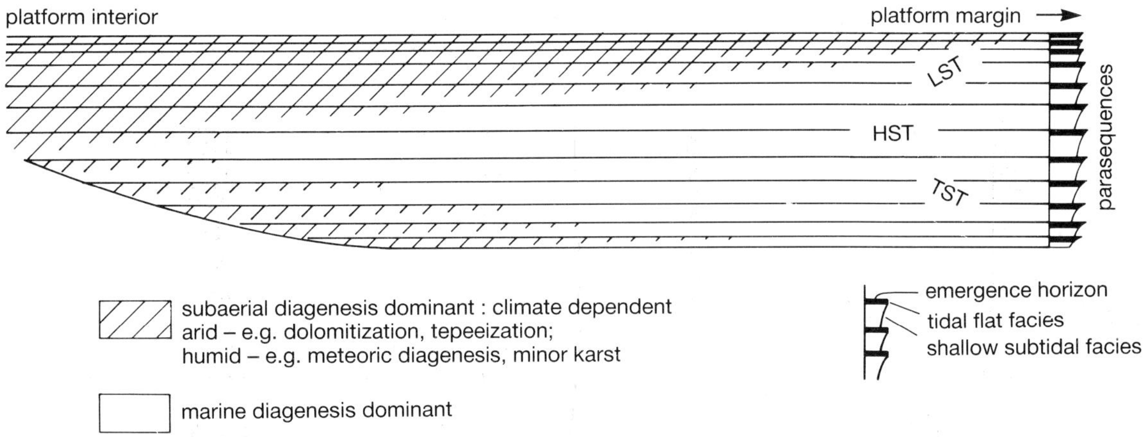

Fig. 8. Generalized scheme indicating the lateral and vertical variations in carbonate diagenesis for parasequences of different systems tracts. The type of subaerial (or surface-related) diagenesis depends very much on the climate, with palaeokarst surfaces and meteoric cements characterizing a humid climate, and supratidal dolomitization typical of an arid climate.

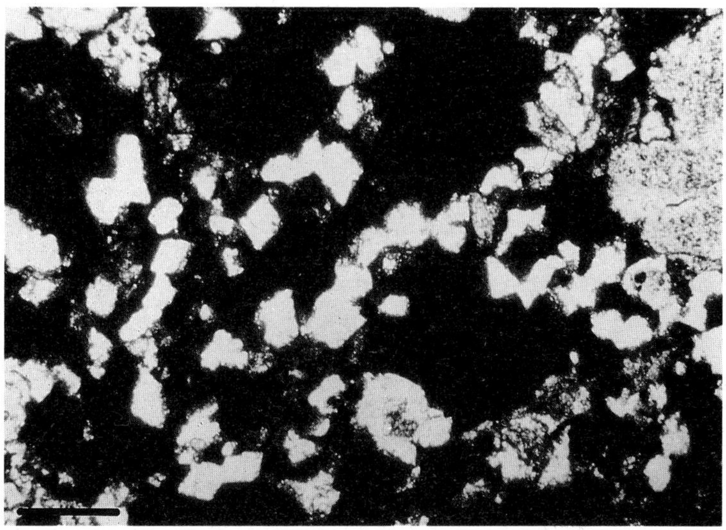

Fig. 9. Postdolomite dissolution: rhombohedral voids in highly porous limestone result from selective dissolution of unstable ferroan dolomite (arrows). Bathonian, Massangis, SE Paris Basin. Scale bar = 50 μm.

dissolution and precipitation, dissolution may theoretically predominate where solutions are highly undersaturated with respect to the precursor carbonate. In this case, dolomitization may have a positive effect on porosity evolution. Also of significance here is the work of Maliva and Siever (1989) proposing the importance of the force of crystallization versus undersaturated porewaters.

The second factor influencing the evolution of porosity in dolomites concerns dolomite cement. Clearly, cement reduces or destroys porosity (see earlier discussion and Fig. 2). Where dolospar lines fractures and vugs its cement origins are evident. However, the same limpid dolomite may constitute part of the matrix dolomite, where it may form the peripheral parts of individual crystals. Crystal growth into voids is cement whatever the nature of the crystal. There are several examples in this volume demonstrating porosity reduction by dolomite cement (Lucia & Major, this volume; Purser *et al.*, this volume).

In summary, there are many different types of

pore space, only some of which are related to the dolomitizing process. Since dolomite tends to resist compaction but may be intensely fractured, it may form porous reservoirs at considerable depth (Schmoker & Halley, 1982) or in strongly tectonized regions (Mountjoy & Amthor, this volume). On the whole, dolomitization probably '*is* a good thing', especially in the deeper burial environments.

CONCLUSIONS

We have discussed briefly a number of concepts considered to be significant in the understanding of dolomite. Looking at the 'dolomite problem' from our editorial viewpoint, we nevertheless make the following observations, taking into account the various presentations and discussions during the Dolomieu Symposium.

Recent advances

Inevitably, these are numerous. Perhaps basic to all discussion concerning the origins of dolomite is that it is a highly variable mineral whose molecular structure, stoichiometry and trace-element composition all express different conditions of formation. Therefore, it is an illusion to seek a common dolomitizing model. We have known for many years that dolomites form in many different surface and burial environments. We may be missing the main point in considering whether seawater, mixed or burial dolomitization best explains most dolomites. Furthermore, the fact that dolomite is associated with normal seawater, brine or meteoric water (all being possible) is not absolute proof of cause and effect. Basically, the fact that dolomite crystals may grow in seawater is not important. What we need to know is which physical and chemical factors within a particular environment determine crystal nucleation and growth.

A point of major importance concerns the evolution and modification of the mineral dolomite, i.e. its stability. We have only recently accepted the well known fact that surface dolomite is metastable and that its subsequent diagenetic evolution may begin soon after its formation. The Dolomieu Symposium confirmed that surface Quaternary dolomite may dissolve selectively, recrystallize and cement prior to burial. Therefore, one must be careful in studying Recent dolomite to make the distinction between primary (initial) and secondary (acquired) properties. With burial, dolomite evolves diagenetically and the final (perhaps there is no *final*) product cannot be readily compared with Recent dolomites.

The fact that much dolomite forms on or near the Earth's surface is no revelation. Deep-burial dolomitization has been increasingly well documented and, in recent years, dolomite formation at intermediate depths has been demonstrated, showing that the mineral can form at all depths. The current debate is whether the volume of deep-burial dolomite is important relative to early near-surface dolomites, or whether the effects of burial are recorded essentially by diagenetic modification of pre-existing dolomite. That some 'new' dolomite is formed at depth is beyond question, but whether most of this dolomite is generated from ions within subsurface waters or is the result of 'remobilization' of a precursor dolomite, is not always clear. Even on a case-by-case basis it may be difficult to distinguish between these two possibilities (e.g. Zenger & Dunham, 1988).

Recent advances in the understanding of dolomite have been the natural consequence of new instruments, techniques and approaches. In addition to microscopy in its various and often sophisticated forms, there has been much research involving fluid inclusions and strontium isotopes, both of which are giving more precise information concerning the nature of dolomitizing fluids. Hopefully, the distinction will be made between the nature of the primary fluids and those responsible for the late stages of crystal growth. Overall precision depends to a great degree on sampling. Although many researchers have been utilizing precision drills for sampling specific dolomite fractions, there are no published results of sampling using laser bombardment, although this has been applied to calcites (Dickson *et al.*, 1990). In spite of increasingly sophisticated techniques, interpretations generally remain unanswered or ambiguous.

Growth histories of individual crystals, often being complicated and of long duration, undoubtedly record the varying chemistries of successive waters. This is demonstrated in part by crystal zoning, commonly imaged by cathodoluminescence. However, the great bulk of geochemical studies of dolomite are based on the analysis of whole crystals: as these are often small, it is generally impossible to do otherwise. It is probable that results would be somewhat different if there were more analyses of successive crystal increments.

Outstanding problems and possible future research

Since there have been over 200 years of research concerning the mineral dolomite, the editors cannot predict with any degree of confidence the nature and direction of future research: we can offer only a biased speculation.

Because dolomite does form under near-surface conditions, we should logically make the most of this natural laboratory. In particular, we should try to look closer at the very first stages of dolomite formation. Although nucleation and growth of dolomite crystals is reputed to be long, one could perhaps place prepared samples of aragonite and calcite at specific sites within sabkhas and other modern dolomitizing environments and monitor any changes.

Clearly, sophisticated instruments and techniques will continue to be used. However, it is hoped that this 'high-tech' approach will not become divorced from the more general techniques, including field-mapping and petrography. Although perhaps obvious, a multiscaled/interdisciplinary evaluation of specific dolomite bodies may improve our overall understanding of their origins. Defining the processes that have created the major dolomite platforms of the past is not a simple matter. However, in so doing, many researchers are turning every stone on the *dolomite* mountain in the hope of finding the truth. Of course, it is vital to examine the *limestone* mountain as well, especially if it is of the same age. Comparative studies may be fruitful.

Many aspects of dolomite are important, irrespective of scale. However, in addition to the careful geochemical and petrographic studies necessary for the improved understanding of dolomite, we would suggest that the more *geological* aspects of dolomite, including basin history (sedimentation, burial and tectonic) also be pursued. The possible climatic, eustatic and global tectonic implications of the temporal and spatial distribution of dolomite are important goals.

ACKNOWLEDGEMENTS

In preparing this introduction the authors have incorporated many remarks formulated during official presentations and informal discussions. In principle these comments are acknowledged, but we apologize to those persons whom we have failed to mention. The choice of points treated in this contribution is based to a considerable degree on the replies to a questionnaire distributed after the symposium. This stimulating help is gratefully acknowledged. The authors especially thank the following persons who carefully reviewed our manuscript: Ian Fairchild, Lynton Land, Eric Mountjoy and Duncan Sibley. Their pertinent suggestions have greatly improved the initial manuscript. To all we extend our sincere thanks.

In addition to the editors, who served as reviewers of all manuscripts, outside referees, selected for their expertise in various aspects of dolomitization, contributed to the quality of this volume. The following persons helped in the preparation of this volume by replying to the questionnaire or by reviewing manuscripts: D.M. Aissaoui, J. Amthor, P.A. Baker, A. Brown, D.M. Bliefnick, F. Bourrouilh-Le Jan, H.S. Chafetz, M. Coniglio, L. Devers, J. Dravis, T.L. Elliott, P. Enos, S. Frisia, J.M. Gregg, R.N. Ginsburg, R.H. Groshong, P. Gutteridge, L.A. Hardie, M. Kastner, C. Kerans, L.S. Land, F.J. Lucia, A.M'Rabet, H.G. Machel, J.D. Marshall, S.J. Mazzullo, J.A. McKenzie, W.J. Meyers, J. Miller, D. Morrow, P. Muchez, M. Mutti, H. Nicholson, R.J. Reeder, A.H. Saller, K.-C. Sam Ng, E. Sass, E.A. Shinn, D.F. Sibley, T. Smith, P. Swart, and W.C. Ward.

REFERENCES

Aissaoui, D.M., Coniglio, M., James, N.P. & Purser, B.H. (1986a) Diagenesis of a Miocene reef-platform: Gebel Abu Shaar, Gulf of Suez, Egypt. In: *Reef Diagenesis* (Ed. Schroeder, J.H. & Purser, B.H.) pp. 112–131. Springer Verlag, New York.

Aissaoui, D.M., Buigues, D. & Purser, B.H. (1986b) Model of reef diagenesis: Mururoa Atoll, French Polynesia. In: *Reef Diagenesis* (Ed. Schroeder, J.H. & Purser, B.H.) pp. 27–52. Springer-Verlag, Heidelberg.

Atwood, D.K. & Bubb, J.N. (1970) Distribution of dolomite in a tidal-flat environment, Sugarloaf Key, Florida. *J. Sedim. Petrol.* **78**, 499–505.

Baker, P.A. & Burns, S.J. (1985) The occurrence and formation of dolomite in organic-rich continental margin sediments. *Bull. Am. Ass. Petrol. Geol.* **69**, 1917–1930.

Banner, J.L., Hanson, G.M. & Meyers, W.J. (1988) Water–rock interaction history of regionally extensive dolomites of the Burlington-Keokuk Formation (Mississippian): isotopic evidence. In: *Sedimentology and Geochemistry of Dolostones* (Ed. Shukla, V. & Baker, P.A.) Spec. Publ. Soc. Econ. Paleont. Mineral. **43**, 97–113.

Bathurst, R.G.C. (1971) *Carbonate Sediments and their Diagenesis*. Elsevier Publishing Co., Amsterdam, 620 pp.

Braithwaite, C.J.R. (1991) Dolomites, a review of origins,

geometry and textures. *Trans. Roy. Soc. Edinburgh. Earth Sci.* **82**, 99–112.

Bullen, S.B. & Sibley, D.F. (1984) Dolomite selectivity and mimic replacement. *Geology* **12**, 655–658.

Bureau, S. (1988) Stratigraphie, Sédimentologie, Diagenèse et Paléogéographie du Madison Group, Mississippian, dans l'Overthrust Belt de l'Ouest du Wyoming, Etats Unis. Doctoral Thesis, Université de Paris Sud, Orsay, 301 pp.

Burns, S.J., Baker, P.A. & Showers, W.J. (1988) The factors controlling the formation and chemistry of dolomite in organic-rich sediments: Miocene Drakes Bay Formation, California. In: *Sedimentology and Geochemistry of Dolostones* (Ed. Shukla, V. & Baker, P.A.) Spec. Publ. Soc. Econ. Paleont. Mineral. **43**, 41–52.

Carballo, J.D., Land, L.S. & Miser, D.E. (1987) Holocene dolomitization of supratidal sediments by active tidal pumping, Sugarloaf Key, Florida. *J. Sedim. Petrol.* **57**, 153–165.

Carpenter, A.B. (1980) The chemistry of dolomite formation, I: the stability of dolomite. In: *Concepts and Models of Dolomitization* (Ed. Zenger, D.H., Dunham, J.B. & Ethington, R.L.) Spec. Publ. Soc. Econ. Paleont. Mineral. **28**, 111–121.

Carriere, D. (1987) Enregistrement sédimentaire et diagénétique et morphologique d'un bombement lithosphérique sur l'atoll soulevé de Maré, Archipel des Loyautés, Nouvelle Calédonia. *Comptes Rendus Acad. Sci. Paris.* **305**, 975–980.

Chilingar, G.V., Zenger, D.H., Bissell, H.J. & Wolf, K.H. (1979) Dolomites and dolomitization. In: *Diagenesis in Sediments* (Ed. Larsen, G. & Chilingar, G.V.) pp. 423–536. Elsevier, Amsterdam.

Choquette, P.W. & Pray, L.C. (1970) Geologic nomenclature and classification of porosity in sedimentary carbonates. *Bull. Am. Ass. Petrol. Geol.* **54**, 207–250.

Choquette, P.W., Cox, A. & Meyers, W.J. (1992) Characteristics, distribution and origin of porosity in shelf dolostones: Burlington-Keokuk Formation (Mississippian), US midcontinent. *J. Sedim. Petrol.* **62**, 167–189.

Compton, J.S. (1988) Sediment composition and precipitation of dolomite and pyrite in the Neogene Monterey and Sisquoc Formations, Santa Maria Basin area, California. In: *Sedimentology and Geochemistry of Dolostones* (Ed. Shukla, V. & Baker, P.A.) Spec. Publ. Soc. Econ. Paleont. Mineral. **43**, 53–65.

Cullis, C.G. (1904) The mineralogical changes observed in the cores of the Funafuti borings. In: *The Atoll of Funafuti*. Royal Society, London, pp. 392–420.

Dickson, J.A.D., Smalley, P.C., Raheim, A. & Stijfhorn, D.E. (1990) Intracrystalline carbon and isotope variations in calcite revealed by laser microsampling. *Geology* **18**, 809–811.

Dunham, J.B. & Olson, E.R. (1980) Shallow subsurface dolomitization of subtidally deposited carbonate sediments in the Hanson Creek Formation (Ordovician-Silurian) of central Nevada. In: *Concepts and Models of Dolomitization* (Ed. Zenger, D.H., Dunham, J.B. & Ethington, R.L.) Spec. Publ. Soc. Econ. Paleont. Mineral. **28**, 139–161.

El-Sayed, A., Fairchild, I.J. & Spiro, B. (1991) Kuwait dolocrete: petrology, geochemistry and groundwater origin. *Sedim. Geol.* **73**, 59–76.

Fairbridge, R.W. (1957) The dolomite question. In: *Regional Aspects of Carbonate Deposition* (Ed. LeBlanc, R.J. & Breeding, J.G.) Spec. Publ. Soc. Econ. Paleont. Mineral. **5**, 125–178.

Fairchild, I.J., Knoll, A.H. & Swett, K. (1991) Coastal lithologies and biofacies associated with syndepositional dolomitisation (Draken Formation, Upper Riphean, Svalbard). *Precambrian Research* **53**, 165–198.

Fowles, J. (1991) Dolomite: the mineral that shouldn't exist. *New Scientist* **132**, 46–50.

Friedman, G.M. & Sanders, J.E. (1967) Origin and occurrence of dolostones. In: *Carbonate Rocks, Part A: Origin Occurrence and Classification* (Ed. Chilingar, G.V., Bissell, H.J. & Fairbridge, R.W.) pp. 267–348. Elsevier, Amsterdam.

Gaines, A.M. (1980) Dolomitization kinetics: recent experimental studies. In: *Concepts and Models of Dolomitization* (Ed. Zenger, D.H., Dunham, J.B. & Ethington, R.L.) Spec. Publ. Soc. Econ. Paleont. Mineral. **28**, 81–86.

Garrison, R.E., Kastner, M. & Zenger, D.H. eds. (1984) *Dolomites of the Monterey Formation and Other Organic-Rich Units*. Soc. Econ. Paleont. Mineral., Pacific Section **41**, 215 pp.

Given, R.K. & Wilkinson, B.H. (1987) Dolomite abundance and stratigraphic age: constraints on rates and mechanisms of Phanerozoic dolostone formation. *J. Sedim. Petrol.* **57**, 1068–1078.

Graf, D.L. & Goldsmith, J.R. (1956) Some hydrothermal syntheses of dolomite and protodolomite. *J. Geol.* **64**, 173–186.

Gregg, J.M. & Sibley, D.F. (1984) Epigenetic dolomitization and the origin of xenotopic dolomite texture. *J. Sedim. Petrol.* **54**, 908–931.

Gregg, J.M., Howard, S.A. & Mazzullo, S.J. (1992) Early diagenetic recrystallization of Holocene (<3000 years old) peritidal dolomites, Ambergris Cay, Belize. *Sedimentology* **39**, 143–160.

Hanshaw, B.B., Back, W. & Deike, R.G. (1971) A geological hypothesis for dolomitization by groundwater. *Econ. Geol.* **66**, 710–724.

Hardie, L.A. (1987) Dolomitization: a critical view of some current views. *J. Sedim. Petrol.* **57**, 166–183.

Hein, J.R., Gray, S.C., Richmond, B.M. & White, L.D. (1992) Dolomitization of Quaternary reef limestone, Aitutaki, Cook Islands. *Sedimentology* **39**, 645–662.

Ingerson, E. (1962) Problems of the chemistry of sedimentary carbonate rocks. *Geochim. Cosmochim. Acta* **26**, 815–857.

Katz, A. & Mathews, A. (1977) The dolomitization of $CaCO_3$: an experimental study at 252–259°C. *Geochim. Cosmochim. Acta* **41**, 297–308.

Land, L.S. (1980) The isotopic and trace element geochemistry of dolomite: the state of the art. In: *Concepts and Models of Dolomitization* (Ed. Zenger, D.H., Dunham, J.B. & Ethington, R.L.) Spec. Publ. Soc. Econ. Paleont. Mineral. **28**, 87–110.

Land, L.S. (1985) The origin of massive dolomite. *J. Geol. Educ.* **33**, 112–125.

Land, L.S. (1991) Dolomitization of the Hope Gate Formation (N. Jamaica) by seawater: reassessment of mixing-zone dolomite. In: *Stable Isotope Geochemistry: a tribute to Samuel Epstein* (Ed. Taylor, H.P., O'Neil, J.R. &

Kaplan, I.R) Geochem. Soc. Spec. Pub. **3**, 121–133.
LAST, W.M. (1990) Lacustrine dolomite; an overview of modern, Holocene and Pleistocene occurrences. *Earth Sci. Rev.* **27**, 221–263.
LUMSDEN, D.N. (1985) Secular variations in dolomite abundance in deep marine sediments. *Geology* **13**, 766–769.
LUMSDEN, D.N. & CHIMAHUSKY, J.S. (1980) Relationships between dolomite stoichiometry and carbonate facies parameters. In: *Concepts and Models of Dolomitization* (Ed. Zenger, D.H., Dunham, J.B. & Ethington, R.L.) Spec. Publ. Soc. Econ. Paleont. Mineral. **28**, 123–137.
MACHEL, H-G. & MOUNTJOY, E.W. (1986) Chemistry and environments of dolomitization: a reappraisal. *Earth Sci. Rev.* **23**, 175–222.
MCKENZIE, J.A. (1981) Holocene dolomitization of calcium carbonate sediments from the coastal sabkhas of Abu Dhabi: a stable isotope study. *J. Geol.* **89**, 185–198.
MCKENZIE, J.A. (1991) The dolomite problem: an outstanding controversy. In: *Controversies in Modern Geology* (Ed. Müller, D.W., McKenzie, J.A. & Weissert, H.) pp. 37–54. Academic Press, London.
MALIVA, R.G. & SIEVER, R. (1989) Diagenetic replacement controlled by force of crystallization. *Geology* **16**, 688–691.
MATTES, B.W. & MOUNTJOY, E.W. (1980) Burial dolomitization of the Upper Devonian Miette Buildup, Jasper National Park, Alberta. In: *Concepts and Models of Dolomitization* (Ed. Zenger, D.H., Dunham, J.B. & Ethington, R.L.) Spec. Publ. Soc. Econ. Paleont. Mineral. **28**, 259–299.
MAZZULLO, S.J. (1992) Geochemical and neomorphic alteration of dolomite: a review. *Carbonates and Evaporites* **7**, 21–37.
MONTANEZ, I.P. & READ, J.F. (1992) Eustatic control on early dolomitization of cyclic peritidal carbonates: evidence from the Early Ordovician Upper Knox Group, Appalachians. *Bull. Am. Ass. Petrol. Geol.* **104**, 872–886.
MORROW, D.W. (1982a) Diagenesis 1. Dolomite – Part 1: The geochemistry of dolomitization and dolomite precipitation. *Geoscience Canada* **9**, 5–13.
MORROW, D.W. (1982b) Diagenesis 2. Dolomite – Part 2: Dolomitization models and ancient dolostones. *Geoscience Canada* **9**, 95–107.
MURRAY, R.C. (1960) Origin of porosity in carbonate rocks. *J. Sedim. Petrol.* **30**, 59–84.
PRAY, L.C. & MURRAY, R.C. eds. (1965) *Dolomitization and Limestone Diagenesis*. Spec. Publ. Soc. Econ. Mineral. **13**, 180 pp.
PURSER, B.H., PHILOBBOS, E.R. & SOLIMAN, M. (1990) Sedimentation and rifting in the NW parts of the Red Sea: a review. *Bull. Soc. Geol. France.* **3**, 371–384.
REEDER, R.J. (1991) Surfaces make a difference. *Nature* **353**, 797–798.
SALLER, A.H. (1984) Petrologic and geochemical constraints on the origin of subsurface dolomite, Enewetak Atoll: an example of dolomitization by normal seawater. *Geology* **12**, 217–220.
SANDBERG, P.A. (1983) An oscillating trend in Phanerozoic non-skeletal carbonate mineralogy. *Nature* **305**, 19–22.
SCHMOKER, J.W. & HALLEY, R.B. (1982) Carbonate porosity versus depth: a predictable relation for South Florida. *Bull. Am. Ass. Petrol. Geol.* **66**, 2561–2570.
SHUKLA, V. & BAKER, P.A. eds. (1988) *Sedimentology and Geochemistry of Dolostones*. Spec. Publ. Soc. Econ. Paleont. Mineral. **43**, 266 pp.
SIBLEY, D.F. (1980) Climatic control of dolomitization, Seroe Domi Formation (Pliocene) Bonaire, N.A. In: *Concepts and Models of Dolomitization* (Ed. Zenger, D.H., Dunham, J.B. & Ethington, R.L.) Spec. Publ. Soc. Econ. Paleont. Mineral. **28**, 247–258.
SIBLEY, D.F. (1990) Unstable to stable transformations during dolomitization. *J. Geol.* **98**, 739–748.
SIBLEY, D.F. (1991) Secular changes in the amount and texture of dolomite. *Geology* **19**, 151–154.
SIBLEY, D.F. & GREGG, J.M. (1987) Classification of dolomite rock textures. *J. Sedim. Petrol.* **57**, 967–975.
SIBLEY, D.F., DEDOES, R.E. & BARTLETT, T.R. (1987) Kinetics of dolomitization. *Geology* **15**, 1112–1114.
SLAUGHTER, M. & HILL, R.J. (1991) The influence of organic matter in organogenic dolomitization. *J. Sedim. Petrol.* **61**, 296–303.
SMART, P.L. & WHITAKER, F.F. (1990) Comment and reply on 'Geologic and environmental aspects of surface cementation, north coast, Yucatan, Mexico'. *Geology* **18**, 802–803.
SONNENFELD, P. (1964) Dolomites and dolomitization: a review. *Bull. Can. Petrol. Geol.* **12**, 102–132.
STEIDTMANN, E. (1911) Evolution of limestone and dolomite. *J. Geol.* **19**, 323–348.
SUN, S.Q. (1992) Skeletal aragonite dissolution from hypersaline seawater: a hypothesis. *Sedim. Geol.* **77**, 249–257.
TUCKER, M.E. (1992) The Precambrian–Cambrian boundary: seawater chemistry, ocean circulation and nutrient supply. In. *Evolution. Extinction and Biomineralization. J. Geol. Soc. London* **149**, 655–668.
TUCKER, M.E. (1993) Carbonate diagenesis and sequence stratigraphy. *Sedimentol. Rev.* **1**, 51–72.
TUCKER, M.E. & WRIGHT, P. (1990) *Carbonate Sedimentology*. Blackwell Scientific Publications, Oxford, 482 pp.
VAN TUYL, F.M. (1916) The origin of dolomite. *Iowa Geol. Surv. Ann. Rep.* **25**, 251–422.
VAHRENKAMP, V.C., SWART, P.K. & RUIZ, J. (1991) Episodic dolomitization of Late Cenozoic carbonates in the Bahamas; evidence from strontium isotopes. *J. Sedim. Petrol.* **61**, 1002–1014.
VON DER BORCH, C.C. (1965) The distribution and preliminary geochemistry of modern carbonate sediments of the Coorong area, South Australia. *Geochim. Cosmochim. Acta* **29**, 781–799.
WARD, W.C. & HALLEY, R.B. (1985) Dolomitization in a mixing zone of near-seawater composition. Late Pleistocene, Northeastern Yucatan Peninsula. *J. Sedim. Petrol* **55**, 407–420.
WEYL, P.K. (1960) Porosity through dolomitization: conservation of mass requirements. *J. Sedim. Petrol.* **30**, 85–89.
ZENGER, D.H. (1972a) Significance of supratidal dolomitization in the geologic record. *Geol. Soc. Am. Bull.* **83**, 1–12.
ZENGER, D.H. (1972b) Dolomitization and uniformitarianism. *J. Geol. Educ.* **20**, 107–124.
ZENGER, D.H. (1983) Burial dolomitization in the Lost Burro Formation (Devonian), east-central California,

and the significance of late diagenetic dolomitization. *Geology* **11**, 519–522.

ZENGER, D.H. (1989) Dolomite abundance and stratigraphic age: constraints on rates and mechanisms of Phanerozoic dolostone formation – discussion. *J. Sedim. Petrol.* **59**, 162–164.

ZENGER, D.H. & DUNHAM, J.B. (1980) Concepts and models of dolomitization: an introduction. In: *Concepts and Models of Dolomitization* (Ed. Zenger, D.H., Dunham, J.B. & Ethington, R.L.) Spec. Publ. Soc. Econ. Paleont. Mineral. **28**, 1–9.

ZENGER, D.H. & DUNHAM, J.B. (1988) Dolomitization of Siluro-Devonian limestones in a deep core (5350 m), southeastern New Mexico. In: *Sedimentology and Geochemistry of Dolostones* (Ed. Shukla, V. & Baker, P.A.) Spec. Publ. Soc. Econ. Paleont. Mineral. **43**, 161–173.

ZENGER, D.H. & MAZZULLO, S.J. eds. (1982) *Dolomitization. Benchmark Papers in Geology*. Hutchinson Ross, Stroudsburg, 426 pp.

ZENGER, D.H., DUNHAM, J.B. & ETHINGTON, R.L. eds. (1980) *Concepts and Models of Dolomitization*. Spec. Publ. Soc. Econ. Paleont. Mineral. **28**, 320 pp.

Spec. Publs Int. Ass. Sediment. (1994) **21**, 21–28

Dolomieu and the first description of dolomite

D.H. ZENGER,* F.G. BOURROUILH-LE JAN† *and* A.V. CAROZZI‡

* *Department of Geology, Pomona College, Claremont, California 91711, USA;*
† *Laboratoire CIBAMAR, Bâtiment de Géologie-Recherche, Université de Bordeaux I, Avenue des Facultés, 33405 Talence Cedex, France; and*
‡ *Department of Geology, University of Illinois at Urbana-Champaign, Urbana, Illinois 61801-2999, USA*

The combined meeting on carbonate platforms and dolomitization held in Ortisei, Italy, 16–21 September 1991 marked the 200th anniversary of the first comprehensive description of the rock dolomite by Déodat de Dolomieu (Fig. 1) in the French *Journal de Physique* in 1791 (Dolomieu, 1791a). This present volume is dedicated to this impressive scientist and man. Dolomieu was an extremely interesting figure, scientifically, personally and politically. We present a brief account of some highlights of his life and contributions to geology, particularly emphasizing dolomite, which is generally removed from much of his other geological work involving mineralogy and volcanic rocks and processes. This paper is by no means a complete coverage of Dolomieu and his accomplishments. For details regarding his life and times the reader is referred to more complete biographies by Lacroix (1921), Moret (1953), Taylor (1971) and Bourrouilh-Le Jan (1982, 1990).

Dieudoné Sylvain Gui Tancrède (following his certificate of baptism), called Déodat de Dolomieu, was a well-informed mind. He was a direct inheritor of the Naturalists who, from their 'salons' (XVIIth and XVIIIth century salons were periodic gatherings of social, artistic, intellectual or scientific people, customarily held at the residence of a well known person) and 'cabinets' (scientific collections of samples and instruments) enlightened western civilization, and whose progressive political ideas helped provoke the French and American Revolutions.

He was born on 23 June 1750, in the parish of Dolomieu, France, in the Dauphiné Province, foreland of the Western Alps. His early education was completely neglected. He had no private tutor as would have befitted a child of a noble family, so he learned to read and count by himself. Yet he soon became acquainted with the Alps through long rambles and observation of the expansive Subalpine and Alpine scenery surrounding the family mansion at Dolomieu, near the small town of La Tour du Pin, 30 km west of Grenoble. He started his military career at the age of 12, in the Order of Malta, because his father was Knight of the Royal and Military Order of St Louis and his godfather belonged to the Order of St Jean-de-Jérusalem. (The Order of Malta was a military religious order founded among European crusaders during the Middle Ages, including the Order of St Louis and the Order of St Jean-de-Jérusalem.) A major event was a duel he fought at the age of 18, killing his adversary. Thanks to the intercession on his behalf by the King of France and Pope Clement XIII, he was sentenced to a 1-year imprisonment only, instead of the death penalty or life sentence.

Not until about 1771, when he reached the age of 21, did he meet the intellectuals of his time, when he was stationed first at Metz (Lorraine, N.E. France) and then in Paris. The Army medical officer began his education in chemistry and physics. Alexandre, Duc de La Rochefoucauld, a member of the Academy of Sciences of Paris, introduced him both to mineralogy and to the Parisian salons, where he established lasting relationships with important European philosophers, scientists and politicians, such as Turgot, Condorcet and H.-B. de Saussure.

His first scientific writings, dating from 1775 when he was 25, appeared in the *Journal de Physique* and dealt with the variations in gravity of different bodies according to their respective distances from the centre of the Earth. This work was based on observations he had made at the mines of Montrelay in Brittany (W. France). Between 1776 and 1779, on the advice of Daubenton (an important co-worker of

Fig. 1. Déodat de Dolomieu. Photograph taken by F.G. Bourrouilh-Le Jan from a portrait in the Archives of l'Académie des Sciences, Institut de France. Courtesy of MM. les Secrétaires Perpétuels de l'Académie des Sciences, Institut de France.

Buffon), Dolomieu took an interest in the geology of Sicily, travelled to the Alps and to Malta, and from there to Portugal as secretary of the 'Grand Maître' of the Malta Order, Prince de Rohan. He observed ancient volcanic formations in the area around Lisbon.

In 1779, at the age of 29, Dolomieu became 'corresponding member' of the Academy of Sciences of Paris and soon left the Order of Malta in preference for geology. He was none the less appointed Commander of the Order of Malta in 1780. During that year he studied the geology of Sicily, and in 1782 the Pyrenees, where he discovered the famous mineral deposits of Barèges Valley. At that time he was also drawn to meteorology and astronomy, and persuaded the Grand Maître, Prince de Rohan, to found an observatory. In 1783, he was appointed Commanding Officer of the Harbour of La Vallette in Malta, and troop commander. As his peers were jealous of him he resigned, and his passion for geology stimulated a journey through Italy. In 1786, his application for membership of the Committee of the Order of Malta was rejected. A 4-year lawsuit ensued during which each party tried bitterly to secure the protection of the Pope in Rome and King Louis XVI in Paris. On the death of the Count of Vergennes, the famous French minister of Foreign Affairs, his patron and protector, he hurried back to Paris on horseback, on one occasion riding for 64 consecutive hours. At Versailles, he clashed with the bailiff de Suffren, lost and won his case three times!

Dolomieu, well known for his liberal opinions, was an enthusiastic supporter of the French Revolution. In 1791 he left Rome and hurried to Paris. He joined the Club des Feuillants, and met again many of his friends who had become members of the Academy of Sciences of Paris and professors at the Jardin des Plantes (the modern National Museum of Natural Sciences). All of them were constitutional monarchists and none of them was to survive the Terror. La Rochefoucauld, his patron and friend, then chairman of the Departmental Council of Paris, was slaughtered before his eyes on 4 September 1792. Dolomieu, as well as the mother and the wife of the Duke, had a somewhat miraculous escape.

During his stay in Paris in those gloomy days, he devoted most of his time and energies to pleading with the Committee of Public Safety for the release of the Duke's wife and mother, as well as for his own mother and sister, who were imprisoned in Grenoble. He also supervised the printing of his numerous works on the 'figured stones' of Florence and on the methodical classification of volcanic materials.

In 1795, during the Directoire (the executive body in charge of the French government from 1795 to 1799), the Ecoles Centrales were founded and Dolomieu was appointed Professor of Natural Sciences. In 1796, the Corps des Mines (Institute of Mines) was established and he was appointed Inspector of Mines. Dolomieu's name as a reputable scientist grew continuously, although his fortune had been lost almost entirely as a consequence of the Revolution. 'Science and the teaching of science', he wrote, 'provide me with an income which makes up for the meagerness of my fortune'. He gave his first lecture on the subject of physical geography and geology at the School of Mines of

Paris (the Ecole Nationale Supérieure des Mines de Paris, where Dolomieu's portrait by Angela Kaufman is kept at the entrance to the Library). His passion for science, his dynamism and his geological discoveries drew to him a company of distinguished scientists, especially members of the Learned Societies of Geneva and Bern. In about 1797 he was appointed Founder Member of the Academy of Sciences of Paris (Institut de France, founded in 1635).

Meanwhile, Napoléon Bonaparte, then First Consul and future Emperor, was secretly preparing a military as well as scientific expedition to Egypt, and Dolomieu was among the first scientists requested to join it. When the ships called at Malta, Bonaparte sent Dolomieu and Junot, his aide-de-camp, to demand the surrender of the city, thus breaking a distinct promise he had made to Dolomieu prior to the expedition. Under such trying circumstances, Dolomieu made every endeavour to help his former comrades-in-arms, and especially his old enemy, the bailiff de Loras. This loyalty to the Knights of Malta is said to have caused a stormy dispute between Dolomieu and Bonaparte.

Arriving in Egypt, Dolomieu set about checking the geographical data he had culled from Greek and Latin writers. He studied the geography and geology of the Nile and its delta. All his samples and observations, except for notes on the preservation of ancient monuments, were lost when the expedition hastily left Egypt.

He sailed from Alexandria on 7 March 1799, in a state of illness. The ship was caught in a storm and found shelter in an Italian harbour where the passengers were considered prisoners of war. During his first two months in prison, Dolomieu read Pliny (killed by the eruption of Vesuvius in 79 AD). Then, denounced by the Knights of Malta and an object of hatred by the Queen of the Two Sicilies, Marie-Caroline of Austria, he was sent to a dungeon in Messina (Sicily) with practically no light and only the few books he had managed to smuggle in. He suffered severe mental torture at the hands of his jailers. He remained thus for 21 months, on the charge of having committed a crime against the State of the Two Sicilies.

In the meantime, the scientific world was moved to action. The Institut de France took measures on his behalf. Sir Joseph Banks, the great English scientist and President of the Royal Society of London, interceded on his behalf, although England was at war with France at the time. Finally, Charles IV, King of Spain, prevailed upon the King of Naples to refrain from delivering Dolomieu into the hands of Tsar Paul I, Grand Maître of the Order of Malta, which had been re-established in Russia, where he would have met his death. The French Foreign Minister Talleyrand wanted the Pope to intervene, which prompted this answer from Napoléon Bonaparte, still First Consul, on 4 June 1800: 'I do not, Citizen Minister, approve of all these petty interventions on behalf of Dolomieu; they debase the Government to no avail; I consider the thought of making the Pope intervene as highly improper and I wish that it should not be carried out'.

It was the victory of Bonaparte at Marengo (Italy) which enabled Dolomieu to recover his freedom. His release was the first condition laid down by Bonaparte to the King of the Two Sicilies. In Paris, Dolomieu received a triumphant welcome by his friends and from Bonaparte. When Daubenton died, Dolomieu delivered the first lecture on mineralogy at the Museum of Natural History of Paris, of which he was in charge from 1800. But he longed to return to the Alps. Dolomieu then retired to the Massif Central, to the home of his sister and brother-in-law, the marquis de Dre. There he prepared for a journey to Saxony to visit Werner, a geologist and the leading neptunist, concerning the origin of basalts.

The main preoccupation of his life, and his pride and joy, was his 'cabinet'. Dolomieu's collections consisted of mineralogical and geological samples, and were kept in Malta. The Order of Malta wanted to sell them to the University of Palermo, but Dolomieu wished to present them either to the Academy of Lyon or the city of Grenoble. Finally, he considered sending them as a gift to the 'Congress of the United States of America, on condition that one hundred years hence the Land of Freedom should send to France a collection of samples from America, which the American people will have had time to know and to collect.' Dolomieu's wish was fulfilled much more than 100 years later, unwittingly, when John Pierpont Morgan, an American citizen, endowed the National Museum of Natural Sciences of Paris with a complete collection of gems from America.

Dolomieu, his health having been undermined by the long months of captivity in Sicily, became ill while staying with his sister and died a few days later, on 26 November 1801, at the age of 51.

Scientifically, Dolomieu was known primarily for his studies of vulcanism, although he was greatly interested in earthquakes, mountain belts, mineralogy and the classification of rocks. In fact, not long before his death he had planned to visit Werner (see above), in part to attempt to reconcile French and German classifications of minerals as well as to discuss the origin of basalts. According to his biographer, K.L. Taylor (1971), Dolomieu was '... esteemed as a judicious inquirer within the existing styles of geological research.' Although interested in theory, he was essentially a pragmatist.

Dolomieu became a proponent of the volcanic origin of basalt and published considerably on this subject. Although he believed that vulcanism had been very important, he felt that water was the outstanding cause of geological change (Taylor, 1971, p. 151). He estimated that about 1/20 of the Earth's surface had been affected by volcanic action. His historical scheme contained many elements of neptunism: he believed that the world's oldest rocks had been precipitated out of a universal fluid. Dolomieu could appropriately be considered a catastrophist, as he opposed the idea of slow and cumulative change (Taylor, 1971). He felt that 10 000 years was a generous length of time since the period of great catastrophic upheavals that resulted in rearrangement of the originally crystallized rocks following precipitation in a universal ocean. But, despite his playing down the significance of vulcanism, he was considered an authority on the subject. He believed the sources of volcanoes to be quite deep below the surface of the Earth and not limited to a particular rock. One of his major concerns was the nature and source of volcanic ejecta, and perhaps his expertise here was what most impressed contemporary vulcanologists. Dolomieu envisaged himself as a keen observer, and he greatly respected others such as Horace-Bénédict de Saussure. He believed that geological disputes could be settled by simple observations. Thus, Dolomieu was best known among the natural scientists of his time as a vulcanologist and mineralogist. Ironically, his brief diversion into the study of a calcareous *rock* that we know today as *dolomite*, is one of the major reasons for his reputation two centuries later.

In 1791 the *Journal de Physique* published Dolomieu's paper 'On a Type of Calcareous Rock that Reacts Very Slightly with Acid and that Phosphoresces on Being Struck' (Dolomieu, 1791a). (For the translation of Dolomieu's paper with detailed comments, see Carozzi & Zenger, 1981.) During some mineralogical excursions in the Tyrolean mountains (used here in the broadest of senses), Dolomieu recognized an abundance of rocks that resembled limestones but were only weakly effervescent with acid. At first he thought his acid had lost its strength, but he noted that, when powdered, these rocks reacted with small effervescence and completely dissolved. Later, he found in Roman ruins a rock that was harder and denser than the more usual, softer Greek marbles; this rock also resisted effervescence on the application of acid.

He made these observations initially in the Stubaier Alps west of the Brenner Pass between Innsbruck and Sterzing (Austria). We must remember that Dolomieu could not have understood the extremely complex structures of the eastern Alps, and that he believed that mountain building resulted from the precipitation of successive formations in a universal ocean, followed by violent upheavals. He described a central core of steeply dipping metamorphics and granites against which were leaning the Secondary and Tertiary Mountains consisting of sandstones and limestones. According to the leading ideas of Dolomieu's time, calcareous rocks occurred only in the Secondary Mountains and were directly related to the occurrence of marine organisms. In this area, in the 'Primitive Mountains', he observed what he believed to be an association of his rock with the crystalline basement. He could not have appreciated at the time that this was a structural relationship owing to the intricate folding of the basement and a younger sedimentary cover within which were dolomites, including some dolomitized reefs and dolomitized deep-water limestones with a marble-like appearance.

On returning to Italy, he noted in the calcareous mountains overlying the basement in the Bolzano-Trento area, beds having similar characteristics to those in the Stubaier Alps, but which were horizontally disposed and contained shell impressions. Also, these rocks were fine-grained and fractured conchoidally, but did not possess the lustre and brilliance that distinguishes dolomites of his 'primitive' type. The Bolzano-Trento area is in the southern Calcareous Alps, where the Mesozoic sequence was not involved in as complicated structures. There the sedimentary cover includes the spectacular Middle and Upper Triassic platform dolomites (Fig. 2) and variegated volcanics of the present Dolomite Mountains, which were to be named after him.

In these less deformed dolomites, Dolomieu also referred to cavities lined with small rhombic crystals characterized by a lustre and curved faces that indicate 'spath perlé' (pearl spar), which dissolves slowly

Fig. 2. View from north of the Sassolungo, near Ortisei, one of the many spectacular Triassic carbonate platforms in the Dolomites.

and without effervescence. This point is of interest: his pearl spar is probably our 'saddle dolomite'. Pearl spar was known already at the time of Dolomieu (having been described by Woulfe in 1779) but neither he nor other naturalists recognized the similar identity of pearl spar and the mineral in the rock he was describing; we shall return to this point.

Interestingly, Dolomieu discovered that the fine-grained 'marble-like' dolomites of the central and northern Tyrolean Alps had the property of phosphorescence when scratched with an iron nail, whereas the reefal dolomites of the Dolomites did not. This phosphorescence, to our knowledge, is not mentioned in modern mineralogy texts, and is not common to all or even most dolomites. Those that possess it do so to different degrees. (D.H.Z. tested several low-temperature dolomites in his collection with no success, the same being true for most of the high-temperature specimens. A few unlabelled coarsely crystalline dolomites in the Pomona College collection showed the phosphorescence, which may possibly be due to some rare earth substituting for calcium or magnesium in the dolomite lattice (Dr Bruce Loeffler, personal communication, 1980).)

Dolomieu recognized that in the Primitive Mountains, carbonates and, among them, the new type were the most resistant and rugged. He referred to the saying 'there is no mountain without a calcareous rock cap'. In the area of the Upper Austro-Alpine nappes of the Stubaier Alps northwest of the Dolomites, the dolomites that contribute to the sedimentary cover are located in the higher elevations for reasons of general structure. Local fan-shaped structures result in dolomites, as well as fissile metamorphic rocks participating in the formation of the jagged summits of the Primitive Mountains. Under such conditions, dolomite is indeed more resistant to differential weathering than the metamorphics. In the Dolomites the massive platform and dolomites are important cliff and summit formers.

Dolomieu referred to locations in his 'Primitive Mountains' where the dolomites can be collected at low elevations. He described their varieties, including beautiful white types that could rival the Greek and Carrara marbles were it not for logistical problems. Again, in what is now known as the Stubaier Alps west of the Innsbruck–Brenner Pass area, folds bring the Triassic dolomites down to the valley floors. He referred to both homogeneous and massive dolomites and those mixed with mica, having a fissile nature. Tectonism, including mylonitization, has contributed to their marble aspect by generating recrystallization, fissility and other types of mechanical deformation.

Based on his observations, particularly in the Tyrolean Alps, and also in the Pyrenees, of both limestones and dolomites associated with the 'heterogeneous rocks' of the 'Primitive Mountains', he introduced the new concept that limestones and dolomites were formed, although in lesser quantities, in the universal ocean before the time of precipitation of the Secondary Mountains, and hence before the occurrence of marine organisms.

The story cannot end here, however, for to complete the picture we must consider a paper, 'Analyse de la Dolomie', published in the same *Journal de Physique* a year later (1792) by Nicolas-Théodore de Saussure, son of the more famous Horace-Bénédict de Saussure. (For a translation of Saussure's paper with comments on some interesting problems associated with, and arising from, this work, see Carozzi & Zenger, 1991.) To begin with, Saussure named Dolomieu's rock 'dolomie' in the latter's honour, as Dolomieu's descriptive term was so unwieldy. Interestingly, he included an unreferenced Latin quotation by Linnaeus, who, we find, apparently in 1768 (12th edition of *Systema Naturae*) first recognized dolomite near Roedburg, Norway, as 'marmor tardum', a white marble that effervesces very slowly. Also, it turns out that Arduino, in 1779, described a 'vein of white brecciated marble' at Lavina, Italy, and from his description and analysis the rock would be termed a dolomite. At this place a faulted and folded sequence of Triassic and Jurassic limestones is intercalated with crystalline dolomites. Arduino assumed, remarkably, that these rocks represented a change from a calcareous precursor through the addition of 'magnesia' by contact metamorphic processes.

Clearly, Dolomieu was the first to describe adequately the new rock. Saussure used for his study a suite of samples from the naturalist Fleuriau de Bellevue. Later, some samples were sent by Dolomieu himself. Saussure described its external characteristics: texture more compact than other limestones; crystalline fabric; harder; and of a higher specific gravity than regular limestone. He also discussed the action of flame, water and acids on dolomite, gave its chemical analysis and discussed at great length its phosphorescence which, as was pointed out earlier, is not common to all dolomites. At that time Saussure also did not connect 'pearl spar' with the constituent of dolomite rock. Most interesting was his chemical analysis of dolomite, as follows:

lime	44.29 grains (1 grain = 0.053 g)
clay (alumina)	5.86 grains
magnesium	1.4 grains
iron	0.74 grains
carbonic acid	46.1 grains
total	98.39 grains
loss	1.61 grains
	100.00 grains

The main point of interest here is the low content of magnesium, which seemed to show that dolomite was, in modern terms, a double carbonate of calcium and aluminium. Its relative stability in acid was believed to be due to the difficulty the acids encountered in attacking the carbonates of clay and calcareous 'earth'. Thus, as interpreted by Carozzi and Zenger (1991), two problems existed: the non-recognition of the common mineralogical identity of pearl spar and the mineral constituent of this new rock, and acceptance by most of the initial analysis of dolomite by Saussure. (F.G. Bourrouilh-Le Jan would like to express her opinion that at that time there were relatively few scientists able to make reliable analytical measurements.)

Many papers, including Kirwan's 2nd edition of *Elements of Mineralogy* (1794) reproduced Saussure's analysis, assuming its correctness. In 1799 and 1800, Tennant, in an analysis of agricultural lime that inhibited the growth of plants, showed that it contained three parts of 'calcareous earth' and two of magnesium. Given the slowness of its dissolution in acids, he suspected that this was the rock Dolomieu had described and that it was likely that Saussure's analyses were erroneous. Examining three specimens, he found them to consist of magnesia and calcareous 'earth'.

Aware of Tennant's work, Saussure, in 1800, redid some of his laboratory work, coupling this with a thorough literature check, and found that alumina does not experimentally form a solid carbonate of alumina with carbonic acid, and that it did not exist as a mineral in nature. Faulty laboratory procedures used in 1792 led to a deceptively high content of alumina and to a simultaneous elimination of much of the magnesium. In this paper, Saussure intimated that pearl spar and the mineral constituent of dolomite are the same.

However, neither Tennant's 1799 and 1800 papers nor that including Saussure's admission of error (1800) were widely known, so the latter's erroneous figures continued in print.

In 1807, Klaproth showed the erroneous analysis of Saussure and produced a reasonable analysis of dolomite, showing in particular a much greater content of magnesium, without the slightest trace of alumina. And, in 1808, Karsten clearly demonstrated that pearl spar and dolomite are mineralogically identical, and in a footnote he called them both 'dolomit'.

Although these publications should have cleared

the matter of the common identity of pearl spar and dolomite as well as that of the chemical composition of dolomite, many authors remained unaware of the problem. For example, von Morlot, 40 years later (1847) referred to the elaborate mineralogical study and definition by chemical analysis by Saussure, obviously without recognizing the erroneous character of the analysis.

Finally, the term dolomite became widely used. Kirwan (1794) had converted 'dolomie' to the English 'dolomite'. In his text on mineralogy, he listed separately the white, rhomboidal crystals (pearl spar) and the rock type of Saussure that he termed dolomite. Readers are aware of the present use by many of 'dolostone' for this rock (as recommended by Shrock, 1948) and dolomite for the mineral. Many choose to use the name dolomite for both the rock and mineral. Dolostone is in the literature and preferred by many, and will surely remain, but etymologically (Vatan, 1958; Zenger, 1981) it is incorrect. Were priority to prevail, because the material examined by these earliest investigators was rock, the name is officially appropriate for the rock.

A.V. Carozzi has examined the Saussure collection at the Museum of Natural History in Geneva for the latter's analysed samples, but to no avail. Neither was he able to locate in the Saussure archives any manuscript on the details of N.-T. de Saussure's experiment. He did, however, uncover an undated notebook in the hand of Saussure containing four pages of carelessly written notes, one page of which has scant comments on 'Analyse de la Dolomie du Tyrol' (Carozzi & Zenger, 1991).

Carozzi located the original letter of October 1791 from Dolomieu to Saussure (Dolomieu, 1791b) expressing his admiration for the work of the latter. Because Saussure had done so much in the way of analysis of the rock, Dolomieu felt it was Saussure's choice to name the rock and he humbly accepted the name proposed in his honour. Dolomieu mentioned that if he had done the analytical work himself, he would have named the rock in honour of the young Saussure's father, the better-known H.-B. de Saussure! (Not long afterwards, H.-B. de Saussure was to be so honoured by the naming of saussurite by his son.)

It is a fitting tribute to the impressive Dolomieu that not only the rock but also the magnificent Dolomite Mountains were named in his honour.

ACKNOWLEDGEMENTS

We would like to thank MM. les Secrétaires Perpétuels de l'Académie des Sciences, Institut de France, for access to material, including the portrait of Dolomieu (see Fig. 1). We are grateful to Jean MacKay for typing the several drafts of the manuscript.

REFERENCES

Arduino, G. (1779) *Osservazioni Chimiche Sopra Alcuni Fossili*. In. 12, appresso Benedetto Milocco, Venezia, 58 pp.

Bourrouilh-Le Jan, F.G. (1982) Historical résumé of French contributions to the study of dolomitization. *In: Dolomitization* (Ed. Zenger, D.H. & Mazzullo, S.J.) pp. 383–393. Benchmark Papers in Geology **65**, Hutchinson and Ross, Stroudsburg, PA.

Bourrhouilh-Le Jan, F.G. (1990) Diagenèse des carbonates de plates-formes, récifs et mangroves en Atlantique et Pacifique. Contrôle de la diagenèse par les variations thermo-glacio-eustatiques d'émersion–submersion. Aragonite, calcite, dolomite. 216 p., 2 documents annexes: Annexe 1: Les plates-formes carbonatées du Centre et Sud Pacifique: stratigraphie, sédimentologie, minéralogie, géochimie, 273 p. avec annexes historiques: Biographie de Dolomieu et deux siècles d'études de la dolomie et dolomitisation. Facsimile articles de Dolomieu et Saussure. Annexe 2: Les carbonates de plates-formes de la mangrove d'Andros, Grand Banc de Bahama, 235 pp., 4 atlas. Thèse d'Etat Paris, Université P.-M. Curie, Paris VI.

Carozzi, A.V. & Zenger, D.H. (1981) On a type of calcareous rock that reacts very slightly with acid and that phosphoresces on being struck (translation, with notes, of Dolomieu's paper, 1791). *J. Geol. Educ.* **29**, 4–10.

Carozzi, A.V. & Zenger, D.H. (1991) The original chemical analysis of dolomite by Nicolas-Théodore de Saussure (1792): a laboratory error and its historical consequences. *Arch. Sci. Genève* **44**, 163–196.

Dolomieu, D. de (1791a) Sur un genre de pierres calcaires très peu effervescentes avec les acides et phosphorescentes par la collision. *J. Physique* **39**, 3–10.

Dolomieu, D. de (1791b) *Autograph letter to N.-T. de Saussure* dated Paris, October 31, 1791. BPU, Genève, Archives Saussure, no. 201, 6 pp.

Karsten, D.L.G. (1808) *Mineralogische Tabellen mit Rücksicht auf die neuesten Entdekkungen ausgearbeitet und mit erläuternden Anmerkungen versehen*. 2 verb und verm. H.A. Rottmann, Berlin, 103 pp.

Kirwan, R. (1794) *Elements of Mineralogy*, 2nd edn. J. Nichols, London, vol. 1, 510 pp.

Klaproth, M.H. (1807) Chemische Untersuchung des Dolomits. In: *Beiträge zur chemischen Kenntnis der Mineralkörper*. H.A. Rottmann, Berlin, vol. 4, pp. 204–233.

Lacroix, A. (1921) *Déodat de Dolomieu, membre de l'Institut National (1750–1801). Sa correspondence – sa vie aventureuse – sa captivité – ses oeuvres*. Libraire Académique Perrin et Cie, Paris, 2 vols.

Linnaeus, C. von (1768) *Systema Naturae*, 12th improved edition. Impensis Laurentii Salvii, Holmiae, 3, 41 pp.

Moret, L. (1953) Un géologue dauphinois préromantique Déodat Dolomieu (1750–1801). *Bull. Soc. Dauphinoise d'Ethnologie et d'Archéologie*. Grenoble, In-8, 15 pp.

Morlot, C.A. von (1847) Ueber Dolomit und seine künstliche Darstellung aus Kalkstein. *Naturwissenschaftliche Abhandlungen gesammelt und durch Subscription herausgegeben von Wilhelm Haidinger*, Bei Braümuller und Seidel, Wien, 1, pp. 305–315.

Saussure, N.-T. de (1792) Analyse de la dolomie. *J. Physique* **40**, 161–173.

Saussure, N.-T. de (1800) Recherches sur l'alumine. *J. Physique* **52**, 280–296.

Schrock, R.R. (1948) A classification of sedimentary rocks. *J. Geol.* **56**, 118–129.

Taylor, K.L. (1971) Dieudoné (called Déodat) de Gratet de Dolomieu. *In: Dictionary of Scientific Biography* (Ed. Gillispie, C.G.) pp. 149–153. Charles Scribner's Sons, New York, 4.

Tennant, S. (1799) Ueber die verscheidenen zum Dünger anwendbaren Kalkarten. *Scherers Z. der allgemeine Chemie* **5**, 423–429.

Tennant, S. (1800) Sur différentes espèces de chaux employées dans l'agriculture, extrait des Transactions philosophiques. *J. Physique* **51**, 156–163.

Vatan, A. (1958) 'Dolostone': discussion. *J. Sedim. Petrol.* **28**, 514.

Woulfe, P. (1779) Experiments on some mineral substance. *Phil. Trans. Roy. Soc. London* **69**, 11–34.

Zenger, D.H. (1981) On the formation and occurrence of saddle dolomite: discussion. *J. Sedim. Petrol.* **51**, 1350–1352.

Spec. Publs Int. Ass. Sediment. (1994) **21**, 29–33

Summary

B.H. PURSER, M.E. TUCKER *and* D.H. ZENGER

SABKHA, EVAPORITIC AND REFLUX DOLOMITIZATION MODELS

Perkins and others report the present formation of dolomite, as well as calcite and gypsum, in East Salina on West Caicos Island, British West Indies. The 2.4 m of Holocene sediments include three dolomite zones composed of microcrystalline poorly ordered calcian dolomite, each of which formed during one of three restricted events. Porewater chemistry suggests that dolomite is presently precipitating in the uppermost gypsum layer and in the interval beneath, but it has also been observed replacing precursor mud in organic-rich zones currently undergoing sulphate reduction. $\delta^{13}C$ values (−1 to −9‰ PDB) reflect the organic origin of the carbon and $\delta^{18}O$ values (−1 to 4.2‰ PDB) are consistent with hypersaline marine-derived fluids, so the dolomitizing solution is interpreted to be seawater modified by evaporation and sulphate reduction. Apparently, the reflux model is not applicable in this case. Probably evaporation, possibly aided by hydraulic head (sea level is higher than the saline), is able to drive water upward through low-permeability sediments that do not permit reflux by gravity alone.

Recognizing that the present is not a perfect key to the past, Frisia profitably makes a comparison between Holocene dolomitization in the Abu Dhabi Sabkha and the Norian (Triassic) Dolomia Principale. In particular, she invokes Holocene sabkha processes to help fill in the early history of dolomitization in the Dolomia Principale, which is overprinted by later diagenetic dolomitization. Not directly observed in the Dolomia Principale are probable early stages of direct precipitation of intertidal, ideal, dislocation-ridden dolomite and microdomains of dolomite in Mg-calcite, but still remaining are subtidal calcian dolomite (possibly having nucleated in the dolomite microdomains) with fine modulated structure and enriched values of $\delta^{18}O$ comparable to those at Abu Dhabi. During burial, progressively ordered and compositionally ideal, less modulated and $\delta^{18}O$-enriched dolomite could have nucleated and grown on the most unstable phases.

Iannace and Frisia contend that Norian and Rhaetian (Triassic) successions in the Mediterranean region show different patterns of dolomitization, probably related to a change in peculiar palaeogeographic, palaeoclimatic and tectonic conditions to which the Norian (primarily Dolomia Principale) carbonates were exposed during deposition and diagenesis. Although burial diagenesis did affect the Norian, the great bulk of dolomitization was early diagenetic and resulted in pervasively replaced peritidal cyclic sequences, leading to massive mimetic fine-crystalline dolomites for peritidal intervals, and coarser, non-mimetic dolomite for subtidal intervals, both with enriched $\delta^{18}O$. Rhaetian dolomites, on the other hand, are coarse and non-mimetic, and possess depleted values of $\delta^{18}O$; field evidence strongly suggests a late origin by large-scale fluid circulation. The authors feel that a uniformitarian approach cannot account for the early yet pervasive dolomitization in the Dolomia Principale.

Mutti and Simo envisage dolomitization by modified seawater in association with cyclic facies patterns related to eustasy in a Permian shelf deposit (Yates Formation, west Texas). Specifically, they attribute dolomitization to periods of shelfward shifting of the shoreline and prolonged subaerial exposure at the top of each cycle: dolomitization correlates with regression. The early pervasive replacement dolomitization was focused in the inner shelf facies, whereas the outer shelf remained largely unaffected. The occurrence of the xenotopic mimetic dolomites in peritidal facies, associated evaporite moulds and enriched $\delta^{18}O$ relative to Permian marine values indicates that dolomitization occurred by evaporation – concentrated seawater in a semi-arid sabkha setting.

MIXING-ZONE AND SEAWATER DOLOMITIZATION MODELS

Whitaker and others claim that the hydrology and aqueous geochemistry of magnesium-depleted (but not calcium) groundwaters under northern Andros Island, Bahamas, indicates replacement dolomitization by present-day groundwaters of slightly elevated salinity derived from the mixture of eastward bank surface flow beneath the island and cold normal-salinity seawater from the adjacent ocean. Stable isotope and trace element geochemistry of the 'dolomites' (generally samples from Stargate Blue Hole, average only 1–2% dolomite) suggests dolomite precipitation at present groundwater temperatures from waters of near-seawater composition under slightly reducing conditions. The dolomites, petrographically typical of many modern dolomites (0.4–0.8 Ma), clearly postdate the host carbonate sediments (1.5–2.5 Ma), but it is not determined whether the replacement occurred in prolonged fashion whenever conditions were favourable for dolomitization, or during one phase.

A relatively thick (80 m) sequence of Late Cenozoic, nearly completely dolomitized carbonates from Little Bahamas Bank (LBB) is proposed by Vahrenkamp and Swart as an analogue of many ancient platform dolomites. There is a full range of fabrics, from preservation of precursor fabric to total obliteration. Although dolomitization is nearly pervasive, the dolomites are petrographically immature, with very high vuggy and/or intercrystalline porosity and permeability. Although Ca-rich and poorly ordered dolomites are the rule, near-stoichiometric and well-ordered dolomites do occur. The isolated setting of the Bahamas, as well as an oxygen isotopic composition in equilibrium with seawater, suggests the precipitation of dolomite from that medium, although slight dilution or slight concentration cannot be negated. These LBB dolomites have the potential, through subsequent diagenesis by burial meteoric water, to stabilize into dolomite comparable to the ancient ones, many of which could therefore have had a similar seawater origin.

Massive dolomitization of Lower Viséan carbonates in northern Belgium beneath the top of a regressive trend is described by Muchez and Viaene. Their paragenetic sequence consists of uniform orange-luminescent dolomite followed by calcite dissolution and the precipitation of zoned dolomite cement and, together with the palaeogeographic and sedimentologic setting, is interpreted as indicating dolomitization by fluids circulating near or in the mixing zone; the orange-luminescent dolomite probably resulted from seawater fluxes near the mixing zone, whereas calcite dissolution and dolomite cementation occurred in the mixing zone that developed in connection with the regression. Typically, their depleted values of $\delta^{18}O$ are interpreted as the result of re-equilibration and stabilization of the marine to shallow-subsurface dolomite during further burial at higher temperature.

BURIAL DOLOMITIZATION MODELS

Simo and others attribute dolomitization (replacement and cements) in the Middle Ordovician Glenwood Formation of the Michigan Basin, USA, to seawater considerably modified by evaporation. Dolomitization occurred in early (near-surface, probably marine), middle (shallow burial, characterized by mouldic porosity, aggrading neomorphism and precipitation of ferroan and non-ferroan calcite cement), and late stages (pervasive dolomitization of precursor carbonate and dolomite cementation in fractures, and intergranular and shelter porosity). Fluid inclusion homogenization temperatures of 185 ± 15°C strongly suggest a late-stage deep-burial origin for the dolomite cement. The authors feel they can explain the wide variations in $\delta^{18}O$ by variations in isotopic ratios over the range of temperatures estimated by the fluid inclusion study. The best estimate of $\delta^{18}O$ for the dolomitizing solution is 5‰, suggesting an evaporated seawater. The $^{87}Sr/^{86}Sr$ values of the dolomite cement appear to incriminate fluids of Upper Silurian and Middle Devonian ages.

Miller and Folk describe two episodes of post-lithification dolomite following an earlier selective dolomitization event in the Triassic Portoro Limestone in the Portovenere area, northwest Italy. Dolomitization in these two later periods (pre-Oligocene–Miocene thrusting and post-Plio-Pleistocene faulting) was controlled by faults, stylolites, bedding and differential cementation along fractures; dolomite fronts and localization along faults, fractures and stylolites attest to the postlithification development. One of the unique aspects of this study is the consideration of the limits of dolomitization distal to the foci of dolomitization. The lateral extent is controlled by factors such as differential cementation along faults and joints, differential permeabilities along fractures, veins and beds, fluid pathway interruptions, and differential accumulation of insoluble residue along stylolites.

Mountjoy and Amthor answer 'yes' to their question 'Has burial dolomitization come of age?', at least for the western Canada Sedimentary Basin. With the exception of minor seawater dolomitization immediately beneath the sediment–water interface, they contend that most dolomites have formed in burial environments, *assuming* that present geochemical compositions have not been reset. Applying the gamut of petrologic and geochemical techniques, the workers consider their burial dolomites to be of two sorts: shallow-burial (early?), fine- to medium-crystalline, selective to obliterative dolomites that formed at depths of 500–1500 m, and deep-burial (late), coarse crystalline, white cements (including saddle dolomite) that fill pores, vugs and fractures formed at intermediate depths of 1500–3000 m or more, and that constitute up to 5% of dolomites in completely dolomitized rocks. They conclude that hot hydrothermal formation waters have risen to higher levels along fractures and conduit systems (e.g. porous and permeable platform margins and reef trends). Most of their late-stage dolomites seem to be related to basinwide fluid flow, driven both by sedimentary and tectonic loading and by topography.

Coniglio *et al.* interpret their findings in Ordovician Trenton carbonates in southwest Ontario (Canada) as indicating a main episode of burial dolomitization controlled by Palaeozoic tectonic fractures, following an earlier dolomitization of a widespread 'cap dolomite' formed by fluids generated from overlying shales. The later and deeper burial replacement dolomitization focused along fractures varies from pervasive near the fractures to selective further away or along less permeable limestone. Dolomitizing solutions were generated by compaction flow from argillaceous units in the more central parts of the Michigan Basin, and by reflux from overlying Silurian evaporites, or possibly from younger fluids also emanating from the Silurian. Fluid inclusion data indicate that the carbonates experienced temperatures exceeding those representing the geothermal gradient, strongly suggesting the effects of hydrothermal fluids.

According to Kupecz and Land, dolomites in the Ellenburger Group (Lower Ordovician) of west Texas that have initially replaced the predominantly muddy facies in that unit, have undergone a complicated diagenetic history. Although the muds were initially replaced syndepositionally by evaporitic or normal marine water, they were progressively modified after burial; petrographic and geochemical data (stable isotope values, trace element content and strontium isotope ratios) as well as covariance between crystal size, degree of stoichiometry, depleted $\delta^{18}O$, $^{87}Sr/^{86}Sr$ ratios, and low trace element concentrations, support the contention. At least two major events are considered responsible for the overprinting: meteoric modification synchronous with early Middle Ordovician regional exposure and karstification, and later and deeper burial recrystallization by hot basinal-derived pore fluids expelled during Ouachita thrusting.

DOLOMITE RESERVOIRS

In a general review of pore systems in dolomites, Purser and others stress their highly variable origins and petrophysical properties, which may not necessarily be related to dolomitization; in many cases, porosity is inherited from the primary sedimentary textures or from early predolomite diagenesis. The authors point out that dolomitization may conserve, rearrange or reduce predolomite porosity. Because dolomitization involves the dissolution of precursor carbonate, they also suggest that the process may increase porosity. However, because of the problem of scale, this increase is not easy to demonstrate: local increase within part of the dolomite body may or may not be balanced (or even exceeded) by a decrease in other parts of the same body. The authors conclude that, because porosity in dolomites has multiple origins, exploration strategy should be adapted to the type of pore system in question.

Interesting 20 m-thick lensoid bodies of dolomite that transect bedding within Upper Cretaceous chalky limestones of central Tunisia are described by Negra *et al.* These bodies of dolomite are centred on probable once-permeable channel fills, but the dolomite also replaces the pelagic muds adjacent to each channel; clearly, dolomitization is controlled by the lime sand channel fillings and formed during burial. The two kinds of dolomite involved are an idiotopic polymodal dolomite with crystals of varying size, and a coarser, more unimodal xenotopic dolomite with cloudy centres and clear rims. This latter dolomite represents more 'mature' dolomitization that has led to destruction of intercrystalline porosity in the channel fills, whereas the impermeable lime muds have been converted to more permeable dolomite. They conclude that the dolomitizing fluids were related to artesian flow from a palaeo-island to the northwest.

The mole-for-mole theory of dolomitization and an associated production of porosity because of the difference in molar volumes of calcite and dolomite is seriously contested by Lucia and Major. Their main focus is the Seroe Domi Formation (Plio-Pleistocene) on Bonaire, Netherlands Antilles, which has never experienced burial compaction and cementation that could result in low-porosity dolomites (mean ϕ of dolomite = 11%, whereas that for limestone = 25%). These workers contend that there is a net addition of carbonate during both the replacement stage and a subsequent pore-filling phase (after all calcite has been replaced), and a net addition of calcium during the pore-filling stage. More dolomite was formed than the amount of carbonate in the precursor limestone would permit; consequently, carbonate as well as magnesium must have been imported. According to Lucia and Major, perhaps the single most important factor in the evolution of porosity during dolomitization is the nature of the limestone precursor. They propose that dolomite porosity will always be equal to or less than that of the precursor.

PETROLOGY AND GEOCHEMISTRY OF DOLOMITES

Usdowski has synthesized dolomite and magnesite at 60° and 90°C by reacting $CaCO_3$ with saturated $CaCl_2-MgCl_2$ solutions for periods up to 7 years. The ionic strength of normal seawater is too low to permit the dehydration of Mg^{2+} necessary for spontaneous nucleation, and so dolomite does not precipitate. Rather, dolomite, or its less stable, less ordered form, protodolomite, develops diagenetically. Given sufficient permeability, the majority of subsurface solutions are within the dolomite field at about 120°C. He proposes that the probability of dolomitization by a 'statistical encounter' between carbonate sediments and migrating pore solutions increases with higher temperatures. He further proposes that the Ca/Mg ratios of the most abundant subsurface solutions suggest that the most frequent temperatures of dolomitization are between 80° and 100°C. Accepting the increase in dolomitization with age, Usdowski suggests that the older a limestone, the higher the probability that it will be dolomitized by these migrating subsurface solutions.

Based on her own analytical work and a literature compilation, Searl has amassed a large body of data concerning Ca/Mg ratios, based largely on point analyses by microprobe. These data result in a newly reported discontinuous polymodal distribution of mole % $CaCO_3$ in dolomite superimposed on the 'normal' bimodality (reflecting initially Ca-enriched compositions and the later recrystallized stoichiometric form). The presence of these apparently sharply bounded preferred dolomite compositions is analogous to the absence of compositional gradients within dolomite crystals, as shown particularly by transmission electron microscopy; there seems to be an underlying mineralogical constraint on preferred levels of Ca uptake in dolomite crystals. She suggests that the preferred compositions may arise through the low development of cation ordering within Ca and Mg layers, in addition to cation ordering between layers. Patterns of substitution in which there is an even distribution of Ca^{2+} within Mg layers are likely to be more stable than those in which lattice strain is unevenly distributed. This possible lattice control on cation uptake into dolomite could be significant in the interpretation of dolomite stoichiometry and growth zonation.

In a follow-up to the experiments of Baker and Kastner (1981), Morrow and Abercrombie investigated the influence of dissolved sulphate on the dolomitization reaction at temperatures above 200°C, using larger solution to solid mass ratios so as to allow more exact characterization of the thermodynamic state of the reaction system than was previously possible. They again documented an inhibitory effect on dolomitization by sulphate, although the reaction was not halted. By geochemical modelling calculations they found that at these *high temperatures* the extent of calcite undersaturation in sulphate-free solutions is much greater than in sulphate-bearing solutions, which in turn may indicate that calcite dissolution is the rate-controlling factor on dolomitization in their experiments; they suggest extrapolation of this relationship to the deep-burial environment.

The massive dolomitization of subtidal open marine carbonate sediments by seawater or slightly modified seawater is proposed for Silurian–Devonian sequences in the Illinois Basin, USA, by Kruger and Simo. Although the dolomites occur above and below a major (pre-Kaskaskia) unconformity, the evidence suggests that subaerial exposure related to the development of the unconformity had little to do with dolomitization of the underlying carbonates.

Stable isotope data, as well as major, minor and trace element contents in the dolomites above and below the break, are similar and strongly suggest a more or less continuous dolomitization. Conventional models seem to fail, and these workers suggest that the relative sea-level changes that controlled sediment type provided the necessary pump; dolomitization was enhanced by mud-rich lithologies common throughout the basin.

DOLOMITIZATION AND ORGANIC MATTER

At the northeastern part of the Abu Dhabi sabkha, Baltzer *et al.* noted a persistent association of a Holocene mangrove palaeosol overlain by a lenticular body of dolomitized (up to 100% 'replacement') lithified aragonitic mud. Seawater, drawn laterally towards the sabkha by evaporative pumping, moves preferentially through the relatively porous and permeable mangrove palaeosol. Measurements of pH and Eh, as well as $\delta^{13}C$, strongly suggest the influence of organic matter and the creation of a local anoxic environment that results in aragonitic dissolution and the precipitation of metastable dolomite. The dolomite is probably less than 1000 years in age.

Soussi and M'Rabet defined three groups of dolomite in the Jurassic Nara Formation of central Tunisia. Group I dolomites (Liassic) are tidalites that are depleted in $\delta^{18}O$ and Sr, with fluid inclusions showing a T_H of 120°C. Group II dolomites (Toarcian) are organic-rich, ferroan and relatively enriched in Sr, and have light $\delta^{18}O$ and $\delta^{13}C$. The latter covaries negatively with TOC, indicating that part of the carbon was derived from organic matter by sulphate reduction or deeper burial thermal decarboxylation. Group III dolomites (Malm), hemipelagites, are depleted in $\delta^{18}O$ and impoverished in Sr. These groups were formed during burial, from shallow to deep, with the associated rise in temperature. The dolomitizing solutions appear to have been warm, saline and multiple in origin, and include connate seawater, waters expelled from shales, and updip-moving basinal waters. Dolomitization in the Nara Formation was largely accomplished by the end of the Cretaceous, although post-Cretaceous fractures are cemented by saddle dolomite.

Sabkha, Evaporitic and Reflux Dolomitization Models

Spec. Publs Int. Ass. Sediment. (1994) **21**, 37–54

Salina sedimentation and diagenesis: West Caicos Island, British West Indies

R.D. PERKINS,* G.S. DWYER,* D.B. ROSOFF,* J. FULLER,* P.A. BAKER* *and* R.M. LLOYD†

* *Duke University Department of Geology, Durham, North Carolina 27708, USA; and*
† *Consulting Geologist, 5925 Kirby Drive, Houston, Texas 77005, USA*

ABSTRACT

East Salina on West Caicos Island, BWI, is the site of active gypsum, calcite and dolomite formation. Up to 2.4 m of Holocene sediment have accumulated in an elongate topographic low (3 km long by 0.5 km wide) bounded on the west by Pleistocene aeolianite and isolated from shallow platform waters on the east by a series of oolitic beach and beach/dune sequences of Holocene age. The salina is fed by marine groundwaters that primarily seep through the underlying Pleistocene bedrock. Hydraulic head created by the elevation difference between the salina surface and mean sea level drives the marine influx.

The stratigraphic succession of the salina indicates a trend towards increasing marine restriction, grading from marine wackestones, packstones and grainstones at the base, upward through microbially laminated mudstones into gypsum mush in the uppermost part of the section. A thin microbial mat encrusted with gypsum and ephemeral halite covers the salina surface. Faunal diversity analysis suggests that two separate cycles of salina sedimentation exist within the overall succession. This cyclicity indicates that the depression occupied by the present-day salina was opened and closed at least twice during the Holocene.

Sedimentary porewaters approach normal marine salinities near the base of the succession, but become more saline upwards. Evaporation near the salina surface elevates porewater salinities to as much as seven times normal marine water. Chlorinity profiles suggest that reflux of dense brines is not presently occurring within the sediments examined. More subtle changes in interstitial water chemistry indicate that the following reactions are presently taking place at various places in the sediment column: precipitation and dissolution of gypsum, formation of dolomite, precipitation of calcite and Mg-calcite, and microbial sulphate reduction.

Carbonate phases present in small quantities in the gypsum layer include aragonite, calcite, Mg-calcite and dolomite. Below the gypsum zone, aragonite and Mg-calcite dominate the mineralogy. Porewater chemistry suggests that dolomite and Mg-calcite are actively forming in this interval. Mg-calcite occurs in the form of sheaves of prismatic crystals; dolomite occurs as rhombohedral and subhexagonal crystals. Dolomite has also been observed to replace precursor carbonate mud in organic-rich zones presently undergoing microbial sulphate reduction. All authigenic carbonates examined have $\delta^{13}C$ values ranging from -1 to -9‰ (PDB), reflecting a contribution from organic carbon. $\delta^{18}O$ values range from -2.6 to $+4.2$‰ (PDB) and are consistent with formation from hypersaline marine-derived fluids.

INTRODUCTION

The Turks and Caicos Islands, BWI, are located approximately 150 km north of Hispaniola near the southeastern end of the chain of Bahamian Banks (Fig. 1). West Caicos Island lies on the western margin of Caicos Bank, an isolated carbonate platform that measures approximately 60 by 100 km and is covered by marine waters that average 4.5 m deep (Fig. 2). The average annual rainfall on West Caicos

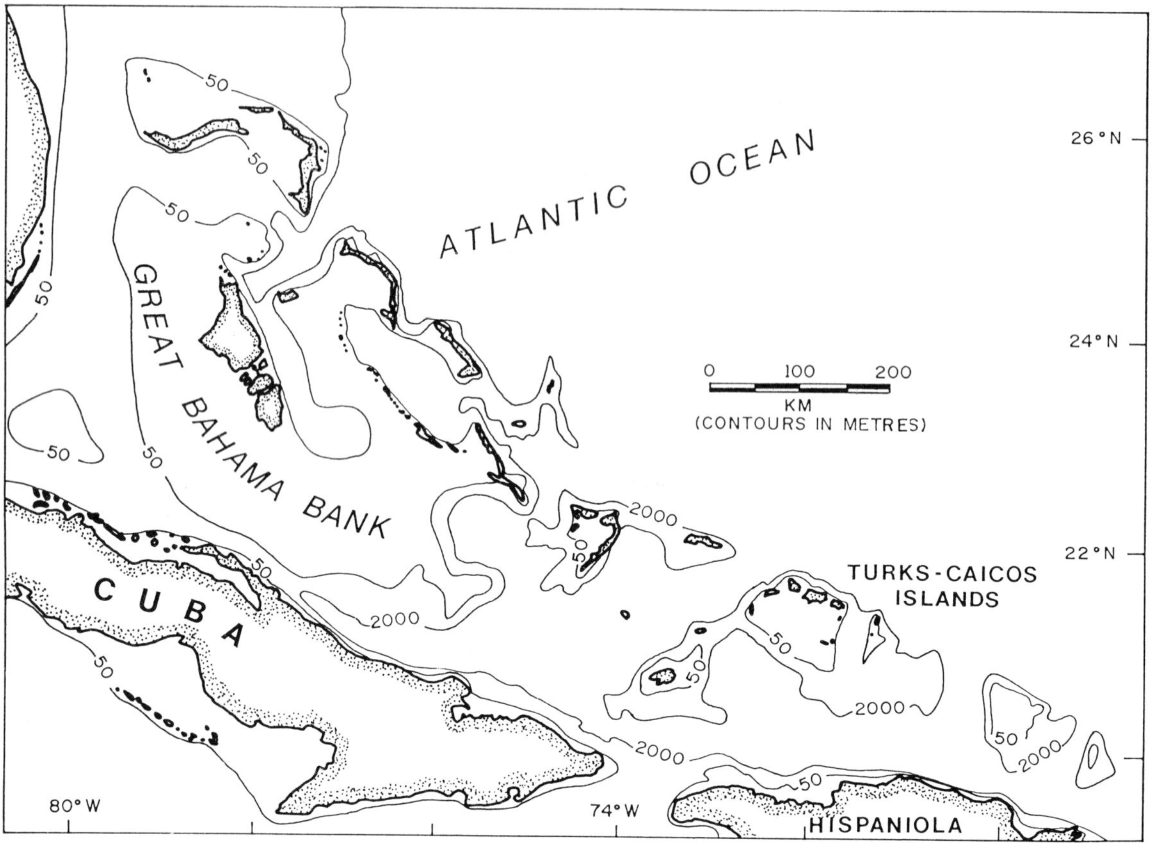

Fig. 1. Map showing regional setting of Turks and Caicos Islands.

is approximately 65 cm, whereas nearby Providenciales averages 112 cm. The climate is mild throughout the year, with an average annual summer temperature of 28.5°C. Easterly trade winds cross the platform at velocities that average 10 knots. Evaporation rates for the Caicos platform are reported as 150 cm year^{-1} (Wanless & Dravis, 1989) to 170 cm year^{-1} (Dwyer, 1991). Broadly speaking, the climate of the Caicos Islands is subtropical, with humidity averaging 75%.

The western two-thirds of West Caicos Island consists of older Pleistocene carbonates, comprising coralgal reefs, prograding beach sequences and dune ridges with elevations as high as 18 m (Fig. 3). Eastward of the most inland Pleistocene dune ridge lies a broad, flat salina (East Salina) that is bounded on its eastern margin by a series of Holocene oolitic dune ridges up to 16 m high. The eastern shore of West Caicos Island consists of oolitic sand that is being generated in the beach and shoreface zone (Lloyd *et al.*, 1987). This series of prograding beach dunes nourished by *in situ* ooid production is responsible for isolating the depression in which East Salina sediments are now accumulating.

SALINA SEDIMENTS AND STRATIGRAPHY

The surface of East Salina is covered by a 5 mm thick rubbery microbial mat encrusted with gypsum and ephemeral halite. Gypsum is also precipitating within this microbial mat, primarily as euhedral lenticular crystals. The mat surface locally exhibits desiccation polygons, but more commonly appears as an irregular bulbous surface with a distinctive pink colour. A few small perennial ponds (visible in the aerial photo in Fig. 3) dot the salina surface and are the sites of active precipitation of gypsum, celestite and Mg-calcite.

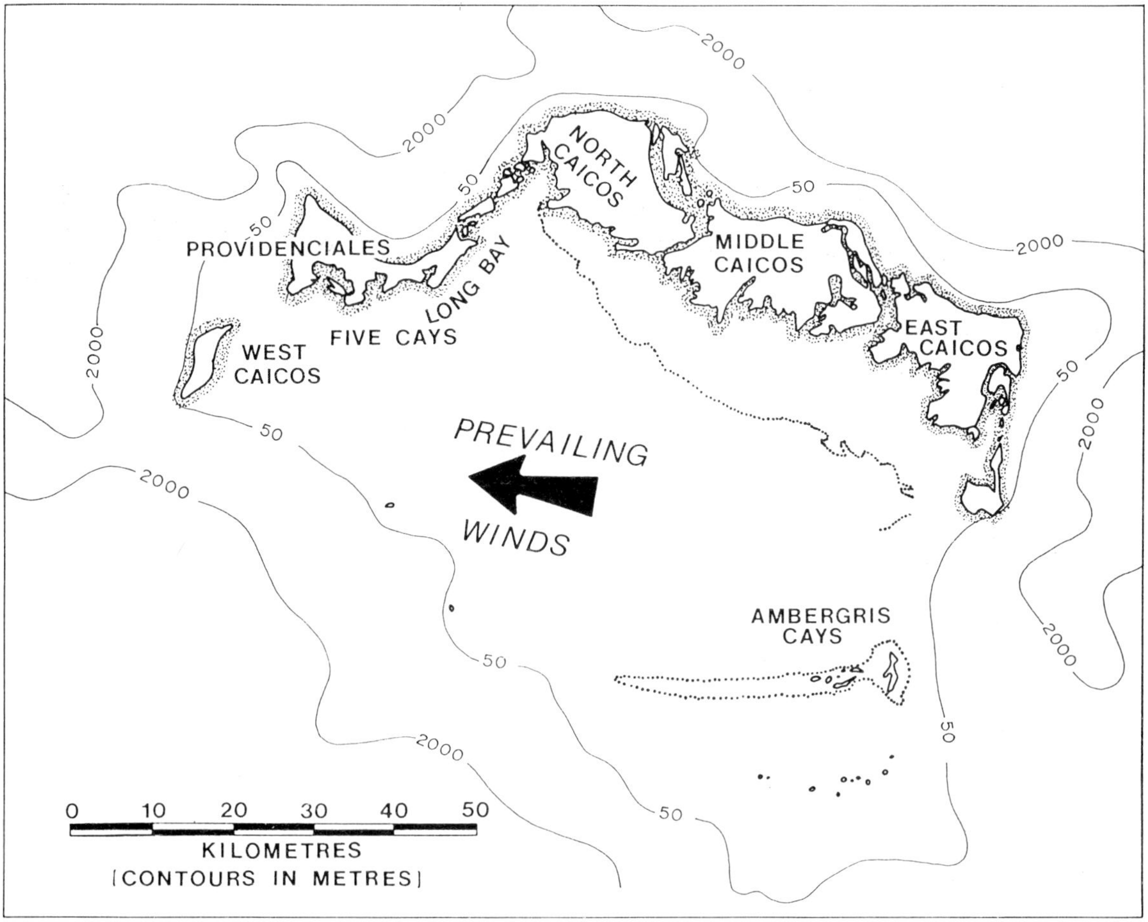

Fig. 2. Map of Caicos Bank showing location of West Caicos.

Occasional rains, often very heavy, inundate the salina surface to a depth of several centimetres. During this time the microbial mat assumes a dark green colour and most of the surficial evaporites are dissolved. The water evaporates quickly (approximately 3 cm week^{-1}) and the salina surface returns to its bulbous character, encrusted with evaporites.

Beneath the salina surface as much as 240 cm of generally unconsolidated carbonate and evaporite sediments have accumulated unconformably on Pleistocene limestone. Twenty-five sediment cores approximately 8 cm in diameter were collected along two east–west transects across the salina. The overall sedimentary record is that of an eastward-thickening wedge comprising six stratigraphically distinct units (Fig. 4 and Fig. 5A) deposited during two restrictive-upward carbonate cycles (Dwyer, 1991). These units can be correlated across the salina with minor thickness variations (Fig. 6). The local absence of some or all units is attributable to non-deposition over Pleistocene rock highs. Exceptions are found in cores taken from the ponds and from dark, organic-rich depressions scattered over the salina surface which contain microbially laminated carbonates in place of the uppermost gypsum mush (Fig. 5B). Along the eastern margin, salina sediments are admixed with skeletal and oolitic sand of adjacent Holocene dunes (Fig. 5C). Here, the gypsum mush interval is also absent, represented instead by an intraclastic packstone cemented by Mg-calcite (Fig. 5C). It is particularly noteworthy that the stratigraphic succession preserved in East Salina sediments is virtually identical to successions that have been described from pro-

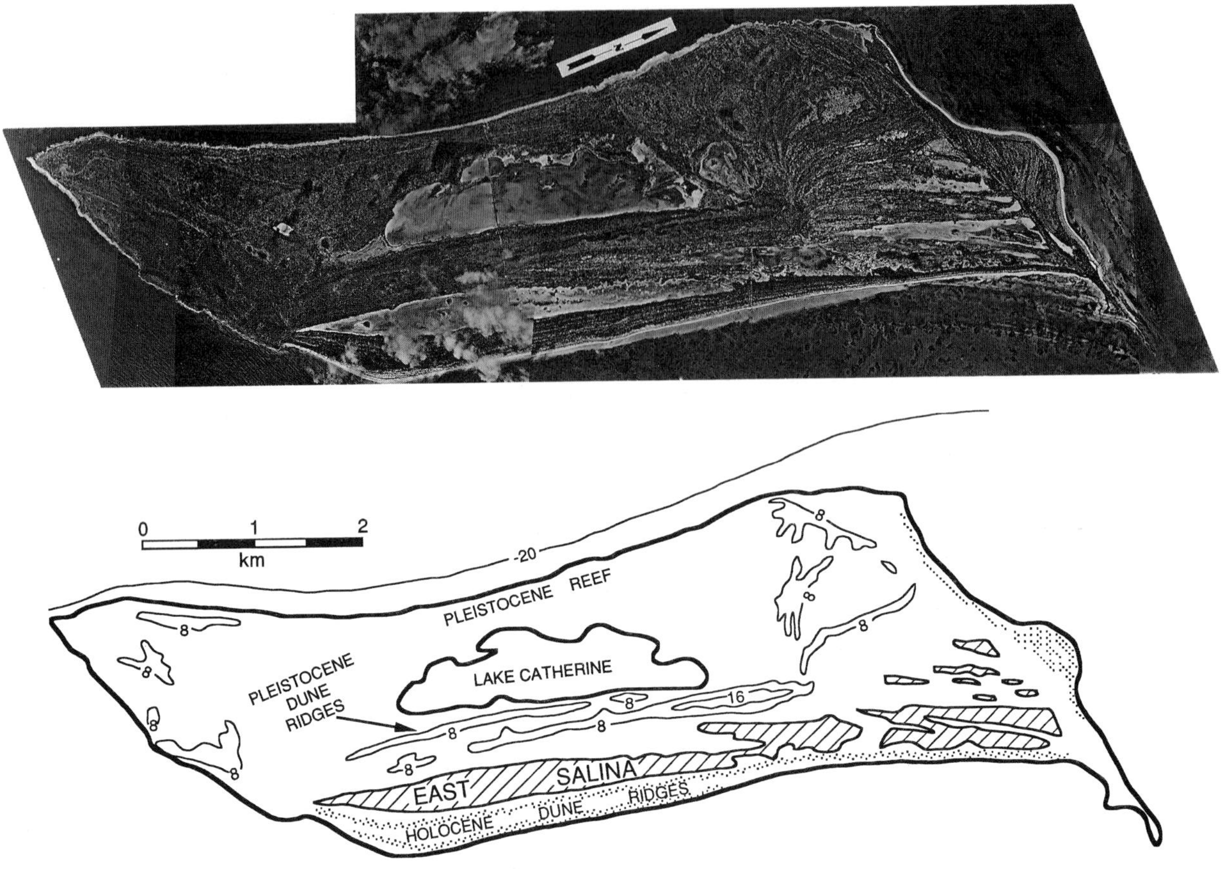

Fig. 3. Aerial photo of West Caicos Island with corresponding map that indicates more salient features. Map contours in metres. Linear feature transecting Lake Catherine and East Salina is an abandoned railroad causeway once used for transporting sisal to the western shore. Modified after Lloyd *et al.* (1987).

grading tidal flats in the Arabian Gulf (Butler *et al.*, 1982; Kendall & Warren, 1988), despite considerable differences in climatic settings.

MINERALOGY OF SALINA SEDIMENTS

Mineral phases identified in the sediment include gypsum, halite, celestite, pyrite, dolomite, aragonite, Mg-calcite and calcite. Dolomite as used herein refers to poorly ordered calcium-rich (52–60 mole % $CaCO_3$) dolomite (protodolomite of Goldsmith & Graf, 1958). Of these, gypsum, celestite, pyrite and dolomite are clearly authigenic, as are some forms of aragonite, Mg-calcite and calcite. These mineral phases are forming in the sediment, either from direct precipitation from porewaters or by replacement of pre-existing carbonate sediments.

Gypsum is the dominant mineral found in the upper salina sediments (generally >90%). A gypsum crystal mush extends downward as much as 70 cm below the salina surface. The gypsum crystals (0.1–2 mm) are mostly discoidal and flattened or lenticular (Fig. 7A). Gypsum also is found in the form of ovoid, hard nodules averaging 0.75 cm in length. These nodules are generally enclosed within microbial layers and are composed of finer-grained, poorly formed crystals. Trace quantities (<1%) of fibrous celestite crystals and crystal aggregates are present locally, commonly incorporated within gypsum clasts. Calcite and Mg-calcite generally make up less than 10% of the gypsum-rich zone.

Most of the Mg-calcite contains less than 9 mole %

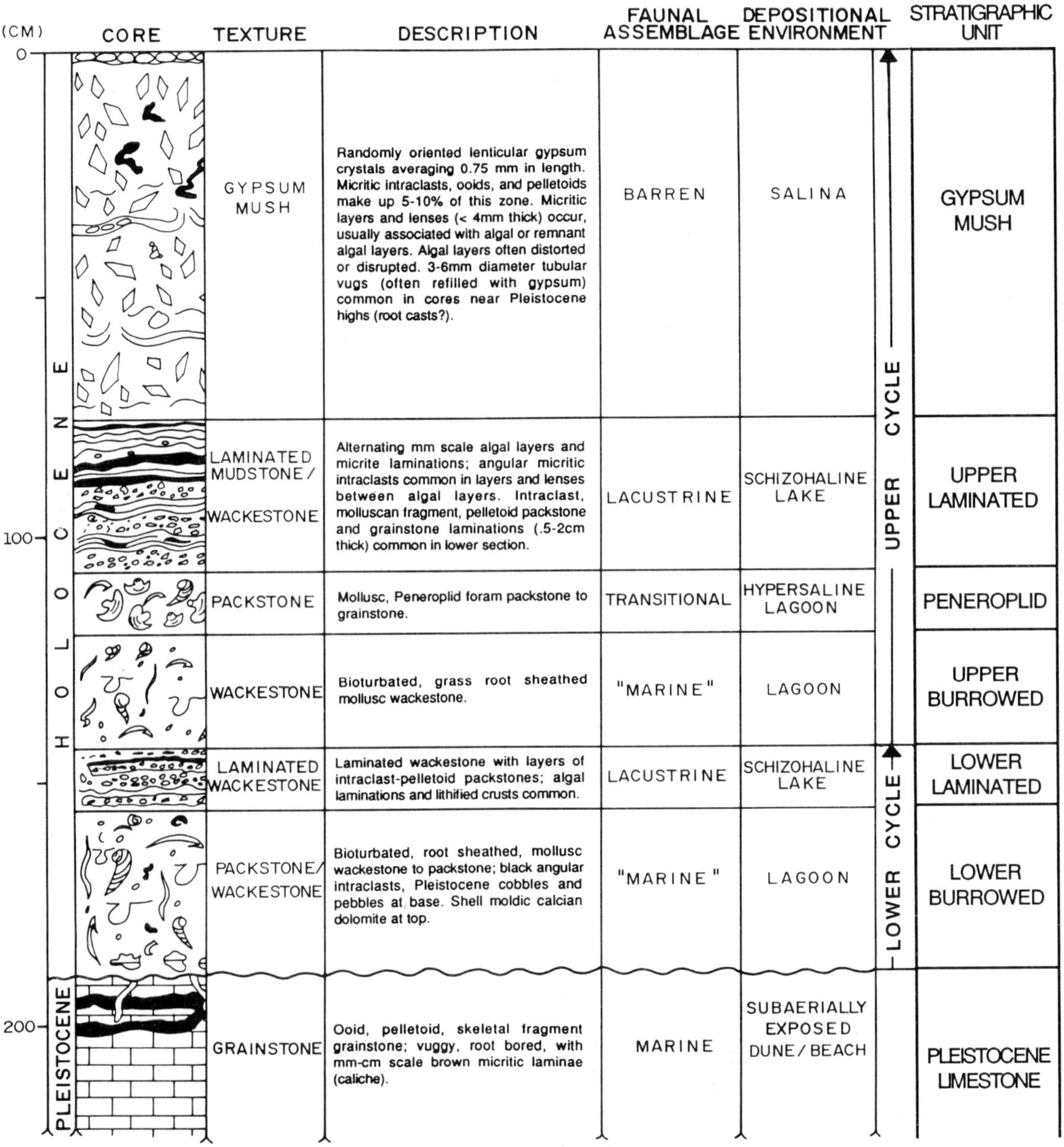

Fig. 4. Idealized salina sedimentary sequence with cycles indicated. After Dwyer (1991).

$MgCO_3$ (determined by X-ray diffractometry), and is generally found as aggregates of poorly formed crystallites or as sheaves of prismatic crystals (Fig. 7B). Calcite is found as aggregates of irregular, poorly formed crystals or as euhedral elongate rhombohedra 5–20 µm in length (Fig. 7C). Both calcite and Mg-calcite in this case are authigenic.

Minor amounts (<1%) of calcian dolomite (57–59 mole % $CaCO_3$) have formed within some of the predominantly calcite clasts through primary pre-

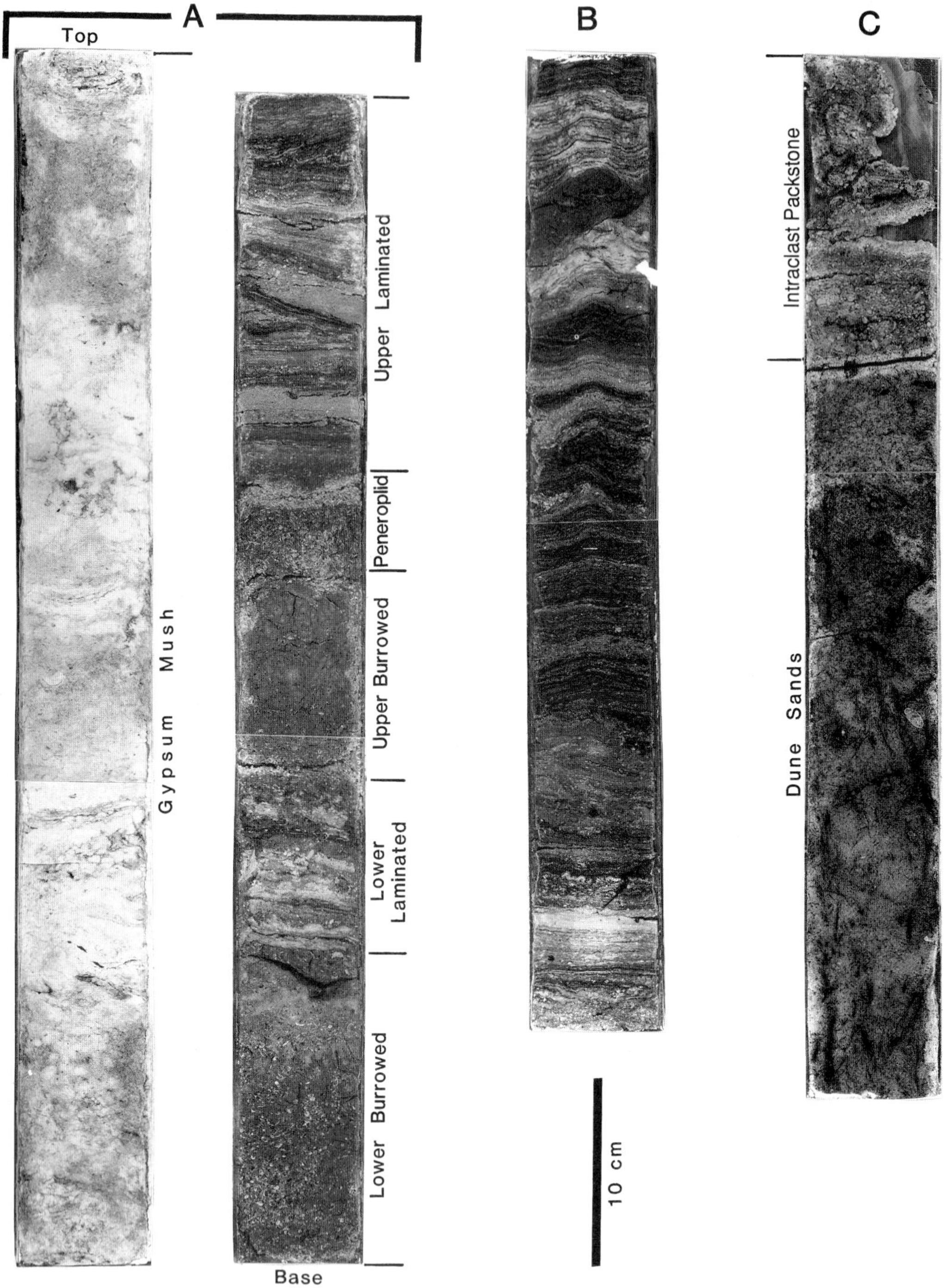

Fig. 5. Plastic-impregnated cores illustrating stratigraphic units recognized in salina sequences. (A) Complete salina sequence from a continuous core illustrating the six stratigraphic units recognized in this study. (B) Typical salina pond core showing microbially laminated carbonates and sparse gypsum. Sediments from these ponds are commonly organic-rich. (C) Core from eastern margin of salina showing rooted back-dune ooid sands overlain by transitional salina sediments comprising organic-rich intraclastic wackestone and packstone. After Dwyer (1991).

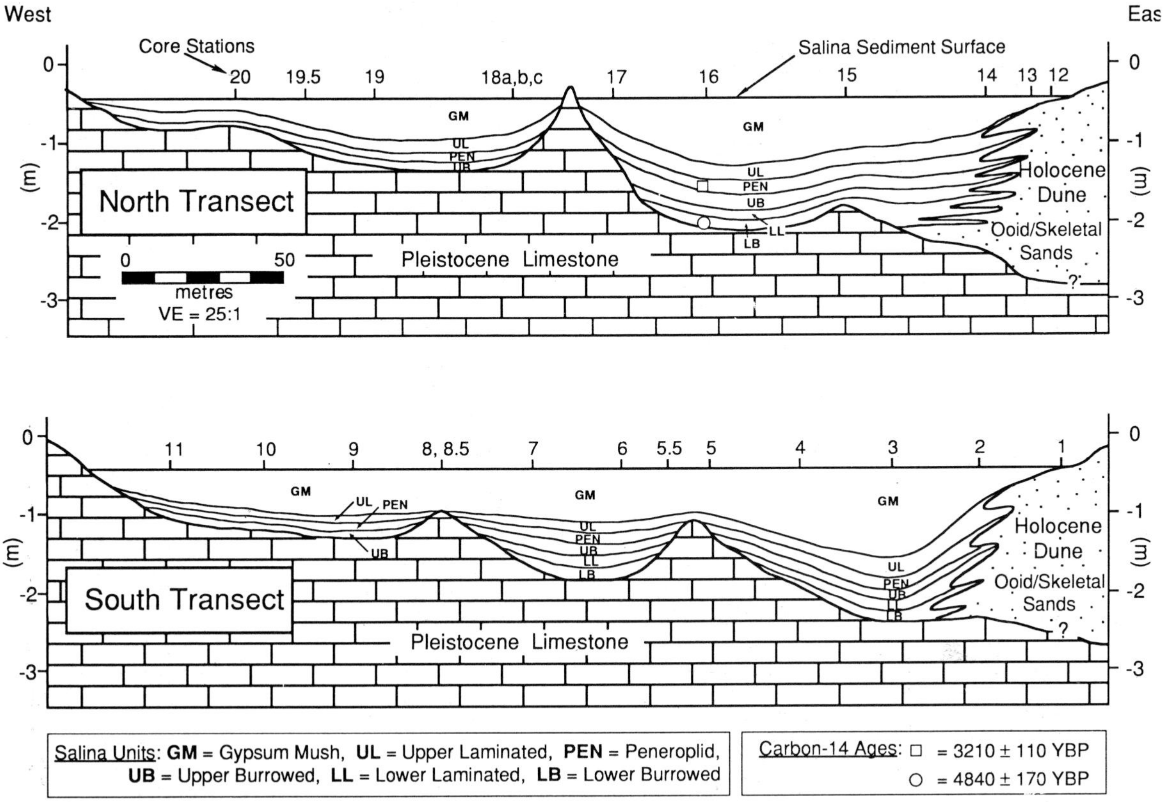

Fig. 6. Stratigraphic cross-sections along transects 150 m north and south of the salina causeway shown in Fig. 3. Vertical scales are in metres relative to sea-level datum.

cipitation or replacement of the precursor calcite phases. Pyrite is also found on the surface of some carbonate grains. The bulk of the dolomite, calcite, pyrite and celestite in the gypsum is concentrated in the $<63\,\mu m$ fraction. Aragonite in the gypsum unit is generally detrital, present in the form of wind-transported ooids and skeletal debris. Less commonly, aragonite precipitates in association with gypsum and other carbonate phases as radiating bundles of small ($<63\,\mu m$) fibrous crystallites (Fig. 7D).

Beneath the gypsum zone the entire section down to bedrock consists almost exclusively of Mg-calcite, calcite, aragonite and dolomite; locally celestite is abundant. Most of the carbonate minerals were deposited during the restricted-marine depositional stages of the salina. These sediments are probably a mixture of detrital carbonate derived from the open bank and beach/dune environments, and *in situ* precipitation of Mg-calcite and aragonite.

Dolomite typically comprises less than 10% of the total carbonate present. However, in individual cores the abundance of dolomite locally increases to 20–95% towards the top of each of the two restrictive-upward carbonate cycles; locally a third dolomitic zone is found as a crust at the base of the Holocene section. In each of these zones, the highest percentage of dolomite is found in grey lithified intervals which are 1–3 cm thick and commonly associated with shell-mouldic porosity. In less dolomitized intervals, dolomite is found as isolated crystals in the mud sized-fraction of unconsolidated sediments. Only the uppermost of these dolomite zones is believed to be forming at present.

In hand samples the lithified intervals from the three dolomitized zones appear similar. Thin-section and scanning electron micrographic examination, however, reveal that these dolomites are different texturally and exhibit different crystal morphologies (Fig. 8). The lower dolomitic crust contains more

Fig. 7. Morphology of salina minerals. (A) Thin-section photomicrograph of gypsum crystals from uppermost gypsum unit. (B) Scanning electron photomicrograph of bundles of prismatic Mg-calcite. (C) Scanning electron photomicrograph of calcite precipitates within the gypsum unit. (D) Scanning electron photomicrograph of bundles of radiating aragonite crystallites surrounded by sheaves of prismatic Mg-calcite.

than 60% dolomite, comprising subhedral crystals averaging 5 µm in size (Fig. 8A) that are associated with irregular fenestral porosity and moulds of gypsum. The middle dolomite zone is most extensively developed near the top of the lower burrowed unit as a grey shell-mouldic packstone containing up to 70% dolomite. These crystals also average 5 µm in size, but exhibit a well-developed rhombohedral habit (Fig. 8B).

The morphology of the upper dolomite is markedly different from that of the lower and middle dolomite zones. Dolomite occurs as granular masses of subhexagonal crystals (Figs 8C and 8D). Recent dolomite of similar morphology and size has been observed in the Coorong Lagoon of Australia (von der Borch & Jones, 1976), in continental saline lakes of western Australia (DeDeckker & Last, 1988), and in Ojo de Liebre, Mexico (Pierre *et al.*, 1984).

Stable isotopes

Carbon and oxygen isotope values were obtained from nearly pure authigenic carbonate phases (aragonite, Mg-calcite, calcite and dolomite) to help determine the conditions under which they formed. Purity of samples was assessed by X-ray diffraction; dolomites were isolated from other carbonates by buffered acetic-acid leaching. The $\delta^{18}O$ values for all carbonates range from −2.6 to +4.2‰ PDB (Fig. 9). These values are consistent with mineral formation from hypersaline marine-derived fluids. Increases in temperature and relative humidity as well

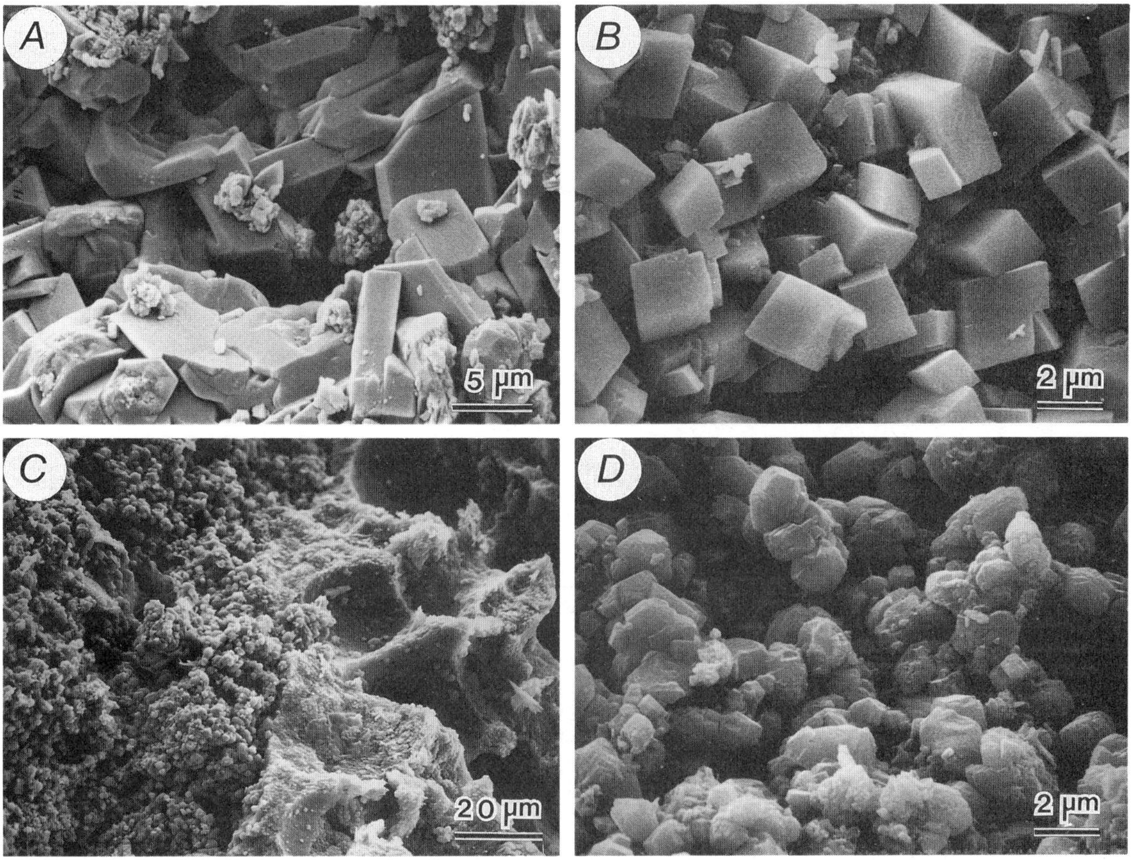

Fig. 8. Variations in dolomite crystal morphology. (A) Subrhombohedral dolomite crystals from lower dolomite zone. (B) Well-developed euhedral rhombic crystals from the middle dolomite zone. (C) Spherical to subhexagonal dolomite crystals (left side of photograph) adjacent to peneroplid foram test from the upper dolomite zone. (D) Higher magnification of dolomite crystals shown in (C) from the upper dolomite zone.

as periodic fresh and/or marine influx could explain the lack of significant oxygen isotopic enrichment normally associated with hypersalinity (Lloyd, 1966). Because the oxygen isotopic values of the pond and porewaters were not measured, equilibrium oxygen isotopic compositions of the carbonate phases can only be estimated.

Lloyd (1966) reported the oxygen isotopic compositions of several pond and porewaters on the nearby island of Inagua, Bahamas. Pondwaters with a salinity range of about 100–350‰ had $\delta^{18}O$ values of 3.5 ± 1‰ SMOW. Porewaters from pits in non-pond areas had values of 2.0 ± 1‰ SMOW over a salinity range of 90–130‰. Water samples from the Pekelmeer Lake on Bonaire, NA, had $\delta^{18}O$ values of 1.5 ± 1‰ SMOW over a salinity range of about 80–250‰ (Lloyd, 1966). In order to calculate equilibrium oxygen isotopic compositions of the carbonate phases, an intermediate value of +2.5‰ SMOW is assumed for the East Salina pond and porewaters. The temperatures (daytime) chosen for these calculations range from 29°C (measured temperature at the base of a core) to 37°C (average measured sediment-surface temperature). Utilizing the fractionation factors reported by O'Neil, Clayton and Mayeda (1969) for calcite-water and by Fritz and Smith (1970) for protodolomite-water, a range of $\delta^{18}O$ values (PDB) of −0.3 to −1.9 for calcite and +2.8 to +1.0 for dolomite is calculated over this temperature range. Some of the measured isotopic values for the calcium carbonate samples are slightly heavier than the calculated values. The pond and

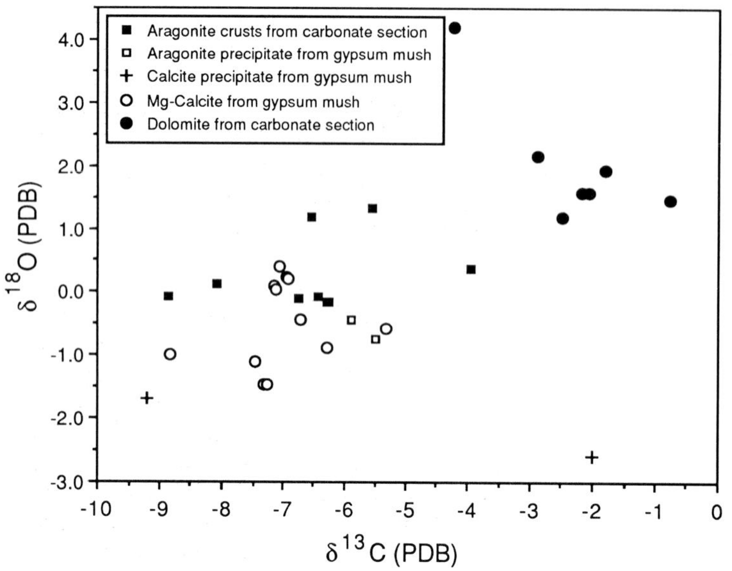

Fig. 9. Oxygen- and carbon stable isotope compositions of mineralogically pure samples of calcite, Mg-calcite, dolomite and aragonite from East Salina sediments. All values are relative to the PDB standard. After Rosoff (1990).

Table 1. Dolomite stable isotopes. All values in ‰ relative to PDB. Samples analysed at North Carolina State University (MEAS). Dolomite was purified by an acetic acid leach of coexistent calcite and aragonite.

Sample no.	Depth (cm)	Stratigraphic unit	Dolomite zone	$\delta^{13}C$	$\delta^{18}O$
18-7 (mud)	47	Upper laminated	Upper	−4.2	+4.2
18B-9 (mud)	75	Lower laminated	Middle	−0.8	+1.5
16-7 (mud)	116	Lower laminated	Middle	−2.2	+1.6
15-9 (mud)	100	Lower burrowed	Middle	−2.1	+1.6
15-9 crust	103	Lower burrowed	Middle	−2.5	+1.7
16-8 crust	122	Lower burrowed	Middle	−1.8	+1.9
16-10 crust	140	Pre-lower burrowed	Lower	−2.9	+2.2

porewaters may have slightly higher $\delta^{18}O$ values than assumed. At isotopic equilibrium aragonite and Mg-calcite are predicted to be slightly enriched isotopically relative to calcite, with Mg-calcite becoming more enriched with increasing Mg (0.06‰ per mole % $MgCO_3$) (Tarutani *et al.*, 1969). Values for the lower and middle dolomites (Table 1) fall into the calculated $\delta^{18}O$ range and could have precipitated in isotopic equilibrium with the assumed fluid under the given conditions. The upper dolomite, however, shows much greater enrichment (Table 1), suggesting formation from waters with a $\delta^{18}O$ value between +3.5 and +5‰ SMOW. This dolomite is likely to be presently forming in porewaters which have a salinity of about 150‰.

The $\delta^{13}C$ values of authigenic carbonates range from −0.8 to −9.2‰ PDB; calcites from −5.3 to −9.2; aragonite from −3.9 to −8.8, and dolomite from −0.8 to −4.2. The range of carbon isotopes is believed to reflect the relative contribution of dissolved CO_2 derived from the microbial oxidation of organic carbon in the microbial mats ($\delta^{13}C$ about −20‰) and from marine biogenic and marine carbonate material ($\delta^{13}C$ of +4.2 to −1.7‰; Keith *et al.*, 1964) and dissolved inorganic carbon. It should be remembered that, at isotopic equilibrium at these temperatures, calcite is enriched about 2‰ and dolomite about 5‰ relative to dissolved inorganic carbon (Emrich *et al.*, 1970; Sheppard & Schwartz, 1970).

Decreases in porewater Mg^{2+} and SO_4^{2-} concentrations (relative to Cl^-) in the upper interval of the carbonate section are consistent with the interpretation of dolomite precipitation accompanied by microbial sulphate reduction. Sulphate reduction is thought to enhance dolomitization (Baker & Burns, 1985). No obvious depletion of Mg^{2+} or $SO_4{}^{2-}$ accompanies the dolomite occurrences in the middle and lower dolomite zones, an indication that these dolomites are not presently forming.

From these observations the following conclusions can be drawn.

1 Three episodes of dolomitization have occurred in East Salina, each associated with a different schizohaline lake/salina depositional event. It is likely that dolomite formed at the expense of precursor $CaCO_3$, most likely aragonite mud.

2 The upper dolomite zone formed from highly evaporated waters (salinity about 150‰) and may presently be active.

3 Less elevated formation salinities are suggested for the lower and middle dolomite zones. However, numerous temperature and $\delta^{18}O_{water}$ combinations could explain their origin and they may actually be hybrids, partially re-equilibrating isotopically and chemically over time, similar to the 'ageing' process proposed by McKenzie (1981) for Recent Abu Dhabi dolomites.

4 Degradation of organic matter by sulphate-reducing bacteria was concurrent with dolomitization in all three zones.

GEOCHEMISTRY OF POND AND POREWATERS

As part of a study of the sedimentology, mineralogy and porewater chemistry of schizohaline pond sediments from the Turks and Caicos Islands, Leaver (1985) studied one core from West Caicos (Fig. 10). Rosoff (1990) and Dwyer (1991) examined East Salina in considerably more detail, utilizing 25 piston cores collected along two east–west transects. With respect to geochemistry, analyses were performed on 160 pond and porewater samples; replicate analyses yielded analytical precisions of +0.58% for Cl^-, +1.01% for Mg^{2+}, +1.03% for $SO_4{}^{2-}$, +1.8% for Ca^{2+}, and +3.3% for Sr^{2+}.

With rare exception, porewater salinities (based on chloride determinations) increase from near-normal marine concentrations at the base of the sediment up to as much as seven times seawater near the sediment surface. Generally, the chloride concentration in the sedimentary porewater also increases from east to west across the salina. Surface water salinities range from 60 to 250‰ during dry periods (isolated ponds and depressions) and between 40 and 50‰ after heavy rains ('lake' conditions).

To examine relationships between major ions in each water sample without the influence of evaporation, ionic concentrations were normalized to the chloride concentration (molar ratio). The Cl ion is considered to be conservative in this environment. Nowhere are the salina porewaters supersaturated with respect to halite. About a tenfold concentration of normal seawater is needed to precipitate halite in these brines (Borchert & Muir, 1964; Sonnenfeld, 1984; Warren, 1989). Dissolution of halite is also not a factor because a solid NaCl phase is not present in any significant quantity within the sediment. Ephemeral halite does precipitate on the salina surface during very dry periods, and is also found crystallizing in and floating on surface waters in some of the shallow ponds and depressions.

Figure 10 summarizes the work of Leaver (1985) on the mineralogy and porewater chemistry of the initial core recovered from East Salina. The profound changes in ion/Cl ratios of the major ions indicated chemical responses of the porewaters beyond that of simple concentration by evaporation. The changes appeared to be related to diagenetic changes in the sediments, principally to gypsum precipitation and dolomitization. The further work of Rosoff (1990) and Dwyer (1991) revealed a more complex relationship between porewater chemistry and mineralogy, adding the processes of gypsum dissolution and sulphate reduction. From their work an idealized porewater profile was generated, as shown in Figure 11.

HYDROLOGY OF EAST SALINA

When this study was initiated, it was hypothesized that downward-refluxing brines would be a major hydrological and geochemical factor in the diagenetic alteration of sediments below the zone of gypsum formation at the sediment surface. The chlorinity profiles determined from porewater studies (Fig. 11), however, indicated that downward reflux of brines through salina sediments was not occurring. It was also observed that porewater in core tubes which penetrated the sediment column below the gypsum layer rose and fell with the tides. More detailed studies were subsequently undertaken to better understand the hydrology of the salina. These

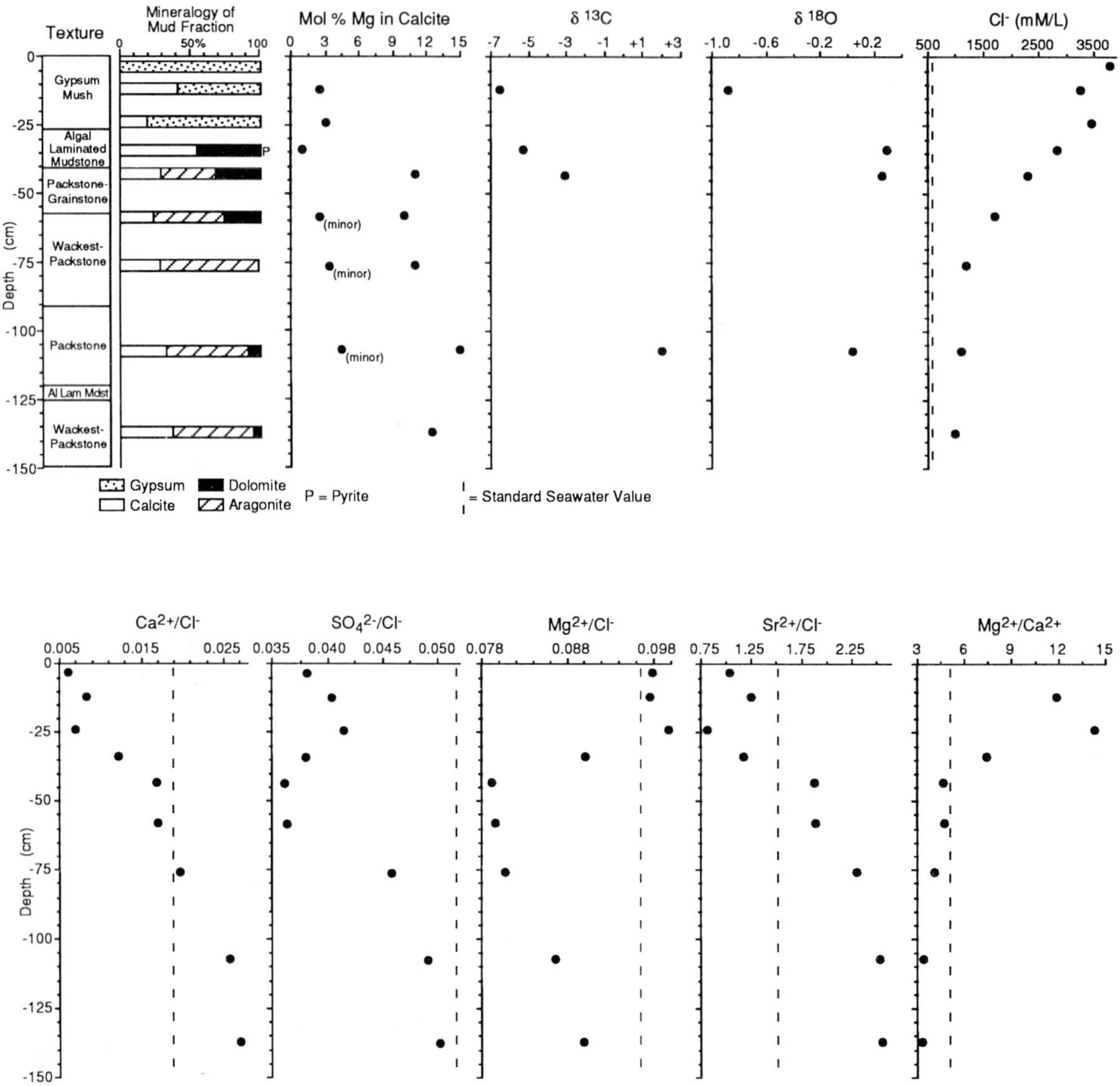

Fig. 10. Sedimentology, mineralogy, stable-isotope and porewater chemistry of core from East Salina. After Leaver (1985). Stable isotopes were measured on the bulk carbonate portion of the mud fraction. In some cases more than one mineral is included in these samples. Data are not corrected for the difference in acid fractionation between calcite and dolomite.

included a stadia rod survey to establish the position of the salina surface relative to sea level, and the installation of tide gauges in the salina.

Four groundwater monitoring wells were contructed along an east–west transect across the salina. Wells were excavated and cased from approximately 60 cm above the salina surface to the top of the Pleistocene with 15 cm PVC pipe. The position of the water level in the wells relative to the salina surface was recorded by hand on an hourly basis during daylight hours. Groundwater levels in the two easternmost wells were monitored continuously over a 2-day period by type F Stevens water-level recorders. To permit comparisons between water-level fluctuations in the salina and those in the nearby platform waters, a water-level recorder also

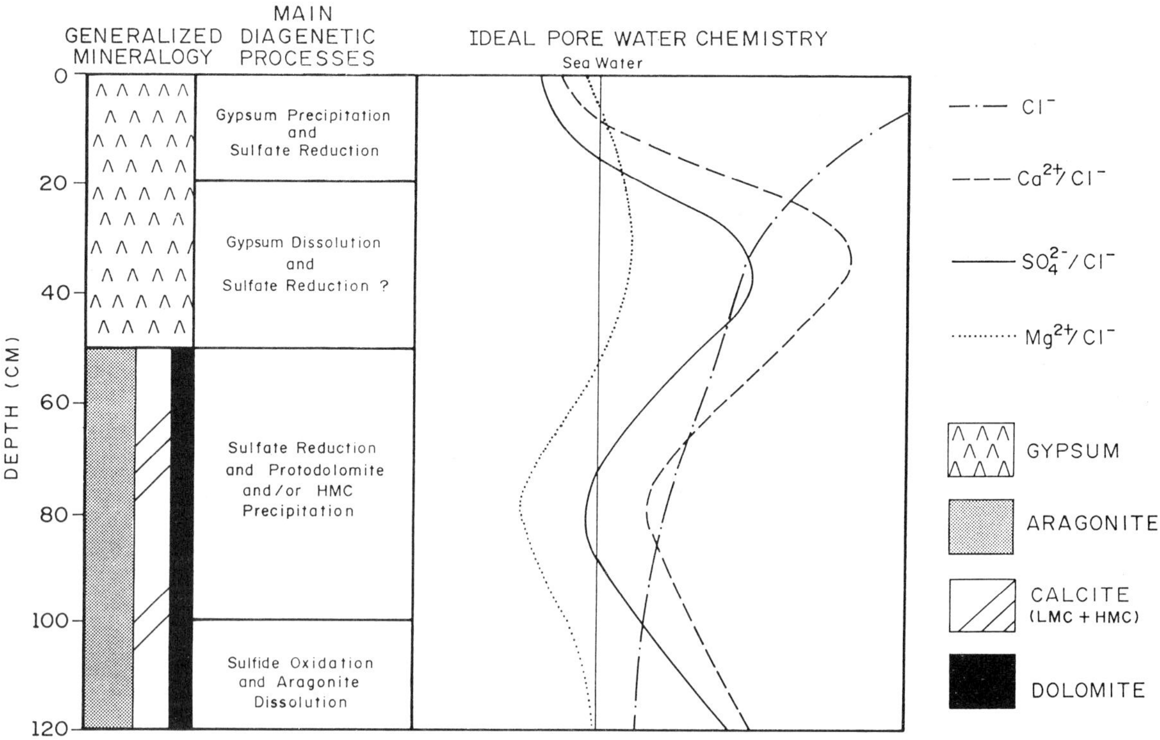

Fig. 11. Idealized mineralogy and porewater chemistry for East Salina sediments. After Rosoff (1990).

was placed in a protected cove on the northeast side of West Caicos.

Comparison between tidal fluctuation and groundwater movement shows that the salina is an artesian groundwater system, where an aquitard (the salina sediments) overlies an aquifer (Pleistocene limestone) fed by marine inflow under hydrostatic head (Figs 12 and 13). When the partially confined aquifer is tapped, the groundwater level in the salina rises and falls with the mixed, semidiurnal ocean tides. Groundwater movement in the salina lags behind the corresponding rising and falling ocean tides by approximately 1 h ± 15 min. Moreover, there is a substantial amplitude reduction in the cycle of groundwater movement compared to the ocean tidal range. The difference in tidal amplitude between wells A and B (Fig. 12) is apparently a function of the distance from the source.

In summary, geochemical and hydrological evidence confirms the existence of a subterranean connection between the marine environment and the salina. Marine-derived inflow seeps or is forced through the underlying karsted Pleistocene limestone, and possibly through the porous and permeable Holocene dunes, into the salina sediments. The marine influx is driven by hydrostatic elevation head (hydrostatic pressure) and, to some extent, by tidal pumping. At maximum recorded low tide, the salina surface lies approximately 8 cm below sea level. Sea level rises approximately 80 cm above the salina surface at maximum high tide. This elevation difference creates a hydrostatic gradient. The potential for a continual supply of seawater is maintained by the hydrostatic differential between the salina and the ocean and the daily process of tidal pumping.

Since the salina is under hydrostatic head (up to −80 cm during normal tidal conditions), the water in the monitored wells rises above the salina surface when the aquitard is penetrated. At maximum high tide the groundwater rose 29 cm above the salina surface in well A and 23 cm above in well B. At maximum low tide the groundwater rose 7 cm above the salina surface in well A and 5 cm above in well B. However, the salina surface is not flooded via groundwater recharge because the rate of recharge

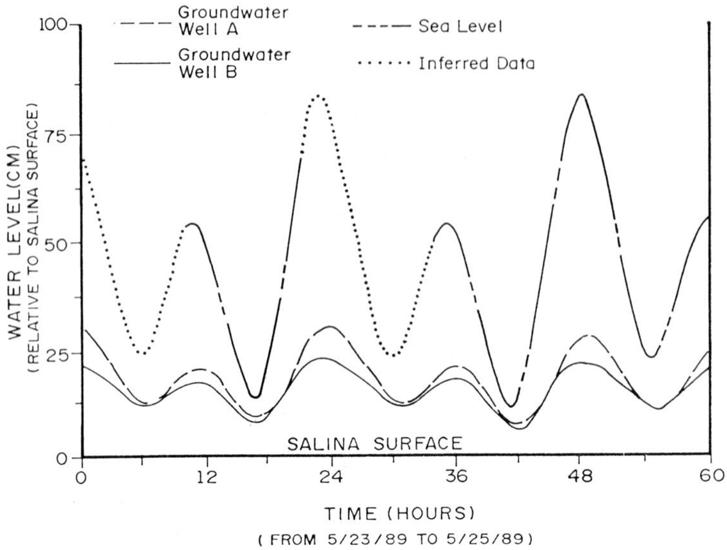

Fig. 12. Comparison of water-level records from well A, well B and sea-level station.

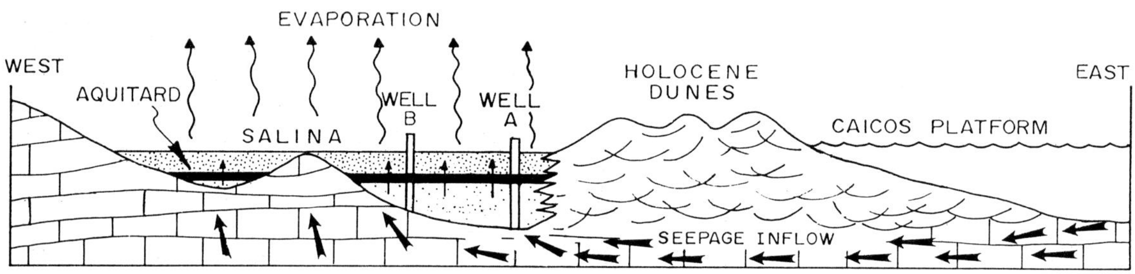

Fig. 13. Hydrological model proposed for eastern shore of West Caicos and East Salina.

is equal to the rate of evaporation. In fact, the evaporitic 'outflow' potential must exceed seepage inflow (the 'phreatic majanna', as described by Logan, 1987). Only small amounts of water move through the aquitard, and these are quickly evaporated near the surface of the gypsum unit. Thus, evaporation in East Salina leads to a concentration of brines near the salina surface and the densest porewaters are suspended at the top of the profile. The non-precipitated solutes remaining in solution after maximum evaporation slowly diffuse downward. No downward reflux of dense brines takes place because of the low permeability of the salina sediments and the high permeability of the underlying Pleistocene limestone. Consequently, seawater recharge can easily keep pace with evaporation. So long as recharge of seawater from below is physically possible, brine reflux should not occur. Seasonal changes in hydrological regime resulting from climatic variations are not dealt with in this study, but should be considered in future work.

Dolomite formation

In order to allow dolomite formation (primary or replacement), a mechanism must exist that can provide magnesium and other reactants to the site of precipitation. Several mechanisms have been proposed to account for the supply of dolomitizing fluids in Holocene carbonate sediments. These include seepage reflux (Adams & Rhodes, 1960; Deffeyes *et al.*, 1965; Muller & Teitz, 1971; Kocurko,

1979), evaporative pumping (Hsu & Siegenthaler, 1969; Hsu & Schneider, 1973; McKenzie *et al.*, 1980), groundwater and seawater mixing (Hanshaw *et al.*, 1971; Land, 1973; Magaritz *et al.*, 1980), tidal pumping (Carballo *et al.*, 1987) and storm recharge (Patterson & Kinsman, 1982).

In East Salina, magnesium is supplied to the sites of dolomite formation by seawater infiltration. Magnesium may also be supplied by dissolution of Mg-calcite phases.

Brine reflux has been proposed as a hydrological mechanism for supplying magnesium in coastal evaporitic settings similar to East Salina (Morrow, 1982a,b; Land, 1983). However, in East Salina the presence of an aquitard (the salina sediments) and a constant landward hydraulic drive prohibits such reflux. As noted previously, dense brine waters sit atop less saline near-normal marine waters.

At present dolomite is forming in the sediment without the reflux mechanism. Seawater is supplied by seepage influx and modified into a dolomitizing solution by heating, evaporation and sulphate reduction. Sulphate reduction has previously been used to help explain deep-water dolomite formation (Kelts & McKenzie, 1982; Baker & Burns, 1985; Compton, 1988). The proposed mechanism of diffusive supply of Mg^{2+} from overlying seawater has been suggested to be incapable of any large-scale dolomitization (Kelts & McKenzie, 1982; Land, 1985). In East Salina, dolomite forms in the zone of microbial sulphate reduction, but magnesium is supplied by seawater recharge from below. Therefore, dolomite formation is not limited by the magnesium supply.

EVAPORITE DISSOLUTION AND STRATIGRAPHIC SIGNIFICANCE

Dissolution of gypsum is presently occurring in the lower part of the gypsum mush, where porewaters are less concentrated than the overlying brine (Rosoff, 1990). Evaporites precipitated during earlier hypersaline stages have been completely dissolved; the only evidence of their former presence is found in the form of gypsum crystal moulds in cemented crusts from the base of the Holocene sediment succession and in the lower and upper laminated units (Fig. 14A). Dissolution of gypsum associated with the lower laminated unit probably occurred during deposition of the overlying upper burrowed unit as the salina was flooded by undersaturated marine waters. Similarly, gypsum precipitated during a proposed salina phase associated with the earliest Holocene transgression would also have been leached during deposition of the overlying lower burrowed unit. Although evaporite dissolution associated with marine flooding of a salina is an obvious mechanism, a contributing factor may also be the dissolution of gypsum by ascending marine groundwater under hydrostatic head.

The preserved stratigraphic record of such stacked restrictive-upward cycles may, therefore, lack any evidence of evaporites other than the distinctive

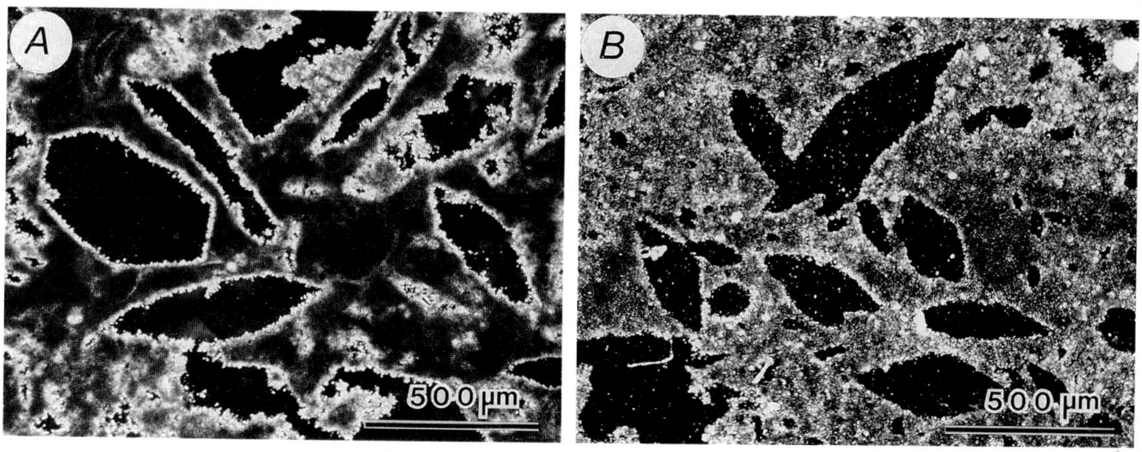

Fig. 14. Holocene gypsum moulds and possible ancient analogue. (A) Gypsum crystal moulds in aragonite-cemented layer from lower laminated unit. (B) Gypsum moulds from peritidal dolomite in the Devonian Jeffersonville Limestone of southern Indiana.

gypsum moulds that may be only locally preserved. Such evidence is commonly preserved in ancient peritidal successions (Fig. 14B), where it is often interpreted to indicate formation in a supratidal setting. Alternatively, such evidence may indicate ephemeral salinas punctuating marine successions in coastal settings where interior lagoons or isolated ponds are periodically invaded by marine waters. Such settings may be far more common in the stratigraphic record than previously suspected.

SUMMARY AND CONCLUSIONS

1 Within East Salina up to 2.4 m of Holocene carbonate and evaporite sediments have accumulated unconformably atop the Pleistocene limestone that floors this intraisland basin. The Holocene sediments contain a succession of six distinct stratigraphic units deposited in two restrictive-upward carbonate cycles, the uppermost of which is capped by an active evaporative gypsum mush. Each restrictive-upward cycle consists of a bioturbated lagoonal carbonate unit followed by highly restricted microbially laminated carbonates and evaporites. Lagoonal units were deposited at salinities that approximated normal marine. The highly restricted units, however, were deposited at salinities ranging from brackish to extremely hypersaline (15–>150‰) in schizohaline lake/salina settings. Although only two restrictive-upward cycles are apparent in the Holocene sediment succession, an additional schizohaline lake/salina setting probably occurred during initial inundation by rising Holocene seas.

2 Three dolomite zones composed of microcrystalline poorly ordered calcian dolomite are recognized in the Holocene sediment succession. Each dolomite zone formed during one of the three restricted events, i.e. the lower dolomite formed during initial Holocene inundation, the middle dolomite during deposition of the lower laminated unit, and the upper dolomite during deposition of the upper laminated and gypsum mush units. The dolomite zones formed through preferential replacement of $CaCO_3$ muds when they were exposed to hypersaline waters.

3 The present porewater chemistry profiles are shaped by a combination of evaporation, mineral precipitation and dissolution, mineral replacement, sulphur redox reactions, fresh and marine water influx, diffusion and advection. Ca^{2+}/Cl^-, Sr^{2+}/Cl^-, and SO_4^{2-}/Cl^- ratios in the porewaters indicate the precipitation of gypsum at the surface and dissolution of gypsum at depth.

4 In the carbonate sediments beneath the gypsum interval, the present formation of dolomite and Mg-calcite is evident from the reduction of Mg^{2+}/Cl^- ratios in the porewaters. SO_4^{2-}/Cl^- ratios also decrease in this zone, suggesting a mechanistic linkage between carbonate precipitation and microbial sulphate reduction. The negative $\delta^{13}C$ values of all the carbonate phases throughout the sediment suggests the incorporation of HCO_3^- derived from the bacterial degradation of isotopically light organic carbon via sulphate reduction. The range of observed $\delta^{18}O$ values is representative of formation in warm, hypersaline waters.

5 The reflux hydrologic mechanism, which is believed by some workers to dominate many modern sabkhas and salinas worldwide, does not apply to the East Salina study area. Brine retention within the salina sediments is caused by seawater recharge from below, due to a landward hydraulic gradient and possibly a minor contribution from capillary pressure resulting from evaporative loss of near-surface waters. The existence of any outflow system for dense brines is as yet undetermined.

6 The preserved stratigraphic record of periodically flooded salinas such as that of West Caicos might well consist of dolomitized marine and microbially laminated couplets containing only associated moulds or calcite pseudomorphs after gypsum.

ACKNOWLEDGEMENTS

Financial support for this research was provided by grants from Amoco Production Company, ARCO Oil and Gas Company, and Shell Development Company. Dr William Showers of North Carolina State University is acknowledged for stable isotope analyses, as well as Dr James Gregory, Chip Cheschiere and Charles Williams for their assistance and equipment use in our water-level studies.

REFERENCES

Adams, J.E. & Rhodes, M.L. (1960) Dolomitization by seepage refluction. *Bull. Am. Ass. Petrol. Geol.* **44**, 1912–1920.

Baker, P.A. & Burns, S.A. (1985) Occurrence and formation of dolomite in organic-rich continental margin sediments. *Bull. Am. Ass. Petrol. Geol.* **69**, 1917–1930.

Borchert, H. & Muir, R.O. (1964) *Salt Deposits: Their Origin, Metamorphism, and Deformation*. Van Nostrand and Co. Ltd., London, 338 pp.

Burns, S.J., Baker, P.A. & Showers, W.J. (1988) *The Factors Controlling the Formation and Chemistry of Dolomite in Organic-Rich Sediments: Miocene Drakes Bay Formation, California*. Spec. Publ. Soc. Econ. Paleontol. Mineral., Tulsa **43**, 41–52.

Butler, G.P., Harris, P.M. & Kendall, C.G.St.C. (1982) Recent evaporites from the Abu Dhabi coastal flats. In: *Depositional and Diagenetic Spectra of Evaporites* (Ed. Handford, C.R., Loucks, R.G. & Davies, G.R.). Soc. Econ. Paleont. Mineral. Core Workshop no. 3, Tulsa, 33–64.

Carballo, J.D., Land, L.S. & Miser, D.E. (1987) Holocene dolomitization of supratidal sediments by active tidal pumping, Sugarloaf Key, Florida. *J. Sedim. Petrol.* **57**, 153–165.

Compton, J.S. (1988) Degree of supersaturation and precipitation of organogenic dolomite. *Geology* **16**, 318–321.

Deffeyes, K.S., Lucia, F.J. & Weyl, P.K. (1965) Dolomitization of Recent and Plio-Pleistocene sediments by marine evaporite waters on Bonaire, Netherlands Antilles. In: *Dolomitization and Limestone Diagenesis – A Symposium* (Ed. Pray, L.C. & Murray, R.C.), Spec. Publs. Soc. Econ. Paleontol. Mineral., Tulsa, **13**, 71–88.

De Deckker, P. & Last, W.M. (1988) Modern dolomite deposition in continental, saline lakes, western Australia. *Geology* **16**, 29–32.

Dwyer, G.S. (1991) *Depositional and Diagenetic Evolution of a Holocene Salina: West Caicos, British West Indies*. Unpublished MS Thesis, Department of Geology, Duke University, 131 pp.

Emrich, K., Ehhalt, D.H. & Vogel, J.C. (1970) Carbon isotope fractionation during precipitation of calcium carbonate. *Earth Planet. Sci. Lett.* **8**, 363–371.

Fritz, P. & Smith, D.G.W. (1970) The isotopic composition of secondary dolomites. *Geochim. Cosmochim. Acta* **34**, 1161–1173.

Goldsmith, J.R. & Graf, D.L. (1958) Structural and compositional variations in some natural dolomites. *J. Geol.* **66**, 678–693.

Hanshaw, B.B., Back, W.E. & Deike, R.G. (1971) A geochemical hypothesis for dolomitization by groundwater. *Econ. Geol.* **66**, 710–724.

Hsu, K.J. & Schneider, J. (1973) Progress report on the dolomitization of Abu Dhabi Sabkhas, Arabian Gulf. In: *The Persian Gulf: Holocene Carbonate Sedimentation and Diagenesis in a Shallow Epicontinental Sea* (Ed. Purser, B.H.) pp. 409–422. Springer-Verlag, New York.

Hsu, K.J. & Siegenthaler, C. (1969) Preliminary experiments on hydrodynamic movement induced by evaporation and their bearing on the dolomite problem. *Sedimentology* **12**, 11–25.

Keith, M.L., Anderson, G.M. & Eichler, R. (1964) Carbon and oxygen isotopic composition of mollusc shells from marine and fresh-water environments. *Geochim. Cosmochim. Acta* **28**, 1757–1786.

Kelts, K. & McKenzie, J.A. (1982) Diagenetic dolomite formation in Quaternary anoxic diatomaceous muds of DSDP Leg 64, Gulf of California. In: *Scientific Party, Initial Reports DSDP, US Government Printing Office, Washington, DC* **64**(2), 553–570.

Kendall, C.G.St.C. & Warren, J.K. (1988) Peritidal evaporites and their sedimentary assemblages. In: *Evaporites and Hydrocarbons* (Ed. Schreiber, B.C.), pp. 66–137. Columbia University Press, 475 pp.

Kocurko, M.J. (1979) Dolomitization by spray-zone brine seepage, San Andres, Columbia. *J. Sedim. Petrol.* **49**, 209–214.

Land, L.S. (1973) Contemporaneous dolomitization of Middle Pleistocene reefs by meteoric water, North Jamaica. *Bull. Marine Sci.* **23**, 64–92.

Land, L.S. (1983) Dolomitization. *Am. Ass. Petrol. Geol. Educ. Course Note Series* **24**, 20 pp.

Land, L.S. (1985) The origin of massive dolomite. *J. Geol. Educ.* **33**, 112–125.

Leaver, J. (1985) *Sedimentology, Mineralogy, and Pore Water Chemistry of Schizohaline Pond Sediments, Turks and Caicos Island, British West Indies*. Unpublished MS Thesis, Department of Geology, Duke University, 75 pp.

Lloyd, M. (1966) Oxygen isotope enrichment of sea water by evaporation. *Geochim. Cosmochim. Acta* **30**, 801–814.

Lloyd, R.M., Perkins, R.D. & Kerr, S.D. (1987) Beach and shoreface ooid deposition on shallow interior banks, Turks and Caicos Islands, British West Indies. *J. Sedim. Petrol.* **57**, 976–982.

Logan, B.W. (1987) The MacLeod Evaporite Basin, Western Australia. *Am. Ass. Petrol. Geol. Memoir, Tulsa, Oklahoma* **44**, 140 pp.

McKenzie, J.A. (1981) Holocene dolomitization of calcium carbonate sediments from the coastal sabkhas of Abu Dhabi, UAE: a stable isotope study. *J. Geol.* **89**, 185–198.

McKenzie, J.A., Hsu, K.J. & Schnieder, J.F. (1980) Movement of subsurface waters under the sabkha, Abu Dhabi, UAE, and its relationship to evaporite dolomite genesis. In: *Concepts and Models of Dolomitization* (Ed. Zenger, D.H., Dunham, J.B. & Ethington, R.L.) Spec. Publs. Soc. Econ. Paleontol. Mineral., Tulsa **28**, 11–30.

Magaritz, M., Goldenburg, L., Kafri, U. & Arad, A. (1980) Dolomite formation in the seawater–freshwater interface. *Nature* **287**, 622–624.

Morrow, D.W. (1982a) Diagenesis 1. Dolomite – Part 1, The chemistry of dolomitization and dolomite precipitation. *Geoscience Canada* **9**, 5–13.

Morrow, D.W. (1982b) Diagenesis 2. Dolomite – Part 2, Dolomitization models and ancient dolostones. *Geoscience Canada* **9**, 95–107.

Muller, S. & Tietz, G. (1971) Dolomite replacing 'cement A' in biocalcarenites from Fuerteventura, Canary Islands, Spain. In: *Carbonate Cements* (Ed. Bricker, D.P.) Johns Hopkins University Press, Baltimore, MD, 376 pp.

O'Neil, J.R., Clayton, R.N. & Mayeda, T.K. (1969) Oxygen isotope exchange between divalent metal carbonates. *J. Chem. Phys.* **51**, 5547–5558.

Patterson, R.J. & Kinsman, D.J.J. (1982) Formation of diagenetic dolomite in coastal sabkhas along the Arabian (Persian) Gulf. *Bull. Am. Ass. Petrol. Geol.* **66**, 28–43.

Pierre, C., Ortlieb, L. & Person, A. (1984) Supratidal evaporitic dolomite at Ojo de Liebre lagoon: mineralogical and isotopic arguments for primary crystallization. *J. Sedim. Petrol.* **54**, 1049–1061.

ROSOFF, D.B. (1990) *Sediment Mineralogy, Pore Water Geochemistry, Groundwater Hydrology, and Hydrocarbon Source Potential of a Holocene Salina, West Caicos, British West Indies.* Unpublished MS Thesis, Department of Geology, Duke University, 171 pp.

SHEPPARD, S.M.F. & SCHWARTZ, H.P. (1970) Fractionation of carbon and oxygen isotopes and magnesium between metamorphic calcite and dolomite. *Contrs. Mineral. Petrol.* **26**, 161–198.

SONNENFELD, P. (1984) *Brines and Evaporites.* Academic Press, New York, 613 pp.

TARUTANI, T., CLAYTON, R.N. & MAYEDA, K. (1969) The effect of polymorphism and magnesium substitution on oxygen isotope fractionation between calcium carbonate and water. *Geochim. Cosmochim. Acta* **33**, 987–996.

VON DER BORCH, C.C. & JONES, J.B. (1976) Spherular modern dolomite from the Coorong area, South Australia. *Sedimentology* **23**, 587–591.

WANLESS, H.R. & DRAVIS, J.J. (1989) *Carbonate Environments and Sequences of Caicos Platform, Field Trip Guidebook T374.* American Geophysical Union, Washington, DC, 75 pp.

WARREN, J.K. (1989) *Evaporite Sedimentology.* Prentice-Hall Inc., Englewood Cliffs, New Jersey, 285 pp.

Spec. Publs Int. Ass. Sediment. (1994) **21**, 55–74

Mechanisms of complete dolomitization in a carbonate shelf: comparison between the Norian Dolomia Principale (Italy) and the Holocene of Abu Dhabi Sabkha

S. FRISIA

Università degli Studi di Milano, Dipartimento di Scienze della Terra, via Mangiagalli 34, I-20133 Milano, Italy

ABSTRACT

The diagenetic history of the Late Triassic Dolomia Principale tidal-flat complex is reconstructed utilizing transmission and analytical electron microscopic techniques and comparison with modern Abu Dhabi dolomite analogues. The Dolomia Principale, which completely dolomitized peritidal and subtidal cycles, shows five dolomite texture types with different stable isotope values. The early products of shallow subsurface dolomitization are preserved, and are calcian and characterized by fine, pervasive modulated microstructures. Texturally, microstructurally and geochemically similar dolomites are observed in subtidal facies of Abu Dhabi, generally replacing aragonites and Mg-calcites. The subsequent dolomitization history of the Dolomia Principale continues with the precipitation of progressively more ideal dolomites, with coarse modulated microstructures or ribbon structures. Ideal dolomites, generally characterizing void-filling crystals or the outer part of matrix-replacive as well as mimetic dolomites, show only a few dislocations. Dislocation networks characterize the highest-temperature dolomites. A considerable spread of $\delta^{18}O$ values (+3.3‰ to below −4‰) characterizes 'matrix' dolomite crystals with different chemical compositions and microstructures coexisting in the same crystal. This spread is considered to be the result of an admixture of signals coming from calcian dolomites and ideal ones precipitated later in the diagenetic history. Calcian dolomites, slightly Ca dolomite, some ideal dolomite 'overgrowths' and void fillings with positive $\delta^{18}O$ still form during the same cycle. $\delta^{18}O$-depleted ideal dolomites formed during burial. The dolomitizing fluids were provided by the cyclic subaerial exposures to which the Dolomia Principale was subjected, allowing saline fluid circulation through pores and fractures.

INTRODUCTION

The formation of massive dolomite in carbonate platforms is a result of several different mechanisms, ranging from single events in the case of hydrothermal circulation of hot seawater (Wilson *et al.*, 1990) or thermal convection of formation fluids, or expulsion of basinal fluids (cf. Machel & Anderson, 1989), to a series of dolomitizing steps that occurred from early, surface to burial diagenetic stages (cf. Land, 1985; Ruppel & Cander, 1988; Grotzinger, 1989). In this last case, the textural and geochemical properties of the earlier precipitates may have been totally or partially obliterated by the last stabilization event (Land, 1985). Up until now, there has been no method of discriminating between all the diagenetic imprints, especially in formations that have complex diagenetic histories. Dolomite textures are not unequivocal: fine-crystalline mimetically replacive dolomite, often associated with algal laminites and considered to be early diagenetic in origin, can be the product of stabilization of earlier, more unstable phases (Land, 1985; Frisia, 1991); coarse-grained fabric-destructive dolomite seems to be more frequently related to burial dolomitization, although the depth of burial may still be shallow, and stabilization a relatively early phenomenon (cf. Sass & Katz, 1982; Land, 1985). Geochemical and structural properties have often been utilized in tracing the dolomitization patterns of carbonate platforms, both ancient and modern. In particular, trace elements, stable isotopes, stoichiometry and cation ordering are used to discriminate 'early' from 'old' dolomites, on the basis of our present knowledge of Holocene dolomitization. In fact, most surface-formed dolomites studied are calcian and characterized by cation disorder (Reeder, 1981, 1983); they are therefore unstable and have a recrystallization potential that increases with increasing burial temperature (Hardie, 1987). Recently, Gregg *et al.* (1992) suggested that dolomite recrystallization occurs in the first few centimetres below the sediment–water interface by surface energy-driven dissolution–reprecipitation, as inferred from the textures observed. The dolomite they studied is calcian and characterized by progressive cation ordering within a depth of 15 cm (unfortunately, microstructures were not observed). However, in arid and strongly evaporative settings dolomite can be ideal (or Mg-rich) and characterized by closely spaced lattice defects: these properties have been considered by Rosen *et al.* (1989) as being indicative of rapid growth from solutions with a high Mg/Ca ratio. In their opinion, based on transmission electron microscopic (TEM) observations and geochemistry, the formation of ideal dislocation-rich dolomite and that of calcian dolomite with modulated microstructures is 'the result of the chemistry of precipitating solutions', and is 'not related to progressive stabilization by dissolution–reprecipitation reactions with increasing age'. Rosen *et al.* (1989) relate cation ordering to relatively slow growth rate. The implication of such a hypothesis is that dolomite composition (and microstructures?) may be related to the chemistry of parent fluids, i.e. original (Sass & Katz, 1982; Bein & Land, 1983; McKenzie, 1985; Sass & Bein, 1988). This is a challenge to the theory of progressive stabilization of dolomites by dissolution–reprecipitation during burial, with subsequent modification of the original texture, isotope, major and trace element signals as dolomite becomes more and more ordered and ideal with burial (Land, 1985; Hardie, 1987). An example against the use of trace elements as evidence of 'purification' by several dissolution–reprecipitation steps (Land, 1985) is given by Vahrenkamp and Swart (1990), who observed a covariance between Sr and the stoichiometry of dolomite without apparent correlation to age, depth, lithofacies and other sedimentological features. Unequivocal answers to the question whether ancient dolomite properties are original or due to several steps of wet recrystallization have not been obtained, even using TEM techniques. Early workers (Wenk *et al.*, 1983) suggested that the modulated microstructure was associated with replacement on the basis of comparison between modern, disordered dolomites with dislocations and ancient, better-ordered calcian dolomites with modulated microstructures. It is now known that it occurs also in modern calcian dolomite, fringing-reef cements (Miser *et al.*, 1987). Furthermore, modulated microstructure occurs in void-filling (although the voids formed after dissolution of allochems and the microstructures are also present in the matrix) Eocene ideal dolomites (Miser *et al.*, 1987), i.e. it is not exclusive to calcian dolomites. It appears that different settings (arid vs. humid; restricted vs. open platforms; shallow subsurface vs. deep burial) and mechanisms (replacement vs. free growth) may give the same microstructures.

In order to understand how, and whether, microstructures and geochemical properties can be utilized in the study of the diagenetic history of carbonate platforms, the products of early surface dolomitization in a modern setting are compared with those of an ancient completely dolomitized carbonate shelf, comparing field and textural observations with geochemical and microstructural analyses. The Norian Dolomia Principale shows a wide variety of dolomite textures and has the potential to yield early to late diagenetic dolomites. In fact, geochemical studies carried out on sediments of the same age and similar facies of the Northern Calcareous Alps and Apennines (Veizer, 1983; Lo Cicero, 1987) show a spread in $\delta^{18}O$ values from positive to negative (below −4‰ PDB), which were explained as the consequence of facies differences and progressive diagenetic stabilization. Previous TEM observations of Dolomia Principale samples by Reeder (1981) and Wenk *et al.* (1983), revealed pervasive modulated microstructures in ordered calcian dolomites, which they interpreted as replacement of previous unstable carbonates. However, in those early TEM studies, facies, stratigraphic level, textural characteristics and isotope data were not considered. Here, a discrimination between different diagenetic phases, with possible recognition of the early products, is attempted by analysing microstructures and the geochemistry of texturally different dolomites, the sedimentological characteristics and isotopic signals of which are known, by means of both transmission and analytical electron microscopy. Abu Dhabi sabkha dolomites are taken as a reference to

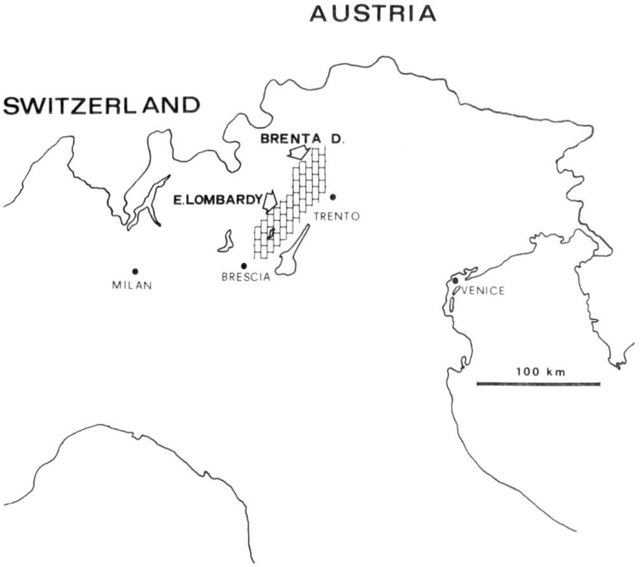

(a)

(b)

Fig. 1. (a) Map showing the location of the areas studied. (b) Brenta Dolomites peritidal and subtidal cycles from Bocca di Brenta.

test the degree of preservation (if any) of surface and shallow subsurface dolomite.

THE NORIAN DOLOMIA PRINCIPALE

The Norian Dolomia Principale consists of pervasively dolomitized sediments deposited on a great tidal flat which characterized the western end of the Pangean Gulf, extending from Spain to Hungary and Greece. The area considered in this study is much smaller (Fig. 1), but representative of the Norian palaeogeography of southern Europe: a carbonate shelf dissected into platforms and intraplatform basins by a series of synsedimentary faults related to the Norian rifting (Jadoul *et al.*, 1992).

The Brenta Dolomites pertain to the structural high domain of the Trento Plateau. Here, the Dolomia Principale is about 1000 m thick, and consists of inner platform facies arranged in shallowing-upwards peritidal and subtidal cycles, commonly grouped into thinning-upwards megacycles capped by flat-pebble breccia, red brecciated horizons and pisoid rudstones (the diagenetic cap of Bosellini & Hardie, 1985). Peritidal cycles consist of (from bottom to top): (a) bioclastic–intraclastic grainstones and fine-grained breccias; (b) massive vuggy dolomites, mostly represented by bioturbated bioclastic packstones and wackestones with gastropod and bivalve moulds; (c) stromatolitic bindstones, with sheet cracks and shrinkage pores, overlying mudstone–wackestone with fenestral and dissolution cavities; and (d) flat-pebble breccias associated with tepee structures and pisoid rudstones commonly developed on the flanks of tepees. These lithofacies are most characteristic of the lower 300 m of the succession, and also intercalate within subtidal cycles. The latter consist of massive vuggy intraclastic bioclastic packstones and wackestones showing bioturbation, with gastropod and bivalve moulds. The top of some cycles is characterized by up to 1 m thick flat-pebble breccias, pisoids, red breccia horizons and subordinate tepee structures, which have been interpreted as diagenetic caps (Bosellini & Hardie, 1985). Red to green clayey layers at the top of some cycles may show erosive lower contacts. Each single subfacies of peritidal and subtidal lithofacies was sampled for the lower, middle and upper stratigraphic levels of the Dolomia Principale.

In eastern Lombardy the inner platform is dissected by synsedimentary faults which determined the development of margins, slopes and intraplatform basins. The latter commonly consist of limestone and dolomitized limestones with interspersed rare chert nodules, intercalated marls and subordinate shales. Dolomitization of the limestones is most pervasive in the proximity of platform margins and below prograding platform complexes (Frisia, 1991). The Dolomia Principale in this area is also arranged into peritidal and subtidal cycles, but here the maximum thickness is about 1800 m. Peritidal cycles show the same sedimentological characteristics as those of the Brenta Dolomites, and are mostly developed in the lower 300 m of the succession. However, the diagenetic caps are lacking. A few embryonic tepees have been observed, as well as sheet cracks, shrinkage pores and dissolution cavities, which may record periods of subaerial exposure. The higher subsidence rate of this area, registered in the considerable thicknesses of both the Dolomia Principale and intraplatform basin facies (Aralalta Group; Jadoul, 1986) had effects on the depositional characteristics of the subtidal cycles. These consist of massive vuggy bioturbated bioclastic packstones, with gastropod and bivalve moulds. The top of these cycles shows a shallowing-upwards trend, as can be observed in the development of fenestral mudstones and stromatolitic bindstones representing a small portion of the cycle, but diagenetic caps are always absent; it is therefore possible that this part of the Dolomia Principale did not undergo prolonged subaerial exposures compared to those documented in the Brenta Dolomites and in the Venetian Alps (Bosellini & Hardie, 1985; Hardie *et al.*, 1986). Field evidence shows only some dissolution cavities filled by fibrous cements or prism cracks, although subaerial exposures are inferred from fossil reptiles, particularly small forms (*Megalancosaurus*; Wild, 1991) adapted to arboreal life, which were recovered from intraplatform basin facies (Renesto, 1993).

The slope facies are of particular interest, being represented by breccias and megabreccias with inner platform and margin-derived clasts embedded in dolomitic marls, marly limestones and shales. The Dolomia Principale clasts embedded in these basin facies of the same Norian age are dolomitized and show the same textural characteristics of the sediment from which they derived. Peritidal, subtidal, margin and slope facies were sampled throughout the succession in order to make textural, geochemical and microstructural comparisons with the Dolomia Principale of the Brenta Dolomites and detect similar or different diagenetic trends in the same formation for the diverse palaeogeographic settings.

ANALYTICAL METHODS

Textural analysis

Five different textures of dolomite were recognized in the Dolomia Principale subtidal and peritidal facies.

1 Type 1 (Figs 2 and 3). Unimodal planar-e (crystal size up to 4 μm), mimetically replacing stromatolites, peloids, forams and other bioclasts, thalli and branches of *Dasycladacean* algae, and the darker laminae in problematica such as *Spongiostromata*. Its distribution relative to the facies is shown in Figure 4.

2 Type 2 (Fig. 2). Unimodal planar-s dolomite (crystal size of about 20 μm), with uniform extinction, forming the light-grey thicker laminae separating the dark organic laminae composed of type 1. The distinction has been made on the basis of coarser size with respect to type 1, poorer preservation of the original textures and the more irregular crystal faces as seen with the scanning electron microscope. Furthermore, in some cases, type 2 may represent a void-filling dolomite. In fact, it has been observed that layers of fibrous cements may coat stromatolites, evidence for *in situ* precipitation of accretionary layers (Grotzinger, 1989). In this case type 2 may have replaced an original fibrous carbonate cement.

3 Type 3 (Fig. 5). Replaces the sediment matrix, especially in subtidal facies. It consists of polymodal dolomite, with crystal sizes ranging from 50 to over 200 μm, planar-s to non-planar, commonly showing

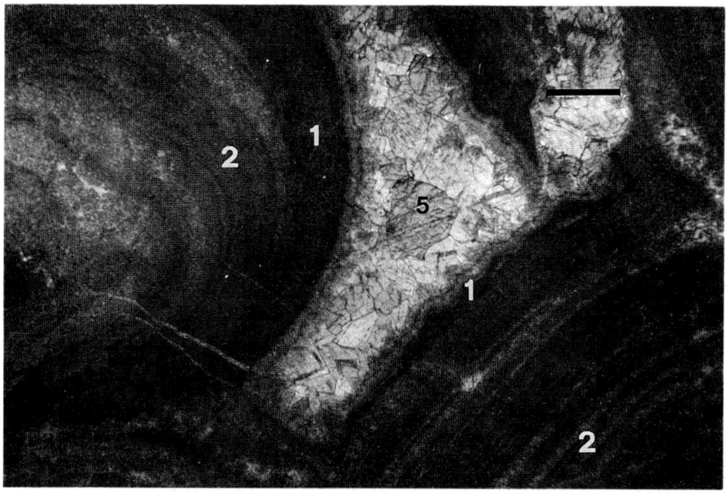

Fig. 2. Completely dolomitized oncoidal rudstone from the intertidal facies of the Dolomia Principale of the Brenta Dolomites. Microbial laminae (dark and irregular) are replaced by type 1 dolomite (1), whereas the more regular, light-grey cortexes showing incipient radial texture are composed of type 2 dolomite (2). Voids between oncoids are here filled by type 5 (5), showing growth bands. The fracture on the lower left of the photo is filled by dolomite and does not cross-cut type 5. Scale bar = 1 mm.

Fig. 3. Scanning electron micrograph of type 1 dolomite which mimetically replaces an allochem surrounded by coarser-grained non-planar type 3 (matrix) dolomite (scale bar = 40 μm). The insert shows the unimodal planar-e texture of type 1 (scale bar = 4 μm).

uniform extinction and subordinate undulatory extinction. This dolomite type may also replace some allochems, as well as types 1 and 2.

4 Type 4 (Fig. 6). Void-filling dolomite precipitated within fenestral dissolution cavities, bivalve and gastropod moulds. Unimodal within the same void (crystal sizes in the range 100–300 μm), planar-e to planar-s, with uniform extinction. It shows a tendency to crystal size coarsening in larger voids (>1 cm).

5 Type 5 (Figs 2 and 7). The least common textural type, mostly present as void filling in the supratidal facies. Rare in subtidal and marginal facies, where it may be associated with bitumen and fractures related to synsedimentary faulting. Type 5 is polymodal (100–400 μm), planar-s to non-planar, with growth lines and showing undulose extinction. It is the only dolomite type showing growth bands and two red-luminescent zones in cathodoluminescence (CL). Type 5 may replace types 1, 2 and, subordinately, 3, at the margin of the void where it has grown.

Types 1, 2 and 4 are cross-cut by millimetre-thick fractures filled with dolomite spar. These fractures

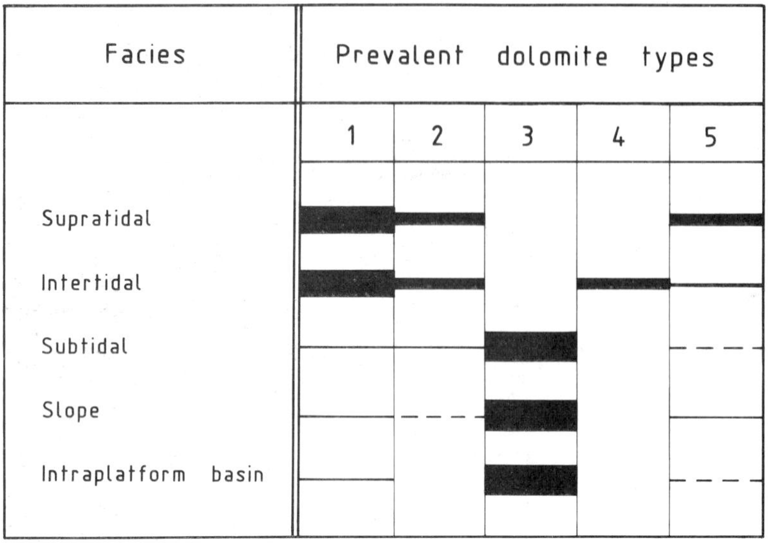

Fig. 4. The distribution of the dolomite textural types recognized in this study relative to the different facies.

also cross-cut the fine crystals of type 3, but stop at the margin of the coarser, non-planar crystals of the matrix dolomite and do not cross-cut type 5. All the recognized dolomite types are cross-cut by stylolites.

Stable isotope analysis

Stable isotope analyses were carried out on samples from supratidal, intertidal and subtidal (including margin) facies at different stratigraphic levels. As can be seen in Figure 8, there is no clear difference in the distribution of $\delta^{18}O$ and $\delta^{13}C$ according to diverse facies, apart from a greater concentration of the intertidal samples in the positive $\delta^{18}O$ field. The most positive $\delta^{18}O$ values were detected in a few subtidal samples; the next more positive dolomites are in stromatolitic bindstones of inter- /supratidal facies from both studied areas. Figure 8 shows the trend towards $\delta^{18}O$ depletion, a characteristic of other Upper Triassic dolomitized carbonate platforms (Veizer, 1983; Lo Cicero, 1987): $\delta^{18}O$ values below −4‰ are not common and pertain to samples showing dissolution phenomena, with reprecipitation of void-filling dolomite commonly associated with bitumen. This trend simply indicates that there occurred a range of diagenetic modifications, first at the surface and in the shallow subsurface (positive $\delta^{18}O$ and $\delta^{13}C$), and later in burial settings ($\delta^{18}O$ below −4‰; $\delta^{13}C$ positive). Even when related to textural characteristics (Fig. 9), stable isotopes show a spread of values, especially wide for type 3 dolomite, which again may be indicative of several diagenetic phases affecting the same textural type. Type 5 is the only one which has a restricted field of values: the $\delta^{18}O$ signal below −4‰ should be considered as due to stabilization of a precursor carbonate during burial at temperatures above 60°C (Land, 1985), an inference that can be supported by the non-planar texture and the undulose extinction of type 5 (Sibley & Gregg, 1987). The $\delta^{18}O$ values of types 1 and 2 are mostly grouped in the positive field but, again, negative signals were recorded, although evidence for strong recrystallization (such as fabric-destructive dissolution and reprecipitation) was not observed with the optical microscope or cathodoluminescence.

The occurrence of $\delta^{18}O$ values more positive than +2.5‰, a characteristic of dolomites forming at the present time in sabkhas, tidal flats and areas with active circulation of marine waters (Land, 1985), suggests that some products of early surface diagenesis in environments dominated by the evaporation of seawater may still be preserved in the Dolomia Principale. Therefore, it was felt necessary to make a comparison with Recent dolomites having isotopic signatures similar to the heaviest ones

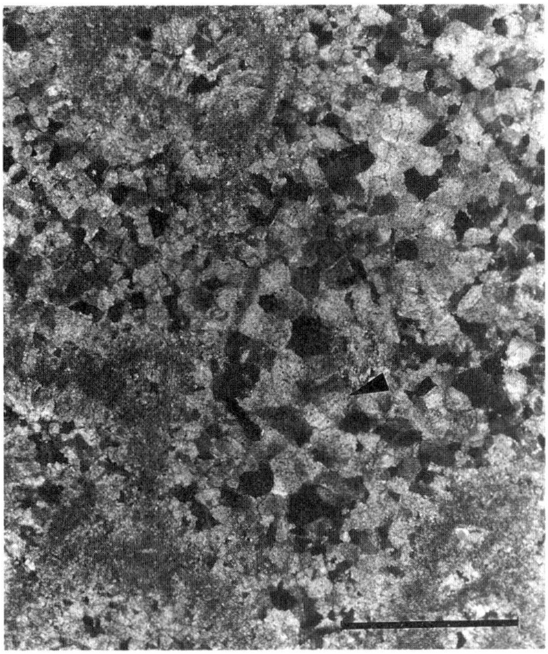

Fig. 5. Packstone with *Dasycladacean* algae from the subtidal facies of the Dolomia Principale of the Brenta Dolomites (crossed polars) composed mostly of type 3 dolomite, which commonly shows undulose extinction (arrow). Scale bar = 1 mm.

Fig. 6. Scanning electron micrograph of the void-filling type 4 dolomite filling a small gastropod mould in the shallow subtidal facies of the Dolomia Principale in Eastern Lombardy (Corna Blacca, Brescia). Scale bar = 100 μm.

recorded in the Dolomia Principale and precipitated in analogous geographic and climatic settings.

Transmission and analytical electron microscopy: experimental procedures

The comparisons discussed above involved utilizing transmission (TEM) and analytical electron microscopy (AEM) techniques which permit the observation of crystal microstructures and ordering reflections, as well as the detection of major and trace elements. The aim of such comparisons was to determine similarities or differences between the recent dolomites and those characterized by the most positive $\delta^{18}O$ in the Dolomia Principale. That is, to understand whether the products of early dolomitization survive burial re-equilibration. Furthermore, TEM has been utilized to trace the subsequent diagenetic modifications in the Triassic samples, responsible for the observed spread of $\delta^{18}O$ values.

Fig. 7. Scanning electron micrograph of the void-filling type 5 dolomite showing growth lines (arrow), from the same sample shown in Figure 2 (Brenta Dolomites). Scale bar = 100 μm.

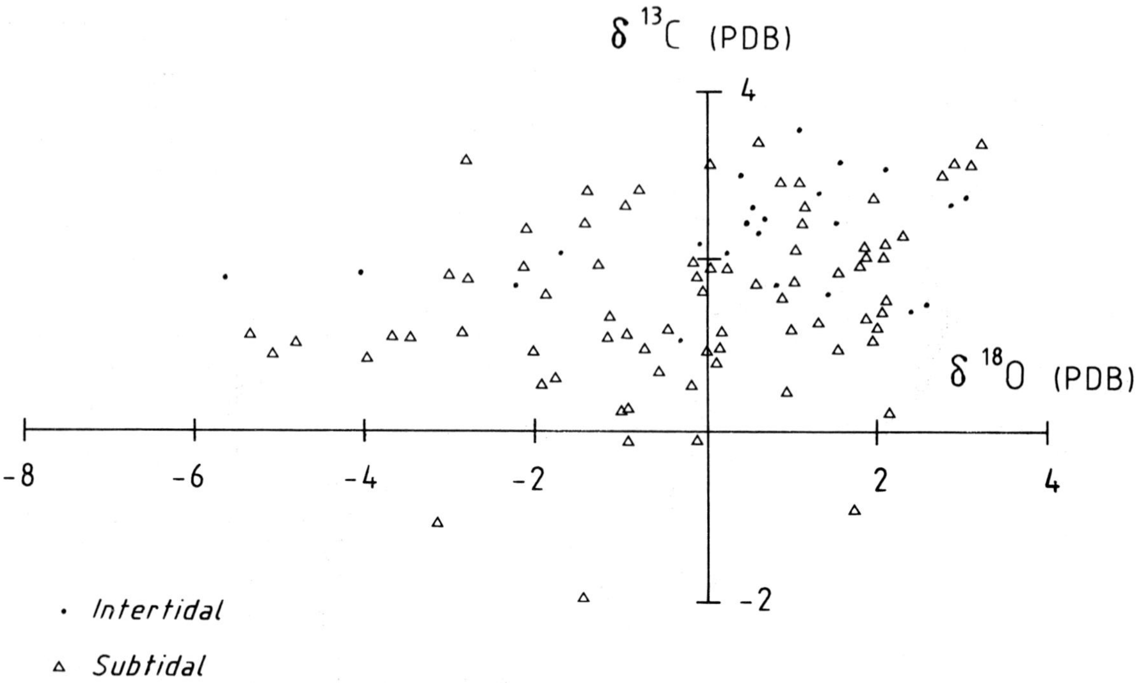

Fig. 8. Stable isotope values distribution of the Dolomia Principale intertidal/supratidal and subtidal samples, showing that intertidal facies values are more concentrated in the positive $\delta^{18}O$ field. However, the isotopic trend only indicates that the dolomites have been subjected to increasing temperature with burial.

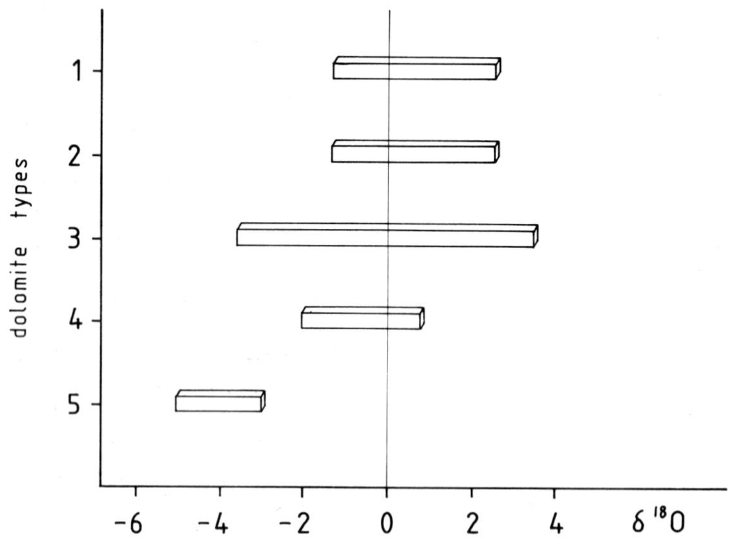

Fig. 9. $\delta^{18}O$ data distribution relative to dolomite types. Note the spread of values, especially in type 3, which may record several steps of diagenetic modification. Type 5 is the only one showing a restricted distribution in the negative field (see text for details).

The modern dolomites come from intertidal and subtidal facies of the Abu Dhabi sabkha, and have been kindly provided by Professor J. McKenzie and Dr D. Müller of the ETH in Zürich.

The Abu Dhabi samples were impregnated with Petropoxy, glued to glass with crystalbond and ground to a thickness of 30 μm. After the TEM mount had been glued on to the area of interest, the thin section was detached from the glass with acetone. A JEOL JEM 200 CX STEM equipped with a LaB6 filament, high-angle X-ray detector and KEVEX EDX analytical system was used. Analyses were carried out on areas with a surface of 0.25 μm^2, by means of a grid where the beam impinges on each spot for 2 s, minimizing radiation damage. For very heterogeneous crystals with respect to microstructures, the spot mode was utilized, with the beam operating for 10 s on each 20 nm wide spot. This method, however, gives mostly qualitative information. The consistency of Ca, Mg, Sr and Fe contents as detected by AEM was subsequently checked by means of an ARL microprobe with beam size of 2 μm. The AEM and microprobe analyses are consistent. Na and Mn were also analysed with the microprobe.

WHY THE COMPARISON WITH THE ABU DHABI SABKHA?

It is impossible to find the exact equivalent depositional and diagenetic environments of the Dolomia Principale because the break-up of the Pangea must have had serious consequences on oceanic circulation and climate (Kutzbach, 1989). Deposition and early dolomitization of the Dolomia Principale tidal flats occurred at the margins of the Pangean supercontinent at the end of one major supercontinent megacycle (Worsley *et al.*, 1984) and at the onset of its incipient fragmentation (Veevers, 1989). The possible implications on depositional and diagenetic environments of the reduced activity of oceanic ridges (which are the present-day major Mg-sink) and conspicuous extraction of Ca by the widespread evaporitic deposits around the Pangean Gulf have yet to be defined. Therefore, one may question the comparison between the Dolomia Principale and the Abu Dhabi sabkha, rather than other sites of modern dolomitization, such as the Bahamas–south Florida–Antilles humid zones, where supratidal flats dolomite crusts are well known (Hardie 1977; Gebelein *et al.*, 1980; Carballo & Land, 1984). These were rejected on the basis of geographic settings and climate. The Bahamas is an isolated platform, developed on a passive margin surrounded by very deep waters, which was not the case for the Dolomia Principale. Furthermore, the Bahamas platform was affected by extensive karst phenomena during major sea-level falls of the Pleistocene sea and, at the present day, meteoric water circulation is documented by strong dissolution of carbonate minerals. In the Dolomia Principale there is evidence of meteoric diagenesis as documented by dissolution cavities, but cave networks comparable to the Blue Holes, solution pipes and fractures typical of warm, wet tropical climates have not been observed. The red horizons that periodically cap subtidal and peritidal cycles of the Dolomia Principale consist of dolomite, detrital quartz and rare mica (possibly wind-blown) and cavernous karst is absent in the underlying sediments. Therefore, the climatic conditions that predominated during the deposition of the Dolomia Principale were more arid. As a consequence, the Dolomia Principale tidal-flat hydrology and diagenetic evolution were controlled by climate and porosity different from those of the Bahamas.

The same geographic and climatic considerations hold true for the Pacific atolls such as Mururoa (Aissaoui, 1988), Enewetak (Saller, 1984), Mataiva and Makatea (Bourrouilh-Le Jan, 1992), strongly affected by tropical karstification, with the development of pinnacles, caves and bauxitic soils.

The dolomitized crusts of Ambergris Cay, Belize (Mazzullo *et al.*, 1987) represent a significant portion of the Holocene supratidal flat sediments, and dolomitization is apparently more widespread than in most other areas of the world. However, the region is located entirely in the humid tropical zone. Desiccation cracks, stromatolites and fenestrae, common in the Dolomia Principale, are not present in Ambergris Cay. Furthermore, the $\delta^{13}C$ values of the dolomites from Belize reflect the influence of fresh water, which has not been observed in the Dolomia Principale. The arid Arabian Gulf shelves do not face an ocean with open circulation and the area is subjected to incipient rifting of the Red Sea and final collision to the northeast (Zagros); therefore, the region may be considered a 'miniature Pangean Gulf'. Dolomitization of both intertidal and subtidal facies is an important phenomenon which can be traced over a relatively wide area along the Arabian coast, north to Kuwait (Purser, 1973; McKenzie, 1981; Gunatilaka *et al.*, 1984), with low average annual rainfalls (30–40 mm/yr in the Trucial Coast, but 100–120 mm/yr in Al-Khiran lagoon, Kuwait, ranging up to 300 mm/yr). Although gypsum or anhydrite are lacking as intercalated beds in the Dolomia Principale, evaporite deposits were extensive in the inner shelf areas of the Pangean Gulf in the Late Triassic (Megard-Galli & Baud, 1977; Ziegler, 1982, 1988). Furthermore, aridity seems to characterize the Upper Triassic, as documented by sedimentology (Tucker & Benton, 1982) and inferred from global circulation models for the Pangea (Kutzbach, 1989). On the basis of the above considerations, the Abu Dhabi sabkha dolomites were chosen as a reference to understand the early mechanisms of dolomitization in the Dolomia Principale, a starting point to unravel the diagenetic history of the Norian carbonate shelf.

RESULTS

Dolomia Principale

Type 1 and Type 2 dolomites

Regardless of the area of provenance, facies and stratigraphic level, these two dolomite types are always characterized by the following (Figs 10a and 10b):

1 fine and pervasive modulated microstructures in calcian dolomites (53–56 mole % Ca);

2 coarser, non-pervasive modulated microstructures in areas with slight excess in calcium (about 52 mole % Ca);

3 areas without modulated microstructures with ideal composition (50 mole % Ca). These latter may show cross-cutting relationships with modulated calcian dolomites. In this case microporosity can be observed at the boundary between the two phases.

The microstructures and different calcium contents coexist within the same dolomite crystal, as shown in Fig. 10b. Ideal dolomite seems to be more common towards the margins of crystals.

Type 3

The matrix dolomite shows a variety of microstructures which, again, are independent of facies, stratigraphic level and area of provenance:

1 fine and pervasive modulated microstructures developed in calcian dolomite (53–56 mole % Ca);

2 coarser, locally pervasive modulated microstructures in slightly calcian dolomites (average 52 mole % Ca), most common in the subtidal part of

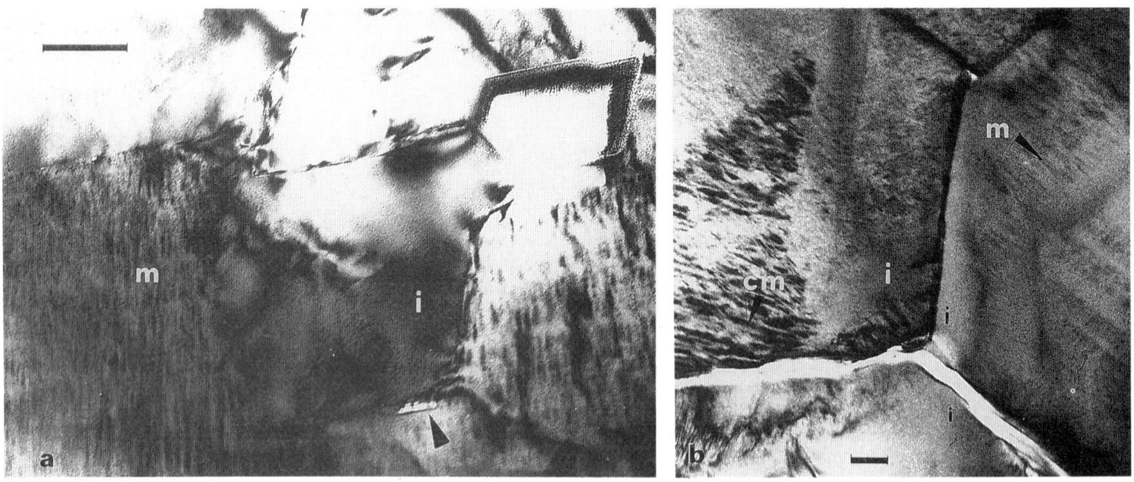

Fig. 10. Transmission electron micrographs (a) of types 1 and 2 dolomites from intertidal facies of Brenta Dolomites, showing pervasive modulated microstructure (m) in calcian dolomite (53–56 mole % Ca). This more unstable phase is cross-cut, with development of microporosity (arrow) by a more stable ideal (i) dolomite showing only a few dislocations. Scale bar = 250 nm. (b) Type 2 dolomite from the oncoidal cortexes of Figure 2. Note the presence of calcian dolomite with fine modulated microstructure (m), slightly calcian dolomite with coarser modulations (cm) and ideal dolomite without modulations (i). Scale bar = 250 nm.

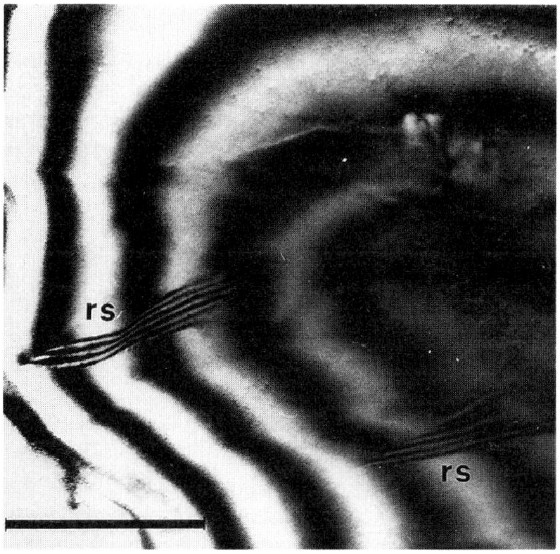

Fig. 11. Ribbon structures (rs), probably due to compositional variations, in type 3 dolomites from the subtidal facies of Eastern Lombardy (Alpo Bondone, Brescia). Scale bar = 250 nm.

the cycles and ribbon structures (Fig. 11). The latter may be related to compositional variations (Barber & Wenk, 1984) developed between calcian dolomite areas and ideal ones. Ribbon structures characterize coarse grains (100 μm or more) of subtidal cycles and slope facies;

3 dislocations and defect-free dolomites in areas with ideal composition, more common than in types 1 and 2.

Type 4

Dolomites filling fenestrae and dissolution cavities have ideal composition and show only a few dislocations (Fig. 12).

Type 5

This dolomite, the most depleted with respect to $\delta^{18}O$, has an almost ideal composition (*ca.* 51–50 mole %) and is characterized by a few dislocations and regular defects parallel to the basal plane (Fig. 13), corresponding to streaking along c* in the diffraction pattern. This phenomenon has been observed by Wenk and Zenger (1983) in burial dolomites formed at temperatures over 60°C.

In most Dolomia Principale dolomites the Sr content is in the range 100–200 ppm, but it is below 80 ppm in type 5. Na is in the range 80–200 ppm in types 1, 2, 3 and 4, whereas it is below microprobe detection limit in type 5.

Abu Dhabi sabkha dolomites

Dolomite rhombs of upper and lower intertidal facies, with crystal sizes ranging from 0.1 to 2 μm, generally develop in small pores (3–4 μm in diameter) among a mesh of micrometre-sized aragonite needles. These surficial dolomites are dislocation-ridden (Fig. 14). The high density of crystal defects has been interpreted by Reeder (1981) and Blake *et al.* (1982) as being due to rapid growth (1000–10 000 years). The composition is mostly ideal; however, there can be a slight excess in Mg. Spot-size analyses

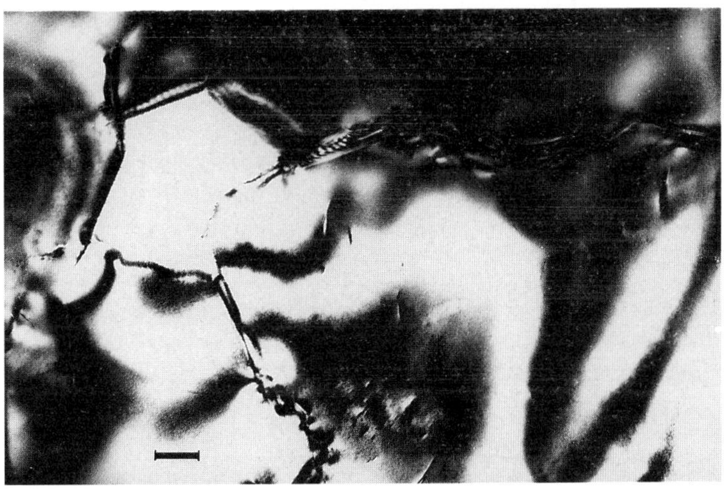

Fig. 12. Transmission electron micrograph of the void-filling type 4 dolomite from fenestral pore (Brenta Dolomites). It has ideal composition and shows only a few dislocations. Scale bar = 250 nm.

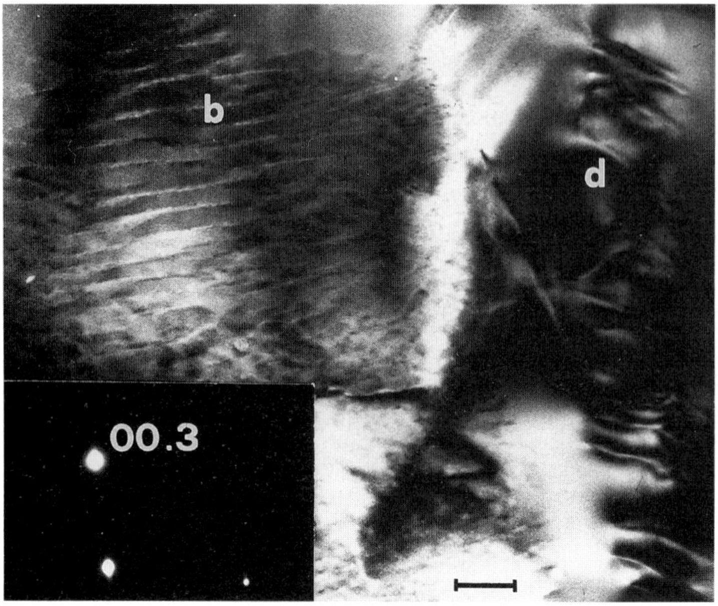

Fig. 13. Transmission electrom micrograph of the void-filling type 5 dolomite, with almost ideal composition, a few dislocations (d) and basal defects (b), as documented by the inserted SAD, showing streaking along 00.3 (supratidal facies, Brenta Dolomites). Scale bar = 250 nm.

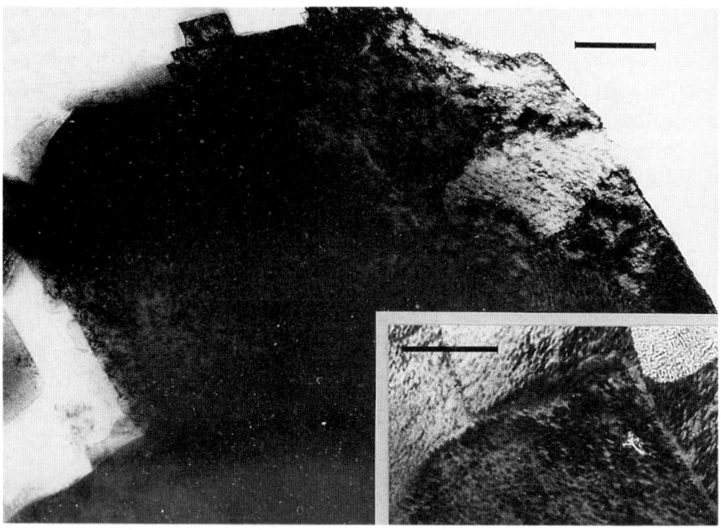

Fig. 14. Transmission electron micrograph of Abu Dhabi dislocation-ridden intertidal dolomite. Scale bar = 100 nm. Inserted for comparison is an 'ancient analogue', a dislocation-ridden ideal dolomite from the Permian Capitan Reef interpreted as being formed from hypersaline fluids. These dolomites have not been observed in the Norian Dolomia Principale. Scale bar = 250 nm.

revealed that some 20 nm areas have 90 mole % Mg, possibly due to magnesite inclusions (McKenzie, 1981), which also precipitate with dolomite in laboratory experiments using hypersaline waters (Morse & Mackenzie, 1990). It is possible that excess Mg is the result of 'averaging in' small Mg-rich crystals (or domains).

In subtidal samples, dolomite rhombs may attain 4 μm in size and show indented irregular boundaries with aragonite needles. Microporosity may characterize the boundaries between the two phases. Dolomite rhombs show aragonite relics about 0.1 μm long (Fig. 15a), a possible evidence of replacement by dissolution–reprecipitation. In some samples,

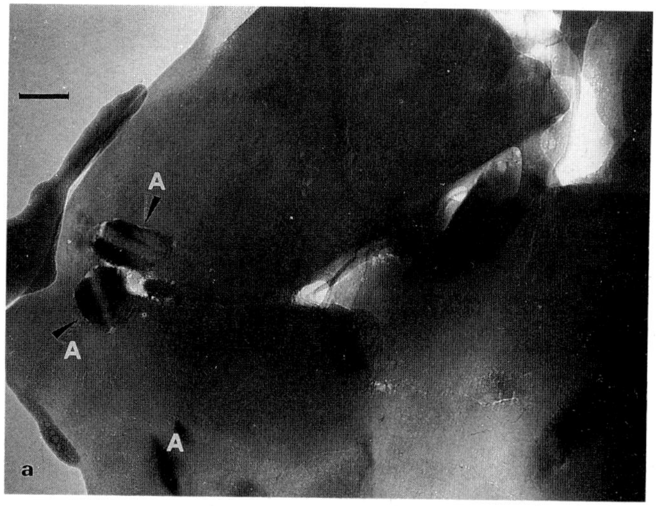

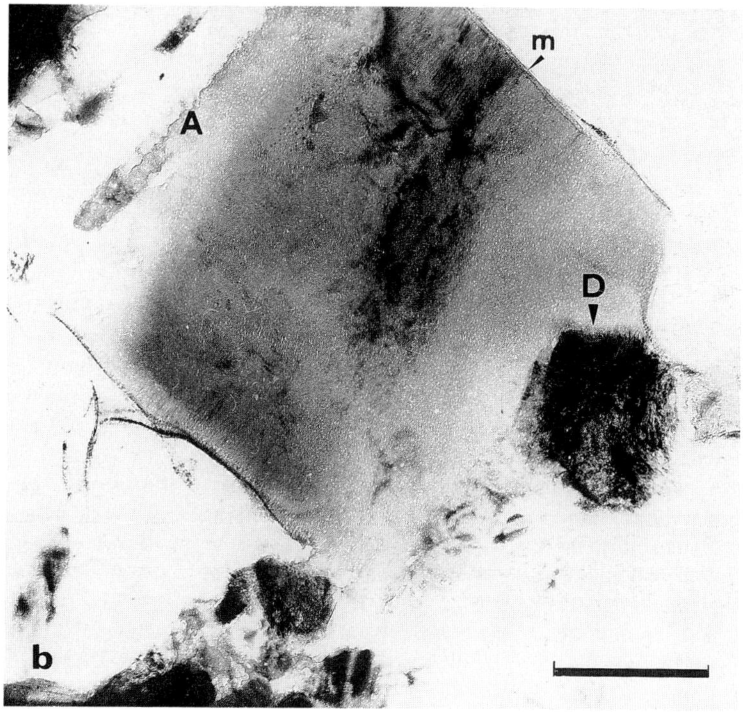

Fig. 15. Transmission electron micrographs (a) of Abu Dhabi subtidal dolomite showing aragonite microinclusions (A) The fuzziness of the picture is due to the instability of the sample under the beam. Scale bar = 100 nm. (b) Documenting the development of modulated microstructures (m) in a calcian dolomite crystal from the Abu Dhabi subtidal facies. (A) indicates an aragonite needle with an irregular surface, possibly due to dissolution; (D) indicates a dislocation-ridden ideal dolomite on which the calcian dolomite may have nucleated. Scale bar = 250 nm.

fine modulated microstructure (similar to that observed in types 1, 2 and 3 of the Dolomia Principale) has been observed in dolomite rhombs with excess calcium (54 mole % Ca) adjacent to aragonite crystals (Fig. 15b). There are smaller rhombs formed in pores and lacking any clear evidence of dissolution–reprecipitation phenomena, and these show the same high crystal defect density as the intertidal specimens, and ideal composition. In the sabkha subtidal facies, dolomite is also present as 20–100 Å

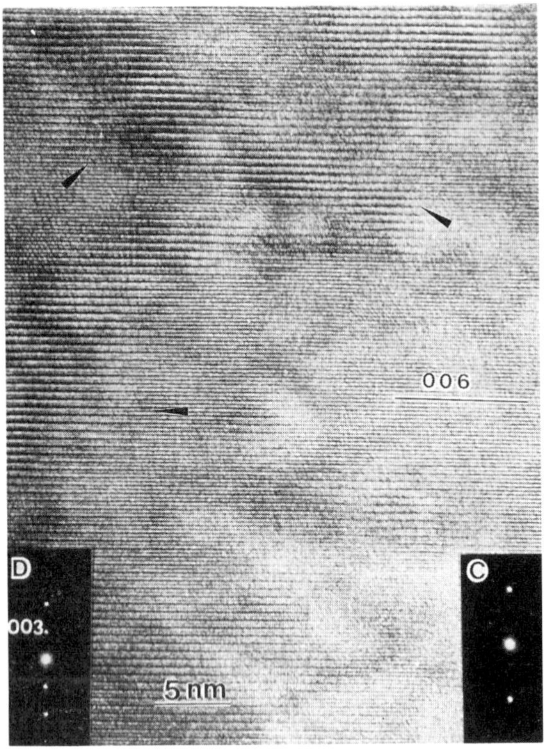

Fig. 16. AREM micrograph of Abu Dhabi magnesian calcite with coherent dolomite domains (arrows) as documented by the presence of the ordering reflections (003) in the diffraction pattern. D = dolomite diffraction pattern; C = calcite diffraction pattern. Scale bar = 5 nm (courtesy of H.R. Wenk).

domains within disordered rhombohedral magnesian calcites, as documented by optical diffractograms obtained in the image negatives with laser diffraction. The crystal structure of these ordered domains is coherent with that of calcite (Fig. 16). These dolomite domains observed in bioclasts consisting of Mg-calcite may represent 'primary dolomite'.

The Sr content of the Abu Dhabi dolomites is biased by aragonite microinclusions.

However, in some subtidal crystals it can be less than 500 ppm, and even reach values similar to those of the Dolomia Principale dolomites (200 ppm). The average Sr content of both intertidal and subtidal sabkha dolomites ranges between 800 and over 2000 ppm, as detected with the microprobe. Sr in single aragonite crystals is over 7000 ppm. Na amounts to 100–300 ppm, values similar to those of the Dolomia Principale.

DISCUSSION: PROPOSED MECHANISMS OF DOLOMITIZATION

This study suggests a rather unexpected reply to the controversy as to whether or not the final chemical and petrographic overprints obscure the original properties. There is a strong diagenetic overprint, but it is still possible to trace the history of dolomitization using what can be termed 'microstructural and compositional' end-members. In this case 'end-members' does not imply a continuum of states: rather, they represent the ends of the diagenetic scale as inferred from textural observations, isotope data, microstructural characteristics, comparison with modern analogues and field observations. In the Dolomia Principale the two end-members are fine-crystalline planar mimetic calcian dolomite (56 mole % Ca), with the most positive $\delta^{18}O$ values and characterized by fine pervasive modulated microstructure; and coarse-crystalline void-filling ideal dolomite with the most negative $\delta^{18}O$ values, and characterized by basal defects.

The first end-member of the Dolomia Principale is here considered to represent the products of the early stages of dolomitization in subtidal settings, on the basis of its analogies with the Abu Dhabi subtidal dolomites. However, TEM observations indicate that the subtidal calcian dolomites of Abu Dhabi both replace aragonite as postulated by Illing *et al.* (1965), Butler (1969), McKenzie (1981), Patterson and Kinsman (1982), and grow (or nucleate) on a more ideal dislocation-ridden dolomite (Fig. 15b) formed in the intertidal and supratidal environments. At the present state of knowledge, this ideal to Mg-rich fine-crystalline dislocation-rich dolomite appears to be related to strongly evaporative settings. In fact, in addition to Abu Dhabi, it has also been observed in lake brines of the Coorong region (Rosen *et al.*, 1989) where it forms laminated units (Warren, 1990). Texturally and microstructurally similar dolomites were not observed in the Dolomia Principale, and therefore either they did not form or they were completely replaced in subtidal settings by less unstable calcian dolomites with modulated microstructures. Sedimentological evidence suggests that, in the Dolomia Principale, primary dolomitization occurred via precipitation of ideal and dislocation-ridden dolomite. First, the aridity of the Dolomia Principale environment should have favoured evaporation of seawater on the platform, and thus the presence of fluids with a high Mg/Ca ratio. These latter decreased the induction stage of dolomitization (Sibley *et al.*, 1987) and probably precipitated dislocation-rich ideal dolomites (Rosen *et al.*, 1989). Secondly, the presence of facies entirely dominated by microbial mats covering millions of square kilometres (considering the whole extension of the Norian carbonate platforms in the Pangean Gulf) may have had conspicuous consequences on the pco_2 of the relatively 'closed' surface-water system of the inner platforms, by increasing bicarbonate concentration (Grotzinger, 1989). The combination of these two factors favours elevated Mg/Ca ratios and bicarbonate enrichment in seawaters, making it possible to precipitate directly ideal to Mg-rich dolomite.

During periodic subaerial exposures the Dolomia Principale tidal flats were thus subjected to conditions favouring the rapid formation of unstable dislocation-rich dolomite, which became the preferred nucleation site for the successive products of diagenetic stabilization (Fig. 17). Other favourable nucleation sites were possibly provided by dolomite microscopic domains in disordered Ca-Mg carbon-

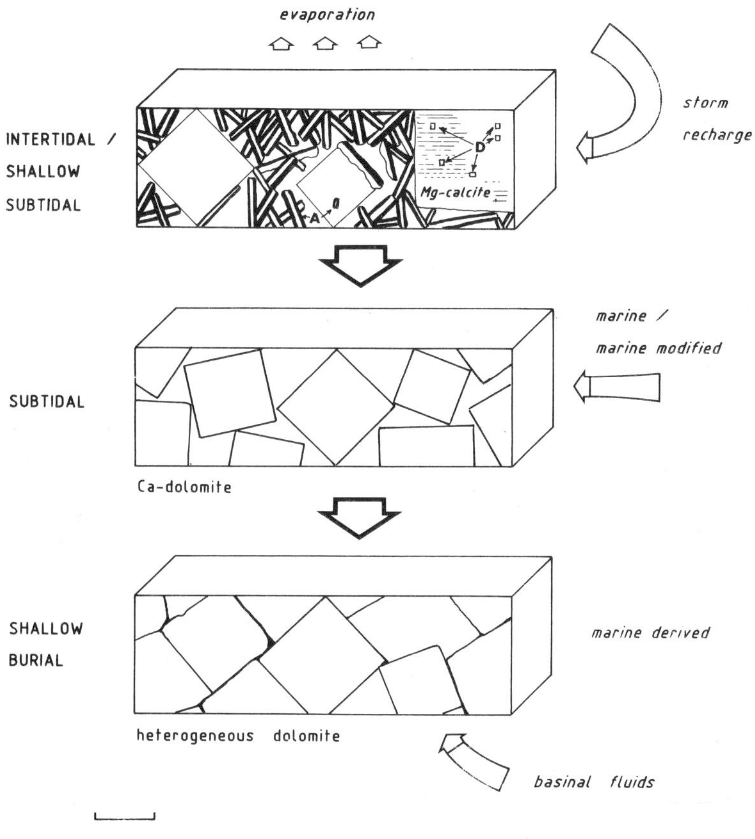

Fig. 17. Proposed dolomitization mechanisms for the Dolomia Principale carbonate shelf in the studied area. The first step represents early superficial diagenetic modifications: Mg-calcites, with coherent dolomite domains, aragonite and ideal dislocation-ridden unstable dolomite are the primary components of the sediment. Dolomite precipitates from marine waters evaporated after storm recharge. The second stage is stabilization of the original sediment in the subtidal setting (very shallow to shallow burial), possibly by evaporated seawater refluxing through the platform: aragonite, Mg-calcite and dislocation-ridden dolomite are dissolved and calcian dolomite precipitates. The unstable ideal dolomite formed in inter-/supratidal settings and microdomains of dolomite in Mg-calcite of subtidal facies provide favourable nucleation sites for the calcian dolomite with modulated microstructure. Continued evaporation–reflux may be favoured by the cyclic subaerial exposures of the Dolomia Principale. The third stage encompasses the shallow burial to deeper burial settings. More ordered and compositionally ideal dolomite precipitates from modified seawater at slower rates under rising temperature. Unstable calcian dolomite may still be preserved when ideal dolomite forms, and 'microstructurally and compositionally heterogeneous dolomite crystals' are the final product. Void-filling dolomite is 'homogeneous'. Scale bar = 1 μm.

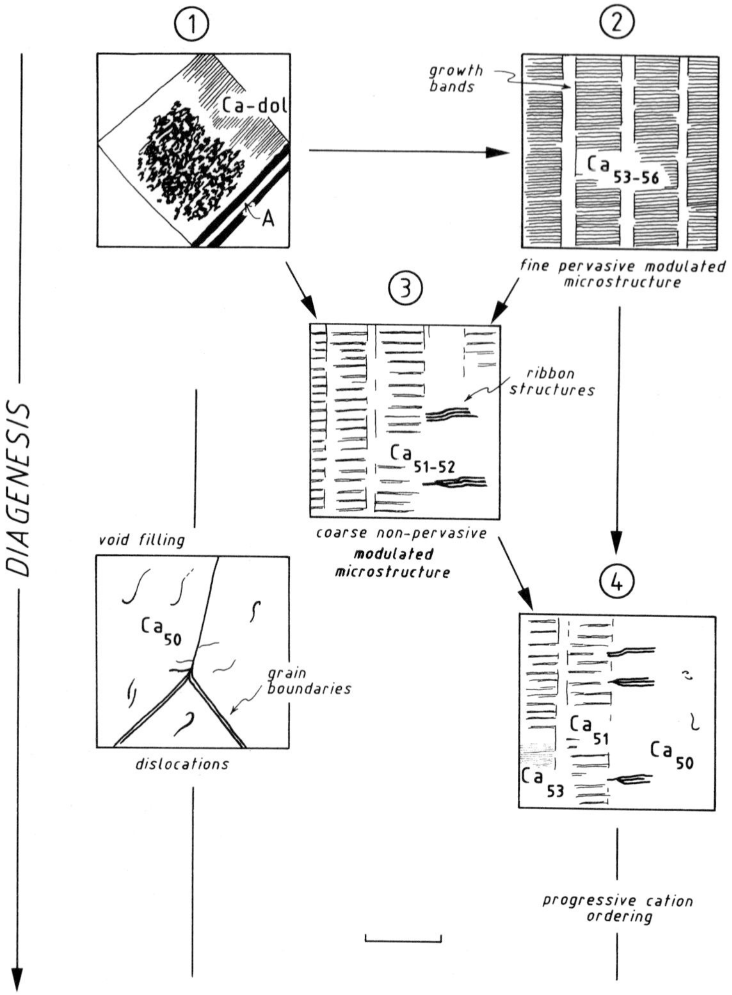

Fig. 18. Proposed microstructural/chemical changes occurring during time in the studied carbonate shelves undergoing progressive dolomitization from surface to deeper burial settings. (1) The first precipitate is dislocation-ridden ideal dolomite which is replaced by (2) calcian dolomite with pervasive fine modulated microstructures. The next step is the stabilization of the latter into a slightly calcian dolomite with non-pervasive coarser modulations (by partial replacement, epitaxy or total replacement). The direct passage from step 1 to step 3 is also envisaged according to rate of nucleation and growth and temperature. The final step is the precipitation of ideal, defect-free dolomite, at higher temperatures. Slow growth rate probably allows for the more stoichiometric characters of void-filling dolomites precipitated in shallow subsurface settings (left part of drawing). Scale bar = 0.5 μm.

ates. Therefore, the calcian dolomites of the Dolomia Principale, with pervasive microstructures observed in types 1, 2 and 3, would be the product of replacement of unstable carbonates in shallow subsurface settings (Figs 17 and 18).

Stable isotope data are consistent with precipitation from marine waters modified by evaporation, which may have been driven through the platform by reflux after storm recharge. The presence of calcian dolomites with modulated microstructures in platform-derived clasts embedded in Norian basin sediments, and the presence of relatively impermeable clayey layers between subtidal cycles in the Brenta Dolomites, suggests that replacement occurred within one 40 000–100 000-year depositional cycle (Hardie *et al.*, 1986). The periodic subaerial exposures possibly supported the evaporation–reflux system for millennia needed to provide the necessary through-flow of dolomite supersaturated waters. The only early diagenetic 'primary' dolomite present in the Dolomia Principale, as inferred from isotope data and from its presence in resedimented clasts, is type 4. The $\delta^{18}O$ and $\delta^{13}C$ values suggest precipitation from marine waters, and their textures indicate that they are not replacive. Thus, in the Dolomia Principale void fillings, marine dolomites that formed in the shallow subsurface are ideal and do not show modulated microstructures.

At the present state of knowledge, little is known concerning the relationships between microstructures, fluid composition and growth processes. Therefore, it may only be inferred, on the basis of textural characteristics, that a slower growth rate with respect to mimetic dolomite, and free growth in voids, could give ideal, better-ordered dolomites. Slightly calcian dolomites with coarse modulated microstructures and ideal dolomites most commonly observed in type 3, probably formed during burial, at progressively rising temperatures. In fact, the lowest $\delta^{18}O$ signals of type 3 come from non-planar coarse crystals, mostly characterized by almost ideal composition and a great number of dislocations. As a consequence, the diagenetic environments where type 3 was formed probably encompassed shallow subsurface to deeper burial. The relationships with fractures suggests that precipitation of the coarser crystals of type 3 occurred (or growth of these crystals continued) when type 5 formed. The latter is considered to be the 'burial end-member' of the Dolomia Principale diagenetic scale on the basis of texture, low $\delta^{18}O$, and the relationships with both other dolomite types and fracture systems. Isotope data are indicative of temperatures of formation over 60°C (Land, 1980). Red-luminescent bands, possibly indicative of the presence of Mn, may support the influence of basinal fluids in the formation of this dolomite. Type 5 is commonly a void filling, mostly ideal with few dislocations. When slightly calcian, it shows basal defects and lacks modulated microstructures. Again, we are confronted with a dolomite which commonly formed in voids and shows textures suggesting a slower rate of nucleation and growth than types 1, 2 and 3.

The coarser crystals of type 3, which are non-planar and not cross-cut by fractures as type 5, suggest that the ideal areas with dislocations observed in these matrix dolomites may also have formed at relatively high sedimentary temperatures. This would explain the observed large spread of $\delta^{18}O$ values characteristic of type 3. This spread is the result of the admixture of signals from compositionally and microstructurally heterogeneous crystals composed of calcian areas with modulated microstructures (early diagenetic), and progressively more ideal areas with coarser modulated microstructures, ribbon structures and dislocations (precipitated during progressive burial). The preservation of the metastable, calcian, early diagenetic dolomites, which 'survived' the burial stabilization observed in types 1, 2 and 3, may be due to a very low reactivity even on the geological timescale. The subsequent transition to other metastable (almost ideal, coarsely modulated dolomite) or stable dolomites may be the result of reaction kinetics (Morse & Casey, 1988), in particular of the rate of nucleation and growth. In fact, the compositional and microstructural characteristics of types 4 and 5 indicate that ideality and lack of modulated microstructures do not depend exclusively on increasing temperature during burial. Elevated temperatures may only 'accelerate' the dolomitizing reaction, regardless of the fluid composition (Hardie, 1987). Secondly, it would be difficult to envisage changes in fluid composition explaining the progressive ideality of dolomite crystals as observed in types 1, 2 and 3. Although stoichiometric dolomite has been related to hypersaline solutions (Sass & Bein, 1988), both the ideal void fillings of the Dolomia Principale have $\delta^{18}O$ values consistent with precipitation from fluids with normal seawater composition (Figs 9 and 18). Personal TEM and AEM observations of fine-crystalline ideal dolomites mimetically replacing pisoids from the Permian Capitan Reef, interpreted as precipitated from hypersaline solutions, show small crystals ($<0.5\,\mu m$ in size) with a great number of dislocations (see Fig. 14). A high Mg/Ca ratio rather than stoichiometry (ordered structure and ideal composition) favours a rapid rate of nucleation and growth (Sibley & Gregg, 1987; Sibley *et al.*, 1987). Ideal composition and ordered structures may be a function of the rate of nucleation and growth, as inferred from textural characteristics of the more ideal Dolomia Principale dolomites: slower rates (longer time) produce stoichiometric dolomite.

Thus, it is the combined action of rising temperature and reaction kinetics which may be responsible for the observed diagenetic pattern. The early calcian dolomite represented a site for nucleation of the more stable phases during the process of wet recrystallization. The possible epitaxy of dolomites precipitated at different diagenetic conditions, now present in variable percentages within types 1, 2 and 3 dolomites, is responsible for the spread of isotope values. Therefore, in the Dolomia Principale there is not a complete resetting of geochemical and petrographic properties with burial, but a 'superposition'. This phenomenon may be a consequence of the length of time required for pervasive dolomitization and the burial temperature. It appears that temperatures of 60°C, characterizing the burial environment when type 5 formed, were not sufficient to cause a complete resetting by wet recrystallization, almost as if there was a sort of 'inability to re-equilibrate' dolomites below and at these temperatures. Higher temperatures may favour this. In fact, a hydrothermal event at temperatures of 120°C (Frisia, 1991), affecting both the Dolomia Principale and the overlying Rhaetian units in the Brenta

Dolomites, formed non-mimetic coarse-crystalline (400 μm) slightly calcian to ideal dolomites, which replace all the dolomite types recognized, regardless of composition and microstructures. It is interesting to note that these high-temperature dolomites are dislocation-ridden. Therefore, this is further proof that temperature alone is not sufficient to 'produce' stoichiometric dolomites. Furthermore, the hydrothermal dolomites clearly replace other dolomites (and limestones), but do not show modulated microstructures. These observations may be further proof that cation ordering may be a function of the length of time taken to form dolomite crystals.

The proposed diagenetic trends for the Dolomia Principale apply only to the studied area, because the effects of hydrothermalism, or more prolonged subaerial exposures in other areas, may have changed the scenario.

CONCLUSIONS

This work further demonstrates that dolomite microstructures cannot be related unequivocally to fluid composition and diagenetic environment parameters. However, this study suggests the following considerations:

1 In the Dolomia Principale there is a relationship between microstructure and chemical composition. In fact, dolomites with 56 mole % excess Ca are always characterized by fine and pervasive modulated microstructures.

2 Dolomites precipitated from high Mg/Ca ratio fluids (Abu Dhabi) or at high temperature (about 120°C) are ideal but dislocation-ridden.

3 Void-filling dolomites precipitated both in the shallow subsurface and in burial settings at temperatures above 60°C are ideal and lack modulated microstructures.

4 Single dolomite crystals, both modern and ancient, may be compositionally and microstructurally heterogeneous. Chemical composition and isotope data obtained from such dolomites are the results of 'averaging in' different domains (or different crystals).

Points 2 and 3 clearly indicate that the same microstructures (or lack of microstructures) characterize dolomites precipitated in different diagenetic environments under the influence of diverse chemical and physical parameters. However, microstructural analyses can still be useful in interpreting the diagenetic history of carbonate platforms when other techniques and field observations are used. In the present work the author has demonstrated that, even if the present is not the key to the past (Zenger, 1972), the modern mechanisms of sabkha dolomitization may still be taken as sound analogues for early diagenetic steps on some Pangean carbonate shelves. If this avenue of approach is valid, more comparisons utilizing textural, chemical and microstructural methods are needed.

ACKNOWLEDGEMENTS

This work was supported by M.P.I. grant 40% to F. Jadoul. Most of the study has been carried out at the Department of Earth Sciences of the University of California, Berkeley. Therefore I would like to thank H.R. Wenk for useful discussions and for providing me with AREM data, C. Ecker for helping with AEM analyses, T. Tigue for TEM sample preparation and J. Donovan for help in obtaining careful microprobe analyses. I am also very grateful to R.J. Reeder and F. Jadoul for helpful suggestions. I also wish to thank B. Purser, M. Tucker and D. Zenger for comments and corrections. Furthermore, I wish to thank L. Mattavelli and T. Ricchiuto for the scientific support of the Geochemical Laboratories of AGIP SpA, San Donato Milanese, where the stable isotopes analyses were carried out.

REFERENCES

Aissaoui, D.M. (1988) Magnesian calcite cements and their diagenesis: dissolution and dolomitization. Mururoa Atoll. *Sedimentology* **35**, 821–841.

Barber, D.J. & Wenk, H.R. (1984) Microstructures in carbonates from the Alno and Fen carbonatites. *Contrib. Mineral. Petrol.* **88**, 233–245.

Bein, A. & Land, L.S. (1983) Carbonate sedimentation and diagenesis associated with Mg-Ca chloride brines: the Permian San Andres Formation in the Texas Panhandle. *J. Sedim. Petrol.* **53**, 243–260.

Blake, D.F., Peacor, D.R. & Wilkinson, B.H. (1982) The sequence and mechanisms of low-temperature dolomite formation: calcian dolomites in a Pennsylvanian echinoderm. *J. Sedim. Petrol.* **52**, 59–70.

Bosellini, A. & Hardie, L.A. (1985) Facies e cicli della Dolomia Principale delle Alpi Venete. *Mem. Soc. Geol. Ital.* **30**, 245–266.

Bourrouilh-Le Jan, F. (1992) Evolution des karst océaniques (karst, bauxites et phosphates). *Karstologia* **19**, 31–50.

Butler, G.P. (1969) Modern evaporite deposition and geochemistry of coexisting brines, the sabkha, Trucial Coast, Arabian Gulf. *J. Sedim. Petrol.* **39**, 70–89.

CARBALLO, J.D. & LAND, L.S. (1984) Holocene dolomitization of supratidal sediments by active tidal pumping, Sugarloaf Key, Florida. *Am. Ass. Petrol. Geol. Bull.* **68**, 459.

FRISIA, S. (1991) *Caratteristiche Sedimentologiche ed Evoluzione Diagenetica della Dolomia Principale (Norico) del Lago d'Idro e delle Dolomiti di Brenta*. PhD Thesis, Università degli Studi, Milano, 156 pp.

GEBELEIN, C.D., STEINEN, R.P., GARRETT, P., HOFFMANN, E.J., QUEEN, J.M. & PLUMMER, L.N. (1980) Subsurface dolomitization beneath the tidal flats of central west Andros Island, Bahamas. In: *Concepts and Models of Dolomitization* (Ed. Zenger, D.H., Dunham, J.B. & Ethington, R.L.) Spec. Publs. Soc. Paleont. Mineral. **28**, 31–49.

GREGG, J.M., HOWARD, S.A. & MAZZULLO, S.J. (1992) Early diagenetic recrystallization of Holocene (<3000 years old) peritidal dolomites, Ambergris Cay, Belize. *Sedimentology* **39**, 143–160.

GROTZINGER, J.P. (1989) Facies and evolution of Precambrian carbonate depositional systems: emergence of the modern platform archetype. In: *Controls on Carbonate Platform and Basin Development* (Ed. Crevello, P.D., Wilson, J.L., Sarg, F.J. & Read, J.F.) Spec. Publs. Soc. Econ. Paleont. Mineral **44**, 79–106.

GUNATILAKA, A., SALEH, A., AL TEMEEMI, A. & NASSAR, N. (1984) Occurrence of subtidal dolomite in hypersaline lagoon, Kuwait. *Nature* **311**, 450–452.

HARDIE, L.A. (1977) *Sedimentation on the Modern Carbonate Tidal Flats of Northwest Andros Island, Bahamas*. Johns Hopkins University Press, Baltimore, 202 pp.

HARDIE, L.A. (1987) Dolomitization: a critical view of some current views. *J. Sedim. Petrol.* **57**, 166–183.

HARDIE, L.A., BOSELLINI, A. & GOLDHAMMER, R.K. (1986) Repeated subaerial exposure of subtidal carbonate platforms, Triassic, Northern Italy: evidence for relative sea-level changes on 10 000–100 000-year scale. *Paleoceanography* **1**, 447–457.

ILLING, L.V., WELLS, A.J. & TAYLOR, J.C.M. (1965) Penecontemporaneous dolomite in the Persian Gulf. In: *Dolomitization and Limestone Diagenesis* (Ed. Pray, L.C. & Murray, R.C.) Spec. Publs. Soc. Econ. Paleont. Mineral. **13**, 89–111.

JADOUL, F. (1986) Stratigrafia e paleogeografia del Norico nelle Prealpi Bergamasche occidentali. *Riv. Ital. Paleont. Strat.* **91**, 479–512.

JADOUL, F., BERRA, F. & FRISIA, S. (1992) Stratigraphic and paleogeographic evolution of a carbonate platform in an extensional tectonic regime: the example of the Dolomia Principale in Lombardy. *Riv. Ital. Paleont. Strat.* **98**, 29–44.

KUTZBACH, J. (1989) *Past Climates and Climatic Change*. University of Wisconsin, Madison, 247 pp.

LAND, L.S. (1980) The isotopic and trace element geochemistry of dolomite: the state of the art. In: *Concepts and models of dolomitization* (Ed. Zenger, D.H., Dunham, J.B. & Ethington, R.L.) Spec. Publs. Soc. Paleont. Mineral. **28**, 87–100.

LAND, L.S. (1985) The origin of massive dolomite. *J. Geol. Educ.* **33**, 112–125.

LO CICERO, G. (1987) Carbon and oxygen isotopic composition of Norian sediments. Panormide carbonate platform, Palermo Mountains, Sicily. *Rend. Soc. Geol. Ital.* **9**(1986), 209–218.

MACHEL, H.G. & ANDERSON, J.H. (1989) Pervasive subsurface dolomitization of the Nisku Formation in Central Alberta. *J. Sedim. Petrol.* **56**, 891–911.

MCKENZIE, J.A. (1981) Holocene dolomitization of calcium carbonate sediments from the coastal sabkha of Abu Dhabi, UAE: a stable isotope study. *J. Geol.* **89**, 185–198.

MCKENZIE, J.A. (1985) Stable isotope mapping in Messinian evaporative carbonates of central Sicily. *Geology* **13**, 851–854.

MAZZULLO, S.J., REID, A.M. & GREGG, J.M. (1987) Dolomitization of Holocene Mg-calcite supratidal deposits, Ambergris cay, Belize. *Geol. Soc. Am. Bull.* **28**, 224–231.

MEGARD-GALLI, J. & BAUD, A. (1977) Le Trias Moyen et Supérieur des Alpes nord-occidentales et occidentales: données nouvelles et corrélations stratigraphiques. *Bull. BRGM, IV*, **3**, 233–250.

MISER, D.E., SWINNEA, J.S. & STEINFINK, H. (1987) TEM observations and X-ray crystal-structure refinement of a twinned dolomite with modulated microstructure. *Am. Mineralogist* **72**, 188–193.

MORSE, J.W. & CASEY, W.H. (1988) Ostwald processes and mineral paragenesis in sediments. *Am. J. Sci.* **288**, 537–560.

MORSE, J.W. & MACKENZIE F.T. (1990) *Geochemistry of Sedimentary Carbonates*. Elsevier, Amsterdam, 707 pp.

PATTERSON, R.J. & KINSMAN, D.J.J. (1982) Formation of diagenetic dolomite in coastal sabkha along Arabian (Persian) Gulf. *Am. Ass. Petrol. Geol. Bull.* **66**, 28–43.

PURSER, B.H. (1973) *The Persian Gulf*. Springer-Verlag, New York, 471 pp.

PEEDER, R.J. (1981) Electron optical investigation of sedimentary dolomites. *Contrib. Mineral. Petrol.* **76**, 148–157.

REEDER, R.J. (1983) Crystal chemistry of the rhombohedral carbonates. In: *Carbonates: Mineralogy and Chemistry: Reviews in Mineralogy* (Ed. Reeder, R.) Vol. 11, pp. 1–48. Mineralogy Society of America.

REEDER, R.J. & PROSKY, J.L. (1986) Compositional sector zoning in dolomite. *J. Sedim. Petrol.* **56**, 237–247.

RENESTO, S. (1993) *Megalancosaurus*: an arboreal archeosauromorph (Reptilia) from the Upper Triassic of Northern Italy. *J. Verteb. Paleontol.* (in press)

ROSEN, M.R., MISER, D.E., STARCHE, M.A. & WARREN, J.K. (1989) Formation of dolomite in the Coorong Region, South Australia. *Geochim. Cosmochim. Acta* **53**, 661–669.

RUPPEL, S.C. & CANDER, H.S. (1988) Dolomitization of shallow water platform carbonates by seawater and seawater-derived brines: San Andres Formation (Guadalupian), West Texas. In: *Sedimentology and Geochemistry of Dolostones* (Ed. Shukla, V. & Baker, P.A.) Spec. Publs. Soc. Econ. Paleont. Mineral. **43**, 245–262.

SALLER, A.H. (1984) Petrologic and geochemical constraints on the origin of subsurface dolomite, Enewetak Atoll: an example of dolomitization by normal sea water. *Geology* **12**, 217–220.

SASS, E. & BEIN, A. (1988) Dolomites and salinity: a comparative geochemical study. In: *Sedimentology and Geochemistry of Dolostones* (Ed. Shukla, V. & Baker, P.A.) Spec. Publs. Soc. Econ. Paleont. Mineral. **43**, 223–233.

SASS, E. & KATZ, A. (1982) The origin of platform dolomites: new evidence. *Am. J. Sci.* **282**, 1184–1213.

SIBLEY, D.F., & GREGG, J.M. (1987) Classification of dolomite rock textures. *J. Sedim. Petrol.* **57**, 967–975.

SIBLEY, D.F., DEDOES, R.E. & BARTLETT, T.R. (1987) Kinetics of dolomitization. *Geology* **15**, 1112–1114.

TUCKER, M. & BENTON, M.J. (1982) Triassic environments, climates and reptile evolution. *Palaeogeog. Palaeoclim. Palaeoecol.* **40**, 361–379.

VAHRENKAMP, V.C. & SWART, P.K. (1990) New distribution coefficient for the incorporation of strontium into dolomite and its implications for the formation of ancient dolomites. *Geology* **18**, 387–391.

VEEVERS, J.J. (1989) Middle/Late Triassic (250 ± 5 Ma) singularity in the stratigraphic and magmatic history of the Pangean heat anomaly. *Geology* **17**, 784–787.

VEIZER, J. (1983) Trace elements and isotopes in sedimentary carbonates. In: *Carbonates: Mineralogy and Chemistry: Reviews in Mineralogy* (Ed. Reeder, R.J.) Vol. 11, pp. 265–300. Mineralogical Society of America.

WARREN, J.K. (1990) Sedimentology and mineralogy of dolomitic Coorong lakes, South Australia. *J. Sedim. Petrol.* **60**, 843–858.

WENK, H.R. & ZENGER, D.H. (1983) Sequential basal faults in Devonian dolomite, Nopah Range, Death Valley, California. *Science* **22**, 502–504.

WENK, H.R., BARBER, D.J. & REEDER, R.J. (1983) Microstructures in carbonates. In: *Carbonates: Mineralogy and Chemistry: Reviews in Mineralogy* (Ed. Reeder, R.J.) vol. 11, pp. 301–367. Mineralogical Society of America.

WILD, R. (1991) *Aetosaurus* (Reptilia, Thecodontia) from the Upper Triassic (Norian) of Cene near Bergamo, Italy, with a revision of the genus. *Riv. Museo Civ. Sci. Nat. 'E. Caffi'* **14**(1989), 1–24.

WILSON E.N., HARDIE L.A. & PHILLIPS, O.M. (1990) Dolomitization front geometry, fluid flow patterns, and the origin of massive dolomite: the Triassic Latemar buildup, Northern Italy. *Am. J. Sci.* **290**, 741–796.

WORSLEY, T.R., NANCE, D. & MOODY, J.B. (1984) Global tectonics and eustasy for the past 2 billion years. *Marine Geology* **58**, 373–400.

ZENGER, D.H. (1972) Dolomitization and uniformitarianism. *J. Geol. Educ.* **20**, 107–124.

ZIEGLER, P.A. (1982) *Geological Atlas of Western and Central Europe*. Elsevier Scientific Publications, Amsterdam, 130 pp.

ZIEGLER, P.A. (1988) Evolution of Arctic–North Atlantic and the Western Tethys. *Am. Ass. Petrol. Geol. Memoir* **43**, 198 pp.

Spec. Publs Int. Ass. Sediment. (1994) **21**, 75–89

Changing dolomitization styles from Norian to Rhaetian in the southern Tethys realm

A. IANNACE* *and* S. FRISIA†

* *Dipartimento di Scienze della Terra, Università di Napoli; and*
† *Dipartimento di Scienze della Terra, Università di Milano, Italy*

ABSTRACT

A comparative study of Upper Triassic carbonate platform successions carried out in the southern Alps and southern Apennines, together with a review of available literature, suggests that two different dolomitization styles characterize respectively the Norian and Rhaetian over most of the peri-Mediterranean region. The Norian is characterized by the dominance of the Dolomia Principale facies, i.e. completely dolomitized peritidal cyclic successions. The bulk of the dolomitization was completed by the end of the Norian and its characteristics are strictly related to this specific stratigraphic interval. Rhaetian facies, on the contrary, are more diversified, typically record a sudden terrigenous input and generally have a much lower dolomite content. Norian dolomites are commonly mimetically replacive and finely crystalline, especially in the inter/supratidal intervals, alternating with coarser, non-mimetic dolomite in the subtidal intervals. Their $\delta^{18}O$ values are typically as high as +3.5 per mil (PDB), especially in the fine-crystalline types. Rhaetian dolomites, on the contrary, are mostly coarse and non-mimetic, with field evidence clearly indicating their origin through large-scale fluid circulations. Earlier diagenetic dolomites are of minor importance and are associated only with the inter-/supratidal facies. Moreover, all these dolomites have $\delta^{18}O$ systematically lower than texturally analogous Norian dolomites.

Two different styles of dolomitization thus pertain to these Late Triassic ages. The Rhaetian appears as a period of transition between the production of large bodies of early diagenetic dolomite of the Norian and the lack of such in the succeeding Jurassic/Cretaceous, at least in the Tethyan domain. The Rhaetian transition is probably related to a change from the very particular palaeogeographic/tectonic/climatic setting in which deposition and diagenesis of the Norian carbonates occurred. Norian carbonate shelves developed in the extensive Pangea Gulf, which was bordered landward by wide evaporitic deposits, indicating arid climatic conditions. The Rhaetian transgression, with terrigenous input and a more humid climate, by terminating these conditions, greatly hindered surface dolomitization.

A strictly uniformitarian approach can be used only in explaining Rhaetian dolomitization, whereas the very thick dolomite deposits of the Dolomia Principale should be viewed as the result of processes not operating, at least on the same scale, in modern environments.

INTRODUCTION

The Dolomia Principale/Hauptdolomit formation of the Alpine–Apennine domain represents a case of the 'Law of Persistence of Facies' (Ager, 1973), i.e. the widespread occurrence of given facies over very large areas during specific periods of geological time.

The Hauptdolomit Formation was defined by Gümbel (1861) in the northern Calcareous Alps on the basis of its particular lithological and morphological characteristics, which are especially striking in the Dolomites. Similar facies, known as the Dolomia Principale formation, are found throughout the whole Italian peninsula, and also in southern Spain, Greece, Turkey, Hungary, Dinarids and Carpathians. The Dolomia Principale consists of a 1000 m thick dolomitized succession of cyclic subtidal and peritidal inner platform facies with

Fig. 1. Geographic location of the investigated areas.

restricted faunas. Sedimentation of the Dolomia Principale took place on a wide pericontinental platform under semiarid conditions from Upper Carnian (?) through the entire Norian and, in some areas, Rhaetian (De Zanche, 1990). This 'homogenized' a Ladinian–Carnian palaeogeography characterized by many carbonate platforms and relatively deep basins. During the Rhaetian the extent of the Dolomia Principale platform facies was conspicuously reduced until its complete disappearance in the Lower Liassic of the Tethys.

The impressive amount of dolomite concentrated in a specific time interval (i.e. the Norian) makes the study of dolomitization in the Dolomia Principale an important challenge towards the understanding of the genesis of massive dolomite in general. On the basis of our studies in the southern Alps and southern Apennines (Fig. 1) and a reappraisal of the literature, we will try to approach the problem by placing the history of the Dolomia Principale platform within the geological evolution of the Tethys during the Upper Triassic to Jurassic transition. Our working hypothesis is that dolomitization of the Dolomia Principale is considered to be the result of particular geologic conditions that terminated following the palaeoceanographic changes which affected the Tethys domain after the Rhaetian.

This work is part of an international joint research programme headed by M. Boni (Università di Napoli, Italy) and J.A. McKenzie (ETH Zurich, Switzerland).

The hypothesis of a link between climatic changes from Late Triassic through Liassic, and a decrease in dolomite formation in the Tethyan realm, is not new (D'Argenio, 1974). However, to date, Upper Triassic dolomites of the peri-Mediterranean region have been traditionally considered as a whole. Our field, petrographic and geochemical approach was first aimed at the distinction of early diagenetic dolomite, which led us to recognize a significant difference between dolomitization patterns in the Norian and the Rhaetian, resulting in more precise scenarios for the genesis of Upper Triassic dolomites.

We wish to stress, in agreement with the criticism formulated by Zenger (1989), that any attempt to evaluate the dolomite problem in a global perspective, such as that proposed by Given and Wilkinson (1987), should be based on, rather than leading to, a precise discrimination of dolomitization styles. The lack of any such preliminary assessment may explain, for example, why the Upper Triassic, a dolomite-rich geological interval in the shallow Tethyan realm, appears unfavourable to dolomitization in Given and Wilkinson's curve.

BRENTA DOLOMITES

The Norian Dolomia Principale is the backbone of the Brenta Dolomites (see Fig. 1), where it consists of a 1000 m thick succession of completely dolomitized inner platform carbonates. Two main lithofacies can be recognized: peritidal cycles, which predominate in the lower 300 m and towards the top of the succession, and subtidal cycles in the middle–upper part of the formation.

The first lithofacies is organized into shallowing-upwards cycles, 1–5 m thick, composed of (a) bioclastic and intraclastic grainstones and fine breccias at the base above an erosive lower limit; (b) massive vuggy dolomites (packstones/wackestones) with dasycladacean algae, Foraminifera, gastropods and bivalves; (c) stromatolitic bindstones and fenestral mudstones/wackestones in the upper part of the cycle (Fig. 2a); (d) flat-pebble breccia commonly associated with tepee structures and vadose pisoids at the top of the cycle. The upper limit may be an irregular surface with pockets of greenish clay or red dolomicrite. The subtidal cycles, 2–5 m thick,

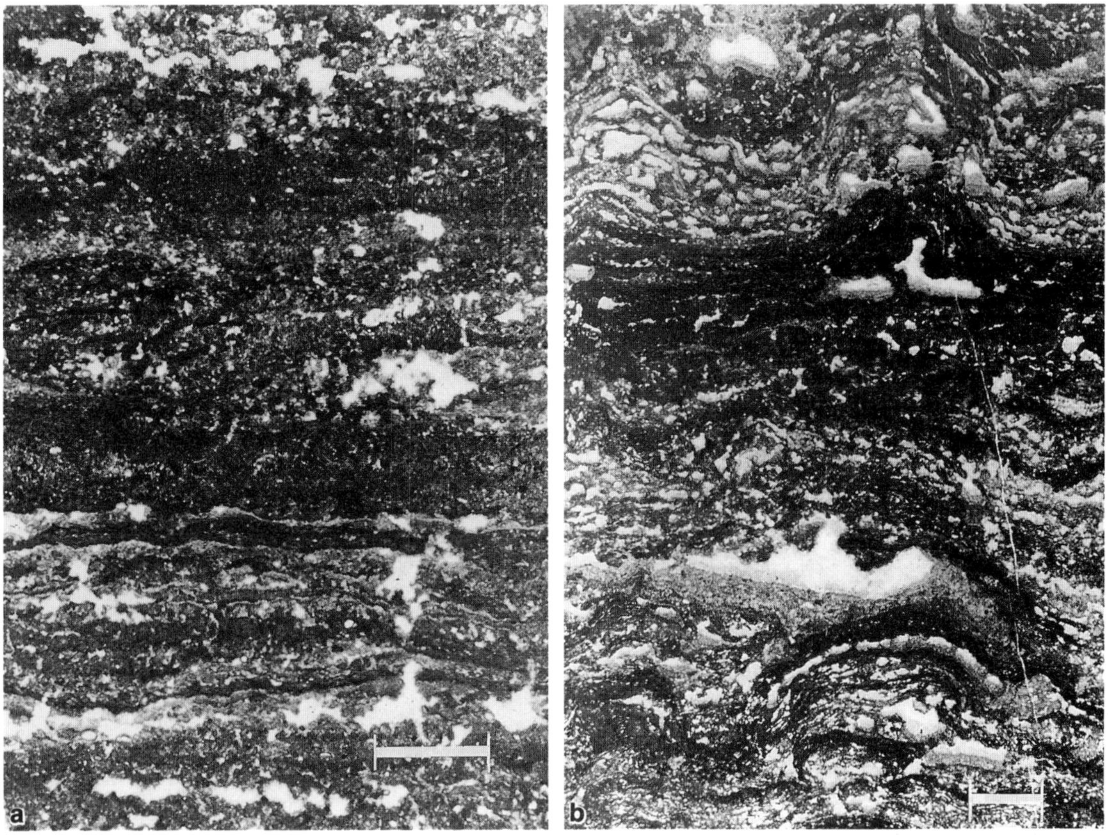

Fig. 2. Dolomite textures in Norian and Rhaetian intertidal facies. (Scale bars = 2 mm). (a) Laminated fenestral bindstone, totally dolomitized. Norian. (b) Same facies as in (a) but with partial dolomitization. Dolomite is finely crystalline and is present only in the micrite. Rhaetian.

consist mostly of intraclastic and peloidal packstones and intraclastic–bioclastic packstone/wackestones (Fig. 3a) with Megalodon and gastropod moulds, dasycladacean algae, Foraminifera and bivalves. Microbial bindstones are rare because cycles are often truncated by subaerial exposure surfaces.

Dolomitization of both types of cycle shows the same pattern. Microbial laminites in both peritidal and subtidal cycles are replaced by 2–6 μm-sized (average 4 μm) unimodal planar-e mimetic dolomite, (Figs 2a and 4a). Pisoids and the outline of vadose cements also occur in this dolomite, which preserves the former shapes. The 'bulk' of dolomitization is expressed by polymodal (20 to well over 100 μm) planar-s to non-planar (subordinate) non-mimetic dolomite, mostly as matrix replacive (Fig. 4c), subordinately as void-filling (Sibley & Gregg, 1987). Former textures are relatively well preserved (Fig. 3a). These dolomites probably replaced less stable carbonates through processes which had already taken place in the Norian. In fact, blocks and clasts produced by synsedimentary faulting of the platform and resedimented in coeval marly basinal facies are already dolomitized, and show the same dolomite textures and stable isotope composition as the Dolomia Principale facies from which they derived (Frisia, 1991 and this volume).

These dolomites may be locally obliterated by a coarser crystalline (up to 400 μm), planar-s to non-planar non-mimetic dolomite, which is most common along bedding planes or along faults. The latter is depleted in ^{18}O ($\delta^{18}O$ as low as −17‰ PDB), shows conspicuous intracrystalline porosity and is remarkably similar to the massive dolomites replacing limestones in the overlying Rhaetian facies.

The Dolomia Principale is overlain by the

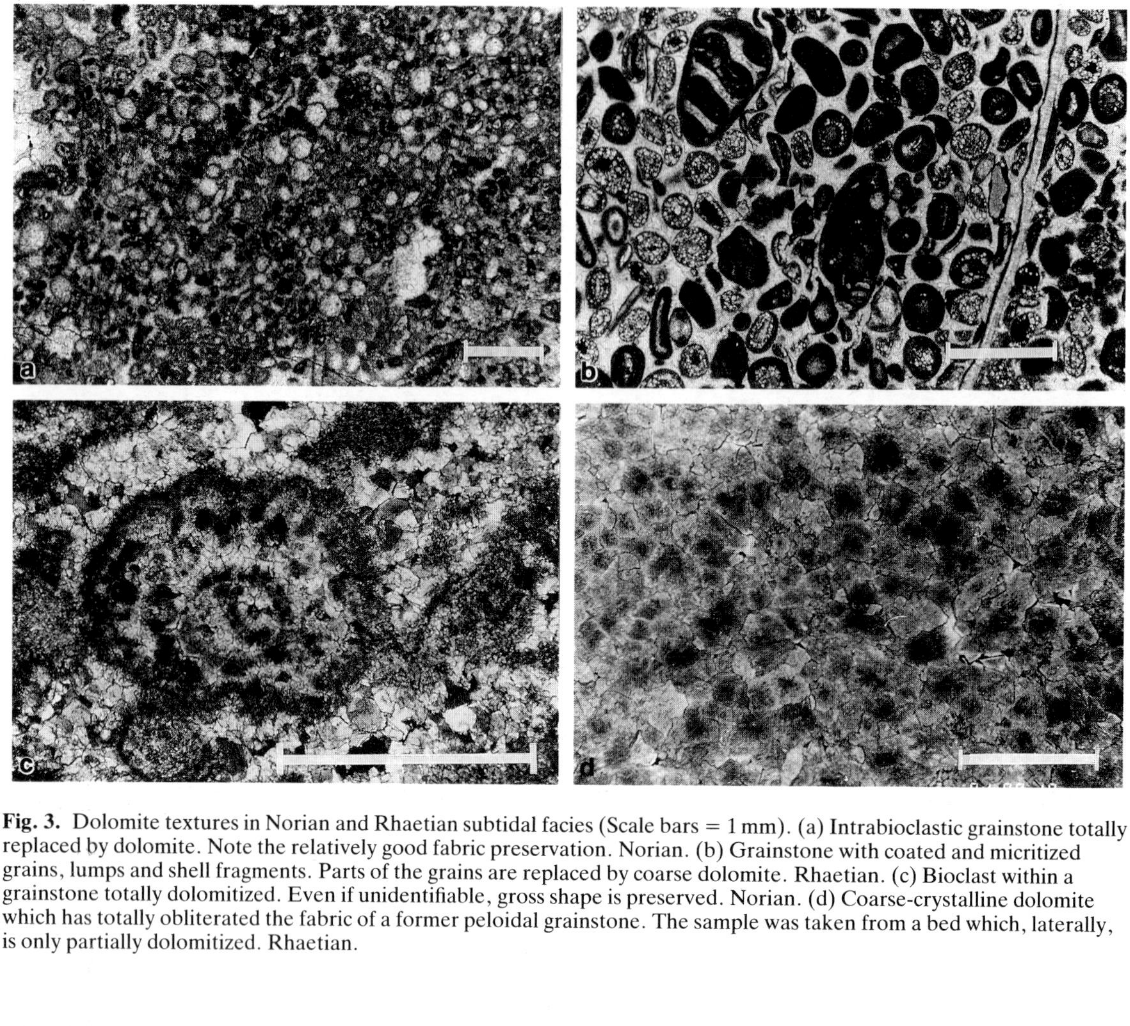

Fig. 3. Dolomite textures in Norian and Rhaetian subtidal facies (Scale bars = 1 mm). (a) Intrabioclastic grainstone totally replaced by dolomite. Note the relatively good fabric preservation. Norian. (b) Grainstone with coated and micritized grains, lumps and shell fragments. Parts of the grains are replaced by coarse dolomite. Rhaetian. (c) Bioclast within a grainstone totally dolomitized. Even if unidentifiable, gross shape is preserved. Norian. (d) Coarse-crystalline dolomite which has totally obliterated the fabric of a former peloidal grainstone. The sample was taken from a bed which, laterally, is only partially dolomitized. Rhaetian.

Rhaetian Calcare di Zu, consisting of shallowing-upwards cycles with shales and marls towards the base, where subtidal facies are prevalent. The subfacies of both subtidal and peritidal cycles are very similar to those of the Dolomia Principale (Fig. 2b), although higher-energy episodes were more frequent, as documented by common bioclastic–intraclastic–oolitic grainstones/packstones (Fig. 3b).

Dolomitization of stromatolitic bindstones in the Calcare di Zu is similar to that of the same facies of the Dolomia Principale. Dolomitization of the subtidal beds is usually incomplete (Fig. 3b), and due to polymodal planar-s to non-planar non-mimetic dolomite showing conspicuous intracrystalline porosity, as revealed by scanning electron microscopy (Fig. 4d). Microstructures, as seen with transmission electron microscopy, are remarkably similar to high-temperature dolomites (Barber & Wenk, 1984). This inference is further supported by $\delta^{18}O$ values as low as −11.8‰ (PDB) (Frisia, 1991).

Thus, regardless of textural characteristics, Norian dolomites are commonly ^{18}O-enriched compared to all Rhaetian ones (see Fig. 6).

Lattari Mountains

The Lattari Mountains (see Fig. 1) are a northeast–southwest-trending 40 km long belt along the southern limit of the Gulf of Naples. They consist mostly of a Mesozoic carbonate platform succession and Tertiary calcareous–siliciclastic cover. During the Miocene the succession was involved in the formation of the Apennine fold and thrust belt.

The oldest part of this succession, Norian–Lower

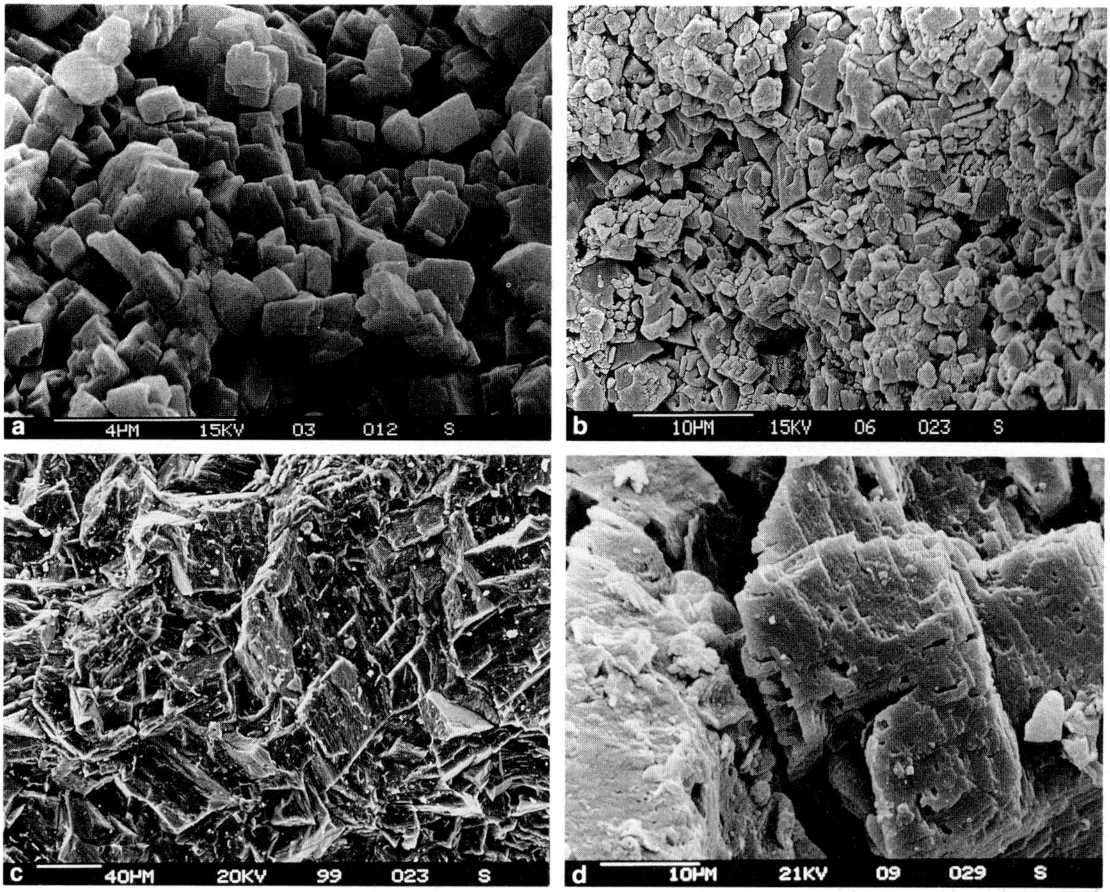

Fig. 4. Scanning electron microphotographs of Norian and Rhaetian dolomites. (a) Micritic dolomite replacing matrix. Norian. (b) Micritic–microsparitic dolomite mimetically replacing a pisoid cortex. Rhaetian. (c) Coarse-crystalline matrix dolomite. Growth zonations are observable. Norian. (d) Coarse-crystalline dolomite with intracrystalline porosity. Rhaetian.

Liassic in age and cropping out in the southeastern part of the belt, has been recently studied by Iannace (1991a). The Norian part of the succession consists of dolomites organized into shallowing-upward cycles. Each ideal cycle consists of basal massive bioclastic–intraclastic packstones/grainstones with gastropods and bivalves (see Fig. 3a) followed by laminated fenestral wackestones (Fig. 2a) overlain by metre-sized tepee structures with spectacular laminated cements of vadose origin (Iannace, 1991b).

The Rhaetian–Liassic succession consists of diverse lithofacies assemblages. In its Rhaetian part a platform margin complex consisted of back-reef partially dolomitized grainstones/packstones with Megalodons and Foraminifera; a periodically emergent barrier with abundant synsedimentary laminated and botryoidal cements; and an upper slope characterized by dolomitic breccias and pseudobreccias.

Different dolomite types may also be recognized in the Norian facies. Although a sharp distinction is not possible, two end-members are represented respectively by an aphanitic to micrite type and a sparitic type. The latter can be further subdivided into mosaic replacive (matrix dolomite) and void-filling. The fine-crystalline type is the most abundant and this results in good textural preservation in most of the Norian facies (Fig. 3a). Less common is a coarser (up to 300 μm), non-mimetic dolomite which,

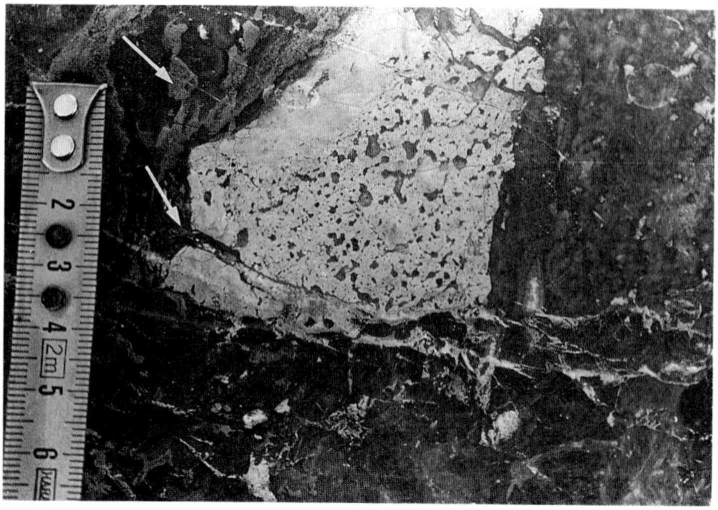

Fig. 5. Clast of fenestral fine-crystalline dolomite surrounded by laminated and fibrous calcite cements, both cut by veins of coarser dolomite (arrows). Lattari Mountains.

however, preserves ghosts of both allochems and cements (Fig. 3c).

Dolomitization patterns are strikingly different in the Rhaetian facies: whereas the Norian facies are completely dolomitized, the Rhaetian contain both limestones and dolomites, making it possible to analyse the geometry of the dolomitization front. Thus, on the basis of field, as well as petrographic and geochemical evidence, it has been possible to distinguish three dolomite types (Iannace, 1990). By far the most abundant type is coarsely crystalline, non-mimetic and almost ideal in composition. This dolomite replaces most of the marginal facies and about 50% of the outcropping back-reef facies along dolomitization fronts, several tens of metres long, which cross-cut bedding planes. Depositional textures are completely obliterated by this dolomite. Much less abundant is a fine-crystalline Ca-rich dolomite present only in inter/supratidal facies (Fig. 2b). Cross-cutting relationships indicate that this dolomite predates laminated and fibrous calcite cements (Fig. 5) related to prolonged Rhaetian subaerial exposures (Iannace, 1991; 1993).

In analogy to what was observed in the Brenta dolomites, all Norian dolomites are ^{18}O-enriched compared to the Rhaetian ones (Fig. 6; Iannace, 1991a), regardless of any differentiation among diverse dolomite types.

Other areas

The sedimentary and diagenetic characteristics of the Upper Triassic–Lower Jurassic of the Brenta and Lattari Mountains can be found in other contemporaneous carbonate successions in the western Mediterranean. Although textural and geochemical data directly comparable to ours are not always available, similar sedimentology, facies distributions, dolomitization patterns etc. greatly help in the comparison with our units, and for the scope of this paper are sufficient to interpret the possible diagenetic history.

Apennines and Sicily

In the southern Apennines continuous successions of Norian to Liassic age have been studied in the Picentini Mountains (De Castro, 1990) and Maratea Mountains (Civita, 1964). In both areas, the evolution from the massive peritidal dolomites of the Norian to the lagoonal dolomitic limestone with *Triasina hantkeni* and Megalodontids of the Rhaetian has been reported. With respect to the dolomitization styles, one of us (A.I.) has observed differentiation similar to that of the Lattari Mountains, i.e. fine fabric-preserving dolomite in the Norian, and mostly saccharoidal non-pervasive and fabric-destructive dolomite in the Rhaetian in irregular bodies that do not follow bedding. In addition to these detailed reports, it is generally accepted that the limestone : dolomite ratio in southern Apennines carbonate successions gradually increases (Carannante *et al.*, 1988) between the Late Triassic and the Early Liassic (often undistinguished and called Infralias).

More to the north, in the Aurunci Mountains,

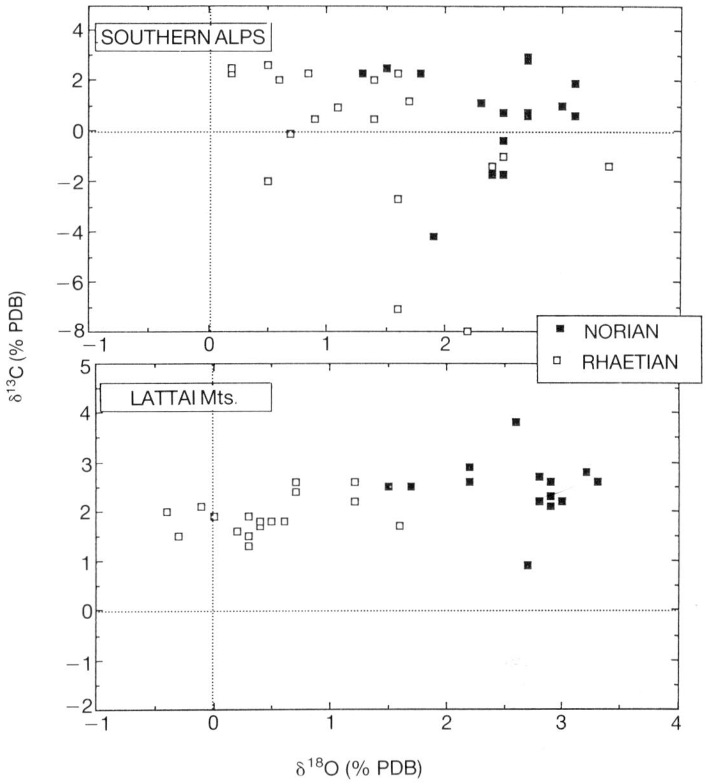

Fig. 6. Carbon and oxygen isotope compositions of Norian and Rhaetian dolomites in southern Alps and southern Apennines. In both cases Norian dolomites are generally[18] O-enriched respect to the Rhaetian, regardless of dolomite type.

Carannante *et al.* (1978) have described a Rhaetian limestone–dolomite succession characterized by subtidal back-reef facies alternating with intertidal facies which commonly show evidence of subaerial exposure. These facies overlie a 200 m thick completely dolomitized succession.

This general trend towards more open and concomitantly less dolomitic facies in the latest Triassic characterizes the entire central Apennines (Accordi & Carbone, 1988; Adamoli *et al.*, 1990). Moreover, during the Hettangian there is a prevalence of limestone facies and the dolomite present has been considered to be of late diagenetic origin (Accordi & Carbone, 1988).

In the northern Apennines, the Norian is characterized by dolomites and evaporites, the dolomites having been deposited in shallow mesosaline seas (Ciarapica, 1990). A drastic change in depositional environments occurred during the Rhaetian: areas of evaporite deposition became sites of clastic sedimentation, whereas the Norian dolomitic platforms evolved towards a basinal trough, where black mudstones were deposited.

In Sicily, during Norian–Rhaetian times, reef limestones gradually prograded over calcareous–dolomitic peritidal facies (Di Stefano, 1990). A very detailed sedimentological study of the latter was carried out by Catalano *et al.* (1974). They recognized that dolomite content decreases towards the top of the section of Upper Norian age, and is more abundant in inter/supratidal rather than in subtidal facies. Although no information is given concerning the type of dolomite present, the high-quality illustrations suggest that most of the dolomite has a mimetically replacive fine-crystalline fabric.

In contrast, Infraliassic breccias, interpreted as the resedimentation product at the foot of this platform (Dolomie Fanusi, in Scandone *et al.*, 1972), consist of coarse-crystalline non-mimetic dolomites. Relic masses of the previous limestones are found within this dolomite.

Alps

The Norian Briançonnais Units of the western Alps consist mostly of facies similar to those of the

Dolomia Principale (Megard-Galli & Baud, 1977; Lualdi, 1983); those in the Piedmont Units are intercalated with basalts indicative of the incipient rifting of the Ligurian Ocean. In the central Alps and in the south Alpine domain the widespread cyclic inner platform facies of the Dolomia Principale intercalate with, or pass laterally into, dark limestones, marly limestones and dolomites, shales of the intrashelf basin facies and deeper lagoon (Furrer, 1985; Jadoul, 1985). One of us (S.F.) observed that Norian intraplatform sediments are dolomitized only adjacent to the platform or immediately below prograding platform complexes, whereas towards the basinal depocentre dolomitization is absent.

Eastward, the Dolomia Principale is transitional into the Dachstein facies, well represented in the northern Calcareous Alps (Furrer, 1985), where the Dachstein Riffkalf reef complex interfingers with the basinal Hallstatt facies. It must be stressed that, adjacent to the open ocean, the Norian platform facies are not dolomitized.

The Rhaetian is mainly represented by limestones in all the south Alpine and Alpine domain, from the Calcaires Gris of the Briançonnais Units to the Calcare di Zu of the southern Alps, to the Kössen Formation, representing deeper lagoonal facies of the Upper Dachstein cyclic subtidal/intertidal platform facies in the northern Calcareous Alps. Dolomitization still characterizes the lower part of the Calcare di Zu and Kössen Beds, although it is mostly fabric-destructive and, where mimetic, restricted to intertidal/supratidal intervals.

Hungary

According to Haas (1988) and Balog *et al.* (1991), a progressive transgression of the Dachstein limestone reefal facies over the Hauptdolomit peritidal dolomites took place during the Norian–Rhaetian. These two formations are separated by a Transitional Unit with diagenetic characters intermediate between the completely dolomitized facies of the Hauptdolomit and the undolomitized Dachstein limestone. As a rule, in the Transitional Unit only the intertidal facies are dolomitized, whereas the subtidal ones are not.

Dinarids

A sedimentation pattern similar to that observed in the Alps is found in the Dinarids. In particular, facies similar to the Dolomia Principale are transitional basinwards into the Dachstein limestone (Cadet, 1975) and subsequently to pelagic facies. The overall dolomite content decreases in the Uppermost Triassic (Radoicic, personal communication). The Rhaetian may be dolomitized, especially over structural highs, but generally it is represented by limestones and dolomitic limestones similar to those of the Zu and Kössen Beds.

Hellenids

The Upper Triassic is characterized by the transition from platform to open-basin facies (Pindo Basin), with subsequent irregular dolomitization of platform facies which becomes gradually more pervasive toward the inner platform (Parnassus-Kiona, Hofbauer, 1985). The Rhaetian seems to be similar to that of the Norian, and a marked change in sedimentation and diagenesis occurs in the Late Lias–Dogger.

Carpathians

Michalik (1980) has shown that, in most of the Carpathian region, the Norian is characterized by a relative uniformity of facies, consisting mainly of Dolomia Principale type and evaporite facies. During the Rhaetian, intensive rifting led to the formation of many shallow elongate troughs, resulting in a more diversified facies distribution. Dolomite drastically diminished in the Rhaetian. The authors consider this evolution to be a local expression of a Tethyan reorganization in terms of geodynamics, seawater salinity, seawater circulation and climate, causing a decrease in cosmopolitanism of facies, floras and faunas.

SUMMARY OF DATA AND DISCUSSION OF THE REGIONAL FRAMEWORK

The comparative study of Upper Triassic dolomites of the Brenta Dolomites (and adjacent areas) and the Lattari Mountains has shown that the Norian and Rhaetian are characterized by two different styles of dolomitization. This seems to be further supported by the data present in the literature concerning the Upper Triassic of the peri-Mediterranean area (Fig. 7).

Norian carbonate shelf successions are pervasively affected by fine-crystalline fabric-preserving dolo-

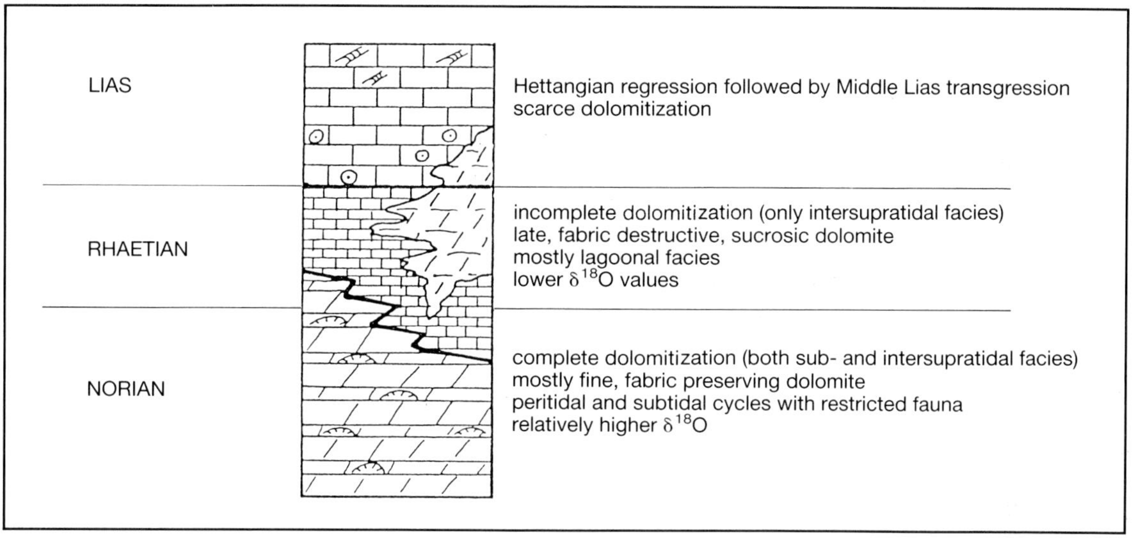

Fig. 7. Synopsis of the main sedimentological features common to most Upper Triassic–Lower Liassic platform carbonates of the peri-Mediterranean region.

mite. The dolomitization is thought, on the basis of field evidence, to have been almost completed in the Norian. Dolomite distribution in intraplatform basin sediments and its relationships with the prograding platform facies, suggest that dolomitizing fluids were produced on the platform and/or coming from the platform. Most probably these were slightly evaporated marine waters, as indicated by high $\delta^{18}O$ values typical of Norian dolomites in both studied areas. Furthermore, transmission electron microscope observations indicate that only a few Norian dolomites show microstructures related to high-temperature processes.

In contrast, Rhaetian carbonates are only partially dolomitized and dolomites are commonly massive and fabric-destructive. They occur in large irregular bodies cross-cutting the stratification. Moreover, some isotopic signals and transmission electron microscope observation (Brenta dolomites) point to relatively high temperatures during replacement. Fine-crystalline dolomites, replacing intertidal to supratidal facies similar to those of the underlying Dolomia Principale, are much rarer, and cross-cutting relationships with synsedimentary cements indicate their earlier origin with respect to the saccharoidal one.

The differences in dolomitization style revealed by our studies in the Alps and southern Apennines become more significant when considered within the Upper Triassic sedimentological scenario of the western Tethys. In fact, the observed overall consistency of the dolomitization style of the Norian sediments over a wide region, implies that this phenomenon is stratigraphically controlled, i.e. related to this specific time interval. This, together with our conclusion that dolomitization was completed during the Norian, means that the diagenesis of the Dolomia Principale platform was controlled by the Norian sedimentological environment.

On the contrary, in the Rhaetian not all the platform successions are dolomitized, and the bulk of the dolomite cuts across bedding. Facies-related earlier dolomite is far less common. These differences suggest that some important environmental changes occurred at the Norian-Rhaetian boundary in the shallow seas of the western Tethys.

In fact, important palaeoenvironmental changes affected all shallow Tethysian seas, from uppermost Norian to Upper Hettangian. The transition from the 'persistence of facies' of the Norian Dolomia Principale to the much more complex Liassic facies distribution is the result of a sudden increase in tectonic activity, which took place in the Upper Norian–Rhaetian and can be regarded as the prelude to the opening of the Ligurian–Piedmont ocean (Dewey *et al.*, 1973; Laubscher & Bernoulli, 1977; Catalano & D'Argenio, 1982; Jadoul, 1985). More globally, this process is part of the Pangea fragmentation, which took place from the Carnian to the Middle Lias, completely changing the pattern of ocean/landmass distribution and thus dramatically affecting the climate and palaeocirculation over the entire Earth (Veevers, 1989).

A global change at the Triassic–Liassic boundary is also postulated on the basis of long-term cyclicity. According to Fischer (1981, 1984) this boundary coincides with a transition from an ice-house to a greenhouse mode, comparable to the transition from the oscillatory to the submergent mode of MacKenzie and Piggott (1981). Based on studies of petrography and the distribution of oolites, Sandberg (1983) claims that the same boundary coincides with a transition from aragonite seas to calcite seas, as is also supported by variation in the mineralogy of calcareous sponges (Cuif & Gautret, 1991). In all these cases, the Triassic–Jurassic boundary is seen to be a time of first-order changes in climate, palaeoceanography, mineralogy of carbonate sediments, atmospheric CO_2 content, etc. All these theories suggest a first-order *eustatic* variation of Vail *et al.* (1977) as the main control of these changes.

The facies changes observed can also be considered as an expression of the worldwide Rhaetian transgression. Palaeontological evidence in northwestern Europe suggests that the transgressive seas were characterized by lower salinity (Hallam & El Shaarawy, 1982). Rapid transgression and clastic input characterize the Rhaetian deposits of the Alps and northern Apennines, whereas in the southern Apennines terrigenous sediments are lacking. This suggests that the clastic input had its source in the northern Laurasia supercontinent, and that it was triggered by increased river discharge following a climatic change. It is historically known that the Rhaetian marks a transition towards a more humid climate following the Norian aridity. The increased river discharge also had the effect of lowering the salinity in the western Tethys. Clastics did not reach its southern margin because they were trapped in basinal troughs.

In conclusion, in the frame of a global change occurring between Upper Triassic and Lower Jurassic, it appears that the Rhaetian is the most critical period for this transition. This is especially true in the Tethyan realm, but is also supported by the classic French and German Triassic stratigraphy in which a marine Rhaetian overlies the evaporitic Keuper (Carnian + Norian). Furthermore, in the stratigraphic record of northwest Europe, the Rhaetian has palaeontological and lithological peculiarities making it 'special' with respect to both Triassic and Jurassic (Wells, 1959).

THE HYPOTHESIS

The basis of our hypothesis is that the Norian Dolomia Principale reflects very special, non-actualistic palaeogeographic and geodynamic settings; this idea should be the basic constraint in any attempt to explain the massive dolomitization of this formation. By way of contrast, the Rhaetian can be seen as a period of transition marking the beginning of a new geological cycle, which will be characteristic of the following Jurassic and Cretaceous of the Tethys. As a consequence, dolomitization of Rhaetian carbonates can be more easily explained by dolomitization models based on modern analogues.

Norian

During the Norian the western Tethys was a large, shallow megagulf, some thousands of kilometres wide, characterized by semiarid climates and bordered by evaporitic belts (Fig. 8). The physicochemical characteristics of the water over the platform were peculiar. First, temperatures should be expected to have been higher than the mean oceanwater temperatures because the shallow depth of the sea over a very large area hindered free exchange with the ocean. The large blanket of water, through the combined effect of prolonged insolation and limited exchange with open seas, warmed easily.

With respect to chemistry, it has been suggested that the Tethys gulf acted in a manner comparable to many modern lagoons, with a zonation of evaporite deposition (Jansa *et al.*, 1980). These authors have observed that Upper Triassic Tethyan evaporitic deposits change from potash salt to halite to gypsum and finally carbonates from west to east. They believe that this was an effect of chemical fractionation occurring over the entire western Tethys, and reflecting differences in solubility. We emphasize here that these carbonates are mostly represented by the dolomites of the Dolomia Principale formation.

We can find no modern analogue of comparable size. Probably a giant Persian Gulf would be the best approximation. We only envisage that dolomitizing fluids were evaporated marine waters, produced on

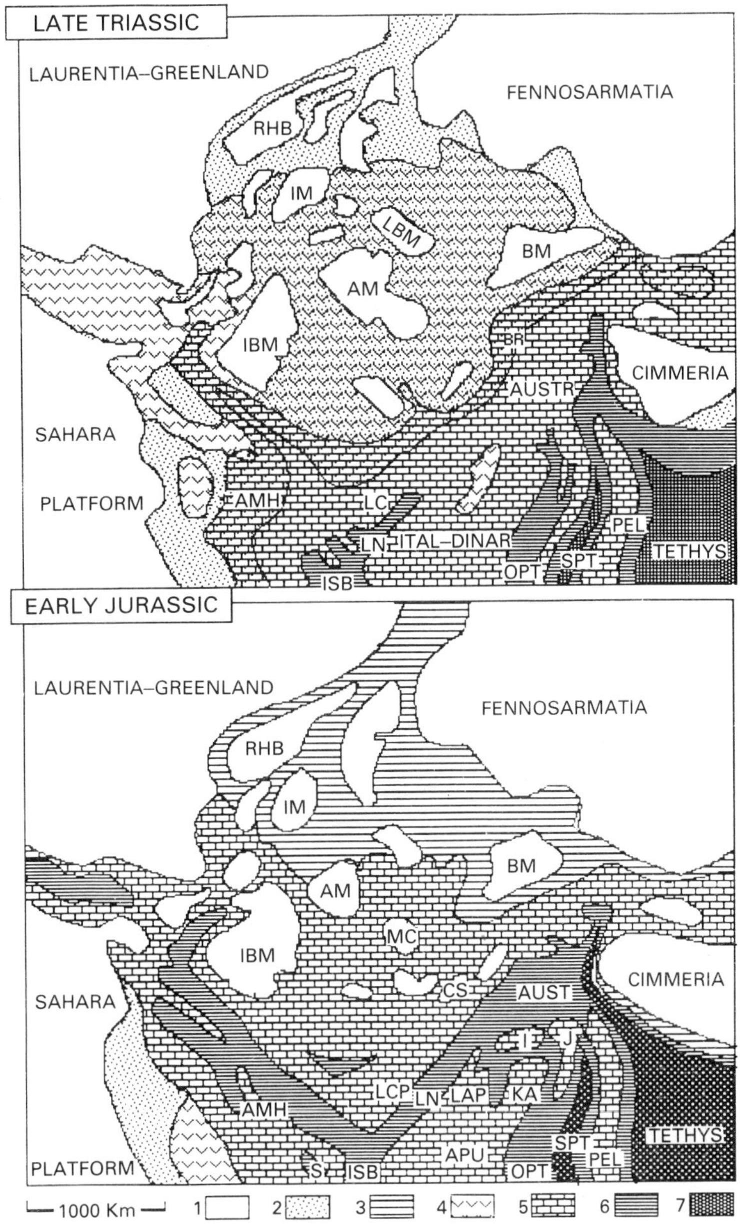

Fig. 8. Simplified palaeoceanographic map of Western Tethys during Late Triassic (Keuper = Carnian + Norian) and Early Jurassic (Pliensbachian–Toarcian). Modified after Ziegler (1988). 1 = emergent lands; 2 = shallow clastics, mainly sands; 3 = shallow clastics, mainly shales; 4 = evaporites; 5 = shallow-water carbonates; 6 = deep-sea pelagic sediments; 7 = oceanized areas; AM, Armorican Massif; AMH, Ain M'lila High; APU, Apulian Block; AUSTR, Austroalpine domain; BR, Brianconnais Domain; BM, Bohemian Massif; CS, Corsica–Sardinia Block; ISB, Ionian Sea Basin; J, Julian platform; KA, Karst Platform; LAP, Latium–Abbruzzi Platform; LCP, Lucania–Compania Platform; LN, Lagonegro Basin; MC, Massif Central; OPT, Olenos-Pindos trough; PEL, Pelagonian Block; SPT, Sub-Pelagonian trough.

a wide platform and over a long period of time, as a consequence of the aforementioned very special environmental conditions. The hydrological system could have been similar to that of the 'reflux' of Adams and Rhodes (1960) or the 'sabkha model' of Müller *et al.* (1990), i.e. a flow of more dense saline waters from the inner platform towards the basins. The persistence in time and space of such special conditions made possible the formation of the gigantic dolomite body of the Dolomia Principale.

Rhaetian

The very specific uniform sedimentological setting during the Norian disintegrated in the Rhaetian. The main interrelated factors contributing to this drastic change were:

1 Reduction in the extent of mesosaline settings and their replacement by normal marine conditions;

2 Change to a more humid climate, with subsequent changing of the evaporation : precipitation ratio;

3 Sudden decrease of evaporite deposition in the marginal marine settings of the western Tethys (Fig. 8).

We further speculate that at least two additional factors concurred to change the sedimentological–diagenetic style of the platform carbonates. It cannot be presumed that these occurred suddenly at the Norian–Rhaetian boundary, in that they probably spanned from Norian to Middle Lias. However, their effect may have been critical at the Norian–Rhaetian transition.

The first factor concerns the connection of western Tethys with the Arctic Ocean. According to presently available data (Ziegler, 1988), these two oceans had no connections during 'Keuper time' (Carnian + Norian), whereas a seaway linked them in the Middle Liassic (Fig. 8). The problem is the precise timing of this connection. According to palaeontological data (Ager, 1956) the first Lower Liassic ammonites and brachiopods had an Arctic affinity. Subsequently, in the Jurassic, there was a well established differentiation between a northern Arctic and a southern Tethyan Province.

However, considering the specificity of Rhaetian lithologies and faunas, it cannot be excluded that the incursion of Arctic ocean waters in the Norian megagulf occurred during the Rhaetian, as suggested by Wicher (1938) on the basis of studies of Foraminifera. In any case, such an oceanographic event, introducing colder and less saline Arctic waters into the tropical Tethyan megagulf, should have had a dramatic effect on the diagenesis of platform carbonates. In particular, most of the aragonitic sediments were probably calcitized and not dolomitized.

The second, much more speculative, factor concerns the Mg recycling at mid-ocean ridges. It has been suggested that submarine hydrothermal activity represents the major sink in the Mg cycle (Wolery & Sleep, 1976; Edmonds *et al.*, 1979). During the Lias the disaggregation of Pangea opened the Atlantic–Pennidic system (Veevers, 1989). We may speculate, therefore, that a complementary factor in reducing the potential for early dolomitization potential may have been the removal of Mg from the 'platform sink' in favour of the 'mid-oceanic ridge sink'.

CONCLUSIONS

Late Triassic carbonates of the peri-Mediterranean Alpine region are represented mainly by great amounts of massive dolomite. In this contribution we have stressed that, within the Upper Triassic carbonates, a difference in dolomitization patterns between Norian and Rhaetian can be recognized. The understanding of this difference stems from field observations and laboratory analyses of two local case histories framed in a wide regional setting provided by previous literature. Although laboratory analyses are essential for the evaluation of dolomitization mechanisms, without such a global approach we would fail to grasp the peculiar character of the Norian Dolomia Principale, that is, the great extent of pervasive dolomite related to a specific chronostratigraphic stage, as opposed to what is observed from both lower and upper parts of the Mesozoic of western Tethys.

In this perspective we postulate that the bulk of the dolomite was formed as a result of processes triggered by very particular geographic and sedimentological settings (Fig. 9) which have no modern counterparts, even though later diagenetic overprints cannot be excluded. The dolomitization mechanisms we envisaged were somewhat comparable to those operating today in the Persian Gulf. However, the scale at which these mechanisms operated was substantially different in terms of time and space. In consequence, Uniformitarianism seems to be insufficient, even in respect of methodology (see Gould, 1988), in providing a modern counterpart fully explaining the genesis of the Dolomia Principale.

By way of contrast, the dolomitization of the Rhaetian does not seem to be stratigraphically constrained and can be more easily explained by means of currently used dolomitization models. For example, the 'Florida–Bahamas' dolomitization model could be applied to the fine-crystalline dolomite observed in decimetre-thick inter/supratidal facies. The irregular dolomitized bodies cropping out as 'patches' in the Rhaetian should be considered each as a single case, and possibly related to large-scale fluid cirulation which may have been active from the Rhaetian onwards.

It is evident from this review that the Rhaetian was a period of transition, radically different from the Norian and sharing many characteristics with the following Jurassic. Thus, we believe that sedimentology and diagenesis of Upper Triassic

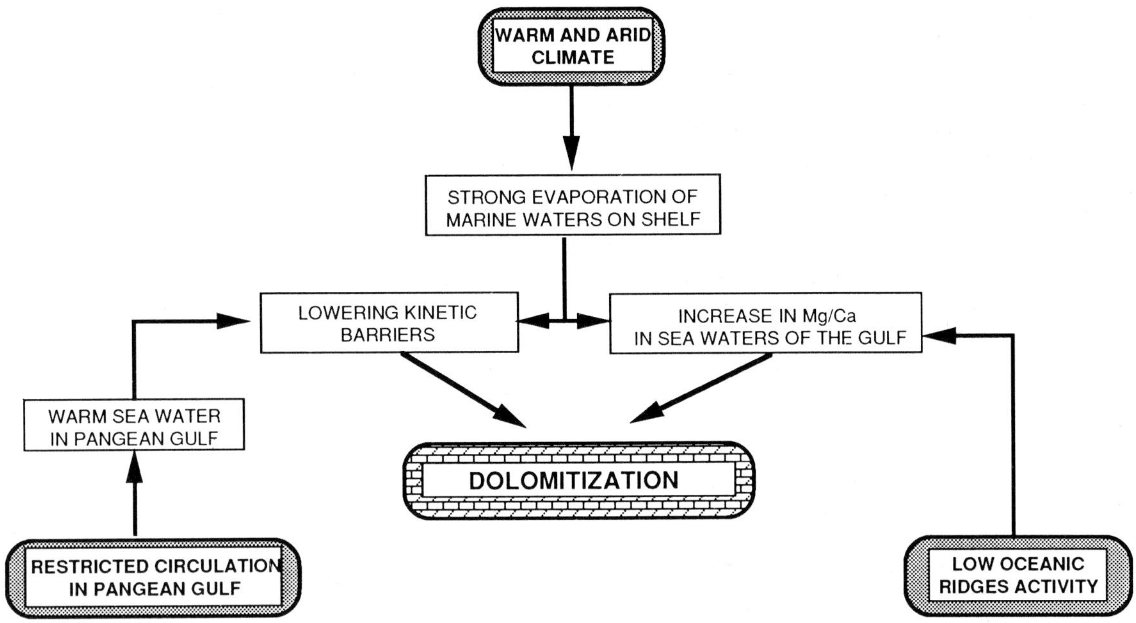

Fig. 9. Palaeoclimatic and geographic factors determining dolomitization in the Tethyan region.

carbonates gives further support to the views of palaeontologists and stratigraphers, such as Ager (1987) and Golebiowski (1990), who wish to maintain the distinction between Norian and Rhaetian, in contrast to a North American tendency to suppress the Rhaetian (Palmer, 1983; Tozer, 1988).

Finally, we wonder if a similar approach may be used in other cases, such as the Precambrian/Cambrian, the Cambro-Ordovician and the Permian–Triassic transitions, where the occurrence of widespread pervasive and fabric-preserving dolomites seems to be stratigraphically related.

ACKNOWLEDGEMENTS

We wish to thank the editors of this volume and the external reviewer L. Hardie for valuable suggestions which helped us in better focusing the hypothesis on Norian–Rhaetian dolomitizing trends. This work benefited from MURST 40% grants to Professor Maria Boni and Professor Flavio Jadoul, who supervised our PhD thesis, and whose constant encouragement we greatly acknowledge.

REFERENCES

Accordi, G. & Carbone, F. (1988) Sequenze carbonatiche meso-cenozoiche. *Quaderni Ric. Sci. Cons. Naz. Ricer., Roma (Italia)* **114**, 11–92.

Adamoli, L., Bigozzi, A., Ciarapica, G., Cirilli, S., Passeri, L., Romano, A., Duranti, F. & Venturi, F. (1990) Upper Triassic bituminous facies and Hettangian pelagic facies in the Gran Sasso range. *Boll. Soc. Geol. Ital.* **109**, 219–230.

Adams, J.E. & Rhodes, M.L. (1960) Dolomitization by seepage refluxion. *Am. Ass. Petrol. Geol. Bull.* **44**, 1912–1921.

Ager, D.V. (1956) The geographical distribution of Brachiopods in the British Middle Lias. *Q. J. Geol. Soc. London* **112**, 157–182.

Ager, D.V. (1973) *The Nature of the Stratigraphical Record*. McMillan, London.

Ager, D. (1987) A defence of the Rhaetian stage. *Albertiana* **6**, 4–12.

Balog, A., Read, J.F. & Haas, J. (1991) Dolomitization of a Late Triassic carbonate platform. *Abst. Dolomieu Conf. on Carbonate Platforms and Dolomitization, Ortisei (Italia)*, 16–21 Sept. 1991, 11.

Barber, D.J. & Wenk, H.R. (1984) Microstructures in carbonates in Alno and Fen carbonatites. *Contrib. Mineral. Petrol.* **88**, 233–245.

Cadet, J.P. (1975) Sur la géologie des confins méridionaux de la Bosnie et de la Serbie: mise en évidence de la nappe de Seurec (région de Visegrad et Rogatica, Yugoslavie). *Bull. Soc. Geol. Fran.* (7) XII, **6**, 967–972.

Carannante, G., Carbone, F., Catenacci, V. & Simone,

L. (1978) I carbonati triassici dei Monti Aurunci: facies deposizionali e diagenetiche. *Boll. Soc. Geol. Ital.* **97**, 687–698.

Carannante, G., D'Argenio, B. & Sgrosso, I. (1988) Le successioni mesozoiche dell'Appennino campano-lucano: inquadramento generale. *74th Nat. Congr. Soc. Geol. Ital., Sorrento (Napoli)*, 13–17 Sept. 1988; *Relazioni*, 25–31.

Catalano, R. & D'Argenio, B. (1982) Infraliassic strike-slip tectonics in Sicily and southern Apennines. *Rend. Soc. Geol. Ital.* **5**, 5–10.

Catalano, R., D'Argenio, B. & Lo Cicero, G. (1974) I ciclotemi triassici di Capo Rama (Monti di Palermo). *Geol. Romana* **13**, 125–145.

Ciarapica, G. (1990) Central and Northern Apennines during the Triassic: a review. *Boll. Soc. Geol. Ital.* **109**, 39–50.

Civita, M. (1964) Osservazioni geologiche nei Monti di Maratea (Lucania Meridionale). *Mem. Note Ist. Geol. Appl. Napoli* **9**, 24 pp.

Cuif, J.P. & Gautret, P. (1991) Etude de la repartition des principaux types de démosponges depuis le Permien. Hypothèse d'une incidence des conditions océanologiques sur la biominéralisation carbonaté des spongiaires. *Bull. Soc. Geol. France* **162**, 875–886.

D'Argenio, B. (1974) Le piattaforme carbonatiche periadriatiche. Una rassegna di problemi nel quadro geodinamico mesozoico dell'area mediterranea. *Mem. Soc. Geol. Ital.* **13**, 137–159.

De Castro, P. (1990) Studies on the Triassic carbonates of the Salerno province (Southern Italy): the Croci di Acerno sequence. *Boll. Soc. Geol. Ital.* **109**, 187–217.

Dewey, J.F., Pitman, W.C., Ryan, W.B.F. & Bonnin, J. (1973) Plate tectonics and the evolution of the alpine system. *Geol. Soc. Am. Bull.* **84**, 3137–3180.

De Zanche, V. (1990) A review of Triassic stratigraphy and paleogeography in the Eastern and Southern Alps. *Boll. Soc. Geol. Ital.* **109**, 59–71.

Di Stefano, P. (1990) The Triassic of Sicily and the Southern Apennines. *Boll. Soc. Geol. Ital.* **109**, 21–37.

Edmonds, J.M., Measures, C., McDuff, R.E., Chan, L.H., Collier, R., Grant, B., Gordon, L.I. & Corliss, J.B. (1979) Ridge crest hydrothermal activity and the balance of the major and minor elements in the Ocean: the Galapagos data. *Earth Planet. Sci. Lett.* **46**, 1–18.

Fischer, A.G. (1981) Climatic oscillation in the biosphere. In: *Biotic Crises in Ecological and Evolutionary Time* (Ed. Nitecki, M.H.) pp. 103–131. Academic Press, New York.

Fischer, A.G. (1984) The two Phanerozoic supercycles. In: *Catastrophes and Earth History* (Ed. Berggrenn, W.A. & Van Couvering, J.A.) pp. 129–149. Princeton, NJ.

Frisia, S. (1991) *Caratteristiche Sedimentologiche ed Evoluzione Diagenetica della Dolomia Principale (Norico) del Lago di Idro e delle Dolomiti di Brenta.* PhD thesis, University of Milan, Italy, 158 pp.

Furrer, H. (Ed.) (1985) *Field Workshop on Triassic and Jurassic Sediments in the Eastern Alps of Switzerland.* ETH Zurich (Switzerland), Neue Folge, **248**, 81 pp.

Given, R.K. & Wilkinson, B.H. (1987) Dolomite abundance and stratigraphic age: constraints on rates and mechanisms of Phanerozoic dolostone formation. *J. Sedim. Petrol.* **57**, 1068–1078.

Golebiowski, R. (1990) The alpine Kössen Formation, a key for European topmost Triassic correlation. A sequence and ecostratigraphic contribution to the Norian–Rhaetian discussion. *Albertiana* **8,** 25–35.

Gould, S.J. (1988) *Time's Arrows, Time's Cycles*. Penguin Books, London, 222 pp.

Gümbel, C.W. (1861) *Geognostiche Beschreibung des Bayrischen Alpengebirges und seines Vorlandes*. Gotha, (Perthes).

Haas, J. (1988) Upper Triassic carbonate platform evolution in the Transdanubian mid-mountains. *Acta Geol. Hungar.* **31**, 299–312.

Hallam, A. & El Shaarawy, Z. (1982) Salinity reduction of the end-Triassic sea from the Alpine region into northwestern Europe. *Lethaia* **15**, 169–178.

Hofbauer, G. (1985) Stratigraphie, Fazies und Tectonik am SW-Rand des Parnassos-Kiona Gebirge (Mittelgriechland). *Erlanger Geol. Abh.* **112**, 11–45.

Iannace, A. (1990) Massive dolomitization in Rhaetian carbonates of Southern Apennines (Italy). *Abstr. 13th Int. Ass. Sedimentology Meeting, Nottingham, UK*, 26–31 August 1990, 235.

Iannace, A. (1991a) *Ambienti Sedimentari e Processi Diagenetici in Successioni di Piattaforma Carbonatica del Trias Superiore dei Monti Lattari e Picentini (Salerno).* PhD thesis, University of Naples, Italy, 221 pp.

Iannace, A. (1991b) Late Triassic of Cetara. *5th Int. Symp. Fossil Algae, Capri (Italy)*, 7–12 April 1991, *Field Trip Guide Book*, pp. 69–71.

Iannace, A. (1993) I caratteri diagenetici delle facies di margiue del Trias Superiore dell'Appennino Meridionale e le loro implicazioni paleogeografiche. *Riv. Ital. Paleont. Stratigr.* **99**, 57–80.

Jadoul, F. (1985) Stratigrafia e paleogeografia del Norico nelle Prealpi Bergamasche occidentali. *Riv. Ital. Paleont. Stratigr.* **91**, 479–512.

Jansa, L.F., Bujak, J.P. & Williams, G.L. (1980) Upper Triassic deposits of the western north Atlantic. *Can. J. Earth Sci.* **17**, 547–558.

Laubscher, H. & Bernoulli, D. (1977) Mediterranean and Tethys. In: *The Ocean's Basins and Margins* (Ed. Nairn, A.E.M., Kanes, W.H. & Stheli, F.G.) vol. 4A, pp. 1–28. Plenum, New York.

Lualdi, A. (1983) Ricerche stratigrafico sedimentologiche nel Prepiemontese Ligure. I: Il Trias dell'Unità Arnasco-Castelbianco. *Atti Ist. Geol. Univ. Pavia (Italia),* **30**, 197–204.

Mackenzie, F.T. & Piggott, J.D. (1981) Tectonic controls of Phanerozoic sedimentary rock cycling. *J. Geol. Soc. London* **138**, 183–196.

Megard-Galli, J. & Baud, A. (1977) Le Trias Moyen et Supérieur des Alpes Nord-Occidentales et Occidentales: données nouvelles et corrélations stratigraphiques. *Bull. BRGM IV* **3**, 233–250.

Michalik, J. (1980) A paleoenvironmental and paleoecological analysis of the west Carpathians part of the Northern Tethyan nearshore region in the latest Triassic time. *Riv. Ital. Paleont. Stratigr.* **85**, 1047–1064.

Müller, D.W., McKenzie, J.A. & Mueller, P.A. (1990) Abu Dhabi sabkha, Persian Gulf, revisited. Application of strontium isotope to test an early dolomitization model. *Geology* **18**, 618–621.

PALMER, A.R. (1983) The decade of the North American Geology, 1983 Geological Time Scale. *Geology* **11**, 503–504.

SANDBERG, P.A. (1983) An oscillating trend in Phanerozoic non-skeletal carbonate mineralogy. *Nature* **305**, 19–22.

SCANDONE, P., RADOICIC, R., GIUNTA, G. & LIGUORI, V. (1972) Sul significato ambientale dell Dolomie Fanusi e dei Calcari ad Ellipsactinie della Sicilia Settentrionale. *Riv. Min. Siciliana* **133**, 51–61.

SIBLEY, D.F. & GREGG, J.M. (1987) Classification of dolomite rock textures. *J. Sedim. Petrol.* **57**, 967–975.

TOZER, E.T. (1988) Rhaetian: a substage, not a stage. *Albertiana* **7**, 9–15.

VAIL, P.R., MITCHUM, R.M., TODD, R.G., VIDMEIR, J.M., THOMPSON, S., SANGREE, J.B., BUBB, J.W. & HATFIELD, W.G. (1977) Seismic stratigraphy – application to hydrocarbon exploration. *Am. Ass. Petrol. Geol. Mem.* **26**, 49–212.

VEEVERS, J.J. (1989) Middle Late Triassic singularity in the stratigraphic and magmatic history of the Pangean heat anomaly. *Geology* **17**, 784–787.

WELLS, A.K. (1959) *Outline of Historical Geology*, Murby & Co, London, 398 pp.

WICHER, C.A. (1938) Microfauna aus Jura und Kreide insbesondere Nordwest Deutschland. I: Lias alpha bis epsilon. *Abh. Pr. Geol. Landesamt, NF* **193**, 1–16.

WOLERY, T.J. & SLEEP, N.H. (1976) Hydrothermal circulation and geochemical flux at mid ocean ridges. *J. Geol.* **84**, 249–275.

ZENGER, D.H. (1989) Dolomite abundance and stratigraphic age: constraints on rates and mechanisms of Phanerozoic dolostone formation – Discussion. *J. Sedim. Petrol.* **59**, 162–164.

ZIEGLER, P.A. (1988) Post-hercynian plate reorganization in the Tethys and Arctic–North Atlantic domains. In: *Triassic Jurassic Rifting: Developments in Geotectonics* (Ed. Manspeizer, W.) pp. 711–755. Elsevier, Amsterdam.

Spec. Publs Int. Ass. Sediment. (1994) **21**, 91–107

Distribution, petrography and geochemistry of early dolomite in cyclic shelf facies, Yates Formation (Guadalupian), Capitan Reef Complex, USA

M. MUTTI* *and* J.A. SIMO†

* *Geologisches Institut, ETH-Z, Sonneggstrasse 5, CH-8092 Zürich; and*
† *Department of Geology and Geophysics, 1215 W. Dayton St., Madison, WI 53706, USA*

ABSTRACT

This paper documents the geometry, distribution, petrography and geochemistry of early diagenetic shelf dolomites within a sequence-stratigraphic framework. The Guadalupian Yates Formation is composed of siliciclastic/carbonate cycles, deposited under subaerial to subtidal and shallow-marine conditions respectively, on a gently basinward-sloping shelf. Each cycle is composed of a lower transgressive siliciclastic and restricted carbonate facies, and an upper open-marine to restricted high-stand carbonate facies, capped by a subaerial exposure surface.

The shelf strata are now composed of dolomite, whereas the basinward equivalent rocks, the massive Capitan Reef and the foreslope facies are mostly limestone. The stratigraphic distribution of dolomite, the landward and basinward migration of the dolomite–limestone interface, and cross-cutting relationships document the occurrence of several early generations of dolomite in each of the Yates cycles. Dolomite clasts ripped from the underlying unit are incorporated together with undolomitized clasts at the base of the cycles. Dolomitization events occur in supra- and intertidal facies during the landward shift of facies belts in the transgressive portion of the cycle, whereas during the late high-stand and subaerial exposure at cycle tops, the dolomite–calcite interface is shifted basinward and dolomitization occurs indiscriminately in supra-, inter- and subtidal facies.

Two types of dolomite are recorded in these deposits. Fine-crystalline dolomite is the most common and replaced carbonate grains, aragonitic and Mg-calcite cements and carbonate mud. Oxygen isotope data ($\delta^{18}O$ is enriched up to 5‰ with respect to Permian marine values) and major element concentrations are consistent with dolomitization from a marine fluid, modified by evaporation, in an oxidizing environment. Coarse-crystalline dolomite, precipitated as a primary dolomite cement in fenestral pores, has a limited areal distribution. Oxygen isotopes ($\delta^{18}O = -1.3 \pm 1.8\ \sigma$) are also consistent with precipitation from seawater, possibly at higher temperatures and salinities. Carbon isotopes ($\delta^{13}C = 5.0 \pm 0.6\ \sigma$) and trace element concentrations suggest precipitation during or above the zone of bacterial sulphate reduction.

Both dolomite types are interpreted to have formed in response to seawater flow through the shallow subsurface driven by tidal pumping, or by storm-recharge flooding and evaporative pumping, and by high evaporation rates on the supra- and intertidal areas. Dolomitization occurs during both deposition and subaerial exposure of each depositional cycle. Dolomitization of supra- and intertidal facies within one cycle occurred approximately during a period less than 100 000 years, whereas dolomitization of subtidal carbonates (and of marine cements) in association with the exposure at cycle tops may be related to a longer time duration (200 000 or 300 000 years).

INTRODUCTION

Many carbonate platforms are characterized by extensively dolomitized cyclic inner shelf facies, whereas outer shelf and basinal facies remain undolomitized (Wilson, 1975; p. 314). One of the problems concerning the origin of ancient massive shelf dolomites is that the volume of dolomite occurring in comparable modern environments is minimal relative to that of ancient massive dolomites. Furthermore, differences exist between the geochemical characteristics of modern near-surface dolomites and ancient dolomites, which commonly record a complex history of stabilization by many generations of fluids (Land, 1980; Gregg & Shelton, 1990; Gao & Land, 1991).

Temporal variations in the relative abundance of dolomite vs. calcite occurs in the geological record and has been related to long-term marine transgressions and regressions (Given & Wilkinson, 1987; Sibley, 1991). The postulated genetic link between dolomite abundance and global eustasy is based mainly on the assumption that most dolomites have formed in the presence of marine or modified marine water at the Earth's surface (Land, 1985; Machel & Mountjoy, 1986). However, dolomite in many ancient sequences is preferentially associated with the regressive portion of depositional cycles, whereas transgressive depositional facies commonly remain limestone (Dunham & Olson, 1978; Kerans & Lucia, 1989; Montañez, 1991). Dolomitization may occur in different stratigraphic frameworks and dolomites formed in different sequence-stratigraphic frameworks may have different characteristics. This paper documents the petrology and geochemical characteristics of several generations of dolomite within a sequence-stratigraphic framework.

The common cyclicity of upward-shallowing strata indicating repeated fluctuations of relative sea level, has been envisaged by Wilson (1975) as a mechanism to expose carbonate shelves either to sabkha, reflux or mixing-zone dolomitization. If relative sea-level variations control dolomitization, we should expect to find a distinct dolomite in each cycle, and the dolomite–calcite interface to shift shelfward or basinward within each cycle, according to the stratigraphic pattern of the depositional facies.

The upper Yates Formation in the Guadalupe Mountains provides an excellent example of cyclic dolomitization (Mutti, 1990) that suggests a genetic relationship between dolomite distribution, stratigraphic patterns and relative sea-level variations. This paper describes the dolomite distribution and geometry across the shelf profile, documents the occurrence of different generations of dolomitization phases within each depositional cycle, describes the petrographical and geochemical characteristics of dolomite types, and discusses dolomitization in terms of relative sea-level variations and explores the potential that repetition of several separate episodes of early dolomitization has to form massive shelf dolomite.

GEOLOGICAL SETTING

The study area is located in the eastern portion of the Guadalupe Mountains (Fig. 1). The Permian Capitan Reef Complex is exposed for 80 km along a northeast–southwest trend in New Mexico and west Texas, and consists of a shelf with an evaporitic lagoon, shelf crest and outer shelf, shelf margin and

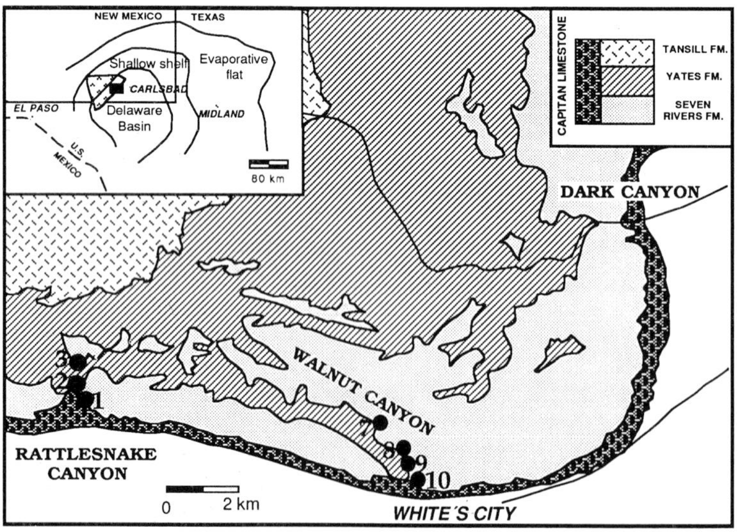

Fig. 1. Geological and palaeogeographic (inset) map of the study area showing the location of the selected measured sections used to construct Figure 3. The triangle in the inset indicates the position of the Guadalupe Mountains, the black rectangle the location of the study area. The geological map is modified after Hayes (1957), the palaeogeographic map after Ward *et al.* (1986).

deep basin (King, 1948; Pray and Esteban, 1977; Pray, 1988). The evolution of the platform during the deposition of the Capitan Reef Complex is characterized by maximum progradation during deposition of the Seven Rivers Formation on a low-relief margin, and aggradation and consequent steepening of the shelf margin during deposition of the Yates and Tansill Formations (Garber *et al.*, 1989). During deposition of the Yates and Tansill, shelf-crest restrictions were effectively creating abrupt shelfward changes in facies types (Pray, 1988). Siliciclastics are commonly interbedded with the shelf carbonates (Candelaria 1982, 1989; Garber *et al.*, 1989; Borer & Harris, 1991a,b). The deposition of the siliciclastic/carbonate packages is controlled by three orders of superimposed low-amplitude relative sea-level variations (Borer & Harris, 1991a,b).

Shelf carbonates of the Capitan Reef Complex are entirely composed of dolomite (Dunham, 1972; Neese, 1979; Schwartz, 1981; Candelaria, 1982; Garber *et al.*, 1989; Mutti, 1990; Borer & Harris, 1989, 1991a,b), whereas the outer shelf, the massive Capitan Reef (Garber *et al.*, 1989) and the foreslope (King, 1948; Mruk, 1985) are composed mainly of calcite.

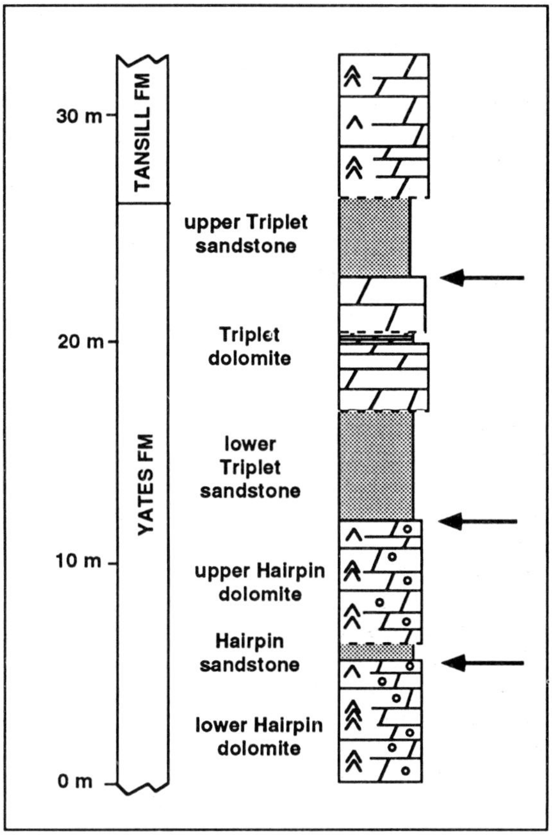

Fig. 2. Informal terminology used for the siliciclastic/carbonate units present within the upper Yates and lowermost Tansill Formations. The base of the siliciclastics is marked by a sharp and planar surface, while the transition from siliciclastics to carbonate is gradual. The inverted V symbols indicate tepees, the circles indicate pisoids.

UPPER YATES DEPOSITIONAL FACIES AND CYCLES

The Yates Formation is characterized by depositional facies belts which remain fairly constant in bands parallel to the Capitan shelf edge, and which grade seaward into massive outer shelf carbonates (Pray & Esteban, 1977; Neese & Schwartz, 1977; Neese, 1979; Schwartz, 1981). Different stratigraphic units have received different names (Fig. 2). The Yates Formation contains depositional cycles characterized by sharp, erosional basal surfaces where sandstone overlies carbonate, and by upward gradations from sandstone to carbonate (Borer & Harris, 1989; Mutti, 1990; Mutti & Simo, 1990) (Fig. 2).

The typical cycle is deposited on a basinward gently sloping shelf and is composed of a lower transgressive siliciclastic and restricted carbonate facies and an upper high-stand carbonate facies (Fig. 3). In the transgressive portion, siliciclastic facies, composed of well-sorted coarse silt to fine sandstone and local dolomite matrix, occur as tabular bodies with sharp, erosive bases (Fig. 4a). The siliciclastics near the shelf margin contain evaporitic moulds, are interbedded with inter- to supratidal marine peloidal packstones and wackestones, and pinch out basinward within 300–500 m from the shelf margin. These characteristics suggest reworking of a probable aeolian sand in a marine environment (Mutti & Simo, 1990; Borer & Harris, 1991a,b). The siliciclastics grade transitionally upward in the Hairpin cycle into restricted inter- to supratidal peloidal and pisolitic packstones, and in the Triplet cycle in the inner shelf into mudstones and ostracod wackestones with evaporite moulds (Fig. 4b), which are progressively replaced basinward by inter- to supratidal peloidal packstones with semirestricted fauna (Fig. 3). The complex relationships between siliciclastics and carbonates near the shelf margin, and cyclic

Fig. 3. Lithofacies in the Yates Formation. (a) Siliciclastic/carbonate cycles in Rattlesnake Canyon (Section 3, Fig. 1). The base of the siliciclastics is characterized by a sharp and planar surface, whereas their top is transitional to carbonates. (b) Mudstones and wackestones with moulds of evaporative minerals (Rattlesnake Canyon, Triplet dolomite, Section 3, Fig. 1). (c) Laminated fenestral peloidal packstone (Walnut Canyon, Hairpin dolomite, Section 8, Fig. 1). The lens cap is 51 mm in diameter. (d) Dolomite rip-up clasts reworked by the siliciclastics (mouth of Walnut Canyon, Triplet dolomite, Section 10, Fig. 1). The lens cap is 51 mm in diameter.

inter- to supratidal vertical trends suggest higher-frequency cyclicity.

During the upper high-stand portion, pisolitic packstone and grainstones were deposited in sub- to supratidal conditions in the Hairpin cycle (Esteban & Pray, 1983; Mutti, 1990). In the Triplet cycle, inter- to supratidal peloidal packstones with semirestricted fauna are widespread in the inner and middle shelves. In both cycles, the facies toward the shelf margin are replaced by open-marine skeletal–peloidal grainstones in thick massive or cross-bedded strata. The upper high-stand portion is capped in the inner shelf by sharp and erosive surfaces and in the outer shelf by subaerial exposure surfaces.

At the base of the cycles facies belts are shifted significantly basinward, where aeolian siliciclastics overlie open-marine facies, and are shifted landward during the transgressive part of the cycles (Fig. 3). Facies stacking in the upper, carbonate part of the cycle indicates a progressive increase in water circulation, with deposition of sub- to supratidal facies.

The upper Yates Formation was probably deposited during the high-stand phase of a third-order sea-level fluctuation and each cycle represents frequencies in the order of 400 000 years (Sarg & Lehmann, 1986; Borer & Harris, 1991a,b). Borer and Harris also suggested that shorter-duration 100 000-year eustatic cycles determine the internal packaging of the larger (400 000 years) and the third-order sequences. Borer and Harris showed that the preservation potential of siliciclastics depends on composite sea-level fluctuations, and suggested that siliciclastics dominated the shelf during the low-stand parts of asymmetric, 400 000-year eccentricity cycles, whereas carbonates were deposited during higher stands of relative sea level. In particular they

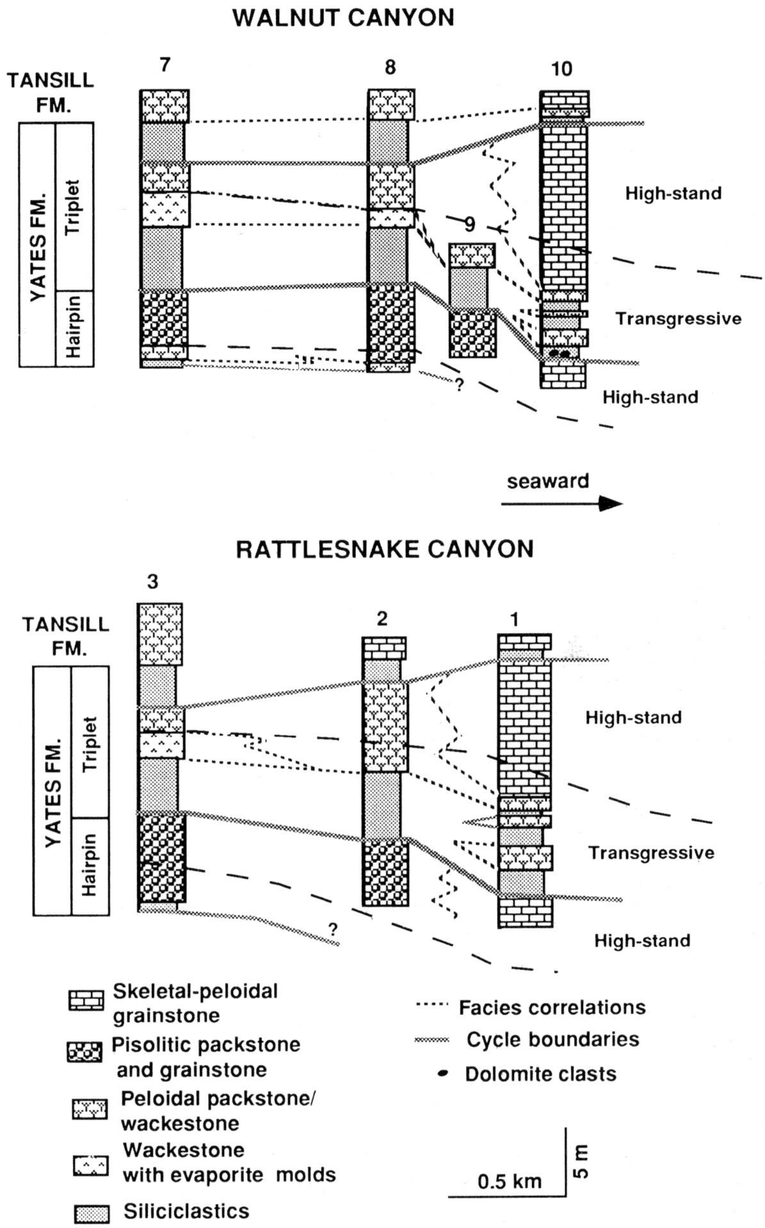

Fig. 4. Distribution of facies types along a dip profile on the shelf in Walnut and Rattlesnake Canyons.

suggested that carbonate sedimentation mainly occurred when 400 000-year high-stands acted in phase with 100 000-year high-stands.

DOLOMITE DISTRIBUTION

Massive dolomites are restricted to shelf strata, whereas the massive Capitan Reef and foreslope facies are mostly limestone (Fig. 5). Dolomite may occur as both a replacive and a primary precipitate in the Capitan Reef (Garber *et al.*, 1989) and in the foreslope (Mruk, 1985), but is never pervasive and has a patchy distribution.

The distribution of dolomite on the shelf, although generally related to facies belts, does not appear to be directly related to depositional facies, degree of cementation and palaeoporosity (Fig. 5). In the

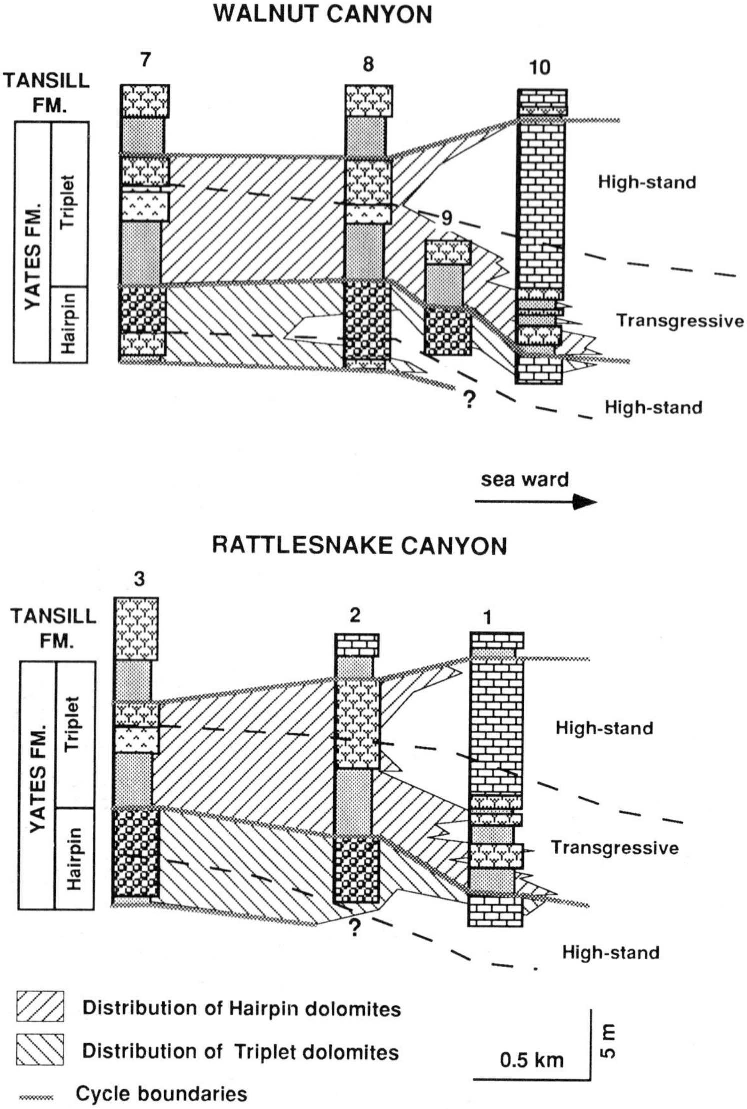

Fig. 5. Distribution of dolomite and its relationship to facies types. Dolomite distribution depends on both the distribution of the depositional facies and the sequence-stratigraphic framework.

inner shelf all facies, from sub- to supratidal, are now composed of dolomite, whereas toward the shelf margin, approaching the dolomite–calcite interface, only the inter- and supratidal facies are dolomitized, whereas subtidal grainstones are limestone (Fig. 5). Dolomite also occurs in subtidal facies only below cycle boundaries (Fig. 5).

The dolomite–limestone interface for Hairpin dolomite in Walnut Canyon shifts progressively shelfward during the transgressive part of the cycle (Fig. 5). No data on dolomite distribution are available for the same interval in Rattlesnake Canyon. However, a reversal of this transgressive trend is observed below the cycle boundaries in both canyons, where the uppermost layer (0.1–0.5 m) is composed of dolomite. At this level the dolomite–limestone interface is shifted significantly basinward (Fig. 5), and dolomite occurs in supra-, inter- and subtidal facies indiscriminately.

The dolomite–calcite interface for Triplet dolomite in both Rattlesnake and Walnut Canyons shifts progressively shelfward during the transgressive portion of the cycle (Fig. 5). Near the shelf margin (Sections 1 and 10), dolomitized supratidal facies cap the inter- and subtidal portions of cycles, suggesting that high-frequency cyclicity also controls dolomitization. Below the cycle boundary, supra-, inter- and subtidal facies are dolomitized indiscriminately, and the dolomite–calcite interface is shifted basinward.

DOLOMITE PETROGRAPHY

Two types of dolomite are distinguished on the basis of crystal size: fine- and coarse-crystalline. Fine-crystalline dolomite is volumetrically dominant and consists of closely packed anhedral crystals <4 μm in size, with irregular intercrystalline boundaries. Dolomite crystals are turbid and yellow-brown in colour under the petrographic microscope, and dull red under cathodoluminescence. Fine-crystalline dolomite replaced both depositional fabrics, such as pisoids, peloids and dasycladaceans (Figs 6a, 7b and 7c), carbonate mud (Fig. 6d), and early syndepositional cements (Fig. 7). Fine-crystalline dolomite shows a variable degree of fabric preservation, ranging from mimetic to destructive fabric (Figs 7a and 7c). In fabric-destructive replacement, original fabrics are commonly outlined by abundant inclusions (Figs 7c and 7d). Fine-crystalline dolomite commonly replaces radial cement fans, in which square terminations (Figs 7a and 7b) suggest a former aragonitic composition (Assereto & Folk, 1980) and isopachous fibrous to bladed cements, whose textural analogies with modern cements suggest a former Mg-calcite composition (James & Choquette, 1983).

Coarse-crystalline dolomite is less common and occurs as pore-lining and pore-filling cement in fenestral pores of inner-shelf facies (Fig. 8a). Coarse-crystalline dolomite consists of subhedral to anhedral dolomite crystals 20–300 μm in size, where crystal size increases towards the centre of the pore (Fig. 8b).

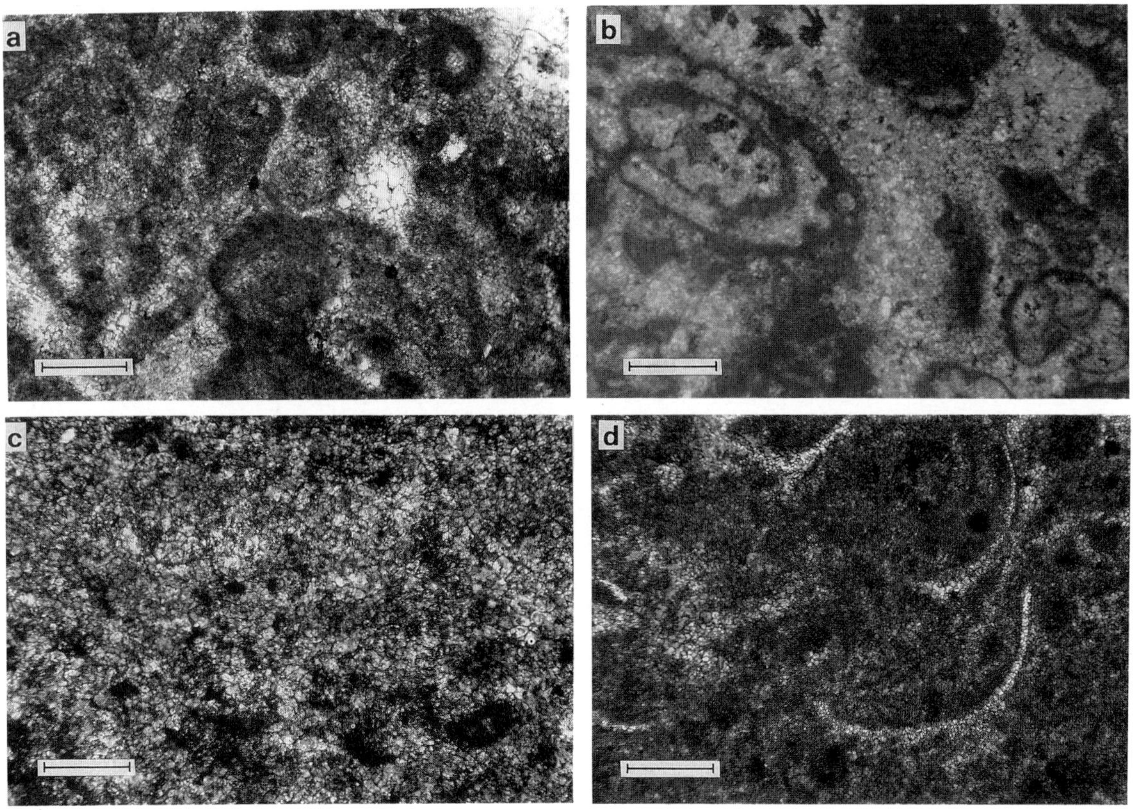

Fig. 6. Fine-crystalline dolomite replacing syndepositional fabrics. (a) The original fabric of this peloidal packstone has been partially destroyed by replacement with fine-crystalline dolomite. (Hairpin Dolomite, Rattlesnake Canyon.) (b) Skeletal grainstone showing mimetic dolomite replacement. (Hairpin Dolomite, mouth of Walnut Canyon.) (c) Non-mimetic dolomite replacement on peloidal packstone. (Triplet Dolomite, Walnut Canyon.) (d) Fine-crystalline dolomite replaces mud-rich facies. (Triplet Dolomite, Walnut Canyon.) Scale bars = 0.2 mm.

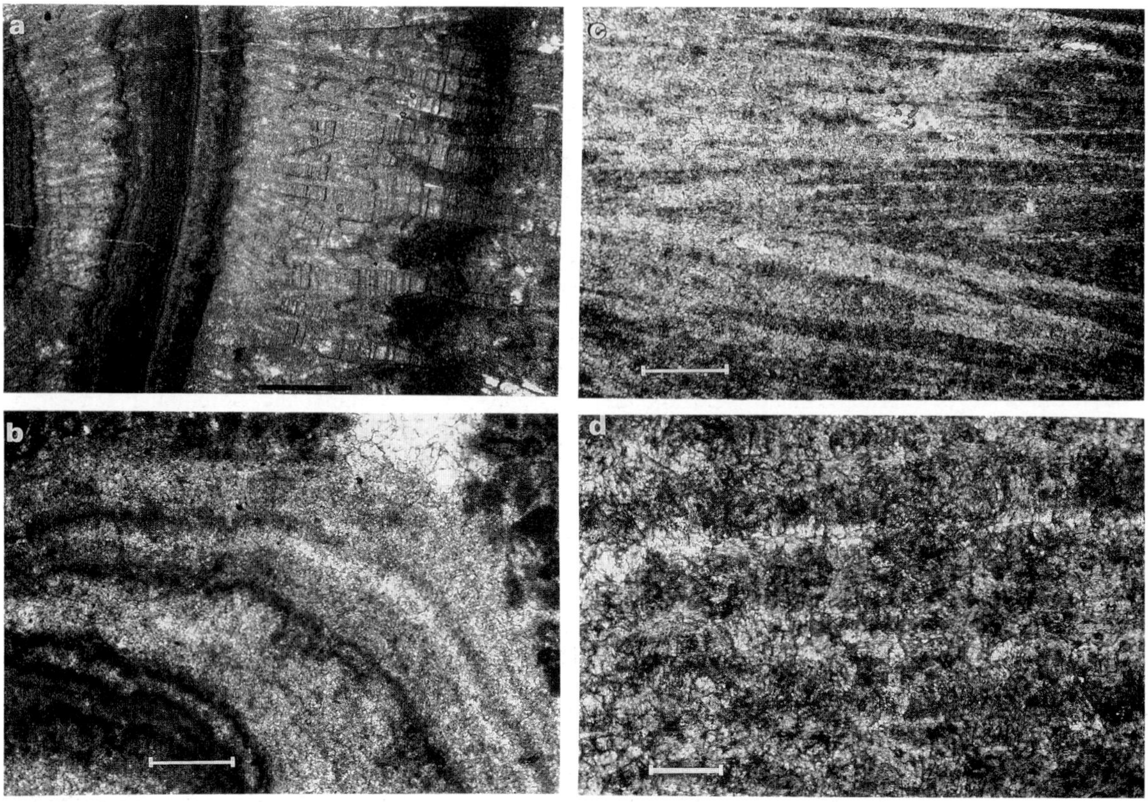

Fig. 7. Fine-crystalline dolomite replacing syndepositional cements. (a) Mimetic replacement of fibrous cements. Crystals clearly show square terminations, indicative of a former aragonitic composition. Scale bar = 0.2 mm. (b) Partially fabric-destructive replacement of fibrous cement. The presence of inclusions delineates former square crystal terminations. Scale bar = 0.2 mm. (c) Non-mimetic replacement of dolomite on fibrous cement. The original morphology is outlined by inclusions. (Hairpin Dolomite, Walnut Canyon.) Scale bar = 0.2 mm. (d) Detail of (c), showing how the original texture is outlined by inclusions. Scale bar = 0.1 mm.

Crystals are clear and transparent under the petrographic microscope, with both planar and straight compromise boundaries. Coarse-crystalline dolomite is generally black under cathodoluminescence, although thin circular zoning (<20 μm) has been observed in the outer crystal boundaries, ranging from bright red to orange in colour. Several features suggest that coarse-crystalline dolomite is a primary cement rather than a former calcitic cement that has been dolomitized:

1 Dolomite crystals are clear, whereas other dolomitized cements are darker and show abundant inclusions.

2 Dolomitized cements are generally finer-grained and have a more heterogeneous texture.

3 Intercrystalline boundaries are planar.

4 Cathodoluminescence colour is either black or zoned, but is never patchy-dull as the replacement.

5 This cement lines primary pores or grows from dolomitized cements.

A Pleistocene textural analogue for this dolomite is described by Kaldi and Gidman (1982) in Great Abaco Island, Bahamas, and it is interpreted to have been precipitated in a mixed marine–meteoric environment.

TIMING OF DOLOMITIZATION

The stratigraphic distribution of dolomites (Fig. 5), cross-cutting relationships and dolomite fabrics allows the identification of separate dolomitization events and an estimation of the relative timing of dolomitization (Mutti, 1990).

Fine- and coarse-crystalline dolomites are intimately related to supra- and intertidal facies. Fine-

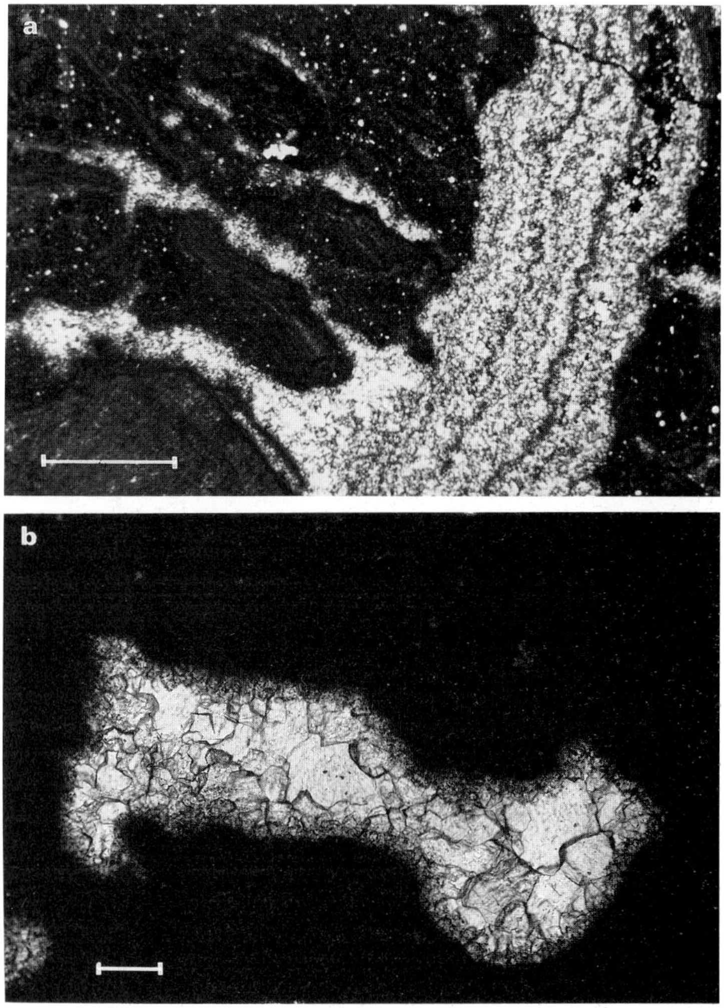

Fig. 8. Coarse-crystalline dolomite. (a) Dolomite cement in syndepositional fenestral pores. The depositional fabrics and the laminated cement to the right have been completely replaced by fine-crystalline dolomite. (Hairpin Dolomite, Walnut Canyon.) Scale bar = 1 mm. (b) Fenestral pore lined and filled by clear dolomite cement. Note that dolomite crystal size increases toward the centre of the pore. (Hairpin Dolomite, Walnut Canyon.) Scale bar = 0.1 mm.

crystalline dolomite also occurs on subtidal facies below cycle boundaries (Fig. 5), suggesting a relationship between depositional environments, cycle boundaries and dolomitization. Fine-crystalline dolomite is incorporated as clasts at the base of the transgressive siliciclastic sandstones, together with non-dolomitized clasts, suggesting dolomitization prior to reworking. Further evidence of several dolomitization events, each one related to the deposition of each depositional cycle, is provided by the seaward and landward shifts of the dolomite–limestone interface. These dolomitization events represent the diagenetic record of higher-frequency depositional episodes within cycles.

The timing of precipitation of coarse-crystalline dolomite cement within each cycle cannot be constrained exactly. Dolomite cement in the Hairpin unit occurs mostly in fenestral pores and is laterally associated with evaporitic moulds now filled by sparry calcite. Sparry calcite precipitates in shallow-burial conditions later in the diagenetic history (Mutti, 1990). The fact that no coarse-crystalline dolomite cement occurs in the evaporite moulds suggests that they were still filled by evaporite minerals during the replacive dolomitization and precipitation of dolomite cements. This relationship indicates that precipitation of dolomite cement occurred in an environment in which fluids were saturated with respect to gypsum-anhydrite.

Figure 9 shows the occurrence of dolomitization

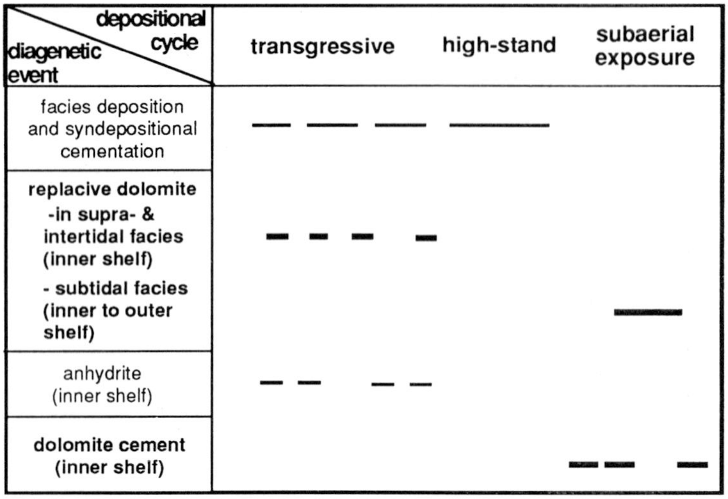

Fig. 9. Generalized paragenetic sequence of the dolomitization phases and some of the depositional and early diagenetic events in the Yates cycles.

events within cycles; several high-frequency dolomitization events occur in supra- and intertidal facies during the landward shift of facies belts in the transgressive portion of the cycle, whereas dolomitization is shifted basinward and occurs also in subtidal facies during the late high-stand and subaerial exposure at cycle tops.

DOLOMITE GEOCHEMISTRY

Microprobe analyses

Concentrations of Mg, Fe, Mn and Sr were determined by electron microprobe analyses with a nine-spectrometer SEMQ electron microprobe at the University of Wisconsin-Madison. Operating conditions were 15 KeV accelerating voltage, 15 mA absorbed current, approximately 10 μm spot size and 80 s counting time for each element. Detection limits at the 95% level of significance for Mn, Fc, Sr and Mg are approximately 110 ppm, 120 ppm, 160 ppm, and 280 ppm, respectively (E. Glover, personal communication).

Minor element concentrations of replacive dolomites, dolomite cement and, for comparison, calcite cement analogous to the cements replaced by dolomite, are listed in Table 1. Mg and Ca concentrations suggest that all dolomites are nearly stoichiometric, as supported by the high degree of crystallographic ordering shown by X-ray diffraction (Mutti, 1990). Fe concentrations display a difference of almost an order of magnitude between fine-crystalline dolomites

Table 1. Summary of major and trace element concentrations

	Dolomite replacing syndepositional cements	Dolomite cement	Bladed and radial-fibrous calcite cements
Fe (ppm)	398	2612	174
Mn (ppm)	129	401	b.d.l.
Sr (ppm)	b.d.l.	128	b.d.l.
$MgCO_3$ (mole %)	49.7	50.7	0.15

(average 398 ppm) and coarse-crystalline dolomites (average 2612 ppm). Coarse-crystalline dolomites are also enriched in Mn (average 401 ppm) with respect to microcrystalline dolomites (average 129 ppm). Differences in element concentrations further support the distinction between fine- and coarse-crystalline dolomite, and suggest a difference in the environment of precipitation and in the relative timing of formation.

Fine-crystalline dolomite is generally characterized by low Fe and Mn concentrations, although iron oxides (haematite) commonly occur in the carbonate matrix. This would suggest an oxidizing environment of formation for fine-crystalline dolomite. If the environment had been reducing, Fe from the oxides could have been reduced and brought into the solution.

Coarse-crystalline dolomites are characterized by high Fe and Mn concentrations, and cathodoluminescence zoning patterns reflect changes in the con-

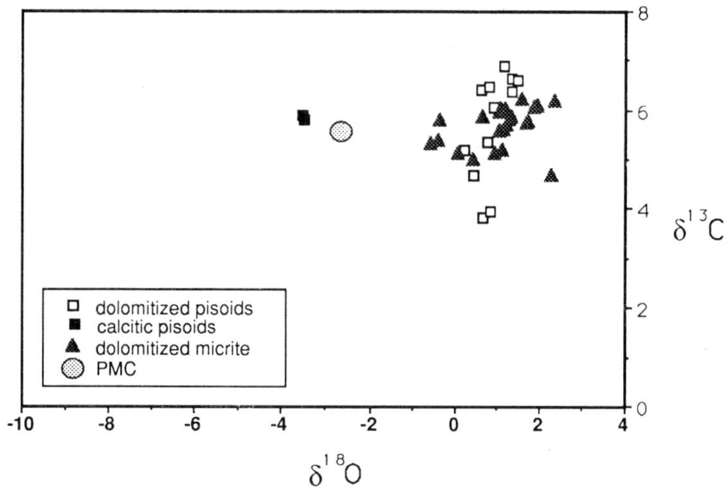

Fig. 10. Carbon and oxygen isotopic composition of replacive dolomites in mudstones (dolomicrite) and dolomite pisoids, compared to the composition of calcitic pisoids. PMC refers to the composition of Permian marine carbonates (PMC), $\delta^{18}O_{PDB} = -2.7‰$ and $\delta^{13}C_{PDB} = +5.3‰$, estimated by Given and Lohman (1985) on multiple covariant analyses of botryoidal and isopachous fibrous cements from the Capitan Reef Complex.

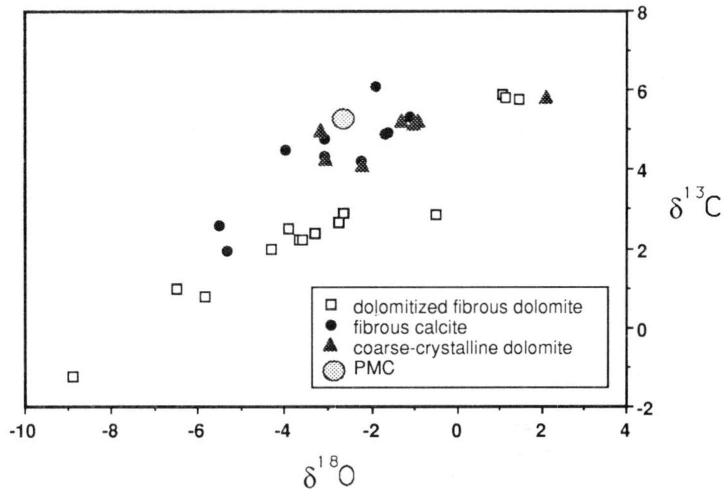

Fig. 11. Carbon and oxygen isotopic composition of replacive dolomites of fibrous cement (dolomitized fibrous cement), of primary dolomite cements and of undolomitized fibrous cement.

centrations of these elements. These variations are related to a reducing environment, with fluctuations either in the supply of Fe (and Mn) or fluctuations in Eh. The Fe and Mn could be provided by the sulphates in the inner-shelf facies, from the oxides and silicates common in the siliciclastic units, or from the red shales in the lagoon. A reducing environment is required to bring these elements into the water and then incorporate them in the coarse-crystalline dolomite.

Stable isotopes

Samples for stable isotope analysis were collected from thin-section chips drilled using a 0.5 mm carbide burr. CO_2 for the analyses was produced by reaction of the powder with H_3PO_4 at 25°C for 8 h, and analysed on a MAT 251 triple-collector mass spectrometer at the University of Wisconsin-Madison. Data were standardized using NBS-16.

Carbon and oxygen isotope data of replacive and primary dolomites of the Upper Yates Formation are shown in Figures 10 and 11, are summarized in Table 2 and listed in Table 3. The measured values are compared to the average composition of Permian marine carbonates (PMC), $\delta^{18}O_{PDB} = -2.7‰$ and $\delta^{13}C_{PDB} = +5.3‰$, estimated by Given and Lohmann (1985) on multiple covariant analyses of marine botryoidal and isopachous fibrous cements from the Capitan Reef Complex.

Table 2. Summary of stable isotopic data

	$\delta^{13}C$	$\delta^{18}O$	Analysed samples
Fine-crystalline dolomite			
Dolomitized micrite	5.6 ± 0.5	1.1 ± 0.8	23
Dolomitized pisoids	5.7 ± 1.0	0.8 ± 0.4	13
Calcitic pisoids	5.8 ± 0.08	−3.5 ± 0.03	2
Dolomitized fibrous cement	2.7 ± 2.0	−3.0 ± 3.0	14
Fibrous calcite cement	4.4 ± 1.2	−3.0 ± 1.5	10
Coarse-crystalline dolomite	5.0 ± 0.6	−1.3 ± 1.8	7

Fine-crystalline dolomites have been analysed in dolomitized micrite, dolomitized pisoids (Fig. 8) and dolomitized fibrous cements (Fig. 9). Dolomicrite samples (Fig. 8) are characterized by a relatively narrow range of $\delta^{13}C$ (average +5.6‰ ± 0.5) corresponding to the PMC value, whereas $\delta^{18}O$ ranges from −0.6‰ to +2.3‰. The average enrichment of $\delta^{18}O$ with respect to PMC is +3.8‰ (ranging between 2.3 and 5‰). No significant differences in isotope composition are apparent in dolomicrite samples from different stratigraphic levels, nor with different positions along the shelf profile. Fine-crystalline dolomite in dolomitized pisoids shows quite homogeneous $\delta^{18}O$ values (average +0.8‰ ± 0.4), with a mean enrichment of 3.5‰ relative to PMC and 4.35‰ relative to their undolomitized equivalents (Fig. 8; Table 2). The $\delta^{13}C$ values of dolomitized pisoids are variable (Fig. 8), with an average of 5.7‰ ± 1.0.

Fine-crystalline dolomites on dolomitized fibrous cements, calcitic fibrous cements and coarse-crystalline dolomite all show a strong covariant trend between $\delta^{13}C$ and $\delta^{18}O$ (Fig. 9), which reflects a differential degree of mixing with isotopically lighter (up to $\delta^{18}O$ = −12‰ and $\delta^{13}C$ = −6.9‰) late calcitic cements (Mutti, 1990). The mixture of the different cements is caused by the small size of the analysed cements that does not allow separate sampling with the dental drill technique. Fractionation factors have been calculated for each sample, taking into account the degree of mixing with sparry calcite cement, estimated from petrographic observations. The most enriched $\delta^{18}O$ values are therefore considered most likely to reflect the actual isotopic composition of the dolomite phases. Dolomitized fibrous cement shows a maximum $\delta^{18}O$ enrichment of 4.2‰ with respect to PMC, whereas the most enriched $\delta^{13}C$ values are very close to the PMC values. Coarse-crystalline dolomite shows a maximum enrichment of 3.4‰ with respect to PMC, and $\delta^{13}C$ values are slightly depleted with respect to the PMC values.

Fine-crystalline dolomite is characterized by $\delta^{18}O$ values enriched with respect to the PMC (average values range between 3.7 and 4.2‰) and by $\delta^{13}C$ values close but slightly enriched with respect to the PMC. The association of dolomite with the depositional facies indicates an early origin, and the only fluid capable of extensive dolomitization during or immediately after deposition is Permian seawater of normal or elevated salinity. The unimodal fine-crystalline dolomite texture is typical of near-surface low-temperature conditions, with a high density of nucleation sites (Sibley & Gregg, 1987), as would be expected for inter- and supratidal mud-rich facies. Facies analysis supports an evaporitic setting in the inner shelf, where anhydrite moulds in supratidal facies testify to high evaporation rates on the Yates shelf. The enrichment of $\delta^{18}O$ is in agreement with this interpretation. Estimates of calcite–dolomite oxygen fractionations predict dolomite to be approximately 3‰ enriched in $\delta^{18}O$ with respect to calcite (Land, 1980). Holocene dolomites from the Abu Dhabi sabkha are enriched in $\delta^{18}O$ by 3.2‰ with respect to the coexisting waters (McKenzie, 1981), whereas Sugarloaf dolomites (Carballo *et al.*, 1987) are enriched by 2.8‰.

Coarse-crystalline dolomite is characterized by $\delta^{13}C$ values both slightly depleted and enriched with respect to the PMC, $\delta^{18}O$ enriched by 3.4‰ with respect to the PMC, and by high zoned concentrations of Fe and Mn. Textural relationships discussed in the petrographic section indicate that the fluids from which coarse-crystalline dolomite precipitated were saturated with respect to evaporites. An additional 3–4‰ $\delta^{18}O$ enrichment is predicted by the non-equilibrium evaporation of seawater to near $CaSO_4$ saturation (Lloyd, 1966; Holser, 1979). The oxygen values are generally more depleted with respect to the fine-crystalline dolomite, and it may reflect precipitation at higher temperatures, different fluid composition or fluid–rock interaction. The depletion in $\delta^{13}C$ values requires input of a light carbon (^{12}C) such as soil gas carbon or carbon liberated during the reduction of sulphates or organic matter. There is no evidence of soil formation during the deposition of the Yates cycles that would justify input of soil carbon. Dolomites formed in or just below the zone of bacterial sulphate reduction should

Table 3. Sample number, geographical location and stable isotope values of analysed samples. The numbers in the location column correspond to the stratigraphic sections indicated in Figure 1.

Sample	Unit	Location (section/canyon)	Facies	$\delta^{13}C$ (PDB)	$\delta^{18}O$ (PDB)
Dolomitized pisoids					
E4-1	Hairpin	3	Pisolitic grainstone and packstone	6.60	1.50
E5-2	Hairpin	3	Pisolitic grainstone and packstone	5.35	0.76
E11-1	Tansill	3	Pisolitic grainstone and packstone	6.06	0.92
T0-1	Hairpin	(Walnut Canyon)	Pisolitic grainstone and packstone	6.40	0.63
T10-1	Tansill	(Walnut Canyon)	Pisolitic grainstone and packstone	6.48	0.82
R3-1	Hairpin	(Walnut Canyon)	Pisolitic grainstone and packstone	6.90	1.18
W2-1	Hairpin	(Walnut Canyon)	Pisolitic grainstone and packstone	6.63	1.35
W13-2	Tansill	(Walnut Canyon)	Pisolitic grainstone and packstone	5.20	0.21
W14-1	Tansill	(Walnut Canyon)	Pisolitic grainstone and packstone	4.70	0.43
Z0-2	Hairpin	(Walnut Canyon)	Pisolitic grainstone and packstone	6.38	1.35
Z10-2	Tansill	(Walnut Canyon)	Pisolitic grainstone and packstone	5.81	1.33
D2-1	Hairpin	(Dark Canyon)	Pisolitic grainstone and packstone	3.96	0.86
D2-1	Hairpin	(Dark Canyon)	Pisolitic grainstone and packstone	3.82	0.67
Calcitic pisoids					
WTC2B-5	Hairpin	8	Pisolitic grainstone and packstone	5.79	−3.47
WTC2B-6	Hairpin	8	Pisolitic grainstone and packstone	5.91	−3.51
Dolomitized fibrous cements					
E5-3	Hairpin	3	Pisolitic grainstone/packstone	0.78	−5.83
E5-4	Hairpin	3	Pisolitic grainstone/packstone	2.89	−2.62
Z5-3	Hairpin	(Walnut Canyon)	Pisolitic grainstone/packstone	5.82	1.15
Z5-4	Hairpin	(Walnut Canyon)	Pisolitic grainstone/packstone	5.76	1.46
Z5-5	Hairpin	(Walnut Canyon)	Pisolitic grainstone/packstone	5.87	1.08
W1A-7	Hairpin	(Walnut Canyon)	Pisolitic grainstone/packstone	2.88	−0.50
D3-4	Hairpin	(Dark Canyon)	Pisolitic grainstone/packstone	1.00	−6.48
D3-5	Hairpin	(Dark Canyon)	Pisolitic grainstone/packstone	1.99	−4.30
D3-6	Hairpin	(Dark Canyon)	Pisolitic grainstone/packstone	2.50	−3.90
E5-5	Hairpin	3	Pisolitic grainstone/packstone	2.24	−3.57
E5-6	Hairpin	3	Pisolitic grainstone/packstone	2.37	−3.30
E5-7	Hairpin	3	Pisolitic grainstone/packstone	−1.26	−8.89
E5-8	Hairpin	3	Pisolitic grainstone/packstone	2.66	−2.74
W5-11	Hairpin	(Walnut Canyon)	Pisolitic grainstone/packstone	2.22	−3.65
Calcitic fibrous cements					
WTC2B-1	Hairpin	8	Pisolitic grainstone/packstone	4.50	−3.99
WTC2B-2	Hairpin	8	Pisolitic grainstone/packstone	6.08	−1.91
WTC2B-3	Hairpin	8	Peloidal packstone/wackstone	4.78	−3.08
WSCT-1	Hairpin	10	Skeletal-peloidal grainstone	2.59	−5.51
WSCT-2	Hairpin	10	Skeletal-peloidal grainstone	4.21	−2.25
WSCT-3	Triplet	10	Skeletal-peloidal grainstone	4.89	−1.71
WSCT-4	Triplet	10	Skeletal-peloidal grainstone	1.94	−5.31
WM3-1	Hairpin	10	Skeletal-peloidal grainstone	4.93	−1.63
WM3-2	Hairpin	10	Skeletal-peloidal grainstone	5.35	−1.12
WM3-3	Hairpin	10	Skeletal-peloidal grainstone	4.32	−3.08

continued on p.104

have depleted $\delta^{13}C$ values, whereas those formed in the zone of methanogenesis should have enriched $\delta^{13}C$ values (Claypool & Kaplan, 1974; Kelts & McKenzie, 1982; Burns *et al.*, 1988). Conditions at the boundary between sulphate reduction and methanogenesis would explain the coexistence in coarse-crystalline dolomite of depleted $\delta^{13}C$ values (with Fe and Mn in solution) and enriched $\delta^{13}C$ values (with incorporation of Fe and Mn in the dolomite).

Table 3. *Continued*

Dolomitized micrite					
E5-1	Hairpin	3	Pisolitic grainstone/packstone	5.12	0.09
E7-1	Triplet	3	Wackestone	4.69	2.31
E7-3	Triplet	3	Wackestone	6.02	1.19
E10-1	Tansill	3	Peloidal packstone/wackestone	4.69	2.31
E11-2	Tansill	3	Peloidal packstone/wackestone	5.88	0.67
E11-3	Tansill	3	Peloidal packstone/wackestone	5.57	1.07
M3-1	Hairpin	2	Pisolitic grainstone/packstone	5.69	1.26
M5-1	Triplet	2	Peloidal packstone/wackestone	6.21	1.61
T2-1	Hairpin	(Walnut Canyon)	Pisolitic grainstone/packstone	5.40	−0.40
T7-1	Triplet	(Walnut Canyon)	Peloidal packstone/wackestone	5.31	−0.57
R5-1	Hairpin	(Walnut Canyon)	Pisolitic grainstone/packstone	6.18	2.35
R8-1	Tansill	(Walnut Canyon)	Peloidal packstone/wackestone	5.76	1.75
W7-1	Triplet	(Walnut Canyon)	Peloidal packstone/wackestone	5.12	0.95
W13-1	Tansill	(Walnut Canyon)	Peloidal packstone/wackestone	5.86	1.35
W14-1-1	Tansill	(Walnut Canyon)	Peloidal packstone/wackestone	5.81	−0.36
Z0-1	Hairpin	(Walnut Canyon)	Pisolitic grainstone/packstone	5.97	1.05
Z0-3	Hairpin	(Walnut Canyon)	Pisolitic grainstone/packstone	6.01	1.11
Z5-1	Hairpin	(Walnut Canyon)	Pisolitic grainstone/packstone	6.06	1.91
Z7-1	Triplet	(Walnut Canyon)	Peloidal packstone/wackestone	5.19	1.14
Z10-1	Tansill	(Walnut Canyon)	Peloidal packstone/wackestone	4.99	0.44
D3-1	Hairpin	(Dark Canyon)	Pisolitic grainstone/packstone	5.62	1.17
D9-1	Triplet	(Dark Canyon)	Peloidal packstone/wackestone	6.09	1.97
D13-1	Tansill	(Dark Canyon)	Peloidal packstone/wackestone	5.75	1.71
Coarse-crystalline dolomite					
W5-1	Hairpin	(Walnut Canyon)	Peloidal packstone/wackestone	4.27	−3.04
W5-2	Hairpin	(Walnut Canyon)	Peloidal packstone/wackestone	4.10	−2.19
W5-3	Hairpin	(Walnut Canyon)	Peloidal packstone/wackestone	5.13	−0.99
W5-4	Hairpin	(Walnut Canyon)	Peloidal packstone/wackestone	5.20	−1.28
W5-9	Hairpin	(Walnut Canyon)	Peloidal packstone/wackestone	4.99	−3.13
W5-10	Hairpin	(Walnut Canyon)	Peloidal packstone/wackestone	5.22	−0.89
W5-13	Hairpin	(Walnut Canyon)	Peloidal packstone/wackestone	5.81	2.11

EARLY DOLOMITIZATION AND RELATIVE SEA-LEVEL CHANGES

The petrographic and geochemical data discussed in the earlier sections indicate that dolomitization both replaced carbonate grains, aragonitic and Mg-calcite cements and fine-grained sediments, and precipitated as pore-filling cement, and occurred early, both during deposition and during subaerial exposure at the cycle top, as fine-grained dolomite was subjected to marine reworking at the base of the next cycle. Oxygen isotopes of both fine- and coarse-crystalline dolomite suggest an original marine fluid, modified by evaporation, and major element concentrations indicate an oxidizing environment for fine-crystalline dolomite and reducing for coarse-crystalline dolomite. Carbon isotopes indicate coarse-crystalline dolomite precipitation during or above the zone of bacterial sulphate reduction.

The fine-grained dolomites are interpreted to have occurred in response to seawater flow through the shallow subsurface driven by tidal pumping (similar to models by Bush (1973) and Carballo *et al.* (1987)) or by storm-recharge flooding and evaporative pumping (McKenzie *et al.*, 1980; Patterson & Kinsman, 1981), and by high evaporation rates on the supra- and intertidal areas. Dolomitization of inter-, supra- and subtidal facies below cycle boundaries occurred during times of relative sea-level falls, probably when the zone of storm flooding and evaporative pumping could expand seaward due to the gentle profile. The environmental conditions of evaporation and sea-level variations required for such models are common during the upper Guadalupian in the Permian basin. Evidence of high evaporation and aridity occurs throughout the section, with common evaporitic moulds, and in the lagoon with precipitation of gypsum and anhydrite (Sarg, 1981). Seawater movement through the intertidal sediment is also common, and probably caused the precipitation of large amounts of fibrous cements associated with the tepee facies (Rosenblum, 1984).

Coarse-crystalline dolomite probably precipitated from a similar, modified marine fluid and by a similar mechanism. However, its distribution limited to fenestral pores, its presence in equilibrium with evaporite minerals and light $\delta^{18}O$ values suggest that high salinities and high temperatures of evaporated seawater (Land, 1985; Sass & Bein, 1988) controlled the precipitation. Depleted to enriched $\delta^{13}C$ values and high Fe and Mn concentrations indicate its precipitation at the boundary between sulphate reduction and methanogenesis.

Dolomitization is interpreted to occur during both deposition and subaerial exposure of each depositional cycle. The assumptions that each depositional cycle has a duration of 400 000 years, that carbonates were deposited within 100 000 years, and that there were high-frequency sea-level changes (Mutti, 1990), suggest that the several episodes of dolomitization of supra- and intertidal facies within one cycle occurred approximately during a period of less than 100 000 years. This result is in agreement with calculated dolomitization rates (Hardie, 1987). Dolomitization of subtidal carbonates (and of marine cements) in association with the exposure at cycle tops, may be related to a longer time duration (200 000 or 300 000 years).

ACKNOWLEDGEMENTS

We thank L.C. Pray and R.H. Dott Jr. (University of Wisconsin), and J.F. Sarg for their interest and support in the project. John Valley provided access to the University of Wisconsin Stable Isotope Laboratory, and Kevin Baker helped to analyse the samples. The Carlsbad National Park, New Mexico, is gratefully acknowledged for giving permission to collect samples. This work benefited from the comments and criticism of P. Enos, B.H. Purser, M.E. Tucker, and D.H. Zenger. This research was supported by Exxon Production Research Company, The University of Wisconsin Graduate School, and GSA Grants-in-Aid.

REFERENCES

Assereto, R.L.A.M. & Folk, R.L. (1980) Diagenetic fabrics of aragonite, calcite and dolomite in an ancient peritidal–spelean environment: Triassic Calcare Rosso, Lombardia, Italy. *J. Sedim. Petrol.* **50**, 371–395.

Borer, J.M. & Harris, P.M. (1989) Depositional facies and cycles in Yates Formation outcrops, Guadalupe Mountains, New Mexico. In: *Subsurface and Outcrop Examination of the Capitan Shelf Margin, Northern Delaware Basin* (Ed. Harris, P.M. & Grover, G.A.) Spec. Publs Soc. Econ. Paleont. Mineral., Tulsa, Core Workshop **13**, 305–317.

Borer, J.M. & Harris, P.M. (1991a) Depositional facies and model for mixed siliciclastics and carbonates of the Yates Formation, Permian Basin. In: *Mixed Carbonate–Siliciclastic Sequences* (Ed. Lomando, A.J. & Harris, P.M.) Spec. Publs. Soc. Econ. Paleont. Mineral., Tulsa, Core Workshop **15**, 1–133.

Borer, J.M. & Harris, P.M. (1991b) Lithofacies and cyclicity of the Yates Formation, Permian Basin: implications for reservoir heterogeneity. *Am. Ass. Petrol. Geol. Bull.* **75**, 726–779.

Burns, S.J., Baker, P.A. & Showers, W.J. (1988) The factors controlling the formation and chemistry of dolomite in organic-rich sediments: Miocene Drakes Bay Formation, California. In: *Sedimentology and Geochemistry of Dolostones* (Ed. Shukla, V. & Baker, P.A.) Spec. Publs Soc. Econ. Paleont. Mineral., Tulsa **43**, 41–52.

Bush, P. (1973) Some aspects of the diagenetic history of the sabhka in Abu Dhabi, Persian Gulf. In: *The Persian Gulf: Holocene Carbonate Sedimentation and Diagenesis in a Shallow Epicontinental Sea* (Ed. Purser, B.H.) pp. 395–408. New York, Springer-Verlag.

Candelaria, M.P. (1982) *Sedimentology and Depositional Environments of Upper Yates Formation Siliciclastics (Permian, Guadalupian), Guadalupe Mountains, Southeast New Mexico.* Unpublished MS Thesis, University of Wisconsin-Madison, 267 pp.

Candelaria, M.P. (1989) Shallow-marine sheet sandstones, Upper Yates Formation, Northwest Shelf, Delaware Basin, New Mexico. In: *Subsurface and Outcrop Examination of the Capitan Shelf Margin, Northern Delaware Basin, New Mexico* (Ed. Harris, P.M. & Grover, G.A.) Spec. Publs Soc. Econ. Paleont. Mineral., Tulsa, Core Workshop **13**, 319–324.

Carballo, J.D., Land, L.S. & Miser, D.L. (1987) Holocene dolomitization of supratidal sediments by active tidal pumping, Sugarloaf Key, Florida. *J. Sedim. Petrol.* **39**, 70–89.

Claypool, G. & Kaplan, I.R. (1974) The origin and distribution of methane in marine sediments. In: *Natural Gases in Marine Sediments* (Ed. Kaplan, I.R.) pp. 99–140. Plenum Press, New York.

Dunham, J.B. & Olson, E.R. (1978) Diagenetic dolomite formation related to Paleozoic paleogeography of the Cordilleran miogeosyncline in Nevada. *Geology* **6**, 556–559.

Dunham, R.J. (1972) *Capitan Reef, New Mexico and Texas: Facts and Questions to Aid Interpretation and Group Discussion.* Spec. Publs Soc. Econ. Paleont. Mineral., Tulsa, Permian Basin Section **72–14**, 294 pp.

Esteban, M. & Pray, L.C. (1983) Pisoids and pisolite facies (Permian), Guadalupe Mountains, New Mexico and West Texas. In: *Coated Grains* (Ed. Peryt, T.M.) pp. 503–537. Springer-Verlag, New York.

Gao, G. & Land, L.S. (1991) Early Ordovician Cool Creek Dolomite, Middle Arbuckle Group, Slick Hills, SW Oklahoma, USA: origin and modification. *J. Sedim. Petrol.* **61**, 161–173.

GARBER, G.A., GROVER, G.A. & HARRIS, P.M. (1989) Geology of the Capitan shelf-margin – subsurface data from the Northern Delaware basin. In: *Subsurface and Outcrop Examination of the Capitan Shelf Margin, Northern Delaware Basin* (Ed. Garber, G.A., Grover, G.A. & Harris, P.M.) Spec. Publs Soc. Econ. Paleont. Mineral., Tulsa, Workshop **13**, 3–269.

GIVEN, R.K. & LOHMANN, K.C. (1985) Derivation of the original isotopic composition of Permian marine cements. *J. Sedim. Petrol.* **55**, 430–439.

GIVEN, R.K. & WILKINSON, B.H. (1987) Dolomite abundance and stratigraphic age – constraints on rates and mechanisms of Phanerozoic dolostone formation. *J. Sedim. Petrol.* **57**, 1068–1078.

GREGG, J.M. & SHELTON, K.L. (1990) Dolomitization and dolomite neomorphism in the back-reef facies of the Bonneterre and Davis Formations (Cambrian), southeastern Missouri. *J. Sedim. Petrol.* **60**, 549–562.

HARDIE, L.A. (1987) Dolomitization: a critical view of some current views. *Perspectives, J. Sedim. Petrol.* **57**, 166–183.

HAYES, P.T. (1957) *Geology of the Carlsbad Caverns East Quadrangle, New Mexico*. US Geol. Survey, Geol. Quadrangle Map GQ-98.

HOLSER, W.T. (1979) Trace elements and isotopes in evaporites. In: *Marine Minerals* (Ed. Burns, R.G.) Mineral. Soc. Am., Reviews in Mineralogy **6**, 295–346.

JAMES, N.P. & CHOQUETTE, P.W. (1983) Diagenesis 6. Limestones – The seafloor diagenetic environment. *Geoscience Canada* **10**(4), 162–179.

KALDI, J. & GIDMAN, J. (1982) Early diagenetic dolomite cements: examples from the Permian Lower Magnesian Limestone of England and the Pleistocene carbonates of the Bahamas. *J. Sedim. Petrol.* **53**, 1073–1096.

KELTS, K. & MCKENZIE, J.A. (1982) Diagenetic dolomite formation in Quaternary anoxic diatomaceous muds of Deep Sea Drilling Project Leg 64, Gulf of California. In: *Initial Reports of the Deep Sea Drilling Project* **64**, 553–570.

KERANS, C. & LUCIA, F.J. (1989) Recognition of second, third and fourth/fifth order scales of cyclicity in the El Paso Group and their relation to genesis and architecture of Ellenburger reservoirs. In: *The Lower Paleozoic of West Texas and Southern New Mexico – Modern Exploration Concepts* (Ed. Cunningham, B.K. & Cromwell, D.W.) Spec. Publs Soc. Econ. Paleont. Mineral., Tulsa, Permian Basin Section **89–31**, 105–110.

KING, P.B. (1948) *Geology of the Southern Guadalupe Mountains, Texas*. USGS Professional paper **215**, 183 pp.

LAND, L.S. (1980) *The Isotopic and Trace Element Geochemistry of Dolomite: the State of the Art*. Spec. Publs Soc. Econ. Paleont. Mineral., Tulsa **28**, 87–110.

LAND, L.S. (1985) The origin of massive dolomite. *J. Geol. Educ.* **33**(2), 112–125.

LLOYD, R.M. (1966) Oxygen isotope enrichment of sea water by evaporation. *Geochim. Cosmochim. Acta* **30**, 801–814.

MACHEL, H. & MOUNTJOY, E.W. (1986) Chemistry and environments of dolomitization: a reappraisal. *Earth Sci. Rev.* **23**, 175–222.

MCKENZIE, J.A. (1981) Holocene dolomitization of calcium carbonate sediments from the coastal sabkhas of Abu Dhabi, UAE: a stable isotope study. *J. Geol.* **89**, 185–198.

MCKENZIE, J.A., HSU, K. & SCHNEIDER, J.F. (1980) Movement of subsurface waters under the sabkha, Abu Dhabi, UAE, and its relation to evaporative dolomite genesis. In: *Concepts and Models of Dolomitization* (Ed. Zenger, D.H., Dunham, J.B. & Ethington, R.L.) Spec. Publs Soc. Econ. Paleont. Mineral., Tulsa **28**, 11–30.

MONTAÑEZ, I.P. (1991) Eustatic control on early dolomitization of cyclic peritidal carbonates: evidence from the Early Ordovician Upper Knox Group, Appalachians and Middle to late Cambrian Bonanza King Formation, Southern Great Basin (abs.). *Am. Assoc. Petrol. Geol. Bull.* **75**, 638.

MRUK, D.H. (1985) *Cementation and Dolomitization of the Capitan Limestone (Permian), McKittrick Canyon, West Texas*. Unpublished MS Thesis, University of Colorado, 153 pp.

MUTTI, M. (1990) *Sedimentology and Diagenesis of Carbonate/Siliciclastic Cycles, Yates Formation, Guadalupian, New Mexico*. Unpublished MS Thesis, University of Wisconsin-Madison, 228 pp.

MUTTI, M. & SIMO, J.A. (1990) Sedimentology and diagenesis of siliciclastic/carbonate cycles, Capitan Reef Complex, USA (abs.). *Int. Ass. Sedimentol., 13th International Sedimentological Congress, Nottingham UK* 376.

NEESE, D.G. (1979) *Facies Mosaic of the Upper Yates and Lower Tansill Formations (Upper Permian), Walnut Canyon, Guadalupe Mountains, New Mexico*. Unpublished MS Thesis, University of Wisconsin-Madison, 110 pp.

NEESE, D.G. & SCHWARTZ, A.H. (1977) Facies mosaic of the upper Yates and the lower Tansill Formations, Walnut and Rattlesnake Canyons, Guadalupe Mountains, New Mexico. In: *Upper Guadalupian Facies, Permian Reef Complex, Guadalupe Mountains, New Mexico and West Texas* (Ed. Hileman, M.E. & Mazzullo, S.J.) *Field Conference Guidebook, 1*. Spec. Publs Soc. Econ. Paleont. Mineral., Tulsa, Permian Basin Section **77–16**, 437–450.

PATTERSON, R.J. & KINSMAN, D.J. (1981) Hydrologic framework of a sabkha along the Arabian Gulf. *Am. Ass. Petrol. Geol. Bull.* **66**, 28–43.

PRAY, L.C. (1988) Geology of the Western Escarpment, Guadalupe Mountains, Texas. In: *Geologic Guide to the Western Escarpment Guadalupe Mountains, Texas* (Ed. Sarg, J.F., Rossen, C., Lehmann, P.J. & Pray, L.C.) Spec. Publs Soc. Econ. Paleont. Mineral., Tulsa, Permian Basin Section **88–30**.

PRAY, L.C. & ESTEBAN, M. (Eds.) (1977) *Upper Guadalupian Facies, Permian Reef Complex, Guadalupe Mountains, New Mexico and West Texas, 1977 Field Conference Guidebook, 2*. Spec. Publs Soc. Econ. Paleont. Mineral., Tulsa, Permian Basin Section **77–16**, 194 pp.

ROSENBLUM, M.B. (1984) *Early Diagenetic Sheet Crack Cements of the Guadalupian (Permian) Shelf, Yates and Tansill Formations, New Mexico, USA*. Unpublished MS Thesis, University of Wisconsin-Madison, 95 pp.

SARG, J.F. (1981) Petrology of the carbonate–evaporite transition of the Seven Rivers Formation (Guadalupian, Permian), southeast New Mexico. *J. Sedim. Petrol.* **51**, 73–95.

Sarg, J.F. & Lehmann, P.J. (1986) Lower-Middle Guadalupian Facies and Stratigraphy, San Andres/Grayburg Formations, Permian Basin, Guadalupe Mountains, New Mexico. In: *San Andres/Grayburg Formations, Guadalupe Mountains, New Mexico and Texas* (Ed. Moore, G.E. & Wilde, G.L.) Soc. Econ. Paleont. Mineral., Permian Basin Section, Annual Field Trip Guidebook 88–25.

Sass, E. & Bein, A. (1988) Dolomites and salinity: a comparative geochemical study. In: *Sedimentology and Geochemistry of Dolostones* (Ed. Shukla, V. & Baker, P.A.) Spec. Publs Soc. Econ. Paleont. Mineral., Tulsa **43**, 223–233.

Schwartz, A. (1981) *Facies Mosaic of the Upper Yates and Lower Tansill Formations (Upper Permian), Rattlesnake Canyon, Guadalupe Mountains, New Mexico*. Unpublished MS Thesis, University of Wisconsin-Madison, 155 pp.

Sibley, D.F. (1991) Secular changes in the amount and texture of dolomites, *Geology* **19**, 151–154.

Sibley, D.F. & Gregg, J.M. (1987) Classification of dolomite textures. *J. Sedim. Petrol.* **57**, 955–963.

Ward, R.F., Kendall, C.G.St.C. & Harris, P.M. (1986) Upper Permian (Guadalupian) facies and their association with hydrocarbons – Permian basin, west Texas and New Mexico. *Am. Ass. Petrol. Geol. Bull.* **70**, 239–262.

Wilson, J.L. (1975) *Carbonate Facies in Geologic History*. New York, Springer-Verlag, 471 pp.

Mixing-Zone and Seawater Dolomitization Models

Spec. Publs Int. Ass. Sediment. (1994) **21**, 111–132

Dolomitization by near-normal seawater? Field evidence from the Bahamas

F.F. WHITAKER,* P.L. SMART,* V.C. VAHRENKAMP,†
H. NICHOLSON‡ *and* R.A. WOGELIUS§¶

** Department of Geography, University of Bristol, Bristol BS8 1SS, UK;*
† Koninklijke/Shell Exploratie en Productie Laboratorium, Volmerlaan 6, 2080AB Rijswijk (Z-H), The Netherlands;
‡ B.P. Research, Sunbury Research Centre, Chertsey Road, Sunbury on Thames, TW16 7LN, UK; and
§ Department of Earth Sciences, University of Oxford, Oxford OX1 3PR, UK

ABSTRACT

Measurements of salinity, temperature and groundwater discharge indicate active circulation of groundwaters of near-normal seawater composition through the Great Bahama Bank beneath Andros Island. Waters of slightly elevated salinity derived from the bank surface flow eastwards beneath the island, and then mix with cold normal-salinity seawaters from the adjacent oceans. Saline groundwaters are significantly depleted in magnesium compared to Great Bahama Bank and open ocean seawaters. In the absence of major calcium depletion, this provides direct evidence for present-day replacement dolomitization. The saline groundwaters are characterized by elevated $p\text{CO}_2$ and, as a consequence, lowered carbonate mineral saturation indices, due to subsurface oxidation of organic matter principally by sulphate reduction under anoxic conditions. In Stargate Blue Hole, South Andros, sparse replacement dolomites and dolomite cements are observed only in rock samples from the depth of the present saline zone. Stable isotope and trace element analyses suggest precipitation from a fluid of near-seawater composition under slightly reducing conditions, and at temperatures in the lower range of those observed for modern saline groundwaters. Strontium isotopes yield a maximum age of 0.4–0.8 Ma for the dolomites, substantially younger than that estimated for initial deposition (1.5–2.5 Ma). Order-of-magnitude calculations based on estimates of groundwater flow combined with the observed magnesium depletion in saline groundwaters indicate dolomitization at a rate of $0.18–3.0 \times 10^{-5}\%\ \text{a}^{-1}$. Over the long term, given the likely effect of Pleistocene sea-level fluctuations on the saline circulation system, this would be sufficient to generate 0.3–4.5% dolomite, a figure comparable to the observed abundance in the Stargate rock samples.

INTRODUCTION

Seawater is the only widely available fluid capable of large-scale dolomitization of carbonate platforms prior to deep burial (Land, 1985). However, despite the considerable thermodynamic drive to precipitate dolomite from seawater, laboratory experiments have consistently failed to synthesize dolomite under reasonable earth surface conditions (Lippmann, 1973). This suggests that either seawater must be modified in some manner to overcome kinetic barrier(s) to dolomitization (reviewed by Machel & Mountjoy, 1986) or, alternatively, the precipitation of dolomite from essentially unmodified seawater must occur at a rate insufficient to produce measurable quantities of dolomite at the timescale of laboratory experiments, but which may be sufficient to generate extensive dolomitization over geological timescales (Land, 1985; Hardie, 1987). In common

¶ Present address: Argonne National Laboratory, University of Chicago, Illinois 60439, USA

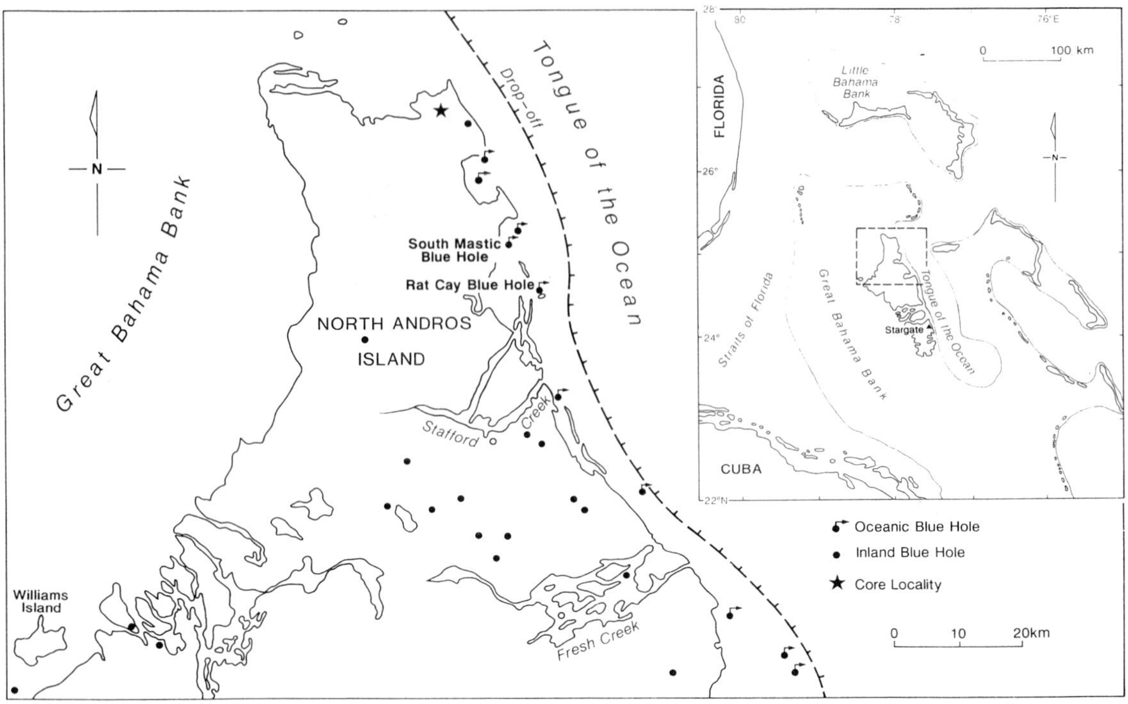

Fig. 1. North Andros study area showing blue hole sampling sites and location in Bahamas (inset).

with all diagenetic reactions dolomitization is dependent both upon fluid circulation to supply the reactants and remove the products of dolomitization, and upon porewater geochemistry to provide the thermodynamic and kinetic controls on precipitation and dissolution of mineral phases. An indication of the potential for dolomitization by seawater-derived fluids within a carbonate platform can thus be obtained from field studies of the groundwater hydrology, whereas its actual occurrence may be directly indicated by magnesium depletion in the saline groundwaters. In this study we combine a modern process-based approach using hydrological and aqueous geochemical techniques with a geological approach to investigate dolomitization in the zone of saline groundwaters beneath Andros Island, Great Bahama Bank.

The Bahama banks comprise the world's largest extant carbonate platform, and provide models for many partially or completely dolomitized platforms in the fossil record. Although their importance as a natural laboratory for the study of carbonate sedimentology and near-surface diagenesis has been recognized for the last 40 years, diagenetic processes beneath the outer 'skin' of the banks remain largely elusive, due to difficulties of subsurface access. Fortunately, North Andros Island on the Great Bahama Bank has a large number of 'blue holes', which are the entrances to flooded cave systems (Fig. 1). Inland blue holes penetrate to depths in excess of 100 m and give access to saline groundwaters in the upper part of the platform (Smart *et al.*, 1988), whereas oceanic blue holes on the east coast function as sites of saline groundwater discharge (Whitaker & Smart, 1990). In this study these sites are used to obtain information on the circulation and aqueous geochemistry of saline groundwaters, and also to sample wall rock.

STUDY AREA

The nature of the Great Bahama Bank is well known from numerous previous studies (e.g. Mullins & Lynts, 1977; Beach, 1982). Briefly, the bank comprises a flat-topped carbonate platform with steep marginal slopes into the Straits of Florida to the west and Tongue of the Ocean to the east. The majority of the platform surface is flooded to shallow depths (<6 m) by seawater, but limited circulation gives rise seasonally to elevated salinities, particularly to the north of Williams Island (see Fig. 1). Only the eastern margin of the bank is presently emergent, where the accumulation of Pleistocene aeolianities gives some relief (<10 m) on North and South Andros Islands and Mangrove Cay. Recharge and area of emergence are sufficient that a thick (30 m) freshwater lens is developed beneath North Andros Island, but to the west low relief gives an extensive area with predominantly brackish waters near the surface. The platform surface was exposed during Pleistocene low sea-stands, and there is considerable evidence of surface and subsurface karstification.

Stabilization of aragonite and high Mg-calcite has been relatively rapid, and calcite is therefore the dominant near-surface carbonate. However, at a relatively shallow depth (>50 m; Beach, 1982) dolomites are present. From studies of comparable dolomites elsewhere in the Bahamas, these are known to be Plio-Pleistocene in age, and trace element and stable isotopic studies infer precipitation from waters of near-seawater composition (Vahrenkamp and Swart, this volume). Detailed petrographic study combined with strontium isotope analysis indicates that up to five generations of dolomite are present, which can be allocated to specific phases (Vahrenkamp *et al.*, 1991).

North Andros Island was selected for this study because of the high density of inland and oceanic blue holes, the former also being accessible via logging roads in the interior. Furthermore, Simms (1984) has previously suggested on theoretical grounds that density reflux of evaporatively concentrated waters may occur on the Great Bahama Bank west of North Andros Island, and that this could drive the circulation of saline groundwater beneath North Andros Island. Some support for the hypothesis of net movement of saline groundwater beneath North Andros is also provided by the distribution of oceanic blue holes (see Fig. 1), which are relatively numerous on the east coast of Andros but essentially absent on the west coast of the island, where the combination of high sediment mobility and lack of active outflow apparently results in choking of the karstic openings.

Three types of blue hole are recognized in this study: inland cenotes, inland fracture and oceanic blue holes (Fig. 2). Inland cenote blue holes are flooded vertical shafts up to 110 m deep which have circular entrances typically 50–150 m in diameter and bell out at depth. A small number have horizontal passages leading off, but these, if present, are usually inaccessible due to accumulation of breakdown material. In contrast to cenotes, which are widely distributed across the island, fracture blue holes are found predominatly near the east coast of the island, and are developed on extensional fractures parallel to the bank margin (Smart *et al.*, 1988). They typically have a linear plan form, and are commonly very deep (>100 m). The vertical morphology of fracture blue holes and the fresh nature of the wall rock due to active spalling and lack of marine organisms, makes these sites ideal for wall-rock collection. Slow groundwater movement is evident in inland fracture blue holes, but the large passage cross-sectional areas give low velocities. Oceanic blue holes open from the shallow flooded banks, and comprise predominantly horizontal passages which connect with vertical shafts in the floor, access to which is often restricted by breakdown. Some of these passages, which generally run parallel or perpendicular to the coast, have been explored for considerable distances by cave divers (Conch Sound Blue Hole on North Andros has been surveyed to over 1.5 km in length). Oceanic blue holes are characterized by strong semidiurnally reversing currents driven by tidally induced differences between sea-surface elevation and inland groundwater levels. The more horizontal oceanic blue holes predominate on North Andros, whereas fracture blue holes are better developed on South Andros. As both are considered to be possible routes for the discharge of saline groundwaters, samples for geochemical analysis were collected from both types of site.

METHODS

Because of the semidiurnal reversal of flow in oceanic blue holes, a continuous record of discharge was needed in order to differentiate between the 'shuttling' of locally derived seawater and any larger-scale net saline groundwater circulation. Aanderaa oceanographic recording current meters (RCMs) were therefore installed in two oceanic blue holes, Rat Cay and South Mastic on the east coast of North Andros (see Fig. 1). Discharge was calculated from passage cross-sectional area and average velocities, the latter derived from the RCM record and measurements of the cross-sectional velocity distribution throughout the tidal cycle obtained *in situ* by

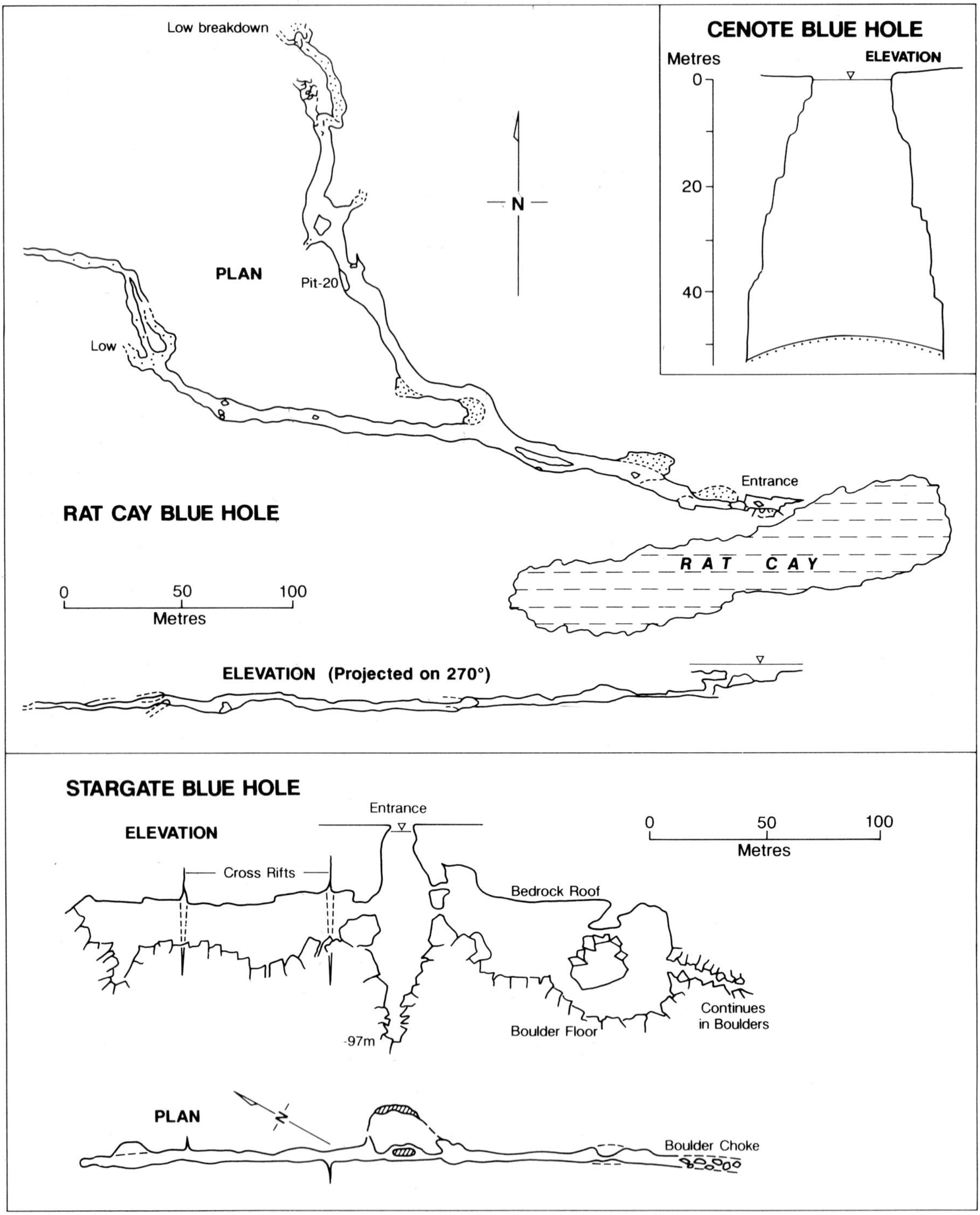

Fig. 2. Surveys of typical blue holes sampled in this study. Rat Cay Blue Hole is an oceanic discharge site on the east coast of North Andros; Stargate Blue Hole is an inland fracture-controlled blue hole adjacent to the east coast of South Andros; the cenote blue hole (inset) is representative of the numerous inland cenotes widely distributed across North Andros Island. See Fig. 1 for location of individual sites.

divers using a hand-held Ott current meter. Data on the local tidal head, water temperature and salinity were also obtained from sensors in the RCM. Further details are given by Whitaker and Smart (1990 and in press a).

Salinity (±0.05‰) and temperature (±0.1°C) profiles were measured in 27 deep inland cenote blue holes using a WTW-LF 191 probe. Water samples were also collected using a Wuidart borehole sampler and by divers working *in situ*. The latter were collected in 1.5 l sample tubes filled by the forward motion of the diver and capped immediately. On return to the surface, samples were flowed through a closed cell in which pH was measured using a digital Walden pH meter standardized with high-precision NBS buffers (±0.02 pH units). In addition, dissolved oxygen was measured in the cell using WTW OXI-91 dissolved oxygen meter (±1% O_2 saturation), but some contamination from atmospheric oxygen (<5% O_2) was evident in waters known to be anoxic. Subsamples were taken for immediate measurement of alkalinity and calcium by titration, and for later analysis of major ions by standard techniques (magnesium, sodium and potassium by spectrophotometry, atomic absorption and flame emission, chloride by Mohr titration, and sulphate by turbidimetry). In all cases dilutions were weighed and analytical precision was better than ±1% (with the exception of sulphate, ±4.5%). Separate subsamples fixed with zinc acetate were also taken for determination of reduced sulphur by iodometric titration (±1.5%). Ion balance errors for all samples were <5%, with an average deviation from zero of 1.4%. Analytical accuracy was confirmed by independent analysis of magnesium (ICPMS) and chloride (Mohr titration) by the laboratories of AMOCO Production Company. Saturation indices with respect to calcite and aragonite, and also ordered and disordered dolomite (K_D $10^{-18.06}$ and $10^{-16.52}$, respectively; Helgeson *et al.*, 1981), and the $p\text{CO}_2$ with which waters were in equilibrium, were calculated using the aqueous speciation model SOLMINEQ.88 (Kharaka *et al.*, 1988). Dissolved organic carbon analyses were performed on selected samples with a Shimadzu TOC-500 total organic carbon analyser, using the high-temperature catalytic oxidation method of Sugimura and Suzuki (1988). Unfortunately, however, precision was low (±28%) due to sensitivity variations between runs. Analytical and derived data are presented in full by Whitaker (1992).

In order to examine the distribution and nature of Pleistocene dolomites within the platform, vertical suites of wall-rock samples were collected from two inland fracture sites on South Andros, Stargate and Evelyn Green's Blue Hole (0–40 m and 0–22 m depths, respectively). The stable isotope geochemistry of the dolomite was measured using two methods. A dolomite concentrate was prepared by dissolution of all calcite from a bulk sample in acetic acid. CO_2 gas derived from the dolomite was analysed by standard techniques using a triple-collector mass spectrometer. Correction was made for isobaric influences and also for fractionation during dolomite dissolution by phosphoric acid using a fractionation factor of 1.0102. Spot analyses were also performed on thin sections using a laser ablation system for stable isotope extraction (LASSIE; Smalley *et al.*, 1992) allowing isotopic analysis of areas as small as 20 μm in diameter. However, due to the high transparency of the sample, a higher laser power was required and the analytical areas were considerably larger (50–70 μm). A scanning proton microprobe (SPM) was used to determine the major and trace element content of the dolomites, as well as to look for trace element zoning patterns. The SPM offers excellent spatial resolution and concentration sensitivity, making it suitable for detecting zoning in samples as small as 50 μm (Fraser, 1990). Maps, linescans and point analyses were carried out on dolomite grains using proton-induced X-ray analysis (PIXE). Details of analytical conditions and deconvolution of the PIXE spectra for quantitative analyses are given by Wogelius *et al.* (1992).

CIRCULATION OF SALINE GROUNDWATER

The discharge and head recorded at South Mastic oceanic blue hole for a 48 h period during July 1988 are shown in Figure 3a. The semidiurnal response to local tidal head can be clearly seen, but the duration of the inflow phase is significantly longer than of the outflow phase. Furthermore, the maximum velocity reached on outflow is much higher than that during inflow. There is clear net discharge of groundwater at this site, which averaged $2.04 \pm 1.35 \times 10^5\,m^3$ per tidal cycle during the 17-day record (outflow $3.75 \pm 0.25 \times 10^5$ compared to inflow $1.71 \pm 1.46 \times 10^5\,m^3$ per tidal cycle). A much longer record is available from Rat Cay oceanic blue hole, spanning 6 months from July 1988. The cumulative net discharge during this time is shown in Figure 3b. Initially, during the

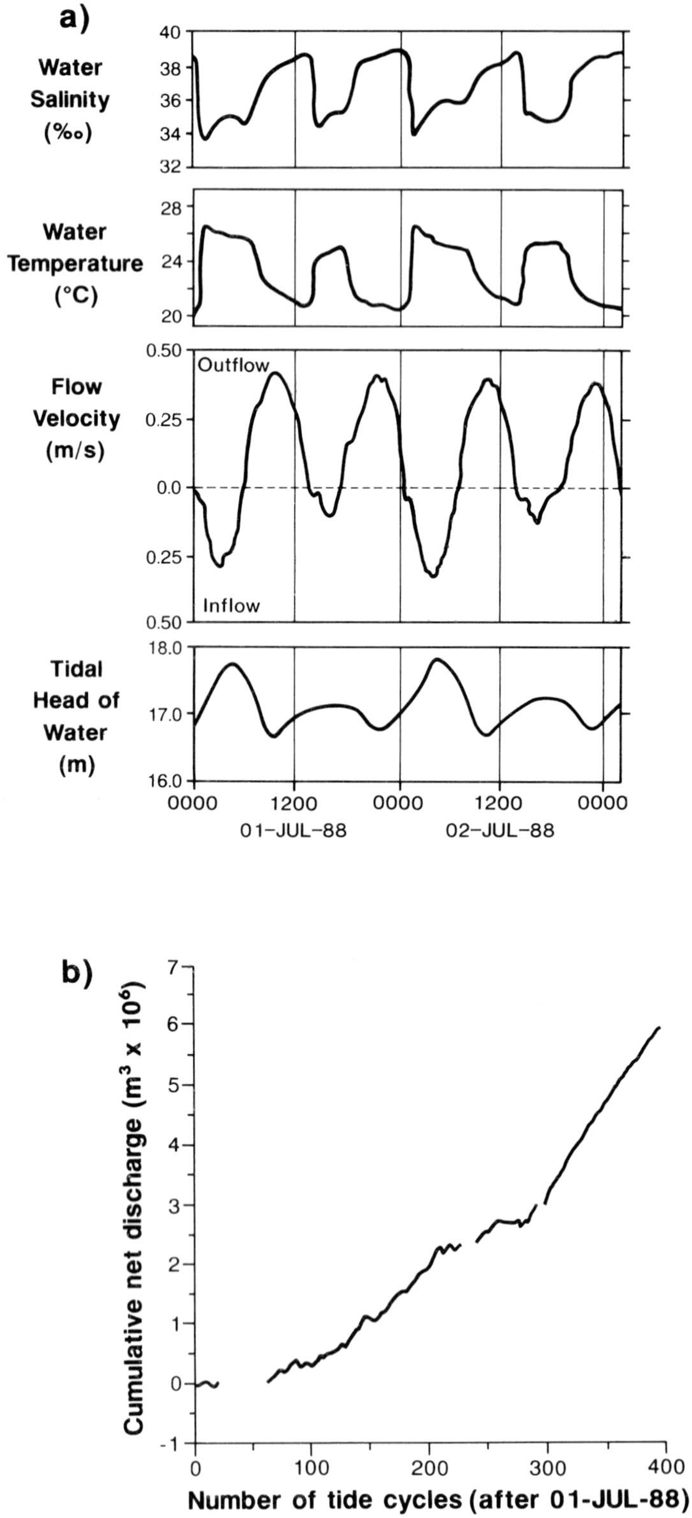

Fig. 3. (a) recording current meter record from South Mastic Blue Hole, North Andros, showing flow velocity and tidal head for a 2-day period in July 1988. (b) Cumulative net discharge from Rat Cay Blue Hole, North Andros over the period July 1988–January 1989 (periods of rotor failure shown by breaks in the line).

summer net discharge was small, but thereafter increased significantly into the autumn and winter. A total of $6.8 \times 10^6\,m^3$ of saline groundwater was discharged from Rat Cay Blue Hole over the observation period (extrapolating over periods of rotor failure), although the average net discharge was less ($1.71 \pm 1.74 \times 10^3\,m^3$ per tidal cycle) than at the much larger South Mastic Blue Hole. Further details and analysis of oceanic blue hole discharge records are given by Whitaker and Smart (in press a).

In order to identify the source(s) of the groundwater discharging from the east coast of North Andros Island, salinity and temperature were used as natural tracers. Representative values of salinity for bank and open ocean waters, discharges from oceanic blue holes and saline groundwaters from inland cenote blue holes are summarized in Table 1. Waters from the Tongue of the Ocean and the Straits of Florida are of comparable salinity (36.6 ± 0.1 and 36.3 ± 0.1‰, respectively), the former being slightly more saline due to its enclosed position (Sverdrup *et al.*, 1946; Busby & Dick, 1964). In comparison, the salinity of groundwaters discharging at the end of the outflow phase from nine oceanic blue holes during July/August was significantly higher (37.70 ± 1.65‰; Table 1). Such elevated-salinity waters can only derive from the Great Bahama Bank on the western side of North Andros Island, where high evaporation rates and long residence times increase the salinity of the shallow waters over large areas to greater than 38‰, and locally off the west coast of the island to in excess of 45‰ (Smith, 1940; Cloud, 1962). Saline groundwaters beneath the fresh/saltwater mixing zone of central and eastern inland cenote blue holes on North Andros occupy an intermediate position between open ocean and bank seawaters, with a mean salinity of 37.20 ± 1.85‰ (Table 1), confirming that eastward flow of bank waters through the platform must occur. The two cenote blue holes sampled on the west side of the island (Fig. 1) have a salinity of 44.45 ± 0.70‰, significantly higher than those of other inland sites and comparable with the most saline Great Bahama Bank waters.

Salinity data therefore suggest that waters from the Great Bahama Bank are involved in a regional-scale groundwater circulation, flowing eastwards beneath North Andros and discharging via oceanic blue holes on the east coast. A plume of water of slightly elevated salinity identified by Busby and Dick (1964) at depths of 140–180 m in the Tongue of the Ocean adjacent to the eastern platform margin may also derive from this circulation system. Clearly, the observed flows cannot be explained by buoyant circulation in the fresh/saltwater mixing zone, which would generate much lower salinities. Rather, these observations appear to support the theoretical calculations of Simms (1984) and indicate that reflux of only slightly elevated-salinity bank waters does occur, with lateral flow at depth in response to the horizontal density gradient generated by the contrast with normal-salinity seawaters in the adjacent ocean basins.

Table 1. Salinity and Mg/Cl ratio of source seawaters and saline groundwaters. GBB, Great Bahama Bank; TOTO, Tongue of the Ocean; East Bank, shallow bank waters within the North Andros barrier reef.

	Salinity (‰)	Mg/Cl (molar ratio $\times 10^{-2}$)
Source seawaters		
GBB ($n = 5$)	40.25 ± 1.85	10.72 ± 0.24
TOTO ($n = 2$)	36.35 ± 0.65	10.04 ± 0.02
East Bank ($n = 4$)	36.00 ± 0.65	10.40 ± 0.30
Saline groundwaters		
Inland cenote (western)	44.45 ± 0.70 ($n = 2$)	9.41 ± 0.04 ($n = 2$)
Inland cenote (central and eastern)	37.20 ± 1.85 ($n = 15$)	9.86 ± 0.33 ($n = 15$)
Inland fracture	36.35 ± 0.85 ($n = 14$)	9.88 ± 0.25 ($n = 12$)
Oceanic discharge	37.70 ± 1.70 ($n = 9$)	10.01 ± 0.46 ($n = 8$)

Elevated-salinity waters discharging from oceanic blue holes are significantly colder than both surface bank waters (annual range 22–33°C) and the mean annual temperature (25°C), with a minimum temperature of 21°C. Saline waters in inland cenote blue holes are also relatively cold: 24.4 ± 0.5°C; at variance with the pattern expected for static groundwaters, the temperature of which normally increases with depth in response to the geothermal heat flux. At a depth of 80 m an average temperature of 27°C would be predicted from the geothermal gradient of nearby Florida. Saline groundwater temperatures are similar to the minimum (February) value for comparable depths in the Tongue of the Ocean, but less than the mean annual surface-water temperature for both the Tongue of the Ocean and the Straits of Florida. The low temperatures within the saline zone of the platform suggest that cold ocean water is involved in the groundwater system. A conservative mixing model, based on a maximum average bank salinity of 42‰ and minimum winter temperature of 22°C, indicates that the temperature of these cold ocean waters must be less than 20.2°C to produce the observed minimum outflow temperature of 21°C and corresponding maximum salinity of 39‰. Such water would derive from a depth of greater than 260 m in the Straits of Florida or Tongue of the Ocean.

The simplest explanation of the discharge, salinity and temperature data is that large-scale thermal ('kohout') convection occurs within the platform, and these waters mix with the denser, slightly elevated-salinity waters derived by reflux from the bank surface (Fig. 4a). A second possibility is that there is a net flow of cold ocean water eastwards from the Straits of Florida beneath Andros Island (Fig. 4b). This provides a rather better explanation of the low temperatures of saline groundwaters, which increase from west to east beneath the island (Whitaker & Smart, in press a). Such a circulation could be driven by a difference in sea-surface elevation across the platform generated by the Florida Current, which is confined and forced northwards through the shallowing straits between the Great Bahama Bank and the Floridian peninsula (Maul, 1986).

In conclusion, large-scale circulation of near-normal seawater occurs within the Great Bahama Bank beneath North Andros Island. Two types of waters are involved: one of slightly elevated salinity from the bank surface and the other of normal salinity but low temperature derived from significant depth in the adjacent ocean. This active circulation, combined with the Mg-rich nature of the fluids, generates significant potential for platform-wide dolomitization.

GEOCHEMICAL EVIDENCE FOR DOLOMITIZATION BY SALINE GROUNDWATERS

Both primary precipitation of dolomite cements result in consumption and replacement dolomitization of magnesium ions from solution (Table 2). Thus magnesium depletion of the seawater-derived saline groundwaters provides a direct indicator of the occurrence and extent of dolomitization, although an alternative magnesium sink, the precipitation of high Mg-calcite cements, must also be considered. In order to account for dilution and, more importantly, evaporation, we consider the Mg/Cl ratios of saline groundwaters from Andros Island, chloride being conserved during both these processes and during rock–water interaction. The mean Mg/Cl ratios of inland cenote and fracture groundwaters are significantly lower (at 95 and 92.5% confidence limits, respectively) than either bank or open ocean waters, which together form their source, indicating loss of magnesium along the flow path (Table 1). Waters from oceanic discharges have significantly lower Mg/Cl ratios than bank waters, but are not statistically different from Tongue of the Ocean seawater.

The evidence from Mg/Cl ratios of inland cenote waters therefore suggests that dolomitization may be occurring in the zone of saline groundwaters, but for fracture and oceanic waters this conclusion is dependent on the relative contributions of open ocean waters from the Straits of Florida and/or Tongue of the Ocean and waters input from the Great Bahama Bank, which are already geochemically evolved (Broecker & Takahashi, 1966; Morse *et al.*, 1985). Bank water samples from this study are depleted in calcium by an average of 16 ± 8 mg/l, but enriched in magnesium by 102 ± 28 mg/l relative to that predicted from evaporative concentration of open ocean waters, indicating stabilization of high

Table 2. Principal dolomitization reactions

1 Primary precipitation
$Ca^{2+} + Mg^{2+} + 2CO_3^{2-} \longrightarrow CaMg(CO_3)_2$
2 Replacement dolomitization
A $2CaCO_3 + Mg^{2+} \longrightarrow CaMg(CO_3)_2 + Ca^{2+}$
B $CaCO_3 + Mg^{2+} + CO_3^{2-} \longrightarrow CaMg(CO_3)_2$
C $(2-x)CaCO_3 + Mg^{2+} + xCO_3^{2-} \longrightarrow CaMg(CO_3)_2 + (1-x)Ca^{2+}$

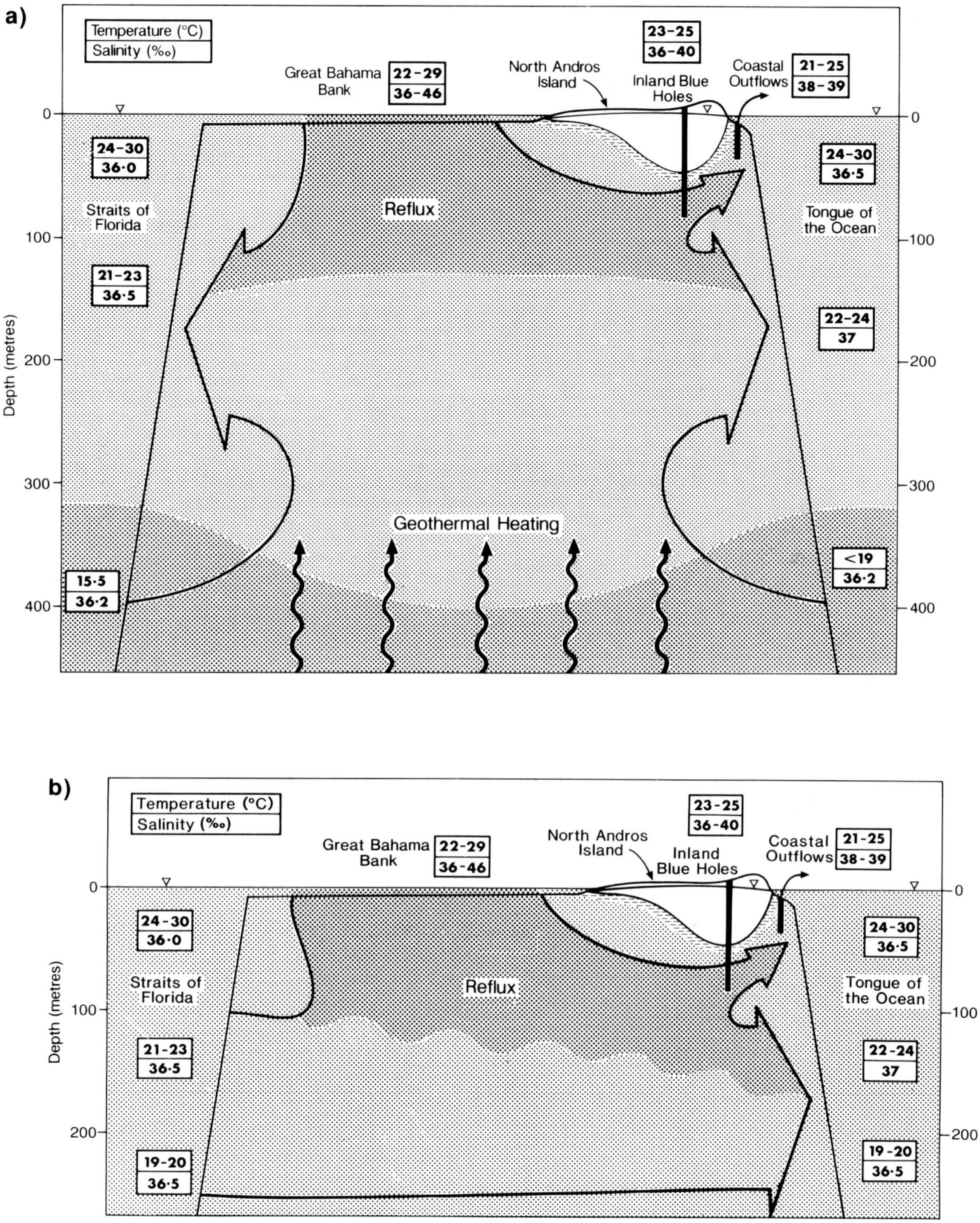

Fig. 4. Summary of temperature and salinity data for ocean, bank and saline groundwaters, showing postulated circulation system beneath North Andros (open arrows), with (a) thermal and reflux drive and (b) reflux with trans-bank difference in sea-surface elevation. Density of stipple is proportional to water density (from Whitaker & Smart, in press).

Mg-calcite to low Mg-calcite. Therefore, to calculate the magnitude of magnesium depletion, and determine whether this is caused by formation of dolomite or high Mg-calcite, the excess amount of magnesium (Mg_{XS}) and calcium (Ca_{XS}) in individual samples has been calculated relative to linear mixing of source waters, using chloride to predict the mix of bank and open ocean water:

$$Mg_P = Mg_O + \left[\frac{Cl_G - Cl_O}{Cl_B - Cl_O}\right] Mg_O - Mg_B \quad (1)$$

and

$$Mg_{XS} = Mg_G - Mg_P \quad (2)$$

where the subscripts G, P, O and B refer to sampled groundwater, predicted groundwater, ocean and bank seawater concentrations of chloride and magnesium. Whereas the chloride concentration of the saline groundwaters is directly measured, and that of open ocean waters has a low variability, there is considerable uncertainty about the salinity of bank-derived source waters. This depends on the relationship between the potential for flow (determined by a combination of elevation head and fluid density, the latter being controlled by salinity and temperature) and the permeability of the Holocene sediments mantling the bank surface. These variables exhibit considerable spatial and, in the case of flow potential, temporal variability. Simms (1984) predicted that waters >38‰ have the potential to reflux, and a figure of 41.05 ± 1.48‰ (the average salinity of bank surface waters between 24°50′N and 25°00′N; Cloud, 1962) is assumed to be representative of input waters. The sensitivity of groundwater geochemistry to variations in bank source-water salinity within the range of observed values is less than the combined 1 δ root mean squared analytical uncertainties of ±28 mg/l for Mg_{XS} and ±8 mg/l for Ca_{XS} (Whitaker, 1992).

Inland cenote blue holes from central and eastern North Andros have a magnesium depletion of 52 ± 12 mg/l (significant at 95% confidence interval), somewhat higher but not statistically separate from the values from South Andros fracture sites (41 ± 15 mg/l, excluding three outliers; Table 3 and Fig. 5). The latter have a much smaller calculated bank contribution, as would be expected given their greater distance from the locus of maximum bank salinities north of Williams Island. Inland fracture and cenote blue hole waters also both show significant increases in calcium concentrations compared to their source waters (Table 3), indicating

Table 3. Summary of differences between observed and predicted values of magnesium (Mg_{XS}) and calcium (Ca_{XS}) calculated from percentage bank-derived input on the basis of salinity.

	% Bank input	Mg_{XS} (mg/l)	Ca_{XS} (mg/l)
Inland cenote (western)	108 ± 2 (n = 2)	−222 ± 9 (n = 2)	−56 ± 33 (n = 2)
Inland cenote (central and eastern)	26 ± 29 (n = 15)	−52 ± 45 (n = 15)	+18 ± 17 (n = 15)
Inland fracture	6 ± 13 (n = 14)	−41 ± 15 (n = 11)	+14 ± 27 (n = 14)
Oceanic discharge	30 ± 32 (n = 9)	+15 ± 20 (n = 7)	−4.5 ± 19 (n = 9)
Oceanic discharges*	56 ± 40 (n = 9)	−36 ± 38 (n = 7)	−15 ± 8.5 (n = 7)

* Recalculated assuming mixing between a subset of inland blue holes from central and eastern North Andros (mean salinity 38.45‰, Mg 1450 mg/l and Mg_{XS} −58 mg/l) and east-coast bank waters.

net dissolution of calcite (strontium : calcium ratios preclude the possibility of aragonite dissolution; Whitaker, 1992). Magnesium depletions therefore cannot be explained by precipitation of high Mg-calcite cements, a conclusion supported by the observed undersaturation of saline groundwaters with respect to high Mg-calcite.

Dolomitization cannot occur via primary precipitation (Reaction 1, Table 2), which would consume both calcium and magnesium. It may, however, occur by replacement, either releasing calcium (Reaction 2A, Table 2) or consuming carbonate (Reaction 2B, Table 2). For inland cenote and fracture blue hole waters, the molar magnesium depletion is, respectively, an average of 4.3 and 3.6 times larger than the calcium release. Thus it can be concluded that dolomitization must be predominantly by replacement with consumption of carbonate (Reaction 2C, Table 2; x = 0.75).

In contrast to the saline groundwaters sampled from inland sites, oceanic discharges are not significantly depleted in magnesium. Indeed, after exclusion of two strongly magnesium-depleted samples which are statistical outliers, these waters are, on average, enriched in magnesium compared to both their source seawaters and inland cenote groundwaters (Table 3). The probable reason is that, although oceanic discharges were sampled at the end of the discharge phase, they still contain a sig-

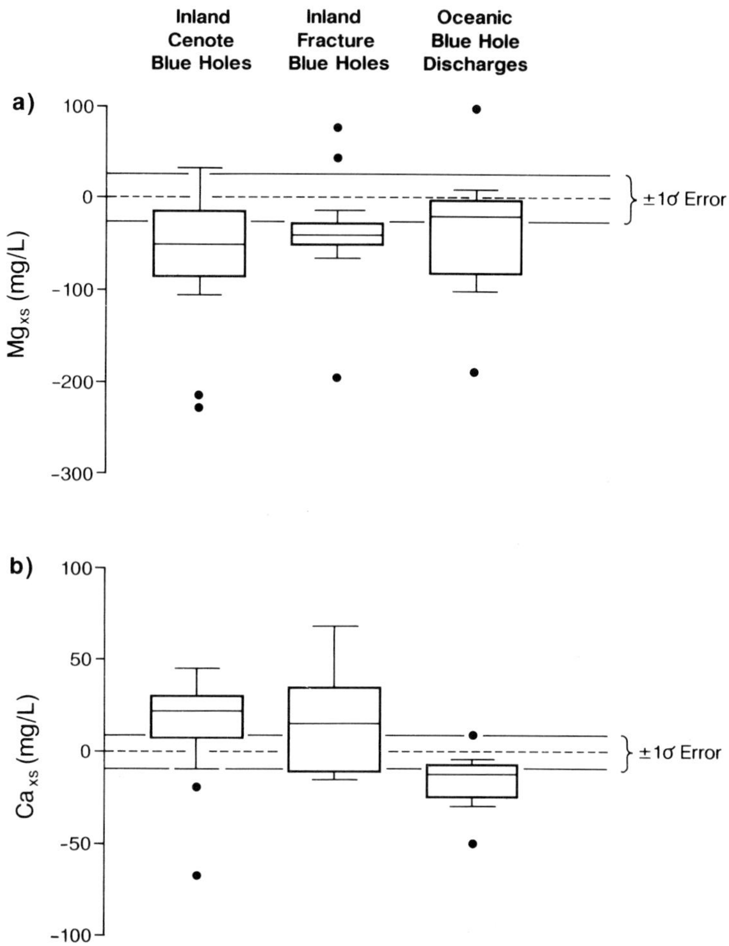

Fig. 5. Boxplots showing median, quartiles, range and statistical outliers of Mg_{XS} and Ca_{XS} for inland cenote and fracture blue holes, and oceanic discharges based on correction for mixing of nondiluted inland cenote blue hole waters and East Bank seawaters.

nificant component of the local bank waters which entered the groundwater circulation during previous high ocean tides when inflow occurred. East-coast bank waters are magnesium-enriched but of comparable salinity to the open ocean end-member used to calculate magnesium excess (Table 1). The simple two-component mixing model based on salinity (Equations 1 and 2) therefore underestimates magnesium depletion. This suggestion was tested by recalculating magnesium excess values for oceanic discharges, assuming that they were derived simply from mixing of east-coast bank waters with inland cenote saline groundwaters, the latter being assumed to represent the average mix of Great Bahama Bank and open ocean waters circulating under North Andros.

This model suggests that, on average, the East Bank component is of the order of 44% in the sampled waters (Table 3), but at some sites this figure is 100% whereas at others it is zero, presumably reflecting both the 'plumbing' of individual sites and the time of sampling during the outflow period. In fact, the two water samples rejected as statistical outliers for oceanic discharges fall within the spread of results for inland cenotes, and may thus represent unmixed discharging groundwaters. The recalculated values of magnesium excess suggest that, once corrected for local mixing, oceanic discharges are also significantly depleted in magnesium (by 36 ± 38 mg/l, n = 7 excluding two outliers; Table 3 and Fig. 5). Furthermore, an average 15 ± 56 mg/l of this depletion is additional to that present in inland cenote waters. This would be equivalent to a total magnesium depletion along the complete groundwater flow path of 67 ± 72 mg/l, excluding the dilution effect of East Bank waters. The loss of magnesium thus appears to be progressive along the groundwater flow path, something which cannot be clearly demonstrated from the geographic distribution of magnesium depletions from individual cenote sample sites. Finally, it should be noted that, unlike inland saline groundwaters, oceanic discharges are significantly depleted in calcium (Table 3 and Fig. 5), probably due to direct precipitation of calcite cements as a result of mixing and degassing of inland groundwaters discharging at the coast (see below).

There is thus good evidence that saline groundwaters circulating beneath Andros Island are depleted in magnesium as a result of replacement dolomitization. In the following section the geochemical nature of these groundwaters is considered further.

AQUEOUS GEOCHEMISTRY OF SALINE GROUNDWATERS

It was suggested above that the dolomitization occurred predominantly by Reaction 2B (Table 2), and would therefore be driven by an increase in the alkalinity of the saline groundwaters. Source bank and seawaters have a low $p\text{CO}_2$ controlled by equilibration with the atmosphere (Table 4). However, an increase in $p\text{CO}_2$ may occur as bank-derived waters pass downwards through the veneer of unconsolidated Holocene sediments, Morse *et al.* (1985) reporting $p\text{CO}_2$ values of up to 0.55% in fine-grained muds. Based on this figure, mixing calculations indicate that cenote groundwaters (26% bank water) might attain a $p\text{CO}_2$ of up to 0.17% from this source. However, this is a maximum estimate as it assumes that all bank-derived waters pass through the finer sediments (high $p\text{CO}_2$), rather than the more transmissive oolites and grainstones (Enos & Sawatsky, 1981), which have a lower $p\text{CO}_2$. The $p\text{CO}_2$ of saline groundwaters from inland cenote blue holes is significantly higher than this predicted maximum, with an average more than 30 times the atmospheric $p\text{CO}_2$ (Table 4). Waters from inland fracture and oceanic blue holes on the east coast have a much lower $p\text{CO}_2$, due to degassing and mixing with low-$p\text{CO}_2$ local seawaters.

Saline groundwaters in both North Andros cenote and South Andros fracture inland blue holes are close to equilibrium with respect to calcite, and almost all undersaturated with respect to aragonite, although oceanic blue holes are slightly supersaturated with respect to both these minerals. All samples are significantly less saturated than source bank surface and open ocean waters, as expected given their increased $p\text{CO}_2$. In fact, undersaturation of these

Table 4. Summary of geochemistry of source seawaters and saline groundwaters. GBB, Great Bahama Bank; TOTO, Tongue of the Ocean; East Bank, shallow bank waters within the North Andros barrier reef. SI_C, SI_A, SI_D and SI_DD, saturation indices (log IAP/K) for calcite, aragonite, dolomite and disordered dolomite; DO, dissolved oxygen; DOC, dissolved organic carbon; $\text{SO}_{4\text{XS}}$, sulphate excess calculated relative to mixing between open ocean and Great Bahama Bank waters (*cenote and East Bank waters).

	$p\text{CO}_2$ (%)	SI_C (SI_A)	SI_D (SI_DD)	DO (%)	DOC (mg/l)	Reduced sulphur species (mg/l)	$\text{SO}_{4\text{XS}}$ (mg/l)
Source seawaters							
GBB	0.05 ± 0.02 (*n* = 5)	+0.66 (+0.52) ± 0.14 (*n* = 5)	+3.31 (+1.08) ± 0.28 (*n* = 5)	96 (*n* = 1)	7.7 ± 2.9 (*n* = 5)	<0.1 (*n* = 4)	—
TOTO	0.05 ± 0.02 (*n* = 2)	+0.80 (+0.66) ± 0.01 (*n* = 2)	+3.52 (+1.99) ± 0.02 (*n* = 2)	96 (*n* = 1)	5.6 (*n* = 1)	<0.1 (*n* = 1)	—
East Bank	0.05 ± 0.03 (*n* = 4)	+0.71 (+0.57) ± 0.17 (*n* = 4)	+3.38 (+1.76) ± 0.38 (*n* = 4)	65 ± 8 (*n* = 2)	12.0 ± 6.8 (*n* = 4)	0.9 (*n* = 2)	—
Saline groundwaters							
Inland cenote (western)	0.34 ± 0.17 (*n* = 2)	−0.07 (−0.21) ± 0.10 (*n* = 2)	+1.68 (+0.14) (*n* = 1)	—	—	−0.58 (*n* = 1)	−32 (*n* = 1)
Inland cenote (central and eastern)	1.04 ± 0.20 (*n* = 15)	−0.10 (−0.24) ± 0.13 (*n* = 15)	+1.70 (+0.16) ± 0.27 (*n* = 15)	4.6 ± 2.5 (*n* = 9)	14.0 ± 2.8 (*n* = 5)	3.8 ± 6.3 (*n* = 13)	−126 ± 133 (*n* = 15)
Inland fracture	0.32 ± 0.08 (*n* = 14)	+0.08 (−0.06) ± 0.07 (*n* = 14)	+2.05 (+0.48) ± 0.15 (*n* = 14)	8.6 ± 7.7 (*n* = 14)	13.6 ± 4.3 (*n* = 8)	0.51 ± 0.47 (*n* = 11)	−14 ± 165 (*n* = 13)
Oceanic discharge	0.20 ± 0.04 (*n* = 9)	+0.29 (+0.15) ± 0.09 (*n* = 9)	+2.53 (+1.01) ± 0.10 (*n* = 9)	—	12.3 ± 4.5 (*n* = 3)	0.58 ± 0.67 (*n* = 5)	+43 ± 120* (*n* = 11)

waters is buffered by calcite dissolution, as indicated by the significant calcium enrichment of inland waters. There is a similar reduction in dolomite saturation in saline groundwaters (compared to source waters), although all waters retain the thermodynamic potential to precipitate ordered and, in most cases, disordered dolomite.

The elevated pco_2 of saline groundwaters appears to be caused by *in situ* oxidation of organic matter along the saline groundwater flow path. Despite analytical uncertainties in the DOC analyses, the majority of groundwaters sampled have a significantly higher dissolved organic carbon concentration than adjacent ocean waters (Table 4), suggesting an inland source of organic matter. Furthermore, all inland saline groundwaters are essentially anoxic and contain measurable concentrations of reduced sulphur species, indicating that further oxidation of organic matter by bacterially mediated sulphate reduction is also occurring. In inland cenote blue holes the average sulphate depletion (SO_{4XS}, calculated in a similar manner to the magnesium depletion) is 126 ± 133 mg/l. Reoxidation of reduced sulphur appears to occur in the oceanic discharges which are not significantly depleted in sulphate (Table 4), as has previously been reported in oxic freshwaters from inland cenotes (Bottrell *et al.*, 1991). Fracture groundwaters occupy an intermediate position and show considerable individual variation.

Although there is considerable scatter, due both to the high combined analytical uncertainty of SO_{4XS} and to reoxidation of reduced sulphur species in the oceanic discharges, there is a statistically significant positive relationship between the magnesium depletion and sulphate depletion of saline groundwaters (Fig. 6a). However, given the relatively small sulphate depletion in these waters (cf. Baker & Burns, 1985), this relationship most probably reflects autocorrelation of these variables, rather than a causal relationship. There does not appear to be any relationship between magnesium depletion and dissolved organic carbon concentration, suggesting that dolomitization is not linked directly to dissolved organic compounds. Nor is there any pattern of increasing magnesium depletion with increasing pco_2 of inland cenote groundwaters, as might be expected if dolomitization is related to elevated alkalinity. However, it could be argued that alkalinity is controlled mainly by conditions in the open-water column at individual sample sites, whereas magnesium depletion is cumulative along the flow path (see Whitaker, 1992 for detailed argument). When the saturation index with respect to calcite is considered in relation to pco_2 (Fig. 6b), a highly significant relationship of increasing calcite saturation with increasing pco_2 is apparent for cenote groundwaters. This is contrary to the pattern expected if degassing of high pco_2 waters is occurring, a trend clearly defined by the fracture and oceanic discharges (Fig. 6b), confirming the earlier suggestion that degassing controls saturation and calcium depletion in the latter waters. Similarly, an increase in pco_2 would be expected to decrease saturation in waters not buffered by the presence of carbonate bedrock, or produce no change in the index where dissolution occurs. The relationship implies the operation of a process that consumes carbonate ions but releases calcium. Morrow (1982) has proposed a general description of the replacement dolomitization reaction (Reaction 2C, Table 2) which conforms with these criteria, and, as suggested by Mg_{XS} and Ca_{XS} data, may provide a description of the dolomitization process beneath North Andros.

In summary, saline groundwater have an elevated pco_2 which drives the dissolution of calcite and may also be of significance in dolomitization. The CO_2 appears to be derived from bacterial oxidation of organic matter, which causes the development and maintenance of anoxic conditions and the depletion of sulphate. Degassing of saline groundwaters and mixing with local bank waters adjacent to the east coast of the island is associated with significant calcium depletion, suggesting precipitation of bank-marginal calcite cements.

ESTIMATION OF THE RATE OF DOLOMITIZATION BY SALINE GROUNDWATERS

The geochemistry of saline groundwaters sampled in inland and oceanic blue holes has provided insights into the nature and controls of dolomitization by near-normal seawater. If we assume that these waters are representative of saline groundwaters throughout the platform, the degree of observed magnesium depletion can be combined with the estimated rate of groundwater flow to derive an order-of-magnitude estimate of the present-day rate of dolomitization of the Great Bahama Bank beneath North Andros. Groundwater flow is estimated from oceanic blue hole discharges, assuming that at the platform margin all flow is concentrated

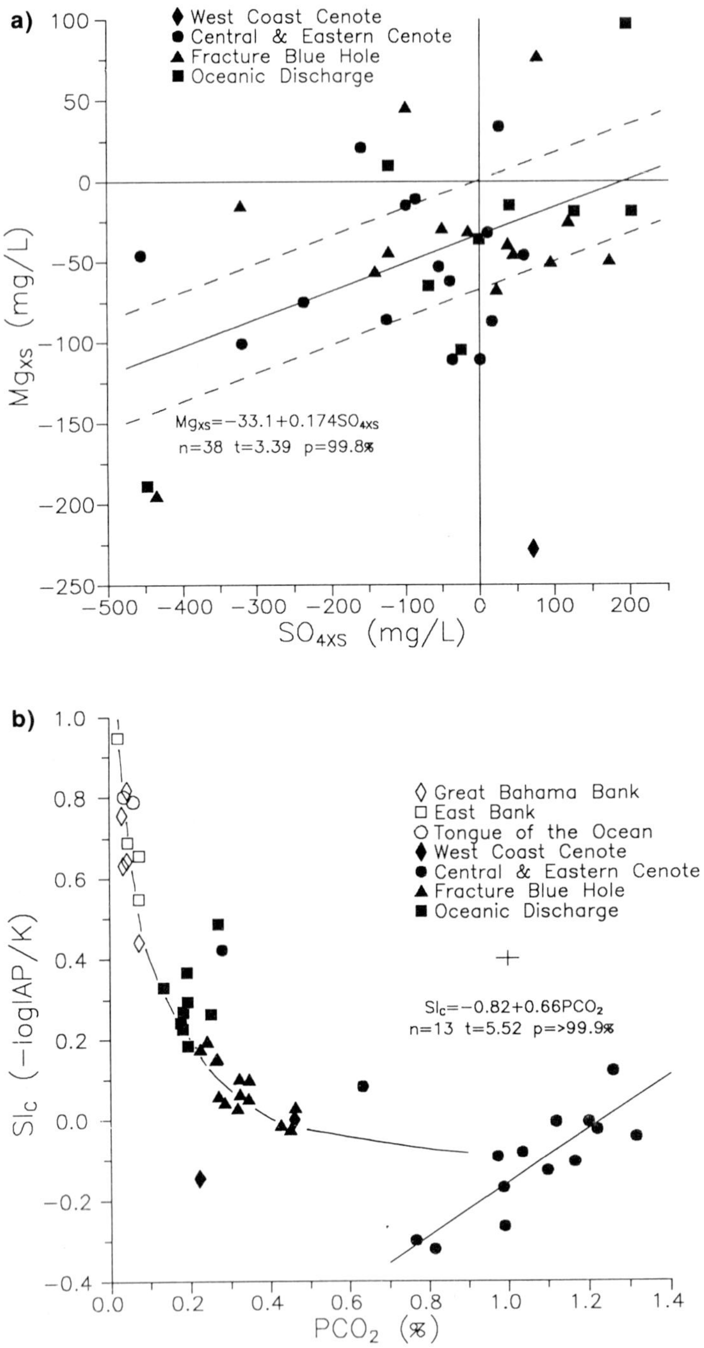

Fig. 6. (a) Positive relationship between Mg_{XS} and SO_{4XS} for inland saline groundwaters and oceanic discharges corrected for mixing with East Bank seawater. (b) Relationship between pco_2 and SI_C for source seawaters and saline goundwaters. Two samples of inland cenote waters (eastern and cental Andros) are statistical outliers and are excluded from the regression.

into major conduits which discharge on to the surface of the east bank, as observed to occur along the major bank-marginal fracture on the east coast of South Andros. The long-term discharge record from Rat Cay Blue Hole, measured by RCM, indicates an average outflow equivalent to a continuous discharge of $0.38\,m^3/s$. The equivalent figure for the shorter record at the much larger South Mastic site is $4.55\,m^3/s$. We use the smaller figure in order to provide a minimum estimate of the dolomitization

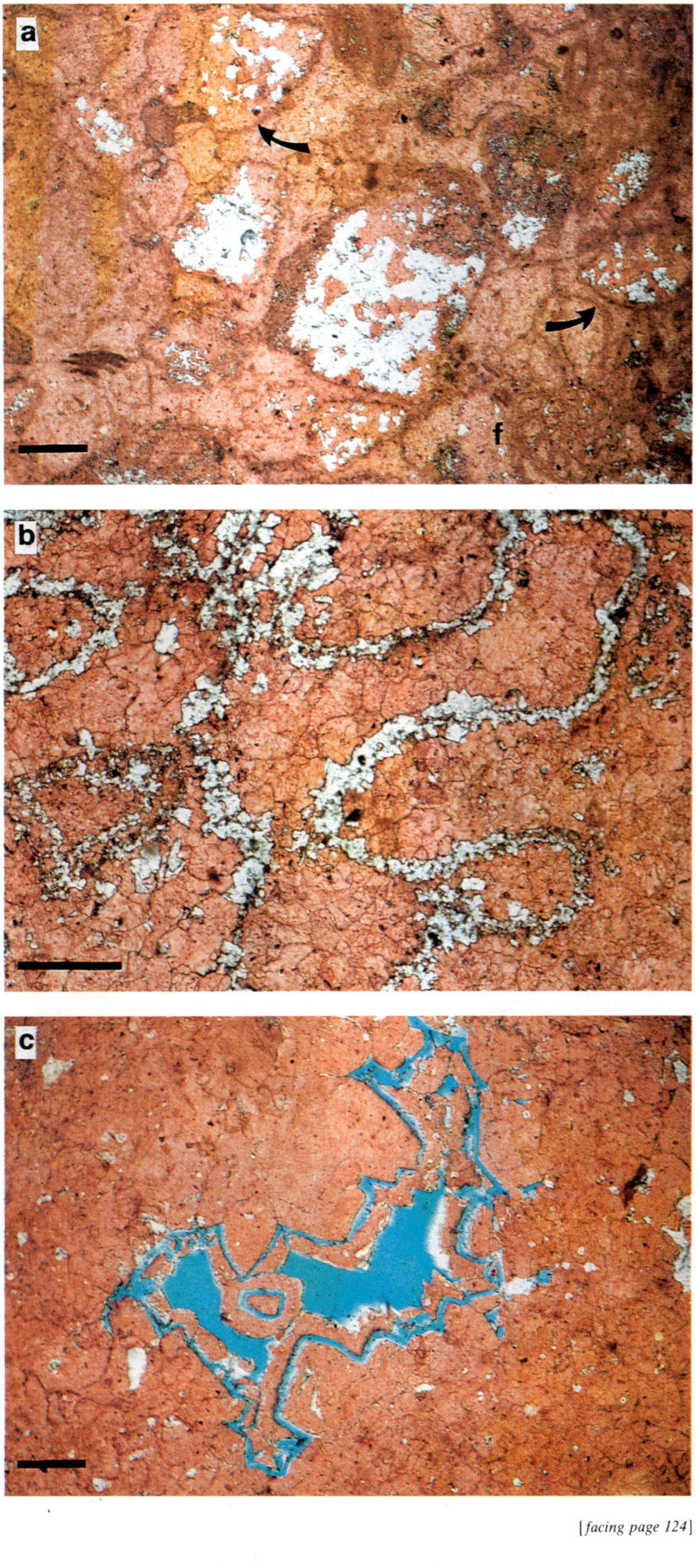

Plate 1 Thin-section photomicrographs from Stargate reefal limestones. Calcite has been stained red with alazarin red-S and pore space is impregnated with blue-dyed plastic resin. (a) Dolomite patches and isolated small rhombs in a well-cemented low Mg-calcite grainstone. Note micritic rinds outlining former grains (arrows) and small foraminifera (F). Scale bar = 200 μm, sample 87–39, 31.5 m depth. (b) Dolomite cement (?) on a micritic rind tracing the outline of a scleractinian coral. Scale bar = 100 μm, sample 87–39, 31.5 m depth. (c) Outline of a former rim cement, possibly of dolomite (*cf*. Ward & Halley, 1985). Scale bar = 200 μm, sample 87–39, 29 m depth.

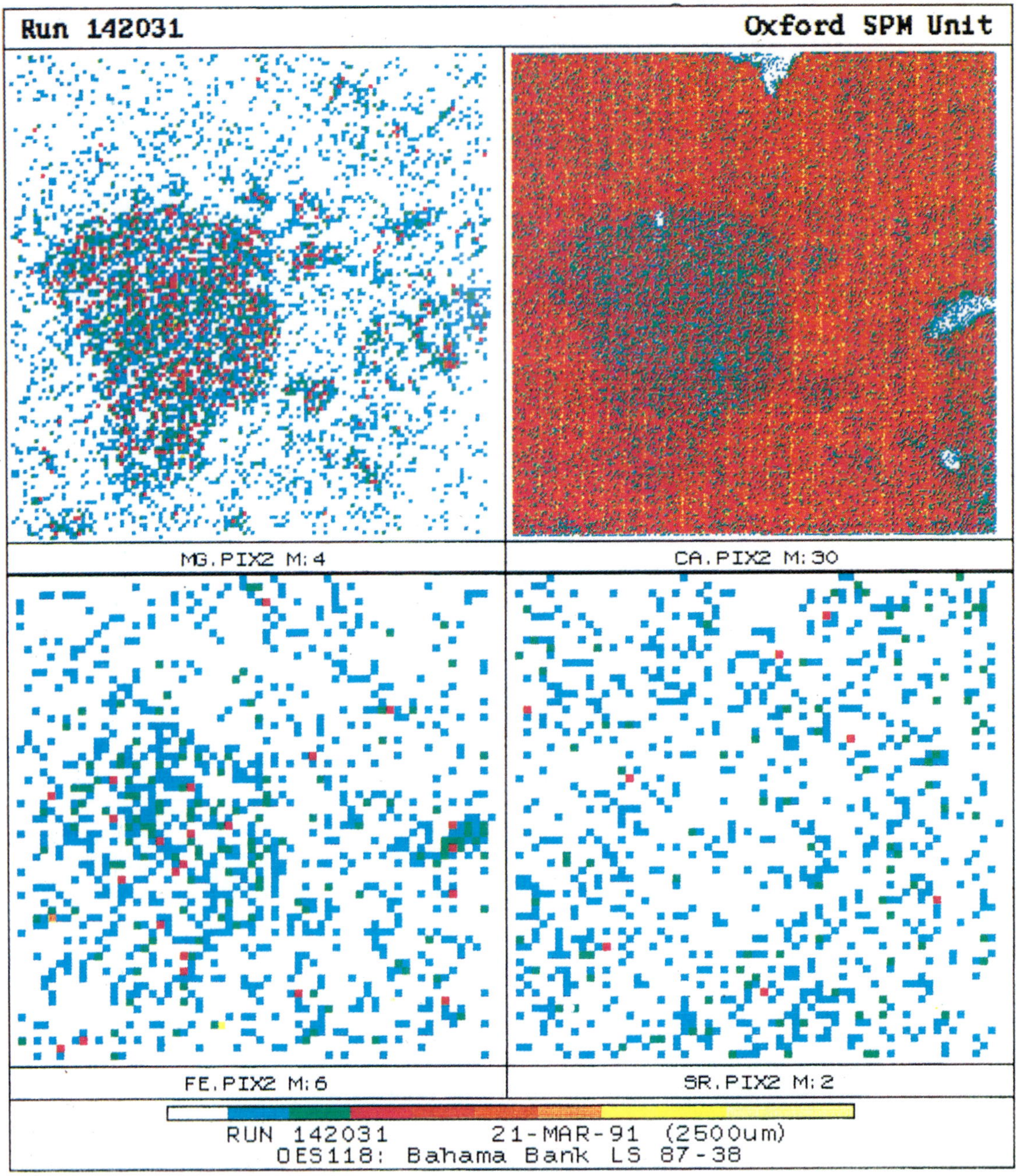

Plate 2 PIXE map (2500 μm^2) of concentrations of magnesium, calcium, iron and strontium in an anhedral aggregate of Stargate dolomite (sample 87–38, 35 m depth) within a low Mg-calcite matrix. Relative concentration intensities are indicated by the colour bar, with white as lowest concentration and yellow as highest. High-magnesium region is a single large patch of dolomite; the remainder of the scan is calcite (higher calcium levels).

rate. On the east coast of the island there are a total of ten known oceanic blue holes (Fig. 1), all of which have tidally reversing currents and are known to discharge cold elevated-salinity saline groundwaters at the end of the outflow phase. In addition there are several smaller openings, which are hydrologically active and may or may not be connected to the large-scale groundwater flow system, plus unobserved discharges directly into the Tongue of the Ocean. At one oceanic blue hole, a karstic collapse more than 30 m in diameter, discharge must be at least an order of magnitude greater than that at South Mastic. Thus, if we assume that an average discharge of 0.38 m^3/s measured at Rat Cay is representative of the long-term discharge of each of these ten oceanic blue holes, the total calculated discharge of 3.3×10^5 m^3 per day is a minimum estimate. This is equivalent to a net outflow of 4.1 m^3/day/m coastline, distributed along the 80 km length of the east coast of North Andros.

Within the platform saline groundwater circulation is constrained beneath an average upper limit of −26 m, which is the average depth of the base of the fresh/saltwater mixing zone beneath North Andros Island (calculated from all inland blue holes penetrating the saline zone). Furthermore, a lower limit of −168 m may be defined for the circulation involving elevated-salinity bank waters, where the average density of saline groundwaters from inland blue holes (δ_t = 26.3, excluding the two west-coast sites) is equivalent to that in the adjacent ocean basins. Thus, saline groundwater flow will occur through an effective aquifer thickness of 142 m, yielding an average flow rate of 2.9×10^{-2} $m^3/day/m^2$. The average rate of dolomitization can now be derived by assuming that the magnesium consumption is evenly distributed along the length of the groundwater flow path, taken to be an average of 100 km from the centre of the flooded bank to the west of the island to the eastern platform margin. A magnesium depletion of 67 mg/l (cumulative total depletion in saline groundwaters from inland cenote and oceanic blue holes, excluding mixing with East Bank waters) is equivalent to precipitation of 0.51 g of dolomite per litre of water. Using a specific gravity of dolomite of 2.85 g/cm^3, and the minimum flow rate of 2.9×10^{-2} m/day, the minimum rate of dolomitization may be calculated as 1.8×10^{-6}%/a. Note that use of the South Mastic average discharge will give a rate 25 times higher, giving an upper probable rate of dolomitization of 3.0×10^{-5}%/a.

Considering carbonates from a depth of 30 m (the shallowest depth at which dolomites were recognized in samples of wall rock from Stargate Blue Hole on South Andros, see below), and assuming an average subsidence rate of 20 m/Ma (McNeil, 1989), the maximum period for dolomitization to occur would be 1.5 Ma. Note that this ignores the initial residence in the freshwater lens and vadose zone, which may be a significant time interval after initial deposition (Matthews & Froelich, 1987). Furthermore, the present groundwater flow system is dependent on large area of the bank surface being flooded to a shallow depth, with the development of restricted seawater circulation on the bank. Given the large-amplitude high-frequency sea-level fluctuations that have characterized the Pleistocene, such conditions will only have occurred for an estimated 10% of the time (Richards *et al.*, in press). Thus, it is unlikely that dolomitization of carbonates by groundwater flow systems similar to that seen beneath Andros at present could have occurred for more than a total of 150 ka. Using this figure, together with the minimum and maximum estimated rates of dolomitization, rocks from 30 m would be predicted to comprise between 0.27 and 4.5% dolomite. In fact, other saline groundwater flow systems, driven by buoyant circulation in the mixing zone and head differences generated, for example, by wind-driven set-up, are likely to operate during periods of emergence when the present reflux system ceases (Whitaker & Smart, in press a). Calculations of the percentage of dolomite generated by the present-day circulation system may thus underestimate the degree of dolomitization.

SHALLOW DOLOMITES FROM SOUTH ANDROS

Dolomite is not present in stained (alizarin red S) thin sections of samples from within the brackish freshwater lens (0–14 m in Stargate; 0–11 m in Evelyn Green's Blue Hole), or the fresh/saltwater mixing zone (14–28 m in Stargate; 11–19 m in Evelyn Green's Blue Hole). However, minor amounts of dolomite (Plate 1, facing p. 124) were found in samples collected from the saline zone in the Stargate Blue Hole below a depth of 29 m. No dolomite was observed in samples from Evelyn Green's Blue Hole, possibly because sampling in the saline zone was limited by the relatively shallow depth of this site.

In the saline zone of Stargate Blue Hole dolomite increases in abundance with depth over an interval of at least 14 m, from 29 to a maximum sampled depth of 43 m. The dolomitization model described above would predict such an increase in dolomite abundance with depth. However, this pattern may also result from the preferential dissolution of dolomite apparent in the upper part of the sampled range (Plate 1, part c). Whereas at present the freshwater lens and saline groundwaters are both supersaturated with respect to disordered dolomite, the base of the fresh/saltwater mixing zone in Stargate Blue Hole (24–28 m) is slightly undersaturated with respect to disordered dolomite (minimum $SI_{DD} = -0.05$), suggesting that dissolution may be a relatively recent phenomenon.

The rock matrix hosting the dolomite consists of a bioclastic grainstone containing coral fragments and other skeletal grains typical of a reefal limestone, and both cement and matrix comprise predominantly stabilized low Mg-calcite. The distribution of the dolomite is highly heterogeneous, on average comprising 1–5% of the wall rock. Dolomite is found both as a matrix replacement and as a primary cement. Four dolomite fabrics can be recognized:

1 Isolated euhedral rhombs, 4–5 μm in diameter, floating within the low Mg-calcite rock matrix (Plate 1, part a);

2 Patches of anhedral crystals, commonly 30–50 μm across, but locally up to a few hundred μm, floating within the rock matrix (Plate 1, part a);

3 Fabric-selective replacement of allochems (Plate 1, part b);

4 Minor rim cements (Plate 1, part c).

Petrographically the dolomite appears to be similar to many recent immature dolomites described from elsewhere in the Bahamas and the Caribbean region (Ward & Halley, 1985; Humphrey, 1988; Vahrenkamp & Swart, this volume) but is distinct from most mature ancient dolomites.

The strontium isotope ratio was determined on a concentrate of dolomite from sample 87-38 from 35 m depth in Stargate Blue Hole. The ratio of 0.709137 corresponds to a seawater strontium isotope age of between 0.4 and 0.8 Ma, using a seawater strontium isotope curve with data combined from De Paolo (1986) and Burke *et al.* (1982) (P.K. Swart, personal communication; NBS987 = 0.710250). Assuming that seawater and the host rock form the principal sources of the strontium in the dolomite, the strontium isotope ratio indicates the oldest possible age of dolomitization (Swart *et al.*, 1987). Given the hydrological setting, it cannot be discounted that strontium with a significantly older isotopic signature has been incorporated into the dolomite. However, in the absence of any evidence for the involvement of an older strontium source, a maximum dolomite age of 0.4–0.8 Ma seems realistic. In comparison, the depositional age of the host limestones can be estimated by correlation with core U-1 on North Andros Island (see Fig. 1). Here, limestones at a depth between 30 and 40 m were estimated to be between 1.5 and 2.5 Ma, based on magnetostratigraphy and biostratigraphic marker horizons (McNeil, 1989). Hence, dolomite from Stargate Blue Hole is significantly younger than the estimated age of the host limestone (1.5–2.5 Ma). Furthermore, it is also younger than the massive Pliocene dolomites found at a depth below 53 m in core U-1 (2.0–4.5 Ma; Vahrenkamp *et al.*, 1991) which are common throughout the subsurface of the Bahamas.

The results of carbon and oxygen stable isotope determinations on calcite and dolomite from sample 87-38 are presented in Table 5a and Figure 7. There appears to be a significant difference between carbon and oxygen isotopic values derived from the bulk and LASSIE spot analyses which may indicate changes in water chemistry and/or temperature. However, the mixing trend in isotopic signatures between matrix calcite and concentrated dolomite evident in Figure 7 suggests that the large LASSIE spot size may have caused a partial admixture of CO_2 gas derived from the matrix calcite.

Assuming that the trend in Figure 7 describes a mixing of isotopes from two end-members, the Stargate dolomites have an isotopic signature of approximately +3.5‰ $\delta^{18}O$ and +2.5‰ $\delta^{13}C$, essentially identical to that of older massive dolomites found elsewhere in the Bahamas (Vahrenkamp and Swart, this volume). Using the approximation of Land (1983) for the equilibrium precipitation of dolomite from seawater with an isotopic composition of 1‰ SMOW (similar to modern Gulf Stream water), the measured oxygen isotopic composition of the dolomite is consistent with precipitation from seawater at 20–22°C. This is somewhat colder than the present saline groundwater temperatures measured in Stargate Blue Hole (24–25°C), but is comparable with minimum temperatures for saline groundwaters discharging from oceanic blue holes on North Andros (see Fig. 4). Note that the minor amounts of evaporitic concentration affecting the bank waters will not be expected to have a significant influence on the isotopic signature.

Major and trace element concentrations of Star-

Table 5. Geochemical composition of Stargate dolomite sample 87–38.
(a) Stable isotopes

	$\delta^{13}C$	$\delta^{18}O$
Lassie dolomite	+0.1	+3.8
Replacive dolomite	+0.9	−0.8
Replacive dolomite	+1.2	+1.1
Replacive dolomite	+1.8	+0.6
Dolomite cement		
Lassie calcite		
Calcite allochem	−0.3	−0.6
Calcite allochem	−1.0	−1.2
Calcite cement	−0.8	−2.8
Dolomite concentrate		
Bulk dolomite	+2.6	+3.2
Bulk dolomite	+2.4	+3.6

(b) SPM point analyses (ppm) of dolomite

Pt. no*	Sr	Zn	Fe	Mn	Cu	Ni	Y	Rb
142028	133 ± 9	<3	185 ± 11	15 ± 5	<4	<6	<5	<4
142030	127 ± 8	<3	166 ± 9	20 ± 6	<5	<8	<5	<4
142044	208 ± 134	±2	104 ± 7	<13	<5	<8	<7	5 ± 3
142052	132 ± 12	<5	167 ± 10	17 ± 7	<6	<8	<9	<6
147009	158 ± 8	<4	116 ± 6	11 ± 4	<4	<6	<5	5 ± 2

* Proton-microprobe analyses have been normalized to the average dolomite Ca concentration (see below) which was measured by electron microprobe.

(c) EMA analyses of dolomite and calcite ([Ca and Mg in element wt %, mole % $MgCO_3$ (Ca/40.08/Mg*24.31]/[1 + Ca/40.08/Mg*24.31]*100), Fe and Mn in ppm)

Ca	Mg	Mole % $MgCO_3$	Fe	Mn
39.55 ± 0.15	0.43 ± 0.01	—	200 ± 100	300 ± 100
24.84 ± 0.12	11.57 ± 0.09	43.4	400 ± 100	0 ± 100
24.51 ± 0.12	11.74 ± 0.09	44.1	100 ± 100	0 ± 100
24.09 ± 0.12	11.77 ± 0.09	44.6	200 ± 100	100 ± 100
23.70 ± 0.12	11.90 ± 0.09	45.3	300 ± 100	0 ± 100

gate dolomites (Tables 5b and 5c) are similar to those of other Bahamian dolomites (Vahrenkamp & Swart, this volume). Dolomites are calcian (43–45 mole % $MgCO_3$) and have relatively low strontium concentrations (127–208 ppm) compared with other dolomites suggested to be of marine origin (e.g. Behrens & Land, 1972). Linescans and maps show no obvious trace element zoning (Plate 2, facing p. 125). Predicted trace element and major element ratios for the dolomitizing fluid at 25°C were calculated by the method of Wogelius *et al.* (1992) and are presented in Figure 8. The molar cation ratios predicted for fluids in equilibrium with the Stargate dolomites are consistent with precipitation from seawater or a slightly reduced pore fluid derived from seawater. The range of fluid composition recorded by the dolomites is relatively narrow compared to the observed range in marine pore fluids. In particular, the limited range of iron and manganese concentrations indicates that redox conditions during the formation of these dolomites were relatively constant (Tables 5b and 5c). The system was probably poised at a slightly reducing Eh, with iron and manganese (hydr)oxides contained within the carbonate sediment matrix providing the Fe^{2+} and Mn^{2+} by reductive dissolution.

In summary, up to 5% of the reefal limestone section immersed in the saline water zone of Stargate

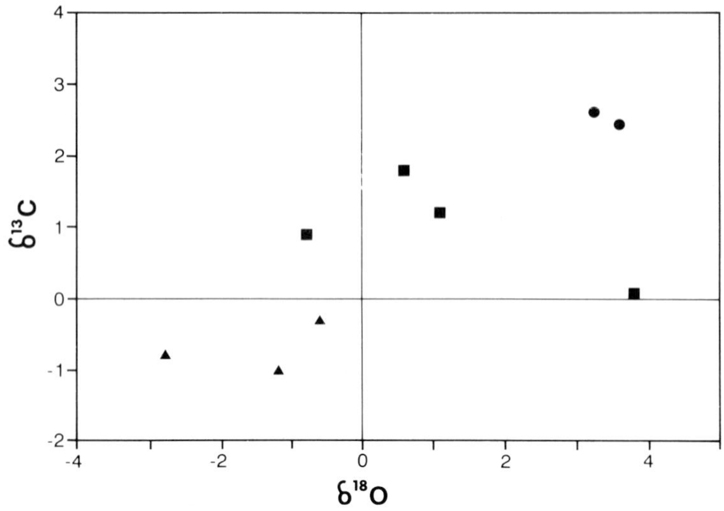

Fig. 7. Cross-plot of carbon and oxygen stable isotopes of Stargate rock sample 87-38, 35 m depth. Calcites and dolomites measured with LASSIE are triangles and squares, respectively, and concentrated dolomite from a bulk sample are circles.

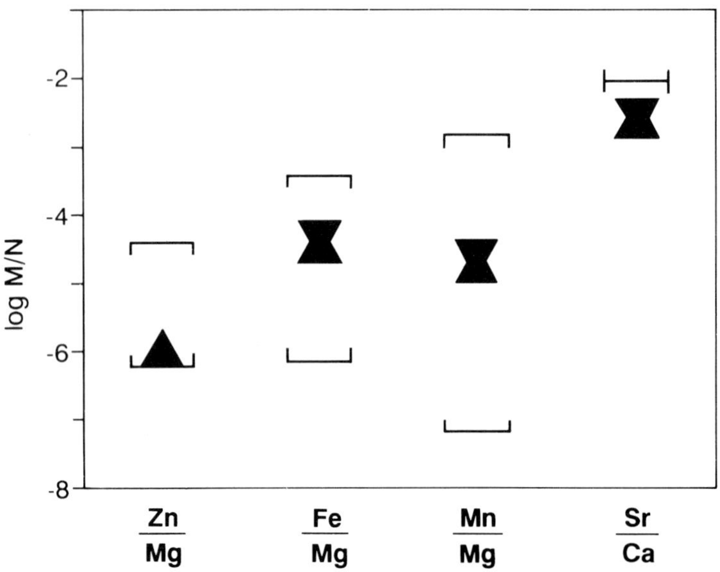

Fig. 8. Range of molar cation ratios of fluids calculated to be in equilibrium with Bahama Bank dolomite (triangles) compared with those measured in seawater (Nordstrom *et al.*, 1979) and marine pore fluids (Froelich *et al.*, 1979; Brooks *et al.*, 1986) (upwards- and downwards-facing brackets, respectively).

Blue Hole comprises dolomite, whereas none is present in the overlying freshwater lens and fresh/saltwater mixing zone (possibly due to dedolomitization). These dolomites are significantly younger than the host rock and mineralogically metastable (non-stoichiometric). The isotopic and trace element composition suggests that the dolomites precipitated from marine waters under slightly reducing conditions.

DISCUSSION

The hydrology and geochemistry of saline groundwaters from North Andros provide clear evidence of dolomitization by present-day saline groundwaters derived from the Great Bahama Bank and adjacent open ocean basin(s). The geochemistry of inland saline groundwaters indicates that dolomitization is predominantly by replacement, and appears to be progressive along the groundwater flow path. A number of 'modifications' to seawater have been suggested in the literature to overcome the apparent barriers to dolomitization (Machel & Mountjoy, 1986). In the Andros groundwaters the Mg/Ca ratios are similar to seawater, and there are insufficient organic acids present to change activities by preferential complexing of calcium (Harrison & Thyne, in press). Sulphate depletions are relatively small compared to those enhancing dolomite precipitation in the laboratory experiments of Baker and Kastner (1981), and may be important only where the carbonate ions are not derived from the precursor calcite (Morrow & Ricketts, 1988). Indeed, other studies suggest that the presence of sulphate enhances dolomitization (Eugster & Hardie, 1978). However, there is a clear association between dolomitization and increased alkalinity, which facilitates carbonate ion penetration of the hydrated magnesium ion barrier on the crystal surface (Lippmann, 1973; Morrow & Ricketts, 1988). The increased alkalinity is associated with bacterial oxidation of organic matter, which generates reducing conditions in the groundwaters, and there is also the possibility of some catalytic bacterial activity (cf. Mansfield, 1980). Finally, dolomite supersaturation is reduced in the saline groundwaters compared to seawater, possibly aiding segregation of magnesium and calcium in the dolomite crystal (Gaines, 1980). It is also noteworthy that many environments known to host modern dolomites are also of this general geochemical character (deep-sea sediments, Baker & Burns, 1985; beachrock, Friedman, 1992; microbial mats, Shinn *et al.*, 1965; saline lagoons, De Deckker & Last, 1988).

Notwithstanding these geochemical arguments, the rate of dolomitization is critically dependent on the rate of groundwater circulation within the platform, which controls the supply of magnesium. The North Andros study suggests that present rates of flow based on groundwater discharge from oceanic blue holes are on average 2.9×10^{-2} m/day. Combined with an average magnesium depletion of 67 mg/l, this indicates a present-day dolomitization rate of $0.18–3.0 \times 10^{-5}$%/a. The derivation of this calculated rate of dolomitization is critically dependent on two assumptions. First, the groundwater discharge from oceanic blue holes is assumed to have been distributed through the whole of the cross-sectional area of the aquifer before integration into the vertically extensive bank-marginal fracture systems feeding the oceanic blue holes. However, we have no direct evidence to indicate the relative proportions of flow through cavernous and non-cavernous porosity, or the degree of interaction between them. In practice, flow appears to occur via a set of ramifying, predominantly bedding-plane controlled 1–10 cm wide dissolutional conduits which exchange with less mobile water within the adjacent porous blocks, possibly assisted by the semidiurnal cyclicity of the pressure regime (Whitaker & Smart, in press b). Secondly, the geochemistry of waters sampled in the blue holes is assumed to be representative of saline groundwaters throughout the aquifer. However, the cavernous environment may be subject to enhanced organic matter input, with marine organic debris drawn into oceanic blue holes by the tidal inflow and, more significantly, surface-derived organic matter and microbial debris generated in the overlying freshwater body input to inland blue holes. Thus, given the aqueous geochemical evidence that organically mediated processes are important in controlling magnesium depletion, a significant overestimate of the present rate of dolomitization may be obtained from these waters. Despite an extensive search, we have not been able to locate any deep boreholes on Andros Island. At this stage we are therefore unable to investigate the degree to which blue hole waters are representative of non-cavernous porewaters.

Sparse Pleistocene dolomites are observed at depths as shallow as 30 m in the wall-rock samples from Stargate Blue Hole, South Andros. They comprise euhedral rhombs, patches of anhedral crystals, fabric-selective replacement of allochems and rim cements, and are petrographically similar to many 'recent' dolomites. In thin section these dolomites typically comprise 1–5% of the wall-rock sample, although their distribution is highly heterogeneous. The stable isotope and trace element geochemistry of the dolomites is compatible with precipitation from waters of near-normal seawater composition under slightly reducing conditions at temperatures in the lower range of those observed for modern saline groundwaters. The Stargate dolomites clearly postdate the carbonate sediments that host them, but two explanations of the strontium isotope 'age' of the dolomites are possible: either the dolomites grew over a prolonged period of time after deposition of the sediments whenever conditions were favourable, so that their age represents a bulked average for this period, or they developed during a single phase of dolomitization, as has been proposed for older Bahamian dolomites (Vahrenkamp *et al.*,

1991). Although the absence of trace element zoning in proton-probe linescans may be interpreted as indicative of a single phase of growth, it is probable that the supply of trace elements to the interior of the platform is limited and relatively constant during all phases of growth, and thus zoning would not be expected. Evidence regarding the 'age' of the dolomite therefore remains equivocal.

If we assume that dolomitization occurs at the estimated present-day rate and during all times of platform flooding (*ca.* 10% of the time) since deposition, the predicted dolomite content of a sample from 30 m depth ranges from a minimum of 0.3% to a maximum of 4.5%. Thus, despite the generalized nature of the calculations, rates of dolomitization observed under present-day conditions are capable of generating the extent of dolomitization observed in the Stargate samples. However, dolomitization is likely to be directly proportional to permeability, and consequently spatially heterogeneous at all scales. We have no information on whether the dolomite observed in the wall rock of the blue holes is generally distributed throughout the aquifer at these depths. Horizontal coring from the walls of the blue holes would be needed to investigate this problem. Alternatively, a detailed inspection of the northwest Great Bahama Bank borehole cores of Beach (1982) may verify the presence of the sparse dolomites at depths above the previously recognized Plio-Pleistocene dolomites (above 53 m in core U-1).

It is important to recognize that over geological timescales carbonate sediments spend the major part of their time bathed in saline groundwaters (Matthews & Froelich, 1987), and thus if circulation is maintained over the longer term complete dolomitization may occur. It can be argued that karstification associated with recent large-amplitude glacio-eustatic sea-level fluctuations has increased rates of circulation, in comparison with less karstified fossil carbonate platforms subject to lower-amplitude oscillations. However, duration of platform flooding is much increased under the latter conditions, and significant dolomitization may occur at the low observed rates over long timescales. Over the short and medium term the saline groundwater circulation system currently active beneath Andros Island appears capable of producing sparse dolomites, which may be important in providing 'seeds' for later more pervasive dolomitization.

ACKNOWLEDGEMENTS

Financial support was provided by Koninklijke/Shell Exploratie en Productie Laboratorium, Shell Bahamas, AMOCO UK Exploration Company and the Royal Society. The Institute of Oceanographic Sciences of the Natural Environment Research Council and Environmental Sciences Division of Wimpol kindly provided, decoded and calibrated the recording current meters. Invaluable field assistance was provided by Steve Hobbs and Rob Palmer, and Neil Sealey and Sue and Jim Phillips provided logistic support. Axel Miller (Plymouth Marine Laboratories) assisted with DOC analyses, Burt Fischer (AMOCO Production Company) provided independent corroboration of magnesium and chloride analyses. Peter Swart (University of Miami) provided stable isotope analyses of the dolomite concentrate, Joaquin Ruiz (University of Arizona) assisted with strontium isotope analyses and Geoff Grime and Mike Marsh (Oxford Scanning Microprobe Unit) assisted with the proton-probe microscopy. Finally Simon Godden drew the diagrams and Liz Owen undertook the preparation of the manuscript. Shell KSEPL and BP Research are thanked for funding the reproduction of Plate 1.

REFERENCES

Baker, P.A. & Burns, S.J. (1985) Occurrence and formation of dolomite in organic-rich continental margin sediments. *Bull. Am. Ass. Petrol. Geol.* **69**, 1917–1930.

Baker, P.A. & Kastner, M. (1981) Constraints on the formation of dolomite. *Science* **213**, 214–216.

Beach, D.K. (1982) *Depositional and Diagenetic History of Pliocene–Pleistocene Carbonates, Northwestern Great Bahama Bank*. Unpublished PhD Thesis, RSMAS, University of Miami, 600 pp.

Behrens, E.Q. & Land, L.S. (1972) Subtidal Holocene dolomite, Baffin Bay, Texas. *J. Sedim. Petrol.* **42**, 155–161.

Bottrell, S., Smart, P.L., Whitaker, F.F. & Raiswell, R. (1991) Geochemistry and isotope systematics in the mixing zone of Bahamian blue holes. *Appl. Geochem.* **6**, 97–103.

Broecker, W.S. & Takahashi, T. (1966) Calcium carbonate precipitation on the Bahama Banks. *J. Geophys. Res.* **71**, 1575–1602.

Brooks, R.P., Presley, B.J. & Kaplan, I.R. (1968) Trace elements in the interstitial waters of marine sediments. *Geochem. Cosmochim. Acta* **32**, 397–414.

Burke, W.H., Denison, R.E., Hetherington, E.A., Koepnick, R.B., Nelson, H.F. & Otto, J.B. (1982) Variation of sea-water ^{87}strontium/^{86}strontium throughout Phanerozoic time. *Geology* **10**, 516–519.

Busby, R.F. & Dick, G.F. (1984) *Oceanography of*

the Eastern Great Bahama Bank, Part 1. Temperature–Salinity Distribution. US Naval Oceanographic Office, 42 pp.

CLOUD, R.E. (1962) *Environments of Carbonate Deposition West of Andros Island, Bahamas*. USGS Professional Paper 350 **43**, 138 pp.

DE DECKKER, P. & LAST, W.M. (1988) Modern dolomite deposition in continental saline lakes, Western Victoria, Australia. *Geology* **16**, 29–32.

DE PAOLO, D.J. (1986) Detailed record of the Neogene strontium isotopic evolution of sea-water from DSDP site 590B. *Geology* **14**, 103–107.

ENOS, P. & SAWATSKY, L.H. (1981) Pore networks in Holocene carbonate sediments. *J. Sedim. Petrol.* **51**, 961–985.

EUGSTER, H.P. & HARDIE, L.A. (1978) Saline lakes. In: *Lakes: Chemistry, Geology, Physics* (Ed. Lerman, A.) pp. 237–293. Springer-Verlag, New York.

FRASER, D.G. (1990) Applications of the high-resolution scanning proton microprobe in the Earth sciences: an overview. *Chem. Geol.* **83**, 27–37.

FREIDMAN, G. (1992) Methane-generated lithified dolostone of Holocene age: Eastern Mediterranean. *J. Sedim. Petrol.* **61**, 188–194.

FROELICH, P.N., KLINKHAMMER, G.P., BENDER, M.L., LEUDTKE, N.A., HEATH, G.R., CULLEN, D., DAUPHIN, P., HAMMOND, D., HARTMAN, B. & MAYNARD, V. (1979) Early oxidation of organic matter in pelagic sediments of the eastern equatorial Atlantic: suboxic diagenesis. *Geochim. Cosmochim. Acta* **43**, 1075–1090.

GAINES, A.M. (1980) Dolomitisation kinetics: some recent experimental studies. In: *Concepts and Models of Dolomitization* (Ed. Zenger, D.H., Dunham, J.B. & Etherington, R.L.) Spec. Publs. Soc. Econ. Paleont. Mineral., Tulsa **28**, 81–86.

HARDIE, L.A. (1987) Dolomitisation: a critical view on some current views. *J. Sedim Petrol.* **57**, 166–183.

HARRISON, W.J. & THYNE, G.D. (in press) Prediction of diagenesis reactions in the presence of organic acids. *Geochim. Cosmochim. Acta*.

HELGESON, H.C., KIKHAM, D.H. & FLOWERS, G.C. (1981) Theoretical prediction of the thermodynamic behaviour of aqueous electrolytes at high temperatures and pressures; IV calculation of activity coefficient, osmotic coefficients and apparent molal properties at 600°C and 5 kb. *Am. J. Sci.* **281**, 1249–1510.

HUMPHREY, J.D. (1988) Late Pleistocene mixing-zone dolomitisation, south-eastern Barbados, West Indies. *Sedimentology* **35**, 327–348.

KHARAKA, Y.K., GUNTER, W.D., AGGARWAL, P.K., PERKINS, E.H. & DEBRAAL, J.D. (1988) *SOLMINEQ.88: A Computer Program for Geochemical Modelling of Rock–Water Interactions*. USGS Water Resources Investigative Report 84-4227, 419 pp.

LAND, L.S. (1983) The application of stable isotopes to studies of the origin of dolomite and to problems of diagenesis of clastic sediments. In: *Stable Isotopes in Sedimentary Geology* (Ed. Arthur, E.A.) Soc. Econ. Paleont. Mineral. Short Course No. 10, **4-1**, 4–22.

LAND, L.S. (1985) The origin of massive dolomite. *J. Geol. Educ.* **33**, 112–125.

LIPPMANN, F. (1973) *Sedimentary Carbonate Minerals*. Springer-Verlag, New York, 228 pp.

MACHEL, H.G. & MOUNTJOY, E.W. (1986) Chemistry and environments of dolomitization – a reappraisal. *Earth Sci. Rev.* **23**, 175–222.

MCNEIL, D.F. (1989) *Magnetostratigraphic Dating and Magnetization of Cenozoic Platform Carbonates from the Bahamas*. Unpublished PhD Dissertation, RSMAS, University of Miami, 210 pp.

MANSFIELD, C.G. (1980) A urolith of biogenic dolomite – another clue to the dolomite mystery. *Geochim. Cosmochim. Acta* **44**, 829–840.

MATTHEWS, R.K. & FROELICH, C. (1987) Forward modelling of bank-margin carbonate diagenesis. *Geology* **15**, 673–676.

MAUL, G.A. (1986) Linear correlations between Florida current volume transport and surface speed with Miami sea-level and weather during 1964–1970. *Royal Astronom. Soc. Geophys. J.* **87**, 55–66.

MORROW, D.W. (1992) Diagenesis 1; Dolomite – Part 1, The chemistry of dolomite precipitation. *Geoscience Canada*, **9**, 5–13.

MORROW, D.W. & RICKETTS, B.D. (1988) Experimental investigation of sulphate inhibition of dolomite and its mineral analogues. In: *Sedimentology and Geochemistry of Dolostones* (Ed. Shukla, V. & Baker, P.A.) Spec. Publs. Soc. Econ. Paleontol. Mineral., Tulsa **43**, 25–39.

MORSE, J.W., ZULLIG, J.J., BERNSTEIN, L.D., MILLERO, F.J., MILNE, P., MUCCI, F.J. & CHOPPIN, G.R. (1985) Chemistry of calcium carbonate-rich shallow-water carbonate sediments in the Bahamas. *Am. J. Sci.* **285**, 147–185.

MULLINS, H.T. & LYNTS, G.W. (1977) Origin of the Northwest Bahama platform: review and interpretation. *Bull. Geol. Soc. Am.* **88**, 1447–1461.

NORDSTROM, D.K., PLUMMER, L.N., WIGLEY, T.M.L., WOLERY, T.J., BALL, J.W., JENNE, E.A., BASSETT, R.L., CRERAR, D.A., FLORENCE, T.M., FRITZ, B., HOFFMAN, M., HOLDREN, G.R., LAFON, G.M., MATTIGOD, S.V., MCDUFF, R.E., MOREL, F., REDDY, M.M., SPOSITO, G. & THRAILKILL, J. (1979) A comparison of computerised chemical models for equilibrium calculations in aqueous systems. In: *Chemical Modelling of Aqueous Systems, Speciation, Sorption, Solubility and Kinetics* (Ed. Jenne, E.A.) pp 857–892. American Chemical Society Series **93**.

RICHARDS, D.A., SMART, P.L., EDWARDS, R.L. (in press) Sea levels for the last glacial period constrained using ^{230}Th ages of submerged speleothems. *Nature*

SHINN, E.A., GINSBURG, R.N. & LLOYD, R.M. (1965) Recent supratidal dolomite from Andros Island, Bahamas. In: *Dolomitisation and Limestone Diagenesis* (Ed. Pray, L.C. & Murray, R.C.) Spec. Publs. Soc. Econ. Paleont. Mineral. **13**, 112–123.

SIMMS, M. (1984) Dolomitization by groundwater flow systems in carbonate platforms. *Trans. Gulf Coast Assoc. Geol. Sci.* **24**, 411–420.

SMALLEY, P.C., MALLE, C.N., COLEMAN, M.L. & ROUSE, J.E. (1992) LASSIE (lasar ablation sampler for stable isotope extraction) applied to carbonate minerals. *Chem. Geol.* **101**, 43–52.

SMART, P.L., PALMER, R.J., WHITAKER, F.F. & WRIGHT, V.P. (1988) Neptunian dykes and fissure fills: an overview and account of some modern examples. In: *Paleokarst* (Ed. James, N.P. & Choquette, P.W.) pp. 149–163. Springer-Verlag, New York.

SMITH, C.I. (1940) The Great Bahama Bank. 1: General

hydrographical and chemical features. *J. Marine Res.* **3**, 147–170.

Sugimura, Y. & Suzuki, Y. (1988) A high temperature catalytic oxidation method for the determination of organic carbon in sea-water by direct injection of a liquid sample. *Marine Chem.* **24**, 105–131.

Sverdrup, H.U., Johnson, M.W. & Fleming, R.H. (1946) *The Oceans; their Physics, Chemistry and General Biology*. Prentice-Hall, New York, 1087 pp.

Swart, P.K., Ruiz, J. & Holmes, C. (1987) The use of strontium isotopes to constrain the timing and mode of dolomitisation of upper Cenozoic sediments in a core from San Salvador, Bahamas. *Geology* **15**, 262–265.

Vahrenkamp, V.C., Swart, P.K. & Ruiz, J. (1991) Episodic dolomitisation of Late Cenozoic carbonates in the Bahamas: evidence from strontium isotopes. *J. Sedim. Petrol.* **61**, 1002–1014.

Ward, W.C. & Halley, R.B. (1985) Dolomitisation in a mixing zone of near-seawater composition, Late Pleistocene, northeastern Yucatan peninsula. *J. Sedim. Petrol.* **55**, 407–420.

Whitaker, F.F. (1992) *Hydrology, Geochemistry and Diagenesis of Modern Carbonate Platforms in the Bahamas*. Unpublished PhD Thesis, University of Bristol, UK, 347 pp.

Whitaker, F.F. & Smart, P.L. (1990) Circulation of saline groundwaters through carbonate platforms: evidence from the Great Bahama Bank. *Geology* **18**, 200–204.

Whitaker, F.F. & Smart, P.L. (in press a) Circulation of saline ground-waters through carbonate build-ups; an overview and case study from the Bahamas. In: *Diagenesis and Basin Development* (Ed. Horbury, A. & Robinson A.) Am. Assoc. Petrol. Geol. Mem.

Whitaker, F.F. & Smart, P.L. (in press b) Hydrogeology, hydrology and water resources of the Bahamian archipelago. In: *Geology and Hydrology of Carbonate Islands* (Ed. Vacher, H.L. & Quinn, T.)

Wogelius, R.A., Fraser, D.G., Feltham, D.J. & Whiteman, M.I. (1992) Trace element zoning in dolomite: proton microprobe data and thermodynamic constraints on fluid compositions. *Geochim. Cosmochim. Acta* **56**, 319–334.

Spec. Publs Int. Ass. Sediment. (1994) **21**, 133–153

Late Cenozoic dolomites of the Bahamas: metastable analogues for the genesis of ancient platform dolomites

V.C. VAHRENKAMP* *and* P.K. SWART†

* *Shell Research, Volmerlaan 6, 2288 GD Rijswijk ZH, Netherlands; and*
† *Marine Geology and Geophysics, RSMAS, University of Miami, Rickenbacker Cswy, Miami, Florida 33149, USA*

ABSTRACT

The petrographic and geochemical characteristics of a more than 80 m thick sequence of shallow-burial Late Cenozoic dolomites from Little Bahamas Bank (LBB), northern Bahamas, indicate that these dolomites are seawater-derived and partially metastable. Dolomites range in composition between calcian and stoichiometric ($Ca_{.60}Mg_{.40}CO_3$–$Ca_{.51}Mg_{.49}CO_3$). Strontium content and oxygen isotopes covary with the major element composition. It is suggested that part of the 2‰ spread in oxygen isotopic composition is a result of natural and laboratory isotope fractionation. After correction for these effects oxygen isotopes are in equilibrium with seawater at 20–22°C. Iron and manganese content of the dolomites is essentially precursor-controlled. The geometry of one of the dolomite bodies suggests that dolomitization occurred in the seawater phreatic zone. Seawater circulation through LBB was probably driven by an overlying freshwater/mixing zone system during (partial?) platform exposure.

Maturation of the metastable dolomites from LBB during future diagenesis may cause evolution of petrographic and geochemical signatures to those typical of many ancient dolomite sequences. Hence, Late Cenozoic dolomites of the Bahamas can be used as analogues for the genesis of many ancient platform dolomites. Their diagenetic potential permits the prediction of secondary diagenesis and the understanding of its impact seen in older platform dolomites.

INTRODUCTION

The formation of apparently minor amounts of dolomite in modern carbonate environments is in sharp contrast to the vast quantities of dolomite in the stratigraphic record (e.g. modern: Shinn *et al.*, 1965; Behrens & Land, 1972; McKenzie, 1981; Mitchell *et al.*, 1987; Carballo *et al.*, 1987; ancient: Given & Wilkinson, 1987). Recent dolomites usually differ both in appearance and composition from their ancient counterparts, and are therefore questionable analogues for the genesis of ancient platform dolomites. Some of the youngest massive dolomite sequences, which are at least by volume comparable with some ancient dolomitized carbonate platforms, occur in Upper Cenozoic strata of the Bahamas (Field & Hess, 1933; Newell & Rigby, 1957; Goodell & Garman, 1969; Supko, 1977; Gidman, 1978; Beach, 1982; Pierson, 1982; Kaldi & Gidman, 1982; Williams, 1985a; Swart *et al.*, 1987; Dawans & Swart, 1988; Vahrenkamp, 1988; Vahrenkamp & Swart, 1990; Vahrenkamp *et al.*, 1991). How do these dolomites compare to their ancient counterparts? In the apparent absence of modern massive dolomites can they be used as analogues for ancient platform dolomites?

We have investigated the petrographic and geochemical characteristics of a more than 80 m thick sequence of Late Cenozoic shallow-burial dolomites from Little Bahama Bank (LBB), northern Bahamas. It is the purpose of this paper to constrain the origin of the LBB dolomites and to determine their potential for diagenetic alteration. Future diagenesis is predicted by comparing the diagenetic potential of LBB dolomites with the composition of ancient dolomites.

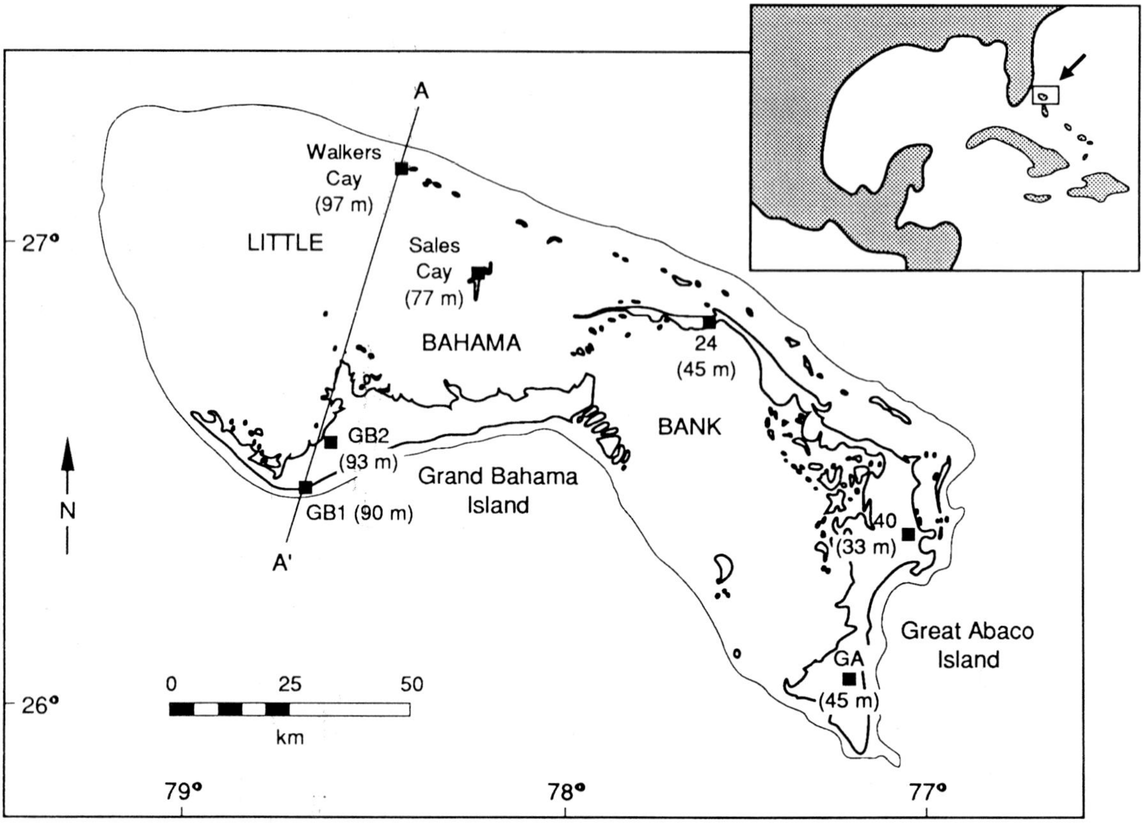

Fig. 1. Little Bahama Bank forms the northernmost platform of the Bahamas. Shown are the locations and core lengths of seven boreholes penetrating pre-Pleistocene sediments. Cores from Walkers Cay, Sale Cay, Grand Bahama Island (GB1, GB2), and Great Abaco Island recovered dolomites.

GEOLOGICAL SETTING

Little Bahama Bank is the northernmost carbonate platform of the Bahamian Archipelago (Fig. 1). Contrary to what its name may suggest, it is a rather large platform that stretches more than 200 km from east to west, reaching a width of more than 100 km. The upper 100 m of the platform have been penetrated by a number of core borings (Figs 1 and 2). A Middle Miocene to Pleistocene stratigraphy has been established here based on biostratigraphy, magnetostratigraphy and strontium isotope data (Williams, 1985a,b; McNeill 1989; Vahrenkamp *et al.*, 1991).

An extensive dolomite body has been cored in five wells at a burial depth as shallow as 20 m (Gidman, 1978; Kaldi & Gidman, 1982; Williams, 1985a; Vahrenkamp, 1988; Vahrenkamp *et al.*, 1991). Strontium isotope signatures separate the dolomitized section into at least three distinct bodies generated during successive diagenetic episodes in the early Late Miocene, Late Pliocene and latest Pliocene to early Pleistocene (Fig. 2; Vahrenkamp *et al.*, 1988, 1991). The timing of dolomitization established that parts of the dolomite formed essentially at the surface, prior to the deposition of the overlying Pleistocene Lucayan limestone (Fig. 2).

METHODS

More than 300 partially stained thin sections covering a total of about 210 m of dolomite core from four wells form the basis for the following petrographic descriptions. Some 429 bulk rock samples were taken from the dolomitized part of the cores for

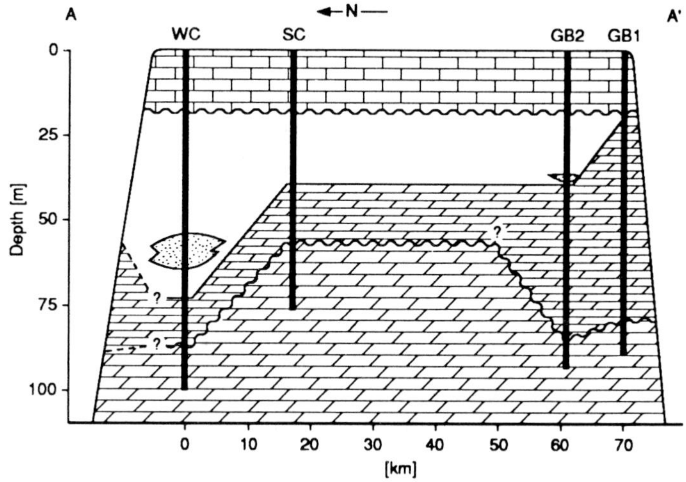

Lithology	Lower Dolomite	Upper Dolomite		Limestone	
Deposition	Middle Miocene (or older)	Late Miocene to Pliocene		Late Miocene ? to Pliocene	Pleistocene
Dolomitization/ Lst. stabilization	e. L. Miocene	L. Pliocene	l. Pliocene/ e. Pleistocene	L.Pliocene ?	Pleistocene
Dolomitization Phase	Phase I	Phase II	Phase III		
Exposure surface					

Fig. 2. North–south cross-section of Little Bahama Bank along line (A–A′) of Fig. 1 (GB2 and SC are projected on to line A–A′). A Pleistocene limestone unit (Lucayan Formation) of almost uniform thickness covers pre-Pleistocene limestones and dolomites. The lower dolomite strata were dolomitized during the early Late Miocene (phase I), the upper dolomite during the Late Pliocene and latest Pliocene/early Pleistocene (phases II and III). The timing of replacement dolomitization suggests a Middle Miocene and a pre-Late Pliocene age for the respective stratigraphic units. The contact between lower and upper dolomite strata indicates significant depositional and/or erosional topography on Little Bahama Bank before deposition of the precursor sediments of the upper dolomite. Note the geometry of the dolomite/limestone contact. Any diagenetic model explaining the regional dolomitization must feature a hydrological system responsible for reactant transport, which can create the outlined geometry (see also Fig. 12). (From Vahrenkamp *et al.*, 1991).

geochemical analysis (sample spacing: min. 4 cm; max. 1.5 m; average one sample per 0.5 m of section). Mineralogy was analysed with X-ray diffraction (XRD). Major and trace element composition was determined using an inductively coupled plasma spectrometer (ICP). Comparison of major element composition determined by XRD (peak shift method of Lumsden and Chimahusky, 1980) with results from ICP analyses ensured compatibility of samples, methods and results. Most bulk rock samples plus some additional micro samples were analysed for their carbon and oxygen stable isotopic composition. Isotope data have been corrected for isobaric interferences following the modified procedures of Craig (1957) for a triple-collector mass spectrometer and for fractionation during the dissolution of dolomite by phosphoric acid using a value of 1.0102. This value is approximately 0.89‰ different from that applied to calcites reacted at the same temperatures. For further details on sample locations and techniques see Williams (1985a) and Vahrenkamp (1988).

RESULTS

Lithological and geochemical characteristics of LBB dolomites

The lithological characteristics of the Late Tertiary section of LBB, including depositional facies, depositional and diagenetic texture, mineralogy and other petrographic features are summarized in Figure 3.

Depositional texture

Despite extensive dolomitization, lithofacies were identified based on their original depositional textures and compositions (Fig. 3). Textural variety is large, ranging from mud-supported over grain-supported textures to boundstones. Common

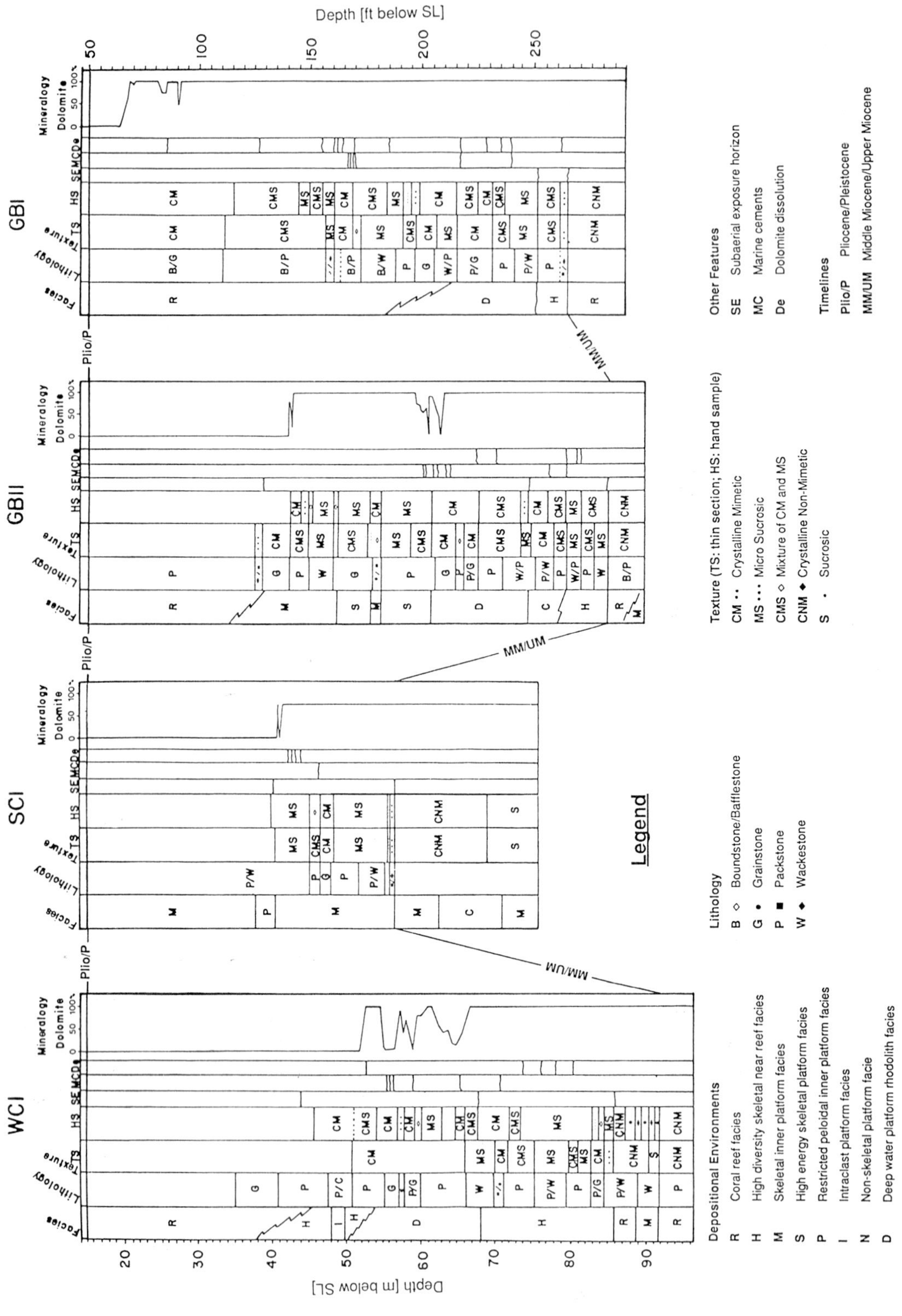

Fig. 3. Summary sheet of LBB petrography.

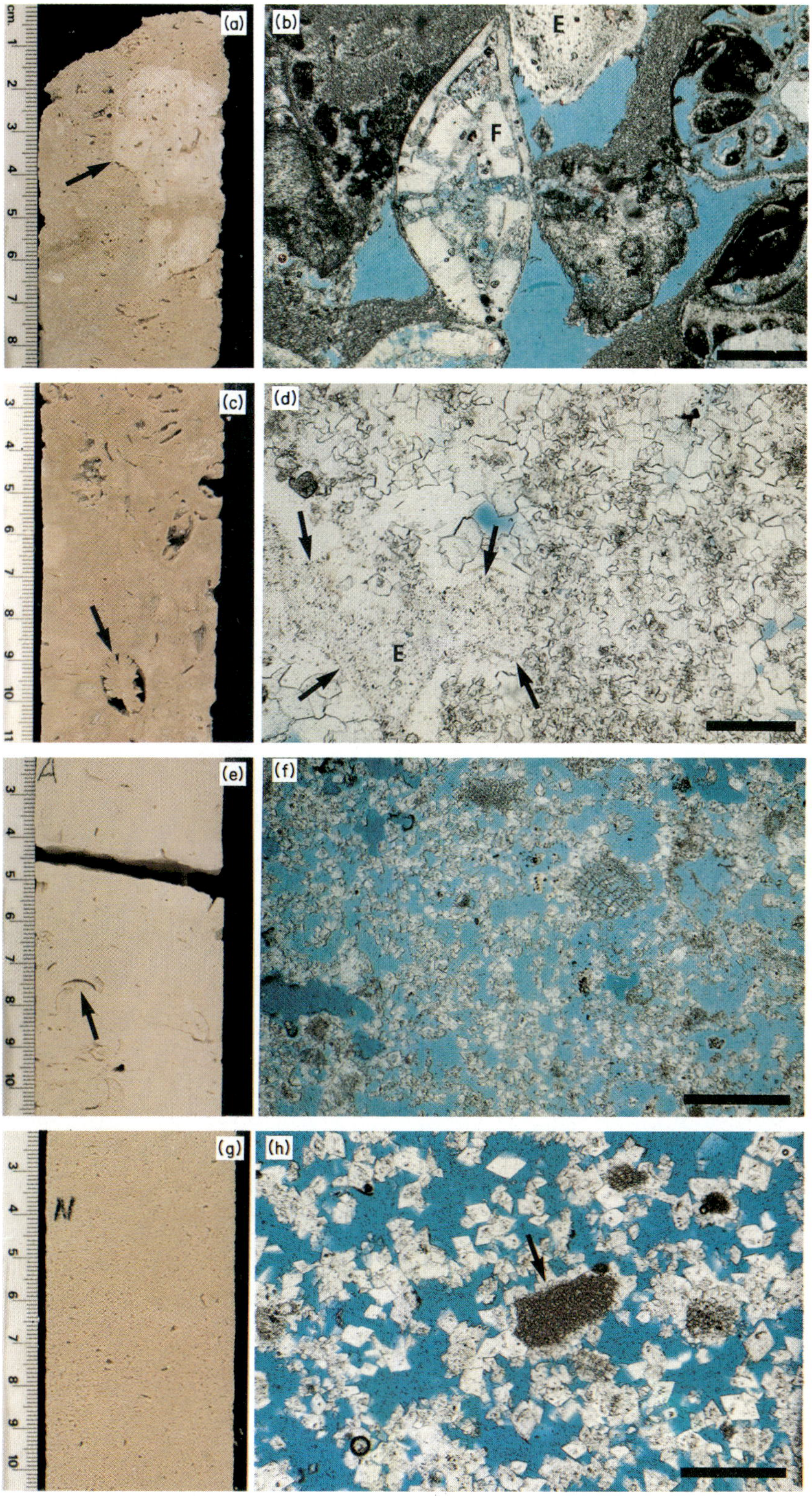

Plate 1 See p. 138, legend to Fig. 4 for caption.

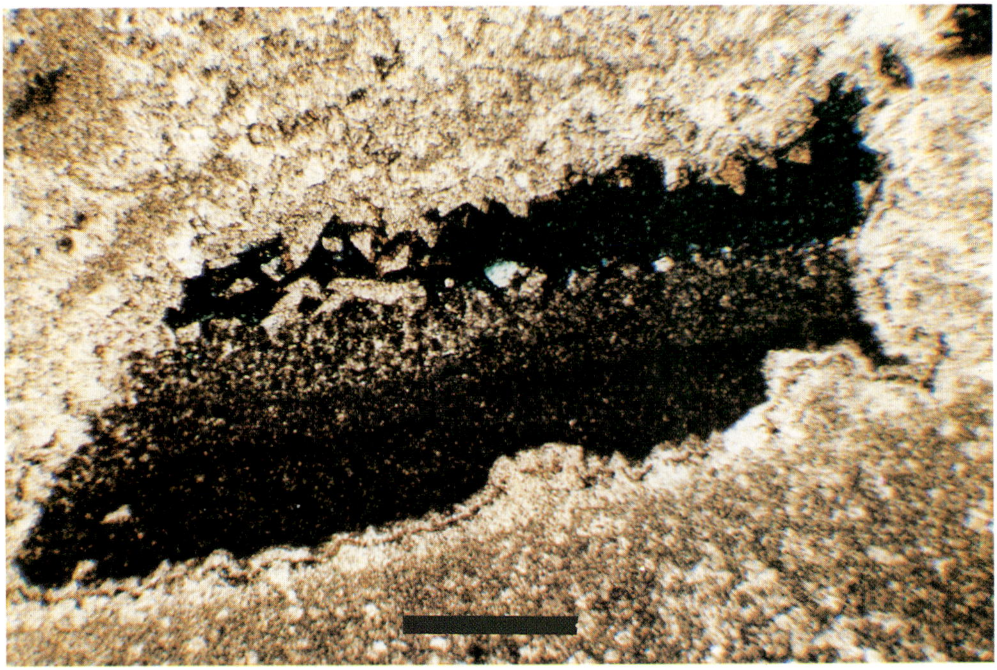

(a)

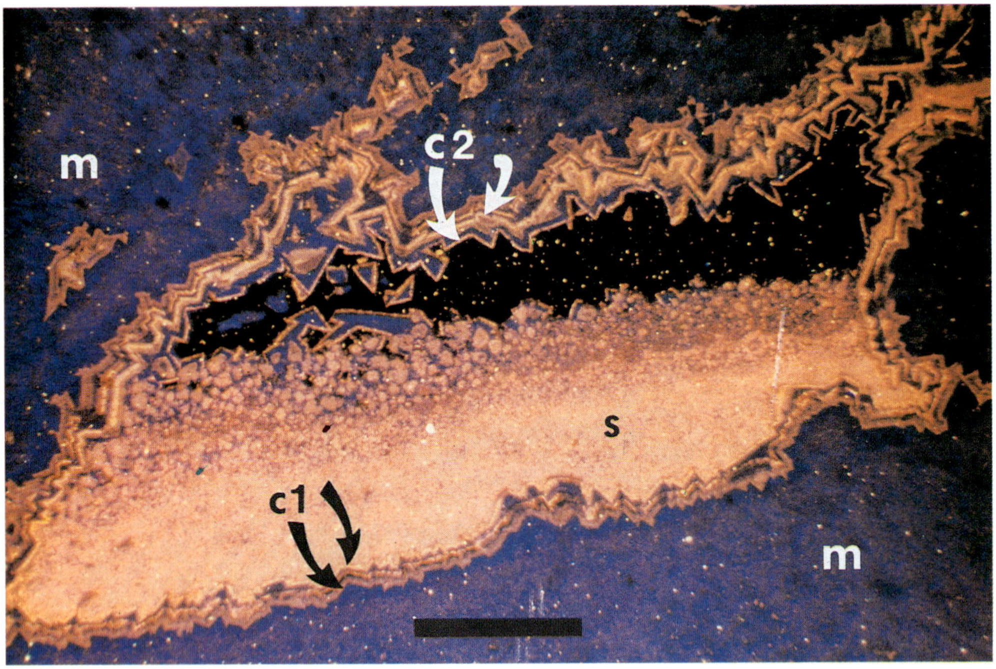

(b)

Plate 2 (a) A vug in CNM dolomite partially filled with internal sediment and cement below the erosional boundary separating early Late Miocene from Late Pliocene dolomites (GB1 63 m; transmitted light; scale bar = 500 μm). (b) Same view as upper photo but in cathodoluminescent light. The matrix of CNM dolomite (m) does not luminesce. First the vug was lined with a zoned dolomite cement (c1; up to a yellow-orange luminescing zone). Subsequently, silt (s) was washed in and dolomitized, probably simultaneously with the precipitation of additional dolomite cementation (c2; non-luminescing and final orange luminescing bands). Hence, at least two phases of dolomitization occurred separated by renewed sediment deposition (scale bar = 500 μm; vacuum: 170 millitorr; current: 0.2 mA at 15 kV).

biogenic components are corals, a variety of foraminifera, echinoderms, red algae, rhodoliths and molluscs (Fig. 4/Plate 1, facing p. 136). For a more complete description of depositional composition and textures see Gidman (1978), Williams (1985a) and Vahrenkamp (1988).

Diagenetic texture

Dolomitization is usually complete (see Fig. 3), with a full range of lithofabrics from precursor fabric preservation to complete fabric destruction. Principally, two dolomite varieties exist: friable to ringing-hard coarse-crystalline tan dolomite, and friable to well-cemented fine-crystalline cream-coloured dolomite (Fig. 4/Plate 1). The former is limited to the Middle Miocene section, whereas the latter occurs in the upper Late Miocene/Pliocene part of the section and is difficult to differentiate from the overlying friable to well-cemented cream-coloured limestones. In analogy to a classification developed for other Bahamian dolomites (Dawans & Swart, 1988), LBB dolomites were classified according to their texture. Variations with depth were logged in detail by visual inspection of both hand samples and petrographic thin sections (Fig. 3). Hand sample and thin-section classification agree reasonably well.

Crystalline mimetic dolomites (CM) (Fig. 4a,b/Plate 1). Crystalline mimetic dolomites are characterized by the mimetic preservation of the precursor fabric and by an abundance of clear void-lining or void-filling cement. Original grains are either mimetically replaced or leached, creating considerable mouldic porosity. Primary and secondary porosity is often occluded by a variety of clear dolomite cements. The most common types are void-lining subhedral to euhedral bladed rim cements (50–150 μm), void-lining anhedral equant cements and anhedral to euhedral syntaxial overgrowth on echinoderm fragments (Fig. 4b/Plate 1).

Crystalline non-mimetic dolomites (CNM) (Fig. 4c,d/Plate 1). Crystalline non-mimetic dolomites are composed of a tight interlocking mosaic of anhedral to euhedral large dolomite crystals, commonly with cloudy inclusions and clear outer rims (Fig. 4d/Plate 1). The fabric suggests severe recrystallization, with often complete obliteration of precursor components and fabric.

Sucrosic dolomites (Fig. 4e–h/Plate 1). Sucrosic dolomites are mosaics of euhedral intergrown dolomite rhombs. The precursor fabric is completely destroyed and original components are largely dissolved, with the exception of red algae and occasional echinoderm fragments. Because of a distinct bimodal size distribution, sucrosic dolomites are subdivided into a microsucrosic (MS; <50 μm; Fig. 4e,f/Plate 1) and a sucrosic (S; >50 μm; Fig. 4g,h/Plate 1) variety. MS dolomites are found in the Late Miocene to Pliocene section only, whereas S dolomites are limited to the Middle Miocene section.

CMS dolomites. CMS dolomites are a transitional category between CM and MS dolomites. This subdivision was used to describe a considerable part of the section which exhibits characteristics of both CM and MS dolomites.

Porosity/permeability

Porosity varies between 5% and 43% (Williams, 1985b; own visual estimates). Average porosity is high, probably between 25% and 30%. Permeability reaches up to 3900 Md (Williams, 1985b).

Cathodoluminescence

Cathodoluminescence is overall dull, yet alternating dull and bright zones may occur associated with subaerial exposure horizons (Plate 2, opposite).

Major elements

LBB dolomites are characterized by a large spread in major element composition, ranging between almost stoichiometric dolomite (49 mole % $MgCO_3$) to calcian (as little as 40 mole % $MgCO_3$; Fig. 5). A correlation exists between stoichiometry and dolomite fabric (MS is often stoichiometric; CM is always Ca-rich) and location with respect to the platform margin (platform edge is Ca-rich as well as stoichiometric dolomite; platform centres are only Ca-rich dolomite; Fig. 5).

Trace elements

The concentration of Sr in the dolomite is relatively low (range 70–300 ppm) and covaries with major element geochemistry (Fig. 6; Vahrenkamp & Swart, 1990). The Fe and Mn concentrations range from 10 to 300 ppm and from 5 to 100 ppm, respectively. Depth plots show large variations in trace element

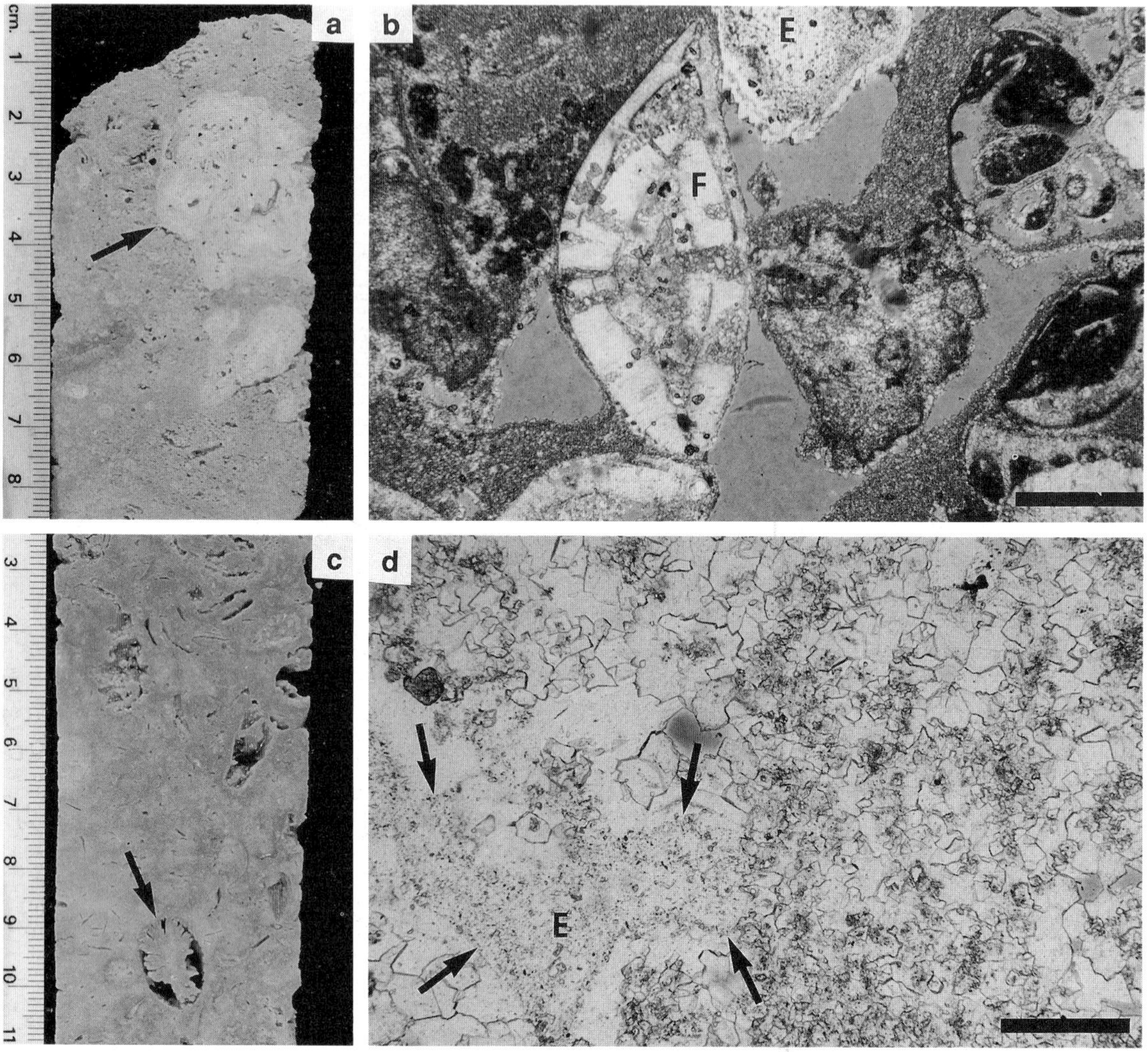

Fig. 4. (*also reproduced in colour, see Plate 1, facing p. 136*) Textures of dolomites from Little Bahama Bank. The classification has been adopted from Dawans and Swart (1988). Crystalline non-mimetic and sucrosic dolomites are found only in the lower dolomite unit, whereas crystalline-mimetic and microsucrosic textures prevail in the upper dolomite lithologies (see also Figs 2 & 3). Note that transitional dolomite textures exist between these four classes. The scale in core-sample pictures a, c, e and g is in cm; porosity in thinsection photomicrographs b, d, f and h is impregnated with blue plastic resin. (a, b) Crystalline mimetic dolomites (a: WC 60 m depth; b: GB1 61.4 m depth, scale bar = 500 μm). The rock is well indurated and white to cream in colour. Mimetic replacement of biogenic components is common: the arrow in (a) points towards a rhodolith nodule; in (b) a large foraminifera (F – *Amphistegina*) and an echinoderm fragment (E) are mimetically replaced. Clear rim cements and secondary porosity are common. (c, d) Crystalline non-mimetic dolomites (c: GB2 87.8 m depth; GB1 80.5 m depth, scale bar = 500 μm). These dolomites are ringing-hard and tan in colour. Biomoulds are common (arrow in (c) points towards a coral mould). However, in thin-sections original components are non-mimetically replaced and difficult to identify. Common exceptions are echinoderm fragments with cloudy outlines and clear syntaxial rim cements (E and arrows in (d)). (e, f) Microsucrosic dolomites (e: SC 56.8 m depth; f: SC 44.8 m depth, scale bar = 400 μm). These rocks are white in colour and usually friable. Biomouldic porosity is common outlining a time-diagnostic mollusc assemblage (arrow in (e); Williams, 1983, 1985a). The dolomite rhombs less than 50 μm in size contain up to 40% intercrystalline porosity. Identifiable biocomponents are red algae and a few echinoderm fragments. (g, h) Sucrosic dolomites (g: SC 75 m depth; h: SC 75 m depth; scale bar = 400 μm). This dolomite type is tan to cream in colour and friable to sandy in consistency. Few biocomponents are identifiable. Dolomite rhombs are usually between 100 and 150 μm in size; porosity may exceed 40%.

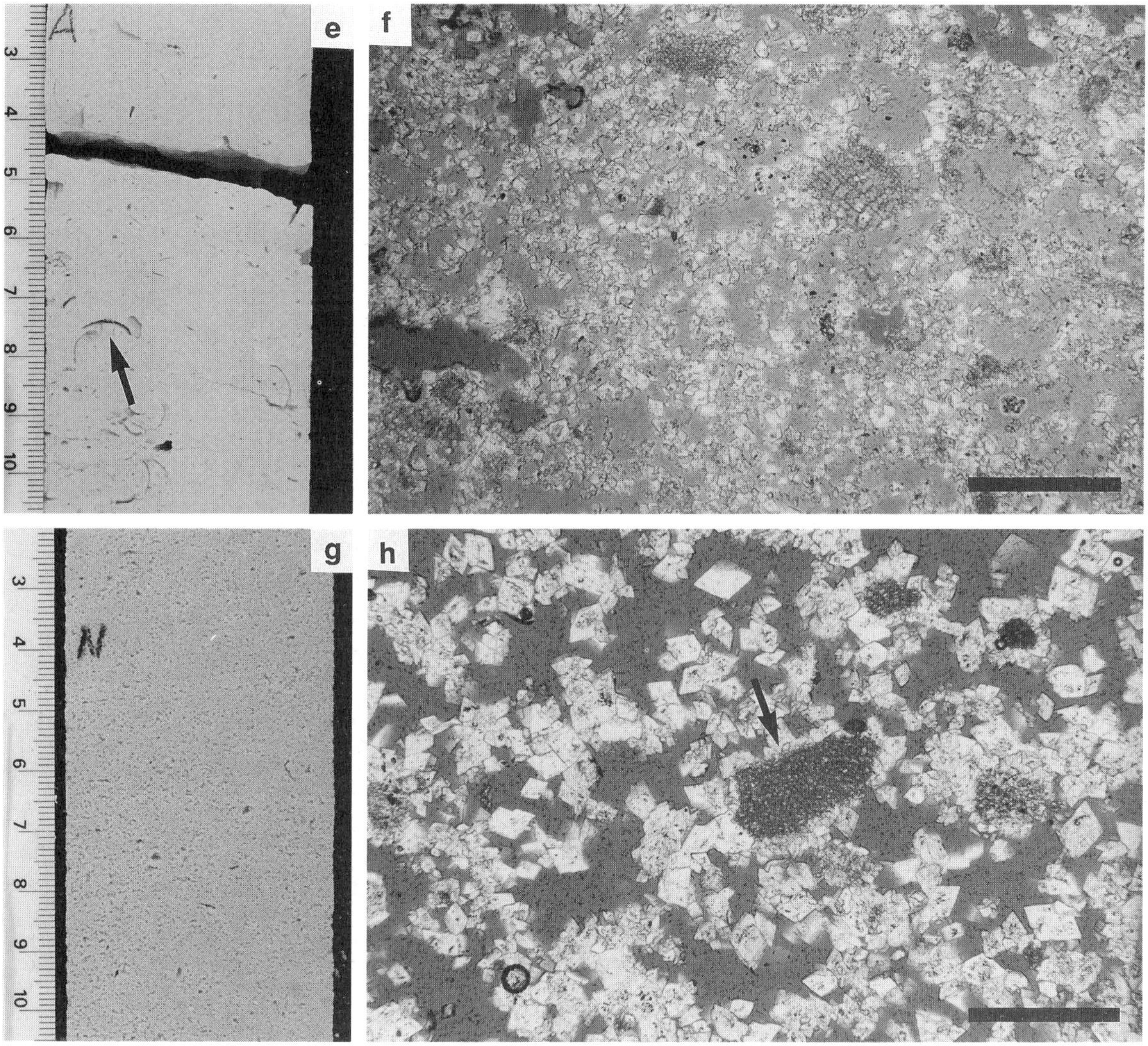

Fig. 4. *Continued.*

content, with an overall increase in Fe and decrease in Mn away from the platform edge (Fig. 7a,b).

Stable oxygen and carbon isotopes

The $\delta^{18}O$ ranges from +1.5‰ to +4‰ PDB (m = +3.2‰, σ = 0.78; Fig. 8) and covaries with the $MgCO_3$ content (Fig. 9). The $\delta^{13}C$ varies between 0‰ and 3.5‰ (m = 2.1).

DISCUSSION

Depositional environment

The Middle Miocene to Pliocene section of LBB is characterized by predominantly skeletal, carbonate platform sediments. Depositional environments range from shallow-water platform interior over platform margin reef to deeper water (30–60 m)

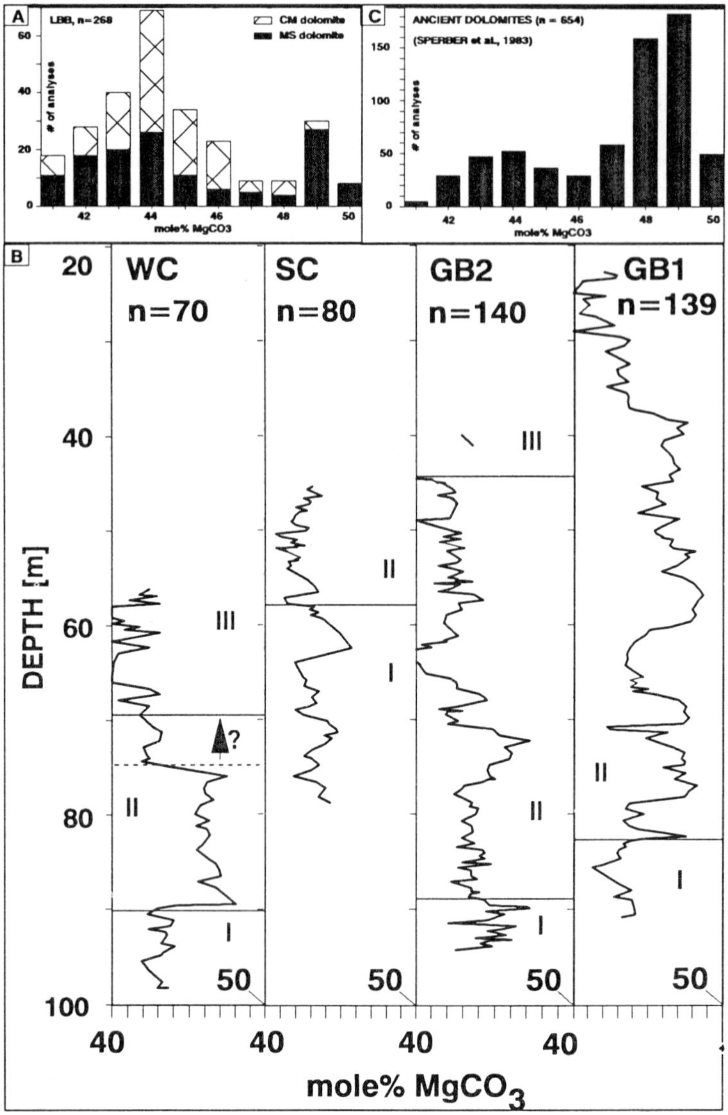

Fig. 5. Mole % $MgCO_3$ composition of Little Bahama Bank dolomites. (A) Histogram of the mole % $MgCO_3$ composition of dolomites with crystalline-mimetic (CM) and microsucrosic (MS) textures. Compositions of these relatively young dolomites range from stoichiometric to strongly magnesium-depleted. Note the bimodality with peaks at 44 and 49 mole % $MgCO_3$. CM textures never reach a stoichiometric composition. (B) Mole % $MgCO_3$ composition of four cores (WC, SC, GB1 and GB2) versus depth. The section has been subdivided into the three dolomite generations differentiated by Sr isotopes (compare with Fig. 2). Note that away from the platform margin (GB1) dolomite units with stoichiometric compositions become less frequent (WC, GB2) and are absent in the centre of the platform (SC). (C) Major element composition of ancient dolomites (Sperber *et al.*, 1984). As for LBB dolomites, bimodality is apparent with peaks at 44 and 49 mole % $MgCO_3$. However, the strongest peak has shifted to a nearly stoichiometric composition. This may indicate progressive evolution of dolomite sequences from an initially partially metastable to a more ideal geochemical composition.

forereef settings. Vertical and lateral facies shifts are indicative of a dynamic shallow-water carbonate system responding to numerous relative sea-level fluctuations. For further details on depositional settings and history see Gidman (1978), Williams (1985a) and Vahrenkamp (1988).

Petrographic maturity

The almost complete dolomitization of an up to 80 m thick, near-surface and relatively young carbonate section in LBB poses the question whether these dolomites are petrographically mature and comparable to ancient dolomites, or whether they are more similar to immature 'protodolomites' described from modern settings (Shinn *et al.*, 1965; Behrens & Land, 1972; McKenzie, 1981; Mitchell *et al.*, 1987; Carballo *et al.*, 1987). One of the criteria considered useful for the differentiation of immature and diagenetically matured dolomites has been their texture (e.g. Sibley & Gregg, 1987; Gregg & Skelton, 1990; Mazullo, 1992). Coarse-crystalline rock textures common in ancient dolostones are often taken as an indication for the maturation of a metastable

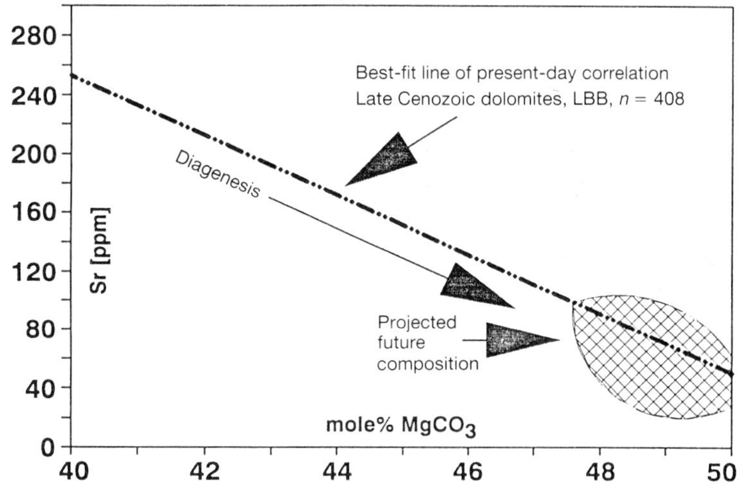

Fig. 6. The Sr composition of LBB dolomites covaries with mole % $MgCO_3$ composition (best-fit line from Vahrenkamp & Swart, 1990). Based on comparison with many ancient dolomites, LBB dolomites are expected to develop into stoichiometric, low-Sr dolomites during future diagenesis. However, in the case of dolomite diagenesis, the marine Sr in those LBB dolomites which are already nearly stoichiometric will probably be preserved, because these dolomites are likely to remain unaltered during dolomite maturation.

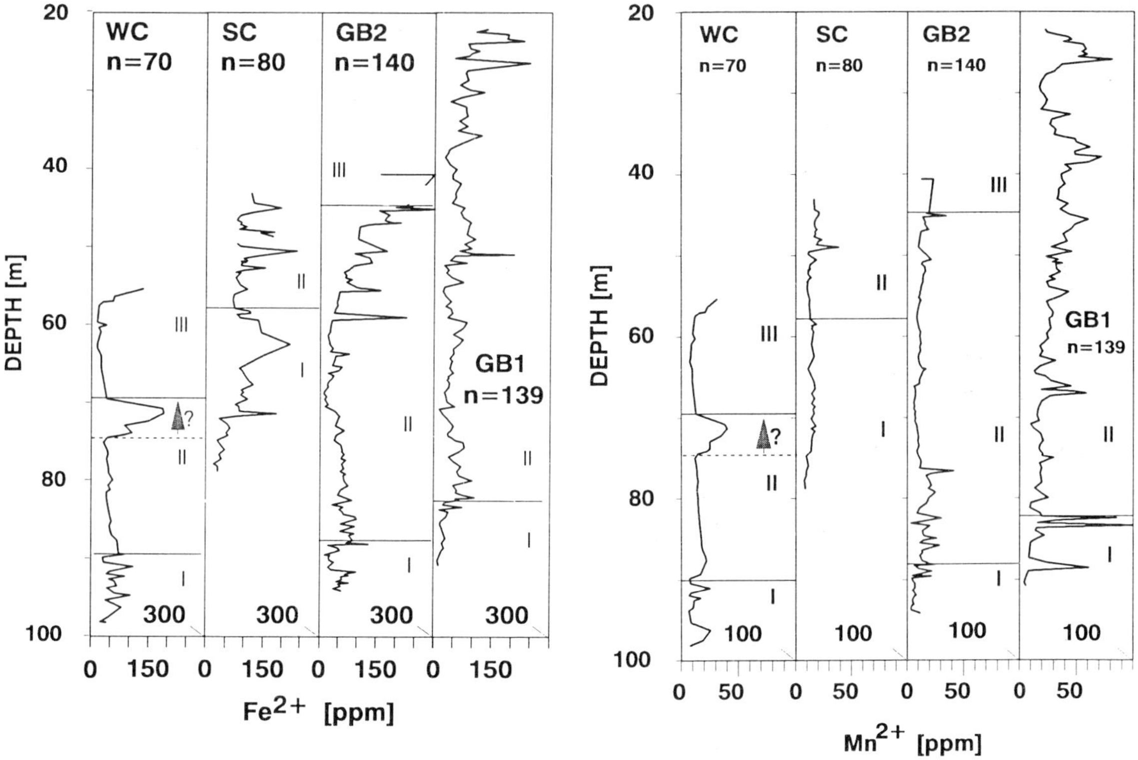

Fig. 7. (a) Iron and (b) manganese compositions of LBB dolomites versus depth. A base composition of 15 ppm for iron and 5 ppm for manganese deviates locally. This is probably related to predolomitization diagenesis in submarine hardgrounds and subaerial exposure horizons. The section has been subdivided into the three dolomite generations differentiated by Sr isotopes (compare with Fig. 2).

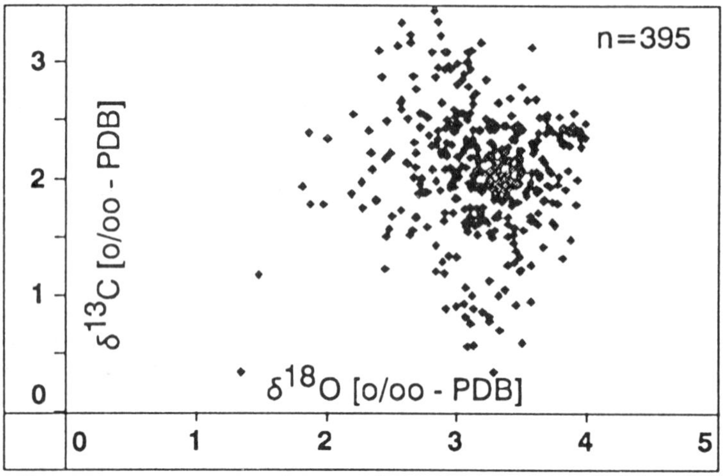

Fig. 8. Cross-plot of oxygen versus carbon isotopic composition of LBB dolomites.

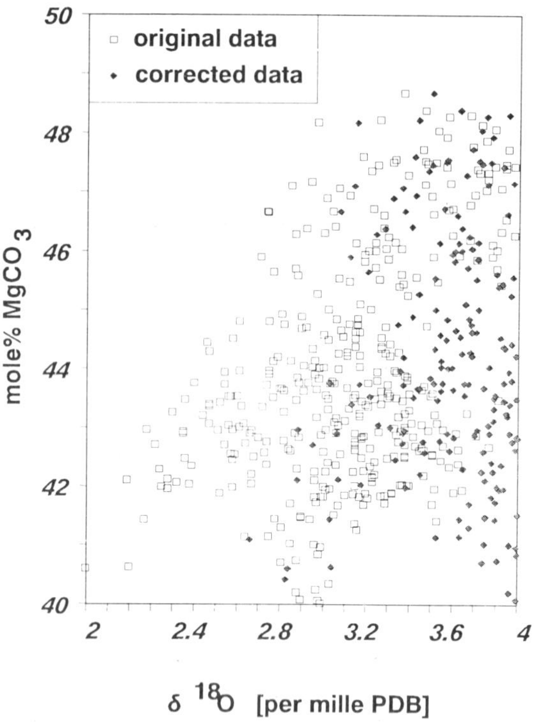

Fig. 9. The relationship between mole % $MgCO_3$ and oxygen isotopic composition for dolomites from LBB. The open symbols show the original data, the closed symbols have been corrected by 0.1‰/mole % $MgCO_3$ deficiency from an ideal dolomite.

dolomite precursor (e.g. Mazullo, 1992). In LBB both fine- and coarse-crystalline rock textures exist (Fig. 4/Plate 1).

The most obvious systematic variation in the petrography of LBB dolomites is related to the boundary between Miocene and Pliocene dolomites (Fig. 2). The former are composed of coarse-crystalline CNM and S types only, whereas the latter are composed exclusively of more fine-crystalline CM, MS and CMS dolomites (Fig. 4/Plate 1). A similar arrangement has been observed in other Bahamian dolomite sections (Dawans & Swart, 1988; Vahrenkamp *et al.*, 1991). An explanation could be that the CM–CMS–MS textures are 'metastable precursor' forms of CNM and S dolomites. Based on strontium isotope signatures and petrographic observations (Vahrenkamp *et al.*, 1991) it is established that the Middle Miocene strata were already dolomitized during dolomitization of the overlying Late Miocene and Pliocene sediments (dolomitization phase I vs. II & III; Fig. 2). Hence, Miocene dolomites were probably exposed to dolomitizing fluids several times during later dolomitization periods. This could have caused dolomite maturation by further textural destruction of CM dolomites and additional growth of MS dolomite crystals to form, respectively, CNM and S textures. However, Miocene dolomites are calcium-rich (average <45 mole % $MgCO_3$) and therefore geochemically as, or even less, stable than the younger dolomite generations (see Fig. 5; Hardie, 1987; see also discussion below). Dolomite maturation implies an increase in stability and organization of the resulting mineral (Land, 1985;

Hardie, 1987). Hence, either CM, CMS and MS precursors of the Miocene CNM and S dolomites were even less stoichiometric than their Pliocene counterparts, or CNM and S dolomites crystallized as their present form and did not experience neomorphism. If this is true, in LBB coarse-crystalline textures are as mature or immature as fine-crystalline textures. In any case, all described dolomite textures (CM, CNM, MS, S, CMS) formed at near-surface conditions under apparently similar chemical conditions (see also discussion below).

A relation appears to exist between texture types and their occurrence within the carbonate platform: the CM variety seems to be more commonly encountered near the bank margins (Fig. 5). This impression could be merely the result of the thicker Pliocene dolomite section near the platform margin (Fig. 2). Alternatively, it may reflect a correlation between dolomite texture and micrite content of the precursor sediment. A gradual change from CM over CMS to MS texture may be related to an increase in the micrite content of the original sediments. CM dolomites are most frequently encountered in grainstone and boundstone deposits with little pore space-occluding micrite, whereas CMS textures are characteristic for packstones and MS types for packstones and wackestones (Fig. 3). Dawans and Swart (1988) propose that the dolomite texture is related to the rate of precipitation as a function of permeability and porosity, with CM dolomite precipitating the fastest and MS dolomite being the product of slow precipitation. Pore space-occluding and permeability-reducing internal micrite may therefore be the governing factor that determines the texture of the dolomites via growth kinetics. Furthermore, a micritic matrix may provide more homogeneously distributed crystallization sites which, given a uniform growth rate, may lead to sucrosic dolomite growth (Sibley & Gregg, 1987; Gregg & Skelton, 1990; Mazullo, 1992). If the dolomite texture is indeed precursor-related, the overall distribution of dolomite textures in a carbonate platform may be predictable, since deposits from the bank interior are commonly more micritic than bank margin sediments (e.g. Wilson, 1975).

Most notable for LBB dolomites is the overall extremely high vuggy and/or intercrystalline porosity and high permeability (up to 43% porosity and 3900 Md permeability; Williams, 1985b; Fig. 4/Plate 1) resulting commonly in a friable to even sandy rock consistency. In some intervals porosity appears to be so high (Fig. 4e–h/Plate 1) that the rock texture may conceivably be not only the result of dolomite precipitation but also of the subsequent dissolution of a (limestone?) matrix. Matrix dissolution could have been caused by aggressive meteoric waters passing through the rock during numerous Pleistocene sea-level low-stands.

In summary, based on high porosities and an often friable consistency, we interprete LBB dolomites to be overall petrographically immature compared to many ancient dolomites. However, petrographic criteria alone do not permit one to discount a maturation step from fine-crystalline CM, CMS and MS textures to the coarse-crystalline CNM and S textures of the early Late Miocene dolomites.

Chemical maturity

Disordered calcian dolomites are as much as 1.5 orders of magnitude more soluble than fully ordered stoichiometric dolomites (Hardie, 1987). Hence, depending on the amount of their excess calcium and degree of disordering, LBB dolomites are to a varying degree more soluble than ordered stoichiometric dolomite. A comparison of stoichiometry histograms (Fig. 5a,c) suggests that LBB dolomites are, on average, also more soluble than most ancient dolomites (Lumsden & Chimahusky, 1980; Sperber *et al.*, 1984). However, it is important to realize that near-stoichiometric and well-ordered dolomites do exist on LBB (e.g. GB1, 56 m; Fig. 5b). These dolomites have a low recrystallization potential and are more likely to be preserved in their present composition than their calcian associates.

In summary, most but not all LBB dolomites are chemically metastable and immature compared to ideal dolomite, as well as most ancient dolomites.

Oxygen isotopes

The variation of approximately 2‰ in the $\delta^{18}O$ data poses the question as to whether such variation reflects changes in the isotopic composition of the water, temperature, or perhaps some other factor. Of particular importance in this study and previous work (Dawans & Swart, 1988) is the positive correlation between dolomite stoichiometry and oxygen isotopic composition (Fig. 9). In this section we will argue that the positive correlation could in fact be an artefact caused by differences in the equilibrium precipitation between oxygen isotopic values of Ca-rich and stoichiometric dolomites; variations in the

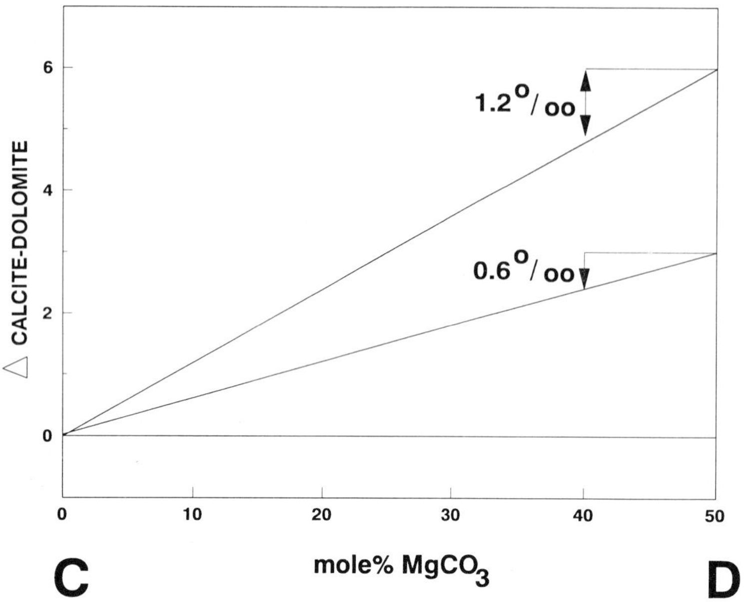

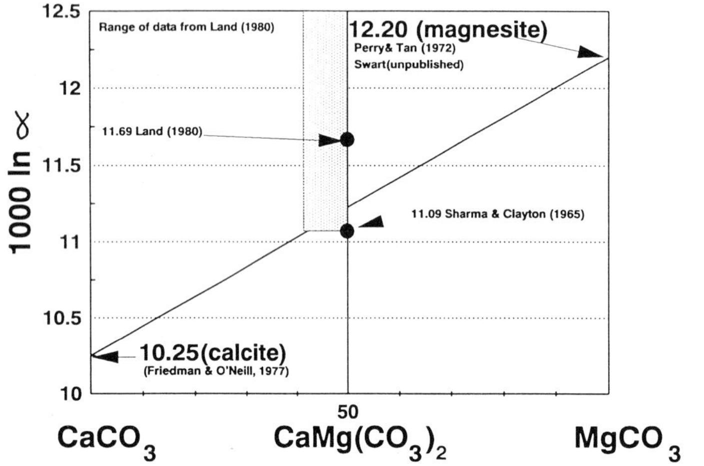

Fig. 10. (a) Theoretical fractionation between calcite and dolomite showing how equilibrium fractionation might vary based on the different estimates of calcite–dolomite. (b) Various fractionation factors for the dissolution of calcite, dolomite and magnesite in phosphoric acid at 25°C. In our studies we have used a value of 1.01109 after Sharma and Clayton (1965).

phosphoric acid fractionation factor with respect to stoichiometry; and/or fractionation related to the rate of dolomitization.

Equilibrium precipitation. Although the magnitude is still uncertain, it is beyond dispute that dolomites are isotopically enriched in $\delta^{18}O$ compared to coprecipitated calcites. The cause of the fractionation is probably the result of the varying strengths of the Mg—O and Ca—O bonds, and for high Mg-calcites, has been measured experimentally to approximate 0.06‰ per mole % change in $MgCO_3$ (Tarutani *et al.*, 1969). Extrapolation of this difference from a low Mg-calcite to a stoichiometric dolomite predicts that coprecipitated dolomites should be 3‰ heavier than calcites, this value being at the lower end of the postulated range of differences between calcite and dolomite (ΔC-D) (Land,

1980). Alternatively, if it is assumed that the ΔC-D is 6‰ (the maximum range of the difference postulated for calcite and dolomite; Land, 1980), then the oxygen isotopic composition will change by 0.12‰ per mole % change in $MgCO_3$. These two extremes are plotted in Figure 10a. Following these arguments, dolomites with more calcian compositions can be expected to have isotopically lighter oxygen isotopic compositions than true stoichiometric dolomites. In the case of Little Bahama Bank, whose dolomites have compositions between $Ca_{.60}Mg_{.40}CO_3$ and $Ca_{.51}Mg_{.49}CO_3$, a discrepancy in the $\delta^{18}O$ of between 1.2 and 0.6‰ can be explained by this mechanism.

Phosphoric acid fractionation. Variation in the measured $\delta^{18}O$ can be induced through changes in the fractionation factor (α) between phosphoric acid and the mineral being dissolved. For calcite, α has been measured as approximately 1.01025 (at 25°C), whereas for dolomite values of between 1.01109 (Sharma & Clayton, 1965; reported in Friedman & O'Neil, 1977) and 1.01169 (Land, 1980) have been reported. In order to obtain a more precise estimate of α, we can add to the database the fractionation factor between magnesite and phosphoric acid as an end-member of a hypothetical solid solution series between calcite and magnesite (Fig. 10b). Extrapolation of the $MgCO_3$ and $CaCO_3$ data to an ideal dolomite composition yields an α value of 1.0111, similar to previously reported experimental data (Sharma & Clayton, 1965; reported in Friedman & O'Neil, 1977; Land, 1980). Based on this extrapolation, dolomites with, for example, a composition of $Ca_{.56}Mg_{.44}CO_3$ will have an α of 1.01095. In other words, if a conventional ∝ correction is applied to these data, dolomites with this composition will be overcorrected with respect to their oxygen isotopic composition by approximately 0.15‰. This difference may, in fact, be an underestimate in so far as experiments by Land (1980) reported a range in α values of 1.2‰. Hence, following our arguments, a variation in $\delta^{18}O$ of up to 0.25‰ may occur for dolomites from LBB with a composition between $Ca_{.60}Mg_{.40}CO_3$ and $Ca_{.51}Mg_{.49}CO_3$ due to differential phosphoric acid fractionation.

Land (1980, p. 107) concluded that, until we can be certain of what fractionation factors to apply, it is better that none are applied at all. Unfortunately, some workers have taken Land's remarks to heart while others have not. Worse, authors frequently fail to mention whether their data have been corrected or not, or what corrections were actually applied. We urge authors to define precisely whether they use fractionation corrections or not.

Kinetic controls. Although there have been no experimental studies to examine the role of kinetics in the control of the fractionation of oxygen isotopes in dolomites, in other mineral systems faster rates of precipitation usually result in lighter isotopic compositions. In the case of the calcian versus the stoichiometric dolomites, this difference in major element chemistry is perhaps the least understood aspect of the geochemistry. One postulated theory is that the calcian dolomites form at faster rates than the more ordered stoichiometric ones. The calcian dolomites might therefore also have lighter isotopic compositions.

The combination of the uncertainties associated with equilibrium precipitation and phosphoric acid fractionation introduces a possible difference of at least between 0.85 and 1.45‰ between dolomites with a composition of $Ca_{.60}Mg_{.40}CO_3$ and $Ca_{.51}Mg_{.49}CO_3$. In addition there may be further non-equilibrium fractionation associated with the rate of dolomite formation, which may make the calcian dolomite isotopically lighter than the stoichiometric variety.

Assuming that the differences between the calcian and stoichiometric dolomites are indeed, at least partially, an artefact, then we can correct for these factors discussed above. In this study we have estimated a correction of approximately 0.1‰ per mole% $MgCO_3$ deficiency. We have applied this correction to the oxygen isotopic data shown in Figure 9. The result shows a narrower range of oxygen isotopic compositions (2.6–4‰) than originally measured, with a smaller standard deviation and a heavier $\delta^{18}O$ (m = +3.8‰). Such a range of values could conceivably be produced from a fluid with a homogeneous isotopic composition considering normal temperature variations of near-surface groundwaters. In the case of the Bahamas the only fluid that satisfies these requirements is pure or slightly modified seawater. Using the approximation of Land (1983) for the equilibrium precipitation of dolomite from a seawater with an isotopic composition of 1‰ SMOW (similar to the $\delta^{18}O$ for modern Gulf Stream water), then the measured oxygen isotopic composition of the dolomite is consistent with precipitation at 20–22°C.

Carbon isotopes

Deposition and dolomitization of the Late Miocene to Pliocene section on LBB were separated by up to several millions of years (McNeill, 1989; Vahrenkamp *et al.*, 1991). Hence, it can be expected that the originally metastable shallow-marine carbonates were at least partially altered diagenetically prior to dolomitization. Negative carbon isotope anomalies at several levels of the stratigraphic section (Fig. 11) are interpreted as the effects of repeated subaerial exposure of the platform. During exposure, freshwater vadose and/or phreatic diagenesis leads to the uptake of light soil carbon and (partial?) stabilization of the metastable marine carbonates to low Mg-calcite prior to dolomitization (Allan & Matthews, 1977). Preservation of negative carbon isotope trends indicates that the carbon system was relatively closed during dolomitization.

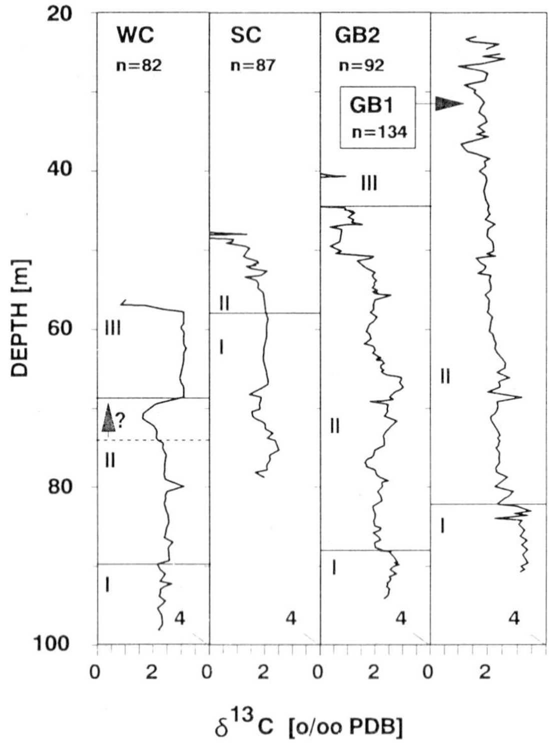

Fig. 11. Carbon isotopes of LBB dolomites versus depth. The section has been subdivided into the three dolomite generations differentiated by Sr isotopes (compare with Fig. 2).

Dolomitizing waters

The isolation from sources of Mg^{2+} other than seawater indicates seawater or a derivative as the diagenetic medium. Oxygen isotope compositions of the dolomite bodies are in equilibrium with ambient seawater, assuming normal temperature and water composition ranges encountered in the Bahamas. However, based on the available data it cannot be entirely discounted that seawater was slightly modified, either by dilution with fresh waters or by mixing with waters of raised salinity.

Manganese and iron

Disregarding local variations, a base composition of about 15 ppm Fe and 5 ppm Mn is typical for LBB dolomites (Fig. 7). Assuming a seawater Fe and Mn content of 2 ppb and 0.3 ppb, respectively (Drever, 1982), hundreds of litres of seawater would be necessary to produce only 1 mole of dolomite with a typical Fe and Mn content, even if all seawater-derived Fe and Mn were to be totally utilized. In view of the vast amounts of dolomite in LBB, clearly other sources for both Fe and Mn must be considered. Pliocene low Mg-calcites from LBB have an average Fe and Mn content of 20–100 ppm and 3 ppm, respectively (Vahrenkamp, 1988). Modern marine carbonate sediments vary widely in their trace element content (Fe = ten to several thousand ppm; Mn = 3–100 ppm; Milliman, 1974). Hence, both the iron and the manganese contents of LBB dolomites are probably derived from precursor sediment. Positive variations with respect to the base composition are probably related to predolomitization processes such as submarine hardground formation and subaerial exposure, and further support relative closed system diagenesis with respect to Fe and Mn during dolomitization.

Environment of dolomite formation

On LBB the delineation of discrete dolomite bodies helps to constrain further possible mechanisms of platform dolomitization and relative sea-level position during dolomitization. Whereas the boundary between early Late Miocene and overlying Pliocene dolomites follows the geometry of depositional units separated by an erosion surface, Pliocene and Pleistocene dolomite bodies cut across time lines and depositional surfaces. For example, during the Late Pliocene, replacement dolomites formed

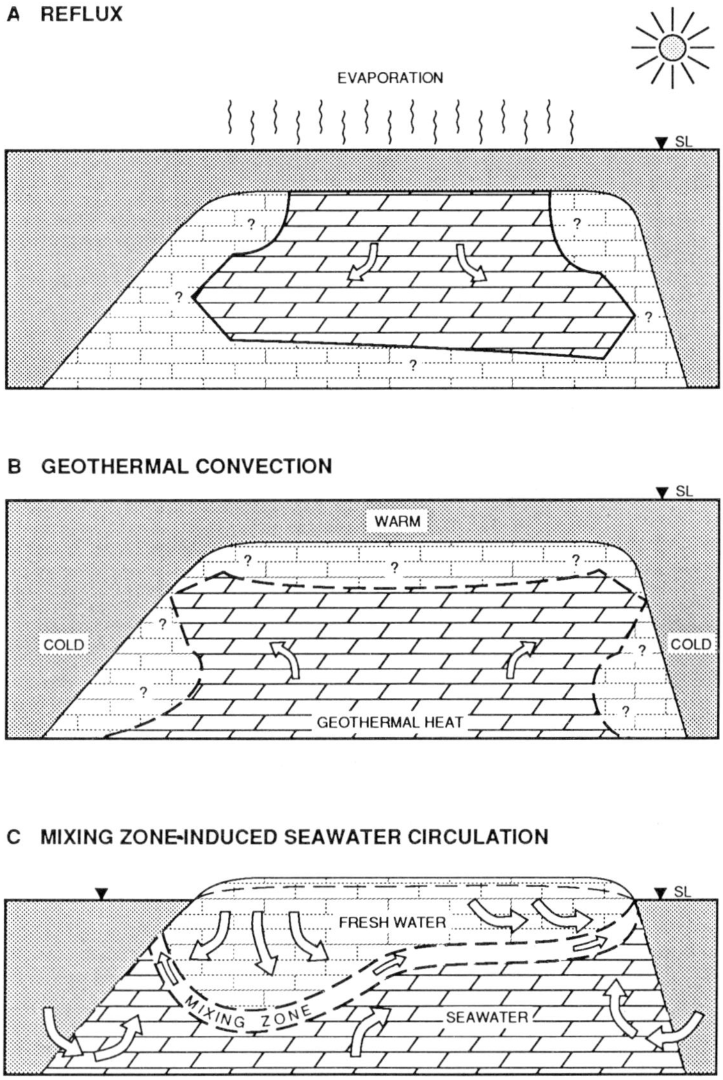

Fig. 12. Three groundwater flow regimes with seawater circulation which have been associated with platform dolomitization (e.g. Simms, 1984; Vahrenkamp, 1988). Seawater refluxing (A) and geothermal convection (B) require (partial) platform submergence for shallow-burial near-surface dolomitization, whereas mixing zone-induced seawater circulation (C) can only develop during (partial) emergence. Asymmetry of flow regimes and hence dolomite bodies may be the result of density gradients, platform shape, topography, permeability differences, etc.

close to the surface of an almost level carbonate platform near the southern edge of the bank (GB1; Fig. 2). To the north (GB2, SC, WC; Fig. 2) dolomites of the same age are found only at depth, leaving an unreplaced limestone cover up to 40 m thick (WC; Fig. 2).

During dolomitization rocks must be near or below sea level in order to physically ensure that seawater can provide the necesary magnesium ions. In addition, any model of dolomitization considered for the Late Pliocene must feature a hydrological system responsible for reactant transport able to create the outlined geometry. The reflux of seawater (Fig. 12), as suggested by Simms (1984), would require water flow through up to 40 m of limestone before the onset of dolomitization. Geothermal convection (or Kohout convection; Simms, 1984; Fig. 12) is unlikely to induce near-surface water circulation as would be required for dolomitization of the rocks recovered from GB1 (Fig. 2).

However, seawater circulation induced by an overlying freshwater lens (Fig. 12) could create a dolomite body of the indicated geometry (compare Fig. 2 and Fig. 12). Exposure of a carbonate platform and rain will cause the creation of a freshwater lens. At the interface between fresh water and seawater, diffusion and physical mixing as a result of energy-level variations (i.e. tides, wet/dry seasons, etc.) will create a zone of mixed-salinity waters. However, mixed-salinity waters are hydrodynamically unstable and will flow towards a lower hydrostatic level (Bear & Todd, 1960). In island systems this level is near the edge of the island, where the outmost extent of the freshwater lens meets seawater. Hence flow in the mixing zone is from the deepest part of the freshwater lens towards its outer limits parallel to the seawater–freshwater interface (Fig. 12). Discharge of water from a mixing zone with a more or less constant volume into the ocean requires recharge from the overlying freshwater lens and in equal parts from the underlying seawater phreatic system. As a result, seawater has to enter the side of carbonate platforms and circulate in the subsurface with a volume equal to half of the mixing-zone discharge into the ocean (Fig. 12). On islands with a relatively constant freshwater aquifer size, freshwater as well as seawater circulation approximates net rain recharge (total rain minus evaporation) and can therefore be substantial. Seawater circulation at depth is an often overlooked aspect of a freshwater/mixed-water lens system. This physical pump system must not be confused with the mixing-zone dolomitization model proposed by Hanshaw *et al.* (1969), Badiozamani (1973) and others, which was based mainly on chemical principles. The type of seawater circulation proposed here requires (partial) exposure of the platform. A dolomite body related to such a flow system should reflect as its upper boundary the outline of the overlying mixing zone and freshwater aquifer (compare Figs 2 and 12). Evidence of prolonged subaerial exposure of LBB during the Late Pliocene/early Pleistocene are numerous exposure horizons (Williams, 1985a; Vahrenkamp, 1988) and a condensed stratigraphic section (McNeill, 1989). Hence, the shape of the Late Pliocene dolomite body in LBB (Fig. 2), in combination with the age of dolomitization, provides additional support for the generation of dolomites from seawater. Seawater circulation through the platform is proposed to have been driven by an overlying freshwater lens and its associated mixing zone during platform exposure. However, the need for seawater-derived Mg for dolomite formation implies that diagenesis occurred below sea level, indicating a minor regression only or dolomitization during the early stages of sea-level fall or shortly before renewed flooding of the platform. Similar flow systems, which are probably related to dolomitization, have been reported from the subsurface of the present-day Bahamas (Whitaker *et al.*, this volume).

Possible future diagenesis

In many respects, dolomites from the Bahamas are already typical platform dolomites. For example, dolomitized limestone sections are tens of metres thick, and replacement has been nearly complete. However, dolomite fabrics and geochemistry are, at least in part, immature compared to many ancient platform dolomites. Differences concern the overall extremely high porosities, the apparent predominance of a single dolomitization phase for a given sequence (Vahrenkamp *et al.*, 1988, 1991) and the significant geochemical metastability (90% of dolomites <47 mole % $MgCO_3$).

The stabilization of marine metastable aragonite and high Mg-calcite to low Mg-calcite is an accepted axiom of calcite diagenesis (e.g. Land, 1967). In agreement with other workers (e.g. Land, 1985; Hardie, 1987; Banner *et al.*, 1988; Gao & Land, 1991; Gao *et al.*, 1992) we concur that a similar concept is applicable for dolomites: given appropriate geochemical conditions, metastable Ca-rich dolomites will stabilize to stoichiometric forms. However, due to the primary range in stoichiometry, mixing will occur between a preserved marine composition of originally (near) stoichiometric dolomites and the signatures of stabilized dolomites that reflect conditions of secondary diagenetic environments (Figs 7 and 13).

The occurrence of hollow dolomite rhombs (Fig. 14), especially in near-platform-margin cores, suggests that the stabilization process has already started. Hollow dolomite rhombs are nearly stoichiometric and exhibit relatively heavy oxygen isotope signatures. Rhomb centres were apparently more soluble, and hence probably less stoichiometric. Dissolution may have occurred in undersaturated waters of freshwater phreatic or mixing-zone environments which passed through the rock column several times during periods of Pleistocene sea-level fluctuation.

Ancient dolomites: a view into the future of LBB dolomites

Oxygen isotope trends, stoichiometry and low Sr concentrations (<100 ppm) are commonly observed in ancient dolomites (e.g. Weber, 1964; Mattes & Mountjoy, 1980; Banner *et al.*, 1988; Wallace, 1990;

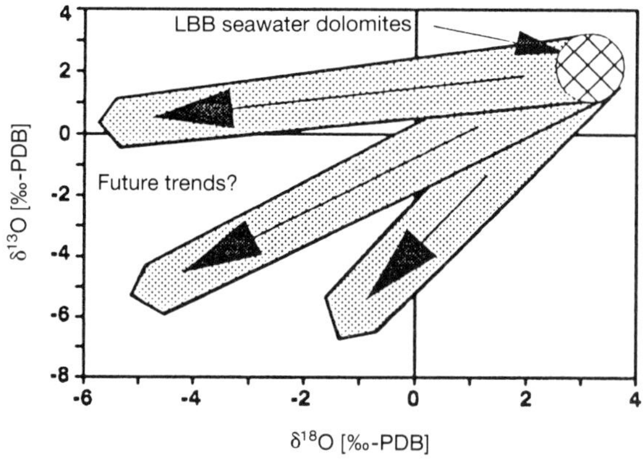

Fig. 13. The development of mixing trends in the bulk rock carbon and oxygen isotopic composition of LBB dolomites during future diagenesis. Progressive recrystallization of metastable dolomites in hot and/or meteoric waters will introduce a significant modification of the oxygen isotopic composition, with a probable negative trend. However, due to the presence of some original stoichiometric dolomites an original marine signature component may be preserved. The modification of the carbon isotopic signature is likely to be less pronounced due to the relative abundance of carbon from the rock system.

Migaszewski, 1991). Assuming a genetic origin of these ancient dolomites similar to that of LBB dolomites, the oxygen isotope trends should contain some isotope values that reflect the preserved isotopic signatures of originally stoichiometric dolomites. Oxygen isotope trends of Devonian dolomites from three different continents (Mattes & Mountjoy, 1980; Wallace, 1990; Migaszewski, 1991) all approach a common origin at about −1 to −2‰ PDB (Fig. 15), and therefore may lend credence to the notion of a depleted oxygen isotopic signature for Devonian seawater. Using a ΔC-D of 3‰, this value is in equilibrium with calcite cements presumably carrying a Devonian seawater signature (Hurley, 1986; Lohmann, 1988). The oxygen isotope signature of Devonian marine calcite was derived by using isotope trends of originally marine calcite cements (e.g. Given & Lohmann, 1986; Lohmann, 1988). This technique implies that marine isotope signatures are partly preserved during diagenetic stabilization (dissolution/reprecipitation) of aragonite and high Mg-calcite to low Mg-calcite (Given & Lohmann, 1986). However, applying the principles of this technique to dolomites should lead to equally good, or even more precise, results, because a preserved direct precipitate from seawater is used instead of a (closed system?) dissolution/reprecipitation product.

Both former marine calcites and dolomites often exhibit negative oxygen isotopes trends away from the common point of original marine signatures (e.g. Mattes & Mountjoy, 1980; Banner *et al.*, 1988; Wallace, 1990; Migaszewski, 1991). For calcites these trends are usually interpreted to be the result of diagenesis in meteoric waters (e.g. Allan & Matthews, 1977; Given & Lohmann, 1986). Diagenesis of dolomites, however, is less well understood and additional data need to be considered.

Fig. 14. Scanning electron photomicrograph of hollow dolomite rhombs (GB2, 72 m). One bar = 3 μm. Corrosion of rhomb centres may indicate the onset of diagenetic 'purification' of the dolomite section. More stable stoichiometric rims of dolomite rhombs and their original marine geochemical signatures are preserved, whereas metastable centres of dolomite rhombs are dissolved. Hollow rhomb centres may be filled again by later dolomite generations. Note, the age of dolomite generations in ancient rhombs with irregular internal generation boundaries is not necessarily from oldest in the centres to youngest at the peripheries.

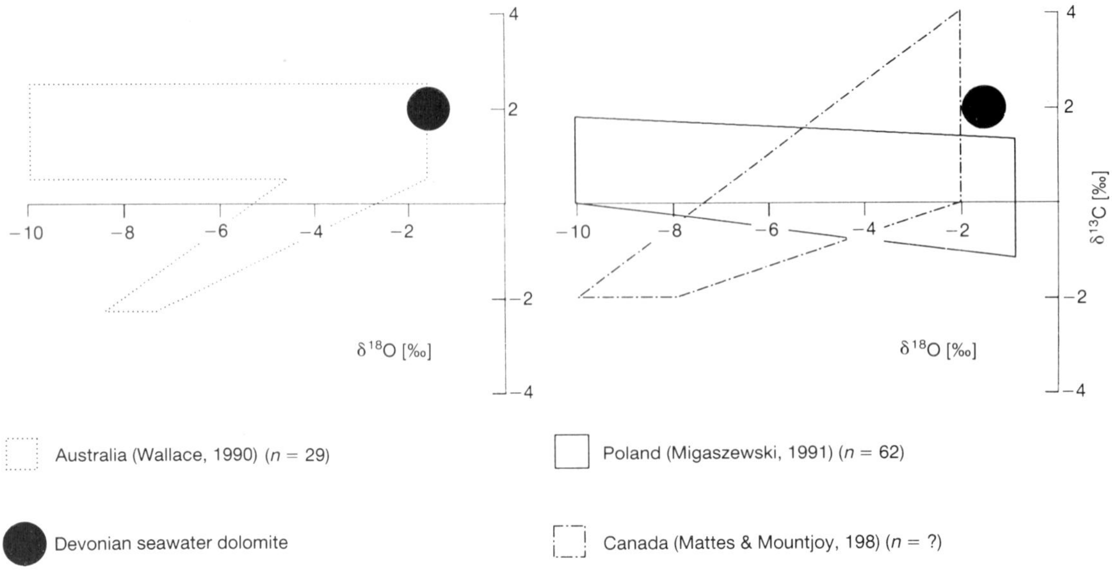

Fig. 15. Oxygen versus carbon isotope composition of three Devonian dolomites compiled from literature. In each case the range of carbon isotopes does not exceed 4‰, whereas oxygen isotopes exhibit a range of 8‰ or more. The heaviest oxygen isotopes in each example approach the hypothetical composition of dolomite in equilibrium with Devonian seawater. This may indicate preservation of original marine dolomite of Devonian age. The Devonian seawater signature is derived from applying a $\Delta_{calcite-dolomite}$ of 3‰ to the composition of Devonian marine $CaCO_3$ cement proposed by Hurley (1986) and Lohmann (1988).

The Sr concentration of ancient stoichiometric dolomites is usually less than 100 ppm, similar to that of stoichiometric dolomites from LBB (e.g. Weber, 1964; Mattes & Mountjoy, 1980; Banner *et al.*, 1988; Wallace, 1990; Migaszewski, 1991). However, the possibility that all dolomites precipitated directly from seawater can be rejected, based on the deviation of the observed oxygen isotopes from marine values (e.g. Fig. 15). Hence, dolomites with non-marine geochemistry are the result of either diagenesis of metastable precursors or of a later genesis of additional dolomite. In either case, waters must have been depleted in $\delta^{18}O$ and/or hot in order to cause negative isotope trends. Furthermore, waters must have had a Sr/Ca ratio as low as or even lower than that of seawater in order to cause Sr concentrations of significantly less than 100 ppm. A compilation of Sr/Ca ratios of natural waters (Vahrenkamp & Keuning, 1990) suggests that burial waters, with the exception of meteoric and mixed marine/meteoric waters, have Sr/Ca ratios commonly far in excess of that of marine waters. Hence, it can be deduced that dolomites with negative oxygen isotope trends and low Sr concentrations were probably formed in low-salinity meteoric waters. Furthermore, precursors for these dolomites were probably metastable dolomites and not limestones, since meteoric waters are unable to provide sufficient Mg for massive replacement dolomitization of limestones (e.g. Given & Wilkinson, 1987). Hence, we propose that many ancient platform dolomites had a seawater origin similar to LBB dolomites. Even though later diagenesis and recrystallization usually overprinted original dolomite composition, careful analyses can reveal primary geochemical signatures of preserved original marine dolomite phases (e.g. Gao & Land, 1991; Gao *et al.*, 1992).

We conclude that diagenesis in meteoric waters is a probable fate for LBB dolomites. Such diagenesis will likely cause an overall decrease in porosity, permeability and Sr content. Dolomites will become more stoichiometric and negative oxygen isotope trends will develop. Preservation of the original composition may occur as a result of primary stoichiometry or in intervals of low porosity and permeability which are essentially isolated from the circulation of diagenetic waters.

CONCLUSIONS

1 Repeated episodes of massive replacement dolomitization in the shallow subsurface of Little Bahama Bank created platform dolomites similar in extent and thickness to many ancient dolomite sequences.

2 Petrographically, LBB dolomites are interpreted to be immature, despite some indications for a relative increase in textural maturity of early Late Miocene dolomites. A correlation seems to exist between precursor micrite content and sucrosic dolomite texture.

3 Geochemically both mature and immature dolomites exist, as indicated by the coexistence of stoichiometric and highly calcian dolomites. A correlation between textural and geochemical maturity could not be established.

4 Carbon isotope signatures suggest that metastable marine precursor carbonates were at least partially stabilized to low Mg-calcite prior to dolomitization.

5 In the isolated setting of the Bahamas, seawater is the only Mg source available for massive replacement dolomitization. This, in combination with an oxygen isotopic composition in equilibrium with seawater, suggests that dolomites precipitated essentially from seawater. However, neither slight dilution of seawater with fresh water nor slight concentration due to evaporation can be ruled out.

6 Based on the geometry of one of the major dolomite bodies, dolomitization occurred in the seawater phreatic zone, with seawater circulation probably driven by an overlying freshwater/mixing-zone system during (partial?) platform exposure.

7 A comparison between the chemical composition of ancient dolomites and the geochemical potential of LBB dolomites suggests that:

(a) many ancient dolomite sequences could have had a seawater origin similar to that of the LBB dolomites;

(b) LBB dolomites have the potential to stabilize into 'typical' ancient dolomites during further diagenesis;

(c) in spite of possible future diagenesis, partial preservation of an early genetic signature is probable since existing stoichiometric dolomites are unlikely to be affected during dolomite maturation;

(d) diagenesis of metastable dolomites into typical ancient dolomites is likely to occur in shallow to deep-burial meteoric-water environments.

ACKNOWLEDGEMENTS

The authors would like to thank Drs R.N. Ginsburg and S.C. Williams for creating the base for this study, and for many stimulating thoughts. This research was funded by NSF grant EAR-86-07688 to P.K. Swart and the Industrial Sponsors of Fisher Island. Support by Shell Research and permission to publish are gratefully acknowledged. Dr Burt Fisher and the AMOCO corporation are thanked for ICP analyses. Critical reviews by S.J. Mazullo, Bruce Purser, Maurice Tucker and Don Zenger significantly improved the manuscript and are gratefully acknowledged.

REFERENCES

Allan, J.R. & Matthews, R.K. (1977) Carbon and oxygen isotopes as diagenetic and stratigraphic tools: surface and subsurface data, Barbados, West Indies. *Geology* **5**, 16–20.

Badiozamani, K. (1973) The Dorag dolomitization model – application to the Middle Ordovician of Wisconsin. *J. Sedim. Petrol.* **43**, 965–984.

Banner, J.L., Hanson, G.N. & Meyers, W.J. (1988) Water–rock interaction history of regionally extensive dolomites of the Burlington-Keokuk Formation (Mississippian): isotope evidence. In: *Sedimentology and Geochemistry of Dolostones* (Ed. Shukla, V. & Baker, P.A.) Spec. Publs Soc. Econ. Paleont. Mineral., Tulsa **43**, 97–115.

Beach, D.K. (1982) *Depositional and Diagenetic History of Pliocene-Pleistocene Carbonates of Northwestern Great Bahama Bank: Evolution of a Carbonate Platform*. PhD Dissertation, University of Miami, Coral Gables, 452 pp.

Bear, J. & Todd, D.K. (1960) The transition zone between fresh and salt waters in coastal aquifers. *Calif. Univ. Water Resources Cent. Contribution* **29**, 156 pp.

Behrens, E.W. & Land, L.S. (1972) Subtidal Holocene dolomite, Baffin Bay, Texas. *J. Sedim. Petrol.* **42**, 155–161

Carballo, J.D., Land, L.S. & Miser, D.E. (1987) Holocene dolomitization of supratidal sediments by active tidal pumping, Sugarloaf Key, Florida. *J. Sedim. Petrol.* **57**, 153–165.

Craig, H. (1957) Isotopic standards for carbon and oxygen for mass-spectrometric analysis of carbon dioxide. *Geochim. Cosmochim. Acta* **12**, 133–149.

Dawans, J.M. & Swart, P.K. (1988) Textural and geochemical alternations in late Cenozoic Bahamian dolomites. *Sedimentology* **35**, 385–403.

Drever, J.I. (1892) *The Geochemistry of Natural Waters*. Prentice-Hall, Englewood Cliffs, 388 pp.

Field, R.M. & Hess, H.H. (1933) A borehole in the Bahamas. *Am. Geophys. Un., Trans. Ann. Mtg.*, 234–235.

Friedman, I. & O'Neill, J.R. (1977) Compilation of stable isotope fractionation factors of geochemical interest. *US Geol. Survey Professional Paper* 440-KK.

Gao, G. & Land, L.S. (1991) Early Ordovician Cool Creek dolomite, Arbuckle group, Slickhills, SW Oklahoma, USA: origin and modification. *J. Sedim. Petrol.* **61**, 161–173.

Gao, G., Land, L.S. & Folk, R.L. (1992) Meteoric modification of early dolomite and late dolomitization by basinal fluids, Upper Arbuckle Group, Slick Hills, Southwestern Oklahoma. *Bull. Am. Ass. Petrol. Geol.* **76**, 1649–1664.

Gidman, J. (1978) *Diagenesis of Cored Pleistocene Carbonates, Great Abaco Island, Little Bahama Bank.* PhD Dissertation, University of Liverpool, 359 pp.

Given, R.K. & Lohmann, K.C. (1986) Isotopic evidence for the early meteoric diagenesis of the reef facies, Permian reef complex of west Texas and New Mexico. *J. Sedim. Petrol.* **56**, 183–193.

Given, R.K. & Wilkinson, B.H. (1987) Dolomite abundance and stratigraphic age: constraints on rates and mechanisms of Phanerozoic dolostone formation. *J. Sedim. Petrol.* **57**, 1068–1078.

Goodell, H.G. & Garman, R.K. (1969) Carbonate geochemistry of Superior deep test well, Andros Island, Bahamas. *Bull. Am. Ass. Petrol. Geol.* **53**, 513–536.

Gregg, J.M. & Skelton, K.L. (1990) Dolomitization and dolomite neomorphism in the back reef facies of the Boneterre and Davis formations (Cambrian), southeastern Missouri. *J. Sedim. Petrol.* **60**, 549–562.

Hanshaw, B.B., Back, W. & Deicke, R.G. (1969) A geochemical hypothesis for dolomitization by ground water. *Econ. Geol.* **64**, 710–724

Hardie, L.A. (1987) Dolomitization: a critical view of the current views. *J. Sedim. Petrol.* **57**, 166–183.

Hurley, N.F. (1986) Diagenesis of Devonian reefs in the Oscar Range, Canning Basin, Western Australia. *Abstracts, 12th International Sedimentological Congress, Canberra, Australia*, 148.

Kaldi, J. & Gidman, J. (1982) Early diagenetic dolomite cements: Examples from the Permian Lower Magnesian Limestone of England and the Pleistone carbonates of the Bahamas. *J. Sedim. Petrol.* **52**, 1073–1085.

Land, L.S. (1976) Diagenesis of skeletal carbonates. *J. Sedim. Petrol.* **37**, 914–930.

Land, L.S. (1980) The isotopic and trace element geochemistry of dolomite: the state of the art. In: *Concepts and Models of Dolomitization* (Ed. Zenger, D.H., Dunham, J.B. & Ethington, R.L.) Spec. Publs Soc. Econ. Paleont. Mineral., Tulsa **28**, 87–110.

Land, L.S. (1983) The application of stable isotopes to studies of the origin of dolomite and to problems of diagenesis of clastic sediments. In: *Stable Isotopes in Sedimentary Geology* (Ed. Arthur, M.A., Anderson, T.F., Kaplan, I.R., Veizer, J. & Land, L.S.) Soc. Econ. Paleont. Mineral. Short Course 10, 4.1–4.22.

Land, L.S. (1985) The origin of massive dolomite. *J. Geol. Educ.* **33**, 112–125.

Lohmann, K.C. (1988) Geochemical patterns of meteoric diagenetic systems and their application to studies of paleokarst. In: *Paleokarst* (Ed. James, N.P. & Choquette, P.W.) pp. 58–80. Springer-Verlag, Berlin.

Lumsden, D.N. & Chimahusky, J.S. (1980) Relationship between dolomite nonstoichiometry and carbonate facies parameters. In: *Concepts and Models of Dolomitization* (Ed. Zenger, D.H., Dunham, J.B. & Ethington, R.L.) Spec. Publs Soc. Econ. Paleont. Mineral., Tulsa **28**, 111–121.

McKenzie, J.A. (1981) Holocene dolomitization of calcium carbonate sediments from the coastal sabkhas of Abu Dhabi, UAE: a stable isotope study. *J. Geol.* **89**, 185–198.

McNeill, D.F. (1989) *Magnetostratigraphic Dating and Magnetization of Cenozoic Platform Carbonates from the Bahamas.* PhD Dissertation, University of Miami, Coral Gables, 210 pp.

Mattes, B.W. & Mountjoy, E.W. (1980) Burial dolomitization of the Upper Devonian Miette buildup, Jasper National Park, Alberta. In: *Concepts and Models of Dolomitization* (Ed. Zenger, D.H., Dunham, J.B. & Ethington, R.L.) Spec. Publs Soc. Econ. Paleont. Mineral., Tulsa **28**, 259–297.

Mazullo, S.J. (1992) Geochemical and neomorphic alteration of dolomite: a review. *Carbonates and Evaporites* **7**, 21–37.

Migaszewski, Z.M. (1991) Devonian dolomites from the Holy Cross MTS, Poland: a new concept of the origin of massive dolomites based on petrographic and isotopic evidence. *J. Geol.* **99**, 171–187.

Milliman, J.D. (1974) *Marine Carbonates.* Springer-Verlag, New York, 375 pp.

Mitchell, J.T., Land, L.S. & Miser, D.E. (1987) Modern marine dolomite cement in a north Jamaican fringing reef. *Geology* **15**, 557–560.

Newell, N.D. & Rigby, J.K. (1957) Geological studies on the Great Bahama Bank. In: *Regional Aspects of Carbonate Deposition* (Ed. Leblanc, R.J. & Breeding, J.G.) Spec. Publs Soc. Econ. Paleont. Mineral., Tulsa **5**, 15–72.

Pierson, P.J. (1982) *Cyclic Sedimentation, Limestone Diagenesis and Dolomitization in Upper Cenozoic Carbonates of the Southeastern Bahamas.* PhD Dissertation, University of Miami, Coral Gables, 312 pp.

Sharma, T. & Clayton, R.N. (1956) Measurement of $^{18}O/^{16}O$ ratios of total oxygen of carbonates. *Geochim. Cosmochim. Acta* **29**, 1347–1353.

Shinn, E.A., Ginsburg, R.N. & Lloyd, R.M. (1965) *Recent Supratidal Dolomite from Andros Island, Bahamas.* In: *Dolomitization and Limestone Diagenesis, a Symposium* (Ed. Pray, L.C. & Murray, R.C.) Spec. Publs Soc. Econ. Paleont. Mineral., Tulsa **13**, 112–123.

Sibley, D.F. & Gregg, J.M. (1987) Classification of dolomite rock textures. *J. Sedim. Petrol.* **57**, 967–975.

Simms, M. (1984) Dolomitization by groundwater flow systems in carbonate platforms. *Trans. Gulf Coast Ass. Geol. Soc.* **34**, 411–420.

Sperber, C.M., Wilkinson, B.H. & Peacor, D.R. (1984) Rock composition, dolomite stoichiometry, and rock/water reactions in dolomitic carbonate rocks. *J. Geol.* **92**, 609–622.

Supko, P.R. (1977) Subsurface dolomites, San Salvador, Bahamas. *J. Sedim. Petrol.* **47**, 1063–1077.

Swart, P.K., Ruiz, J. & Holmes, C. (1987) The use of strontium isotopes to constrain the timing and mode of dolomitization of Upper Cenozoic sediments in a core from San Salvador, Bahamas. *Geology* **15**, 262–265

Tarutani, T., Clayton, R.C. & Mayeda, T.K. (1969) The effect of polymorphism and magnesium substitution on oxygen isotope fractionation between calcium carbonate and water. *Geochim. Cosmochim. Acta* **33**, 987–996.

Vahrenkamp, V.C. (1988) *Constraints on the Formation of Platform Dolomites: a Geochemical Study of Late Tertiary Dolomite from Little Bahama Bank, Bahamas.* PhD Dissertation, University of Miami, Coral Gables, 434 pp.

Vahrenkamp, V.C. & Keuning, F. (1990) Sr/Ca ratios of natural waters: implication for dolomite genesis. In: *Abstracts, 13th International Sedimentological Congress, Nottingham, England*, 565.

Vahrenkamp, V.C. & Swart, P.K. (1990) New distribution coefficient for the incorporation of strontium into dolomite and its implication for the formation of ancient dolomites. *Geology* **18**, 387–391.

Vahrenkamp, V.C., Swart, P.K. & Ruiz, J. (1988) Constraints and interpretation of $^{87}Sr/^{86}Sr$ ratios in Cenozoic dolomites. *Geophys. Res. Lett.* **15,** 385–388.

Vahrenkamp, V.C., Swart, P.K. & Ruiz, J. (1991) Episodic dolomitization of Late Cenozoic carbonates in the Bahamas: evidence from strontium isotopes. *J. Sedim. Petrol.* **61**, 1002–1014.

Wallace, M.W. (1990) Origin of dolomitization on the Barbwire Terrace, Canning Basin, Western Australia. *Sedimentology* **37**, 105–123.

Weber, J.N. (1964) Trace element composition of dolostones and dolomites and its bearing on the dolomite problem. *Geochim. Cosmochim. Acta* **28**, 1817–1868.

Williams, S.C. (1985a) *Stratigraphy, Facies Evolution and Diagenesis of Late Cenozoic Limestones and Dolomites, Little Bahama Bank, Bahamas.* PhD Dissertation, University of Miami, Coral Gables, 462 pp.

Williams, S.C. (1985b) Late Cenozoic subsurface dolomites of Little Bahama Bank. *Soc. Econ. Paleont. Mineral. Annual Midyear Meeting* **2,** 97.

Wilson, J.L. (1975) *Carbonate Facies in Geologic History.* Springer-Verlag, New York, 471 pp.

Spec. Publs Int. Ass. Sediment. (1994) **21**, 155–166

Dolomitization caused by water circulation near the mixing zone: an example from the Lower Viséan of the Campine Basin (northern Belgium)

P. MUCHEZ *and* W. VIAENE

Fysico-chemische Geologie, K.U. Leuven, Celestijnenlaan 200C, B-3001 Heverlee, Belgium

ABSTRACT

Massive Lower Viséan dolomites are present in the subsurface of the Campine Basin (northern Belgium) and border the palaeocontinental margin of the Brabant–Wales Massif. Their occurrence is restricted to sediments below the top of the regressive trend that has been recognized within the Lower Viséan. Evaporites or their relics are absent from the Lower Viséan strata of the Campine Basin. The paragenetic sequence in the Lower Viséan is very characteristic: the formation of uniform orange-luminescent dolomites is followed by a period of calcite dissolution and the precipitation of zoned dolomite cement. After this cementation, dolomite crystals were partly dissolved and a calcite cement filled primary and secondary pores.

The palaeogeography, sedimentological setting and paragenetic sequence are interpreted to be indicative of a dolomitization related to fluid circulation near the coastal mixing zone. The uniform orange-luminescent dolomites were formed by the migration of seawater near this mixing zone. Subsequent calcite dissolution and dolomite cementation occurred in the mixing zone, which was formed in response to the lowering of relative sea level. Dolomite dissolution and calcite cementation took place in the meteoric phreatic environment in the final stage of this lowering.

Stable oxygen isotope measurements of the uniform luminescent dolomites (−6‰ to −12‰ PDB) do not support a shallow-marine origin, but represent re-equilibration of the dolomite during further burial. The calculated $\delta^{18}O$ composition of the zoned dolomite cement (+0.8‰ PDB) is comparable with that of Viséan dolomites with a seawater origin (*ca.* +1‰ PDB). This strongly suggests that the mixing zone in which dolomite cementation took place was dominated by a marine fluid.

INTRODUCTION

During the last three decades several dolomitization models have been proposed. One of the most extensively applied to ancient rocks is the mixed-water model (e.g. Land, 1973; Badiozamani, 1973; Choquette & Steinen, 1980; Dunham & Olson, 1980; Churnet *et al.*, 1982). However, this model has several serious weaknesses (Machel & Mountjoy, 1986; Hardie, 1987). Limestone replacement by dolomite has rarely been observed in modern coastal mixing zones. In addition, in a number of dolomites interpreted to be of mixing-zone origin, dolomite has precipitated without dissolution of the calcite substrate, evidence that negates the fundamentals of the mixing-water model (Hardie, 1987). It is therefore unlikely that widespread massive dolomitization occurred in the mixing zone of sea- and fresh water.

However, a characteristic feature of the mixing zone which has received little attention, is that high fresh- and seawater fluxes occur near the mixing zone (Simms, 1984). Coastal groundwater circulation (Bear, 1979) consists of flow in the freshwater lens, flow in a zone of dispersion between the fresh water and the underlying and adjacent seawater, and circulation in the underlying seawater (see Fig. 5). The fresh- and seawater mix is due to hydrodynamic dispersion. This causes a lowering of density compared with seawater and the mixed water moves upward. Seawater flowing from the shelf downwards into the subsurface replaces that lost by mixing. Flow patterns near the mixing zone can be calculated by the model of Lebbe (1983), which is based on the mathematical model of solute transport and dispersion of Konikow and Bredehoeft (1978) and which incorporates the effect of fluid density on the velocity distribution. The seawater circulation, induced by the flow in the mixing zone, can provide the magnesium required for dolomitization (Simms, 1984).

In this paper we will show that the formation of massive dolomites may be related to fluid flow occurring near the mixing zone. The interpretation is based on the palaeogeographical and sedimentological setting, and on the typical paragenetic sequence present in the Lower Viséan carbonates in the deeper subsurface of northern Belgium. Geochemistry supports the proposed model, although stable isotope data must be interpreted carefully. The criteria discussed in this paper can be used to identify dolomitization related to fluid flow near the mixing zone in other areas.

METHODS

The carbonates were examinated by conventional and cold cathodoluminescence (CL) petrography carried out with a Technosyn Cold Cathodo Luminescence Model 8200 MkII (16–20 kV, 420 μA gun current, 0.05 mmHg vacuum, 5 mm beam width).

The dolomite crystals were analysed for manganese and iron on an Applied Research Laboratories model 'AMX' electron microprobe. The detection limits for $FeCO_3$ and $MnCO_3$ are 0.11 mole % and 0.09 mole %, respectively, using siderite and rhodochrosite as standards.

A dental drill was used to obtain carbonate samples for oxygen and carbon isotope analyses. Before drilling, photographs of the selected area were taken under cathodoluminescence for point-counting. The carbon and oxygen isotopic compositions of dolomite and calcite crystals were measured on a Finnigan-Mat Delta-E stable isotope ratio mass spectrometer at the Free University of Brussels, Belgium. All samples were powdered and 1 mg aliquots were reacted with 100% orthophosphoric acid under vacuum. Reaction time for all the calcite samples was about 12 h at 25°C. The dolomite reaction time was about 84 h at 25°C. The isotopic compositions are expressed as δ values in per mil (‰) difference from the PDB standard. Corrections for mass spectrometric analyses are based on the procedures of Craig (1957). Reproducibilities are better than 0.1‰ for oxygen and carbon at the 2σ level based on repeated analyses.

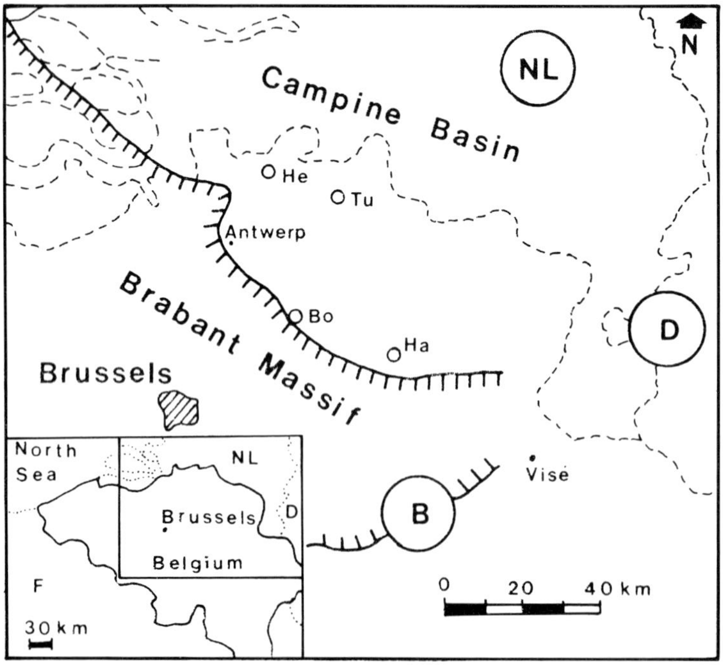

Fig. 1. Location of the wells penetrating Lower Viséan strata in northern Belgium. Dinantian strata are present to the north and south of the Brabant Massif. The area outside the hatched line represents Viséan strata at outcrop and in the deeper subsurface. Bo, Booischot; Ha, Halen; He, Heibaart; Tu, Turnhout; NL, Netherlands; D, Germany; B, Belgium.

Dolomite $\delta^{18}O$ values have not been further corrected for a dolomite phosphoric acid fractionation factor, according to the recommendation of Land (1980).

SEDIMENTOLOGICAL AND PALAEOGEOGRAPHICAL SETTING

Description

The dolomites investigated are found to the north of the Brabant–Wales Massif in northern Belgium (Fig. 1) and typically border the continental margin of this palaeohigh. They are absent farther to the north. The lower part of the dolomite replaces limestones of Tournaisian or Lower Viséan age, and the upper part is of Lower Viséan age (Bless *et al.*, 1976). The total thickness is more than 200 m in the Halen well (Fig. 2). The dolomites replaced open-marine bioclastic and peloidal packstones to grainstones, as shown by relic textures and by incompletely dolomitized intercalations. Relics of peloids, crinoids, brachiopods and corals have been recognized.

Near the palaeocontinent (Booischot borehole), sandy bioclastic and peloidal wackestones and packstones (Fig. 3A), calcareous sandstones and dolo-

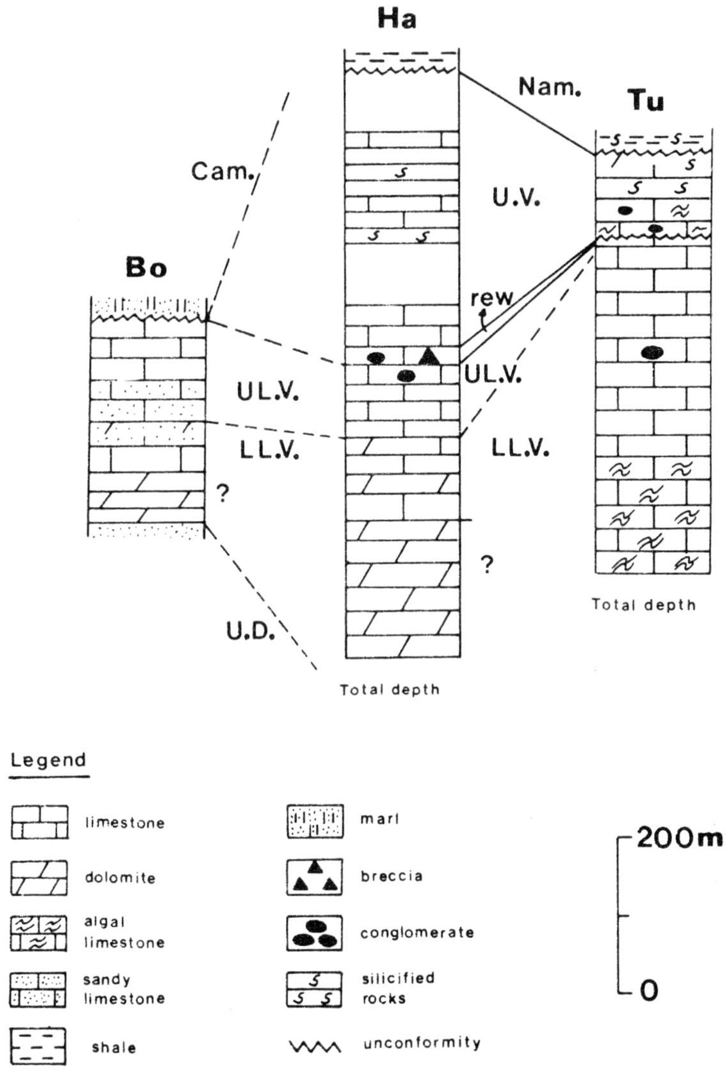

Fig. 2. Logs of the Viséan carbonates in the Booischot (Bo), Halen (Ha) and Turnhout (Tu) wells (U.D., Upper Devonian; LL.V., Lower part of the Lower Viséan; UL.V., upper part of the Lower Viséan; rew, reworked sequence; U.V., Upper Viséan; Nam, Namurian; Cam, Campanian).

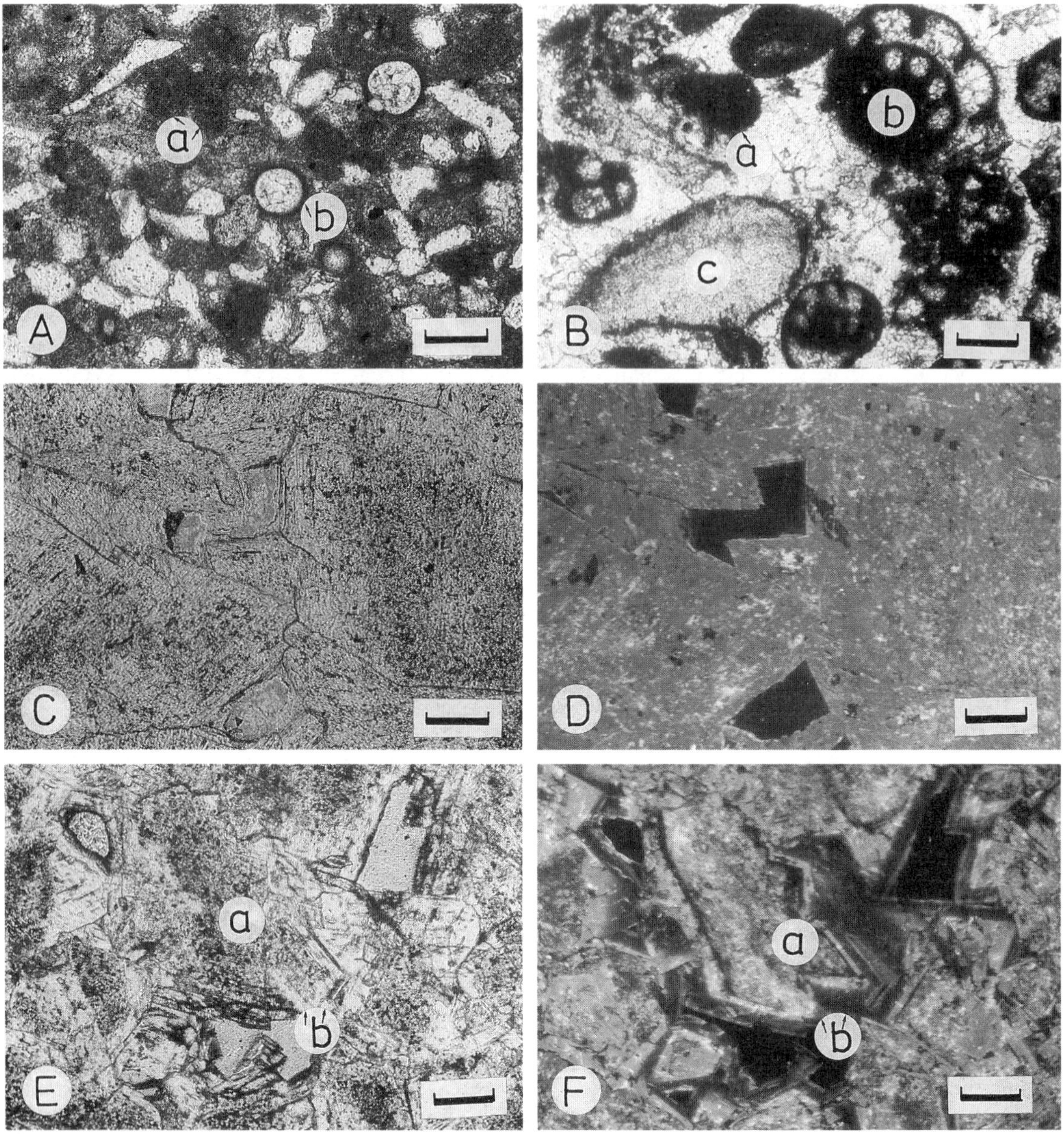

Fig. 3. Thin-section photographs of Lower Viséan carbonates. (A) Booischot, 587 m. Sandy packstone with peloids (a) and calcispheres (b). (B) Halen, 1021 m. Pack- to grainstone with peloids (a), foraminifera (b) and crinoids (c), which may be micritized. (C) Booischot, 659 m. Massive dolomite with large anhedral dolomite crystals. (D) Cathodoluminescence (CL) petrography of (C). The dolomite crystals possess a uniform moderate-orange luminescence. (E) Halen, 1331 m. Cloudy dolomite core (a) followed by a clear rim (b). (F) CL petrography of (E). The core (a) is orange luminescent. The rim (b) contains a bright yellow, a non- and a green-luminescent zone. Scale bars = 100 μm.

microsparite layers are present above the dolomites (Fig. 2; Muchez *et al.*, 1987). The biota consists mainly of ostracods, calcispheres and algal tubes. Within these carbonates and sandstones seven dolocrete horizons, formed in a brackish supratidal marsh environment, have been described by Muchez and Viaene (1987). Peloidal bioclastic packstones and grainstones with an open-marine biota (Fig. 3B) and bioclastic wackestones with a restricted fauna occur above this sandy sequence in the upper part of the Lower Viséan. However, they have only been sporadically dolomitized. In the Halen well, further to the north of the palaeohigh, partly dolomitized bioclastic packstones and grainstones are followed by an alternation of bioclastic packstones and grainstones and bioclastic wackestones situated in the upper part of the Lower Viséan (Fig. 2).

Interpretation

The limestone intercalations in the dolomite and the relic textures indicate a replacive origin of the dolomites and an open-marine subtidal depositional environment for most of the limestone precursor. The sandy bioclastic wackestones and packstones with a low biotic diversity indicate sedimentation in a restricted lagoon. The evolution in sedimentation from open-marine towards restricted and supratidal conditions reflects a lowering of relative sea level within early Viséan times. Following this regressive trend, sedimentation occurred on a shallow open-marine shelf. Massive dolomitization is restricted to the carbonates below the top of the regressive trend, represented by the sandy facies and the dolocrete horizons. In the Halen well this lowering of relative sea level has not been observed; however, this might be due to the block-faulted character of the Campine Basin and a high subsidence rate in the Halen area (Muchez *et al.*, 1987).

The typical occurrence of the dolomites near the continent and their restriction to the lower part of the Lower Viséan indicates that the dolomitizing fluids were restricted in space and time. Furthermore, the dolomites appear to be related to the lowering of relative sea level. Since several exposure surfaces have been recognized at Booischot, no distinct sequence boundary at which dolomitization took place can be distinguished. In southern Belgium, however, the top of the Lower Viséan dolomites is intensely karstified (Pirlet, 1970). Karst pockets are filled with non-dolomitized crinoidal limestones of the overlying unit, confirming a Lower Viséan age of the dolomites. In addition, no evaporites or relics thereof are present in the Lower Viséan of northern Belgium. A working model is based on these observations and will be critically evaluated in the light of the petrographic data. Such a model suggests that dolomitization was caused by water circulation near the mixing zone in the coastal area (Simms, 1984; Vahrenkamp *et al.*, 1991).

PETROGRAPHY AND PARAGENETIC SEQUENCE

Description

The dolomites consist of anhedral to euhedral crystals ranging in size between 45 and 1500 μm (Fig. 3C), and exhibit a uniform dull to moderate orange luminescence (Fig. 3D). At a very few intervals in the Booischot well, but more commonly in the Halen well, an orange-luminescent core is followed by a dolomite rim (Fig. 3E,F). This rim may make up 40% of the dolomite. Large oversized pores of several millimetres existed before the precipitation of the rim. The latter are thought to be the result of the dissolution of the calcite between the dolomite crystals. The rim consists of multiple yellow to green and non-luminescent dolomite zones (Fig. 3F).

The core and the rim are irregularly truncated by a calcite cement (Fig. 4A,B). The rim is completely dissolved where the dolomite crystals have been in contact with the pores, which subsequently became filled with calcite (Fig. 4A,B). These observations clearly demonstrate that calcite cementation postdated dolomitization and dolomite cementation, and that dolomite dissolution occurred before calcite precipitation. No dedolomite crystals are present.

The calcite cement shows a complex pattern of non-, bright- and dull-luminescent zones (Fig. 4C,D). The blocky calcites that fill the pores between the allochems and the dolomite crystals are dull-luminescent. In the dolomites (Fig. 4B) very irregular cavities, which have been cemented by the dull-luminescent calcites, are sometimes present. The allochems themselves show a moderate to dull luminescence. Compaction is of minor importance in the lower part of the Lower Viséan limestones, suggesting that pore occlusion occurred early in diagenesis; sporadic stylolites cross-cut both dolomite and calcite cement. In the limestones in the upper part of the Lower Viséan, however, a stylolitic contact is present between the peloids and a fitted fabric results. This fitted fabric indicates that mechanical and chemical compaction preceded calcite cementation, and that cementation was not an early diagenetic process (see also Muchez, 1988). These textural criteria indicate that calcite cementation in the lower part of the Lower Viséan strata predated the deposition of the overlying Lower Viséan strata. Since dolomitization predated calcite cementation, this process occurred during or at the end of the lower part of the Lower Viséan.

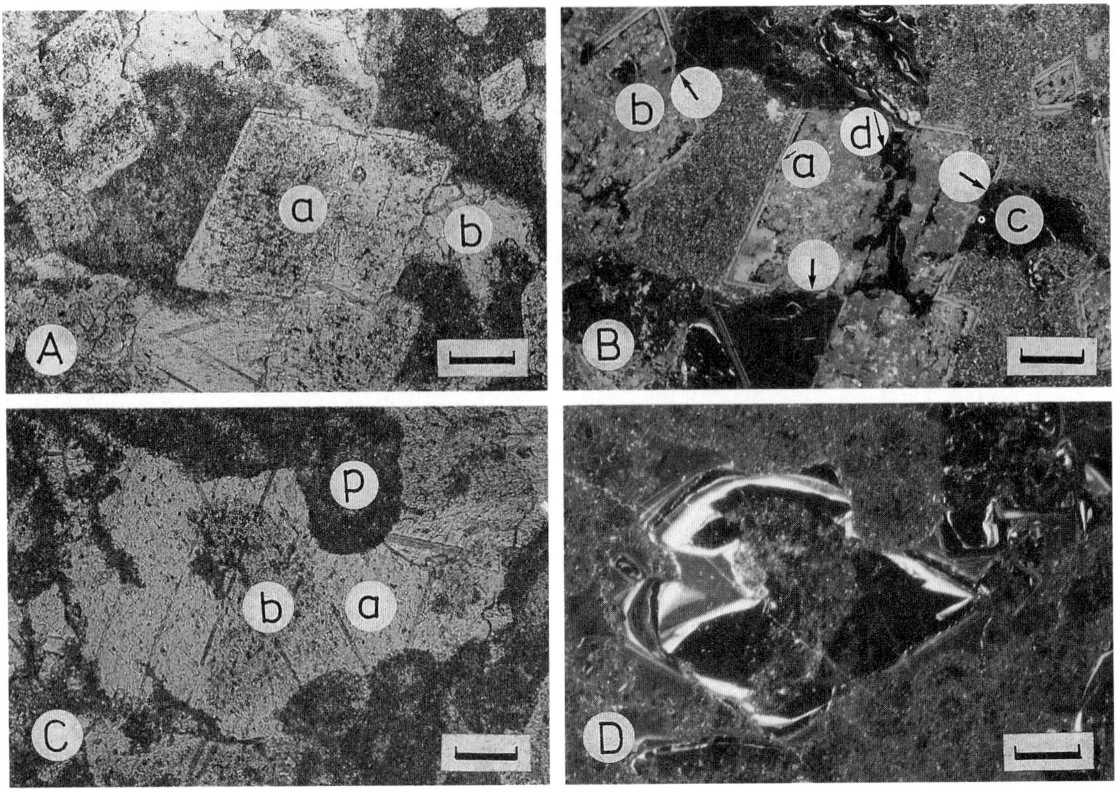

Fig. 4. Thin-section photographs of the dolomites and the subsequent calcite cements. (A) Halen, 1167 m. Dolomite rhombohedra (a) in a peloidal grainstone. The grainstone is cemented by blocky calcite cement (b). Scale bar = 100 μm. (B) Cathodoluminescence petrography of (A). The zoned dolomite cement (a) and the orange dolomite core (b) have been dissolved before the precipitation of the dull-luminescent blocky calcites (c). The dolomite rim is dissolved where it was in contact with a pore (large arrows). The dull-luminescent calcites fill cavities within the dolomite core (d). (C) Halen, 1105 m. Rim cements (a) developed around crinoid ossicles (b) in peloidal packstone (p). Scale bar = 90 μm. (D) CL petrography of (C). The rim cement shows a complex pattern of non-, bright- and dull-luminescent zones.

Interpretation

The uniform luminescence of the dolomite crystals suggests that the major part of the dolomite was formed from a fluid of nearly constant chemical composition. Subsequently, fluids undersaturated in calcite appear to have migrated through the carbonates, causing the development of large irregular oversized pores within the replacive dolomite. Dolomite cements formed after this dissolution, which occurred during one or several periods. The zoned pattern in the luminescence of the rim suggests that porewater chemistry varied during dolomite precipitation. A drastic change in the chemistry of the porewater occurred after the precipitation of the dolomite cement, resulting in partial dissolution of the dolomite rim and of the core. This dissolution was followed by the precipitation of a calcite cement.

As already noted, the uniform luminescent dolomites in the Campine Basin formed in a fluid with a nearly constant chemical composition, and replaced open-marine carbonates. This first dolomitization occurred due to seawater circulating near the mixing zone during the early Viséan (Fig. 5). The humid climate present during the early Viséan (Peeters *et al.*, 1992) contributed to water circulation within and around the mixing zone. Early Viséan mixing-zone dolomitization under humid conditions has been described by Hird *et al.* (1987).

Secondly, during the lowering of relative sea level, the mixing zone moved seaward (for this process see also Kaldi and Gidman (1982) and Humphrey

(1988), and the remaining calcium carbonate dissolved in this zone undersaturated in calcium carbonate and a dolomite cement precipitated. Undersaturation in calcium carbonate and supersaturation in dolomite can be caused by the mixing of fresh- and seawater (Plummer, 1975; Wigley & Plummer, 1976). The zoned pattern of the Lower Viséan dolomite cement suggests chemical variations in the waters. Zoned dolomite cements formed in the mixing zone of the Yucatan Peninsula (Ward & Halley, 1985) and of the Florida aquifer (Randazzo & Cook, 1987).

Thirdly, when lowering of the relative sea level continued during early Viséan times, the mixed water was replaced by fresh waters. The latter are undersaturated in dolomite and dolomite dissolution may have taken place. This process may have occurred in the Lower Viséan dolomites where the dolomite cores and rim were partly dissolved early in diagenesis. The non-, bright- and dull-luminescent calcites occurring in the pores could represent a meteoric phase (Fig. 5).

The blocky calcite cements in the Upper Viséan strata of the Campine Basin precipitated from fresh waters probably at the end of the Viséan and certainly during the Namurian (Muchez *et al.*, 1991). A similar time period may be tentatively proposed for the calcite cements in the upper part of the Lower Viséan.

GEOCHEMISTRY

Trace elements

Microprobe analyses of the dolomites are summarized in Table 1. Although average $FeCO_3$ and $MnCO_3$ composition of crystal cores is 0.47 mole % and 0.21 mole %, respectively, the $FeCO_3$ content can locally be high (1.48 mole % in the core and 1.69 mole % in the uniform luminescent dolomites). The iron and manganese concentrations in the different cement zones are variable, but generally higher than in the associated core. The highest iron concentrations occur in the non-luminescent zone. The $FeCO_3$ content can be as high as 2.89 mole %.

The highly variable iron and manganese concentrations in the zones support the interpretation of fluctuating chemical conditions during the formation of the dolomite rim. High iron and manganese concentrations in the porewaters can be due to the reduction of iron and manganese oxides in the Lower Viséan siliciclastics near the Brabant Massif (Fig. 2). From the present data it is not clear whether the local high concentrations of iron in the dolomite cores are due to neomorphism or whether they represent original values. Iron concentrations up to 9160 ppm have been reported from marine or hypersaline Miocene dolomites (Coniglio *et al.*, 1988). However, these data do not exclude an enrichment caused by recrystallization.

Stable isotopes

Results

Dolomites with uniform orange luminescence have $\delta^{18}O$ values between −6‰ and −12‰ and $\delta^{13}C$ values between +1‰ and +2‰ (Fig. 6). The dolomites that consist of a core and a rim show a variation in the $\delta^{18}O$ composition between −4‰ and −7‰. $\delta^{13}C$ values are slightly higher for the zoned dolomites (+3‰ to +4‰) relative to the homogeneous dolomites. The volume percent of the

Table 1. Microprobe analyses of $FeCO_3$ and $MnCO_3$ content of uniformly luminescent dolomites, dolomite cores and zoned dolomites in the Halen and Booischot wells.

	$FeCO_3$ range (mole %)	$FeCO_3$ average (mole %)	$MnCO_3$ range (mole %)	$MnCO_3$ average (mole %)
Uniform orange luminescent dolomites ($n = 23$)	b.d.–1.69	0.37	b.d.–0.59	0.20
Orange luminescent core ($n = 23$)	b.d.–1.48	0.47	b.d.–0.32	0.21
1st bright yellow zone ($n = 21$)	0.46–0.85	0.64	0.46–1.95	0.91
2nd non-luminescent zone ($n = 22$)	1.91–2.89	2.14	0.57–0.92	0.72
3rd green luminescent zone ($n = 18$)	1.00–1.56	1.24	0.40–0.67	0.55

(b.d., below detection limit).

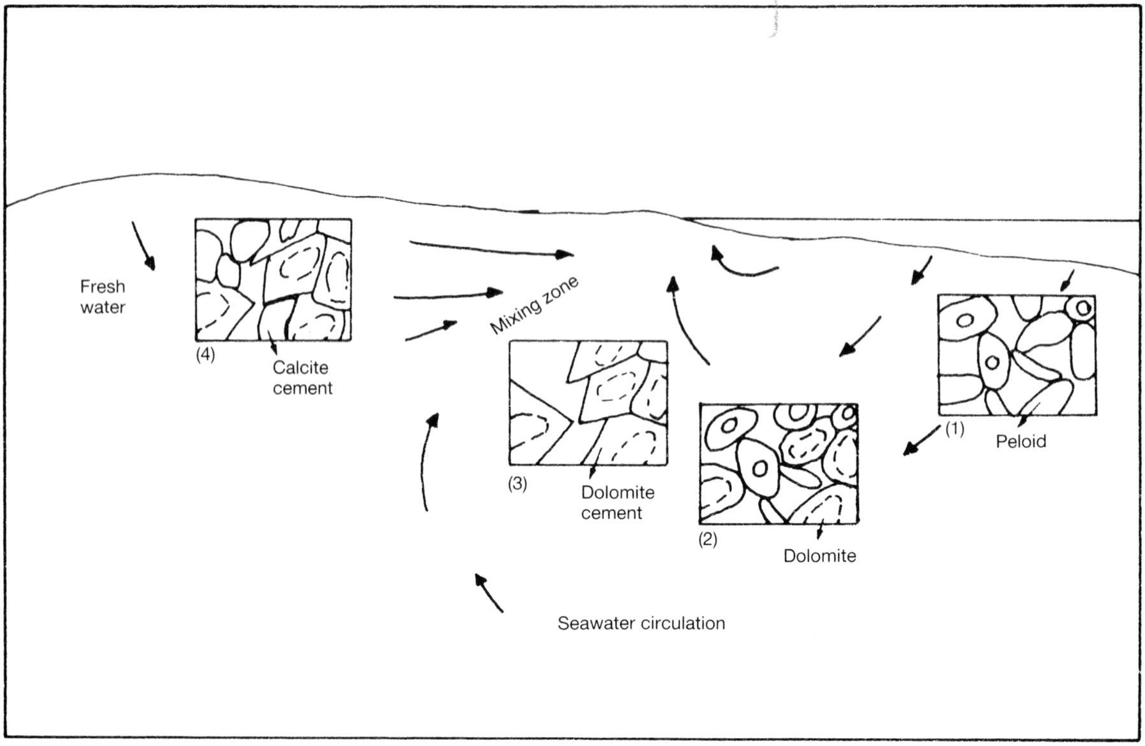

Fig. 5. The diagenetic processes that occurred in the carbonates during the early Viséan. Flow pattern after Simms (1984). (1) sedimentation; (2) dolomitization; (3) calcite dissolution and dolomite cementation; (4) dolomite dissolution and calcite cementation.

dolomite core and rim has been measured by point-counting photographs taken under cathodoluminescence before sampling. Figure 7, showing the volume percent of the dolomite rim versus the $\delta^{18}O$ composition of the dolomites, suggests a strong correlation (r = 0.89) between the two variables. Extrapolation of the data indicates a $\delta^{18}O$ value of +0.8‰ for the composition of the rim dolomites.

Furthermore, the stable isotopic composition of Lower Viséan neomorphosed yellow luminescent wackestones in the Booischot well, and of moderate to dull-luminescent allochems and blocky calcites in the Lower Viséan of the Halen well, have been analysed. The $\delta^{18}O$ composition of the wackestones lies between −10.0‰ and −7.9‰, and carbon values vary between −1.6‰ and +1.1‰. The oxygen isotopic composition of the allochems and blocky calcites ranges between −9.8‰ and −7.3‰, whereas carbon values cluster around +1.5‰.

Interpretation

The carbon isotopic composition of the zoned dolomites in the Halen well is similar to that of the Lower Carboniferous marine carbonates ($\delta^{13}C$ = +4‰; Meyers & Lohmann, 1985). The carbon isotopic composition of the uniform luminescent dolomites in the Booischot well indicates a slight depletion compared with this value.

The negative oxygen isotope values of the uniform luminescent dolomites do not support an origin from marine fluids. They are more compatible with a burial origin, or with a re-equilibration during further burial. Petrographic evidence, however, indicates an early diagenetic origin and a re-equilibration origin is proposed. The latter process has been widely recognized and reflects a common discrepancy between petrographic and geochemical data (see also Land, 1980; Hardie, 1987; Peeters *et al.*, 1993).

Based on the literature, an average $\delta^{18}O$

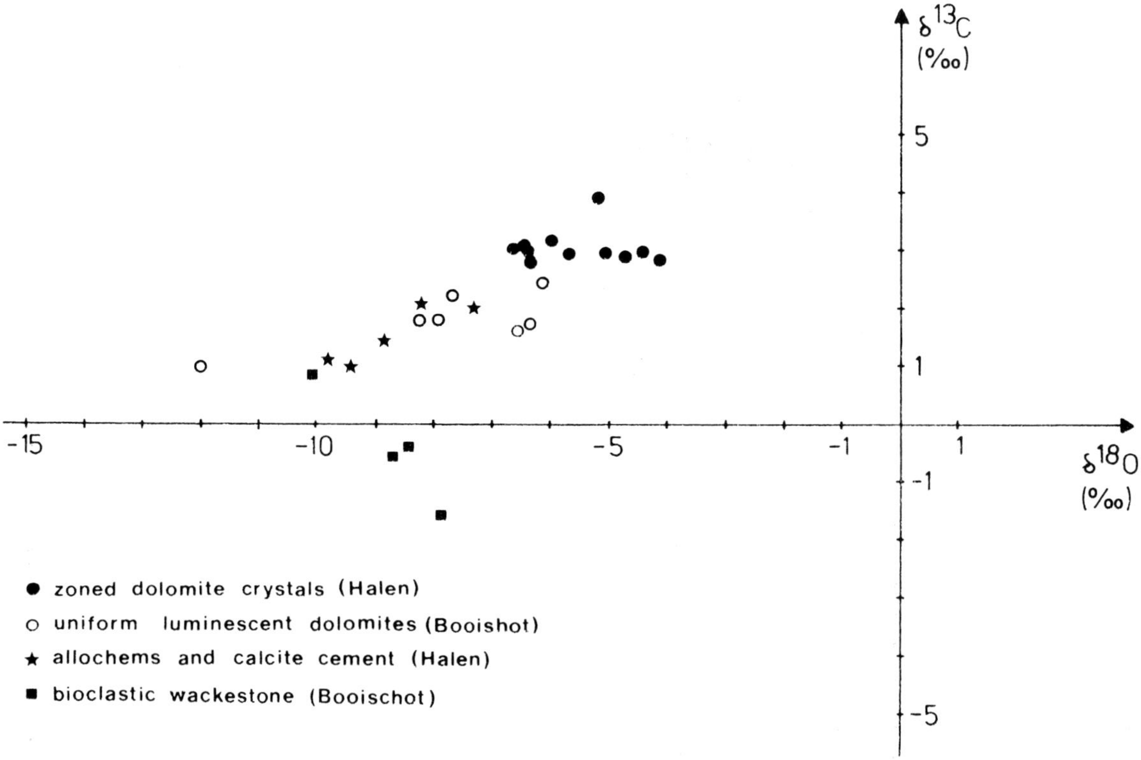

Fig. 6. Carbon and oxygen isotopic composition of uniformly luminescent dolomites, zoned dolomite crystals, yellow luminescent wackestones, dull to moderate luminescent allochems and blocky calcites in the Lower Viséan.

Table 2. $\delta^{18}O$ values for Lower Carboniferous marine limestones and calculated oxygen isotopic composition of associated marine dolomites.

$\delta^{18}O$ composition of Lower Carboniferous marine limestones	$\delta^{18}O$ composition of Lower Carboniferous marine dolomites
−1.5 to −2‰ PDB (Brand, 1982)	+1.5 to +1.0‰ PDB
−1.5‰ PDB (Meyers & Lohmann, 1985)	+1.5‰ PDB
−4.0‰ PDB (Hudson & Anderson, 1989)	−1.0‰ PDB

composition of Viséan marine dolomites has been calculated (Table 2), taking into account the different fractionation factors of calcite and dolomite ($\delta^{18}O_{dolomite} - \delta^{18}O_{calcite}$ at 25°C = 3‰; Matthews & Katz, 1977; McKenzie, 1981). If we accept a value of *ca.* +1‰ for Viséan marine dolomites, the dolomite rim with a value of +0.8‰ (Fig. 7) is very likely to have been precipitated in a fluid dominated by marine water. This value too may represent a re-equilibrated one. However, the fact that the rim still contains its typical zonation and that there is a clear linear relationship between the relative volume of the rim and the oxygen isotopic composition, tends to exclude re-equilibration of the rim and the value of 0.8‰ most likely approximates the original value in the dolomites. We suggest that the core, formed in marine waters, re-equilibrated because it was initially less stoichiometric and that the rim, precipitated in mixed waters, still contains its primary stoichiometry. Holocene and modern marine dolomites are calcium-rich, with a composition around $Ca_{1.2}Mg_{0.8}(CO_3)_2$ (Carballo *et al.*, 1987; Mitchell *et al.*, 1987). In contrast, dolomite rims that precipitated in the mixing zone in the Floridan aquifer have a much higher stoichiometry: their average composition is $Ca_{1.04}Mg_{0.96}(CO_3)_2$ (Randazzo & Cook, 1987).

If we accept the above values and a mixing-zone origin of the rim, this mixing zone was dominated by

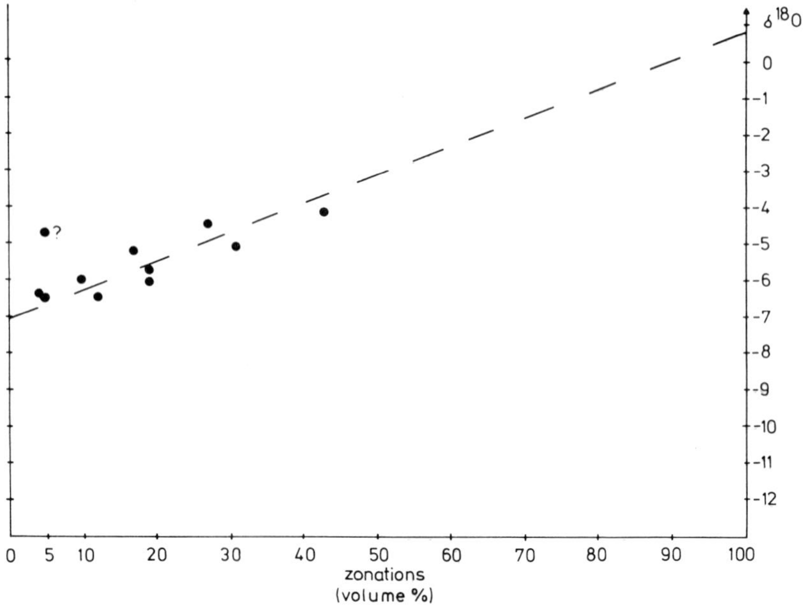

Fig. 7. Oxygen isotopic composition of the zoned dolomite crystals versus the volume of the rim or zones.

marine fluids, a conclusion also formulated by Ward & Halley (1985) for the Late Pleistocene mixing-zone dolomites in the northeastern Yucatan Peninsula. The $\delta^{18}O$ composition of the Yucatan dolomite and of modern groundwater suggests dolomite precipitation from mixed water containing at least 75% seawater. Indeed, supersaturation with respect to dolomite is maintained over this range of salinities (Stoessell *et al.*, 1989). The fact that calcite dissolution preceded the dolomite cement in the early Viséan of the Campine Basin does not conflict with high salinities: in the Yucatan Peninsula, mixtures of as much as 90% seawater are undersaturated with respect to calcite (Stoessell *et al.*, 1989).

Two aspects of the isotope data merit further comment. First, a linear relation exists between the relative volume of the dolomite rim and the oxygen isotopic composition of the total rock in the Halen well, although the core of the dolomites has been recrystallized. This relation strongly suggests that recrystallization occurred under uniform conditions in the Lower Viséan dolomites of the Halen well.

Secondly, the considerable spread in the oxygen isotopic composition of the dolomites in the Booischot well indicates that recrystallization in this well was not uniform. The difference between the two wells cannot readily be explained, but is probably related to the general geological setting. In the Halen well, the Upper Viséan strata are overlain by a very thick sequence of Upper Carboniferous shales, whereas in the Booischot well the Upper Carboniferous is absent and the Cretaceous occurs on top of the Lower Viséan. Between the dolomites and the Cretaceous marls, only 200 m of limestone are present (Fig. 2). The Booischot well is situated close to the Brabant Massif and fresh waters from this palaeohigh could have penetrated the subsurface during several periods since the early Viséan. The top of the Lower Viséan is intensely mineralized. Sphalerite and galena occur in veins and in recrystallized limestones associated with a karstic zone (Muchez & Viaene, 1990), and mineralizing fluids migrated through the Lower Viséan of the Booischot area. We believe that the dolomites in the Booischot and Halen well underwent different evolutions and that different fluids have circulated through them since their formation, resulting in different recrystallization conditions in the two areas.

The carbon isotopic composition of the neomorphosed wackestones in the Booischot well and of the allochems and the blocky calcites in the Halen well are more negative than those of the marine carbonates (Meyers & Lohmann, 1985). The range of the oxygen isotopic values is similar to that reported for Lower Viséan carbonates in southern Belgium, which had been neomorphosed in a meteoric pore fluid (Peeters *et al.*, 1993; Muchez *et al.*, 1992) and which show an inverted J trend (see also Meyers & Lohmann, 1985). Neomorphism of the host limestone and cementation in a meteoric phreatic environment is therefore consistent with the proposed evolution from marine to mixing zone and to meteoric porewaters in the Lower Viséan carbonates.

CONCLUSIONS

Thick massive dolomites occur in the Lower Viséan of the Campine Basin (northern Belgium). In contrast with the classical model of mixing-zone dolomitization, we suggest that the major part of the dolomite body formed due to the seawater circulation adjacent to and induced by the mixing zone. Such dolomites would normally border palaeocontinental margins, or develop beneath and adjacent to islands.

In the Lower Viséan of northern Belgium, and probably at many other places, such marine dolomite formation was followed by calcite dissolution and dolomite cementation in the mixing zone and calcite cementation in the meteoric phreatic environment. This evolution occurs in response to a lowering of relative sea level, which is most pronounced in shallow coastal areas.

The use of stable isotopes in the reconstruction of the origin of dolomites has been questioned by many authors. Also, in this study stable isotope data must be interpreted carefully. However, the combination of detailed petrography, including point-counting, and stable oxygen isotopes supports the proposed model. No dolomitization model may be based exclusively on isotope analyses, since this would lead to speculative interpretations.

ACKNOWLEDGEMENTS

We sincerely thank Jos Bouckaert, Geological Survey of Belgium, for permission to study the cores and for his stimulating interest. Rudi Swennen provided constructive criticism of the manuscript. The constructive reviews by Peter Gutteridge, Bruce Purser, Maurice Tucker and Donald Zenger improved the manuscript and are appreciated. We thank Eddy Keppens and Peter Nielsen for the stable isotope measurements. Jacques Wautier kindly carried out microprobe analysis on a Camebax at the Centre d'Analyse par Microsonde pour les Sciences de la Terre at Louvain-la-Neuve. The thin sections were carefully prepared by Cyriel Moldenaers. Philippe Muchez thanks the National Fund for Scientific Research of Belgium for their financial support.

REFERENCES

Badiozamani, K. (1973) The Dorag dolomitization model – application to the Middle Ordovician of Wisconsin. *J. Sedim. Petrol.* **43**, 965–984.

Bear, J. (1979) *Hydraulics of Groundwater*. McGraw-Hill, New York, 569 pp.

Bless, M.J.M., Bouckaert, J., Bouzet, Ph., Conil, R., Cornet, P., Fairon-Demaret, M., Groessens, E., Longerstaey, P.J., Meessen, J.P.M.Th., Paproth, E., Pirlet, H., Streel, M., Van Ameron, H.W.J. & Wolf, M. (1976) Dinantian rocks in the subsurface north of the Brabant and Ardenno-Rhenish massifs in Belgium, the Netherlands and the Federal Republic of Germany. *Meded. Rijks Geol. Dienst* **27**, 81–195.

Brand, U. (1982) The oxygen and carbon isotope composition of Carboniferous fossil components: seawater effects. *Sedimentology* **29**, 139–147.

Carballo, J.D., Land, L.S. & Miser, D.E. (1987) Holocene dolomitization of supratidal sediments by active tidal pumping, Sugarloaf Key, Florida. *J. Sedim. Petrol.* **57**, 153–165.

Choquette, P.W. & Steinen, R.P. (1980) Mississippian non-supratidal dolomite, Ste Genevieve Limestone, Illinois basin: evidence for mixed-water dolomitization. In: *Concepts and Models of Dolomitization* (Ed. Zenger, D.H., Dunham, J.B. & Ethington, R.L.) Soc. Econ. Paleont. Mineral. Spec. Publ. 28, 163–196.

Churnet, H.G., Misra, K.C. & Walker, K.R. (1982) Deposition and dolomitization of Upper Knox carbonate sediments, Copper Ridge district, East Tennessee. *Geol. Soc. Am. Bull.* **93**, 76–86.

Coniglio, M., James, P. & Aissaoui, D.M. (1988) Dolomitization of Miocene carbonates, Gulf of Suez, Egypt. *J. Sedim. Petrol.* **58**, 100–119.

Craig, H. (1957) Isotopic standards for carbon and oxygen correction factors for mass spectrometric analysis for carbon dioxide. *Geochim. Cosmochim. Acta* **12**, 133–149.

Dunham, J.B. & Olson, E.R. (1980) Shallow subsurface dolomitization of subtidally deposited carbonate sediments in the Hanson Creek Formation (Ordovician-Silurian) of Central Nevada. In: *Concepts and Models of Dolomitization* (Ed. Zenger, D.H., Dunham, J.B. & Ethington R.L.) Soc. Econ. Paleont. Mineral. Spec. Publ. 28, 139–161.

Hardie, L.A. (1987) Dolomitization: a critical view of some current views. *J. Sedim. Petrol.* **57**, 166–183.

Hird, K., Tucker, M.E. & Waters, R. (1987) Petrology, geochemistry and origin of Dinantian dolomites from southeast Wales. In: *European Dinantian Environments* (Ed. Adams, A.E., Miller, J. & Wright, V.P.) pp. 359–377. Wiley, New York, 402 pp.

Hudson, J.D. & Anderson, T.F. (1989) Ocean temperatures and isotopic compositions through time. *Trans. Roy. Soc. Edinburgh, Earth Sci.* **80**, 183–192.

HUMPHREY, J.D. (1988) Late Pleistocene mixing zone dolomitization, southeastern Barbados, West Indies. *Sedimentology* **35**, 327–348.
KALDI, J. & GIDMAN, J. (1982) Early diagenetic dolomite cements: examples from the Permian Lower Magnesian Limestone of England and the Pleistocene carbonates of the Bahamas. *J. Sedim. Petrol.* **52**, 1073–1085.
KONIKOW, L.F. & BREDEHOEFT, J.D. (1978) *Computer Model of Two-Dimensional Solute Transport and Dispersion in Ground-Water*. US Geological Survey Techniques of Water-Resources Inv., Book 7, Chap. C2, 90 pp.
LAND, L.S. (1973) Contemporaneous dolomitization of Middle Pleistocene reefs by meteoric water, North Jamaica. *Bull. Marine Sci.* **23**, 64–92.
LAND, L.S. (1980) The isotopic and trace element geochemistry of dolomite: the state of the art. In: *Concepts and Models of Dolomitization* (Ed. Zenger, D.H., Dunham, J.B. & Ethington, R.L.) Soc. Econ. Paleont. Mineral. Spec. Publ. 28, 87–110.
LEBBE, L.C. (1983) Mathematical model of the evolution of the fresh water lens under the dunes and beach with semi-diurnal tides. *Geol. Appl. Idrogeol.* **18**, 211–226.
MACHEL, H.G. & MOUNTJOY, E.W. (1986) Chemistry and environments of dolomitization – a reappraisal. *Earth Sci. Rev.* **23**, 175–222.
MCKENZIE, J.A. (1981) Holocene dolomitization of calcium carbonate sediments from the coastal sabkhas of Abu Dhabi, UAE: a stable isotope study. *J. Geol.* **89**, 185–198.
MATTHEWS, A. & KATZ, A. (1977) Oxygen isotope fractionation during the dolomitization of calcium carbonate. *Geochim. Cosmochim. Acta* **41**, 1431–1438.
MEYERS, W.J. & LOHMANN, K.C. (1985) Isotope geochemistry of regionally extensive calcite cement zones and marine components in Mississippian limestones, New Mexico. In: *Carbonate Cements* (Ed. Schneidermann, N. & Harris P.M.) Soc. Econ. Paleont. Mineral. Spec. Publ. 36, 223–239.
MITCHELL, J.T., LAND, L.S. & MISER, D.E. (1987) Modern marine dolomite cement in a north Jamaican fringing reef. *Geology* **15**, 557–560.
MUCHEZ, PH. (1988) *Sedimentologische, Diagenetische en Geochemische Studie van de Dinantiaan Strata ten Noorden van het Brabant Massief (Bekken van de Kempen)*. PhD thesis, Katholieke Universiteit Leuven, 311 pp.
MUCHEZ, PH. & VIAENE, W. (1987) Dolocretes from the Lower Carboniferous of the Campine–Brabant Basin, Belgium. *Pedologie* **37**, 187–202.
MUCHEZ, PH. & VIAENE, W. (1990) Lithogeochemistry of the Dinantian strata of the Campine–Brabant Basin (northern Belgium). *Ann. Soc. Géol. Belg.* **113**, 341–358.
MUCHEZ, PH., VIAENE, W., WOLF, M. & BOUCKAERT, J. (1987) Sedimentology, coalification pattern and paleogeography of the Campine Brabant Basin during the Visean. *Geol. Mijnbouw* **66**, 313–326.
MUCHEZ, PH., VIAENE, W. & MARSHALL, J.D. (1991) Origin of shallow burial cements in the Late Viséan of the Campine Basin, Belgium. *Sedimentol. Geol.* **73**, 257–271.
MUCHEZ, PH., PEETERS, C., VIAENE, W. & KEPPENS, E. (1992) Stable isotopic evolution of an evaporitic dissolution breccia in the Lower Viséan of SE Belgium. *Chem. Geol.* **102**, 119–127.
PEETERS, C., MUCHEZ, PH. & VIAENE, W. (1992) Paleogeographic and climatic evolution of the Moliniacian (Lower Viséan) in southeastern Belgium. *Geol. Mijnbouw* **71**, 39–50.
PEETERS, C., SWENNEN, R., NIELSEN, P. & MUCHEZ, PH. (1993) Sedimentology and diagenesis of the Visean carbonates in the Vesder area (Verviers synclinorium, E-Belgium). *Zentralbl. Geol. Paläont.* **Teil 1** (5), 519–547.
PIRLET, H. (1970) L'influence d'un karst sous-jacent sur la sédimentation calcaire et l'intéret de l'étude des paléokarsts. *Ann. Soc. Géol. Belg.* **93**, 247–254.
PLUMMER, L.N. (1975) Mixing of sea water with calcium carbonate ground water. *Geol. Soc. Am. Mem.* **142**, 219–236.
RANDAZZO, A.F. & COOK, D.J. (1987) Characterization of dolomitic rocks from the coastal mixing zone of the Floridan aquifer. *Sediment. Geol.* **54**, 169–192.
SIMMS, M. (1984) Dolomitization by groundwater flow systems in carbonate platforms. *Trans. Gulf Coast Ass. Geol. Soc.* **34**, 411–420.
STOESSELL, R.K., WARD, W.C., FORD, B.H. & SCHUFFERT, J.D. (1989) Water chemistry and $CaCO_3$ dissolution in the saline part of an open-flow mixing zone, coastal Yucatan Peninsula, Mexico. *Geol. Soc. Am. Bull.* **101**, 159–169.
VAHRENKAMP, V.C., SWART, P.K. & RUIZ, J. (1991) Episodic dolomitization of Late Cenozoic carbonates in the Bahamas: evidence from strontium isotopes. *J. Sedim. Petrol.* **61**, 1002–1014.
WARD, W.C. & HALLEY, R.B. (1985) Dolomitization in a mixing zone of near-seawater composition, late Pleistocene, north-eastern Yucatan Peninsula. *J. Sedim. Petrol.* **55**, 407–420.
WIGLEY, T.M.L. & PLUMMER, L.N. (1976) Mixing of carbonate waters. *Geochim. Cosmochim. Acta* **40**, 989–995.

Burial Dolomitization Models

Spec. Publs Int. Ass. Sediment. (1994) **21**, 169–186

Burial dolomitization of the Middle Ordovician Glenwood Formation by evaporitic brines, Michigan Basin

J.A. SIMO, C.M. JOHNSON, M.R. VANDREY,* P.E. BROWN, E. CASTROGIOVANNI, P.E. DRZEWIECKI, J.W. VALLEY *and* J. BOYER

Department of Geology and Geophysics, University of Wisconsin, 1215 West Dayton Street, Madison, Wisconsin 53706-1692, USA

ABSTRACT

The Middle Ordovician Glenwood Formation in the Michigan Basin represents a transitional stratigraphic unit from siliciclastic- to carbonate-dominated sedimentation. Deposition occurred in a distal shallow-marine mixed carbonate–siliciclastic setting relative to the shallow beach-to-bar shoreface deposits of the underlying St Peter Sandstone and the open-marine limestones of the overlying Black River Formation. The Glenwood dolomites form the upper seal of a gas compartment in the deep Michigan Basin.

Diagenesis of the Glenwood Formation carbonate lithologies can be divided into three stages, each occurring in progressively deeper burial diagenetic environments. Early diagenesis occurred in a near-surface, probably marine, setting. Middle diagenesis occurred during shallow burial and is characterized by mouldic porosity, aggrading neomorphism and by precipitation of ferroan and non-ferroan equant calcite cement. Late diagenesis is characterized by deep-burial dolomitization. Dolomite replaced most of the precursor carbonate sediment and occluded fracture, shelter and intergranular porosity. Late postdolomite diagenesis included clay cementation and localized anhydrite formation.

Variations of $\delta^{18}O$ and $\delta^{13}C$ in carbonate phases define four distinct groups.

1 Middle diagenetic stage neomorphic calcites ($n = 6$) and calcite cements ($n = 3$) have $\delta^{18}O$ values that vary from 17.4 to 23.6‰ SMOW, and relatively constant $\delta^{13}C$ values (−3.5 to −1.3‰ PDB).

2 Most late-stage fine-crystalline replacive dolomites have a narrow range of $\delta^{18}O$ values (19.4 ± 0.5‰ SMOW), but $\delta^{13}C$ values range from −4.5 to −1.1‰.

3 Late-stage coarse-crystalline replacive dolomites ($n = 6$) have distinctly lower $\delta^{18}O$ values (14.8 to 17.6‰ SMOW), compared to fine-crystalline replacive dolomite, and mainly constant $\delta^{13}C$ (−1.5 ± 0.3‰ PDB).

4 Late-stage dolomite cements ($n = 17$) have $\delta^{13}C$ values of −4.4 to −1.3‰ PDB, and all but two $\delta^{18}O$ values have a narrow range (18.3 ± 0.7‰ SMOW).

Fluid-inclusion homogenization temperatures of 175 ± 25°C support a late-stage origin of the dolomite cements. At 175°C, the range of $\delta^{18}O$ values for dolomite cements suggests precipitation from a water with a constant $\delta^{18}O$ value of 6 ± 1‰ vs. SMOW. These data suggest dolomitization from an evaporitic fluid.

Replacive dolomites and dolomite cements have low Sr concentrations (typically <100 ppm). Replacive dolomites have high Rb concentrations (up to 12 ppm) and $^{87}Sr/^{86}Sr$ ratios, which reflect contamination by radiogenic clay. In contrast, dolomite cements have low Rb contents, indicating no clay contamination, and lower $^{87}Sr/^{86}Sr$ ratios of 0.7084–0.7086, similar to the Sr isotopic composition of Ordovician and Silurian seawater. The homogeneous Sr isotope composition for the dolomite-forming fluids does not allow basinwide circulation of diagenetic fluids, which would probably have interacted with radiogenic sediments, and indicates that the fluids maintained their original chemical signatures for long periods of time, probably in a closed system.

The proposed dolomitizing fluid is middle Palaeozoic evaporitic brines of seawater origin that interacted little with radiogenic sediments and maintained the original $^{87}Sr/^{86}Sr$ ratios until dolomit-

* Present address: Exxon Co. USA, PO Box 60626, New Orleans, LA 70160-0626, USA.

ization. The timing of dolomitization is uncertain, but the high homogenization temperatures and age of the clay postdating dolomite cements suggests an early Carboniferous age, when the Glenwood Formation reached maximum burial.

INTRODUCTION

The Michigan Basin is an intracratonic basin believed to be the result either of a rift zone partially filled with dense Keweenawan basalts (Hinze *et al.*, 1975; Fisher *et al.*, 1988) or the result of crustal thermal contraction (Nunn & Sleep, 1984). There is evidence of faulting and folding throughout the history of the basin, although the lateral continuity and homogeneity of the Middle Ordovician strata suggest a relatively flat depositional topography at that time (Barnes *et al.*, 1992b). The Middle Ordovician sandstones and limestones extend across the entire basin, and reach a maximum thickness of 360 m in central Michigan. The Glenwood Formation occurs at depths between 1700 and 3230 m. Upper Ordovician carbonates in the Michigan Basin are up to 320 m thick, and thicken toward the east. Silurian and Devonian sediments attain a maximum thickness of 3100 m, and are characterized by shelf carbonates and basinal evaporites. The Middle Ordovician is a deep, structurally controlled natural gas play, with an average in-place estimate of 10–30 bcf per 640-acre unit in producing fields (Harrison, 1987). The diagenetic history of the Glenwood Formation is relevant to the oil and gas industry because it forms the top seal of an overpressured and a normally pressured gas compartment in the deep part of the Michigan Basin (Bahr, 1989; Moline *et al.*, 1992).

With the exception of some centimetre-scale intervals, the Middle Ordovician Glenwood Formation carbonates consist mostly of dolomites. Dolomite precipitated in large fractures, vugs and mouldic porosity, or replaced depositional carbonates and earlier cements. Replacive dolomites are fine-crystalline or coarse-crystalline. Coarse-crystalline replacive dolomite occurs as bands that appear to parallel bedding and have gradational (1–2 cm) contacts with the surrounding sediments. Dolomite cements are coarse-crystalline (up to 2 cm) and have non-planar crystal faces.

Stratigraphic and paragenetic interpretation of the Glenwood Formation was based on optical and cathodoluminescent petrography. Small (5 mg) homogeneous samples provided information concerning the source of the diagenetic fluid and diagenetic environments. These samples were evenly divided into two splits, one for analysis of $\delta^{13}C$ and $\delta^{18}O$, and the other for Rb and Sr concentration and $^{87}Sr/^{86}Sr$ ratios. The $^{87}Sr/^{86}Sr$ ratios and Rb and Sr concentrations (analysed by isotope dilution) were determined using a VG Sector 54 multicollector mass spectrometer and a mixed $^{87}Rb/^{84}Sr$ spike. Rb/Sr ratios are precise to 1–2%, although individual Rb and Sr concentrations have greater uncertainty due to the small amount (<<10 mg) of carbonate analysed. Carbonates for radiogenic isotope split were dissolved in 6 M HCl. Fifty-two analyses of NBS-987 during the study produced an $^{87}Sr/^{86}Sr$ ratio of 0.710273 ± 11 (2σ); additional details of the radiogenic isotope analytical methods can be found in Johnson and Thompson (1991). Random replicate analyses of samples indicate an average precision of ±0.00005 for the $^{87}Sr/^{86}Sr$ ratio. The stable isotope split was reacted for 12 h with concentrated phosphoric acid at 50°C (McCrea, 1950; Sharma & Clayton, 1965). Carbon and oxygen isotope ratios were determined on a Finnigan MAT/251 multicollector mass spectrometer. Results are presented in the standard per mil notation relative to SMOW and PDB for oxygen and carbon, respectively. Precision and accuracy are better than 0.1‰. Thermometric data were obtained using a gas-flow heating–freezing system calibrated with commercially available chemical standards and synthetic fluid inclusions. Accuracy of the temperature measurements obtained in this study ranges from ±0.2°C (below room temperature) to ±3°C for higher temperatures.

STRATIGRAPHY AND SEDIMENTOLOGY

In the Upper Mississippi Valley and the Michigan Basin, the Middle Ordovician (Champlainian) Glenwood Formation is a transitional unit between clean quartz sandstones of the underlying St Peter Sandstone and overlying marine limestones of the Black River Formation (Ostrom, 1969). As such, the Glenwood Formation contains both sandstone and carbonate facies (Figs 1 and 2). In Michigan, the

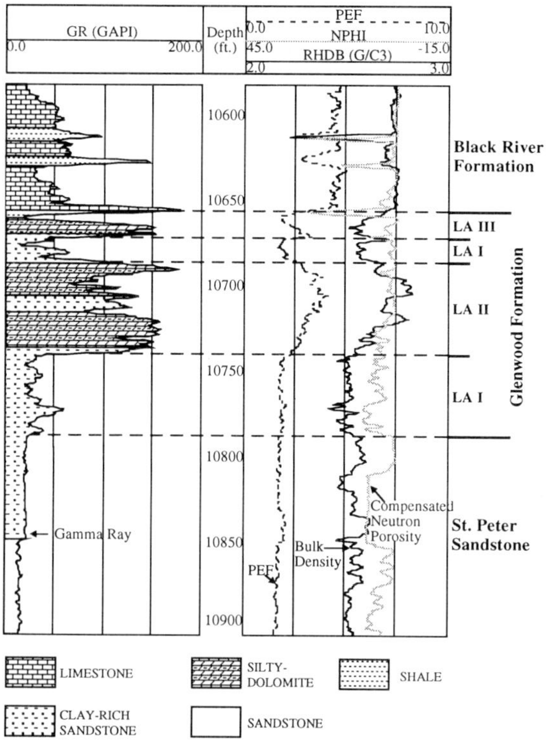

Fig. 1. Stratigraphic section of the Glenwood Formation, showing geophysical log responses of the Glenwood contacts and different lithofacies. JEM Workman 10–31. Location is shown in Figure 3 and Lithofacies (LA I, II and III) are described in Table 1. Modified from Vandrey, 1991.

Glenwood is present only in the subsurface and its contacts are based on geophysical logs and limited core and sample descriptions. The maximum thickness of 80 m decreases towards the basin margin (Fig. 3).

There is little consensus concerning nomenclature for the Glenwood Formation (Bricker *et al.*, 1983; Fisher & Barratt, 1985; Wheeler, 1987; Brady & DeHaas, 1988; Fisher *et al.*, 1988; Barnes *et al.*, 1988; Catacosinos & Daniels, 1991). In this paper the formational boundaries for the Glenwood are Vandrey's (1991). The base of the Glenwood Formation is characterized by the appearance of green to brownish detrital clay in fine- to medium-grained sandstones that overlie the fine- to coarse-grained, clean sand-stones of the St Peter Sandstone. This contact has a distinctive wireline log appearance (Fig. 1). A slight increase on the gamma-ray log and neutron-porosity log produce a shoulder that is distinguishable on most logs. The contact with the overlying Black River Formation occurs at the base of the first massive (bedding typically >3 m) limestone, and is distinguished in geophysical logs based on PEF (photoelectric factor), bulk density and neutron porosity (Fig. 1).

There are three distinct lithofacies within the Glenwood Formation in the Michigan Basin, which represent various combinations of the five different rock types described by Vandrey (1991) and summarized in Fig. 1 and Table 1. These three lithofacies are distinguished in both cores and geophysical logs. Correlations across the basin by the use of wireline logs indicate that interfingering and lateral thinning of the lithofacies are common (Fig. 2).

From bottom to top, the lithofacies (Figs 1 and 2, Table 1) are as follows: (a) Lithofacies I is characterized by densely bioturbated clay-rich sandstone; (b) Lithofacies II is dominated by cyclic carbonate sediments, including mudstones, fossiliferous wackestones to packstones, oolitic wackestones to grainstones and microbial bindstones; and (c) Lithofacies III is characterized by black fissile shales with phosphate nodules, and bioturbated clay-rich sandstones and mud-rich carbonates. The dolomites of Lithofacies II are the subject of this study.

The Glenwood Formation was deposited during an overall transgression, although repetitive small-scale shallowing-upward cycles suggest that sedimentation kept pace with the relative sea-level rises of short-term cycles. Lithofacies I is transitional between the St Peter Sandstone and the Glenwood Formation, and is interpreted as reflecting deepening of the basin (Vandrey, 1991). Transgression continued in the Middle Ordovician and trapped most siliciclastics on the margins of the basin, permitting initiation of carbonate production. Siliciclastic influx was from the northeast and north, and mudstones and wackestones accumulated in more distal lower-energy settings to the southwest (Fig. 3). The packstones to grainstones and bindstones reflect shoaling to sea level. A brief transgressive event drowned the carbonate platform and deposited a shale with phosphatic nodules and localized hardgrounds across the central basin (Lithofacies III). Transgression proceeded in the Late Ordovician, and the shallow subtidal open-marine sediments of the Black River Formation were deposited on the Glenwood Formation (Taylor & Sibley, 1986; Budai & Wilson, 1991).

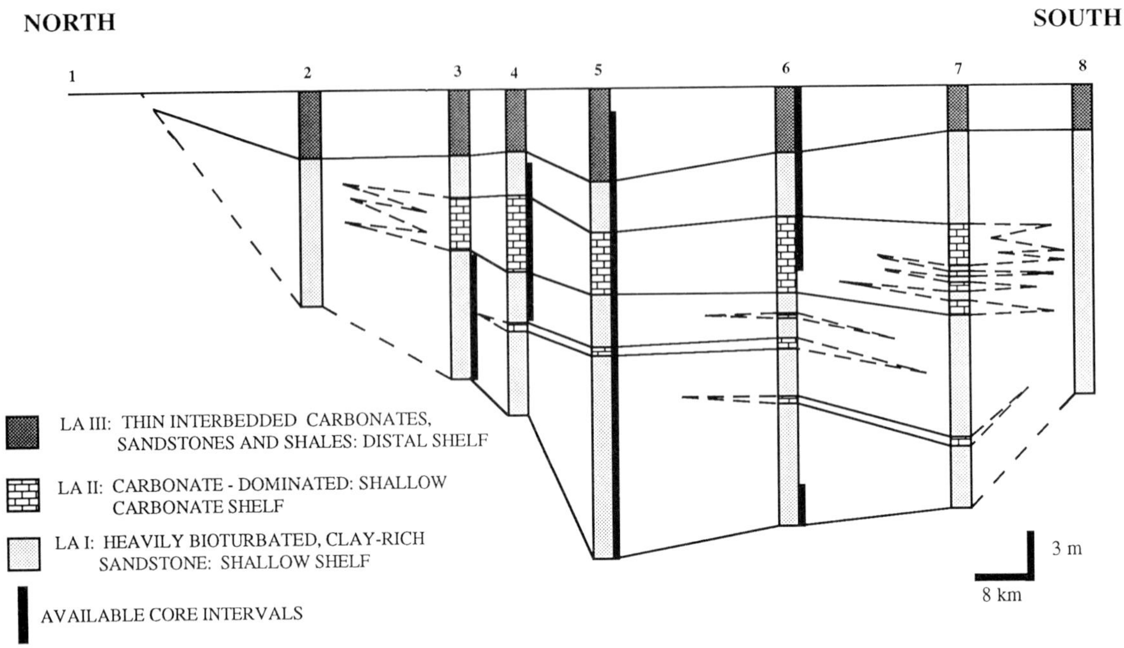

Fig. 2. Stratigraphic north–south cross-section of the Glenwood Formation. Datum is the base of the Black River Formation. Location shown in Figure 3. Lithofacies (LA I, II and III) are described in Table 1. Modified from Vandrey, 1991.

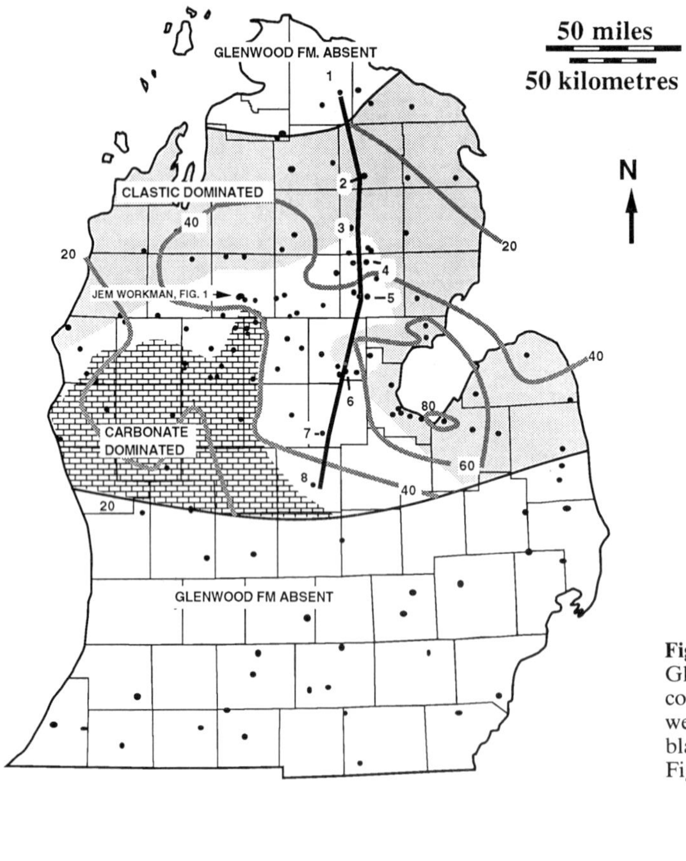

Fig. 3. Isopach (metres) and facies map of the Glenwood Formation (G. Nadon, personal communication, 1992). Dots indicate geophysical well studied and the approximately N–S thick black line is the line of cross-section shown in Figure 2.

Table 1. Lithofacies associations and rock types (after Vandrey, 1991).

Lithology	Grain size or carbonate rock type	Lithofacies occurrence	Framework mineralogy	Colour	Cements	Relative clay abundance	Major depositional features	Major diagenetic features
Sandstone	Fine to coarse sand	I, III	Quartz, K-spar	Green to tannish white	Silica, clay, dolomite	Common to abundant	Heavily bioturbated; rare cross-bedding; coarse burrow fill	Qtz, K-spar overgrowths and dissolution; occasional dissolution seams; pyrite
Conglomerate	Coarse clasts in granule to medium sand matrix	I	Quartz, K-spar, pyrite	White	Silica	None to minor	Graded bedding; restricted thin layer	Quartz overgrowths
Shale	Clay and silt with fine sand stringers	I, II, III	Quartz, K-spar, clay, pyrite, phosphate	Dk. grey to black, red?	Clay, dolomite	Abundant	Heavily bioturbated to unburrowed and fissile	Common dissolution seams; pyrite
Sand-rich carbonate	Silt to fine sand, in mudstone matrix	II, III	Quartz, K-spar, dolomite, clay, pyrite	Dk. grey to white	Clay, dolomite	Common	Heavily bioturbated; thin interbedded sandstone lenses	Common to abundant dissolution seams; pyrite
Carbonate	Mudstones	II, III	Dolomite, clay, quartz, K-spar, pyrite, phosphate	Lt. to dk. grey to black	Clay, dolomite	Common	Heavily bioturbated; intervals of increased siliciclastic content	Common dissolution seams where very fine grained siliciclastics present; fracturing; dolomitization
	Wackestone to grainstones	II, III	Dolomite, calcite Ooids, peloids, fossil fragments	Lt. to dk. grey to brown	Calcite, dolomite	Minor	Horizontal fossil orientation; no visible bedding forms	Minor dissolution seams; fracturing; dolomitization
	Boundstones	II, III	Dolomite Microbial stromatolites	Light and dark grey	Calcite, dolomite	Minor	Scoured surfaces; siliciclastic-rich storm layers	Minor dissolution seams; dolomitization

PARAGENETIC SEQUENCE

The paragenetic sequence of the Lithofacies II carbonates (Fig. 4) is subdivided into early, middle and late stages (Vandrey, 1991). These subdivisions are characterized by porosity evolution, cement type and recrystallization events, and are separated by periods of dissolution and secondary porosity development. The early stage of diagenesis occurred in a near-surface setting, and is characterized by reduction of primary porosity, micritization, and marine and vadose cementation. The middle and late stages reflect postdepositional modification in shallow-meteoric to deep-burial settings. Middle diagenesis is characterized by dissolution, formation of equant calcite spar cements, and neomorphism. The late stage of diagenesis occurred in a deep-burial environment, and is characterized by quartz overgrowths followed by pervasive dolomitization and cementation, dolomite dissolution and, finally, authigenic clay and anhydrite precipitation.

Early diagenesis

Early diagenesis is characterized by bladed (crystal size 20–70 μm × 10–20 μm) and acicular (needle-like crystals up to 30 μm × 3 μm) isopachous cements, and micritic envelopes on skeletal allochems. Only rarely are the micritic rims fractured. Early cements fill primary and, more rarely, mouldic porosity. Crystal forms and terminations suggest an original aragonitic and calcitic mineralogy. Rare meniscus cements occur within some oolitic grainstones (Fig. 5).

Middle diagenesis

Middle-stage diagenesis is characterized by the formation of neomorphic calcite, and of significant secondary (mouldic and fracture) porosity partially filled by equant calcite cements. The equant calcite cement consists of ferroan and non-ferroan precipitates that have crystal sizes up to 0.05 mm and a

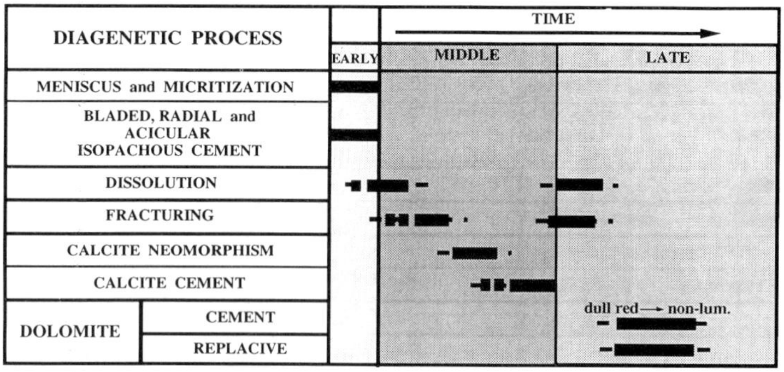

Fig. 4. Paragenetic sequence of the carbonate lithofacies of the Glenwood Formation (after Vandrey, 1991).

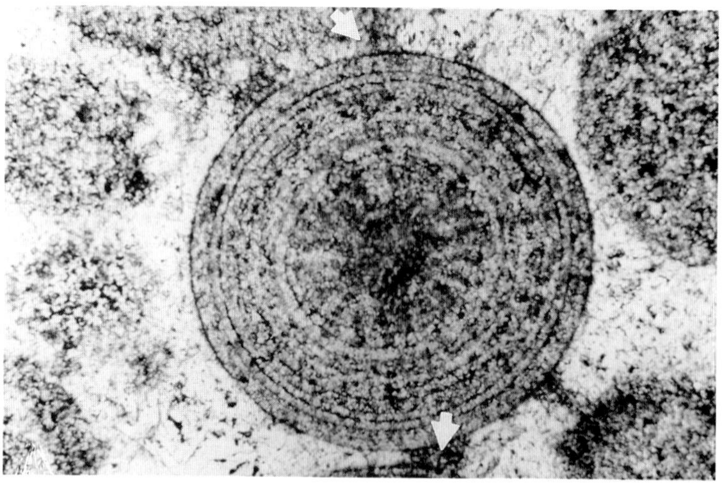

Fig. 5. Plane-light photomicrograph of mimetic fine-crystalline replacive dolomite, and probable meniscus cement (arrow). Field of view is 1.3 mm.

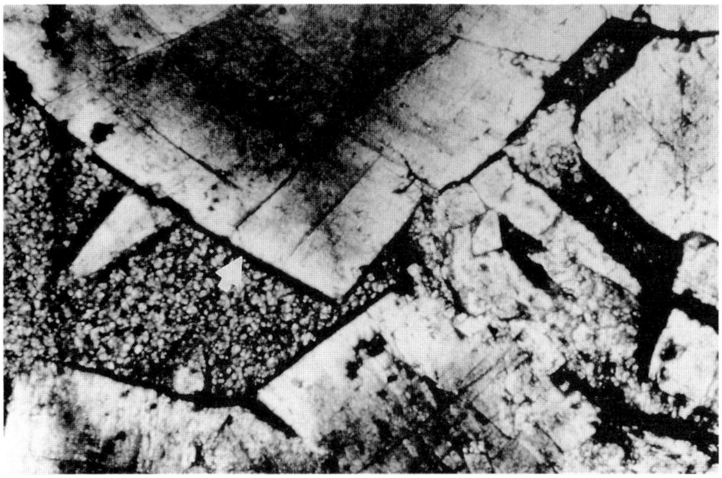

Fig. 6. Coarse-crystalline replacive dolomite displaying typical curved crystal faces of saddle dolomite, inclusion-rich core, clear rim and Fe-rich coating (arrow). It is important to notice that the saddle dolomite is replacing neomorphic calcite. Width is 1.3 mm.

complex cathodoluminescent zonation. Aggrading neomorphism of original micritic sediment is rare, but occurs in two wells as beds 2–0.3 m thick. Neomorphic calcite consists of equant crystals (range 10–100 μm, average 40 μm), that typically stain red and have bright luminescence, but may have patches that stain purple and have a dull-brown luminescence. Neomorphic calcite is situated in the middle diagenetic stage because it is postdated by coarse-crystalline late-diagenetic replacive dolomite (Fig. 6), and there is a close spatial relationship between the neomorphic pseudospar and equant calcite cement. The equant calcite cements were precipitated in fracture and mouldic pores only in beds where the matrix consists of neomorphic pseudospar. These beds typically contain ooids and fossils, and are bounded by thin-bedded, more argillaceous dolomitized sediments.

Diagenetic processes such as mouldic porosity, aggrading neomorphism and ferroan to non-ferroan equant calcite cementation all suggest a meteoric or modified seawater environment during shallow burial for this middle stage (Vandrey, 1991).

Late diagenesis

The volumetrically most important diagenetic phases were acquired during the late diagenetic stage (Vandrey, 1991). Porosity (in the form of vugs and fractures) increased, but was later occluded by pervasive pore-filling dolomites and minor anhydrite and authigenic quartz. The emphasis of this paper is on these late diagenetic dolomites. Late-diagenetic postdolomite cements include clay, rare anhydrite, quartz and pyrite, and are not discussed in this paper. Clays postdating dolomite cements, thus constraining the age of the dolomites, are 346 ± 11 Ma (Barnes *et al.*, 1992a).

Three types of dolomite have been recognized: coarse-crystalline dolomite cement and two replacive dolomite phases, which include coarse- and fine-grained dolomite (Figs 5–7). Replacive dolomite ranges from mimetic (Fig. 5) to non-mimetic. Crystal size (10–2000 μm rhombs) varies according to the original carbonate that was replaced. For example, the non-mimetic dolomites that replaced microbial material are coarser than those in the adjacent micritic matrix. The coarse-crystalline replacive dolomites may have inclusion-rich cores and clear rims, and in some cases are coated by a thin opaque Fe-rich layer (Fig. 6). Coarse-crystalline replacive dolomites (crystal size 0.2–2.0 mm) have curved crystal faces and undulose extinction similar to saddle dolomite cements (Fig. 6), whereas finer-crystalline dolomites (crystal size 0.1–0.06 mm) may appear planar or curved. Like the dolomite cements, the replacive dolomites are light to dark blue when stained. In cathodoluminescence, most of the replacive dolomites are typically non-luminescent, but coarse-crystalline replacive dolomite may be dull-red or zoned dull-red to non-luminescent from core to rim.

Coarse-crystalline dolomite cement greatly reduced or completely filled the available primary and secondary porosity in the Glenwood carbonates. Crystals range in size from 0.2 to 3.5 mm, and com-

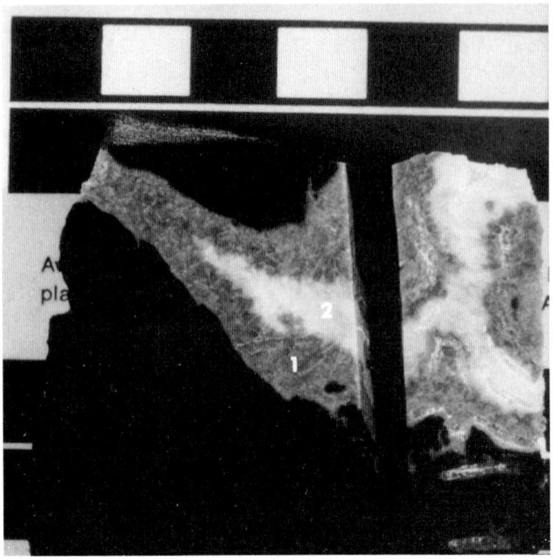

Fig. 7. Fracture-filling dolomite cements. Notice the contrasting colour (light grey, 1, and white, 2) and the drill marks where samples were collected for geochemical analysis. Scale in centimetres.

monly have curved crystal faces typical of saddle dolomite. Rhombs are typically inclusion-rich, but may have clear rims. The dolomites appear white to tan in thin section and hand specimen, dark to light blue with dual staining, and are non-luminescent. Dolomite cements are typically 40–42 mole % $MgCO_3$ and have very high Fe concentrations (2–7 mole % $FeCO_3$, or 7000–41 000 ppm; see Table 3). There may be two generations of dolomite cement in the fractures and vugs (Fig. 7). The boundary between these dolomites is sharp, with no evidence of dissolution.

GEOCHEMISTRY

The carbonate mineral phases of the Glenwood Formation are divided into four groups for comparison: middle-diagenetic calcites (cements and neomorphic), late-diagenetic dolomite cements, and late-diagenetic coarse- and fine-crystalline replacive dolomites. The geochemical properties of the different phases (Table 2 and 3) are utilized to infer the relative timing and possible sources for the fluids that were responsible for late-stage pervasive dolomitization (see below).

Fluid inclusions

Fluid inclusions were analysed from late-diagenetic fracture-filling dolomite cements. Homogenization temperatures (uncorrected for pressure) range from 98 to 189°C (n = 69), with the bulk of the data lying between 120 and 150°C (n = 46; Fig. 8). Halite daughter salts are present in some of the fluid inclusions, suggesting that the dolomitizing fluid was hypersaline. All fluid inclusion data come from the same core and depth interval (Table 2).

The data in Figure 8 show a pronounced peak in the 120–150°C range, with a tail towards higher temperatures. Such a histogram can be interpreted in at least two ways. Accepted at face value, the data might suggest widely varying temperatures during

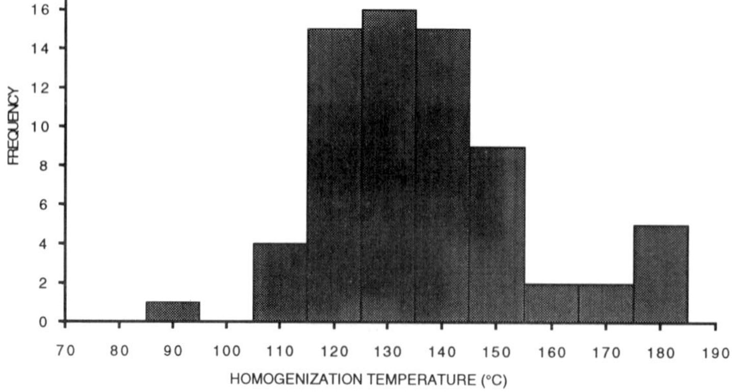

Fig. 8. Fluid-inclusion homogenization temperatures for fracture-filling dolomite cements. Data are not corrected for pressure, which increases temperatures 40–50°C.

Table 2. Interpretation and geochemical data of calcite cements and dolomite types.

Classification	$\delta^{18}O$ (SMOW) average (range)	$\delta^{13}C$ (PDB) average (range)	Measured $^{87}Sr/^{86}Sr$ average (range)	Sr Conc. (ppm) average (range)	Rb Conc. (ppm) average (range)	Fe Conc. (ppm) average (range)	Mn Conc. (ppm) average (range)	Fluid inclusion homogenization temperatures	Interpretation
Calcite Calcite cement	18.4 (17.4–19.8) $n = 3$	−1.9 (−2.2 to −1.3) $n = 3$	0.7087 (0.7084–0.7089) $n = 3$	305.5 (179.7–528.9) $n = 3$	1.48 (0.0–2.93) $n = 3$	5018 (3746–6573) $n = 3$	1002 (582–1317) $n = 3$	—	Shallow burial
Neomorphic calcite	20.5 (17.8–23.6) $n = 6$	−2.6 (−3.5 to −1.5) $n = 6$	0.7092 (0.7091–0.7094) $n = 2$	—	—	2453 (489–4417) $n = 2$	1015 (996–1035) $n = 2$	—	Neomorphism, shallow burial
Dolomite Dolomite cements	18.4 (16.1–22.6) $n = 17$	−2.3 (−4.4 to −1.3) $n = 17$	0.7085 (0.7084–0.7086) $n = 12$	54.1 (31.8–12.2) $n = 13$	0.1 (0.0–0.2) $n = 13$	18 743 (7205–41 386) $n = 13$	1234 (616–1605) $n = 13$	120–190°C	Deep burial, evapoconcentrated marine brine
Replacive Coarse-crystalline	15.4 (14.8–17.6) $n = 6$	−1.6 (−1.0 to −1.7) $n = 6$	0.7092 (0.7090–0.7096) $n = 3$	60.5 (42.6–90.7) $n = 4$	3.0 (0.9–6.5) $n = 4$	28 129 (19 639–38 089) $n = 4$	1459 (1374–1604) $n = 4$	—	Deep burial, evapoconcentrated marine brine
Fine-crystalline	19.6 (17.5–24.0) $n = 23$	−2.0 (−4.5 to −1.1) $n = 23$	0.7106 (0.7084–0.7133; $n = 19$	61.4 (46.7–73.8) $n = 16$	5.8 (1.8–11.9) $n = 16$	11 681 (8319–27 001) $n = 16$	1250 (1047–1562) $n = 16$	—	Deep burial, evapoconcentrated marine brine

Table 3. Isotopic and trace element data for the Glenwood carbonates.

	Depth										
Well	m	ft	Sample number	Measured $^{87}Rb/^{86}Sr$	Measured $^{87}Sr/^{86}Sr$	[Sr] ppm	[Rb] ppm	[Fe] ppm	[Mn] ppm	$\delta^{18}O$ (SMOW)	$\delta^{13}C$ (PDB)
FINE-CRYSTALLINE REPLACIVE DOLOMITE											
State Foster 1–12	3140	10 302.0	SF-18	0.1538	0.709543	53.93	2.87	8352	1333	19.6	−1.7
	3141	10 303.8	SF-7	0.4398	0.711477	67.40	10.24	12 508	1195	19.5	−2.0
	3141	10 304.0	SF-12	0.8307	0.712255	69.32	19.89	10 291	1350	19.4	−1.9
	3141	10 304.0	SF-13	0.2703	0.710770	60.22	5.62	8629	1257	19.4	−1.9
	3142	10 307.1	SF-44	—	—	—	—	—	—	19.7	−1.6
	3142	10 309.1	SF-19	0.4104	0.711714	61.49	8.72	10 878	1239	19.8	−1.4
	3142	10 309.1	SF-20	0.5079	0.712496	58.26	10.22	8142	1048	19.7	−1.5
	3142	10 309.1	SF-21	0.2336	0.710385	46.70	3.77	10 262	1389	19.8	−1.5
	3143	10 311.0	SF-8	0.1657	0.709625	52.19	2.99	9370	1563	19.6	−1.3
	3143	10 311.0	SF-9	0.2538	0.711635	70.67	6.20	8406	1123	19.6	−1.4
	3143	10 311.0	SF-10	0.4672	0.713292	73.81	11.91	9159	1284	19.5	−1.3
	3143	10 312.7	SF-14	0.1499	0.709863	59.45	3.08	8319	1177	19.5	−1.4
	3143	10 312.7	SF-15	0.4241	0.712755	72.60	10.64	9835	1318	19.4	−1.4
	3144	10 314.9	SF-4	0.0696	0.709137	72.79	1.75	10 258	1466	19.4	−1.5
	3144	10 314.9	SF-5	0.3245	0.711242	73.35	8.22	9176	1075	19.2	−1.5
	3146	10 322.5	SF-2	0.0065	0.708430	64.50	0.14	12 144	617	19.9	−3.0
JEM Workman	3256	10 681.3	WM-4	0.0187	0.709447	435.11	2.81	890	1372	—	—
	3258	10 690.1	WM-9	0.2072	0.710619	90.70	6.49	19 640	1442	17.5	−1.1
Joulet Brinks	3270	10 729.7	JB-2	0.1052	0.709460	53.46	1.94	12 794	1123	18.9	−1.9
Hunt Martin	3393	11 132.2	HM-12	—	—	—	—	—	—	18.1	−1.5
	3411	11 191.2	HM-15	—	—	—	—	—	—	19.5	−4.5
	3411	11 191.4	HM-16	—	—	—	—	—	—	24.0	−2.6
	3414	11 200.0	HM-19	—	—	—	—	—	—	19.7	−3.8
Amoco Roscommon	3437	11 277.0	AR-2	0.1207	0.709233	48.73	2.03	23 819	1109	—	—
	3437	11 277.0	AR-3	0.1114	0.709205	57.65	2.22	27 001	1313	19.2	−4.2
COARSE-CRYSTALLINE REPLACIVE DOLOMITE											
Nomeco Gernaat	3158	10 361.5	NG-4	0.0852	0.709102	42.85	1.26	38 089	1374	14.9	−1.6
	3158	10 361.5	NG-8	—	—	—	—	—	—	15	−1.6
	3158	10 361.5	NG-9	—	—	—	—	—	—	14.8	−1.7
	3158	10 361.5	NG-10	—	—	—	—	—	—	14.8	−1.7
JEM Workman	3256	10 682.2	WM-10	0.1500	0.709549	63.96	3.32	34 781	1604	15.4	−1.6
	3258	10 690.1	WM-8	0.0602	0.708967	44.34	0.92	20 008	1419	17.6	−1.0
DOLOMITE CEMENTS											
State Foster 1–12	3140	10 302.0	SF-16	0.0037	0.708426	31.80	0.04	41 386	770	16.1	−2.9
	3140	10 302.0	SF-17	0.0100	0.708544	38.44	0.13	7205	1149	19.6	−1.6
	3141	10 303.8	SF-6	0.0097	0.708506	31.81	0.11	12 508	1376	17.7	−3.0
	3141	10 304.0	SF-11	0.0161	0.708557	34.55	0.19	9700	1333	18.7	−2.3
	3142	10 307.0	SF-45	—	—	—	—	—	—	18.0	−2.4
JEM Workman	3256	10 681.3	WM-2	0.0034	0.708546	73.34	0.09	20 183	1605	17.7	−1.7
	3256	10 681.3	WM-3	0.0034	0.708596	122.94	0.14	21 758	1425	17.6	−1.6
Joulet Brinks	3270	10 729.7	JB-3	0.0114	0.708513	40.50	0.16	15 864	1223	18.5	−1.4
	3270	10 729.7	JB-4	0.0116	0.708432	39.50	0.16	14 389	1197	19.8	−1.3
	3270	10 729.7	JB-5	0.0107	0.708519	43.72	0.16	16 777	1283	18.9	−2.2
	3270	10 729.7	JB-6	0.0118	0.708360	38.96	0.16	9896	1149	18.5	−1.6
Hunt Martin	3393	11 132.4	HM-13	—	—	—	—	—	—	18.0	−2.0
	3393	11 132.6	HM-14	—	—	—	—	—	—	18.1	−1.5
	3411	11 191.6	HM-17	—	—	—	—	—	—	22.6	−2.8
	3411	11 191.8	HM-18	—	—	—	—	—	—	18.2	−4.4
Amoco Roscommon	3437	11 277.0	AR-4	0.0109	0.708370	46.82	0.18	36 469	1546	17.9	−3.1
	3437	11 277.0	AR-5	0.0024	0.708505	95.78	0.08	25 382	1378	17.5	−3.8

Continued

Table 3. *Continued*

Well	Depth m	Depth ft	Sample number	Measured $^{87}Rb/^{86}Sr$	Measured $^{87}Sr/^{86}Sr$	[Sr] ppm	[Rb] ppm	[Fe] ppm	[Mn] ppm	$\delta^{18}O$ (SMOW)	$\delta^{13}C$ (PDB)
CALCITE CEMENTS											
JEM Workman	3256	10681.3	WM-1	0.0033	0.708410	179.68	0.20	4736	1109	19.8	−2.2
	3256	10681.7	WM-7	0.0181	0.708856	207.78	1.30	3746	582	17.4	−2.1
	3256	10681.7	WM-11	0.0160	0.708866	528.92	2.93	6573	1317	18.0	−1.3
NEOMORPHIC CALCITE											
Nomeco Gernaat	3158	10361.1	NG-5	—	—	—	—	—	—	21.5	−3.0
	3158	10361.2	NG-6	—	—	—	—	—	—	23.6	−3.0
	3158	10361.3	NG-7	—	—	—	—	—	—	18.6	−3.5
	3158	10361.7	NG-11	—	—	—	—	—	—	17.8	−2.9
JEM Workman	3256	10681.7	WM-5	—	0.709056	—	—	4417	996	20.2	−1.9
	3256	10681.7	WM-6	—	0.709362	—	—	489	1035	21.4	−1.5

dolomite formation. Variable temperatures are consistent with the strontium and oxygen data (see below). Alternatively, the higher temperature data may have resulted from posttrapping modification of inclusions originally trapped at temperatures near the peak of the histogram (130° ± 15°C). Changes such as necking down, leakage and stretching tend to produce inclusions with lower densities than the originals, and thus higher homogenization temperatures. For either interpretation of the data, a 0.5 kilobar pressure correction would probably be of the order of 40–50°C (Potter, 1977). Consideration of the effect of a pressure correction on the data lends support to the second interpretation: it is not geologically reasonable to suppose temperatures of the order of 240°C, given the burial history of the Michigan Basin. Thus, it is concluded that 175 ± 25°C is the best estimate for temperature of precipitation of the fracture-filling dolomite. These relatively high homogenization temperatures are consistent with dolomitization in a deep-burial setting, but are higher than the temperatures reached at maximum burial depth (Cercone, 1984) with a geothermal gradient of 30°C. These higher temperatures may reflect a heating event.

Stable isotopes

Fifty-five measurements of $\delta^{13}C$ and $\delta^{18}O$ values are shown in Figure 9 and Tables 2 and 3. The $\delta^{18}O$–$\delta^{13}C$ values of the neomorphic calcite (n = 6, $\delta^{18}O$ = 17.8–23.6‰, and $\delta^{13}C$ = −3.5 to −1.5‰) and calcite cements (n = 3, $\delta^{18}O$ = 17.4–19.8‰, and $\delta^{13}C$ =

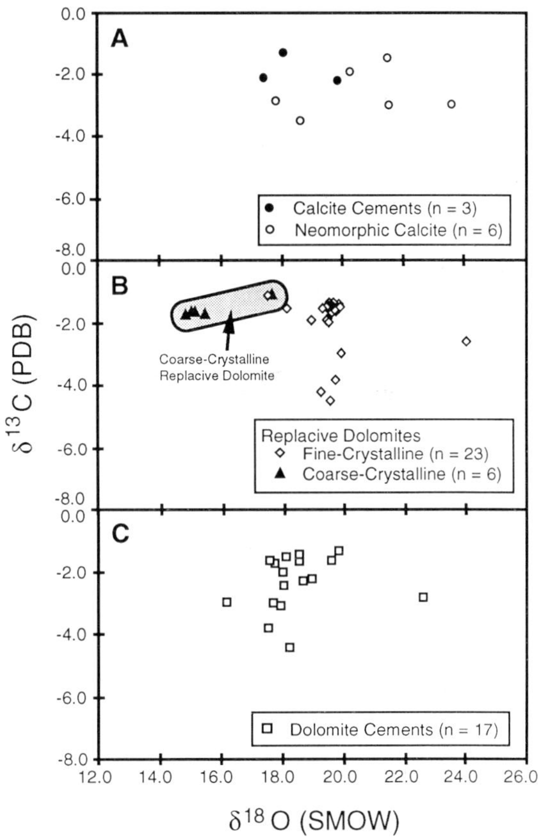

Fig. 9. $\delta^{18}O$–$\delta^{13}C$ composition for (A) calcites, (B) replacive dolomites (shaded field = coarse-crystalline, unshaded field = fine-crystalline), and (C) dolomite cements. Data shown in Table 3.

-2.2 to -1.3‰) show a trend with a large variation in $\delta^{18}O$ and a small variation in $\delta^{13}C$ (Fig. 9A, and Tables 2 and 3). This trend can be explained by variations in temperature and a fluid with uniform $\delta^{18}O$, or by variable fluid–rock interaction.

The $\delta^{18}O$ and $\delta^{13}C$ values of replacive dolomites generally cluster in two distinct groups (Fig. 9B), fine-crystalline and coarse-crystalline (shaded field). Most of the $\delta^{18}O$ values of the fine-crystalline replacive dolomites (n = 21) have a small range of $\delta^{18}O$ values (18.9–19.9‰, mean 19.4 ± 0.4‰) relative to the coarse-crystalline dolomites (14.8–17.6‰, mean 15.4 ± 1.2‰). However, coarse-crystalline dolomites have a small variation in $\delta^{13}C$ values (-1.0 to -1.7‰) compared to fine-grained dolomites (-4.5 to -1.1‰). Dolomite cements have C and O isotope compositions that overlap those of the replacive dolomites (Fig. 9C).

The range of $\delta^{18}O$ values in Glenwood dolomites can be explained by precipitation from a common fluid at variable temperatures, or variable fluids due to mixing or partial water–rock exchange. Most $\delta^{18}O$ values (n = 15 out of 17) of the dolomite cement cluster around 17.5 and 19.6‰ (Table 3), with a range of 2.1‰ (Fig. 9C), and an interpreted constant temperature of 175 ± 25°C (see above), which is consistent with a constant water composition of +6‰ ± 1‰ (Matthews & Katz, 1977) and constant high fluid : rock ratio. This water isotopic composition suggests a high $\delta^{18}O$ evaporitic Palaeozoic seawater (Sheppard, 1986). Comparable changes in $\delta^{18}O$ occur in the fine-crystalline replacive dolomite, which shows a 6.5‰ range in $\delta^{18}O$ values, but 87% of the data cluster around 19.4 ± 0.4‰ (Fig. 9B, Table 3), consistent with a constant water composition and high fluid–rock interaction. A minor portion of the dolomites, two dolomite cement outliers (Fig. 9C) and three fine-grained replacive dolomites are interpreted to have precipitated at the same temperature, but from a water with variable $\delta^{18}O$ or variable amounts of exchange with fluids. The coarse-crystalline replacive dolomite may have precipitated at higher temperatures, or from a water with different initial $\delta^{18}O$.

Sr isotopes

Rb and Sr contents and $^{87}Sr/^{86}Sr$ ratios were measured for 40 carbonate samples (calcite and dolomite; Tables 2 and 3). Two distinctive fields can be defined based on measured $^{87}Sr/^{86}Sr$–$^{87}Rb/^{86}Sr$ variations: neomorphic calcite, calcite and dolomite cements, which have very low $^{87}Rb/^{86}Sr$ ratios, and replacive dolomite, which generally has very high $^{87}Rb/^{86}Sr$ and $^{87}Sr/^{86}Sr$ ratios (Fig. 10). The very low Rb contents and $^{87}Rb/^{86}Sr$ ratios (<0.2 ppm and <0.05, respectively) for the calcites and dolomite cements indicate that these samples are free of silicate contamination, and probably directly reflect the Sr isotope composition of the fluids from which they precipitated. All but three of the 17 calcite and dolomite cements analysed have measured $^{87}Sr/^{86}Sr$ ratios between 0.7084 and 0.7086, and the very low $^{87}Rb/^{86}Sr$ ratios indicate that the measured Sr isotope ratios may be taken as 'initial' Sr isotope compositions in the Middle Ordovician (Fig. 10); these ratios are identical to those of Middle Ordovician and Silurian seawater (e.g. Burke *et al.*, 1982).

$^{87}Sr/^{86}Sr$–$^{87}Rb/^{86}sr$ variations for all but two of the replacive dolomites indicate significant contamination with a high $^{87}Sr/^{86}Sr$ and high $^{87}Rb/^{86}Sr$ component (Fig. 10). This contaminant is most probably small amounts of clay in the sampled carbonate, given the high Rb contents measured, up to 12 ppm, and high $^{87}Rb/^{86}Sr$ ratios, up to 0.50 (Table 3). The large range in measured $^{87}Sr/^{86}Sr$ ratios for the replacement dolomites does not reflect the Sr isotope ratios of the 'initial' carbonate material, and highlights the importance of measuring Rb and Sr contents in Sr isotope studies of carbonates, as noted by earlier workers (e.g. Banner *et al.*, 1988a). The high $^{87}Rb/^{86}Sr$ ratios of the replacive dolomites are similar to those determined by Banner *et al.* (1988a) for dolomites in the Mississippian Burlington-Keokuk Formation in the Illinois basin. Through a series of leaching tests, Banner *et al.* (1988a) determined that a minor (<5%) submicroscopic clay component was present in the dolomites. Similarly, Kralik (1984) determined that dissolution of 'nearly pure' limestone in 1 M HCl leached radiogenic ^{87}Sr from silicate components, which constituted a few percent of the samples.

Dolomites analysed in this study were examined for clay contamination prior to isotopic analysis by examination using the petrographic microscope and X-ray diffraction (XRD). No clay was revealed at the detection limits of XRD (generally 2–5%). Dissolution of dolomites was achieved using 6 M HCl, and residues measured after dissolution for all samples were negligible, within errors of such determinations. We have since tried leaching experiments with weaker HCl and HOAc solutions, and have found that clays can release both Rb and radiogenic

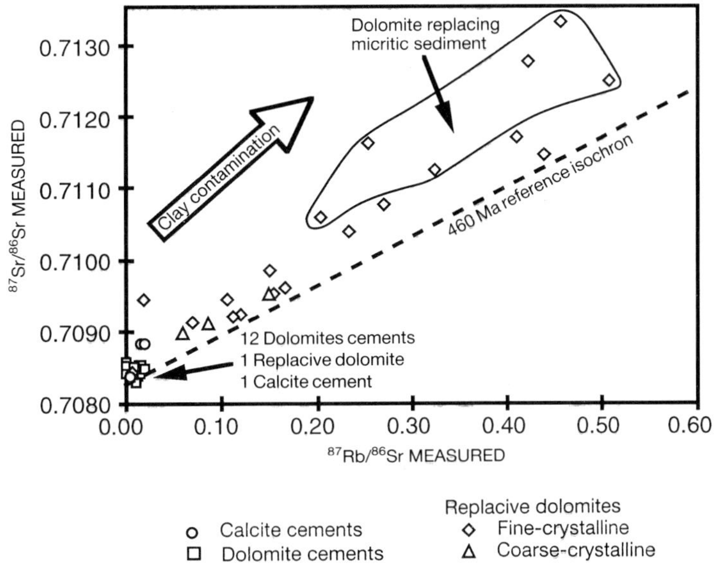

Fig. 10. Measured $^{87}Sr/^{86}Sr$–measured $^{87}Rb/^{86}Sr$ ratios for Glenwood Formation carbonates. The trend of the replacive dolomite samples reflects contamination with clay, as highlighted by the high $^{87}Rb/^{86}Sr$ ratios, although some have similar ratios to the 'initial' Sr isotope composition.

Sr in as weak a solution as 10% HOAc. The positive correlation of measured $^{87}Sr/^{86}Sr$ and $^{87}Rb/^{86}Sr$ ratios for the replacive dolomites (Fig. 10) indicates that both Rb and radiogenic Sr (^{87}Sr) were leached from clays during sample preparation. If only Rb were leached, the data for the replacive dolomites would lie on a horizontal trend in Figure 10. The range of $^{87}Sr/^{86}Sr = 0.7090–0.7133$ in the high-Rb ($^{87}Rb/^{86}Sr > 0.05$) replacive dolomites can be explained by contamination of the sample by 0.2–1.7% detrital Proterozoic clay, or by 0.8–6.5% authigenic Middle Ordovician clay. This level of clay contamination will have no significant effect on the Sr concentration measurements, and stable isotope ratios will not be affected because there is no appreciable carbon in clay, and silicate-bound oxygen does not exchange with phosphoric acid at 50°C over 12 h (the conditions of dissolution).

Projection of the $^{87}Sr/^{86}Sr$–$^{87}Rb/^{86}Sr$ variations of the replacive dolomites to the low $^{87}Rb/^{86}Sr$ ratios that should be appropriated for pure carbonate suggests that the Sr isotope composition that is intrinsic to the carbonate in the replacive dolomite is similar to that of the calcites and dolomite cements ($^{87}Sr/^{86}Sr = 0.7084–0.7086$; Fig. 10). One low $^{87}Rb/^{86}Sr$ sample has an anomalously high $^{87}Sr/^{86}Sr$ ratio of 0.7095 (WH-4, Table 3) which cannot be explained by clay contamination; this sample, in addition to high $^{87}Sr/^{86}Sr$ ratios for two calcite cements (Fig. 10) indicates some Sr isotopic heterogeneity in the diagenetic fluids, but this is conclusively demonstrated by only these three samples. It is not appropriate to project the high $^{87}Rb/^{86}Sr$ replacive dolomite samples back to Middle Ordovician 'initial' $^{87}Sr/^{86}Sr$ ratios, which, if done, might be interpreted as indicating a wide range in $^{87}Sr/^{86}Sr$ ratios of the carbonate material. Thus it is important to report measured and not initial ratios (Figs 10–12). Replacive dolomites with the highest $^{87}Sr/^{86}Sr$ replaced micritic sediment (Fig. 10), which would probably have the greatest clay content. Measured $^{87}Sr/^{86}Sr$ ratios of the replacement dolomites lie above a 460 Ma reference isochron that is fixed by the low $^{87}Rb/^{86}Sr$ samples (Fig. 10), and suggests that they are contaminated by an ancient high-^{87}Sr component such as detrital clay. The data are not consistent with an entirely authigenic clay contaminant that had a middle Ordovician initial $^{87}Sr/^{86}Sr$ ratio that was the same as that of the dolomite cements.

Sr concentration – Sr isotope variations

The Glenwood Formation dolomites generally have low Sr contents (typically <100 ppm; Table 3). The majority of replacive dolomites have Sr contents that overlap those of the dolomite cements (Fig. 11), supporting the extrapolation of Sr isotope compositions of the replacement dolomites to low Rb/Sr (clay-free) compositions. We interpret the $^{87}Sr/^{86}Sr$–Sr content variations for the replacive dolo-

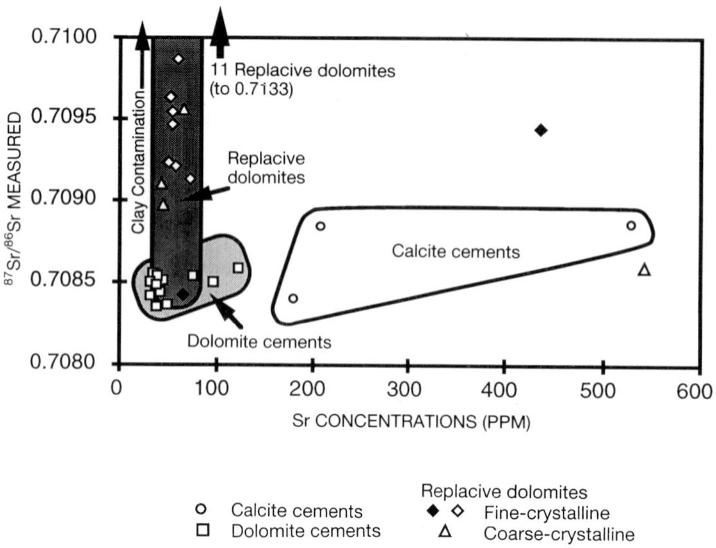

Fig. 11. Sr concentration–$^{87}Sr/^{86}Sr$ ratio variations for the Glenwood carbonates. Solid diamonds are fine-crystalline replacive dolomites with a measured $^{87}Rb/^{86}Sr$ value below 0.05. Data shown in Table 3.

mites as reflecting carbonate precipitation from the same fluid that produced the dolomite cements ($^{87}Sr/^{86}Sr$ = 0.7084–0.7086), followed by incorporation of clay that produced high present-day $^{87}Sr/^{86}Sr$ ratios in the analysed samples. One low-Rb (clay-free) replacive dolomite has an anomalously high Sr content and $^{87}Sr/^{86}Sr$ ratio (Fig. 11), which may suggest interaction with a radiogenic, high-Sr fluid that was a minor component in the diagenetic history of the dolomites. Calcite cements have $^{87}Sr/^{86}Sr$ ratios that overlap those of the low majority of low-Rb dolomites, although the calcites have relatively high Sr contents (Fig. 11; Table 3), consistent with observations in other calcite–dolomite sequences.

Low Sr concentrations in carbonates have been attributed to meteoric–marine mixing prior to carbonate precipitation (McNutt *et al.*, 1987; Banner *et al.*, 1988b), modification during diagenesis (Kinsman, 1969; Morrow & Mayers, 1978; Land, 1985), and variations in the distribution coefficient for the incorporation of Sr into dolomite ($D_{Sr}^{dolomite}$) (Veizer, 1978; Dawans & Swart, 1988; Vahrenkamp & Swart, 1990). In the Glenwood dolomite cements, the presence of halite daughter salts and the high temperatures of homogenization in fluid inclusions might indicate that deep hypersaline brines rather than shallow marine–meteoric mixing waters are responsible for dolomitization. Modification of the trace element composition of dolomite during burial requires flushing large amounts of fluid through the sediments (Morrow & Mayers, 1978), which would probably result in changing $^{87}Sr/^{86}Sr$ ratios in the fluid and precipitated dolomites. The variations in the Sr concentrations of the Glenwood dolomites can be explained by expected variations in $D_{Sr}^{dolomite}$ relative to mole % $MgCO_3$ variations measured by electron microprobe (e.g. Banner *et al.*, 1988b; Vahrenkamp & Swart, 1990; Vandrey, 1991). The low Sr contents do not require precipitation from a dilute fluid, and the salt contents of the fluid inclusions and relatively high $\delta^{18}O$ values indicate instead a saline water as the fluid source.

Sr–O isotope variations

Variations of $\delta^{18}O$ and $^{87}Sr/^{86}Sr$ in the Glenwood dolomites highlight three distinct groups that are also distinguished by petrographic relations: fine-crystalline replacive dolomites, coarse-crystalline replacive dolomites, and fracture-filling coarse-crystalline dolomite cements (Fig. 12).

The fine-crystalline replacive dolomites in Figure 12 lie in a field that is constrained by a narrow range of $\delta^{18}O$ values (18.9–19.9‰) and a broad range of $^{87}Sr/^{86}Sr$ ratios. The nearly vertical trend of these replacive dolomites may be specific to one well, because 13 of the 15 samples in the trend are from one core, and nine of these samples occur within a 2.5 m interval. The relatively restricted range of $\delta^{18}O$ values for most of the fine-crystalline replacive dolomites suggests that they precipitated from a

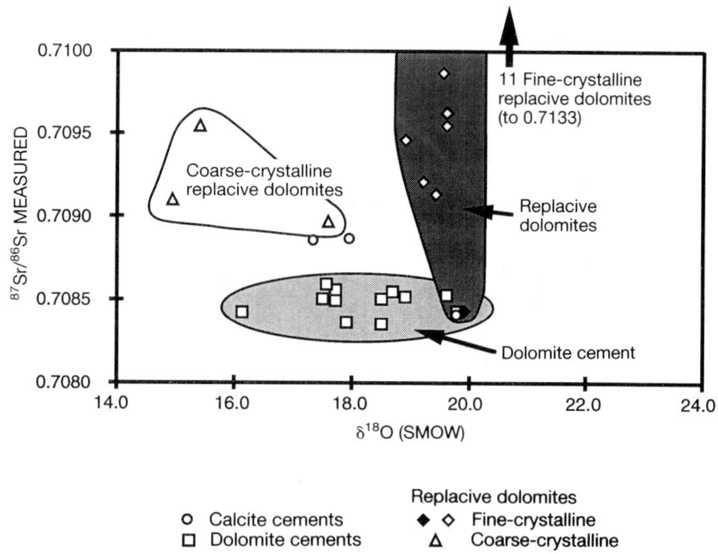

Fig. 12. $\delta^{18}O$–$^{87}Sr/^{86}Sr$ ratio variations of the Glenwood dolomites. Solid diamond is a fine-crystalline replacive dolomite with a measured $^{87}Rb/^{86}Sr$ value below 0.05. Data shown in Table 3.

common fluid, consistent with the interpretation of $^{87}Sr/^{86}Sr$–Sr content variations discussed above.

The coarse-crystalline replacive dolomites have lower $\delta^{18}O$ values (14.8–17.6‰) than the fine-crystalline replacive dolomite, although they overlap in Sr concentrations and $^{87}Sr/^{86}Sr$ ratios (Figs 11 and 12). The $\delta^{18}O$ values of these replacive dolomites are also low relative to the range in the dolomite cements (Fig. 12).

In contrast to the Sr–O isotope variations in fine- and coarse-crystalline replacive dolomite, dolomite cements have uniform $^{87}Sr/^{86}Sr$ ratios (0.7084–0.7086) over a broader range of $\delta^{18}O$ values (Fig. 12; Table 3). All cements have low Rb contents (<0.2 ppm), indicating that they do not contain clay, and this is consistent with the relatively uniform Sr isotope compositions.

SOURCES OF DOLOMITIZING BRINE

Petrographic and geochemical study of the Glenwood dolomites suggests that one fluid of constant Sr concentration and $^{87}Sr/^{86}Sr$ ratio and slightly variable C and O isotopic composition effected the replacement of the bulk of the initial mineralogy and cemented secondary pores with dolomite during deep burial. The initial $\delta^{18}O$ value of the water can best be estimated as +6‰ ± 1‰, using the fluid inclusion homogenization temperatures (175° ± 25°C) and $\delta^{18}O$ (range 17.5–19.6‰) data from the dolomite cement of the same sample. The dolomite–water fractionation curve for oxygen of Matthews and Katz (1977) was used. The high $\delta^{18}O$ values of the initial water and the halite daughter salts within these dolomite cements indicate a saline brine associated with dolomite cementation. The $^{87}Sr/^{86}Sr$ ratios of the majority of carbonate phases in the Glenwood Formation overlap that of seawater during the Middle Ordovician, Upper Silurian and Middle Devonian, indicating the potential for an 'original' marine source. The low and generally homogeneous $^{87}Sr/^{86}Sr$ ratios of the Glenwood carbonates, as measured for the cements, preclude significant interaction with detrital sediments or radiogenic clays. In contrast, modern brines from Ordovician to Devonian strata in the Michigan Basin have highly variable Sr isotope ratios ($^{87}Sr/^{86}Sr$ = 0.7077–0.7103; Wilson, 1989), reflecting interaction with diverse lithologies, including Rb-rich ancient (Precambrian) alumino-silicates, clays and K-feldspar.

Comparison of diagenesis in the Glenwood with that in the underlying St Peter Sandstone places limits on the source of the fluid. The paragenetic sequence of the Middle Ordovician sandstone (Vandrey, 1991; Drzewiecki, 1992) indicates that, prior to dolomitization, a meteoric/marine fluid precipitated silica cements on quartz grains (Winter *et al.*, 1992), in a shallower environment. This meteoric/marine fluid may be similar to the hypothesized fluid responsible for the middle-diagenetic stage. Considering this scenario, Ordovician 'seawater' composition was undoubtedly modified in the Glenwood Formation. It is reasonable to argue that the dolomitizing fluids originated from, or interacted with, Silurian and/or Devonian strata. Silurian and Devonian anhydrites contain fluid inclusions with $^{87}Sr/^{86}Sr$ ratios that are similar (0.7086; Das *et al.*, 1990) to the range observed for the Glenwood dolomite cements. A Silurian/Devonian source, however, requires a downward fluid movement along faults, that can be density- (hypersaline fluids) or pressure-driven. Circulation through faults is required because the Upper Ordovician dolomites have different stable isotope ($\delta^{18}O$ = 17–26.2‰, and $\delta^{13}C = -2.8$ to -1.2‰) compositions (Taylor & Sibley, 1988; Drzewiecki, 1992), indicative of a different dolomitizing fluid. A Silurian marine source is preferred over a Middle Devonian source, because the Devonian seawater would have had to circulate through tight Silurian salts. An alternative scenario to Silurian 'seawater' is that Middle Ordovician 'seawater' remained in the sediments below and migrated into the Glenwood during dolomitization.

Most of the replacive dolomites, although they have a range of replacive textures and radiogenic isotope ratios (see discussion above), have stable isotope compositions that are similar to those of dolomite cements, and thus are interpreted to have precipitated from the same fluid (Figs 9–12). However, the coarse-crystalline replacive dolomites have different $^{87}Sr/^{86}Sr$, $^{87}Rb/^{86}Sr$ and $\delta^{18}O$ ratios (Figs 10–12; Table 3) indicating the influence of a second fluid.

SUMMARY

Diagenesis of the Middle Ordovician shallow-marine Glenwood Formation in the Michigan Basin can be divided into early, middle and late stages. Each stage is characterized by a unique combination of porosity development, cementation and recrystallization. Early diagenesis occurred in a near-surface setting, where primary porosity was partially filled with marine and vadose cements. Middle-stage diagenesis is recognized by calcite cementation (both non-ferroan and ferroan) filling primary intergranular and secondary mouldic pores. In addition, these cements have a close spatial relationship with neomorphic calcite pseudospar. Late-stage diagenesis is characterized by pervasive dolomitization of precursor carbonate sediments and dolomite cementation filling fracture, intergranular and shelter porosity.

Two-phase fluid inclusions from dolomite cements homogenize between 98° and 189°, with the mode of the data around 135 ± 15°C. Pressure correction increases the temperature 40–50°C, thus 175 ± 25°C is the inferred temperature of precipitation of the dolomite cement, supporting a deep-burial setting.

Stable isotope analyses of the dolomite phases indicate up to a 9‰ range of $\delta^{18}O$ values, but most of the data have a much narrower range of 2.0‰. This tight clustering of values suggests constant temperature and water composition. The outlying $\delta^{18}O$ values can be explained by variations in isotopic ratios of water, temperature or fluid–rock interaction. This is probably the case for the coarse-crystalline replacive saddle dolomite. The best estimate for the constant $\delta^{18}O$ value of the water is +6‰ ± 1, which suggests an evaporitic seawater source.

Replacive dolomites and dolomite cements have low Sr concentrations (typically <100 ppm), and the majority of dolomites are inferred to have precipitated from a fluid that had $^{87}Sr/^{86}Sr$ = 0.7084–0.7086. This overlaps the Sr isotope composition of Middle Ordovician, Silurian and Middle Devonian seawater. The most likely scenarios involve either a Middle Ordovician or a Silurian seawater source. A Middle Devonian source is improbable because the fluids would have had to move through low-permeability Silurian evaporites. The Middle Ordovician scenario requires that sufficient seawater was stored in the underlying Ordovician strata, whereas the Silurian scenario requires downward movement along faults. High and variable $^{87}Sr/^{86}Sr$ ratios of the replacive dolomites relative to the dolomite cements are attributed to incorporation of variable amounts (<6%) of detrital clay into the dolomites during dolomitization of the precursor sediments, and are not interpreted to reflect fluid interaction with radiogenic sediments.

The geochemical characteristics of the Glenwood dolomites highlight the importance of circulation at depth of isotopically uniform modified seawater. It also emphasizes that 'seawater' in the subsurface can maintain a relatively uniform isotopic composition, despite the fact that dolomitization occurred at high temperatures and well after deposition. Dolomitization and dolomite cementation by the proposed mechanism are an effective means to form hydrocarbon reservoir seals at depth and maintain overpressures.

ACKNOWLEDGEMENTS

Acknowledgement is made to the Gas Research Institute, grant 5089–260–1810, for support of this research. Jennifer Cappel is thanked for keeping the radiogenic isotope chemistry laboratory running smoothly. Valuable assistance with stable isotope analysis was provided by Kevin Baker, Mike Spicuzza and Bryce Winter. This paper benefited from the reviews of R.H. Dott Jr, T. Elliot, B. Purser, G. Nadon, M. Tucker, B. Winter, and D. Zenger.

REFERENCES

Bahr, J.M. (1989) Evaluation of pressure distribution in the St Peter Sandstone, Michigan basin. *EOS, Trans. Am. Geophys. Union* **70**(43), 1097 (abs.).

Banner, J.L., Hanson, G.N. & Meyers, J.W. (1988a) Determination of initial Sr isotopic compositions of dolostones from the Burlington-Keokuk Formation (Mississippian): constraints from cathodoluminescence, glauconite paragenesis and analytical methods. *J. Sedim. Petrol.* **58**, 673–687.

Banner, J.L., Hanson, G.N. & Meyers, W.J. (1988b) Water–rock interaction history of regionally extensive dolomites of the Burlington-Keokuk Formation (Mississippian): isotopic evidence. In: *Sedimentology and Geochemistry of Dolostones* (Ed. Shukla, V. & Baker, P.A.) Spec. Publs Soc. Econ. Paleont. Mineral. **43**, 97–113.

Barnes, D.A., Girard, J.P. & Aronson, J.L. (1992a) *K/Ar Dating of Illite Diagenesis in the Middle Ordovician St Peter Sandstone, Central Michigan Basin, USA: Implications for Thermal History*. Spec. Publs Soc. Econ. Paleont. Mineral. **47**, 35–48.

Barnes, D.A., Lundgren, C.E. & Longman, M. (1992b) Sedimentology and diagenesis of the St Peter Sandstone, Central Michigan basin, USA. *Am. Assoc. Petrol. Geol.* **76**, 1507–1532.

Barnes, D.A., Harrison, W.B. III, Lundgren, C.E. & Wieczorek, L.M. (1988) *The Lower Paleozoic of the Michigan Basin, a Core Workshop*. Western Michigan University Core Research Laboratory, Kalamazoo, 65 pp.

Brady, R.B. & Dehaas, R. (1988) The 'deep' (pre-Glenwood) formations of the Michigan basin: Part 1 – The Goodwell Unit. *Michigan's Oil and Gas News* **26**, 32–38.

Bricker, D.M., Milstein, R.L. & Reszka, C.R. Jr. (1983) *Selected Studies of Cambro-Ordovician Sediments Within the Michigan Basin*. Michigan Geol. Surv. Rep. Invest. **26**, 54 pp.

Budai, J.M. & Wilson, J.L. (1991) Diagenetic history of the Trenton and Black River Formations in the Michigan basin. In: *Early Sedimentary Evolution of the Michigan Basin* (Ed. Catacosinos, P.A. & Daniels, P.A. Jr.) Geol. Soc. Am. Spec. Paper 256, 73–88.

Burke, W.R., Denison, R.E., Hethenrington, E. A. Koepnick, R.B., Nelson, H.F. & Otto, J.B. (1982) Variations in seawater $^{87}Sr/^{86}Sr$ throughout Phanerozoic time. *Geology* **10**, 516–519.

Catacosinos, P.A. & Daniels, P.A. Jr. (1991) Stratigraphy of Midle Proterozoic to Midddle Ordovician formations of the Michigan basin. In: *Early Sedimentary Evolution of the Michigan Basin* (Ed. Catacosinos, P.A. & Daniels, P.A. Jr.) Geol. Soc. Am. Spec. Paper **256**, 53–71.

Cercone, K.R. (1984) Thermal history of Michigan basin. *Am. Assoc. Petrol. Geol. Bull.* **68**, 130–136.

Das, N., Horita, J. & Holland, H.D. (1990) Chemistry of fluid inclusions in halite from the Salina Group of the Michigan basin: implications for late Silurian seawater and the origin of sedimentary brines. *Geochim. Cosmochim. Acta* **54**, 319–327.

Dawans, J.M. & Swart, P.K. (1988) Textural and geochemical alteration in late Cenozoic Bahamian dolomites. *Sedimentology* **35**, 385–403.

Drzewiecki, P.A. (1992) *Sedimentology, Diagenesis and Geochemistry of the Middle Ordovician St Peter Sandstone of the Michigan basin*. Unpublished MS Thesis, University of Wisconsin, Madison, Wisconsin, 215 pp.

Fisher, J.H. & Barratt, M.W. (1985) Exploration in Ordovician of Central Michigan basin. *Am. Assoc. Petrol. Geol. Bull.* **69**, 2065–2076.

Fisher, J.H., Barratt, M.W. Droste, J.B. & Shaver, R.H. (1988) Michigan basin. In: *Sedimentary Cover – North American Craton, US* (Ed. Sloss, L.L.) Geol. Soc. Am., The Geology of North America **2**, 361–382.

Friedman, I. & O'Neil, J.R. (1977) Compilation of stable isotope fractionation factors of geochemical interest. In: *Data of Geochemistry*, 6th edn (Ed. Fleicher, M.). US Geol. Surv. Professional Paper **440-KK**, 12 pp.

Harrison, W.B. III (1987) Michigan's deep St Peter gas play continues to expand. *World Oil* **204**, 56–61.

Hinze, W.J., Kellogg, R.L. & O'hara, N.W. (1975) Geophysical studies of basement geology of Southern Peninsula of Michigan. *Am. Assoc. Petrol. Geol. Bull.* **59**, 1562–1584.

Johnson, C.M. & Thompson, R.A. (1991) Isotopic composition of Oligocene mafic volcanic rocks in the northern Rio Grande rift: evidence for contributions of ancient intraplate and subduction magmatism to evolution of the lithosphere: *J. Geophys. Res.* **96**, 13593–13608.

Kinsman, D.J.J. (1969) Interpretation of Sr^{2+} concentrations in carbonate minerals and rocks. *J. Sedim. Petrol.* **39**, 486–508.

Kralik, M. (1984) Effects of cation-exchange treatment and acid leaching on Rb-Sr system of illite from Fithian, Illinois. *Geochim. Cosmochim. Acta* **48**, 527–533.

Land, L.S. (1985) The origin of massive dolomite. *J. Geol. Educ.* **33**, 112–125.

McCrea, J.M. (1950) On the isotopic chemistry of carbonates and a paleotemperature scale. *J. Chem. Phys.* **18**, 849–857.

McNutt, R.H., Frape S.K. & Dollar, P. (1987) A strontium, oxygen, and hydrogen isotopic composition of brines, Michigan and Appalachian basins. *App. Geochem.* **2**, 495–505.

Matthews, A. & Katz, A. (1977) Oxygen isotope fractionation during dolomitization of calcium carbonate. *Geochim. Cosmochim. Acta* **41**, 1431–1438.

Moline, G.R., Bahr, J.M., Drzewiecki, P.A. & Shepherd, L.S. (1992) Identification and characterization of pressure seals through the use of wireline logs: a multivariate statistical approach. *Log Analyst* **33**, 362–372.

Morrow, D.W. & Mayers, I.R. (1978) Simulation of limestone diagenesis – a model based on strontium depletion. *Can. J. Earth Sci.* **15**, 376–396.

Nunn, J.A. & Sleep, N.H. (1984) Thermal contraction and flexure of intracrational basins: a three-dimensional study of the Michigan basin. *Geophys. J. Roy. Astronom. Soc.* **76**, 587–635.

Ostrom, M.E. (1969) Champlain Series (Middle Ordovician) in Wisconsin. *Am. Assoc. Petrol. Geol. Bull.* **53**, 672–678.

Potter, R.W. III (1977) Pressure corrections for fluid-inclusion homogenization temperatures based on the volumetric properties of the system NaCL-H_2O. *J. Res. US. Geol. Surv.* **5**(5), 603–607.

Sharma, T. & Clayton, R.N. (1965) Measurements of $^{18}O/^{16}O$ ratios of total oxygen in carbonates. *Geochim. Cosmochim. Acta* **36**, 129–140.

Sheppard, S.M.F. (1986) Characterization and isotopic variations in natural waters. In: *Stable Isotopes in High Temperature Geochemical Processes* (Ed. Valley, J.W., Taylor, H.P. Jr. & O'Neil, J.R.) Mineral. Soc. Am., Rev. Mineral. 16, 165–183.

Taylor. T.R. & Sibley, D.F. (1986) Petrographic and geochemical characteristics of dolomite types and the origin of ferroan dolomite in the Trenton Formation, Ordovician, Michigan basin, USA. *Sedimentology* **33**, 61–86.

Vahrenkamp. V.C. & Swart, P.K. (1990) New distribution coefficient for the incorporation of strontium into dolomite and its implications for the formation of ancient dolomites. *Geology* **18**, 387–391.

Vandrey, M.R. (1991) *Stratigraphy, Diagenesis and Geochemistry of the Middle Ordovician Glenwood Formation, Michigan Basin.* Unpublished MS Thesis, University of Wisconsin, Madison, Wisconsin, 239 pp.

Veizer, J. (1978) Simulation of limestone diagenesis – a model based on strontium depletion: discussion. *Can. J. Earth Sci.* **15**, 1683–1685.

Wheeler, C.T. Jr. (1987) Subsurface study of the Lower Ordovician Prairie du Chien and underlying Cambrian Formations and their relations to Pre-Glenwood unconformity in southern Peninsula of Michigan. *Am. Assoc. Petrol. Geol. Bull.* **77**, 1112 [abs.].

Wilson, T.P. (1989) *Origin and Geochemical Evaluation of the Michigan Basin Brine*. Unpublished PhD Thesis, Michigan State University, East Lansing, Michigan, 272 pp.

Winter, B.L., Castrogiovanni, E., Simo, T., Johnson, C.M., Drzewiecki, P.E. & Brown, P. (1992). Geochemistry of the Middle Ordovician St Peter Sandstone, Michigan basin–Wisconsin Arch: implications for pore water history. *Geol. Soc. Am., North-Central Section, Abstract with Programs*, p. 72.

Spec. Publs Int. Ass. Sediment. (1994) **21**, 187–202

Petrographic, geochemical and structural constraints on the timing and distribution of postlithification dolomite in the Rhaetian Portoro ('Calcare Nero') of the Portovenere Area, La Spezia, Italy

J.K. MILLER* *and* R.L. FOLK

University of Texas at Austin, Austin, Texas 78713, USA

ABSTRACT

The Uppermost Triassic Portoro Limestone of Liguria, Italy, shows a striking structural control of late dolomitization. Large white fronts and irregular pods consisting of coarse, commonly anhedral, dolomite crystals replace jet-black limestone. Dolomitizing fluids were fed along faults, dolomitizing laterally until they contacted joints, fractures or stylolites. The fluids then either migrated elsewhere if the joint or stylolite was open, or were stopped by vein-filling cements of less permeable sparry calcite.

The Portoro underwent three major episodes of dolomitization. The first occurred soon after deposition and preferentially dolomitized certain beds and fabrics. This was followed by two fault- and fracture-related postlithification dolomitization events. The genesis of these two later, petrographically indistinguishable, dolomites was determined using field relations and Sr isotopes. The first of the later events occurred after complete lithification of the rock but prior to a major Oligocene–Miocene thrusting event. The dolomites associated with it can be identified in the field by the shearing of the more ductile microsparry calcite around the dolomite masses, and in the laboratory by their $^{87}Sr/^{86}Sr$ ratio of 0.7078. This ratio falls within the range of values for late Triassic seawater. Identification of the dolomites associated with the second postlithification event is more difficult. These dolomites can be distinguished only where field evidence shows that the dolomite is associated with a Plio-Pleistocene fault or replaces sheared limestone. These latest fault-related dolomites give $^{87}Sr/^{86}Sr$ ratios of 0.7080–0.7090, which we interpret as representing various degrees of buffering of relatively recent seawater by Triassic host rock. $^{87}Sr/^{86}Sr$ ratios thus are useful in determining the timing and genesis of diagenetic events. Radiogenic Sr decreases laterally away from the Plio-Pleistocene faults until it reaches a value equal to that of the earlier dolomites, suggesting that dolomitizing fluids were buffered after only a few metres of transport away from the faults.

INTRODUCTION

The latest Triassic (Rhaetian) Portoro limestone crops out on the Portovenere Peninsula and on the adjoining islands of Palmaria and Tino (Fig. 1) in northwest Italy. Spectacular fronts and aureoles of massive white dolomite are exposed on the sea cliffs. These emanate from and stop at fractures, faults and joints. This massive dolomite is the most conspicuous result of a long series of complicated structural and diagenetic events (Folk & Tiezzi, 1985; Miller, 1988; Folk *et al.*, 1989; Miller & Folk, 1988, 1991).

Over 15 distinct types of dolomite are recognized petrographically and geochemically in the Portoro (Miller, 1988; Miller & Folk, 1988). In this paper, however, we focus attention upon the coarsely crystalline late dolomite that formed after an early synsedimentary dolomitization event, lithification of the precursor muds and tectonic fracturing.

The structural history of the Portoro is complex. The limestone has been thrust, folded, sheared and

* Present address: Exxon Company USA, Houston, Texas 77210-4698, USA.

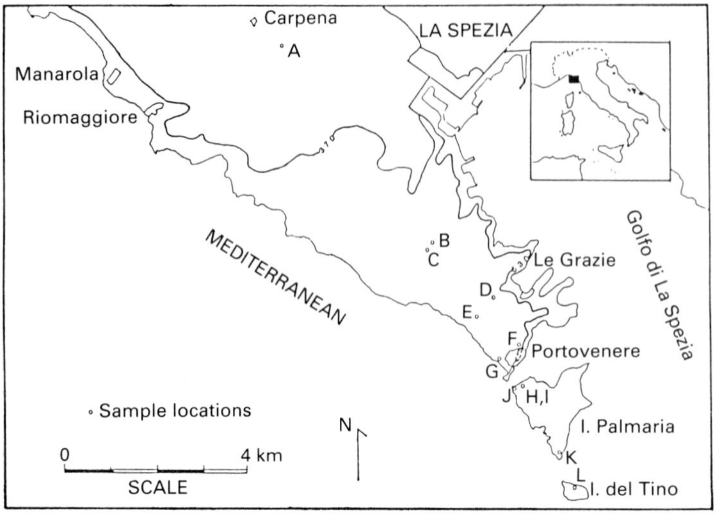

Fig. 1. Location map. Letters represent important sampling localities. A, Carpena quarry; B, Portoro mine; C, quarry; D, mushroom-shaped dolomite mass; E,F,G, quarries; H, quarry opposite Portovenere; I, tidal shelf; J, Plio-Pleistocene fault; K, S. Palmaria quarry; L, Tino quarry.

faulted. Most of this occurred during the extensive Oligocene–Miocene shearing associated with the formation of the Apuanian Alps (Federici & Raggi, 1975; Carmignani *et al.*, 1978; Carter, 1990). During this orogeny, the Portoro was buried under a pile of thrust sheets to a depth of 5–10 km, heated to 200–250°C, sheared and overturned. Faulting and jointing occurred both before and after the shearing and continued into the Quaternary.

Field, petrographic and geochemical evidence support the premise that the massive dolomites of the Portoro formed after complete lithification of the precursor sediments. Other workers have used relationships between dolomites, limestones and faults to postulate a late postlithification dolomitization. Provana di Collegno (1835), Harkness (1859), Spurr (1898), Van Tuyl (1916), Hewett (1928), Ohle (1951), Ham (1955), Miller (1966), Chilingar *et al.* (1979), Nichols and Silberling (1980), Black *et al.* (1981), Zenger (1983), Gregg (1985), Schofield and Adams (1986), Taylor and Sibley (1986), and Mountjoy and Halim-Dihardja (1991) have noted that zones of dolomitization sometimes follow and radiate out from faults, joints and fractures. They also noticed that the dolomite issuing from the faults preferentially replaced certain beds over others. Spurr (1898), Hewett (1931) and Newton and Hardie (1986) found that the more permeable limestone beds were more likely to be dolomitized away from the faults. Many of these same relationships are seen in the Portoro.

THE PRECURSOR ROCK

The undolomitized limestone consists of a jet-black 10–20 μm calcite microspar which is intersected by white and black calcite veinlets, rubble-filled fractures and gold-coloured stylolites. It is interbedded with beds of finely crystalline gold-coloured dolomite which contain pale-grey calcitic nodules. Deposited originally in a somewhat restricted reducing lagoonal environment, the precursor muds were largely aragonitic, as suggested by a strontium trace element content of 1000–7500 ppm (Pavlicek, 1990). The palaeontology and stratigraphy were described by Capellini (1862) and Ciarapica and Passeri (1980).

Widely used as an ornamental stone, the microspar takes a splendid polish and is now a thoroughly lithified 'marble.' Its porosity has been measured by absorption testing to be 0.08% (Italian Institute of Foreign Trade, 1982), a value lower than that of granite. The crystalline structure consists of a polyhedral 'soap-foam' mosaic of closely fitted 10–20 μm calcite microcrystals. Direct scanning electron microscopic (SEM) measurements of broken Portoro surfaces at 100 000× reveal intercrystal boundary spaces of 100–200 Å in width. Calculating porosity based on these spaces and the average crystal size yields an estimated porosity of about 0.1% (Fig. 2). Permeabilities are probably less than 0.001 mD, comparable to those of the nearby Carrara marble.

The early synsedimentary dolomite in the Portoro

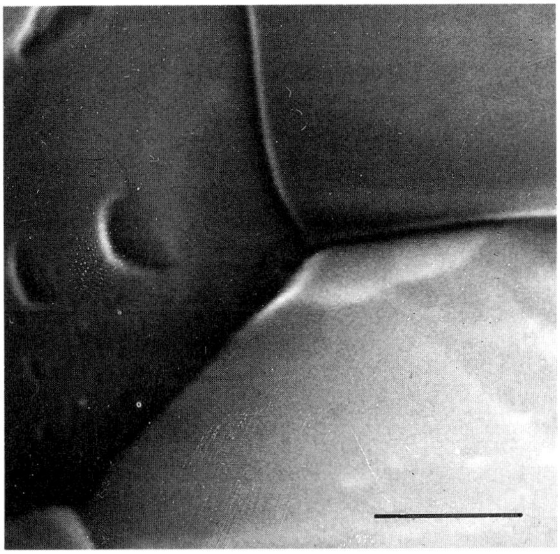

Fig. 2. Polyhedral closely fitted microcrystals of calcite. Intercrystalline boundaries are less than 0.02 μm wide. Scale bar = 1 μm.

occurs in irregular masses and in bedding-parallel gold to grey-coloured ptygmatic-like seams. These dolomites consist of fine 0.06–0.15 mm idiotopic to anhedral crystals. When viewed with a petrographic microscope using oblique reflected light and a white card (Folk, 1987), many of the crystals are seen to contain crystallographically controlled zones of amorphous black organics. These zones make the rhombs appear to have black crosses bisecting them from corner to corner. Folk (1987) calls this zoned dolomite 'St Andrew's cross' dolomite, after the X-shaped cross upon which St Andrew was martyred. This dolomite formed from seawater at subsediment depths of less than a few metres before the precursor aragonite sediment was lithified (Miller, 1988; Miller & Folk, 1988).

POSTLITHIFICATION DOLOMITES

There are two major types of massive postlithification dolomite in the Portoro: a dolomite that formed prior to the Oligocene–Miocene thrusting, overturning and shearing, and a later dolomite associated with late Tertiary or Quaternary normal faulting. Both types typically consist of milky-white coarsely crystalline (0.15–0.5 mm) dolomite. This is in contrast to the more gold to grey finely crystalline early dolomites. Both postlithification dolomites form bedding-constrained replacement fronts or aureoles, which start and stop at faults and joints (Folk & Tiezzi, 1985). These fronts range in lateral extent from 5 cm to 100 cm or longer.

Although they have similar morphology, controls and composition, the early preshearing dolomites and the Pliocene fault-related dolomites can be distinguished using field relations. Limestone is commonly sheared around the coarse massive dolomite fronts in the grotto of Arpaia, in the sea-cliffs of the north coast of Palmaria, and in the quarries further north on the peninsula (Fig. 3). In addition, the dolomite is often fractured and smeared along shear

Fig. 3. Limestone is sheared around a pod of white preshearing late dolomite.

planes. The dolomite thus formed prior to the major mid-Tertiary shearing event. In contrast, the limestone has not been sheared around the dolomite associated with the Pliocene to Pleistocene-aged normal faults. These faults must be of Plio-Pleistocene age because they formed after the upper Miocene shearing events, they parallel regionally documented normal faults, they are open or only partially filled with brecciated clasts, and they are oriented right-side-up and so formed after beds were overturned (Folk *et al.*, 1989).

Petrography of postlithification dolomites

The postlithification dolomite crystals are coarsely crystalline, anhedral to euhedral, have a slight undulatory extinction of 4–9°, and range in size from 0.1 to 0.9 mm. The crystals most commonly fall within a range of 0.3–0.5 mm. Although the size distribution is somewhat patchy, within any one fracture-bound unit most of the crystals lie within a size range of 0.2 mm. The local, approximately unimodal size distribution of the crystals in the massive dolomite suggests nucleation during a single event on homogeneously distributed nucleation sites with a uniform growth rate (Gregg, 1985). In many of the Portoro limestone beds, the microsparite also contains scattered 0.1–0.3 mm replacement rhombs of dolomite. These rhombs are generally euhedral, but share all the other characteristics of the massive postlithification dolomite. Clusters of fine crystals frequently form around these scattered crystals, suggesting that they served as nucleation sites. Whereas the free-floating rhombs are generally euhedral, those within the massive dolomites are subhedral to anhedral and often zoned.

Abundant fluid inclusions within the crystals give them a brown, cloudy appearance in transmitted light and render them white in reflected light. Their rims are clear. These features are commonly observed in massive postlithification dolomites and have been noted by Ohle (1951), Schofield and Adams (1986), and Budai *et al.* (1987). The fluid inclusions are approximately 1 μm in size and appear pink at high magnification in transmitted light. The inclusions' pink colour, lack of birefringence and lower index of refraction than the crystal indicate that they contain fluid. Rare two-phase inclusions are also present.

In any given crystal about 3–5% of the total volume consists of 10–15 μm birefringent carbonate inclusions. When viewed in polarized light these inclusions have the same shape and size as microspar polyhedra (Fig. 4). In areas of coarser microsparite the dolomite contains similarly coarser carbonate inclusions. The inclusions are hence undolomitized remnants of host rock. Because the inclusions are microsparite-sized, the dolomite formed after recrys-

Fig. 4. Euhedral rhomb of late preshearing dolomite. Note similarity in size and shape between the inclusions in the rhomb and the surrounding microsparite groundmass. Field of view is 1.23 mm.

tallization of the host rock. In rare cases some of the massive dolomites also contain some 10–50 μm rhombic inclusions, some of which contain the black organic crosses that are unique to the early dolomites. They are hence believed to represent engulfed relicts of early dolomites The postlithification dolomite crystals also contain many 1–1.5 μm haematite framboids and a few, rare, 30–50 μm inclusions of albite.

Former voids (1–8 mm in radius) cemented with sparry calcite and baroque dolomite are locally present within the massive dolomite. As porosity is commonly associated with massive dolomitization, the voids possibly formed during the dolomitization event. Therefore, prior to the healing of the voids, the dolomite was more permeable and would have formed ideal conduits for additional dolomitizing fluids to move through the rock and affect limestone.

The Portoro contains evidence of pyrite precipitation before, during and after the dolomitization events. All the pyrite is now converted to haematite, which is present both within and between crystals. Interestingly, the massive dolomites appear pinkish to lavender after rain. The moisture adds to the translucency of the rock and pink light is transmitted from the abundant minute haematite framboids. Haematite after pyrite seems to be ubiquitously associated with massive postlithification replacement dolomites. Spurr (1898), Parsons (1918), Wong and Oldershaw (1981), Taylor and Sibley (1986) and Friedman (1987) noted the presence of pyrite and haematite in massive fault-related dolomites. Authigenic albite is also associated with the massive postlithification dolomites in the Portoro. Black *et al.* (1981) and Ricketts (1983) have also noted the association of authigenic albite with fault-related dolomites. Within the Portoro it is most commonly found within the microsparite at the peripheries of fractures and dolomite fronts.

Fig. 5. White dolomite fronts stop at a joint (arrow) to the left of the major fault. Scale is 5′4″ tall.

Fracture-controlled dolomitization

Observations

Although constrained by bedding, vertical and subvertical veins, fractures and joints control the lateral extent of the dolomite fronts. Following a dolomitized bed from a fracture, one typically finds that the bed is completely dolomitized (at least megascopically) until it contacts a barrier. The dolomite then disappears abruptly (Fig. 5). In other occurrences the dolomite ends more gradationally. On the northernmost tip of Palmaria the dolomite bordering a Plio-Pleistocene normal fault preferentially follows certain beds. This dolomite diminishes away from the fault in a stepwise manner across each succeeding fracture or vertical stylolite until it ends at a fracture. Locally, thin stringers of dolomite will follow bedding laminae for a short distance after crossing a fracture (Fig. 6). Kushnir and Kastner (1984) described similar coalescing microdolomite layers within the Monterey Formation of California. The dolomite fronts thus sometimes end abruptly at a fracture and sometimes diminish gradually.

On the far northwest tip of Palmaria dolomitization was controlled by 'breadcrust' fracture patterns (so called because of their near-random orientation and similarity to fractures in hard rolls). Dolomite 'realms' are bounded by fractures; some fracture-bounded polyhedra are partially dolomite, some completely dolomite, and some pristine limestone (Fig. 7). Each polyhedron apparently acted as its own compartmentalized diagenetic capsule.

Fig. 6. Dolomite front associated with a Plio-Pleistocene normal fault partially halts at first fracture. White 'fingers' of dolomite continue through, halting at final fracture (arrow). Dolomite is bounded at its base by a stylolite.

Fig. 7. The distribution of dolomite is controlled by fine fractures. Some fracture polyhedra, such as the one to the right-centre, remain undolomitized; this is probably the result of differential cementation along the fractures.

Discussion

Surprisingly, although many workers have reasoned that fractures, faults, breccia zones and more permeable beds served as conduits for dolomitizing fluids, few have discussed the role of fractures in halting dolomitization fronts. Field evidence from the Portoro indicates that faults and fractures often served as barriers. Harkness (1859) first observed that some dolomite bodies stop at fractures, and thus must have formed later than jointing. Hewett (1931, p. 60) remarked that the outer boundaries of dolomite masses were 'generally very obscure, but in places they are marked by fractures.' Also, Ohle (1951) noted that Mississippi Valley-type ore bodies associated with dolomites were frequently bounded by fractures.

How can fractures serve as dolomitization barriers? In the Portoro there are several alternatives. If a fracture was still open when dolomitizing fluids moved through a bed, the fluids would have flowed into the fissure rather than dolomitizing the neighbouring block of relatively impermeable limestone. A fracture would also halt fluid flow if it were cemented with calcite. On weathered Portoro outcrops, fractures filled with sparry calcite commonly

Fig. 8. Fronts of white dolomite issue from joint. Note the sharp basal contacts of the dolomite fronts, the diffuse upper contacts and the presence of dolomite fronts in different beds on each side of the fracture.

protrude from the surrounding microsparite. The microsparite weathers more rapidly because of its finer crystal size and greater number of intercrystalline boundaries; it must therefore also transmit fluids more easily. Thus, if a fracture had been cemented prior to the dolomitization event it could easily have served as a barrier. And, if it had been cemented and reopened, it might have served as both barrier and conduit.

This relationship might explain the asymmetry of the dolomite fronts that sometimes issue from one side of a fracture but not the other. One would assume that, if a given bed of limestone was slightly more conducive to dolomitization, it would be dolomitized on both sides of a fracture or fault even if the original bed had been displaced. Although this sometimes is true, often a bed will be dolomitized on one side of a fault and the same bed will not be dolomitized on the opposite side (Figs 5, 8 and 9). Discontinuous mineralized coatings on the fracture walls may explain this situation. Fractures may have opened, been cemented with calcite, and then reopened. A reopened fracture would tend to follow the plane of greatest weakness: the boundary between vein calcite and host limestone. It might also switch from side to side of the original fracture, exposing different beds to dolomitizing fluids on either side of the original fracture.

The question arises as to whether the dolomitizing fluids are sourced from the fractures or from the host rock. To supply the necessary quantity of Mg ions for large-scale dolomitization, an outside source of fluids must be invoked. Hence, it makes

Fig. 9. Quarry block from Portoro Mine. Block is completely dolomitized to one side of joint, only partially on other side.

Fig. 10. The tree-shaped white dolomite body (outlined in black) illustrates one mechanism of fluid migration through the Portoro. Fluids moved down the thin, barely discernible dolomitized fault zone to the right of the mushroom-shaped mass (arrow). They then migrated laterally to form the mass. The stratigraphic top is to the right of the photo. Quarry is to the east of Le Grazie.

sense to assume that the fluids migrated first along the paths of easiest migration – the fractures – and then selectively into the less permeable limestones. Figure 10 confirms this assumption. Dolomitizing fluids apparently percolated either up or down a fracture or fault and spread out laterally to form the tree-shaped dolomite body to the right and slightly above the deep quarry. The beds are now vertical, but at the time of dolomitization the 'up' direction would have been to the right of the figure. This suggests that the fluids in this case migrated down and then laterally.

The reason for the 'all or none' dolomitization of separate fracture-bounded limestone blocks is obscure. Apparently, Mg saturation must build up to a certain concentration within one block, and then dolomitization occurs 'explosively'. Experimental work by Sibley *et al.* (1987) demonstrated that dolomite has a long induction stage during which no products form, followed by a relatively rapid replacement stage which does not take place until all the requisites for dolomitization are ideal. Thus, once dolomitization commenced it usually went to completion within one bounded unit. The abundant inclusions within the dolomite, the uniform crystal sizes and the disordered lattice structure (the individual crystals are calcium-rich and elongated along the *c*-axis) suggest a rapid crystallization (Miller, 1988). Perhaps it is not until the first block is completely dolomitized that Mg is available to travel on to the next block and become concentrated.

Bedding controls

Observations

The fronts of both postlithification dolomites have sharp basal limits but upper gradational limits. This suggests that the dolomitizing fluids migrated along the lower bedding plane boundary and migrated upward (Figs 6 and 8).

Discussion

Fluids migrating along a bedding surface might tend to migrate up rather than down because the upper boundary of a bed probably represents a slight hiatus in sedimentation. During this hiatus the sediment surface might become slightly lithified or mineralized, causing a decrease in permeability. Fluids would thus tend to migrate upward from the bedding surface as permeability would be greater upward than downward through the early lithified layer. Changes in grain or crystal size within the limestone beds might also exert some control on the dolomite distribution. However, as the limestone had already been neomorphosed to a consistent crystal size, it is unlikely that crystal size had much effect.

Differences in permeability between beds might also affect dolomitization. The distribution of earlier generations of dolomite might encourage front migration. The coarse replacement dolomite was at one time more permeable than the microsparite. This is evidenced by the presence of zoned pore-filling dolomite cements in the massive preshearing dolomite. Fluids travelling along a conduit would therefore preferentially feed into an already dolomitized bed rather than one which still remained microsparite. Once dolomitization of a bed had commenced, it would continue until it met a barrier (Miller, 1988).

Stylolite-controlled dolomitization

Observations

On the south side of Palmaria many of the dolomite fronts are constrained by bedding-parallel stylolites. Because these stylolites have the same amplitude in the limestone and the dolomite, even though the relative hardness and permeability are different between the two lithologies, they predate the dolomitization. Narkiewicz (1979) described similar stylolites that modify the geometry of limestone–dolomite boundaries in the Upper Devonian of Poland.

On North Palmaria, vertical stylolites commonly served as 'leaky' boundaries to the dolomite fronts. A few stringers of dolomite will leak through the first stylolite only to be halted at the next (Fig. 11).

Most stylolites in the Portoro contain euhedral rhombs of replacement dolomite. Although in some cases such dolomite may be insoluble residue, in others the dolomite must have precipitated as there is no dolomite in the surrounding host rock. Dolomite is commonly found on only one side of a stylolite. However, in some cases where a stylolite bends sharply or has less insoluble residue, dolomite is present on both sides.

Fig. 11. Dolomite front mostly halts at a vertical stylolite (arrow). Some patchy dolomite formed to the right of the stylolite, indicating that the stylolite served as a 'leaky' seal.

Discussion

Stylolites serve not only as barriers to dolomitization but also as conduits for dolomitizing fluids. Ohle (1951), Freeman (1965), Logan and Semeniuk (1976), Wanless (1979) and Zenger (1983) have noted dolomite growing along and out from stylolites. The many rhombs lining stylolites in the Portoro bear this out.

The presence of dolomite along only one side of stylolites indicates that dolomitizing fluids flowed along one side of the stylolite but were unable to communicate with the limestone on the other side. In areas where dolomitization occurred on both sides, the stylolite probably served as a leaky barrier; in the troughs and on the pillars where insoluble residue had accumulated, the stylolite served as a barrier to fluids, but in the areas lacking insoluble residue the dolomitizing fluids passed through.

Breccia-controlled dolomitization

Observations

The preshearing and postshearing dolomites are also associated with breccias. Dolomitized breccia matrix and partially dolomitized breccia occur in subparallel breccia zones on the tidal shelf opposite the town of Portovenere on Palmaria, in the low-angle fault at the northwest tip of Palmaria, along the vertical breccia zone in the cliff below San Pietro, and in the fractures and faults on the south side of Palmaria. Both the dolomitized matrix and the breccia are sheared in the breccia zones of the tidal shelf and below the church of San Pietro, indicating that these breccias formed and were dolomitized prior to shearing and overturning.

Discussion

Breccia zones were probably areas of greater permeability, and thus were more susceptible to dolomitizing fluids. Since the matrix was dolomitized more thoroughly than the coarser breccia fragments, the uncemented breccia matrix was the most likely fluid conduit. Permeability was the major control of dolomitization. Many workers have found late dolomites to be preferentially associated with breccia zones (Hewett, 1931; Ohle, 1951; Narkiewicz, 1979; Taylor & Sibley, 1986; Morrow *et al.*, 1986).

Stable carbon and oxygen analysis

The $\delta^{13}C$ of the postlithification dolomites ranges between 3.4 and 4.4‰ PDB (Table 1), which falls within the range of Portoro microsparites (4.2–5.6‰), the upper range of modern normal marine carbonates (4 to −2‰; Gross, 1964) and most ancient platform dolomites (4 to −3‰; Land, 1980). As the $\delta^{13}C$ of biogenic and inorganic calcites precipitated in seawater is generally enriched by 1–2‰ relative to mean ocean water, the carbon values probably represent those of normal marine water. The upper range of $\delta^{13}C$ of Portoro dolomites is slightly enriched compared to modern examples; this may be due to either modification by organic processes or a more enriched Triassic seawater. Veizer and Hoefs (1976) suggested that the oceanic carbonate reservoir was slightly heavier (2–4‰) with respect to modern seawater during the Triassic. These heavy values of carbon strongly suggest precipitation by brines or normal seawater rather than by meteoric waters. They are also most likely inherited from the original sediment. Fritz and Smith (1970) found that the $\delta^{13}C$ of secondary dolomites is uniform within individual deposits and falls within the range of values for the associated limestones.

The massive postlithification dolomites of the Portoro have a $\delta^{18}O$ range of −1.1 to −4.2‰ (Table 1). Again, this is within the range of normal marine carbonate (Mattes & Mountjoy, 1980). No trends could be found in the oxygen values relating to distance from source or to preshearing versus Plio-Pleistocene fault-related dolomites.

In order to get oxygen values as heavy as −1 to −4‰, the dolomites must have formed from heavy waters at high temperatures, or from seawater at only slightly elevated temperatures. Figure 12 plots $\delta^{18}O$ (PDB) dolomite for various $\delta^{18}O$ (SMOW) water values. Since evaporitic brines and deep basinal fluids rarely become more enriched than +6‰ (Taylor, 1974), this puts an upper limit to the temperature of formation of the dolomite. If the dolomites of −4‰ compositions formed from a water of +6‰ composition, they would have formed at a temperature of 85°C, much lower than that which the Portoro experienced at its maximum depth of burial (200–250°C). If the dolomite had formed from water of normal seawater composition, it would have formed over a range of temperature from 40 to 55°C. Therefore, the dolomite must have formed at

Table 1. Stable isotope (‰, PDB) data.

Sample	$\delta^{13}C$	$\delta^{18}O$	Location
Microsparite–Black mother rock			
2/11JCAL	5.04 ± 0.01	−5.79 ± 0.02	Carpena quarry
6/20MCAL	4.58 ± 0.09	−3.19 ± 0.05	N. Palmaria
9/13M	4.55 ± 0.03	−3.46 ± 0.07	S. Palmaria
V25RB	5.59 ± 0.01	−3.15 ± 0.08	Portovenere
V25VB	4.33 ± 0.03	−3.23 ± 0.06	Portovenere
V25YB	5.41 ± 0.02	−2.64 ± 0.07	Portovenere
F3B	4.86 ± 0.10	−3.24 ± 0.09	Portovenere
F4B	5.05 ± 0.06	−3.05 ± 0.01	Portovenere
TGB	4.15 ± 0.02	−2.25 ± 0.03	Portovenere
Early dolomites			
4/13M	4.31 ± 0.03	−2.06 ± 0.06	S. Palmaria
9/24M	3.81 ± 0.04	−1.91 ± 0.08	N. Palmaria
7/28M	4.14 ± 0.01	−2.00 ± 0.02	N. Palmaria
7/24MS	4.27 ± 0.05	−2.01 ± 0.06	N. Palmaria
2/24MS	4.11 ± 0.01	−2.08 ± 0.01	N. Palmaria
27/25M	4.20 ± 0.04	−1.99 ± 0.02	Stone cutter's
Replacement dolomites associated with Plio-Pleistocene faults			
1/13J	4.11 ± 0.04	−2.40 ± 0.02	N. Palmaria
2/13J	4.15 ± 0.04	−2.71 ± 0.06	N. Palmaria
4/19M	3.88 ± 0.04	−3.29 ± 0.06	N. Palmaria
5/19M	3.98 ± 0.15	−4.22 ± 0.06	N. Palmaria
6/19M	3.90 ± 0.04	−4.12 ± 0.04	N. Palmaria
7/19M	3.76 ± 0.01	−2.50 ± 0.03	N. Palmaria
Preshearing massive dolomites			
2/10M	3.60 ± 0.03	−3.86 ± 0.02	Below cemetery
3/10M	3.43 ± 0.04	−4.12 ± 0.04	Below cemetery
1/12M	3.68 ± 0.05	−2.26 ± 0.06	NE Portovenere
8/13M	3.58 ± 0.05	−1.20 ± 0.07	S. Palmaria
7/13M	4.42 ± 0.11	−2.99 ± 0.07	S. Palmaria
1/13M	3.75 ± 0.06	−2.48 ± 0.04	S. Palmaria
1/26M	3.41 ± 0.05	−1.10 ± 0.04	Portoro mine
2/26M	2.03 ± 0.03	−2.01 ± 0.03	Portoro mine
16/25M	4.29 ± 0.09	−2.66 ± 0.04	Portoro mine
2/25M	3.55 ± 0.06	−2.91 ± 0.06	Mean dog quarry

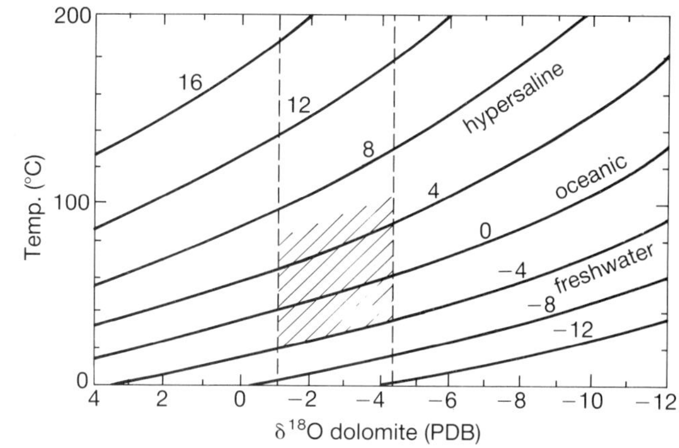

Fig. 12. Graphical representation of the oxygen isoptopic equilibrium between dolomite, the water it precipitated from (SMOW scale) and temperature. The area between the dotted lines represents the range in oxygen values for Portoro values. The hachured region represents the possible range of fluid compositions. (Modified from Land, 1985).

non-metamorphic temperatures at relatively shallow burial depths.

Surprisingly, the range of oxygen isotopic compositions for the microsparite and the dolomite replacing it is almost identical. This, however, only indicates that the dolomite must have formed from a fluid of different composition or temperature than that responsible for the neomorphism of the micrite to microsparite. Had they formed from the same fluid, the dolomite would be heavier (Land, 1980).

The scatter of values for late dolomites from the same localities and even the same beds (all the data for the Plio-Pleistocene dolomites came from one bed) may represent either a slight local buffering or changing fluid temperatures. It is most likely, however, that this variation records diffential alteration and stabilization of the first precipitated dolomite. Land (1985) advocated progressive stabilization of the first precipitated disordered dolomite as it is buried deeper and exposed to higher temperatures. Petrographic properties reveal that the massive dolomites do not consist of a simple phase, and that much zoning is present. Therefore, the isotopic signature may vary according to the degree of alteration subsequent to its first precipitation.

The relatively heavy carbon (2.0–4.3‰) and oxygen (−1.1 to −4.2‰) (Table 1), as well as non-radiogenic Sr values, suggest that the most likely fluid to have produced the dolomites is seawater or modified seawater at less than 60°C (Fig. 12).

Sr analysis

$^{87}Sr/^{86}Sr$ ratios have commonly been used to date marine limestones. We think that they can also be used to pinpoint the timing and origin of diagenetic fluids. The results of Sr analyses on 21 samples are summarized in Table 2. The two samples from the microspar give a ratio of approximately 0.7076. All the preshearing dolomites and the early dolomites give ratios of approximately 0.7077–0.7078. The ratios of both the calcites and the dolomites fall within the range of the ratio for late Triassic seawater. Burke *et al.* (1982) gave a range of 0.7076–0.7078 for the Late Triassic and Faure *et al.* (1978), Haq *et al.* (1987) and Koepnick *et al.* (1990) give a range of 0.7075–0.7079. Our values suggest that both the early and the later preshearing dolomites were precipitated from fluids of very similar Sr composition to the host rock. This similarity in fluid composition may be the result of rock buffering.

In contrast to the early dolomites and the limestone, the fault-related dolomites contain a range of strontium values. Miller (1988) collected samples along a bed of massive replacement dolomite which emanates from an open-gaping Pleistocene fault (fault 'A') located at the northwest tip of Palmaria. Extending into the ocean, this fault contains loose breccia as well as some dolomite-cemented breccia and modern horizontally bedded travertine (Fig. 13). Figure 7 displays one of the dolomite beds that radiates from the fault. The change in $^{87}Sr/^{86}Sr$ ratio in the dolomite bed versus distance away from the fault in a similar bed is plotted in Figure 14. The most radiogenic values (0.7090) of replacement dolomite were found 2 m from the fault. This ratio is very close to that of modern seawater: 0.7091 (Burke *et al.*, 1982). The ratio within the dolomite bed decreases steadily away from the fault: at about 5 m from the fault it has a value equivalent to that of Triassic seawater. This trend may represent the constant buffering of the dolomitizing fluid as it

Table 2. Strontium isotopic data (by comparison with NBS SRM 987).

Sample	$^{87}Sr/^{86}Sr \pm 2s$	Location
Calcite microsparite values		
CAR-2†	0.70758 ± 0.00006	Carrara marble cutters
CAR-2G†	0.70753 ± 0.00005	Carrara marble cutters
Early dolomite		
2/24M†	0.70777 ± 0.00011	N. Palmaria
7/16M†	0.70786 ± 0.00003	Tino
6/17M†	0.70768 ± 0.00003	S. Palmaria
Preshearing massive dolomite		
8/13M†	0.70782 ± 0.00008	S. Palmaria
1/26M†	0.70782 ± 0.00005	Portoro mine
Dolomite associated with Plio-Pleistocene fault		
7/19M†	0.70811 ± 0.00003	N. Palmaria—in fault
6/19M†	0.70824 ± 0.00003	N. Palmaria—in fault breccia
5/19M†	0.70908 ± 0.00003	N. Palmaria—1.5 m from fault
4/19M†	0.70902 ± 0.00003	N. Palmaria—1.8 m from fault
1/13J†	0.70820 ± 0.00004	N. Palmaria—3.9 m from fault
2/13J†	0.70782 ± 0.00003	N. Palmaria—5.0 m from fault
4/13J†	0.70765 ± 0.00005	N. Palmaria—6.4 m from fault
Dolomite associated with fault 'A'		
D1*	0.70829 ± 0.00012	N. Palmaria—0.3 m from fault
D3*	0.70886 ± 0.00013	N. Palmaria—0.9 m from fault
D8*	0.70782 ± 0.00014	N. Palmaria—2.5 m from fault
Dolomites from extensively faulted area		
A-1‡	0.70799 ± 0.00014	N. Palmaria—0.05 m from fault
A-2‡	0.70878 ± 0.00007	N. Palmaria—0.08 m from fault
A-3‡	0.70804 ± 0.00015	N. Palmaria—0.15 m from fault
A-4‡	0.70802 ± 0.00009	N. Palmaria—0.6 m from fault
A-5‡	0.70782 ± 0.00013	N. Palmaria—0.46 m from fault
D-1‡	0.70922 ± 0.00014	N. Palmaria—0.25 m from fault
F-1‡	0.70870 ± 0.00011	N. Palmaria—1.0 m from fault
F-2‡	0.70810 ± 0.00013	N. Palmaria—2.0 m from fault

* Data collected by JKM in 1986.
† Data collected by RLF in 1988.
‡ Data collected by RLF and GG in 1991.

migrated through and replaced the microspar; after passing through 4–6 m of sediment the fluid was possibly entirely rock-buffered.

The values from the dolomite filling of the fault zone are lower than the most radiogenic values. Although surprising at first glance, these values from the fault gouge represent the last crystallization that sealed the conduit. This cement would thus have precipitated from less rapidly moving fluids, which would have had more time to be buffered by the limestone.

That the limestone should have such a strong buffering capacity is not surprising. The initial microspar contains 1500–7500 ppm Sr (Pavlicek, 1990), whereas the dolomite that replaced the microspar has less than 200 ppm Sr (Miller, 1988). Seawater contains only 8 ppm Sr (Krauskopf, 1979). Thus the thousands of ppm of non-radiogenic Sr released during dolomitization, as well as any Sr released by the dissolution of microspar along fractures, would easily mask the signature of the highly radiogenic seawater. In addition, high Sr concentrations of Triassic isotopic composition would probably also be present in the connate waters as a result of the conversion of the originally aragonitic muds to microspar (Folk, 1988). Faure *et al.* (1978) also

Fig. 13. Plio-Pleistocene fault gapes (arrow), but is locally filled with partially dolomitized breccia.

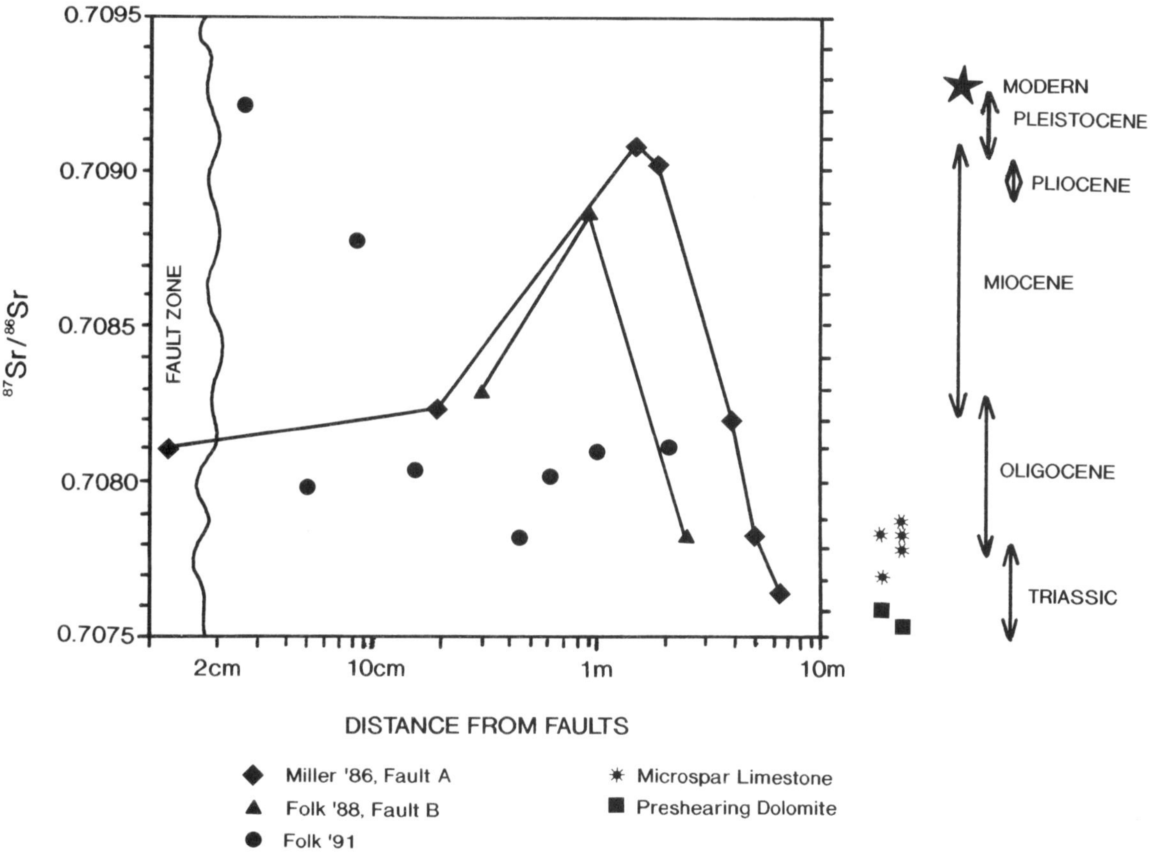

Fig. 14. Transects away from normal faults. The sampling from 1986 and 1988 shows a decrease in radiogenic strontium in dolomites located further away from the fault. Samples taken by Folk in 1991 from a number of faults give more ambiguous results.

found that Sr isotopic ratios may remain unaffected by dolomitization, even though dolomitization causes a drastic decrease in total Sr content. They, too, suggested that this is the result of rock-buffering.

Later sampling by R.L.F. of an extensively faulted area just east of the Pleistocene fault discussed above suggests a more complex picture. Samples taken in 1988 from one fault, named fault 'B,' gave similar results to those from the first fault sampled by J.K.M. Samples taken in 1991 from this same area, however, did not show the same trend: the Sr ratios were all higher than those of the surrounding Triassic rocks, but did not match modern seawater compositions. These values are also shown in Figure 14.

The strontium data can be interpreted as follows:

1 Some of the dolomite corresponds in $^{87}Sr/^{86}Sr$ ratio with late Miocene to modern seawater, agreeing with the field evidence for a very young age of faulting and dolomitization.

2 After some 5 m of percolation into the host limestone, the dolomite has picked up enough old Sr to be totally buffered by the host rock, and gives late Triassic values.

3 There is a complex group of intermediate values, with $^{87}Sr/^{86}Sr$ of 0.7078–0.7083 that may represent various degrees of buffering between host rock and seawater of late Oligocene to Holocene age.

4 It is unlikely that the values higher than 0.7080 came from fluids derived from the weathering of feldspars within other formations, because no ratios higher than those of modern seawater were observed.

CONCLUSIONS

Postlithification dolomitization in the Triassic Portoro limestone was controlled by bedding, stylolites, breccias and differential cementation along joints and fractures (Fig. 15). It occurred after complete lithification of the limestone and neomorphism of the original micritic muds to 10–20 μm microspar. The dolomites are characterized by their milky white colour, coarse crystallinity, cloudy appearance in transmitted light, and microspar inclusions. The first postlithification dolomites underwent brittle fracturing and are recognized by the ductile deformation of the surrounding limestone beds. The sheared limestone enwrapping the dolomite bodies indicates that this dolomite formed prior to the extensive Miocene shearing associated with the formation of the Apuanian Alps.

During the last episode of dolomitization, fluids migrated along Plio-Pleistocene normal faults, dolomitizing fault breccia and forming radiating dolomite fronts. Controlled by permeability differences along the fault surfaces, different beds are dolomitized on either side of the faults. Both field observations and geochemistry constrain the timing of this last event. The host rock is unsheared around

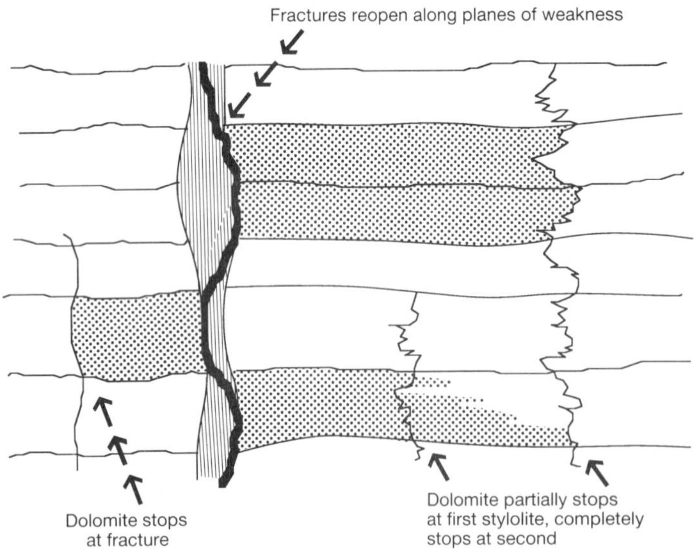

Fig. 15. Schematic drawing showing the various controls on the emplacement of dolomite fronts. Dolomitizing fluids were fed up or down the fracture, issuing laterally to dolomitize where they came into direct contact with the limestone. The vertically hachured area represents an earlier sparry calcite fracture fill that locally serves as a barrier to dolomitizing fluids.

the dolomite fronts; thus the dolomite formed in the post-Miocene and after overturning. Unlike the earlier replacement dolomites, which have Sr isotopic ratios of Triassic seawater composition, $^{87}Sr/^{86}Sr$ ratios from the fault-related dolomite yield Pleistocene seawater ratios at some points near the faults. Strontium ratios may thus serve to identify different episodes of dolomitization. An unsolved problem is how the latest dolomitizing fluids were able to permeate rocks that are more impermeable than even Carrara marble.

ACKNOWLEDGEMENTS

We would like to thank L.S. Land for his assistance in analysing and interpreting our isotope samples, G. Gao and J. Kupecz for their assistance in running Sr samples, and J. Greenberg and K.E. Carter for field assistance. We would also like to thank J.M. Gregg, B.H. Purser, M.E. Tucker, and D.H. Zenger for their help in preparing the manuscript for publication.

REFERENCES

BLACK, D.F.B., MACQUOWN, W.C. Jr. & DEHANS, R.J. (1981) *The Relation of Dolomite Associated with Faults to the Stratigraphy and Structure of Central Kentucky*. USGS Professional Paper 1151A, A1–A17.

BUDAI, J.M., LOHMANN, K. & WILSON, J.L. (1987) Dolomitization of the Madison Group, Wyoming and Utah overthrust belt. *Bull. Am. Assoc. Petrol. Geol.* **71**(8), 909–924.

BURKE, W.H., DENISON, R.E., HETHERINGTON, E.A., KOEPNICK, R.B., NELSON, H.F. & OTTO, J.B. (1982) Variation of seawater $^{87}Sr/^{86}Sr$ throughout Phanerozoic time. *Geology* **10**, 516–519.

CAPPELLINI, G. (1862) Studi stratigrafici e paleontologici sull'Infralias del Golfo della Spezia: *Mem. Acc. della Sci. dell'Ist. Bologna* Ser. 2, **1**, 247–318.

CARMIGNANI, L., GIGLIA, G. & KLIGFIELD R. (1978) Structural evolution of the Apuane Alps: an example of continental margin deformation in the Northern Appenines, Italy. *J. Geol.* **86**, 487–504.

CARTER, K.E. (1990) *Construction and Collapse of an Orogen: Tectonic Strain and Fluid History of the Tuscan Nappe, Northern Appenines, Italy*. PhD Thesis, University of Texas, Austin, 188 pp.

CHILINGAR, G.V., ZENGER, D.H., BISSELL, H.J. & WOLF, K.H. (1979) Dolomites and dolomitization. In: *Diagenesis in Sediments and Sedimentary Rocks* (Ed. Larson, G. & Chilingar, G.) pp. 423–536. Elsevier, Amsterdam, Developments in Sedimentology 25A.

CIARAPICA, G. & PASSERI L. (1980) La lithostratigrafia della Serie Triassica del Promontorio Occidentale del Golfo di La Spezia. *Mem. Soc. Geol. Ital.* **21**, 51–61.

FAURE, G., ASSERETO, R. & TREMBA, E.L. (1978) Strontium isotope composition of marine carbonates of Middle Triassic to Early Jurassic age, Lombardic Alps, Italy. *Sedimentology* **25**, 523–453.

FEDERICI, P.R. & RAGGI, G. (1975) Una nuova interpretazione della tettonica dei Monti della Spezia. *Boll. Soc. Geol. Ital.* **94**, 945–960.

FOLK, R.L. (1987) Detection of organic matter in thin-sections of carbonate rocks using a white card. *Sedim. Geol.* **54**, 193–200.

FOLK, R.L. (1988) Discovery in Columbus' land: carbonate mud diagenesis to microspar or metamicrite? *Soc. Econ. Paleont. Mineral. Midyear Meeting Abstracts* **5**, 19(abstract).

FOLK, R.L. & TIEZZI P.A. (1985) Diagenesis and dolomitization in Triassic (Rhaetian) 'Portoro' Limestone, Portovenere, Liguria, Italy. *Bull. Am. Assoc. Petrol. Geol. Meetings with Abstracts, New Orleans* **69**, 256(abstract).

FOLK, R.L., PURSELL, V.J., GREENBERG, J., MOSHER, S., HELPER, M. & CARTER, K.E. (1989) Inverted tectonic veins in the Triassic Portoro Limestone, Portovenere area (La Spezia). Italy. *Annales Tectonicae* **3**, 25–33.

FREEMAN, T. (1965) *Post-Lithification Dolomite in the Joachim and Plattin Formations (Ordovician), Northern Arkansas*. Geol. Soc. Am. Ann. Mtg., Kanses City, Spec. Paper **87**, 58(abstract).

FRIEDMAN, G.M. (1987) Deep-burial diagenesis: its implications for vertical movements of the crust, uplift of the lithosphere and isostatic unroofing: a review. *Sedim. Geol.* **50**, 67–94.

FRITZ, P. & SMITH, D.W. (1970) The isotopic composition of secondary dolomites. *Geochim. Cosmochim. Acta* **43**, 1357–1365.

GREGG, J.M. (1985) Regional epigenetic dolomitization in the Bonneterre dolomite (Cambrian), southeastern Missouri. *J. Sedim. Petrol.* **54**, 908–931.

GROSS, M.G. (1964) Variation in the $^{18}O/^{16}O$ and $^{13}C/^{12}C$ ratios of diagenetically altered limestones in the Bermuda Islands. *J. Geol.* **72**, 170–194.

HAM, W.E. (1955) Field conference on geology of the Arbuckle Mountain region April 22–23, 29–30, 1955. *Guidebook 3. Okla. Geol. Survey*, 61 pp.

HAQ, B.V., HARDENBOL, J. & VAIL, P.R. (1987) Chronology of fluctuating sea levels since the Triassic. *Science* **235**, 1156–1164.

HARKNESS, R. (1859) On the jointings in the Carboniferous and Devonian rocks in the district around Cork; and on the dolomites of the same district. *Q. J. Geol. Soc. London* **15**, 86–104.

HEWETT, D.F. (1928) Dolomitization and ore deposits. *Econ. Geol.* **23**, 821–863.

HEWETT, D.F. (1931) *Geology and Ore Deposits of the Goodsprings Quadrangle, Nevada*. US Geol. Surv. Prof. Paper **162**, 172 pp.

ITALIAN INSTITUTE OF FOREIGN TRADE (1982) *Marmi Italiana*. Fratelli Vallarde, Milano.

KOEPNICK, R.B. DENISON, R.E., BURKE, W.H., HETHERINGTON, E.A. & DAHL, D.A. (1990) Construction of the Triassic and Jurassic portion of the Phanerozoic curve of seawater $^{87}Sr/^{86}Sr$. *Chem. Geol.* **80**, 327–329.

KRAUSKOPF, K.B. (1979) *Introduction to Geochemistry*.

McGraw-Hill, New York, 263 pp.

KUSHNIR J. & KASTNER, M. (1984) Two forms of dolomite occurrences in the Monterey Formation, California: concretions and layers – a comparative mineralogical, geochemical and isotopic study. In: *Dolomites of the Monterey Formation and Other Organic-Rich Units* (Ed. Garrison, R.E., Kastner, M. & Zenger, D.H.) Pacific Section Soc. Econ. Paleont. Mineral. **41**, 171–183.

LAND, L.S. (1980) The isotopic and trace element geochemistry of dolomite: the state of the art. In: *Concepts and Models of Dolomitization* (Ed. Zenger, D.H., Dunham, J.B. & Etherington, R.L.) Soc. Econ. Paleont. Mineral. Spec. Publ. **28**, 87–110.

LAND, L.S. (1985) The origin of massive dolomite. *J. Geol. Educ.* **33**, 112–125.

LOGAN, B.W. & SEMENIUK, V. (1976) *Dynamic Metamorphism: Processes and Products in Devonian Carbonate Rocks, Canning Basin, Western Australia*. Geol. Soc. Australia, Spec. Pub. **6**, 138 pp.

MATTES, B.W. & MOUNTJOY, E.W. (1980) Burial dolomitization of the Upper Devonian Miette buildup, Jasper National Park, Alberta. *In: Concepts and Models of Dolomitization* (Ed. Zenger, D.H., Dunham, J.B. & Etherington, R.L.) Soc. Econ. Paleont. Mineral. Spec. Publ. **28**, 259–297.

MILLER, D.N. JR. (1966) Diagenesis of sedimentary rocks. In: *Symposium on Recently Developed Geologic Principles and Sedimentation of the Permo-Pennsylvanian of the Rocky Mountains*. Wyoming Geol. Assoc., 12th Annual Conference.

MILLER, J.K. (1988) *Multistage Dolomitization of the Portoro Limestone, Liguria, Italy*. MA Thesis, University of Texas at Austin, 215 pp.

MILLER, J.K. & FOLK, R.L. (1988) Petrographic, geochemical, and structural constraints on the timing and emplacement of dolomite in the Portoro Limestone, Liguria, Italy. In: *Soc. Econ. Paleont. Mineral. Mid year Meeting Abstracts* **5**, 36 (abstract).

MILLER, J.K. & FOLK, R.L. (1991) Post-lithification dolomitization in the Rhaetian Portoro ('Calcare Nero') of the Portovenere Area, La Spezia, Italy. In: *Dolomieu Conference on Carbonate Platforms and Dolomitization, Abstracts* (Ed. Bosellini, A., Brandner, R., Flugel, E., Purser, B., Schlager, W., Tucker, M. & Zenger D.), 168(abstract).

MORROW, D.W., CUMMING, G.L. & KOEPNICK, R.B. (1986) Manetoe facies – a gas-bearing, megacrystalline Devonian dolomite, Yukon and Northwest Territories, Canada. *Bull. Am. Assoc. Petrol. Geol.* **70**, 702–720.

MOUNTJOY, E.W. & HALIM-DIHARDJA, M.K. (1991) Multiple phase fracture and fault-controlled burial dolomitization, Upper Devonian Wabamun Group, Alberta. *J. Sedim. Petrol.* **61**(4), 590–612.

NARKIEWICZ, M. (1979) Telo- and mesogenetic dolomites in subsurface Upper Devonian to Lower Carboniferous sequences of Southern Poland. *Neues Jahrb. Geologie u. Palaeontologie Abh.* **158**, 180–208.

NEWTON, E. & HARDIE, L.A. (1986) Dolomitization front geometry, Triassic laternar buildup, North Italy: a different approach to the problem of the origin of massive dolomite. *Soc. Econ. Paleont. Geol. Midyear Meeting* **3**, 83(abstract).

NICHOLS, K.M. & SILBERLING, N.J. (1980) Eogenetic dolomitization in the pre-Tertiary of the Great Basin. In: *Concepts and Models of Dolomitization* (Ed. Zenger, D.H., Dunham, J.B. & Etherington, R.L.) Soc. Econ. Paleont. Mineral. Spec. Publ. **28**, 237–246.

OHLE, E.L. (1951) The influence of permeability on ore distribution in limestone and dolomite. *Econ. Geol.* **46**, 667–706, 871–908.

PARSONS, L.M. (1918) Dolomitization and the Leicestershire dolomites. *Geol. Magazine* **5**, 246–258.

PAVLICEK, D.J. (1990) *Petrography and Geochemistry of the Upper Triassic Portoro Limestone, Liguria, Italy*. MA Thesis, University of Texas at Austin, 91 pp.

PROVANA DI COLLEGNO, G. (1835) Notes sur quelques points des Alpes Suisses. *Bull. Soc. Geol. France* **6**, 106–114.

RICKETTS, B.D. (1983) The evolution of a Middle Precambrian dolostone sequence – a spectrum of dolomitization regimes. *J. Sedim. Petrol.* **53**, 565–586.

SCHOFIELD, K. & ADAMS, A.E. (1986) Burial dolomitization of the Woo Dale Limestone Formation (Lower Carboniferous) Derbyshire, England. *Sedimentology* **33**, 207–219.

SIBLEY, D.F., DEDOES, R.E. & BARTLETT, T.R. (1987) Kinetics of dolomitization. *Geology* **15**, 1112–1114.

SPURR, J.E. (1898) *Geology of the Aspen Mining District, Colorado*. US Geol. Surv. Monograph 31, 260 pp.

TAYLOR, H.P. (1974) The application of oxygen and hydrogen isotope studies to problems of hydrothermal alteration and ore deposition. *Econ. Geol.* **69**, 843–883.

TAYLOR, T.R. & SIBLEY, D.F. (1986) Petrographic and geochemical characteristics of dolomite types and the origin of ferroan dolomite in the Trenton Formation, Ordovician, Michigan Basin, USA. *Sedimentology* **33**, 61–86.

VAN TUYL F.M. (1916) The origin of dolomite. *Iowa Geol. Surv.* **25**, 251–422.

VEIZER, J. & HOEFS, J. (1976) The nature of $^{18}O/^{16}O$ and $^{13}C/^{12}C$ secular trends in sedimentary carbonate rocks. *Geochim. Cosmochim. Acta* **40**, 1387–1397.

WANLESS, H.R. (1979) Limestone response to stress: pressure solution and dolomitization. *J. Sedim. Petrol.* **49**, 437–462.

WONG, R.K. & OLDERSHAW, A. (1981) Burial cementation in the Devonian Kaybob reef complex, Alberta, Canada. *J. Sedim. Petrol.* **51**, 507–520.

ZENGER, D.H. (1983), Burial dolomitization in the Lost Burro Formation (Devonian), east-central California and the significance of late diagenetic dolomitization. *Geology* **11**, 519–522.

Spec. Publs Int. Ass. Sediment. (1994) **21**, 203–229

Has burial dolomitization come of age? Some answers from the Western Canada Sedimentary Basin

E.W. MOUNTJOY *and* J.E. AMTHOR*

Department of Earth and Planetary Sciences, McGill University, 3450 University St., Montreal, Quebec, H3A 2A7, Canada

ABSTRACT

Many, if not most, of the dolomites in the Western Canada Sedimentary Basin appear to have formed in the subsurface during two or more stages: intermediate (between 500 and 1500 m) and deep (1500–3000 m or deeper). These subsurface dolomites exhibit similar textural characteristics, isotopic and geochemical trends, and paragenetic relationships in different parts of the basin. Dolomitization varies from partial to complete, and is texturally preserving or destructive, although most is destructive. Significant calcite dissolution is often coincident with, or postdates, dolomitization. Seafloor dolomitization, occurring centimetres below the sediment–water interface, or very shallow-burial dolomitization (less than 500 m) is suggested in some cases by microcrystallinity (5–20 μm), mimetic textures, marine carbon, oxygen and strontium isotopic compositions.

Massive replacement dolomites form between 50 and 90% of all dolomites, yet their origin is still problematic. Oxygen isotopic compositions of these dolomites suggest that they formed at between 50° and 70°C in the shallow to intermediate subsurface. Strontium isotopic compositions are slightly more radiogenic than Devonian seawater. A key problem is explaining how the vast quantities of formation waters (probably modified seawater) required for dolomitization were pumped through intermediately buried sediments.

Late-stage deeper-burial coarse-crystalline dolomite cements, including saddle dolomites, occur as vug and fracture fillings, and generally constitute between 1% and 5% of the rock. Temperatures derived from fluid-inclusion analysis range from 80°C to 200°C, and salinity estimates range from 18 to 31 wt/% NaCl equivalent, respectively. Given the temperature, the depth can be determined assuming normal geothermal gradients and, in turn, the approximate geological time of dolomitization. Generally, fluid inclusions indicate higher temperatures than expected from maximum burial, suggesting that hydrothermal formation waters moved to higher levels along fractures and conduit systems. Indeed, saddle dolomites are often, but not always, associated with Mississippi Valley-type ore deposits, which probably resulted from the circulation of hydrothermal fluids. Locally, dissolution preceded or is associated with this late phase of dolomitization. In some petroleum reservoirs and many Pb-Zn deposits (e.g. Pine Point), solution by hydrothermal fluids was extensive and overprinted many of the earlier dolomites. These examples of extensive solution and diagenetic overprinting also appear to be related to the upward movement of deep-basin brines through conduit systems.

The latest carbonates to precipitate in the Western Canada Sedimentary Basin tend to have the most radiogenic Sr compositions. High $^{87}Sr/^{86}Sr$ ratios indicate that some radiogenic Sr was introduced to these carbonates from adjacent and underlying clastic strata and/or the Precambrian basement. The saline formation waters that precipitated these late dolomites and calcites were radiogenic and similar in terms of their Sr isotope ratios to the present-day deep-basin formation waters in the Western Canada Sedimentary Basin, supporting the idea of their late timing. There is little or no direct evidence for suitable heat sources in the Western Canada Sedimentary Basin to drive thermal convection. Therefore, most late-stage dolomites appear to be related to basinwide fluid flow, probably caused by a combination of sedimentary and tectonic loading and topographically driven fluid flow during the Cretaceous and Early Tertiary. Some other basins also exhibit two or more phases of dolomitization, except that the early replacement dolomites are interpreted to have formed much earlier during the seaward progradation of tidal flats, or in a marine–meteoric mixing zone.

* Current address: Koninklijke/Shell Exploratie en Produktie Laboratorium, Volmerlaan 6, 2288 GD Rijswijk, The Netherlands.

INTRODUCTION

Massive replacement dolomites are generally considered to be difficult to form in subsurface settings (Land, 1985; Given & Wilkinson, 1987). However, textural and geochemical data indicate that many replacement dolomites and dolomite cements of the Western Canada Sedimentary Basin formed in the subsurface in two or more stages: intermediate (between 500 and 1500 m) and deep (1500–3000 m or deeper; these definitions of intermediate and deep burial are used throughout this paper). In some cases dolomitization overlaps with hydrocarbon generation (e.g. Machel, 1987a; Mountjoy & Halim-Dihardja, 1991). Accounts of burial dolomitization in various basins have appeared in Freeman (1972), Mattes and Mountjoy (1980), Mountjoy and Krebs (1983), Zenger (1983), Gregg (1985), Land (1985), Machel and Mountjoy (1986, 1987), Gawthorpe (1987), Zenger and Dunham (1988), Choquette and James (1990), Gregg and Shelton (1990), Morrow (1990a), Morrow *et al.* (1990), Narkiewicz (1990), Wallace (1990), Gao and Land (1991), Gao *et al.* (1992), Montañez (1992), Montañez and Read (1992), Barnaby and Read (1992) and many others. For overviews of Devonian sedimentation and the stratigraphy of western Canada see reviews in Moore (1988) and Morrow and Geldsetzer (1988).

In the Western Canada Sedimentary Basin, replacement dolomitization of many Devonian carbonate platforms and buildups ranges from absent to pervasive. Many buildups and some platform margins now consist of extensive replacive dolomites. Dolomites occur in many different parts of the basin and at several stratigraphic levels: Middle Devonian (Fig. 1) Keg River, Pine Point, Slave Point, Swan Hills; and Upper Devonian (Figs 2 and 3) Leduc, Nisku and Wabamun. The problem of the origin of the well known Leduc dolomites in the

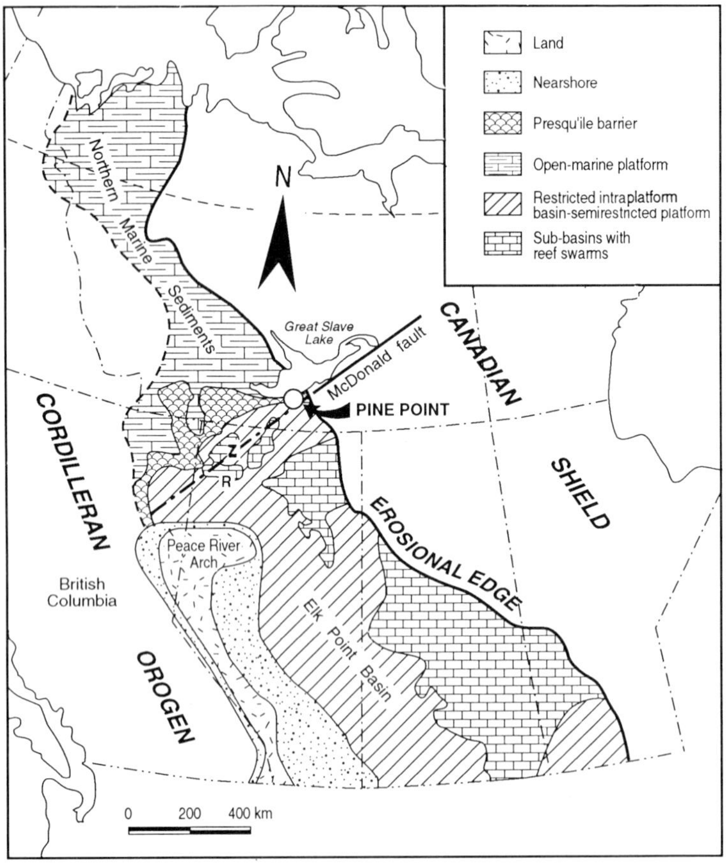

Fig. 1. Simplified regional geological map of the Western Canada Sedimentary Basin during Middle Devonian (after Moore, 1988). The Presqu'ile barrier crops out in the Pine Point area west of the Canadian Shield. It extends westward into the subsurface of the Northwest Territories and northeastern British Columbia, where its present burial depth is between 2000 and 2300 m. The McDonald-Hay River fault bounds two major geological provinces of the Canadian Shield and appears to have been active during parts of the Phanerozoic. R, Rainbow; Z, Zama subbasins.

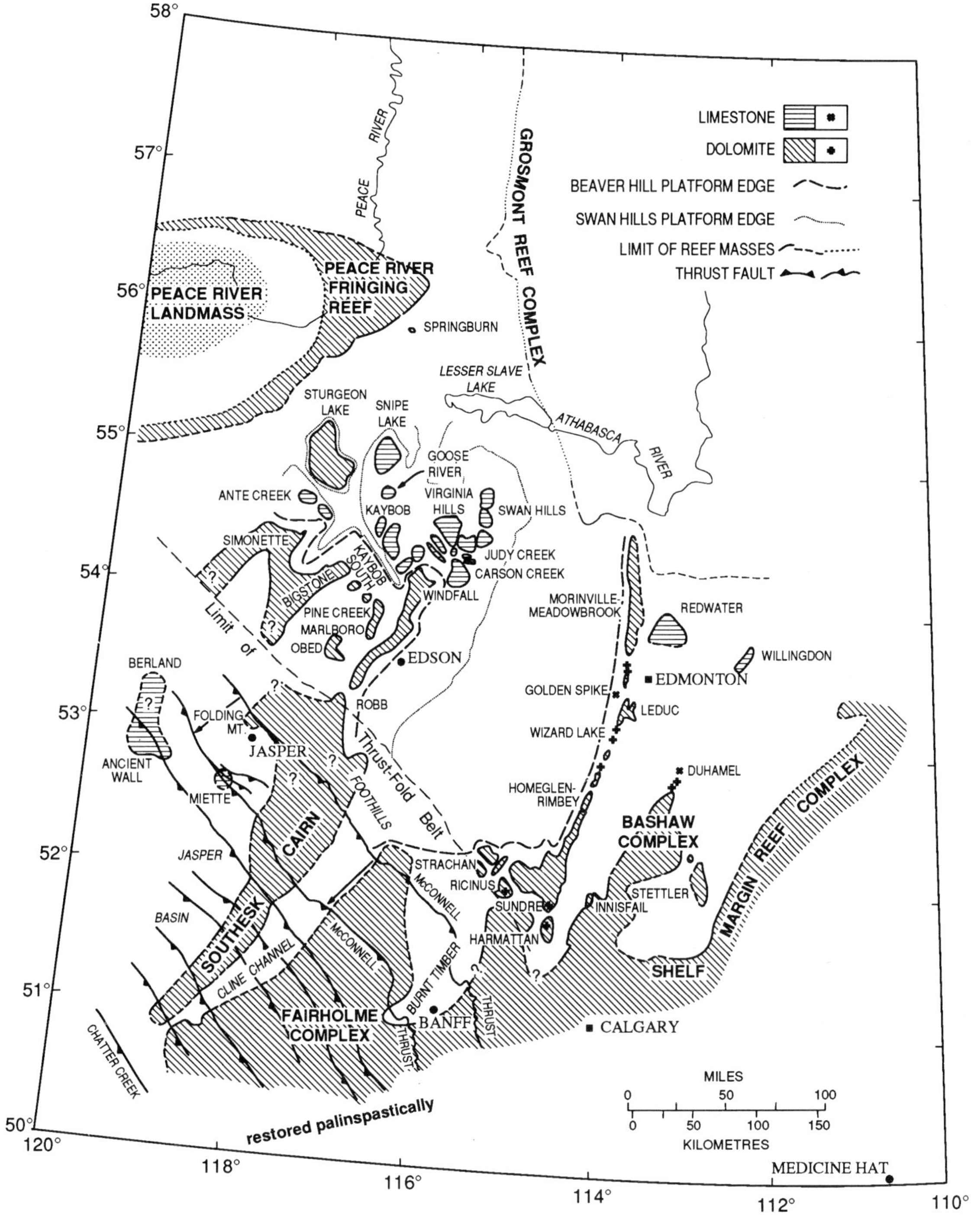

Fig. 2. Distribution of Upper Devonian Leduc and Swan Hills reefs and carbonate platforms, Alberta (updated from Mountjoy, 1980). Most reefs and platforms are dolomitized.

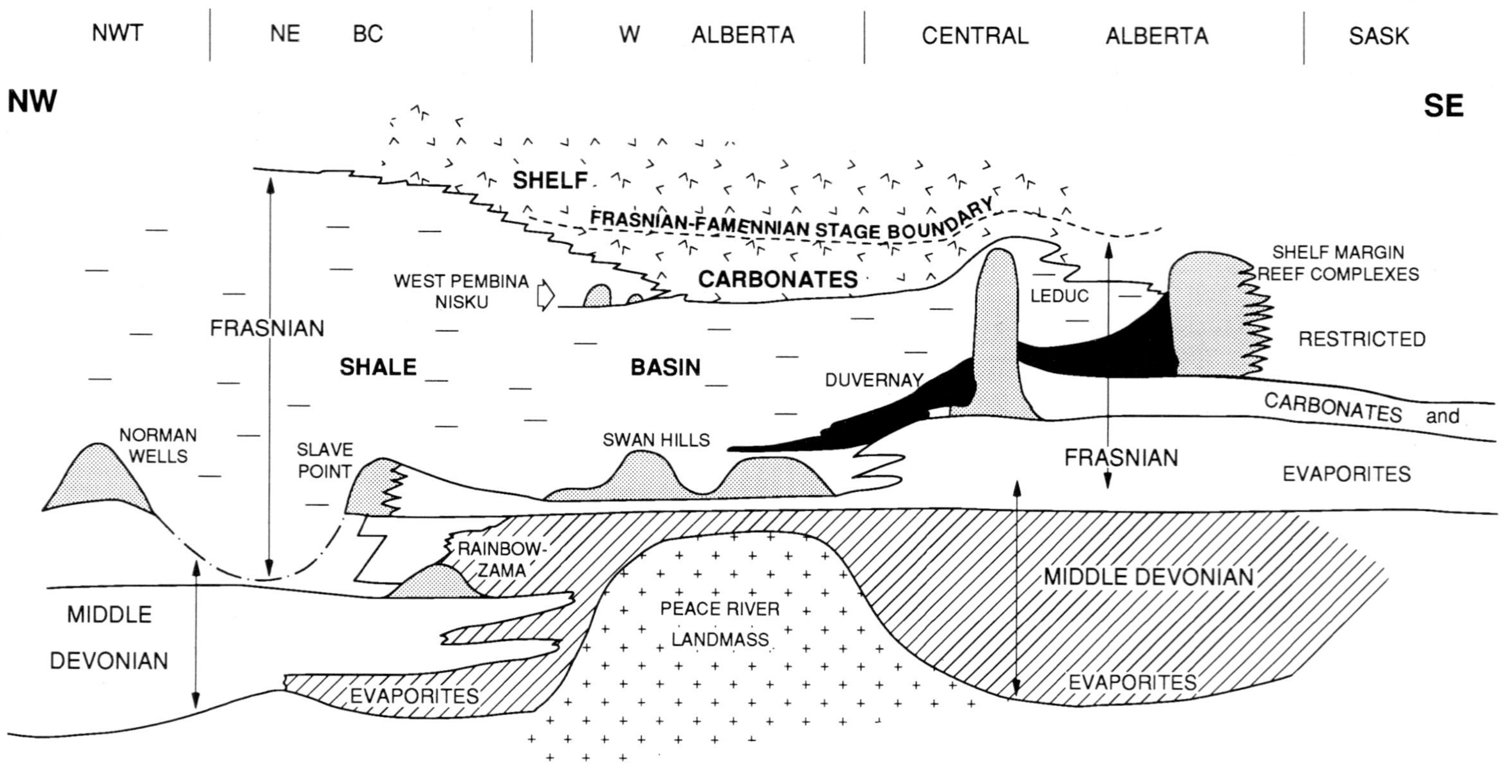

Fig. 3. Schematic northwest–southeast cross-section of the Middle and Upper Devonian from the Northwest Territories to southern Alberta and adjacent Saskatchewan, illustrating five levels of reef development, Keg River (Rainbow–Zama), Slave Point, Swan Hills, Leduc and Nisku (after Bassett & Stout, 1967).

Rimbey-Meadowbrook reef trend (Figs 2, 3 and 4) was first discussed by Illing (1956, 1959). He observed that the spatial distribution of massive replacive dolostones was compatible only with intermediate subsurface to deep-burial dolomitization. Illing suggested that some of the dolomitizing fluids were derived from the compaction of shales surrounding the buildups, but the majority of fluids came from underlying strata, with the chain of reefs acting as a conduit for water escaping from the shelf carbonates to the south. This model has also been applied to other examples, notably the Miette reef complex in the Rocky Mountain Front Ranges, by Mattes and Mountjoy (1980, 1989); for a more complete review see Machel and Mountjoy (1987). Until recently, the petrography, geochemistry and diagenesis of the Leduc dolomites had not been documented, and some of our current research is included in this paper (see also Amthor *et al.*, 1993a,b). In the past few years several important papers concerning Devonian dolomites of the Western Canada Sedimentary Basin have been published: Machel (1986, 1987a, 1988), Machel and Anderson (1989), Qing and Mountjoy (1989), Packard *et al.* (1990), Walls and Burrowes (1990), Kaufman *et al.* (1990, 1991), Mountjoy and Halim-Dihardja (1991), Mountjoy *et al.* (1992), Qing and Mountjoy (1992; in press), or are available in theses (Laflamme, 1990; Teare, 1990; Qing, 1991) or short course notes (Martindale & McDonald, 1990;

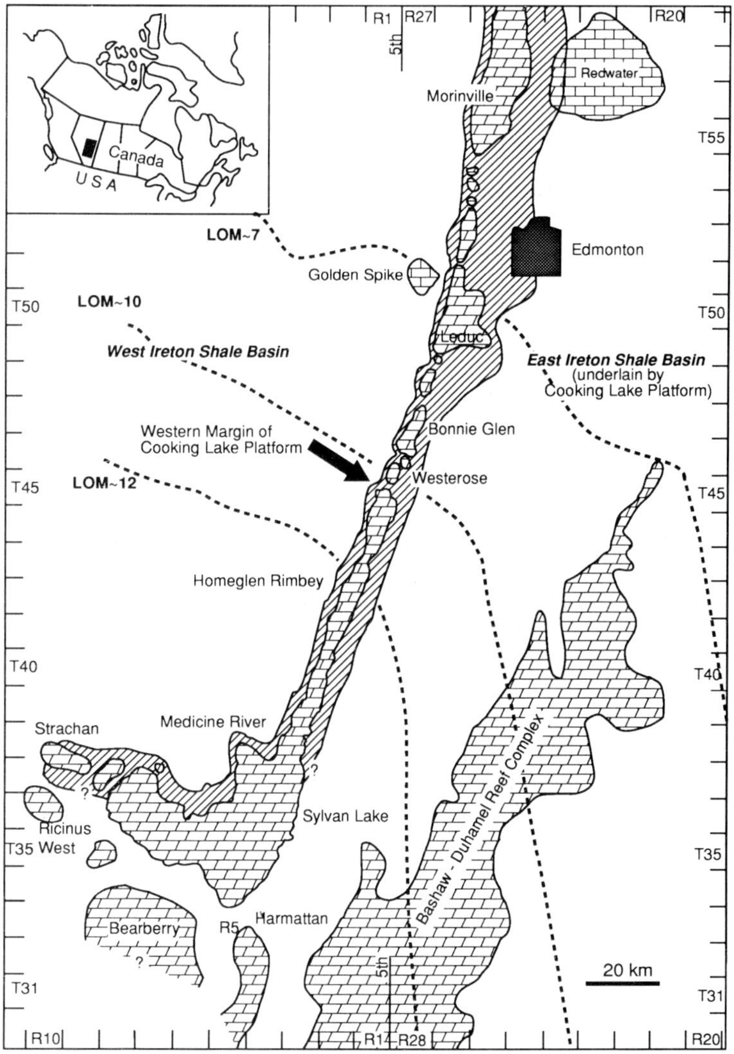

Fig. 4. Map of Upper Devonian Rimbey-Meadowbrook Leduc reef trend showing buildups and extent of dolomitization in the Cooking Lake platform (hachured pattern). Buildups are dolomitized where the margin of the Cooking Lake platform is dolomitized. Golden Spike situated off the platform margin and Redwater located above non-dolomitized platform are not dolomitized, except for the westernmost part of Redwater. Also shown is the maturation (LOM) of the adjacent Duvernay Formation source beds of the Leduc crude oils. Source beds are immature updip from Golden Spike and Leduc buildups, and oil pooled updip must have utilized the Cooking Lake dolomitized platform as a conduit system during secondary migration. (From Amthor *et al.*, 1993)

Walls & Burrowes, 1990; Dravis & Muir, 1992). These studies have shown that dolomitization in western Canada occurred in several stages. Generally, the most extensive dolomites are replacement dolomites which formed early (e.g. very shallow-burial dolomitization) or later during intermediate to deep burial. Whereas late-stage dolomite cements that fill pores, vugs and fractures are relatively well understood, the origin of the medium to coarse-crystalline replacement dolomites is less clear. Some workers (Machel & Anderson, 1989; Qing & Mountjoy, 1989; Amthor *et al.* 1993a) interpret them as primary subsurface replacement dolomites, but others would invoke recrystallization of earlier dolomites for much burial dolomite (Mazzullo, 1992; Montañez & Read, 1992; Gao *et al.*, 1992; Gregg *et al.*, 1992).

Thus the question is not whether dolomites can form in the subsurface, but how extensive are subsurface dolomites and under what conditions do they form? Another key problem is whether earlier dolomites (sea floor, mixing zone etc.) were later overprinted and/or neomorphosed in such a way as to completely obliterate earlier textures and geochemical signatures (Mazzullo, 1992; Montañez & Read, 1992). Morrow (1990b) provided an overview of dolomitization models using Canadian examples.

In this paper, recent petrographic and geochemical data from Middle and Upper Devonian dolomites of western Canada are assessed and integrated with the various concepts and models of dolomitization in order to elucidate these points. Also discussed are the problems of timing of dolomitization, sources of magnesium and transport conduits of the dolomitizing fluids. We begin with the origin of very early dolomites, since their characteristics are important in addressing the problem of dolomite recrystallization, and then discuss the various subsurface dolomites.

SEA-FLOOR OR VERY SHALLOW-BURIAL DOLOMITIZATION

In different parts of the basin, and at most Devonian stratigraphic levels, limestone buildups lacking significant dolomitization are present (lack of pervasive dolomitization model, Machel & Mountjoy, 1987). Examples of sea-floor or very shallow-burial dolomitization (<500 m) are rare in the Western Canada Sedimentary Basin. The best examples of sea-floor-related dolomitization are restricted to sabkha-type dolomites and Middle Devonian Winnipegosis reefs in the Elk Point Basin in Manitoba and Saskatchewan (Teare, 1990; Martindale & MacDonald, 1990). Some sabkha-related dolomites have also been studied immediately adjacent to and southeast of the Presqu'ile Barrier (Qing, 1991; Qing & Mountjoy, 1990), in the Grosmont Formation (Theriault & Hutcheon, 1987; Machel, 1992, personal communication), and in the Wabamun (Mountjoy & Halim-Dihardja, 1991). Fine-crystalline dolomite in some Rainbow reefs has been interpreted as having formed during subaerial exposure (Schmidt *et al.*, 1985; Qing & Mountjoy, 1989). These types of dolomite have not been observed in the Leduc Formation buildups.

Example 1: Winnipegosis buildups

The Winnipegosis Formation buildups that crop out at Dawson Bay, Manitoba, consist of a series of reef complexes that are extensively dolomitized and only locally contain calcite. The reef complexes are 75–100 m thick and consist of a series of individual buildups and associated flank and interbuildup rocks. Similar dolomitized Winnipegosis reefs are scattered in the adjacent subsurface throughout southwestern Manitoba and southern Saskatchewan. The dolomites are all microcrystalline replacement dolomites in which primary textures are well preserved (Martindale & Orr, 1987; Teare, 1990; Martindale & MacDonald, 1990). Oxygen isotope values range from $\delta^{18}O = -3.0$ to $-4.5‰$ PDB. Four dolomite samples have $^{87}Sr/^{86}Sr$ ratios between 0.70795 and 0.70804, and include two matrix dolomites and two mimetically dolomitized submarine cements. These data support the interpretation that these dolomites formed either on the sea floor or during earliest burial from the circulation of Middle Devonian Elk Point seawater through the buildups, and/or from early compaction fluids from the overlying Elk Point evaporites that were funnelled through the buildups during shallow burial (Teare, 1990). On the other hand, Kendall (1989) interpreted these dolomites and the pattern of anhydrites that encircle the individual buildups as resulting from topographically driven fluids derived from the margins of the Elk Point Basin during times of drawdown and evaporation.

Example 2: Pine Point fine-crystalline dolomites

Micro- to very fine-crystalline dolomites occur southeast of the Presqu'ile Barrier in the back-

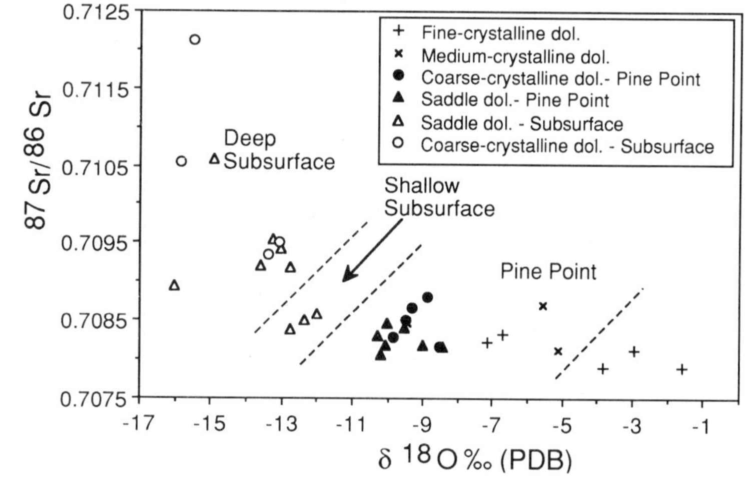

Fig. 5. The $^{87}Sr/^{86}Sr$ vs $\delta^{18}O$ for Middle Devonian Presqu'ile dolomites from along the 400 km Presqu'ile barrier, showing the progressive decrease in $\delta^{18}O$ values from fine-crystalline to saddle dolomites from Pine Point and their clear separation from subsurface coarse-crystalline and saddle dolomites (from Mountjoy *et al.*, 1992, with permission, Pergamon Press).

barrier facies of the Muskeg Formation, and become interbedded with evaporites of the Elk Point Basin (Skall, 1975; Rhodes *et al.*, 1984; Qing, 1991). Three subsurface samples of very fine-crystalline dolomites have $^{87}Sr/^{86}Sr$ ratios from 0.70790 to 0.70812 and $\delta^{18}O$ values from −1.58 to −3.84‰ PDB (Fig. 5). These values are close to those estimated for dolomites precipitated from Middle Devonian seawater, suggesting that very fine-crystalline dolomite probably formed penecontemporaneously at, or just below, the Devonian sea floor. Two fine-crystalline dolomites sampled near Pine Point ore bodies have slightly increased $^{87}Sr/^{86}Sr$ ratios (0.70822–0.70833) and depleted $\delta^{18}O$ values (−6.73 and −7.14‰ PDB), indicating alteration, probably by hydrothermal fluids during later dolomitization and mineralization (Qing, 1991; Mountjoy *et al.*, 1989, 1992).

Example 3: Wabamun Group

Very fine-crystalline sabkha dolomites interbedded with anhydrites from the Wabamun Group of the Okotoks gas field near Calgary were analysed. One sample gave an Sr isotope ratio of 0.70832 (Fig. 6). This and other sabkha dolomites from this well yielded heavy $\delta^{18}O$ values (−3.5 to −4‰ PDB, average $\delta^{18}O$ = −3.9‰ PDB). The relatively heavy oxygen isotopes and the Sr isotope ratios that fall on the Famennian part of the Sr seawater curve of Burke *et al.* (1982) (0.7082–0.7083) indicate that these dolomites were probably derived from Famennian seawater (Mountjoy & Halim-Dihardja, 1991; Mountjoy *et al.*, 1992).

Characteristics of early dolomites and implications for possible neomorphism

Dense mosaics of fine-crystalline (5–60 μm) planar-s dolomite are typical of dolomites formed under near-surface low-temperature saline to hypersaline conditions (Gregg & Sibley, 1984; Machel & Mountjoy, 1987; Sibley & Gregg, 1987; Zenger & Dunham, 1988; Gregg & Shelton, 1990). They are commonly mimetic, preserving textural details of the precursor limestones as well as primary depositional fabrics. Where these are not overprinted, their isotopic composition is close to those values estimated for dolomites precipitated from seawater. That neomorphism/recrystallization can modify or destroy original dolomite textures has been shown (e.g. Gregg & Sibley, 1984; Gregg & Shelton, 1990; Montañez & Read, 1992), but the question is whether such neomorphism can completely obliterate all earlier textures and drastically change their geochemistry. Several factors suggest that this is not the case: many ancient dolomites have a large range of crystal sizes (e.g. Amthor & Friedman, 1991; Gregg & Shelton, 1990) and neomorphism would probably eliminate the finer crystals (Sibley, 1991). Also, it has been shown by studies discussing possible neomorphic recrystallization (e.g. Zenger & Dunham, 1988; Gao, 1990; Gao & Land, 1991; Kaufman *et al.*, 1991; Gregg & Shelton, 1990; Amthor & Friedman, 1992; Montañez & Read, 1992) that dolomite neomorphism is not formation-wide, but associated with either meteoric modification and/or hot, mineralizing brines. Thus, it appears from these examples, as well as the examples of the Western

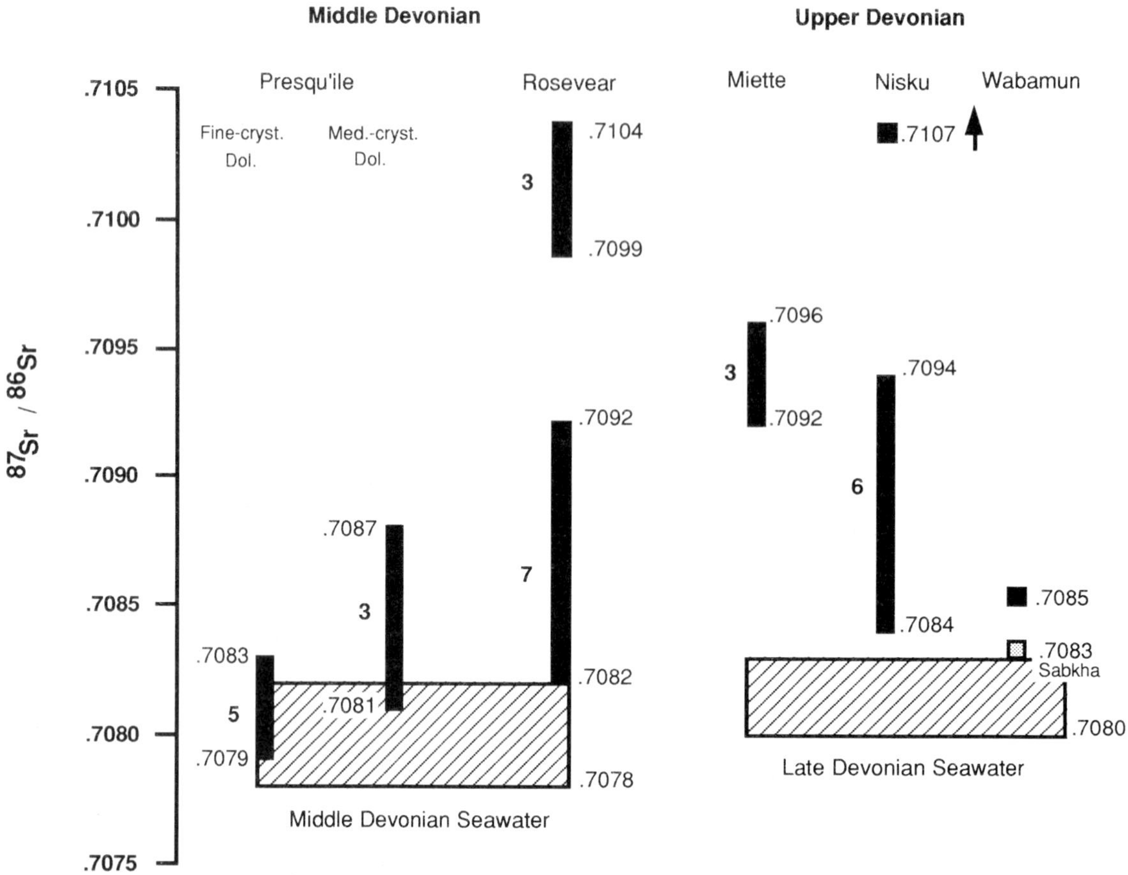

Fig. 6. The $^{87}Sr/^{86}Sr$ values of early replacive matrix dolomites, Middle and Upper Devonian of western Canada, showing ranges for Middle and Upper Devonian seawater. Presqu'ile dolomites are divided into fine- and medium-crystalline types (from Mountjoy *et al.*, 1992, with permission, Pergamon Press).

Canada Sedimentary Basin, that whenever there has been extensive early dolomitization, neomorphism is not sufficiently prevalent in ancient dolomites to completely obliterate the textures of these early dolomites. When overprinted during burial (e.g. Pine Point and Wabamun fine-crystalline dolomites) the isotope geochemistry may be modified but the distinct fine-crystalline texture is, in many instances, more or less preserved, leaving a textural record of the very early stages of dolomitization.

MASSIVE REPLACEMENT DOLOMITES

Massive replacement dolomites are ubiquitous throughout the Upper and Middle Devonian of the Western Canada Sedimentary Basin. Most carbonates have been thoroughly and completely dolomitized, generally leaving few clues as to how and when the dolomitization occurred. To understand this form of dolomitization we are dependent on the few buildups and areas where partial dolomitization can be studied (Mattes & Mountjoy, 1980; Machel & Anderson, 1989; Qing & Mountjoy, 1989, 1990; Qing 1991; Kaufman *et al.*, 1991; Mountjoy & Halim-Dihardja, 1991). Typically, massive replacement dolomites are medium- to coarse-crystalline, subhedral to euhedral planar mosaic dolomites, and coarse-crystalline anhedral non-planar dolomite (Table 1). Coarse-crystalline mosaic dolomites are the most abundant (≈ 90%) and show relatively uniform petrography, with homogeneous red-mottled cathodoluminescence (Fig. 7). Both texturally pre-

Table 1. Petrographic and geochemical characteristics of replacement dolomites in Devonian buildups, western Canada.

	Miette*	Bearberry†	Swan Hills‡	Westerose-Leduc
Petrography	m.-c.x (60–110 μm, 300–600 μm) pl-s(e); non-luminescent	m.-c.x (62–700 μm) pl-s(e); homogeneous to slightly mottled red cathodoluminescence	m.-c.x. (100–600 μm) pl-s(e), and npl-a; red cathodoluminescence	m.-c.x (100–800μm) pl-s(e), and npl-a; homogeneous red to orange-red mottled cathodoluminescence
Geochemistry	$\delta^{13}C$: $\bar{x}$ = 3.0	1.77	2.5	3.0
	$\delta^{18}O$: $\bar{x}$ = −5.0	−4.1	−7.5	−5.98
	$^{87}Sr/^{86}Sr$: $\bar{x}$ = 0.70924	0.7082	0.70890	0.7083
	Fe: $\bar{x}$ = 197	1797	700–4500	1182
	Mn: $\bar{x}$ = 59	189	<200	409
	Sr: $\bar{x}$ = 57	84	23–44	—
	$CaCO_3$: $\bar{x}$ = —	49.9	50.1–51.1	49.3
	$MgCO_3$: $\bar{x}$ = 46.2	50.2	48.3–49.8	50.0
Distribution	Buildup margin; overlying whole buildup strata		Margin of marine channel, capping and underlying platform	Whole buildup, underlying platform and overlying strata
Present depth	5000 m max	3200–4300 m	3050–3450 m	1600–2500 m

* From Mattes and Mountjoy (1980) and Mountjoy *et al.* (1992).
† From Laflamme (1990).
‡ From Kaufman *et al.* (1991).

serving and texturally destructive dolomites are present, but most have textures destroyed. Massive replacement dolomites postdate early-diagenetic features such as micritization, submarine cementation and low-amplitude wispy stylolites. They crosscut depositional facies boundaries and extend into overlying limestones, suggesting that pervasive replacement dolomitization postdates the deposition of adjacent basin limestones and shales. These features are compatible only with subsurface dolomitization.

Two stages of replacement dolomitization are recognized in many areas, based on petrographic and geochemical analyses. In this review we present the example of Leduc dolomites from the Rimbey-Meadowbrook reef trend, and include some comparisons with other examples of replacement dolomites.

Example: Rimbey-Meadowbrook reef trend

Within the Cooking Lake platform and the overlying Leduc Formation buildups, three types of replacement dolomite, mostly coarse-crystalline subhedral to euhedral planar mosaic, form 90% of these dolomites, the remainder consisting of two types of later dolomite cements (Fig. 7) (Amthor *et al.*, 1993a). The mosaic dolomites form unimodal mosaics that show a relatively uniform petrography and homogeneous red-mottled cathodoluminescence (Table 1). One way to explain the coarse-crystalline planar–subhedral texture of mosaic dolomite is that it may have resulted from the growth of a few large crystals in a homogeneous calcite matrix. For example, in stromatoporoid wackestones of the Cooking Lake Formation, the finer-grained matrix is preferentially dolomitized, but the fossils remain undolomitized. If the dolomite in the matrix continued to grow until the crystals met, it would result in a unimodal planar-s dolomite mosaic with calcite · allochems. Alternatively, if any of the remaining undolomitized matrix and fossils were dissolved, the resultant dolomite would be a unimodal planar-s(e) dolomite with intercrystalline and vuggy porosity, a common texture in the completely dolomitized strata of the Leduc Formation. For the diagenesis of the undolomitized Golden Spike buildup (Figs 2 and 4) see McGillivary and Mountjoy (1975), Walls *et al.* (1979) and Carpenter *et al.* (1991).

Stable isotope values of replacement dolomites from widely separated buildups in the central part of the Rimbey-Meadowbrook reef trend range from $\delta^{13}C$ = 1.5 to 6.0‰ PDB, and $\delta^{18}O$ = −7.5 to −5.0‰ PDB (Fig. 8). The composition of hypothetical Upper Devonian marine dolomites, calculated assuming that dolomite is enriched in $\delta^{18}O$ by about 2–4‰ (Land, 1980) over coexisting Upper Devonian marine calcite values, is shown for com-

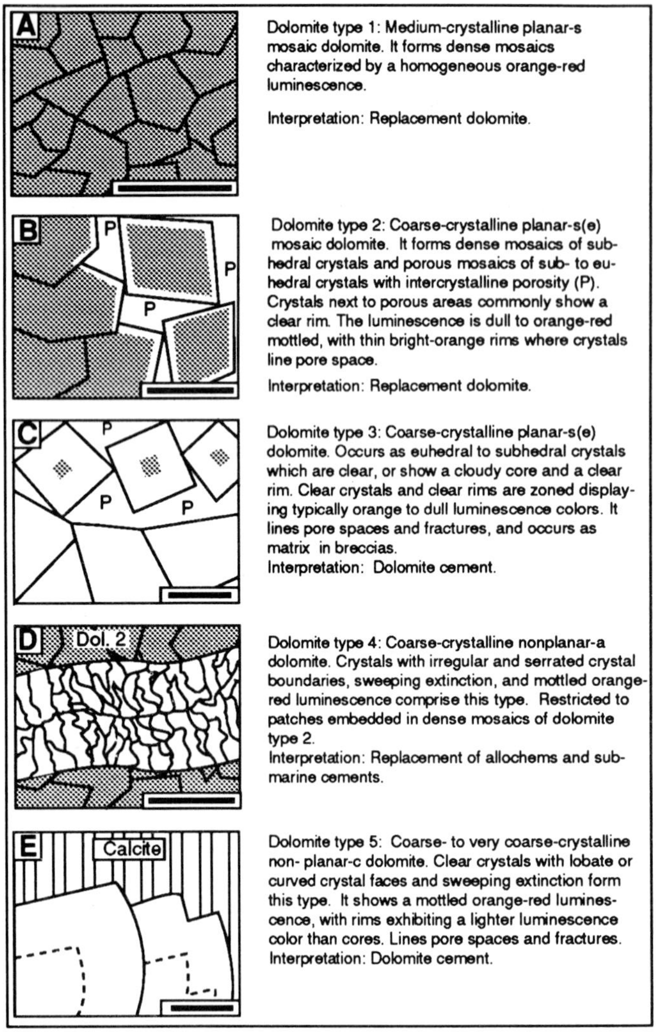

Fig. 7. The five dolomite textures observed in Upper Devonian Cooking Lake/Leduc Formation carbonates of the Rimbey-Meadowbrook reef trend. Scale bars = 500 μm (from Amthor *et al.*, 1993).

parison purposes. The replacement dolomites are significantly depleted in $\delta^{18}O$ relative to hypothetical Upper Devonian marine dolomites (Fig. 8). The three replacement dolomite textures (types 1, 2 and 4, Figs 7 and 10; Amthor *et al.*, 1993a) overlap (Fig. 8). Given the large number of samples of replacement dolomites (n = 102) and the large spatial distribution of the sample localities, there is a remarkably narrow range of isotopic compositions. Samples from replacement dolomites (types 1 and 2) in breccias are enriched up to 2‰ in $\delta^{13}C$ (breccia field in Fig. 8). The later planar and non-planar dolomite cements (types 3 and 5, Fig. 7) have the most negative $\delta^{18}O$ values (average = −7.0 and −8.2‰ PDB, respectively; Fig. 8). Similar isotope values have been reported by others, with $\delta^{18}O$ values as low as −8.0% PDB in the Miette buildup (Fig. 9; Mattes & Mountjoy, 1980) and −9.0‰ PDB in Nisku reefs (Machel & Anderson, 1989). Microprobe analyses indicate that coarse-crystalline replacement dolomites are nearly stoichiometric, with intermediate Mn (average = 409 ppm) and Fe concentrations (average = 1182 ppm; Table 1). Microprobe traverses across individual crystals of all replacement dolomites show no systematic variations in Ca, Mg, Fe and Mn contents (Amthor *et al.*, 1993a). No covariant, regional or stratigraphic trends are apparent. The geochemical compositions of replacement dolomites from several different buildups (Table 1), indicate that these dolomites from widely separated parts of the basin have similar composition.

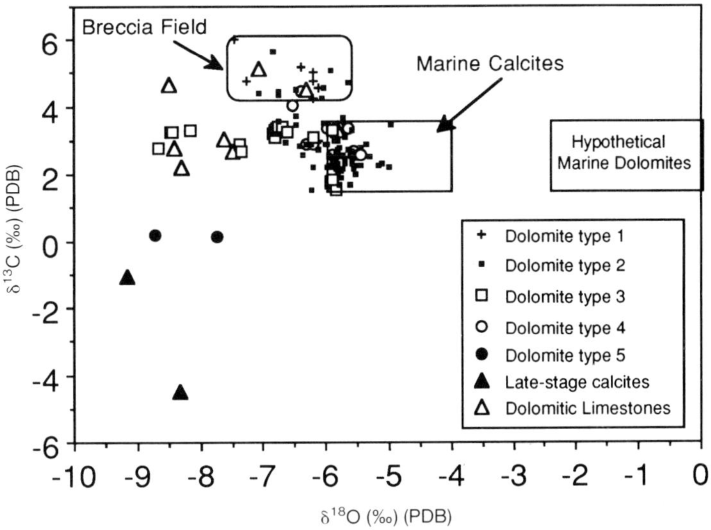

Fig. 8. Plot of $\delta^{13}C$ versus $\delta^{18}O$ isotope values for the five dolomite types and late-stage calcite cements from different Leduc buildups, stratigraphic levels and positions within buildups, and for dolomite samples from dolomitic Cooking Lake limestones in the Rimbey-Meadowbrook reef trend. Replacement dolomites (types 1, 2 and 4) form a relatively tight cluster, whereas later dolomite cements (types 3 and 5) are lighter in both $\delta^{13}C$ and $\delta^{18}O$. Samples from Cooking Lake dolomitic limestones overlap with both replacement dolomites and void-filling dolomites. Late-stage calcites have the most negative isotopes values. Massive replacement dolomites (types 1, 2 and 4) are significantly more negative than hypothetical dolomites precipitated in equilibrium with Upper Devonian seawater. Replacement dolomites (types 1 and 2) from breccias (breccia field) show the most positive $\delta^{13}C$ values (Amthor *et al.*, 1993).

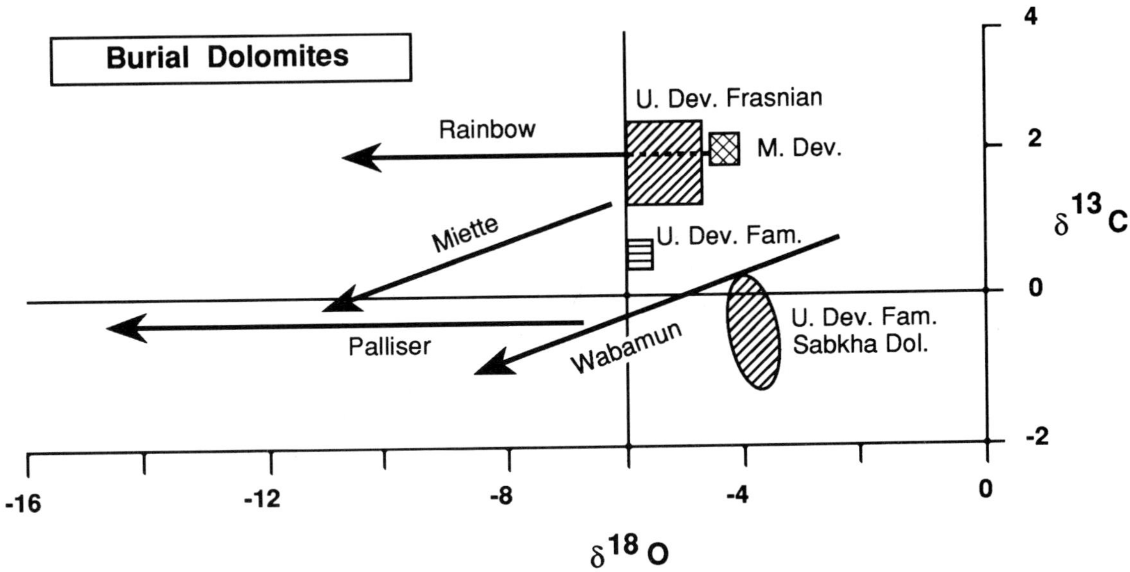

Fig. 9. Stable isotope trends of burial dolomites from Upper and Middle Devonian fields and outcrops in Western Canada (Mattes & Mountjoy, 1980; Qing & Mountjoy, 1989; Mountjoy & Halim-Dihardja, 1991; unpublished data). Boxes are values from calcite fossils and marine cements based on our data and those of others.

Discussion

Several lines of evidence suggest that the medium to coarse-crystalline replacement dolomites do not represent the neomorphic modification of sea-floor-related fine-crystalline dolomites, because (a) evidence for early seawater-related dolomitization has not been found in limestones and limestone buildups, and in incompletely and completely dolomitized buildups (except possibly for Middle Devonian Winnipegosis reefs); (b) reworked clasts in debris-flow deposits of buildup-slope facies are limestone, indicating a lack of early dolomitization; (c) coarse-crystalline replacement dolomites grade directly into limestones and not into fine-crystalline dolomites; (d) dolomitization cross-cuts facies boundaries and members of the underlying Cooking Lake platform; (e) there is no evidence for either regional meteoric recharge or hot mineralizing fluids in the Leduc Formation buildups and underlying Cooking Lake platform, which could have resulted in neomorphic alteration; and (f) the diagenetic paragenesis indicates that dolomitization took place after submarine cementation, and overlaps with low-amplitude wispy stylolites. Theoretically, the oxygen isotopic values of dolomites that precipitated from seawater should be 2–4‰ heavier than those of calcites precipitated in marine seawater (Land, 1980). None of the replacement dolomites reported here comes close to these values (Figs 7 and 8). Thus, except for a few fine and microcrystalline or sabkha dolomites, none of the replacement dolomites appears to represent precipitates from seawater. This isotopic discrepancy could be explained if these dolomites did represent sea-floor precipitation but had undergone complete exchange of oxygen isotopes during neomorphism. Such an exchange would have involved considerable amounts of pore fluids that would have differed isotopically from Devonian seawaters that presumably were buried with these sediments. There is little or no evidence for these fluids and, therefore, it seems unlikely that the oxygen isotopic composition of the replacement dolomites resulted from such an exchange. Thus, we conclude that these dolomites represent replacement of precursor limestones during burial.

Replacement dolomites in the Rimbey-Meadowbrook reef trend formed at 50–70°C (Fig. 10), assuming that the dolomitizing fluids had a $\delta^{18}O$ value near −2.5‰ SMOW, as estimated for Late Devonian seawater, and assuming the $\delta^{18}O$ values of the formation waters were not modified significantly during early burial. Using a geothermal gradient of 30.5°C km^{-1} and a surface temperature of 30°C (Hudson & Anderson, 1989), temperatures of 50–60°C could have been attained when the Leduc Formation was buried to 600–1200 m, depths reached towards the end of the Late Devonian to Early Carboniferous (Fig. 11a). Little significant change in burial temperatures occurred until the Cretaceous, when the Leduc Formation was buried to maximum depths of 2800–4500 m, corresponding to burial temperatures ranging between 110° and 180°C (Fig. 11a). Similar temperatures (35–75°C) have been calculated for early replacement dolomites from other parts of the basin (Machel & Anderson, 1989; Qing & Mountjoy, 1989; Kaufman *et al.*, 1991; Mountjoy & Halim-Dihardja, 1991).

In most areas the formation waters responsible for the replacement matrix dolomites were slightly

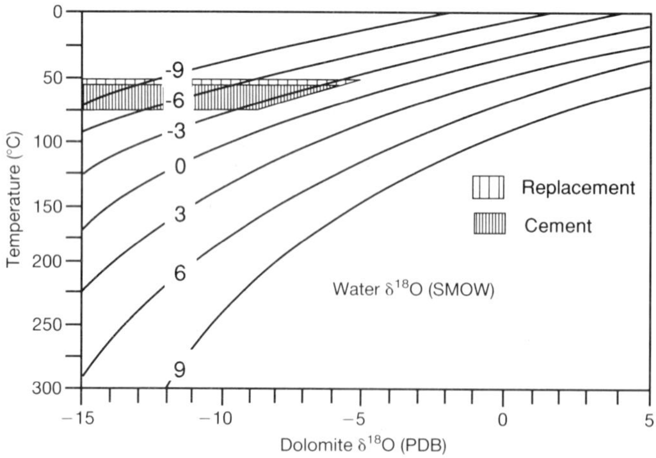

Fig. 10. Plot of equilibrium relationships between temperature and isotope composition of water and dolomite (Land, 1985), assuming a $\delta^{18}O$ of the dolomitizing fluid of −2.5 (SMOW), a range of $\delta^{18}O$ = −5.0 to −7.5 PDB for replacement dolomites and $\delta^{18}O$ = −8.0 PDB for dolomite cements. Replacement dolomites and cements could have formed at minimum temperatures ranging from 50° to about 75°C. If the $\delta^{18}O$ of the dolomitizing fluids increases, higher temperatures are required (from Amthor *et al.*, 1993).

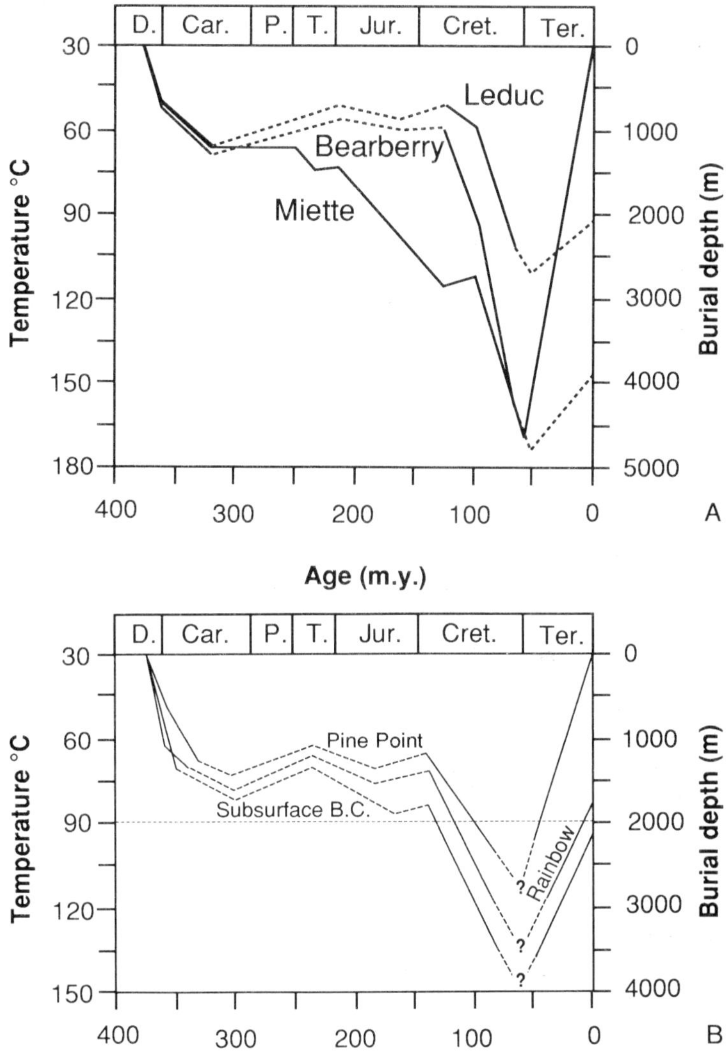

Fig. 11. Representative burial-history plots. (A) Upper Devonian Miette (Mattes & Mountjoy, 1980), Bearberry (Laflamme, 1990), and Leduc buildups (Amthor *et al.*, 1993), constructed using a geothermal gradient of 30°C km^{-1} and a surface temperature of 30°C. (B) Middle Devonian Presqu'ile barrier (top of the Sulphur Point Formation) at Pine Point, Rainbow basin subsurface (Qing & Mountjoy, 1989), and subsurface of northeast British Columbia.

enriched in ^{87}Sr, but not strongly radiogenic. Replacement matrix dolomites from the Presqu'ile barrier, Rainbow, Rosevear (except for three samples), Miette, Leduc, Nisku and Wabamun all have slightly higher $^{87}Sr/^{86}Sr$ ratios (0.7081–0.7094) than the corresponding Devonian seawater (Fig. 6; Table 1). Dolomitizing fluids were probably modified from Devonian seawater, with small amounts of ^{87}Sr being added during early compaction from adjacent or underlying clastics or older carbonates (Mountjoy *et al.*, 1992). Clearly, these formation waters differed significantly from the strongly saline formation waters responsible for the late-stage saddle dolomites, and probably represent seawater that was entrained with the Devonian sediments deposited in the Western Canada Sedimentary Basin.

Porosity development in replacement dolomites

In many regions dolomitization has strongly controlled the origin and quality of carbonate reservoirs, but the role of dolomitization with respect to the enhancement or reduction of porosity in dolomites is poorly understood (Purser *et al.*, this volume). The reasons for this appear to be the many different ways in which dolomite may form and/or be modified during later diagenesis and burial. With respect to porosity, it is important to know what chemical reactions were involved (Machel & Mountjoy, 1986;

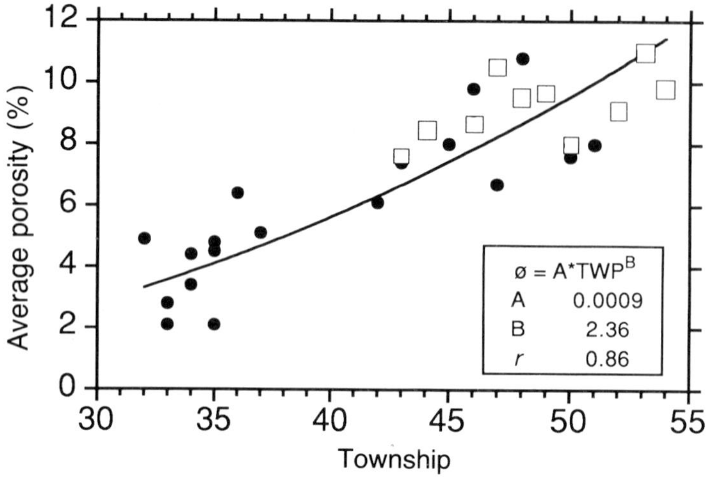

Fig. 12. Average porosity plotted versus north–south townships along the Rimbey-Meadowbrook reef trend. Data (ERCB) obtained from averaging all core porosity analyses from the Leduc and Cooking Lake formations in a series of wells. Curve represents best fit using a statistical computer program. Township 55 represents maxiumum burial depths of about 2500 m and Township 30 burial depths >3500 m (see Fig. 4 for township locations). Some buildup interior wells (filled circles) have the same porosity as buildup flank wells (open squares). Formula provides equation of best fit curve. *r*, correlation coefficient.

Morrow, 1990b), which governs whether or not porosity increases, remains about the same or decreases (Machel & Mountjoy, 1986, 1987).

In completely dolomitized Upper Devonian reefs, moulds and vugs with some intercrystalline porosity are the most prevalent pore types, accounting for 50–100% of the porosity in the studied cores. In many ancient examples, secondary porosity appears to have developed either during dolomitization, or later by a separate dissolution event (or events) that dissolved some or all of the remaining calcite (Choquette *et al.*, 1992; Amthor *et al.*, 1993a). What controls these porous zones is uncertain, but they may reflect areas of higher primary porosity in the precursor limestones.

In the shallow part of the basin the Devonian limestone buildups of western Canada have retained a higher porosity than their dolomitized counterparts (McGillivary & Mountjoy, 1975; Walls & Burrowes, 1985, 1990). For example, the average porosity of the Golden Spike limestone buildup (see Fig. 4) is about 7% for the buildup margins and 15% for the buildup interior (Energy Resources Conservation Board of Alberta (ERCB) data from core analyses). For dolomites along the central part of the Rimbey-Meadowbrook reef trend, porosities average between 4% and 7% (Fig. 12, based on ERCB data, representative of dolomites in many different parts of the basin; Amthor *et al.*, 1993b; Drivet & Mountjoy, 1993), clearly somewhat lower than for limestone buildups such as Golden Spike. Because this dolomitization took place during shallow to intermediate burial, when the limestones were relatively porous, and because dolomites undergo less pressure solution than limestones during burial, replacement dolomitization must have significantly reduced the original primary porosity from something in the order of 25–30% to about 10%. The porosities in limestone buildups for this shallower part of the reef trend were reduced more slowly and progressively during burial from between 20–35% to their present values of 7–15%.

Further downdip in the deeper part of the basin, the Strachan and Ricinus buildups (Figs 2 and 4) provide excellent examples of the differences between limestone and dolomite reservoirs (Marquez, 1993). Dolomites are similar in texture, geochemistry and porosity type to dolomites updip along the reef trend. Porosities vary between 1.2 and 20.7%, averaging 8.4%, and permeabilities between K_h 14 to 100 mD, and K_v 8 to 40 mD. The Strachan gas field forms two separate pools. The western D3-A pool is partly to completely dolomitized, and the eastern D3-B pool consists of partially dolomitized limestones. The D3-A pool, with more extensive replacement dolomitization, is the better reservoir, with porosities ranging from 2% to 10% (average 4%), and permeabilities from 1 to 40 mD. The higher porosities in the D3-A pool appear to be due to less chemical compaction in the dolomites than in the limestones, and also perhaps in part to more solution having occurred in the dolomites (Marquez, 1993; personal communication), a characteristic noted for South Florida dolomites (Schmoker & Halley, 1982).

Origin of widespread dolomitization

Why were the Devonian strata in this basin so extensively dolomitized during shallow to intermediate burial? Devonian rocks in other basins, for example in Europe and Australia, are predominantly limestones (Mountjoy & Krebs, 1983; Wallace, 1990). As noted by many workers (see review in Machel & Mountjoy, 1987), there is something special or unusual about the Western Canada Sedimentary Basin. The reasons for this are not clear, except that pervasive replacement dolomitization is related to the presence of extensive conduit systems along carbonate platform margins at several stratigraphic levels (Presqu'ile barrier, Swan Hills, Cooking Lake and Nisku), as well as along certain faults and fractures (Presqu'ile, Wabamun).

The role of basement faulting with respect to replacement dolomitization is difficult to discern. There is at least an indirect relationship to basement structure because basement faults and trends appear to control the linear platform and reef trends of the Middle Devonian Presqu'ile barrier (MacDonald-Hay River fault system) and the Upper Devonian Rimbey-Meadowbrook (Ribey Arc), Bashaw (Red Deer trend) and southeast Alberta shelf margin reef complex (Mountjoy, 1978, 1980; Ross & Stephenson, 1989; Ross, 1991). Faults do control dolomitization in the case of the Wabamun east of the Peace River Arch (Jones, 1980; Dix, 1990, 1993; Packard *et al.*, 1990; Mountjoy & Halim-Dihardja, 1991), and have been inferred in the Swan Hills field (Viau & Oldershaw, 1984), the Rosevear field (Walls & Burrowes, 1990; Kaufman *et al.*, 1991) and in Nisku reefs (Machel & Anderson, 1989). Now that the basement structure is better understood (Ross & Stephenson, 1989; Ross, 1991), the relationships of these structures to the sedimentary history and subsidence of the basin can be better established. Probably subtle adjustments during sedimentary loading above these major crystalline basement structures led to fracturing that enhanced the permeability in the porous carbonate conduits and, perhaps, the underlying strata. These conduit systems possibly carried the waters responsible for the formation of massive replacement dolomites during intermediate burial, and later, during deeper burial, transported diagenetic fluids that precipitated small amounts of dolomite cements, including saddle dolomites and late-stage calcite cements, and locally metallic sulphides.

Two driving mechanisms for regional basinwide migration of pore fluids are tectonic and sedimentary loading and squeezing of the stratigraphic sequence during continental collision (Oliver, 1986), and topographically driven fluid flow caused by uplifted mountains bordering a sedimentary basin (Garven & Freeze, 1984; Garven 1989). Modelling suggests that gravity-driven flow could drive large amounts of fluids out of sedimentary basins. In the case of the Western Canada Sedimentary Basin, there is conclusive evidence for the concurrent tectonic and sedimentary loading squeezing the sedimentary wedge during tectonic telescoping, and the establishment of regional topographically driven flow systems affecting the basin, beginning in the Late Jurassic and Early Cretaceous and continuing until the Palaeocene (Hitchon, 1984; Qing & Mountjoy, 1992). Earlier tectonic loading in relation to the Antler orogeny (Late Devonian–Early Carboniferous) could have been responsible for driving deep-seated formation fluids, including residual evaporitic brines, northeastward up along porous carbonate conduit systems such as the Rimbey-Meadowbrook reef trend, and for dolomitizing the western margin of the Cooking Lake platform and the overlying Leduc buildups.

The Devonian of western Canada also contains more evaporites than carbonates of the same age in other basins in Europe and Australia. Similarities in the sulphur isotopic composition of anhydrite from Late Devonian evaporites and anhydrite cements in carbonate units of western Canada (Nahnybida *et al.*, 1982; Machel, 1986; Ueda *et al.*, 1987; Krouse *et al.*, 1988; Kaufman *et al.*, 1990; Machel & Burton, 1991) suggest a possible link between evaporites and carbonate diagenesis, including dolomitization (e.g. Spencer, 1987).

DEEPER-BURIAL (LATE) DOLOMITES

Late-stage dolomites occur as coarsely crystalline cements in pores, vugs and fractures, forming up to 5% of completely dolomitized rocks. They also occur in similar situations in limestones (Mountjoy & Halim-Dihardja, 1991). Dolomite cements appear to be less common in the updip shallower parts of the basin (above maximum burial depths of about 3 km), as observed in the Leduc and Grosmont formations along the Rimbey-Meadowbrook reef trend (Fig. 2; Amthor *et al.*, 1993a). In the Presqu'ile barrier, late-stage dolomite cements are abundant in Pine Point ore bodies that occur along three trends and that were buried only to depths of about 2.5 km (Fig. 11b; Qing, 1991; Qing & Mountjoy, 1992). Only recently have detailed isotopic and fluid-inclusion data become available for these dolomites. Dolomites in many parts of the basin have yet to be studied, so that the information reported here may be superseded. Petrographic and geochemical data

Table 2. Petrographic and geochemical characteristics of coarse-crystalline and saddle dolomites in Devonian buildups, western Canada.

	Miette*	Bearberry†	Swan Hills‡	Leduc
Petrography	Several mm, euhedral–subhedral —	Several mm, euhedral–subhedral, homogeneous to zoned, variable fluorescence	Several mm, euhedral–subhedral, CL	Several mm, euhedral–subhedral
Geochemistry ($\bar{x}$)				
$\delta^{13}C$	−0.6	−3.05	+1.77	+0.17 § 3.08¶
$\delta^{18}O$	−9.3	−6.41	−7.84	−8.22, −6.97
$^{87}Sr/^{86}Sr$	0.7107	0.7104	0.7109	—
Fe	841	1490	100–10000	615
Mn	74	165	<300	196
Sr	96	235	67	—
$MgCO_3$	46.8	46.6	—	49.0
Average depth	5000 m max	4115 m	3050–3450 m	1600–2500 m

* From Mattes and Mountjoy (1980) and Mountjoy *et al.* (1992).
† From Laflamme (1990).
‡ From Kaufman *et al.* (1990)
§ Saddle.
¶ Dolomite cements.
CL, cathodoluminescence.

Table 3. Summary of fluid-inclusion data, late-stage dolomites and calcites, Devonian western Canada.

	T_h°C	T_m°C	salinity (wt.% NaCl equiv.)
Presqu'ile barrier (Mid-Dev), Qing (1991)			
Saddle dolomite			
Pine Point	85–115	−7–22	10–24
Shallow subsurface	95–135	−7–34	10–31
Deep subsurface	105–210	−2–22	10–24
Late calcites			
Pine Point	70–130	−3–5	6–8
Shallow subsurface	80–150	−7–10	10–13
Deep subsurface	130–176	−10–16	13–19
Rosevear (Mid-Dev), Kaufman *et al.* (1990)			
Saddle dolomite	125–160	−18–22	21–24
Late calcites	90–126	−6–17	9–21
	148–170	−6–10	9–13
Nisku reefs (Upper Dev), Anderson (1985)			
Saddle dolomite	99–175	−6–25	9–28
Wabamun (Upper Dev), Mountjoy & Halim-Dihardja (1991)			
Saddle dolomite	90–130	−13–24	17–24
	170–210	−21–23	22–24

are shown in Table 2, data from fluid inclusions for saddle dolomites and an even later generation of calcite are summarized in Table 3. In many cases, fluid inclusions in saddle dolomites yield temperatures of homogenization higher than those estimated to have occurred in the rocks during maximum burial (compare with burial curves, Fig. 11), whereas inclusions in the later calcites yield somewhat lower homogenization and less negative final melting temperatures (Table 3), suggesting that the calcites formed from cooler and less saline formation waters. Both the late-stage dolomite (except at Pine Point)

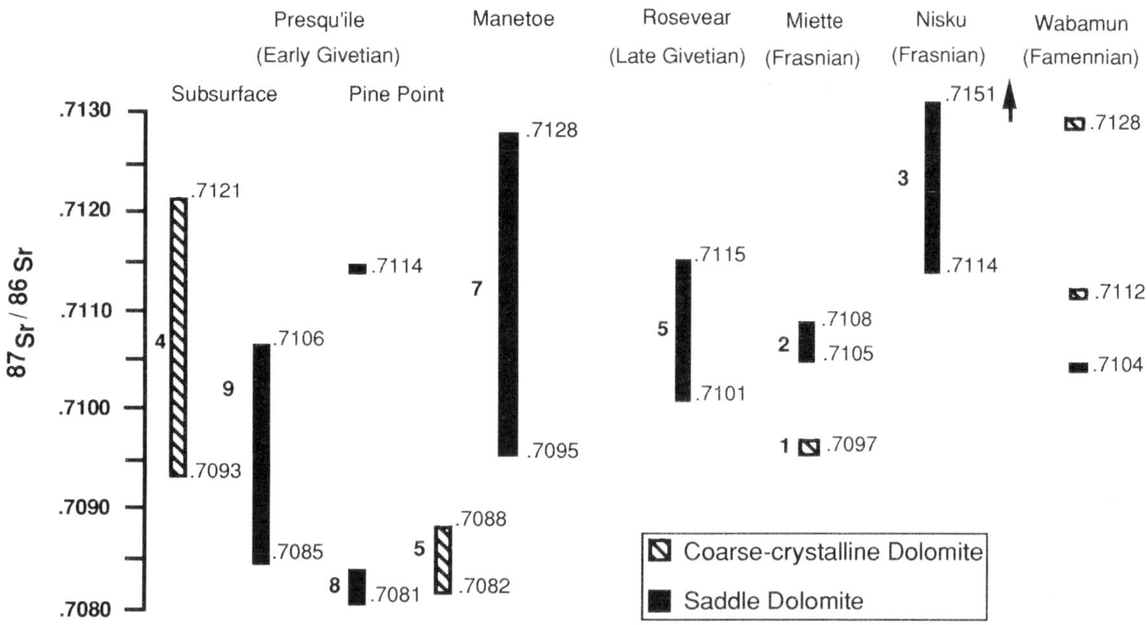

Fig. 13. The $^{87}Sr/^{86}Sr$ values of late-stage burial dolomites, Middle and Upper Devonian, western Canada. Presqu'ile dolomites are divided into samples from Pine Point and the subsurface (from Mountjoy *et al.*, 1992, with permission, Pergamon Press).

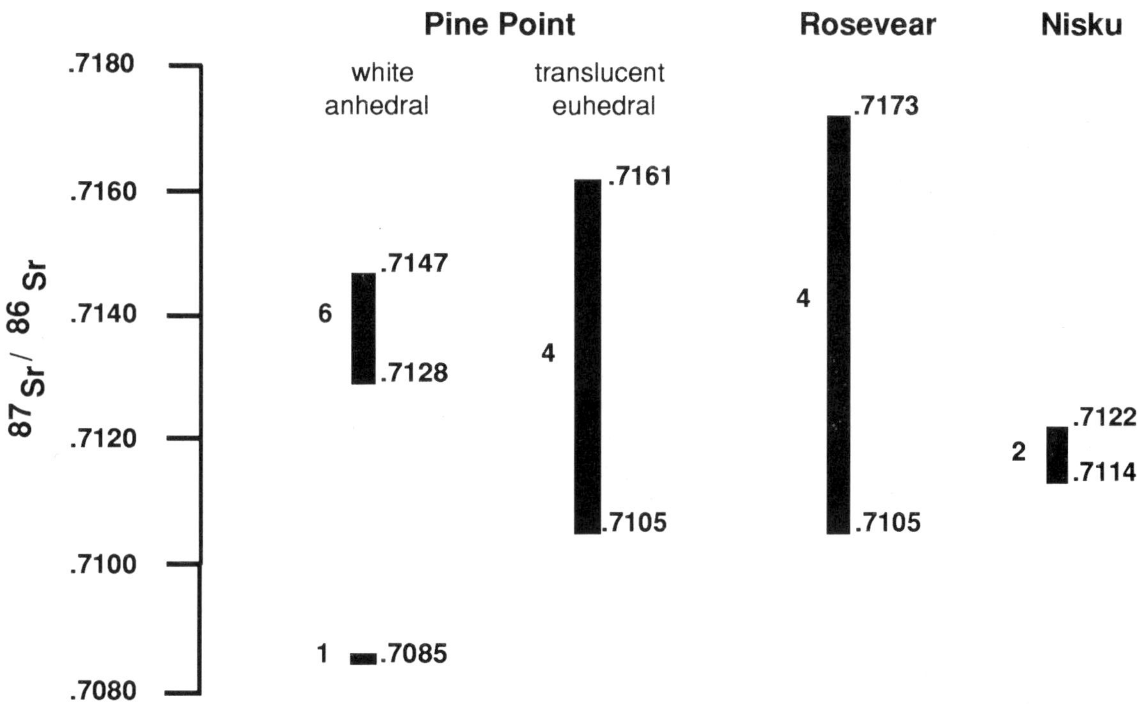

Fig. 14. The $^{87}Sr/^{86}Sr$ ratios of late-stage coarse-crystalline calcites from Pine Point, Rosevear (Kaufman *et al.*, 1990), and Nisku (Anderson, 1985) (from Mountjoy *et al.*, 1992, with permission, Pergamon Press).

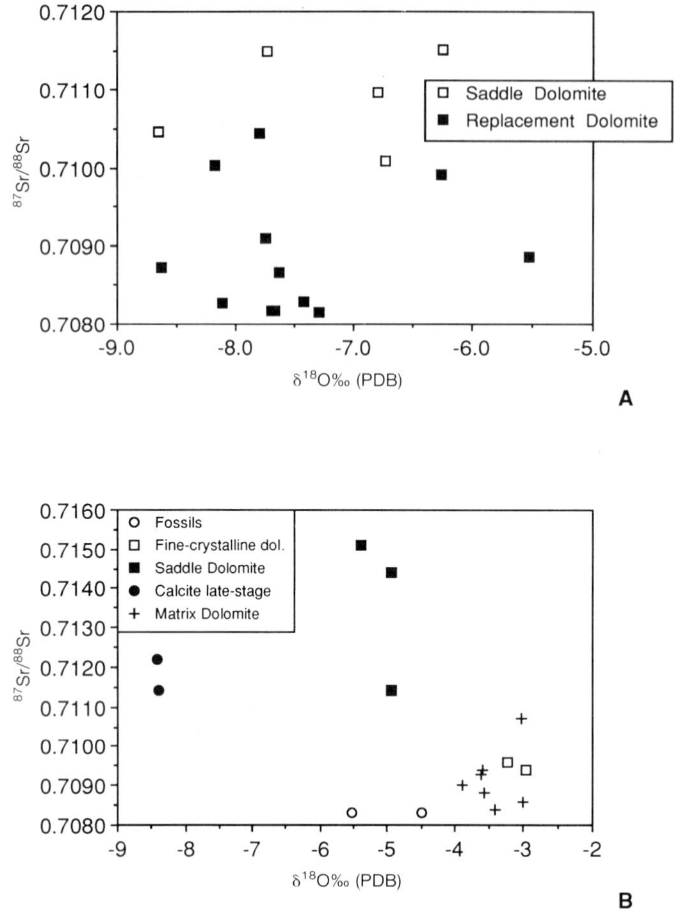

Fig. 15. $^{87}Sr/^{86}Sr$ ratios vs $\delta^{18}O$. (A) Swan Hills Formation, Rosevear Field (Kaufman, 1989), illustrating the scatter of data, except for the more radiogenic saddle dolomites. (B) Upper Devonian Nisku Formation, West Pembina field (Anderson, 1985), showing strongly radiogenic saddle dolomites and late-stage calcites (from Mountjoy *et al.*, 1992, with permission, Pergamon Press).

and calcite cements contain abundant radiogenic Sr (Figs 13 and 14; Mountjoy *et al.*, 1992), similar to present-day formation waters in the basin (Connolly *et al.*, 1990a,b; Machel, 1992, personal communication). Similar relationships occur in the Swan Hills Rosevear field (Kaufman *et al.*, 1990) and the West Pembina Nisku reefs (Machel & Anderson, 1989; Fig. 15). Fluid-inclusion temperatures in conjunction with radiogenic Sr isotopic compositions and other geochemical data suggest that dolomite cements were precipitated from hydrothermal waters that moved updip from deeper in the basin during maximum burial. There is little direct evidence for earlier thermal events occurring throughout the basin that would account for such high temperatures at widely scattered locations in this basin. Hence it is reasonable to infer that, during maximum burial in the Late Cretaceous and Early Tertiary, hot fluids from deeper in the basin to the southwest moved laterally or upwards along conduit systems and fractures and faults (Mountjoy & Halim-Dihardja, 1991; Qing & Mountjoy, 1992). Following are some specific examples of late-stage coarse-crystalline and saddle dolomites.

Example 1: Presqu'ile Barrier

The only regional study of late-stage dolomites is from the 400 km-long Middle Devonian Presqu'ile barrier that forms a northeast–southwest conduit system across the northern part of the Alberta Basin (Qing, 1991; Qing & Mountjoy, 1992 and in press). These dolomite cements show general northeastward trends of decreasing $^{87}Sr/^{86}Sr$ ratios (from 0.7106 to 0.7081) and decreasing homogenization temperatures (from 178° to 92°C) with corresponding heavier $\delta^{18}O$ values from −16 to −7‰ PDB (Fig. 5; Qing & Mountjoy, 1992). These regional trends suggest that hot and radiogenic basinal fluids moved northeastward updip along the Presqu'ile barrier. Such large-scale migrations of basinal fluids were probably related to tectonic thrusting and compression, sedimentary loading and tectonic uplift on the western margin of the Western Canada Sedimentary Basin that took place during the Late Cretaceous and Palaeocene. The Presqu'ile barrier probably acted as a restricted regional conduit system that apparently focused and channelled the basin fluids responsible for extensive local dolomitization, Mississippi Valley-type mineralization and secondary migration of hydrocarbons in the Western Canada Sedimentary Basin (Qing & Mountjoy, 1992 and in press). Similar dolomites occur in the Manetoe facies, which crops out about 100 km northwest of the most northwesterly occurrence of the subsurface Presqu'ile barrier, but have been interpreted very differently by Morrow *et al.* (1986, 1990) to be part of the same conduit system as the Presqu'ile dolomitized barrier, and as forming in the late Devonian. The data of Qing (1991) and Qing and Mountjoy (1992 and in press) do not support this interpretation.

Example 2: Rimbey-Meadowbrook reef trend

Late-stage diagenetic phases that line and fill moulds, vugs and fractures include two types of dolomite cement, anhydrite, bitumen, minor sulphides and calcite. In many samples bitumen and coarse-crystalline calcite cement overlie these dolomite cements. Anhydrite also postdates both replacement dolomites and dolomite cements, filling moulds, vugs and intercrystalline pores, especially in wells from the deeper-buried part of the Rimbey-Meadowbrook reef trend (e.g. south of Homeglen-Rimbey, Figs 2 and 4). Dolomite cements are characterized by two textural types: coarse-crystalline planar cements and non-planar cements. They can be distinguished only in thin sections, which reveal a close association of both cement types, with non-planar cements postdating planar cements in several occurrences. Generally, both types are minor and only form about 5–10% of any one buildup. Both cement types postdate all replacement dolomites. Coarse-crystalline dolomite cements show slightiy higher $CaCO_3/MgCO_3$ ratios relative to replacement dolomites (Tables 1 and 2). Fe and Mn contents are similar, and no systematic variations have been observed within individual dolomite crystals. Some planar dolomite cements display compositional zonation under cathodoluminescence, and have lighter oxygen isotope values ($\delta^{18}O$ = −8.7 to −5.8‰ PDB) than replacement dolomites (Fig. 8). The increasingly negative $\delta^{18}O$ values of later diagenetic phases most likely result from increasing temperature during burial, and changes in $\delta^{18}O$ values of the formation waters (Fig. 16; Amthor *et al.*, 1993a).

Example 3: Wabamun Group east of Peace River Arch

Late-stage sucrosic and saddle dolomites locally fill fractures and cavities in subtidal Famennian limestones on the southeastern side of the Peace River Arch (Fig. 2) in northwest Alberta (Packard *et al.*, 1990; Mountjoy & Halim-Dihardja, 1991). The fractures are interpreted to have formed during

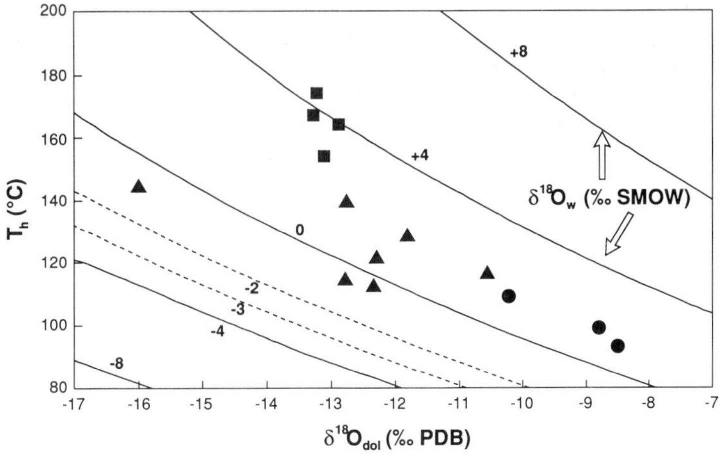

Fig. 16. Homogenization temperature (T_h) and $\delta^{18}O$ values of saddle dolomite, Presqu'ile barrier. Values for $\delta^{18}O$ calculated using equation in Land (1985). Average homogenization temperature is plotted against $\delta^{18}O$ value of dolomite to yield $\delta^{18}O$ values of dolomitizing fluids. Dashed lines are estimated $\delta^{18}O$ values of Middle Devonian seawater. Dots, Pine Point; triangles, NWT; squares northeast BC samples (from Qing & Mountjoy, 1992, with permission, *Geology*).

three different stages, ranging from Late Devonian to Cretaceous and Early Tertiary. Sucrosic and saddle dolomites postdate replacement dolomites, occurring in vugs and late-stage fractures (possibly of Cretaceous age) within replacement dolomites in subtidal limestone sequences (Mountjoy & Halim-Dihardja, 1991). Some saddle dolomite fillings were broken and rotated and coated with a second generation of saddle dolomite, indicating that faulting and fracturing overlapped dolomite precipitation in the eastern part of the Peace River Arch. Both ferroan and non-ferroan saddle dolomites are non-luminescent in their early stages, with several bright-orange luminescing bands in the outer zones of crystals. Sucrosic and saddle dolomites form a group ($\delta^{18}O = -5.7$ to -12.8‰ PDB and $\delta^{13}C = +0.5$ to -4.6‰ PDB) that overlaps with ferroan and non-ferroan replacement matrix dolomites. Primary aqueous fluid inclusions from six saddle dolomites yielded homogenization temperatures between 90° and 130°C, with two samples having much higher temperatures ranging from 170° to 210°C (Table 3). The inclusions have very low eutectic (initial) temperatures (−38° to −56°C) indicating that the fluids are not pure NaCl brines. Final melting temperatures indicate that the fluids have salinities ranging from 17 to 24 equivalent wt% NaCl. These dolomites have $^{87}Sr/^{86}Sr$ ratios ranging from 0.7104 to 0.7128, indicating formation from variably radiogenic fluids. Fluid-inclusion studies indicate that most saddle dolomites formed at around 100°C, but some formed near 200°C from radiogenic saline formation waters. They thus formed from warm to hot highly saline basin formation waters at depths of 2300–3000 m or more (assuming a geothermal gradient of 30°C km^{-1}), or from formation waters derived from even deeper in the basin. This most probably occurred during deepest burial in the Late Cretaceous and Early Tertiary (Fig. 11). These fluids probably moved updip along sedimentary and fault conduit systems and other stratigraphic conduits from deeper parts of the basin and/or the underlying crystalline basement during tectonic loading of the western side of the Western Canada Sedimentary Basin. The late dolomitization appears to have overlapped hydrocarbon generation and migration in western Canada.

Other examples

Other examples of deep-burial dolomite cements have been reported from the Swan Hills Formation in the Rosevear Field (Kaufman, 1989; Kaufman *et al.*, 1990) and from the younger Nisku reefs (Machel, 1986, 1987a,b). In the case of the Rosevear Field, fluid inclusions (T_h = 125–159°C; T_m = −8 to −22°C; Table 3), $^{87}Sr/^{86}Sr$ ratios (0.7101–0.7115), enrichment in Ba and Zn, and light $\delta^{18}O$ values (−6 to −11.2‰ PDB; Fig. 15a) suggest that saddle dolomite precipitated from hot radiogenic brines with salinities six times greater than modern seawater. Anderson (1985) and Machel (1986, 1987a) reported similar data from the Nisku saddle dolomites (Table 4; Fig. 15b) and Machel's (1988) data support updip fluid flow.

Late-stage calcites

Commonly, the latest carbonates to precipitate from highly saline basin brines of the Western Canada Sedimentary Basin were coarse-crystalline, generally euhedral calcites that are locally anhedral. In all cases the calcites are medium- to very coarse-crystalline, with euhedral crystals often >1 cm and, in the case of Pine Point, reaching 20 cm in size. They occur mainly in secondary pore spaces, vugs and fractures. In general, their $\delta^{18}O$ values are slightly heavier than the preceding late-stage burial dolomites. At Pine Point, Rosevear and West Pembina, the highest Sr isotope ratios ratios occur in these calcites, and range from 0.7105 to as high as 0.7173 (Fig. 14). The brines from which these calcites precipitated tend to be the most ^{87}Sr-enriched fluids that developed in the Alberta Basin, more radiogenic than those from which saddle dolomites precipitated. They also are somewhat more radiogenic than the present-day brines of the Edmonton area, whose highest Sr isotope ratio is 0.7129 (Connolly *et al.*, 1990a,b). Relating these calcites to their burial histories and the fact that some may be related to thermochemical sulphate reduction, they reflect the chemistry of the basin fluids at maximum burial or shortly afterwards. The association of some of these calcites with anhydrite indicates that they were precipitated from basin brines. However, present-day river waters and surface waters of the Canadian Shield can be strongly radiogenic (Wadleigh *et al.*, 1985; McNutt *et al.*, 1990). If Shield-derived river waters could be introduced to the deeper levels of the basin, then perhaps ^{87}Sr-enriched meteoric waters may have mixed with basin brines and formed the calcites. However, these fluids contain low $\delta^{18}O$ values (Frape & Fritz, 1987). More late-stage calcite fluid inclusions need to be studied.

Late-stage burial dolomites and calcites (along

Table 4. Potential sources of Sr isotopes in subsurface brines, Western Canada Sedimentary Basin. (From Mountjoy *et al.*, 1992)

	$^{87}Sr/^{86}Sr$
'Connate' seawater	
Mid-Devonian	0.7080
Silurian	0.7080–0.7088
Ordovician	0.7077–0.7090
Cambrian	0.7090+
Mid-Devonian evaporite brines	0.7080
Pressure solution of underlying strata	
Ordovician	0.7070–0.7090
Cambrian	0.7090+
Late Proterozoic	0.7075
Adjacent illite shales	Slightly radiogenic? Higher in older rocks
Cretaceous clastic wedge	Volcanic and igneous detritus highly radiogenic
Metamorphic fluids	Variable—moderate to highly radiogenic?
Precambrian basement (McNutt *et al.*, 1990; Franklyn *et al.* 1991)	Radiogenic 0.7112–0.7136 Yellowknife 0.7188–0.7279 Thompson 0.7105–0.7171 Sudbury I 0.7250–0.7400 Sudbury II 0.7080–0.7103 Norita 0.7065–0.7278 Eye-Dashwa
Present-day western Canada subsurface fluids	
Edmonton area, Connolly *et al.* (1990)	
Tertiary and Upper Cretaceous	0.7057–0.7090
Lower Cretaceous	0.7075–0.7103
Jurassic and Palaeozoic	0.7088–0.7129
Nisku reefs, Machel (pers. comm., 1991)	
Upper Devonian, Frasnian	0.7122–0.7128

with anhydrite and sulphides) occlude pores, thus reducing porosity. However, there is increasing evidence that locally extensive burial dissolution of earlier dolomites has taken place (Amthor & Friedman, 1991; Dravis & Muir, 1992; Mazzullo & Harris, 1992; Amthor & Mountjoy, 1992; Qing & Mountjoy, 1990 and in press; Drivet & Mountjoy, 1993).

Possible origins

Most coarse dolomite cements precipitated from warm (hydrothermal) basin fluids. Sources of Mg for the dolomite cements may well have resulted from pressure solution of earlier replacement dolomites. In many dolomite samples in the Rimbey-Meadowbrook reef trend, some stylolites connect pores which are lined by dolomite cement (Amthor *et al.*, 1993a), indicating a possible relationship between pressure solution and dolomite cementation. McNamara and Wardlaw (1991) observed that flank wells in the Westerose buildup have the largest number of stylolites and the highest amount of dolomite and calcite cements, suggesting a similar relationship. A similar process has been suggested by Machel (1987a) to explain coarse-crystalline (saddle) dolomites in Nisku reefs in the subsurface of Alberta. Non-planar dolomite cements (saddle dolomites; Amthor *et al.*, 1993a), which are more common in the deeper parts of the basin, and late-stage calcite cements are more depleted in $\delta^{13}C$ than planar dolomite cements (Fig. 8), indicating that organically derived carbonate and/or bicarbonate was available in addition to dissolved material from pressure solution. There is scatter in the O and Sr isotope cross-plots from the Presqu'ile Barrier and the Rosevear field (Kaufman, 1989), and no clear trends are evident in Figs 5 and 15. In general, late saddle dolomites and later calcites have more radiogenic Sr compositions than massive replacement dolomites. The Presqu'ile dolomites show two trends: Pine Point coarse-crystalline and saddle dolomites (filled circles and triangles, Fig. 5) shift toward lighter $\delta^{18}O$ values with no, or only a slight, increase in radiogenic Sr compared to the fine-crystalline dolomites; and subsurface Presqu'ile barrier dolomites (open symbols) change progressively with increasing depth in both isotopes, towards lighter $\delta^{18}O$ values concurrent with more radiogenic Sr isotopic ratios (Fig. 5). These are both very different trends from the one inferred by Kaufman *et al.* (1991). Caution in interpreting these data is indicated by oxygen isotope and fluid-inclusion data from coarse-crystalline dolomites which show that these dolomites precipitated from formation waters with variable $\delta^{18}O$ composition (Fig. 16; Qing & Mountjoy, 1992). If dolomitization had occurred earlier in all of the above examples, it would have taken place at shallower burial depths with corresponding lower ambient temperatures. This would create an additional problem of providing suitable heat sources for these dolomitizing fluids in different parts of the basin. The Western Canada Sedimentary Basin rests on continental crust of normal thickness, and therefore probably had geothermal gradients in the order of $30°C\,km^{-1}$. There are no known widespread igneous intrusions of late Palaeozoic or of Triassic and Jurassic ages in the Western Canada Sedimentary Basin. The overlap of saddle dolomites with hydrocarbon generation in some examples and fluid inclusions containing hydrocarbons (Machel, 1987a; Qing & Mountjoy, 1992 and in press) indicates that saddle dolomites occurred late in the basin history, near maximum burial some time during the Late Cretaceous or Early Tertiary. Thus the most likely sources of hot fluids would be formation waters that moved upwards from deeper in the basin, during deep burial (Fig. 11). Saddle dolomites rarely exceed a few percent of all dolomites. The apparent increase of saddle dolomites downdip in the Rimbey-Meadowbrook reef trend (Amthor *et al.*, 1993a; Drivet & Mountjoy, 1993; Marquez, 1992, personal communication) suggests that these dolomites appear to be more abundant in the more deeply buried part of the basin below present depths of about 3 km. As Eliuk (1984) and Machel (1987a,b) have suggested, these dolomites may be in part related to thermal sulphate reduction, which appears to occur mainly at temperatures above 100°C (these temperatures would represent maximum burial depths of about 2.5–2.7 km; Fig. 11).

The basin formation waters did not become markedly radiogenic until relatively deep burial, as only the coarse-crystalline and saddle dolomites and late-stage calcites are strongly radiogenic (Mountjoy *et al.*, 1992). This trend of Sr isotope evolution is similar to that in the midcontinent, southern Appalachian and Gulf Coast basins, where two episodes of dolomitization were regionally extensive. There, late vug-filling calcite and dolomite cements have high $^{87}Sr/^{86}Sr$ ratios, whereas earlier calcites and dolomites have near-marine Sr isotope compositions (Kessen *et al.*, 1981; Grant & Miranda, 1983; Banner *et al.*, 1988, their Fig. 11; Moore, 1985; Moore *et al.*, 1988; Montañez & Read, 1992). This trend is also similar to that of five ore districts in other sedimentary basins (Cambrian, Sardinia; Triassic, Spain; Jurassic, Peru and France; Cretaceous, Argentina) in which Gorzawski *et al.* (1989) showed that the $^{87}Sr/^{86}Sr$ ratio increases with advancing diagenetic stage.

On the basis of burial history and fluid-flow arguments (Qing & Mountjoy, 1989, 1992 and in press; Mountjoy & Halim-Dihardja, 1991), saddle dolomites in the Western Canada Sedimentary Basin do not appear to have been precipitated until deepest burial, although others interpret a Devonian or Carboniferous age for them (Packard *et al.*, 1990; Dravis & Muir, 1992). Potential formation waters available at this time in the Devonian succession could have resulted from compaction by sedimentary and tectonic loading of the western side of the basin (and in part driven by elevated topography). Possibly some formation waters came from underlying and downdip Cambrian and Precambrian clastic successions and Precambrian basement, respectively, and/or from pulses of hydrothermal waters derived from metamorphism in the Late Jurassic and mid-Cretaceous that moved updip and mixed with cooler formation waters. To ascertain the different contributions of each of these potential fluid sources, more research is needed into the analysis of formation waters and rocks in different parts of the basin.

CONCLUSIONS

Data from the Devonian of the Western Canada Sedimentary Basin indicate that dolomitization occurred in both the intermediate and the deep subsurface. The origin of the earlier massive replacement dolomites that form 70–90% of the Devonian dolomites is still uncertain. They cross-cut depositional facies boundaries and postdate overlying basin sediments, early submarine cements and low-amplitude wispy stylolites. These medium- to coarse-crystalline replacement dolomites are significantly depleted in $\delta^{18}O$ relative to hypothetical Devonian marine dolomites. The temperature of formation of these dolomites is calculated to be about 50–70°C, assuming that the dolomitizing fluids had isotopic compositions close to that of Devonian seawater. These temperatures could have been reached when these strata were buried to depths of 600–1200 m, assuming a surface temperature of 30°C and a geothermal gradient of 30°C km^{-1}.

Deep-burial dolomite cements are much better understood than the massive replacement dolomites, because of their distinctive textures and abundant chemical and fluid-inclusion data. These data indicate that, in most cases, the late dolomite cements precipitated from strongly saline formation waters having high $^{87}Sr/^{86}Sr$ ratios at high temperatures, in some cases exceeding estimated maximum burial temperatures. Thus, late dolomite cements probably precipitated in deep-burial settings or at somewhat shallower depths from hydrothermal waters that moved up along stratigraphic and fracture conduit systems from deeper in the basin. The Mg sources are uncertain, but are most likely related to pressure solution of older dolomites or evaporites. The most likely pumping mechanisms for moving these formation waters up the flanks of the basin are sedimentary and tectonic loading of the western side of the basin, in conjunction with topographically driven flow that resulted from uplift of the mountains to the west. In addition, pulses of hot formation waters derived from metamorphism downdip to the west may have moved updip and mixed with cooler waters.

Burial dolomitization has come of age. The preceding examples and discussion demonstrate that burial environments are important settings for dolomitization and the modification of earlier dolomites. This is not to say that these dolomites formed from burial processes (e.g. compaction, mineral transformations, increasing temperatures and pressures) *because* they were buried: rather, they formed in subsurface environments in the presence of dolomitizing fluids (e.g. modified seawater, basin brines, evaporite brines). This requires that special local factors such as the availability of Mg-rich fluids, the presence of a conduit system and an active pumping mechanism control the amount and distribution of burial dolomites. These factors may have been more important than either age- or sea-level related parameters in determining rates of dolomite formation. The Devonian of western Canada contains more dolomites (and more evaporites) than other basins in Europe and Australia, for example. There is conclusive evidence for squeezing of the sedimentary wedge during tectonic telescoping by concurrent tectonic and sedimentary loading, and the establishment of regional topographically driven flow systems affecting the basin, beginning in the Late Jurassic and Early Cretaceous and continuing until the Palaeocene. Also, earlier tectonic loading in relation to the Antler orogeny (Late Devonian–Early Carboniferous) could have been responsible for driving deep-seated formation fluids (including residual evaporitic brines) northward up along porous carbonate conduit systems (such as the Rimbey-Meadowbrook reef trend) and for dolomitizing the margin of the Cooking Lake platform and the overlying Leduc buildups. The Western Canada Sedimentary Basin serves as an example that burial dolomitization is an important process in basins where extensive conduit systems existed and suitable Mg-bearing fluids were present at different times.

ACKNOWLEDGEMENTS

This research was supported by funds from a Natural Science and Engineering Research Council (NSERC) operating grant to Mountjoy and a strategic grant to Mountjoy and Machel. Some funding for part of this research was provided by Chevron, Home Oil, Mobil Oil, Norcen, Petro Canada and Shell Canada. We are grateful for unpublished data, discussions and comments from many colleagues, especially J. Andrichuk, G. Burrowes, H. Chouinard, G. Davies, E. Drivet, P. Gretener, H. Machel, X. Marquez, J. Packard, H. Qing, G. Tebbutt, J. Thiessen and J. Wendte. Earlier versions of this paper were critically read by Eva Drivet, Xiomara Marquez and Jeanne Paquette. We appreciate the editorial comments of M. Coniglio, H.G. Machel, B. Purser, M. Tucker and S. Whittaker. A very special thanks to Don Zenger for his meticulous and careful review and helpful suggestions.

REFERENCES

Amthor, J.E. & Friedman, G.M. (1991) Dolomite-rock textures and porosity development in Ellenburger Group carbonates (Lower Ordovician), Westrn Texas and Southeastern New Mexico. *Sedimentology* **38**, 343–362.

AMTHOR, J.E. & FRIEDMAN, G.M. (1992) Early- to late-diagenetic dolomitization of platform carbonates: Lower Ordovician Ellenburger Group, Permian Basin, West Texas. *J. Sedim. Petrol.* **62**, 131–144.

AMTHOR, J.E. & MOUNTJOY, E.W. (1992) Subsurface dolomite-solution breccias in Upper Devonian Leduc buildups, central Alberta, Western Canada Sedimentary Basin. *Am. Ass. Petrol. Geol. Ann. Mtg. Abstracts*, p. 2.

AMTHOR, J.E., MOUNTJOY, E.W. & MACHEL, H.G. (1992) Burial dissolution of dolomites in Upper Leduc buildups, Western Canada Sedimentary Basin. *Am. Ass. Petrol. Geol. Ann. Mtg. Abstracts*, p. 3.

AMTHOR, J.E., MOUNTJOY, E.W. & MACHEL, H.G. (1993a) Subsurface dolomites in Upper Devonian Leduc Formation buildups, central part of Rimbey-Meadowbrook reef trend, Alberta, Canada. *Bull. Can. Soc. Petrol. Geol.* **41**, 164–185.

AMTHOR, J.E., MOUNTJOY, E.W. & MACHEL, H.G. (1993b) Regional-scale porosity and permeability variations in Upper Devonian Leduc buildups: implications for origin and distribution of porosity in carbonates. *Carbonate Petroleum Reservoirs: models for exploration and development, Geol. Soc. London* (abstract).

ANDERSON, J.H. (1985) *Depositional Facies and Carbonate Diagenesis of Downslope Reefs in the Nisku Formation, Central Alberta, Canada*. Unpublished PhD Thesis, University of Texas at Austin, 393 pp.

AULSTEAD, K.L., SPENCER, R.J. & KROUSE, H.R. (1988) Fluid inclusion and isotopic evidence on dolomitization, Devonian of Western Canada. *Geochim. Cosmochim. Acta* **52**, 1027–1035.

BANNER, J.L., HANSON, G.N. & MEYERS, W.J. (1988) Water–rock interaction history of regionally extensive dolomites of the Burlington-Keokuk Formation (Mississippian): isotopic evidence. In: *Sedimentology and Geochemistry of Dolostones* (Ed. Shukla, V. & Baker, P.A.) Spec. Publ. Soc. Econ. Paleont. Mineral. **43**, 97–113.

BARNABY, R.J. & READ, J.F. (1992) Dolomitization of a carbonate platform during late burial: Lower to Middle Cambrian Shady Dolomite, Virginia Appalachians. *J. Sedim. Petrol.* **62**, 1023–1043.

BASSETT, H.G. & STOUT, J.G. (1967) Devonian of western Canada. *Proceedings International Devonian System, Can. Soc.Petrol. Geol.* **1**, 717–725.

BURKE, W.H., DENISON, R.E., HETHERINGTON, E.A., KOEPNICK, R.B., NELSON, H.F. & OTTO, J.B. (1982) Variations of seawater $^{87}Sr/^{86}Sr$ throughout Phanerozoic time. *Geology* **10**, 516–519.

CARPENTER, S.J., LOHMANN, K.C., HOLDEN, P., WALTER, L.M., HUSTON, T. & HALLIDAY, A.N. (1991) $\delta^{18}O$ values, $^{87}Sr/^{86}Sr$ and Sr/Mg ratios of Late Devonian abiotic marine calcite: implications for the composition of ancient seawater. *Geochim. Cosmochim. Acta* **55**, 1991–2010.

CHOQUETTE, P. & JAMES, N.P. (1990) The burial diagenetic environment In: *Diagenesis* (Ed. McIlreath, I.A. & Morrow, D.W.) Geosci. Can. Reprint series 4, 75–111.

CHOQUETTE, P.W., COX, A. & MEYERS, W.J. (1992) Characteristics, distribution and origin of porosity in shelf dolostones: Burlington-Keokuk Formation (Mississippian), US Mid-continent. *J. Sedim. Petrol.* **62**, 167–189.

CONNOLLY, C.A., WALTER, L.M., BAADSGAARD, H. & LONGSTAFFE, F.J. (1990a) Origin and evolution of formation waters, Alberta Basin, Western Canada Sedimentary Basin. 1. Chemistry. *Appl. Geochem.* **5**, 375–395.

CONNOLLY, C.A., WALTER, L.M., BAADSGAARD, H. & LONGSTAFFE, F.J. (1990b) Origin and evolution of formation waters, Alberta Basin, Western Canada Sedimentary Basin. 2. Isotope systematics and water mixing. *Appl. Geochem*. 5, 397–413.

DIX, G.R. (1990) Stages of platform development in the Upper Devonian (Frasnian) Leduc Formation, Peace River Arch, Alberta In: *Geology of the Peace River Arch* (Ed. O'Connell, S.C. & Bell. J.S.) *Bull. Can. Soc. Petrol. Geol.* **38A**, 66–92.

DIX, G.R. (1993) Patterns of burial and techtonically controlled dolomitization in an Upper Devonian fringing-reef complex: Leduc Formation, Peace River Arch area, Alberta, Canada. *J. Sedim. Petrol.* **63**, 628–641.

DRAVIS, J.J., & MUIR, I.D. (1992) Case study of burial dissolution – Dolomites Devonian Upper Elk Point Group, Western Canada In: *Subsurface Dissolution Porosity in Carbonates–Recognition, Causes and Implications* (Ed. Bloy, G.R.) Can. Soc. Petrol. Geol., Short Course Notes, pp. 55–117.

DRIVET, E. & MOUNTJOY, E.W. (1993) Porosity variations in Upper Devonian Leduc dolomites, central Rimbey Meadowbrook reef trend, Alberta. *Am. Ass. Petrol. Ann. Mtg. Abstracts*.

ELIUK, L.S. (1984) A hypothesis for the origin of hydrogen sulfide in the Devonian Crossfield Member dolomite, Wabamun Formation, Alberta, Canada. Carbonates. In: *Subsurface and Outcrop* 1984 (Ed. Eliuk, L.S.) Core Conference Can. Soc. Petrol. Geol., 245–290.

FRANKLYN, M.T., MCNUTT, R.H., KAMINENI, D.C., GASCOYNE, M. & FRAPE, S.K. (1991) Groundwater $^{87}Sr/^{86}Sr$ values in the Eye-Dechur Lakes pluton, Canada: evidence for plagioclase–water reaction. *Chem. Geol.* (Isotope Geol. Section), **86**, 11–122.

FRAPE, S.K. & FRITZ, P. (1987) Geochemical trends from groundwaters from the Canadian shield. In: *Saline Waters and Gases in Crystalline Rocks* (Ed. Fritz, P. & Frape, S.K.) Spec. Paper Geol. Assoc. Can. 33, 19–38.

FREEMAN, T. (1972) Sedimentology and dolomitization of Muschelkalk carbonates (Triassic), Iberian Range, Spain. *Am. Ass. Petrol. Geol. Bull.* **56**, 434–453.

GAO, G. (1990) Geochemical and isotopic constraints on the diagenetic history of a massive stratal, late Cambrian (Royer) dolomite, lower Arbuckle Group, Slick Hills, SW Oklahoma, USA. *Geochim. Cosmochim. Acta* **54**, 1979–1989.

GAO, G. & LAND, L.S. (1991) Early Ordovician Cool Creek dolomite, middle Arbuckle Group, Slick Hills, SW Oklahoma, U.S.A.: origin and modification. *J. Sedim. Petrol.* **61**, 161–173.

GAO, G., LAND, L.S. & FOLK, R.L. (1992) Meteoric modification of early dolomite and late dolomitization by basinal fluids, Upper Arbuckle Group, Slick Hills, southwestern Oklahoma. *Am. Ass. Petrol. Geol. Bull.* **76**, 1649–1664.

GARVEN, G. (1989) A hydrologic model for the formation of giant oil sands deposits of the western Canada sedi-

mentary basin. *Am. J. Sci.* **289**, 105–166.
GARVEN, G. & FREEZE, R.A. (1984) Theoretical analysis of the role of groundwater flow in the genesis of stratabound ore deposits: 2. Quantitative results. *Am. J. Sci.* **284**, 1125–1174.
GAWTHORPE, R.L. (1987) Burial dolomitization and porosity development in a mixed carbonate–clastic sequence: an example from the Bowland Basin, northern England. *Sedimentology* **34**, 533–558.
GIVEN, R.K. & WILKINSON, B.H. (1987) Dolomite abundance and stratigraphic age: constraints on rates and mechanisms of Phanerozoic dolostone formation. *J. Sedim. Petrol.* **57**, 1060–1078.
GORZAWSKI, H., FONTBOTE, L., SUREAU, J.F. & CALVEZ, J.Y. (1989) Strontium isotope trends during diagenesis in ore-bearing carbonate basins. *Geol. Rund.* **78**, 269–290.
GRANT, N.K. & MIRANDA, C.B. (1983) Strontium isotope and rare earth element variations in non-sulphide minerals from the Elmwood-Gordonsville Mines, central Tennessee. *International Conference on Mississippi Valley-Type Lead-Zinc Deposits, Proceedings.* University Missouri-Rolla, pp. 206–210.
GREGG, J.M. (1985) Regional epigenetic dolomitization in the Bonneterre dolomite (Cambrian), southeastern Missouri. *Geology* **13**, 503–506.
GREGG, J.M. & SHELTON, K.L. (1990) Dolomitization and dolomite neomorphism in the back-reef facies of the Bonneterre and Davis Formations (Cambrian), southeastern Missouri. *J. Sedim. Petrol.* **60**, 549–562.
GREGG, J.M. & SIBLEY, D.F. (1984) Epigenetic dolomitization and the origin of xenotopic dolomite texture. *J. Sedim. Petrol.* **54**, 908–931.
GREGG, J.M., HOWARD, S.A. & MAZZULLO, S.J. (1992) Early diagenetic recrystallization of Holocene (<3000 years old) peritidal dolomites, Ambergris Cay, Belize. *Sedimentology* **39**, 143–160.
HARDIE, L.A. (1987) Dolomitization: a critical view of some current views. *J. Sedim. Petrol.* **57**, 166–183.
HITCHON, B. (1984) Geothermal gradients, hydrodynamics, and hydrocarbon occurrences, Alberta, Canada. *Bull. Am. Ass. Petrol. Geol.* **68**, 713–743.
HUDSON, J.D. & ANDERSON, T.F. (1989) Ocean temperatures and isotopic compositions through time. *Trans. Roy. Soc. Edinburgh: Earth Sci.* **80**, 183–192.
ILLING, L.V. (1956) Dolomitization in relation to porosity in carbonate rocks. *Am. Ass. Petrol. Geol., Program 41st. Ann. Mtg. Abstracts.* Chicago, 13–14 (abstract).
ILLING, L.V. (1959) Deposition and diagenesis of some Upper Paleozoic carbonate sediments in western Canada. *Fifth World Petroleum Congress, New York, Proceedings Section 1*, 23–52.
JONES, R.M.P. (1980) Basinal isostatic adjustment faults and their petroleum significance. *Bull. Can. Petrol. Geol.* **28**, 211–251.
KAUFMAN, J. (1989) *Sedimentology and Diagenesis of the Swan Hills Formation (Devonian), Rosevear Field, Alberta, Canada.* PhD Thesis, State University of New York at Stony Brook, Stony Brook, New York, 412 pp.
KAUFMAN, J., HANSON, G.N. & MEYERS, W.J. (1991) Dolomitization of the Swan Hills Formation, Rosevear Field, Alberta, Canada. *Sedimentology* **38**, 41–66.
KAUFMAN, J., MEYERS, W.J. & HANSON, G.N. (1990) Burial cementation in the Swan Hills Formation (Devonian), Rosevear field, Alberta, Canada. *J. Sedim. Petrol.* **60**, 918–939.
KENDALL, A.C. (1989) Brine mixing in the Middle Devonian of western Canada and its possible significance to regional dolomitization. *Sediment. Geol.* **64**, 271–285.
KESSEN, K., WOODRUFF, M.S. & GRANT, N.K. (1981) Gangue mineral $^{87}Sr/^{86}Sr$ ratios and the origin of Mississippi Valley-type mineralization. *Econ. Geol.* **76**, 913–920.
KROUSE, H.R., VIAU, C.A., ELIUK, L.S., UEDA, A. & HALAS, S. (1988) Chemical and isotopic evidence of thermochemical sulphate reduction by light hydrocarbon gases in deep carbonate reservoirs. *Nature* **333**, 415–419.
LAFLAMME, A.K. (1990) *Replacement Dolomitization in the Upper Devonian Leduc and Swan Hills Formations, Caroline Area, Alberta, Canada.* Unpublished MSc Thesis, McGill University, Montréal, 138 pp.
LAND, L.S. (1980) The isotopic and trace element geochemistry of dolomite: the state of the art. In: *Concepts and Models of Dolomitization* (Ed. Zenger, D.H., Dunham, J.B. & Ethington, R.L.) Soc. Econ. Paleont. Mineral. Spec. Publs. 28, 87–110.
LAND, L.S. (1985) The origin of massive dolomite. *J. Geol. Educ.* **33**, 112–125.
MACHEL, H.G. (1986) Early lithification, dolomitization, and anhydritization of Upper Devonian Nisku buildups, subsurface of Alberta, Canada. In: *Reef diagenesis* (Ed. Schroeder, J.H. & Purser B.H.) pp. 336–356. Berlin, Springer-Verlag.
MACHEL, H.G. (1987a) Saddle dolomite as a by-product of chemical compaction and thermochemical sulfate reduction. *Geology* **15**, 936–940.
MACHEL, H.G. (1987b) Some aspects of diagenetic sulphate-hydrocarbon redox reactions. In: *The Diagenesis of Sedimentary Sequences* (Ed. Marshall, J.D.) Geol. Soc. Lond. Spec. Publs. 36, 15–28.
MACHEL, H.G. (1988) Fluid flow direction during dolomite formation as deduced from trace element trends. In: *Sedimentology and Geochemistry of Dolostones* (Ed. Shukla, V. & Baker, P.A.) Soc. Econ. Paleont. Mineral. Spec. Publs. 43, 115–125.
MACHEL, H.G. & ANDERSON, J.H. (1989) Pervasive subsurface dolomitization of the Nisku Formation in Central Alberta. *J. Sedim. Petrol.* **59**, 891–911.
MACHEL, H.G. & BURTON, E.A. (1991) Burial-diagenetic sabkha-like gypsum and anhydrite nodules. *J. Sedim. Petrol.* **61**, 394–405.
MACHEL, H.G. & MOUNTJOY, E.W. (1986) Chemistry and environments of dolomitization – a reappraisal. *Earth Sci. Rev.* **23**, 175–222.
MACHEL, H.G. & MOUNTJOY, E.W. (1987) General constraints on extensive pervasive dolomitization and their application to the Devonian carbonates of western Canada. *Bull. Can. Soc. Petrol. Geol.* **35**, 143–158.
MCGILLIVARY, J.G. & MOUNTJOY, E.W. (1975) Facies and related reservoir characteristics Golden Spike reef complex, Alberta. *Bull. Can. Soc. Petrol. Geol.* **23**, 753–809.
MCNAMARA, L.B. & WARDLAW, N.C. (1991) Geological description of the Westerose reservoir, Alberta. *Bull. Can. Soc. Petrol. Geol.* **39**, 332–351.
MCNUTT, R.H., FRAPE, S.K., FRITZ, P., JONES, M.G. & MACDONALD, J.M. (1990) The $^{87}Sr/^{86}Sr$ values of Canadian Shield brines and fracture minerals with application to groundwater mixing, fracture history and

geochronology. *Geochim. Cosmochim. Acta* **54**, 205–215.

Marquez, X. (1993) Sedimentological and diagenetic controls on pore systems in a dolomitized reservoir; a case study: the Ricinus West gas field, central Alberta. *Am. Ass. Petrol. Geol. Abstracts.*

Martindale, W. & Orr, N.E. (1987) Middle Devonian Winnipegosis reefs in the Tablelands area, S.E. Saskatchewan. *Second International Symposium on the Devonian System, Can. Soc. Petrol. Geol.*, Program with Abstracts, 156 (abstract).

Martindale, W. & MacDonald, R.W. (1990) Sedimentation and diagenesis of the Winnipegosis Formation Tableland area, southeast Saskatchewan. In: *The Development of Porosity in Carbonate Reservoirs* (Ed. Bloy, G.R. & Hadley, M.G.) Can. Soc. Petrol. Geol., Short Course Notes, Sect. 6, 6.1–6.19.

Mattes, B.W. & Mountjoy, E.W. (1980) Burial dolomitization of the Upper Devonian Miette buildup, Jasper National Park, Alberta. In: *Concepts and Models of Dolomitization* (Ed. Zenger, D.H., Dunham, J.B. & Ethington, R.L.) Soc. Econ. Paleont. Mineral. Spec. Publs. 28, 259–297.

Mazzullo, S.J. (1992) Geochemical and neomorphic alteration of dolomite: a review. *Carbonates and Evaporites* **7**, 21–37.

Mazzullo, S.J. & Harris, P.M. (1992) Mesogenetic dissolution: its role in porosity development in carbonate reservoirs. *Am. Ass. Petrol. Geol. Bull.* **76**, 607–620.

Montañez, I.P. (1992) *Controls of Eustasy and Associated Diagenesis on Reservoir Heterogeneity in Lower Ordovician, Upper Knox Carbonates, Appalachians.* Soc. Econ. Paleont. Mineral., Permian Basin section, Sp. Publs. **92-33**, 165–181.

Montañez, I.P. & Read, J.F. (1992) Fluid–rock interaction history during stabilization of early dolomites, Upper Knox Group (Lower Ordovician), US Appalachians. *J. Sedim. Petrol.* **62**, 753–778.

Moore, C.H. (1985) Upper Jurassic subsurface cements: a case history. In: *Carbonate Cements* (Ed. Schneidermann, N. & Harris, P.M.) Soc. Econ. Paleont. Mineral., Spec. Publs. 36, 291–308.

Moore, C.H., Chowdhury, A. & Chan, L. (1988) Upper Jurassic platform dolomitization, northwestern Gulf of Mexico. A tale of two waters. In: *Sedimentology and Geochemistry of Dolostones* (Ed. Shukla, V. & Baker, P.A.) Soc. Econ. Paleont. Mineral. Spec. Publs. 43, 175–189.

Moore, P.F. (1988) Devonian geohistory of the western interior of Canada. In: *Devonian of the World* (Ed. McMillan, N.J., Embry, A.F. & Glass, D.J.) 1, 67–83.

Morrow, D.W. (1990a) Dolomite – Part 1: The chemistry of dolomitization and dolomite precipitation. In: *Diagenesis* (Ed. McIlreath, I.A. & Morrow, D.W.) Geol. Ass. Can. Reprint series 4, 113–123.

Morrow, D.W. (1990b) Dolomite – Part 2: Dolomitization models and ancient dolostones. In: *Diagenesis* (Ed. McIlreath, I.A. & Morrow, D.W.) Geol. Ass. Can. Reprint series 4, 125–139.

Morrow, D.H. & Geldsetzer, H.H.J. (1988) Devonian of the eastern Canadian Cordillera. In: *Devonian of the World* (Ed. McMillan, N.J., Embry, A.F. & Glass, D.J.) 1, 85–121.

Morrow, D.W., Cumming, G.L. & Koepnick, R.B. (1986) Manetoe facies – a gas-bearing, megacrystalline, Devonian dolomite, Yukon and Northwest Territories, Canada. *Am. Ass. Petrol. Geol. Bull.* **70**, 702–720.

Morrow, D.W., Cumming, G.L. & Aulstead, K.L. (1990) The gas-bearing Devonian Manetoe Facies, Yukon and Northwest Territories. *Geol. Surv. Can. Bull.* **400**.

Mountjoy, E.W. (1978) Upper Devonian reef trends and configuration of the western portion of the Alberta Basin. In: *The Fairholme Carbonate Complex* (Ed. McIlreath, I.A. & Jackson, P.C.) Can. Soc. Petrol. Geol. Guidebook, pp. 1–30.

Mountjoy, E.W. (1980) Some questions about the development of Upper Devonian carbonate (reefs), western Canada. *Bull. Can. Petrol. Geol.* **28**, 315–340.

Mountjoy, E.W. (1989) Miette reef complex (Frasnian), Jasper National Park, Alberta. In: *Reefs, Canada and Adjacent Areas* (Ed. Geldsetzer, H.J.J. *et al.*) Can. Soc. Petrol. Geol., Mem. 13, 497–505.

Mountjoy, E.W. (1990) Use of ^{18}O for determining temperature in secondary dolomites. *Geol. Ass. Can. and Min. Assoc. Can.*, Program with Abstracts, 15, A 92 (abstract).

Mountjoy, E.W. & Halim-Dihardja, M. (1991) Multiple phase fracture and fault-controlled burial dolomitization, Upper Devonian Wabamun Group, Alberta. *J. Sedim. Petrol.* **61**, 590–612.

Mountjoy, E.W. & Krebs, W. (1983) Diagenesis of Devonian reefs and buildups, western Canada and Europe – a comparison. *Z. deutsche geologische Gesellschaft* **134**, 5–60.

Mountjoy, E.W., Qing, H. & McNutt, R. (1989) Sr isotopes in Devonian Dolomites, Western Canada: significance regarding sources of dolomitizing fluids. *Geol. Ass. Can. and Min. Assoc. Can.* Program with Abstracts 14, A 64 (abstract).

Mountjoy, E.W., Qing, H. & McNutt, R.H. (1992) Strontium isotopic composition of Devonian dolomites, Western Canada Sedimentary Basin: significance of sources of dolomitizing fluids. *Appl. Geochem.* **7**, 59–75.

Nahnybida, C., Hutcheon, I. & Kirker, J. (1982) Diagenesis of the Nisku Formation and the origin of late-stage cements. *Can. Min.* **20**, 129–140.

Narkiewicz, M. (1990) Mesogenetic dolomitization in the Givetian to Frasnian of the Holy Cross Mountains, Poland. *Bull. Polish Acad Sci., Earth Sci.* **38**, 101–110.

Oliver, J. (1986) Fluids expelled tectonically from orogenic belts: their role in hydrocarbon migration and other geologic phenomena. *Geology* **14**, 99–102.

Packard, J.J., Pellegrin, G.J., Al-Aasm, I.S., Samson, I. & Gagnon, J. (1990) Diagenesis and dolomitization associated with hydrothermal karst in Famennian upper Wabamun ramp sediments, northwestern Alberta. In: *The Development of Porosity in Carbonate Reservoirs* (Ed. Bloy, G.R. & Hadley, M.G.) Can. Soc. Petrol. Geol., Continuing Education Short Course, Section 9, 9.1–9.19.

Qing, H. (1991) *Diagenesis of Middle Devonian Presqu'ile Dolomite, Pine Point NWT and Adjacent Subsurface.* PhD Thesis, McGill University, 287 pp.

Qing, H. & Mountjoy, E.W. (1989) Multistage dolomitization in Rainbow buildups, Middle Devonian Keg River Formation, Alberta, Canada. *J. Sedim. Petrol.* **59**, 114–126.

QING, H. & MOUNTJOY, E.W. (1990) Petrography and diagenesis of the Middle Devonian Presqu'ile barrier: implications on formation of dissolution vugs and breccias at Pine Point and adjacent subsurface, District of Mackenzie. In: *Current Research, Part D, Geol. Surv. Canada* Paper 90-1D, 37–45.

QING, H. & MOUNTJOY, E.W. (1992) Large-scale fluid flow in the Middle Devonian Presqu'ile Barrier, Western Canada Sedimentary Basin. *Geology* **20**, 903–906.

QING, H. & MOUNTJOY, E.W. (in press) Hydrothermal origin of dissolution vugs and breccias in Middle Devonian Presqu'ile Barrier, host of Pine Point MVT deposits. *Econ. Geol.*

RHODES, D., LANTOS, E.A., LANTOS, J.A., WEBB, R.J. & OWENS, D.C. (1984) Pine Point orebodies and their relationship to the stratigraphy, structure, dolomitization, and karstification of the Middle Devonian barrier complex. *Econ. Geol.* **79**, 991–1055.

ROSS, G.M. (1991) Precambrian basement in the Canadian Cordillera: an introduction. *Can. J. Earth Sci.* **28**, 1133–1139.

ROSS, G.M. & STEPHENSON, R.A. (1989) Crystalline basement: the foundations of Western Canada Sedimentary Basin. In: *Western Canada Sedimentary Basin, a Case History* (Ed. Ricketts, B.D.) Chapter 3, pp. 33–45. Can. Soc. Petrol. Geol.

SCHMIDT, V., MCILREATH, I.A. & BUDWILL, A.E. (1985) Middle Devonian cementation reefs encased in evaporites, Rainbow field, Alberta. In: *Carbonate Petroleum Reservoirs* (Ed. Roehl, P.O. & Choquette, P.W.) pp. 141–160. Springer-Verlag, Berlin.

SCHMOKER, J.W. & HALLEY, R.B. (1982) Carbonate porosity versus depth: a predictable relation for South Florida. *Am. Ass. Petrol. Geol. Bull.* **66**, 2561–2570.

SIBLEY, D.F. (1991) Secular changes in the amount and texture of dolomite. *Geology* **19**, 151–154.

SIBLEY, D.F. & GREGG, J.M. (1987) Classification of dolomite rock textures. *J. Sedim. Petrol.* **57**, 967–975.

SKALL, H. (1975) The paleoenvironment of the Pine Point lead-zinc district. *Econ. Geol.* **70**, 22–45.

SPENCER, R.J. (1987) Origin of Ca-Cl brines in Devonian formations, Western Canada Sedimentary Basin. *Appl. Geochem.* **2**, 373–384.

TEARE, M. (1990) *Sedimentology and Diagenesis of Middle Devonian Winnipegosis Reef Complexes, Dawson Bay, Manitoba.* MSc Thesis, McGill University, Montreal, 150 pp.

THERIAULT, F. & HUTCHEON, I. (1987) Dolomitization and calcitization of the Upper Devonian Grosmont Formation, Northern Alberta. *J. Sedim. Petrol.* **57**, 955–966.

UEDA, A., CAMPBELL, F.A., KROUSE, H.R. & SPENCER, R.J. (1987) $^{34}S/^{32}S$ variations in trace sulphide and sulphate in carbonate rocks of a Devonian reef, Alberta, Canada, and the Precambrian Siyeh Formation, Montana, USA. *Chem. Geol.* (Isotope Geoscience Section) **65**, 383–390.

VIAU, C.A. & OLDERSHAW, A.E. (1984) Structural controls on sedimentation and dolomite cementation in the Swan Hills field, central Alberta. In: *Devonian Lithofacies and Reservoir Styles in Alberta* (Ed. Krause, F.F. & Burrowes, G.) 13th Can. Soc. Petrol. Geol. Core Conference, 201–239.

WADLEIGH, M.A., VEIZER, J. & BROOKS, C. (1985) Strontium and its isotopes in Canadian rivers: fluxes and global implications. *Geochim. Cosmochim. Acta* **49**, 1727–1736.

WALLACE, M.W. (1990) Origin of dolomitization in the Devonian carbonates on the Barbwire Terrace, Canning basin, Western Australia. *Sedimentology* **37**, 105–122.

WALLS, R.A. & BURROWES, G. (1985) The role of cementation in the diagenetic history of Devonian reefs, Western Canada. In: *Carbonate Cements* (Ed. Schneidermann, N. & Harris, P.M.) Soc. Econ. Paleont. Mineral., Spec. Publs. 36, 185–219.

WALLS, R.A. & BURROWES, G. (1990) *Diagenesis and Reservoir Development in Devonian Limestone and Dolostone Reefs of Western Canada*. Can. Soc. Petrol. Geol., Short Course Notes, Section 5, 5.1–5.17.

WALLS, R.A., MOUNTJOY, E.W. & FRITZ, P. (1979) Isotopic composition and diagenetic history of carbonate cements in Devonian Golden Spike reef, Alberta. *Geol. Soc. Am., Bull.* **90**, 963–982.

ZENGER, D.H. (1983) Burial dolomite in the Lost Burro Formation (Devonian), east-central California, and the significance of late diagenetic dolomitization. *Geology* **11**, 519–522.

ZENGER, D.H. & DUNHAM, J.B. (1988) Dolomitization of Siluro-Devonian limestones in a deep core (5350 meters), southeastern New Mexico. In: *Sedimentology and Geochemistry of Dolostones* (Ed. Shukla, V. & Baker, P.A.) Soc. Econ. Paleont. Mineral., Spec. Publs. 43, 161–173.

Spec. Publs Int. Ass. Sediment. (1994) **21**, 231–254

Burial and hydrothermal diagenesis of Ordovician carbonates from the Michigan Basin, Ontario, Canada

M. CONIGLIO,* R. SHERLOCK,* A.E. WILLIAMS-JONES†
K. MIDDLETON* *and* S.K. FRAPE*

* *Department of Earth Sciences, University of Waterloo, Waterloo, Ontario, N2L 3G1, Canada; and*
† *Department of Earth and Planetary Sciences, McGill University, Montreal, Quebec, H3A 9Z9, Canada*

ABSTRACT

Ordovician carbonates in the subsurface of southwestern Ontario and exposed in the Manitoulin Island area are located on the margins of the Michigan Basin. In the subsurface, a widespread ferroan 'cap dolomite' at the top of the Trenton sequence most likely formed from fluids generated from compaction of the overlying shales of the Blue Mountain Formation (Utica Shale equivalent). Other dolomite in the sequence, both in the subsurface and in outcrop, is interpreted to be controlled by tectonic fractures which were intermittently active throughout the Palaeozoic. Saddle dolomite cement occurs in secondary and primary intraparticle pores. Pervasive replacive dolomite and selectively dolomitized beds, fossils and burrows also occur. In the subsurface, fracture-related dolomite is responsible for generating hydrocarbon reservoirs in otherwise impermeable and non-porous limestones.

Based on field, petrographic and oxygen isotopic data, the lithification of these Ordovician carbonates was protracted and can be largely explained as having occurred during burial, probably to depths of 1500–2000 m. Fluid inclusions, however, indicate that these carbonates experienced temperatures that were considerably higher than those attributable to burial alone. In general, fluid inclusions in early calcite cement (predolomitization), dolomite and late calcite cement (postdolomitization) homogenize at temperatures ranging from approximately 100 to 200°C. Dolomitizing fluids were generated through compaction flow from the more central parts of the basin, reflux from the overlying Silurian strata or invasion of younger fluids. Dolomitizing fluids travelled along fractures in the Ordovician sequence, pervasively altering permeable limestones in the immediate vicinity of fractures. Further away from the fractures, or along less permeable limestones, individual beds, fossils and burrows were selectively dolomitized. The fluid-inclusion temperatures can be explained in the context of a burial diagenetic system in which hydrothermal effects were important, although the heat source is uncertain.

INTRODUCTION

Hydrocarbon reservoirs hosted by Middle Ordovician carbonates are currently attractive exploration targets in the Michigan Basin. These structurally controlled fields are interpreted to have resulted from a regional fracture framework which, in addition to assisting hydrocarbon migration, also provided conduits for dolomitizing and other late diagenetic fluids (Sanford *et al.*, 1985; Taylor & Sibley, 1986; Hurley & Budros, 1990; Budai & Wilson, 1991; Middleton, 1991; Colquhoun, 1991). The linear fields range from a few hundreds of metres in width and several kilometres in length, as in the case of most of the Ontario fields, to an order of magnitude larger, as exemplified by the Albion-Scipio and Stoney Point fields in central Michigan. Similar dolomitized and fractured carbonates have also been described from New York, Ohio and Indiana (Keith, 1988). Despite the historical importance of the Michigan Basin, which produced the first commercial well in North America, in 1858 at Oil

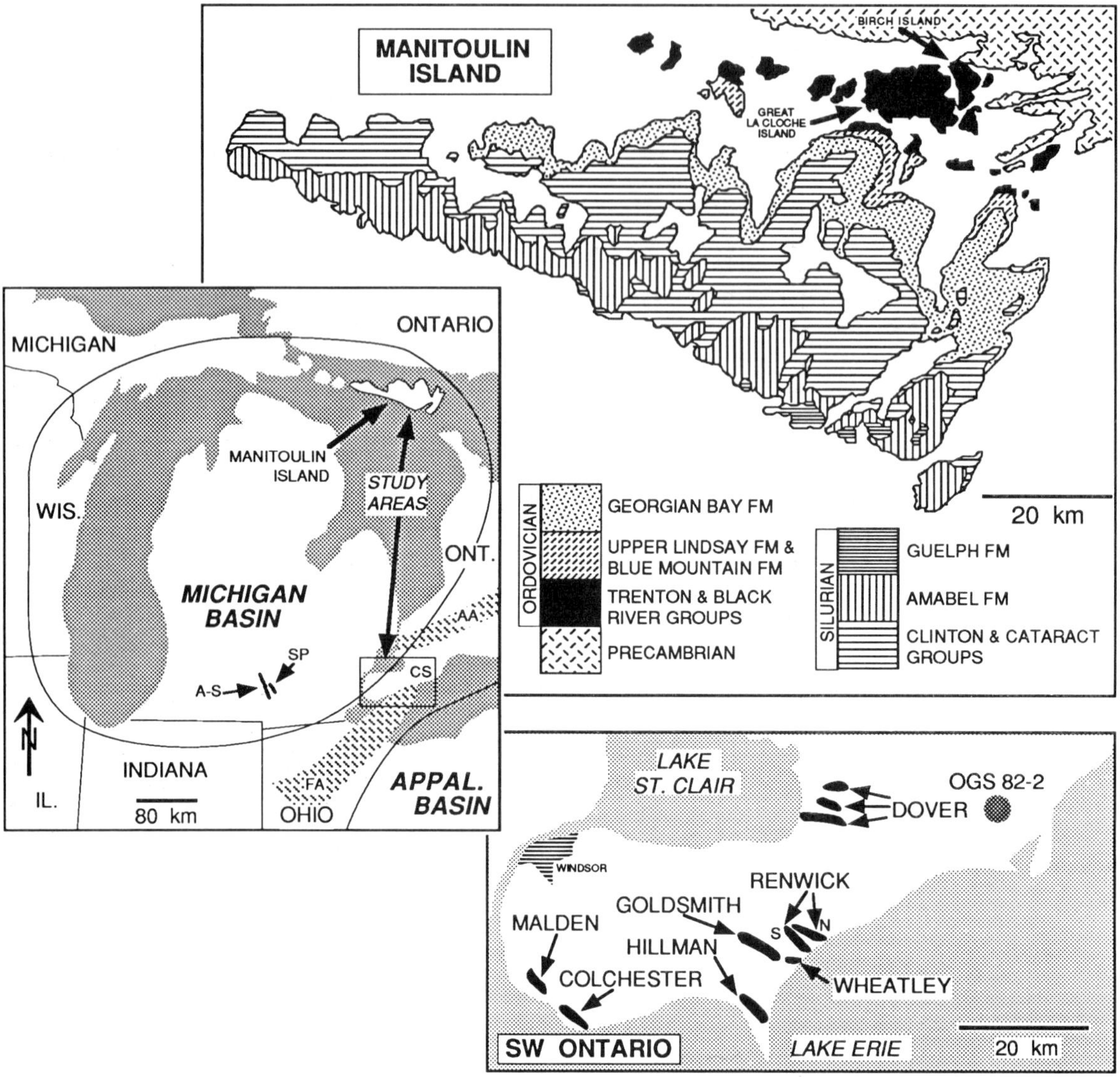

Fig. 1. Maps showing the outline of the Michigan Basin, distribution of Middle Ordovician hydrocarbon fields in southwestern Ontario, and the bedrock geology of the Manitoulin Island area. A-S, Albion-Scipio field; SP, Stoney Point field; FA, Findlay Arch; CS, Chatham Sag; AA, Algonquin Arch. Geological and field maps are modified from Freeman (1979) and Phillips *et al.* (1989), respectively.

Springs, Ontario, it is mainly in the last decade that the diagenetic history of its strata has received any significant amount of attention.

This study is a synthesis and expansion of two recent studies concerning the diagenetic history of Ordovician carbonates from the Michigan Basin in Ontario. Data are derived from the outcrop belt in the area of Manitoulin Island on the northeastern flank of the basin (Coniglio & Williams-Jones, 1992), and from the subsurface of southwestern Ontario (Middleton *et al.*, 1993). Cores from the following fields were examined: Malden, Colchester, Hillman, Wheatley, Renwick, Goldsmith and Dover (Fig. 1). The predominantly limestone core of the OGS-82-2 (Harwich 25-IECR) borehole was examined to assess the range of predolomitization alteration of these rocks in southwestern Ontario (Fig. 1). Through a combination of field and core observations, petrographic study, stable isotope (C, O, S) geochemistry and fluid-inclusion analysis, we will demonstrate that these carbonates clearly preserve a record of burial diagenetic effects which were in part controlled by fractures and hydrothermal circulation. Dolomitization is the most significant of these effects, and is the subject of this investigation. A detailed documentation of the variety of petrofabrics encountered in these rocks is found Coniglio & Williams-Jones (1992) and Middleton *et al.* (1993).

GEOLOGICAL SETTING

Palaeozoic reservoirs in southwestern Ontario occur within and along the southern margin of an area known as the Chatham Sag, a basement low that separates the Algonquin Arch in southwestern Ontario from the Findlay Arch in Ohio (Fig. 1). Together, the Algonquin and Findlay arches and the Chatham Sag separate the Michigan Basin to the northwest from the Appalachian Basin to the southeast (Sanford, 1961). The arches are thought to have been initiated in the Late Precambrian, and they remained intermittently active during the Palaeozoic (Sanford *et al.*, 1985). In southwestern Ontario, Cambrian, Ordovician, Silurian and Devonian reservoirs can be understood in the context of regional fault and fracture systems, expressed through their control on facies distribution and diagenesis (Fig. 2). These faults and fractures resulted from a wrenching tectonic system that originated from regional stresses associated with plate motions and orogenic events at the craton margins (Sanford *et al.*, 1985; Prouty, 1988). Reservoirs in the Middle Ordovician succession, which are aligned approximately northwest–southeast to east–west, are permeability pinchout traps generated along the downdropped side of rotated blocks (Fig. 2). The faults and associated fractures are interpreted to have been pathways that allowed migrating subsurface fluids to dolomitize limestones along and peripheral to the fault systems, thus forming the porous and permeable reservoir rocks (Sanford *et al.*, 1985).

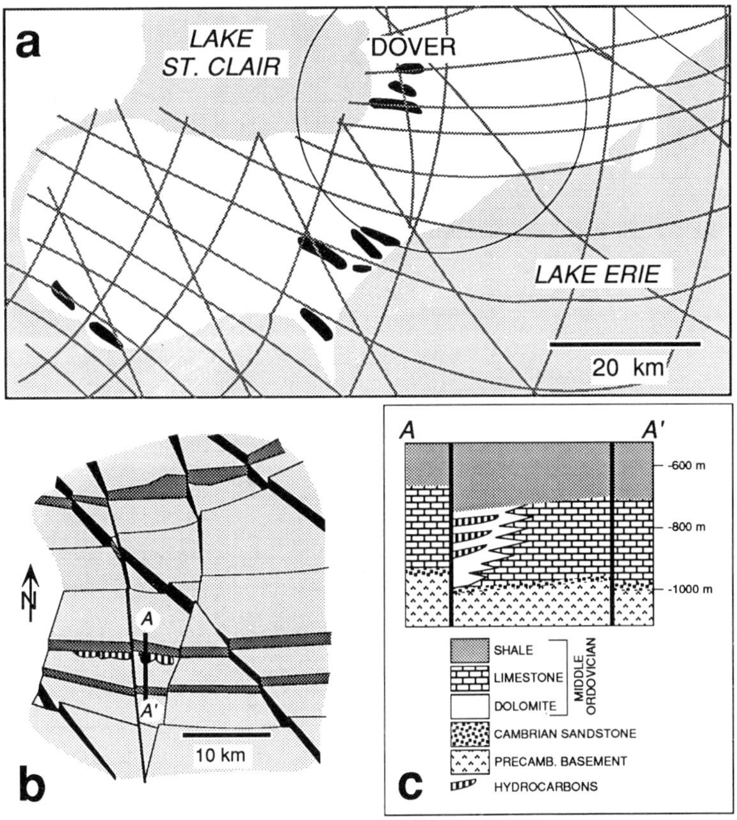

Fig. 2. Fault and fracture systems affecting Middle Ordovician carbonates in southwestern Ontario (a). Hydrocarbon fields from Fig. 1 are also shown. A sketch map of the circled area around the Dover fields is shown in (b). Here, the tilted fault blocks (b) produce permeability pinchout plays where limestones have been dolomitized in proximity to fractures and faults (c). Modified from Sanford *et al.* (1985).

The Palaeozoic succession on Manitoulin Island occurs as a gently southwestward-dipping Middle Ordovician to Middle Silurian carbonate-dominated wedge, which overlies deformed Precambrian metasediments of the Canadian Shield (Fig. 1). Four sets of vertical joints, including northwest–southeast and east–west-trending sets, characterize the Palaeozoic strata (Holst, 1982). Whether any of these joint sets relate to the fracture systems described by Sanford *et al.* (1985) is undetermined, although this seems likely given the regional controls on structure in the Michigan Basin. Minor quantities of hydrocarbons have been encountered on the island, but the controls on migration and trapping remain unknown.

On a regional scale the lithological characteristics of these Ordovician carbonates are remarkably constant. In detail, however, local variations have, in part, been responsible for difficulties in correlating subsurface and outcrop strata. A general stratigraphic correlation chart and lithological summary based on strata from the Manitoulin Island area are shown in Figure 3. These Middle and Late Ordovician beds were deposited on an extensive low-latitude shelf in the area now located between the Appalachians and the Transcontinental Arch (Wilson & Sengupta, 1985; for an alternative view of palaeolatitude, see Brookfield, 1988). The base of the sequence comprises supratidal and tidal-flat clastics and carbonates passing upward into lagoonal carbonates and then into offshore carbonates. The top of the Trenton Group in southwestern Ontario, as well as in much of the Michigan Basin, is a hardground that exhibits a sharp irregular contact with the overlying dark-coloured shales of the Blue Mountain Formation (Utica Shale equivalent; Keith, 1988; Hurley & Budros, 1990; Middleton, 1991). In the northeast to eastern side of the Michigan Basin, however, the transition is more gradual and is represented by interbedded calcareous black shales and the bioclastic limestones of the Collingwood Member (Russell & Telford, 1983), also referred to as the Upper Member of the Lindsay Formation (Johnson & Telford, 1985). The sequence of deep shelf shales of the Blue Mountain and lower Georgian Bay formations gradually shallowed upward to a carbonate tempestite-dominated succession (upper Georgian Bay Formation), terminating with emergence and formation of the disconformable Ordovician–Silurian boundary (Copper, 1978).

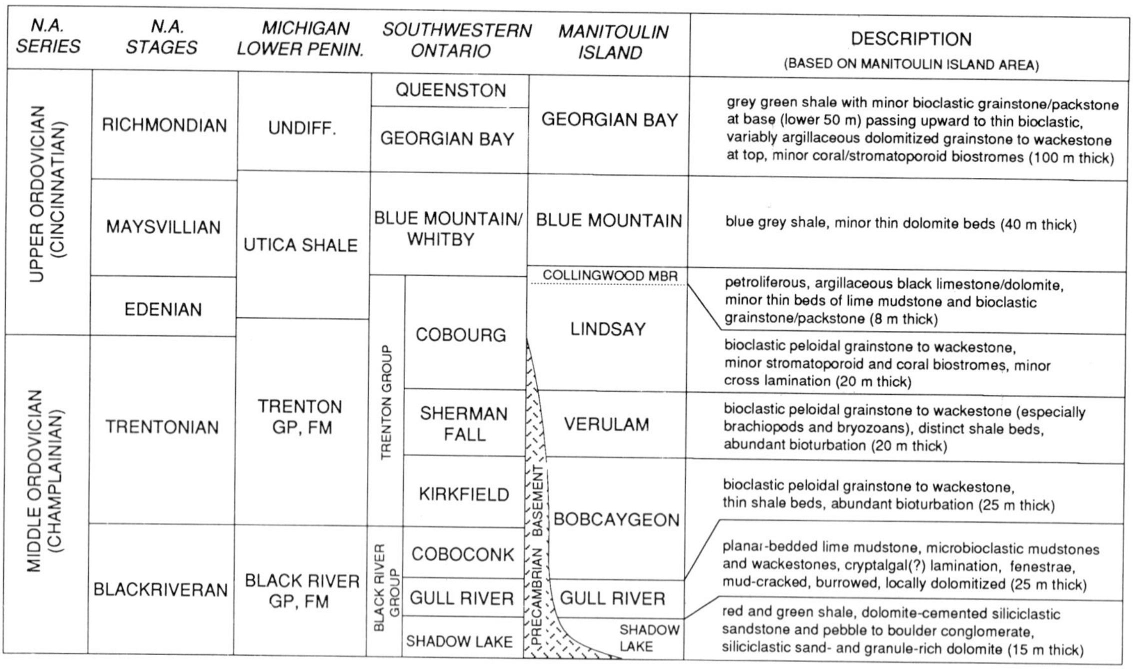

Fig. 3. Correlation chart for the Manitoulin Island area, southwestern Ontario and the Lower Peninsula of Michigan (compiled from Fisher *et al.*, 1988 and Barnes *et al.*, 1981). Lithological descriptions are based on strata from the Manitoulin Island area and are summarized from sources listed in the text.

In southwestern Ontario, the Black River and Trenton strata are up to 280 m thick. The top of the Trenton Group lies at depths varying from 665 m in the southwest part of the study area to 950 m just north of Lake St Clair. The Late Ordovician Georgian Bay and Queenston formations in southwestern Ontario are not considered in the present study, as there is insufficient core. In the Manitoulin Island area, however, the Georgian Bay Formation is well exposed and is therefore included. The entire Ordovician sequence in the Manitoulin Island area is estimated to be approximately 250 m thick (Copper,1978).

DIAGENESIS

Predolomitization diagenetic history

Early lithification of limestone is suggested by the numerous hardgrounds scattered throughout these carbonates and by the lack of obvious mechanical compaction. Vugs and interparticle, intraskeletal and biomouldic pores in these limestones are typically cemented by non-ferroan calcite crystals less than 100 μm in diameter. Ferroan calcite characterizes the last stage of growth of some crystals, and may also fill small discontinuous fractures. Lime mudstones, wackestones and packstones, including hardgrounds, are typically composed of non-ferroan microspar. Calcite cement and microspar generally exhibit dull to weak orange luminescence, with minor zoning that is not correlatable either regionally or locally. Petrographic evidence, such as partially dolomitized calcite cements, clearly indicates that dolomitization postdated the lithification of these carbonates.

Calcite cement and neomorphic crystals generally extinguish uniformly and contain few obvious fluid inclusions. However, fluid inclusion-rich coarsely crystalline calcite cements partially occlude some large vugs developed in leached coral or stromatoporoid heads in otherwise non-porous limestones in the Manitoulin Island area. These distinctive coarse calcite cements postdate the bulk of calcite cementation in these rocks and, in places, are directly overlain by saddle dolomite cement. For convenience, we refer subsequently to all of the predolomitization calcite cement and neomorphic calcite as 'early', in deference to their precipitation prior to dolomitization.

Dolomitization

Replacive dolomite

The scattered nature of Ordovician outcrops in the Manitoulin Island area and the limited number of cores from the subsurface precludes any realistic estimate of the volume or the three-dimensional geometry of dolomitized rock in the succession. Nevertheless, several distinctive styles of replacive dolomitization occur. Common to limestones in the subsurface and outcrop are disseminated idiotopic crystals of dolomite, usually less than 40 μm in diameter. These dolomite crystals, some of which demonstrate cathodoluminescence zoning, may be incorporated into later replacive mosaics that form the predominant dolomite phase in these rocks (see below). The intensity and succession of cathodoluminescence zones visible in the disseminated crystals are not correlatable from one outcrop to the next. Similar floating dolomite rhombs were interpreted by Mountjoy and Halim-Dihardja (1991) to have resulted from the movement of seawater through semilithified sediment near the sediment–water interface, or during early compaction. This volumetrically minor phase is not considered further.

Southwestern Ontario. In the subsurface, two principal types of replacive dolomite are recognized: 'cap' and 'fracture-related'. Figure 4 shows the distribution of these dolomite types for the Hillman field. Cap dolomite typically replaces the upper 1–3 m of the Trenton Group. The strongly ferroan non-luminescent cap dolomite typically occurs as dense xenotopic to hypidiotopic mosaics (Fig. 5a) with variable preservation of original wackestone fabric. In thin section, crystals range from 100 to 500 μm in diameter, have cloudy cores, clearer rims and slightly undulose extinction. A similar cap dolomite is not observed in outcrop.

A more coarsely crystalline, variably ferroan dolomite phase which conspicuously cross-cuts rock units in boreholes is considered to be fracture-related (Fig. 4), an interpretation that is consistent with that of similar fracture-related reservoirs associated with large- and small-scale faults elsewhere in the Michigan Basin (e.g. Taylor & Sibley, 1986; Hurley & Budros, 1990; Budai & Wilson, 1991). Petrographically similar dolomite selectively replaces single grainstone beds in otherwise limestone sequences. Replacive crystals range in diameter from 250 μm to 2 mm, have inclusion-rich cores and rel-

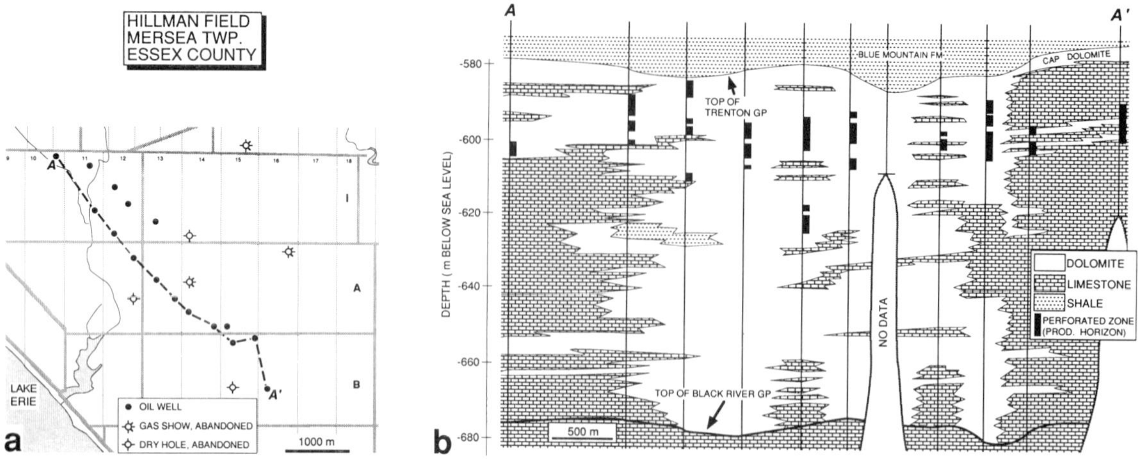

Fig. 4. Location (a) and cross-section (b) of the Hillman field. Lithologies were determined from photoelectric logs. All dolomite, except perhaps for the upper part at the southeastern end of the cross-section, is interpreted to be fracture-related. Producing units are the Cobourg and Sherman Fall Formations. Porosity ranges from 1 to 19%, averaging 8%, and permeability ranges from 0.01 to 1000 mD (Trevail, 1991). Data and interpretation provided courtesy of R.A. Trevail and the Ontario Ministry of Natural Resources.

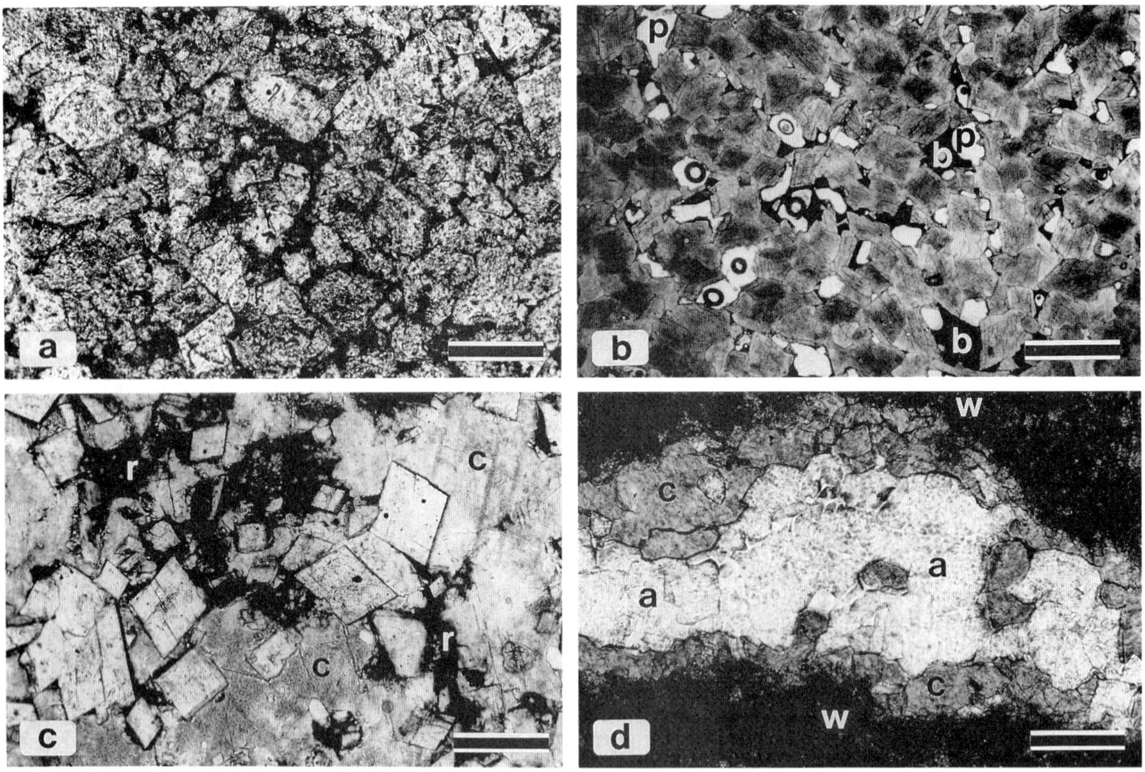

Fig. 5. Thin-section photomicrographs in plane light, southwestern Ontario. (a) Cap dolomite consists of a xenotopic to hypidiotopic mosaic of inclusion-rich ferroan crystals. OGS 82-2 Harwich 25-IECR, 901 m depth. Scale bar = 200 μm. (b) Fracture-related non-ferroan sucrosic dolomite contains solid hydrocarbon (b) in intercrystalline pores (p). Cons *et al.* 33821 Mersea 3-12-I, 876 m depth. Scale bar = 1 mm. (c) Fracture-related non-ferroan dolomite rhombs are mantled by dissolution residue (r) containing siliciclastic silt. Rock is cemented by ferroan late calcite cement (c). Cons *et al.* 33823 Mersea 1-12-A, 793 m depth. Scale bar = 200 μm. (d) Anhydrite (a) and ferroan late calcite cement (c) in veinlet in microbioclastic wackestone (w). Curved embayments of anhydrite penetrating into calcite seen elsewhere in this thin section suggest that anhydrite replaces calcite. OGS 82-2 Harwich 25-IECR, 977 m depth. Scale bar = 200 μm.

atively clear rims, display undulose extinction and are dully luminescent to non-luminescent. In thin section, preservation of depositional fabrics is poor (Fig. 5b,c) but in cores, depositional fabric ranges from well-preserved to poorly preserved (Fig. 6a–d). Cores that contain widespread fracture-related dolomite do not preserve a clearly identifiable cap dolomite. From this we infer that the cap dolomite was overprinted by fracture-related dolomite, producing a fractured cap dolomite. Overprinting is also supported by $\delta^{18}O$ analyses (discussed below) and other studies in the Michigan Basin (Taylor & Sibley, 1986; Prouty, 1988).

The fracture-related replacive mosaics are generally non-porous and xenotopic, and rarely porous and sucrosic. Rare intercrystalline pore space may contain ferroan saddle dolomite cement, late calcite or minor hydrocarbon (Fig. 5b). Some sucrosic samples, when viewed in plane light, contain a nondescript brown substance which irregularly coats dolomite crystals (Fig. 5c). The fine siliciclastic silt identified within this substance indicates it to be a dissolution residue from the original limestone. The remainder of the pore space is filled with late calcite cement.

Manitoulin Island area. In dolomites of the Manitoulin Island area, fossils and sedimentary structures are commonly visible on weathered outcrop surfaces, but in thin section fabric preservation is poor (Fig. 7a,b). All of the petrographic (including luminescence) properties of these replacive dolomite crystals are the same as the fracture-related dolomite from southwestern Ontario. Some portions of the Ordovician sequence, especially near the top, are pervasively dolomitized. Elsewhere, single beds, specific fossil types (e.g. stromatoporoids) and burrows are dolomitized. Other beds contain discontinuous elliptically shaped pods of replacive dolomite up to several decimetres in length. Exposures occasionally show limestone beds which can be traced laterally into pervasive dolomite, over a distance of a few centimetres or less (Fig. 8a). Except for the typical tan-coloured weathering of the dolomite and the light to medium-grey weathering of the limestone in outcrop, the transition is inconspicuous. Dolomite also pervasively replaced coarse grainstones that overlie the Precambrian basement. Tidal-flat sediments in the lower part of the sequence are characterized by beds, up to 50 cm thick, of finely crystalline non-porous dolomite interbedded with shales. Except for these dolomitized tidal-flat sediments, replacive dolomite is characterized by conspicuous biomouldic porosity, intercrystalline porosity, solution-enhanced biomoulds and vugs.

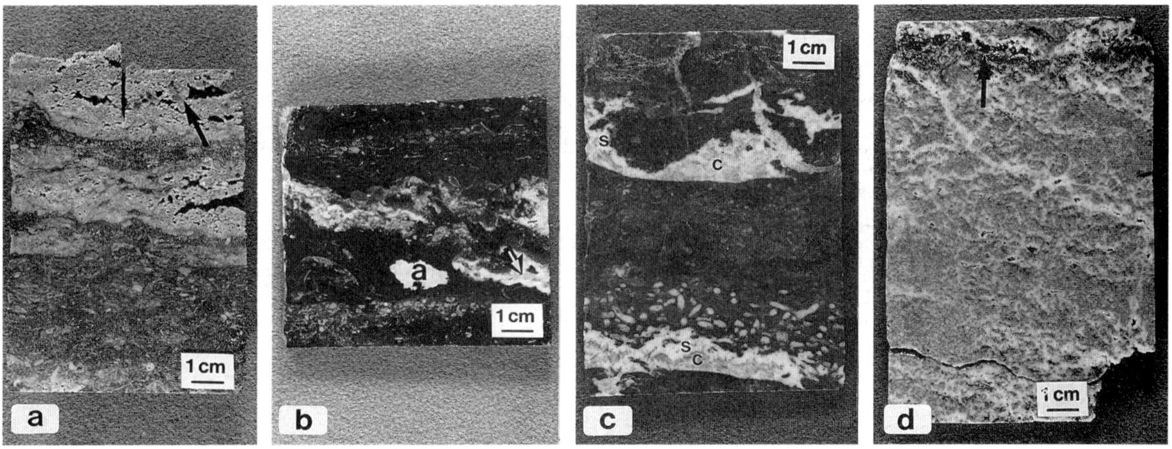

Fig. 6. Core photographs of fracture-related dolomite, southwestern Ontario. (a) Bioclastic dolomite with intercrystalline to vuggy porosity developed within packstone–grainstone nodules. Arrow points to large late-stage sparry calcite cement crystal. Cons *et al.* 33821 Mersea 3-12-I, 791 m depth. (b) Dolomitized bioclastic wackestone contains vugs lined with saddle dolomite cement and occluded by late calcite cement (arrow) and anhydrite (a). Cons *et al.* 33823 Mersea 1-12-A, 813 m depth. (c) Dolomitized bioturbated bioclastic wackestone contains stromatactoid cavities lined with saddle dolomite cement (s) and occluded by late calcite cement (c). Circular to rod-shaped grains near bottom of sample are dolomitized bryozoans. Cons *et al.* 33821 Mersea 3-12-I, 957 m depth. (d) Porosity is almost completely occluded by saddle dolomite cement (white), imparting a mottled white-brown colour in this fracture-related dolomite. Primary fabrics are not preserved. A minor amount of fracture porosity (arrow) remains. Cons *et al.* 33822 Mersea 5-11-I, 764 m depth.

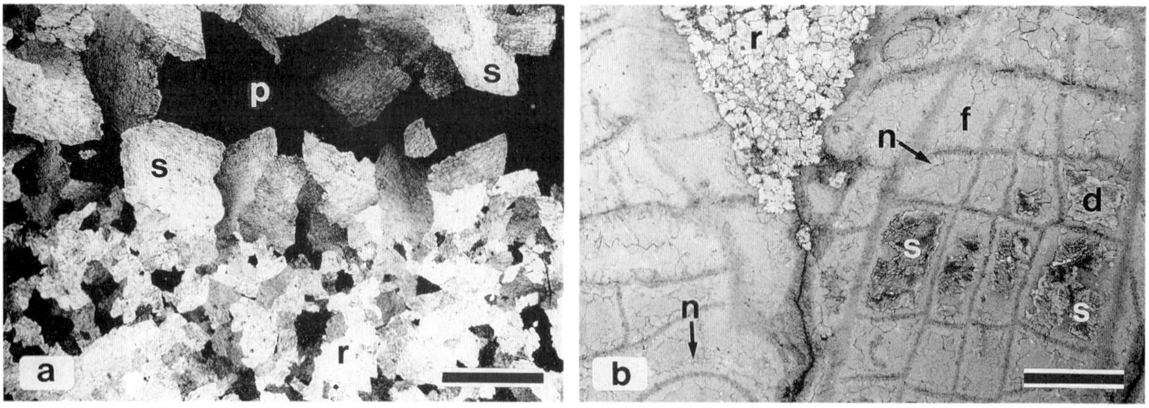

Fig. 7. Thin-section photomicrographs, Manitoulin Island. Scale bar = 1 mm. (a) Crossed polars view of xenotopic fabric-obliterating non-ferroan replacive dolomite (r) and associated pore-lining, very slightly ferroan saddle dolomite cement (s). Pore space is shown by (p). Lower Lindsay Formation. (b) Plane-light view of stained thin section showing coral filled with a fringe of early non-ferroan calcite cement (n) followed by early ferroan calcite cement (f). In some pores the non-ferroan calcite is followed by non-ferroan (d) or ferroan (s) saddle dolomite cement. There is also partial replacement by non-ferroan dolomite (r). Georgian Bay Formation.

Fig. 8. Outcrop photographs, Manitoulin Island area. (a) Sharp lateral transition of tan-weathering dolomite (d, light-grey) to limestone (c, medium-grey). Contact is just to right of hammer. Bobcaygeon Formation, Birch Island. (b) Limestone with vug filled with coarsely crystalline late calcite cement (c) overlying isopachous fringe of ferroan saddle dolomite cement (s). Coin scale is 21 mm in diameter. Lower Lindsay Formation, Manitoulin Island.

Saddle dolomite cement

Saddle dolomite is the most abundant porosity-occluding cement in southwestern Ontario and the Manitoulin Island area. Crystals range from 0.2 to 5 mm in diameter, have curved faces, strongly undulose extinction, inclusion-rich cores and relatively clear margins. Staining with potassium ferricyanide reveals that many crystals are more ferroan on their margins. In general, these crystals are very dully luminescent to non-luminescent. Saddle dolomite cement occurs in replacive dolomite mosaics, where it represents continued enlargement of replacive crystals on the margins of vugs, solution-enhanced fractures and intercrystalline pores (Fig. 7a). It also occurs in intraparticle and biomouldic pores and vugs in limestones, where the dolomite cement commonly overlies an earlier stage of calcite cement (Fig. 7b). Locally in southwestern Ontario, angular fragments of replacive dolomite are partially to completely cemented by saddle dolomite (and less commonly anhydrite cement), resulting in a breccia having a distinctive mottled appearance. This fabric is not observed in outcrop.

Postdolomitization diagenetic history

Late calcite

Coarsely crystalline late calcite cement occurs on fracture surfaces and in vugs and intercrystalline pores in dolomites (Figs 5c and 6a–c). Similar calcite crystals also occur in limestones (Figs 5d and 8b), but in most cases the timing of the precipitation of this phase with respect to dolomite is uncertain. Crystals may attain 3 cm in diameter, but most range from 1 to 10 mm. Late calcite also occurs as minor 'dedolomite', replacing the margins or centres of dolomite crystals. Stained thin sections demonstrate that ferroan, non-ferroan and zoned crystals are equally common. Extinction is typically sharp, and crystals generally contain numerous fluid inclusions. The cathodoluminescence of these crystals is typically uniform and ranges from bright-orange to dull. Discernible zoning is rare, and examples are not correlatable beyond their respective outcrops.

Other diagenetic phases

Several other late diagenetic phases occur, although all are volumetrically minor (<1%) based on visual estimate. Anhydrite is the most common non-carbonate mineral in the subsurface, occurring as millimetre-sized equant and lath-like crystals, and rarely attaining lengths up to 2 cm. It partially replaces limestone (grains and early calcite cement), late calcite cement (Fig. 5d), replacive dolomite and saddle dolomite cement. Anhydrite also occurs as cement within fractures and vugs (Fig. 6b) and between crystals in sucrosic dolomite. Some thin sections show anhydrite replaced by ferroan dolomite or by late calcite cement. In most cases, the relative order of carbonate and anhydrite precipitation is ambiguous.

In the outcrops of the Manitoulin Island area, gypsum postdates saddle dolomite cement typically as selenite crystals or splays of millimetre-wide fibres up to 2 cm in length. Gypsum also occurs as a non-porous white alabaster, composed of equant crystals up to 1 mm in diameter.

Other diagenetic phases include pyrite, authigenic quartz and feldspar, fluorite, barite, celestite and a yellow-brown amorphous material interpreted to be a ferric oxyhydroxide related to modern surface exposure. A similar suite of late-stage non-carbonate phases has also been described from elsewhere in the Michigan Basin (e.g. Budai & Wilson, 1991).

Solid hydrocarbon is observed in only a few samples from the subsurface, occurring in the intercrystalline pores of sucrosic dolomite mosaics (Fig. 5b) or postdating the late calcite cement which fills some of these pores.

Pressure dissolution and fracturing

Well-defined bedding-parallel stylolites truncate both carbonate grains and interparticle calcite cement in limestones. Poorly developed stylolites in both cap dolomite and replacive dolomite suggest that dolomitization followed pressure dissolution in limestone. Rare examples of stylolites cutting single isolated replacive dolomite crystals indicate pressure dissolution during or after dolomitization. Rare stylolites also occur in late calcite cements. Pressure dissolution thus appears to have been a protracted process that occurred before, possibly during, and after dolomitization.

Some stylolites in cores are oriented approximately parallel to vertical fractures, and probably resulted from being deflected near fractures. These stylolites thus postdated fracturing. Other stylolites are cut by fractures. The occurrence of fracturing and brecciation after the onset of pressure dissolution is indicated by rare examples of breccia fragments composed of what appear to be stylocumulate.

STABLE ISOTOPE DATA

Samples of carbonate for stable carbon and oxygen isotopic analysis were extracted using a modified dental drill. Approximately 10 mg of untreated powder were reacted with 100% phosphoric acid at 50°C, 15 min for calcite and 48 h for dolomite, and the evolved CO_2 gas was analysed for $^{18}O/^{16}O$ and $^{13}C/^{12}C$ using standard methods. Results are reported in conventional per mil (‰) notation relative to the PDB standard, using standard correction procedures. No corrections for acid fractionation were made for dolomite. Precision, based on selected replicate analyses, is better than $\pm 0.2‰$ (1σ) for both $\delta^{18}O$ and $\delta^{13}C$ values.

The ^{18}O contents of anhydrite were determined on CO_2 prepared from $BaSO_4$. Dissolution of approximately 100 mg of powdered anhydrite in deionized water was followed by filtering and precipitation of $BaSO_4$ through the addition of $BaCl_2 \cdot 2H_2O$. The precipitate was rinsed with 10% HCl solution and then deionized water to remove any $BaCO_3$ that might have coprecipitated with the $BaSO_4$. The $BaSO_4$ was dried at 80°C prior to mixing with carbon, and then combusted to CO_2 for ^{18}O analysis using the method of Sakai and Krouse (1971). $\delta^{18}O$ values are reported in per mil relative to SMOW, and the reproducibility of an internal $BaSO_4$ standard is better than ±0.5‰ (1 σ). $\delta^{34}S$ values were obtained from $BaSO_4$ using the method described by Yanagisawa and Sakai (1983). The reproducibility of an internal $BaSO_4$ standard is better than ±0.3‰ (1 σ), and results are expressed in per mil relative to the CDT standard.

Calcite

Fifty-two analyses of calcite were carried out on bulk limestone, early (i.e. predolomite) calcite cements, fossils (rugose corals, brachiopods and bryozoans with intrazooecial cement), and late calcite cement (Fig. 9a–c). In general, except for some samples of late calcite cement, $\delta^{13}C$ values are restricted, varying mostly between −3.5‰ and 2.8‰. In contrast, $\delta^{18}O$ values are more widespread, ranging from −11.2‰ to −2.9‰. Late calcite cements tend to be more ^{18}O- and ^{13}C-depleted than other calcites analysed. The most negative $\delta^{13}C$ value measured was −31.2‰ (Fig. 9c).

Most of the bulk limestone and fossil samples and more than half of the early calcite cements have $\delta^{13}C$ and $\delta^{18}O$ values that are close to those of estimated Ordovician low-latitude, marine calcites (Fig. 9a–c; Lohmann, 1988, Fig. 2.8). Assuming that these marine calcites precipitated at 25°C, and using the calcite–water palaeotemperature equation in Friedman and O'Neil (1977), the $\delta^{18}O$ compositions of Middle and Late Ordovician seawater would have been −4.5‰ and −2.5‰ (SMOW), respectively. Several early calcite $\delta^{18}O$ values are, however, more ^{18}O-enriched than the above marine compositions. For example, the highest Late Ordovician $\delta^{18}O$ value of −2.9‰ is from an early intraparticle calcite cement that would have precipitated from seawater at 18°C (Fig. 9c). Similarly, the highest Middle Ordovician $\delta^{18}O$ value of −3.6‰ from an early calcite cement in the Bobcaygeon Formation would have precipitated at 13°C (Fig. 9b). Cooler marine temperatures are a distinct possibility, particularly in the light of recent work suggesting that the Black River and Trenton sequences of southern Ontario represent temperate rather than tropical water conditions (Brookfield, 1988). If this is correct, then the more positive $\delta^{18}O$ values of some early calcites and fossils can be straightforwardly explained. Alternatively, if the temperature of precipitation was similar to that of the low-latitude marine calcites, early diagenetic and possibly marine porewaters were ^{18}O-enriched by up to a few per mil, and therefore closer to modern marine $\delta^{18}O$ values than generally accepted by carbonate workers (see Gregory, 1991).

The large spread in $\delta^{18}O$ values in early calcite is typical of carbonates altered at increasingly higher temperatures encountered during progressive burial diagenesis (e.g. Choquette & James, 1990). Evolving porewater compositions from progressive water–rock interaction and intracrystalline isotopic variations are additional factors that complicate the interpretation of $\delta^{18}O$ values. In light of the previously discussed uncertainties, establishing precipitation temperatures through oxygen isotope palaeothermometry appears to be questionable.

The wide-ranging and negative $\delta^{13}C$ values of late calcite from the Manitoulin Island area contrast sharply with all other carbonate phases analysed in this study, including late calcite cements from southwestern Ontario. The ^{13}C-depleted values strongly suggest that porewater bicarbonate was affected by an organic carbon source, perhaps bacterial or thermochemical sulphate reduction of hydrocarbons in interbedded shales, or from older, contemporaneous or younger strata from more basinal locations (e.g. Machel, 1989).

Dolomite

Sixty-three samples of dolomite were analysed (Fig. 9d–f). Saddle dolomite cement and replacive dolomite have a limited range of $\delta^{13}C$ values, from −1.1 to +1.9‰, and a wider range of $\delta^{18}O$ values from −10.4 to −5.4‰. Although there is clear overlap in the southwestern Ontario samples (Fig. 9d), in the Manitoulin data set replacive dolomite is more ^{18}O-enriched than saddle dolomite cement (Fig. 9e,f). The stable isotopic compositions of saddle dolomite cement and replacive dolomite in the Manitoulin Island area and in southwestern Ontario are comparable to those of burial dolomites reported by Taylor and Sibley (1986) and Granath (1991) from elsewhere in the Michigan Basin. Their burial origin is also supported by the presence of characteristics commonly accepted as criteria for burial dolom-

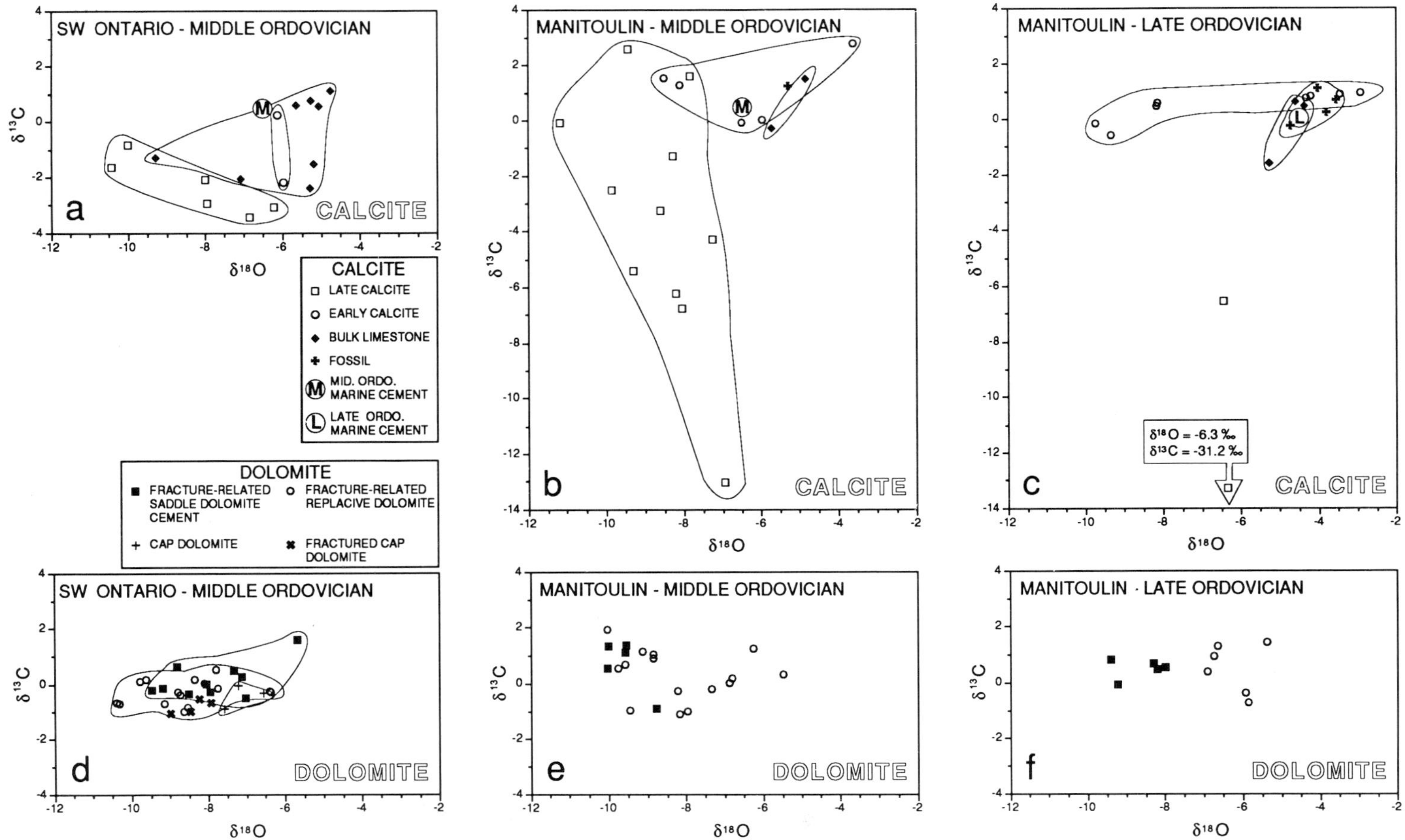

Fig. 9. $\delta^{13}C$ (PDB) versus $\delta^{18}O$ (PDB) for calcite (a–c) and dolomite (d–f) from southwestern Ontario and the Manitoulin Island area. Values of Middle and Late Ordovician marine calcite are from Lohmann (1988).

itization, such as coarse crystal size, abundant xenotopic texture, variable fabric preservation and associated saddle dolomite cement.

Only three samples of cap dolomite from southwestern Ontario were analysed, and their $\delta^{13}C$ values are similar to those of saddle dolomite cement and replacive dolomite (Fig. 9d). Cap dolomite $\delta^{18}O$ values, however, tend to be more ^{18}O-enriched, suggesting that precipitation of this phase might have occurred at temperatures lower than those that characterized precipitation of the other replacive and cement phases. This interpretation is consistent with the previously discussed petrographic evidence. Fractured cap dolomite $\delta^{13}C$ and $\delta^{18}O$ values (n = 4) are more similar to those of saddle dolomite cement and replacive dolomite, supporting the idea that subsequent recrystallization of cap dolomite occurred during the later episode of replacive dolomitization and precipitation of saddle dolomite cement.

Anhydrite

Fifteen samples of anhydrite from the subsurface of southwestern Ontario were analysed for $\delta^{18}O$ and $\delta^{34}S$ (Fig. 10). $\delta^{18}O$ values vary from 11.5 to 19.8‰ and $\delta^{34}S$ values range from 21.8 to 34.3‰. These anhydrite $\delta^{18}O$ and $\delta^{34}S$ values span a range similar

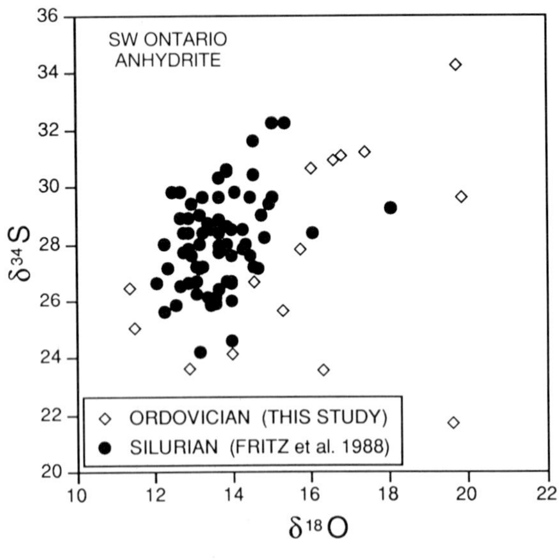

Fig. 10. $\delta^{34}S$ (CDT) versus $\delta^{18}O$ (SMOW) for anhydrite from southwestern Ontario. Ordovician values are from this study. Silurian values are from the Salina evaporite sequence of southwestern Ontario.

to anhydrite from the overlying evaporites of the Salina Formation of southwestern Ontario, which were interpreted by Fritz *et al.* (1988) to reflect closely Late Silurian oceanic sulphate. This idea is returned to later.

FLUID INCLUSIONS

Fluid inclusions were investigated in cleavage fragments of calcite and in 100 μm-thick doubly polished sections prepared from samples containing coarsely crystalline (>1 mm) calcite and/or dolomite. Samples were prepared using standard cold techniques. Microthermometric measurements were made at the University of Waterloo (southwestern Ontario samples) and McGill University (Manitoulin Island area samples) using FLUID INC.-adapted USGS gas-flow heating–freezing stages, calibrated using SYNFLINC synthetic fluid-inclusion standards. Reported temperatures from McGill University are accurate to ±0.1°C in the subzero range and ±2°C at the highest temperatures measured. At the University of Waterloo, accuracy is considered to be ±0.5°C over the range of temperatures measured. Ultraviolet light fluorescence microscopy was used to identify fluid inclusions containing hydrocarbons. Further details on sample preparation and analytical procedures are given by Coniglio and Williams-Jones (1992) and Middleton *et al.* (1993).

Petrography

Southwestern Ontario

Samples of saddle dolomite cement (n = 6) and late calcite cement (n = 3) contain large (6–20 μm) isolated fluid inclusions, characterized by an aqueous liquid and a small vapour bubble. Inclusions displaying consistent microthermometric properties are commonly concentrated within the cloudy cores of dolomite crystals, suggesting a primary and coeval origin. Cubic daughter crystals, tentatively identified as halite, were observed in several inclusions hosted within dolomite, but none were observed within calcite. No fluorescing inclusions were observed.

Manitoulin Island area

Samples analysed for fluid inclusions comprise early calcite cement (n = 3), saddle dolomite cement

(n = 2), replacive dolomite (n = 1) and late calcite cement (n = 3). The only early calcites containing fluid inclusions large enough to analyse were coarsely crystalline pore-lining crystals immediately overlain by saddle dolomite cement and representing the latest phase of early calcite precipitation.

Two types of inclusion were recognized. The first constitutes 90–95% of the fluid inclusion population in most samples, and is characterized by an aqueous phase and a small vapour bubble (type 1). A small number of inclusions are liquid-only. Size is typically less than 10 μm in diameter. The second type of inclusion is also observed in all phases of carbonate, but is more adundant in late calcite. These inclusions are vapour-only or vapour-rich, and 10–15 μm in diameter (type 2). The vapour-rich inclusions contain a minor amount of liquid, probably hydrocarbons, that fluoresces green or orange under ultraviolet light illumination.

Most fluid inclusions from the Manitoulin Island area are considered to be secondary or indeterminate in origin. Fluid inclusions of clearly primary or pseudosecondary origin, recognized by their occurrence in crystals with cores and rims rich in and devoid of fluid inclusions, respectively, were observed in only one sample of early calcite and two samples of dolomite. Unfortunately, most of these inclusions are too small for microthermometric investigation. Significantly, however, they are not noticeably different in other respects from most of the secondary inclusions or isolated fluid inclusions of indeterminate origin in the same samples.

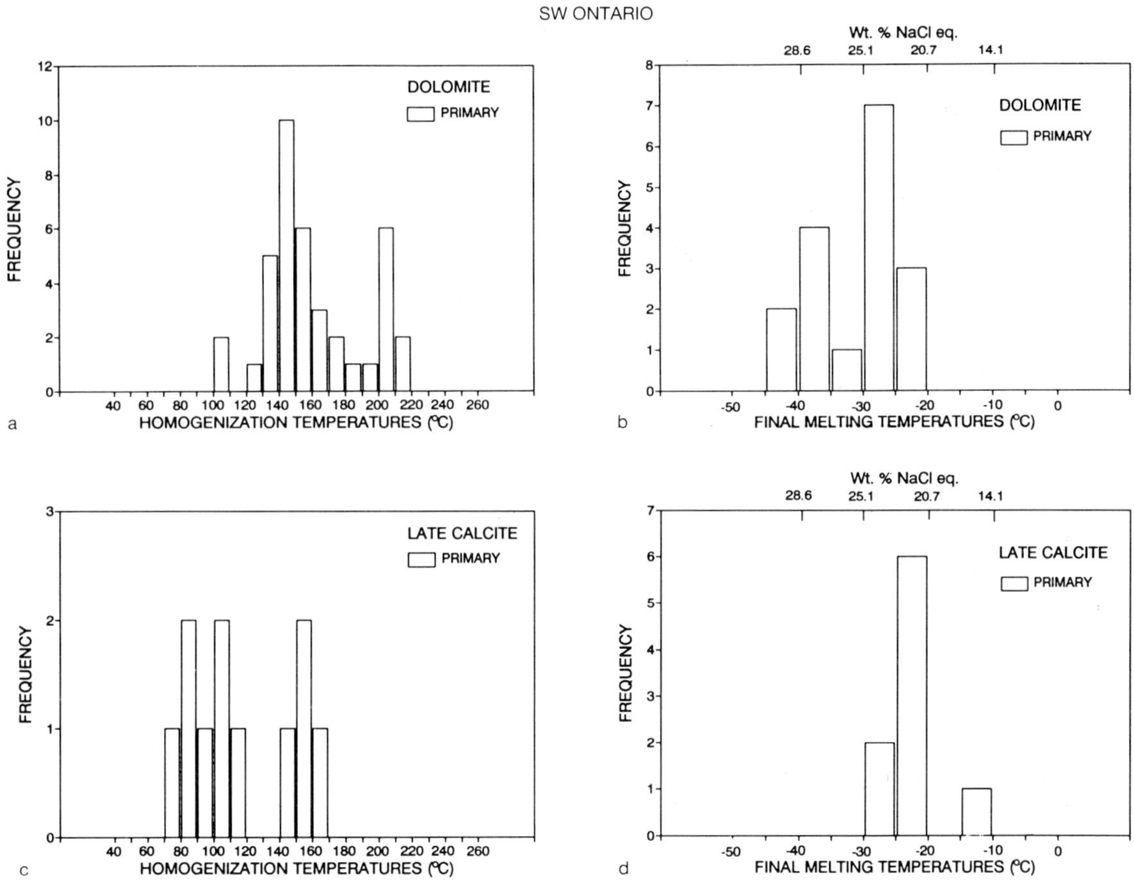

Fig. 11. Histograms showing the distribution of homogenization temperatures (a, c) and final melting temperatures (b, d) of fluid inclusions in saddle dolomite cement and late calcite cement in southwestern Ontario. Fluid salinities were calculated using the method of Oakes *et al.* (1990) and X_{CaCl_2} of 0.90.

Microthermometry

Southwestern Ontario

Fluid inclusions hosted within dolomite have homogenization temperatures displaying two distinct peaks, with modes at 149°C and 206°C (Fig. 11a). Individual samples may display a bimodal temperature distribution. In one sample it was possible to measure variations in homogenization temperature across three growth zones, with the highest temperature (average 201°C, $n = 2$) in the oldest zone, 179°C in an intermediate zone and 148°C in the youngest. These results indicate that precipitating fluids cooled with time. Fluid inclusions hosted within late calcite also display a bimodal distribution in homogenization temperatures, with peaks at 96°C and 156°C (Fig. 11c).

The average initial melting temperatures for fluid inclusions hosted within dolomite and calcite are −73°C and −59°C, respectively. These temperatures are interpreted to reflect eutectic melting in the system NaCl-$CaCl_2$±$MgCl_2$-H_2O. Although they are much lower than the eutectic temperatures in the systems NaCl-$CaCl_2$-H_2O and NaCl-$CaCl_2$-$MgCl_2$-H_2O (−52°C and −55°C, respectively; Crawford, 1981), these temperatures are within the range of initial melting temperatures obtained by Davis *et al.* (1990) on synthetic inclusions with compositions in the systems NaCl-$CaCl_2$-H_2O and NaCl-$MgCl_2$-H_2O. These temperatures are interpreted by them to reflect the melting of metastable phases formed during freezing. Fluids hosted within dolomite have final melting temperatures between −45°C and −23°C (Fig. 11b). The final phase to melt when the temperature is at the low end of the range (−45°C) is an acicular solid, possibly antarcticite ($CaCl_2 \cdot 6H_2O$); at higher temperatures the final solid phase to melt is ice. The most reasonable explanation for this is that the lower temperature represents melting of a fluid with a $CaCl_2$ concentration greater than or equal to the eutectic composition, in the system NaCl-$CaCl_2$-H_2O, and the higher temperatures represent melting on the H_2O-dominated liquidus. This interpretation suggests that $CaCl_2$ was the principal salt in the fluid, i.e. that X_{CaCl_2} may have been close to 0.94, the eutectic composition of the system. Salinities of the fluid inclusions in dolomite calculated from the above final melting temperatures, assuming this value of X_{CaCl_2}, range from 30.3 to 22.1 wt. % NaCl eq., respectively (Oakes *et al.*, 1990). Final melting temperatures for fluid inclusions in calcite range between −28°C and −12°C (Fig. 11d), and the last solid to melt in all cases is ice. The final ice melting temperature of −28°C reflects $X_{CaCl_2} > 0.64$ (Crawford, 1981); the range of salinities calculated using this value is 25.0–15.8 wt. % NaCl eq., and from 24.3 to 15.8 wt. % NaCl eq., if X_{CaCl_2} is assumed to be the same as in the dolomite-hosted inclusions, i.e. 0.94 (Oakes *et al.*, 1990).

Manitoulin Island area

Histograms of homogenization temperatures in type 1 inclusions in early calcite and dolomite display similar distributions, having single well-defined peaks corresponding to modes of 95° and 101°C, respectively (Fig. 12a,c). Late calcites, on the other hand, contain type 1 inclusions that demonstrate a much wider range of homogenization temperatures, including the well-defined peak seen in the early calcite and dolomite data (mode = 107°C) and a second, higher temperature population of fluid inclusions homogenizing mainly between 150° and 210°C (mode = 183°C) (Fig. 12e). The distribution of homogenization temperatures varies greatly among samples of late calcite, which is not unexpected given their heterogeneous nature.

Ice in type 1 inclusions in early and late calcite, and in dolomite, typically begins to melt at about −45°C, although in a few cases ice was observed to begin to melt at temperatures as low as −70°C. These temperatures are consistent with the presence of NaCl−$CaCl_2$±$MgCl_2$ in the fluid (Crawford, 1981; Davis *et al.*, 1990). The temperatures at which the ice in type 1 inclusions in early calcite finally melts are, with two exceptions, <−30°C, and range down to −46.8°C; in dolomite these temperatures are all <−30°C, and are as low as −44.5°C (Fig. 12b,d). Such low temperatures reflect X_{CaCl_2} values >0.90 and salinities between 25 and 31 wt. % NaCl eq. (Oakes *et al.*, 1990). In contrast, the ice in type 1 inclusions in late calcite, with two exceptions, melts at temperatures >−15°C (Fig. 12f). The corresponding salinity is <17.8 wt. % NaCl eq., if X_{CaCl_2} is assumed to be the same as in the early calcite- and dolomite-hosted type 1 inclusions, or <18.6 wt. % NaCl eq. if $X_{CaCl_2} = 0.0$ (Oakes *et al.*, 1990).

The liquid in a small number of type 2 inclusions, several of which are primary, froze in the range −143°C to −175°C. The final melting temperatures, which ranged from −77°C to −88°C, are consistent with light liquid hydrocarbons (*cf.* Weast, 1986).

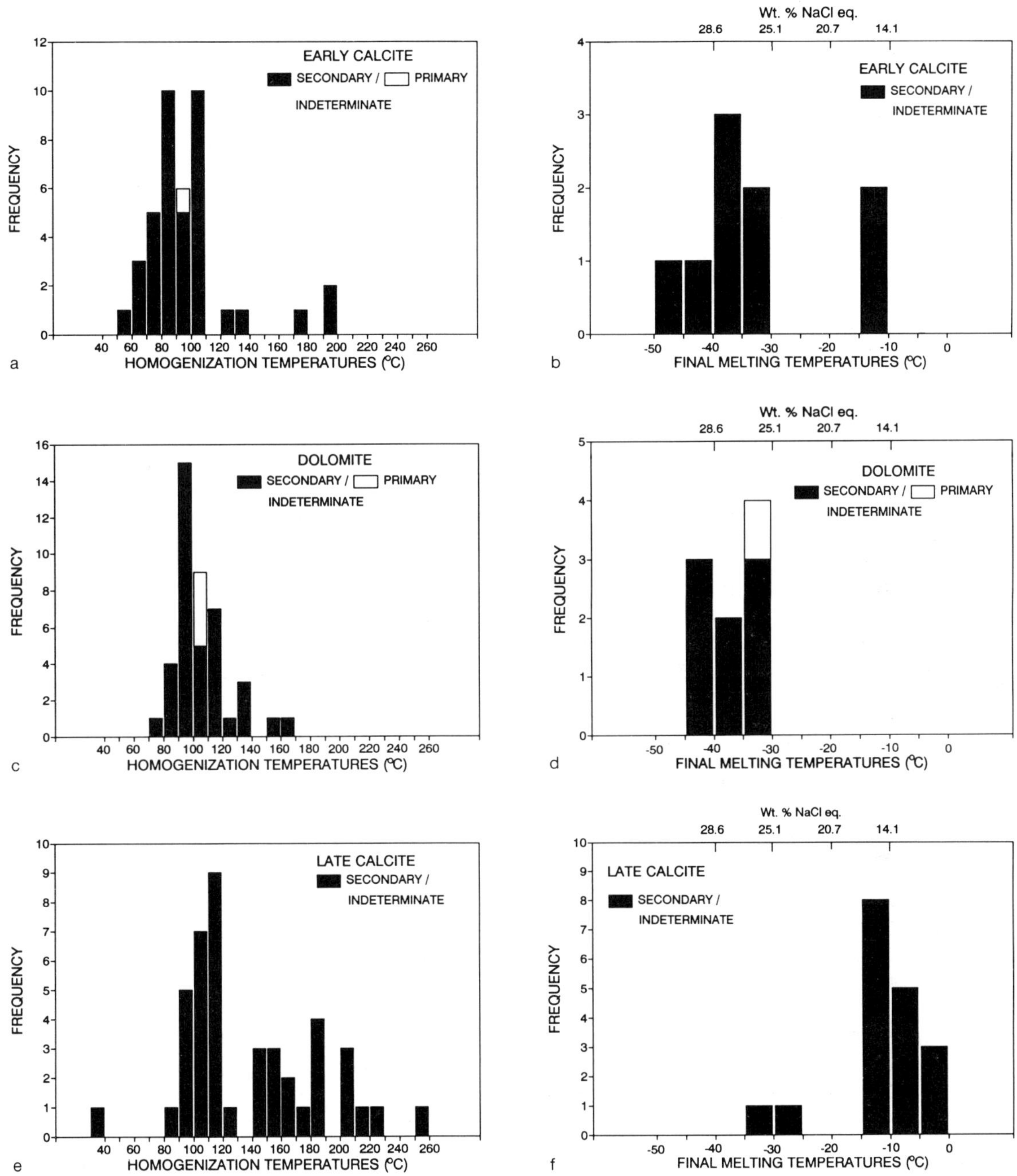

Fig. 12. Histograms showing the distribution of homogenization temperatures (a, c, e) and final melting temperatures (b, d, f) of type 1 fluid inclusions in early calcite cement, dolomite and late calcite cement in the Manitoulin Island area. Fluid salinities were calculated using the method of Oakes *et al.* (1990) and X_{CaCl_2} of 0.90.

DISCUSSION

Introduction

The focus of all other major diagenetic studies of the Ordovician carbonates from the Michigan Basin has been primarily dolomitization, because of the close association of dolomite with hydrocarbon reservoirs (e.g. Taylor & Sibley, 1986; Keith, 1988; Hurley & Budros, 1990; Budai & Wilson, 1991). Previous workers recognized three types of dolomite, based on petrographic and geochemical criteria. A non-ferroan 'regional dolomite', confined to the western and southwestern parts of the basin, is volumetrically the most important and is interpreted to have resulted from the mixing of meteoric and seawater during early diagenesis (Taylor & Sibley, 1986; Budai & Wilson, 1991). A texturally similar ferroan 'cap dolomite' characterizes the uppermost 0.3–15 m of the Trenton sequence, and occurs throughout most of the southern, southwestern (including southwestern Ontario) and some of the northern parts of the Michigan Basin. Equivalent strata of the Lower Member of the Lindsay Formation on Manitoulin Island, although commonly dolomitized, do not show the persistent ferroan composition nor strong stratigraphic control demonstrated by the cap dolomite in the subsurface. Fracture-related epigenetic dolomite, characteristic of Ordovician reservoirs such as the Albion-Scipio trend in central Michigan, is the third and economically most important type recognized (Taylor & Sibley, 1986; Hurley & Budros, 1990; Budai & Wilson, 1991). As will be shown, most if not all dolomite occurring in the Manitoulin Island area can be assigned to this third category.

Cap dolomite

The ferroan composition, relatively uniform distribution at the top of the Trenton sequence, and $\delta^{18}O$ values of the cap dolomite can be explained by compactional dewatering of the overlying thick shale sequence during burial diagenesis. This interpretation agrees with that offered in the comprehensive treatment by Taylor and Sibley (1986) as well as Budai and Wilson (1991). Ferrous iron was derived from the overlying shale sequence (from detrital iron oxides or clay minerals, clay mineral transformations or organic matter) and connate water in pores was expelled during compaction, providing the necessary Mg^{2+} for dolomitization. That this occurred at elevated temperatures accompanying burial is indicated by the $\delta^{18}O$ compositions that are too negative (^{18}O-depleted) to have formed directly from Middle Ordovician seawater.

Fracture-related dolomite

In southwestern Ontario, replacive dolomite and saddle dolomite cement are considered to be fracture-related, based on borehole data showing lateral limestone to dolomite transitions and cementation of breccia fragments. Replacive dolomite and saddle dolomite cement from outcrops of the Manitoulin Island area are similarly interpreted to be fracture-related. The evidence for this, however, is not as convincing. In the field the lateral transition from limestone to dolomite is clearly observed. A fracture-related origin also conveniently explains the apparently erratic distribution of pervasively dolomitized beds or groups of beds to selectively dolomitized macrofossils or portions of individual beds. In these cases preferential dolomitization affected either the beds closest to the fractures or the most permeable lithologies or components. The common dolomitization of grainstone beds adjacent to the relatively impermeable Precambrian basement could have occurred through lateral fluid flow from cross-cutting fractures. In addition, a fracture origin can account for the similar petrographic and stable isotopic characteristics of dolomite throughout the Ordovician sequence in the outcrop belt. The availability of fracture pathways explains why the dolomitization of Georgian Bay Formation carbonates is similar to that of the Trenton and Black River Groups, even though they are separated by a thick shale. Similar fracturing followed by dolomitization is described from several areas in the Devonian of the Western Canada Sedimentary Basin (Stoakes, 1987; Qing & Mountjoy, 1989; Mountjoy & Halim-Dihardja, 1991).

Mechanism for fracture-related dolomitization

The distribution of replacive dolomite and saddle dolomite cement can be explained by the flow of dolomitizing fluids through fractures, as discussed above. Precipitation during burial accounts for the coarse crystal size, abundant xenotopic texture, saddle dolomite cement, negative $\delta^{18}O$ compositions, and the warm homogenization temperatures and high salinities of the primary fluid inclusions analysed both from the outcrop and the subsurface

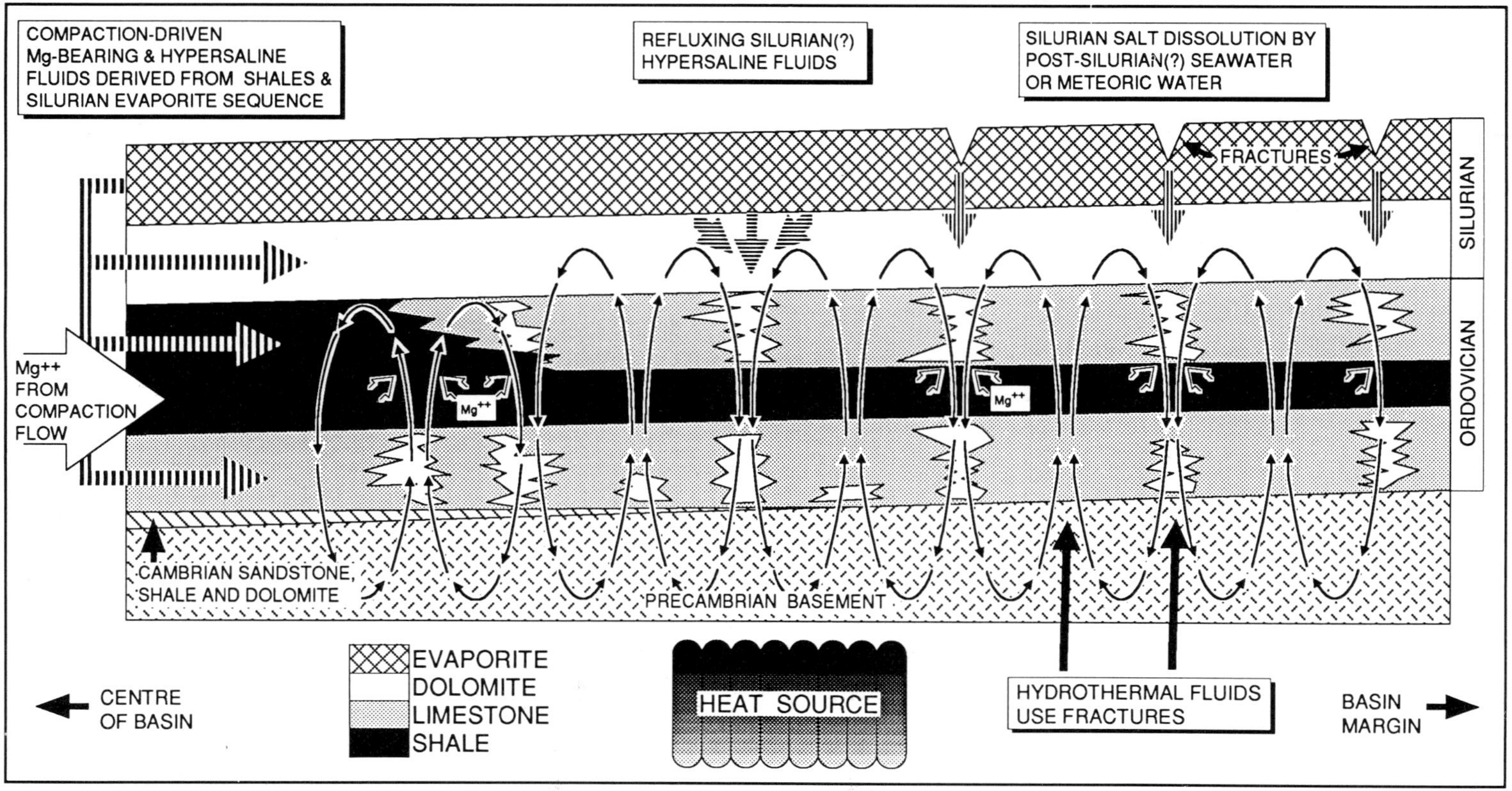

Fig. 13. Schematic representation of origin of fracture-related dolomitization of Ordovician sequence in Ontario. Brines entered the Ordovician sequence mainly through fractures. Hypothetical heat sources in Precambrian basement provide drive for free convection. See text for explanations.

(*cf.* Gregg & Sibley, 1984; Taylor & Sibley, 1986; Lee & Friedman, 1987).

The limited amount of dolomite in the succession avoids the principal weakness of the compaction model, i.e. an insufficient amount of Mg^{2+} from burial porewaters or generated during clay mineral diagenesis (e.g. Land, 1985; Hardie, 1987). Younger, older or correlative argillaceous sediments from deeper in the Michigan and Appalachian basins could be the source of Mg^{2+}-bearing fluids which migrated upwards and laterally towards the basin margins ('oblique compaction flow' of Machel & Mountjoy, 1987; Fig. 13). Compaction flow alone, however, does not adequately account for the high salinities of the fluid inclusions. The most obvious source of saline water in this sequence is the Late Silurian Salina Group, a sequence of carbonates, evaporites and shales that reaches a thickness of up to 1000 m in the central Basin (Fisher *et al.*, 1988). Devonian strata contain smaller volumes of evaporite. In more central parts of the Michigan and Appalachian basins, these evaporitic strata occur structurally lower than the Middle and Late Ordovician succession on the basin margins (e.g. Clifford, 1973; Lilienthal, 1978). Migrating Mg^{2+}-bearing compaction waters could have dissolved salts, or combined with hypersaline pore fluids to form the high salinities of the fluid inclusions in the carbonate minerals. However, a problem of the compaction flow hypothesis is that, if dolomitization occurred during peak burial during the Late Palaeozoic, as suggested by Budai and Wilson (1991), the sequence may have been too deeply buried for compaction flow to have been an effective mechanism of fluid transport. Machel and Mountjoy (1986) suggested that most burial compaction flow probably occurs up to depths of approximately 1000 m. Pressure dissolution, which can occur at depths as shallow as a few hundreds of metres (Choquette & James, 1990), remains a possible mechanism to compact further largely lithified strata. In this case, the regional fracture system affecting these rocks may have been necessary to generate the fluid pathways for lateral and updip migration of saline dolomitizing solutions.

Although compaction flow is a plausible mechanism to explain the fracture-related dolomitization of these rocks, a genetic connection with the overlying Silurian sequence composed of dolomites and evaporites appears to be better supported. Regional studies have shown that the Silurian carbonates, although dolomitic on the basin margins, grade into limestones towards the basin centre (e.g. Sonnenfeld & Al-Aasm, 1991). The reasons for extensive dolomitization of Silurian carbonates around the rim of the Michigan Basin are uncertain, however. Sears and Lucia (1980) explained the dolomitization of Silurian reefs in the northern Michigan Basin as having resulted in part from the reflux of hypersaline brines originating in supratidal areas. Cercone (1988) proposed that the preferential dolomitization of reefal carbonates in the Michigan Basin was due to subsurface seawater flow in response to evaporative drawdown in the basin. Sonnenfeld and Al-Aasm (1991) suggested that brines seeped out from the evaporating pan, carrying with them the requisite Mg^{2+} for dolomitization. Regardless of the details of the pervasive dolomitization of the Silurian strata on the basin margins, reflux of Mg^{2+}-bearing seawater or seawater-derived brines or formation waters is likely to have been involved (Fig. 13; for an alternative view, see Kendall, 1989). The high salinities could be explained by fluid interaction with evaporitic strata or the incorporation of subaerially produced brines (Hanor, 1987). The principal weakness of this hypothesis is that, if dolomitization occurred during the Silurian, the evidence indicating burial dolomitization in the underlying Ordovician must be reconciled in the light of the relatively shallow burial depths implied (probably a few hundred metres).

The connection between dolomitization, halite dissolution and fractures was recently examined by Carter (1991), who noted that the Late Silurian Salina A-1 and A-2 carbonates were preferentially dolomitized where the overlying salt layers (B-Salt, A-2 Salt) thinned or were absent, possibly the result of dissolution by seawater or meteoric water migrating along fractures. These solutions could have travelled into the underlying Ordovician strata to locally dolomitize them (Fig. 13). In this explanation, however, the age of the dolomitizing solution (seawater or meteoric water) could have been post-Silurian, but would have almost certainly been modified substantially from interaction with the thick sequence of Late Silurian evaporites.

The genetic connection between Silurian and Ordovician dolomite finds some support in the study of Late Silurian dolomite by O'Shea *et al.* (1988), in which $\delta^{13}C$ and $\delta^{18}O$ values span a range similar to those of the Ordovician dolomites analysed in this study. Similarly, the strontium isotope results reported by Granath (1991) support the idea that Ordovician and Silurian dolomitization is related. She found that the $^{87}Sr/^{86}Sr$ ratios of matrix (replacive) dolomite and dolomite cement from the Albion-Scipio area were more radiogenic than those expected from Middle Ordovician seawater, and interpreted a Late Silurian seawater influence in the dolomitizing brines. In contrast, McNutt *et al.* (1987) examined strontium isotopes in eight 'matrix' (replacive) dolomites and saddle dolomite cements from the Hillman field of southwestern Ontario, and indicated that some $^{87}Sr/^{86}Sr$ ratios were consistent with Middle Ordovician seawater values. More radiogenic values were suggested to have resulted from the derivation of dolomitizing brines from associated shales or the underlying Precambrian crystalline basement. Alternatively, Cambrian and Lower Ordovician seawater $^{87}Sr/^{86}Sr$ ratios (see Burke *et al.*, 1982) overlap the $^{87}Sr/^{86}Sr$ ranges reported by Granath (1991) and McNutt *et al.* (1987), suggesting that the dolomitizing solutions may have been derived from the compaction of Cambrian and Lower Ordovician strata downdip in the Michigan Basin, supporting the burial compaction hypothesis. The interpretation of the strontium isotope data thus remains equivocal.

The connection between the Ordovician and Silurian diagenesis can also be examined in the light of $\delta^{34}S$ and $\delta^{18}O$ values of anhydrite. Those from the subsurface Ordovician sequence are widely scattered, but their range is only slightly wider than that of the Salina anhydrites from southwestern Ontario, interpreted by Fritz *et al.* (1988) to reflect the composition of oceanic sulphate during the Late Silurian (Fig. 10). The cause of the widespread $\delta^{34}S$ and $\delta^{18}O$ values in the Ordovician sequence remains problematic but, based on the variety of petrographic relations observed with other late phases in these rocks, multiple episodes of anhydrite precipitation (replacement and cement) very likely occurred.

Alternative explanations for fracture-related dolomite in the Michigan Basin have also been proposed. Budai and Wilson (1991), based on $\delta^{18}O$ composition of dolomite from various localities in the Michigan Basin, suggested that mixing of meteoric water recharged from the basin margins with deeply sourced brines, possibly expelled upward through fractures, could explain the origin of fracture-related dolomite. Hurley and Budros (1990) suggested that the principal dolomitizing fluid could have been hypersaline Silurian–Devonian seawater travelling down fractures mixing with warm limestone-dissolving fluids generated deeper in the stratigraphic succession, or possibly in the underlying basement.

Timing of fracture-related dolomitization

Although it is tempting to suggest that fracture-related dolomitization in the Ordovician of the Manitoulin Island area and southwestern Ontario is related to the pervasive dolomitization of the Silurian, details of timing are uncertain. Based on comparison of fracture-related dolomite in the Michigan Basin with Mississippi Valley-type deposits in the central and eastern United States, previous workers have generally agreed on a Late Palaeozoic–Early Mesozoic timeframe for fracture-related dolomitization and hydrocarbon migration, coincident and probably genetically related to the Alleghanian deformation in the east (Prouty, 1988; Hurley & Budros, 1990; Budai & Wilson, 1991). Because southwestern Ontario is located on the margin of the Appalachian Basin (Fig. 1), which is a foreland basin, this interpretation is especially attractive in view of the possibility that foreland basins create favourable conditions for the migration of mineralizing fluids (Leach & Rowan, 1986). In addition to dolomite and late calcite cement, minor late-stage diagenetic minerals, which include anhydrite, barite, fluorite, celestite and sulphides, further enhance the possible relation with Mississippi Valley-type mineralization. The existence of non-economic Mississippi Valley-type deposits in Silurian strata in southern Ontario (Farquhar *et al.*, 1987) and upper New York State (Friedman, 1989) also lends support to the above hypothesis. If the Late Palaeozoic–Early Mesozoic timeframe is correct, the importance of fracture-related dolomitization by reflux from the overlying Silurian sequence could be overestimated. Burial compaction flow and infiltration of dolomitizing fluids along fractures where evaporites were dissolved nevertheless remain possible explanations within this timeframe.

In southwestern Ontario, solid hydrocarbons coating saddle dolomite and late calcite cement suggest that these late-stage diagenetic phases were closely associated with hydrocarbon migration into reservoir rocks, a conclusion also reached by Budai and Wilson (1991). Hydrocarbon-bearing inclusions in cements from the Manitoulin Island area also support this interpretation. Although hydrocarbon-bearing fluid inclusions were not encountered in the subsurface of southwestern Ontario, they have been observed in dolomite and late calcite cements from reservoir rocks elsewhere in the Michigan Basin (e.g. Shaw, 1975; Budai & Wilson, 1991; Granath & McLimans, 1991).

Hydrothermal diagenesis

Although most of the fluid inclusions analysed from the Manitoulin Island area are secondary or indeterminate in origin, the concurrence of the temperature and salinity data from the few available clearly primary inclusions suggest that for the ≈100°C mode, the measured temperatures and salinities are probably close to those that characterized mineral precipitation. The modal homogenization temperature of 100°C for early calcite and dolomite from the Manitoulin Island area is too high to be explained by burial depths alone, especially if these minerals precipitated during the Late Silurian or shortly thereafter. Extrapolation of stratigraphic data from elsewhere in the Michigan Basin (Cercone, 1984) cannot give the thicknesses required to generate these temperatures at peak burial, assuming a 'normal' intracratonic geothermal gradient, such as the 23°C/km suggested by Hogarth and Sibley (1985) and a 20°C surface temperature. Cercone and Pollack (1991) suggested that a thick thermally resistant sequence of Permo-Carboniferous strata, since eroded, could have caused the elevated temperatures responsible for the anomalous maturity of most strata in the Michigan Basin. However, evidence for this is lacking. The best reconstruction of burial depth offered by Coniglio and Williams-Jones (1992) is approximately 1500 m (compacted) for strata of the Manitoulin Island area, yielding a peak burial temperature of 55°C. Doubling burial depth still does not produce peak burial temperatures of 100°C. The problem is exacerbated in trying to explain the high temperature mode of the fluid inclusions seen in late calcite cements.

The fluid-inclusion data from southwestern Ontario, with homogenization temperature modes for saddle dolomite cement at 149°C and 206°C, more clearly demonstrate the contrast with temperatures that logically would have existed during peak burial under a normal palaeogeothermal gradient. The above temperatures are higher than those reported by Granath (1991), who interpreted dolomite fluid-inclusion homogenization temperatures in the Albion-Scipio area to be consistent with deep burial conditions (115°C for matrix and vug-lining dolomite; 133°C for saddle dolomite cements). Like the Manitoulin Island area, the burial history of strata from southwestern Ontario is poorly constrained but may have been as much as 2000 m, based on data extrapolated from elsewhere in the Michigan Basin. Maximum temperatures achieved at peak burial, assuming the same conditions stated above, would have been 66°C. Although not as high, most of the homogenization temperatures of late calcite cements from southwestern Ontario also are not readily explained by burial alone.

A substantially higher palaeogeothermal gradient for the Michigan Basin could have produced higher burial temperatures at the reconstructed depths. Cercone (1984), using Lopatin's method to model the thermal history of the Michigan Basin, proposed that palaeogeothermal gradients in the Michigan Basin probably ranged from 35 to 45°C/km. This would have yielded maximum burial temperatures of 88°C and 110°C for Manitoulin and southwestern Ontario, respectively. Vugrinovich (1988) estimated palaeogeothermal gradients of 50°C/km or more, based on saddle dolomite and fluorite from the Ordovician and Devonian sequences from Michigan's Lower Peninsula and supporting data from coal maturation and oil source studies. Even though the above studies suggest higher temperatures than those indicated by the current knowledge of the burial history of the Michigan Basin, it is uncertain whether these data, although derived from numerous localities, represent more than localized anomalies (Vugrinovich, 1988). The high temperatures (and deduced palaeogeothermal gradients) reported may not be representative of the basin as a whole (see additional discussion in Nunn, 1986).

We propose that the involvement of hydrothermally influenced burial fluids more readily accounts for the high homogenization temperatures of fluid inclusions, without calling upon burial depths which cannot be substantiated or equally inexplicable palaeogeothermal gradients. The late calcite cements that reach temperatures of approximately 200°C in the Manitoulin Island area suggest that the hydrothermal component became more important with time, whereas the generally cooler homogenization temperatures of late calcite cements from southwestern Ontario suggest the opposite, as does the single example of a dolomite crystal with progressively cooler homogenization temperatures in successive growth zones. Note, however, that there is no way of knowing the relative timing of precipitation of phases or hydrothermal diagenesis between southwestern Ontario and the Manitoulin Island area. These variations in temperatures are consistent with the hypothesis that the intermittent reactivation of basement fractures, probably controlled by plate motions and orogenic events at the craton margin (Sanford *et al.*, 1985), may have opened and closed fractures at different times, altering the pathways of hydrothermal circulation. The existence of fracture pathways further allows for a variety of fluids to have circulated during repeated submergence or emergence of the region throughout the Phanerozoic. Halite dissolution along fractures, as described by Carter (1991), would have substantially increased the salinity of dolomitizing solutions. On the other hand, the incorporation of meteoric water during time of emergence could explain the lower-salinity fluid inclusions in late calcite cements.

Regardless of which hypothesis best explains the origin of the dolomitizing fluids, the fluid-inclusion homogenization temperatures strongly suggest a hydrothermal influence (Fig. 13). If convection cells were established and if descending brines were the principal source of Mg^{2+} for dolomitization, these fluids could have circulated in the fractured Ordovician host limestone and shale, with Mg^{2+} replenishment occurring in the upper parts of the convection cell (Fig. 13). The existence of a heat source in the Michigan Basin is completely speculative, however. Carter and Easton (1990) point out that there is no evidence of Phanerozoic igneous rocks in any of the more than 600 wells that penetrate the Precambrian basement in southwestern Ontario. Nevertheless, fractures that penetrate the basement could have been proximal to deeper intrusions.

The fact that most of the fluid inclusions analysed for the Manitoulin Island area are secondary or indeterminate does not alter the basis for involving a hydrothermal influence, but the possibility exists that these inclusions could provide little indication of the physicochemical conditions of mineral precipitation. The precipitation of the latest phase of early calcite cement and subsequent dolomitization could have occurred during burial of this sequence at normal burial temperatures, prior to later involvement of hydrothermal fluids. The few primary inclusions analysed could have been stretched or refilled by later hydrothermal fluids that also filled most of the secondary or indeterminate inclusions (*cf*. Goldstein, 1986; Prezbindowski & Larese, 1987; Barker & Goldstein, 1990). Alternatively, the fluid inclusions may have formed from hydrothermal fluids considerably after carbonate precipitation.

CONCLUSIONS

Our study has focused strictly on the diagenesis of the Ordovician sequence in the Ontario portion of the Michigan Basin. Petrographic, geochemical and fluid-inclusion data indicate that diagenetic alteration of these carbonates, particularly dolomitization, occurred mainly during burial and that fractures provided pathways for dolomitizing fluids. In addition, hydrothermal effects were also important. We have not been able to interpret these rocks unequivocally, however, for the reasons discussed earlier. What is clear from our work is that a better understanding of the diagenesis of these strata will be possible only by examining across the stratigraphic boundaries that typically compartmentalize diagenetic studies.

The situation in southwestern Ontario is an especially good illustration. Almost all of the producing carbonates in the Ordovician, Silurian and Devonian sequences are dolomitized – as dolomitized zones in domes and anticlines and along faults and fractures, as dolomitized reef and reef-associated carbonates, and as dolomitized non-reefal stratigraphic traps (Trevail *et al.*, 1987). In addition, all of these reservoirs can be explained in the context of the fracture model developed by Sanford *et al.* (1985), regardless of whether the first-order controls on reservoir rocks are facies-related, diagenetic or structural in origin. Accordingly, it is tempting to link the origin of dolomite in these various units. If this connection is later shown to be justified through petrographic, geochemical, structural and other studies, explaining the dolomitization of these rocks will almost certainly force us to more rigorously test current burial, seawater and evaporite-related models for regional dolomitization, and perhaps also consider dolomitization models that are hybrids of the currently accepted ones. For southern Ontario, a comprehensive model would need to explain partial dolomitization of the Ordovician and Devonian sequences and almost complete dolomitization in the Silurian, which is also associated with a thick evaporite succession.

ACKNOWLEDGEMENTS

This paper benefited from the critical reviews of W.J. Meyers, B.H. Purser, M.E. Tucker and D.H. Zenger. This research was supported by individual operating grants from Canada's Natural Sciences and Engineering Research Council (NSERC) awarded to M.C., A.E.W.-J. and S.K.F. and by Ontario Geoscience Research Grant No. 353 awarded to M.C. and S.K.F. R.S. acknowledges support from an Ontario Graduate Scholarship. K.M. acknowledges a NSERC Postgraduate Scholarship and additional funding from a Texaco Geological Research Grant. Stable isotope analyses were provided by the Environmental Isotope Laboratory at the University of Waterloo.

We are grateful to the various property owners of the Manitoulin Island area, and especially the First Nations of Cockburn Island, Sheguiandah, Sheshegwaning, West Bay, Whitefish River and Wikwemikong for providing ready access to their land. T.R. Carter (Ontario Ministry of Natural resources (MNR)) and R.A. Trevail (formerly of MNR) provided additional insight and logistical support for the southwestern Ontario portion of this project.

REFERENCES

Barker, C.E. & Goldstein, R.H. (1990) Fluid-inclusion technique for determining maximum temperature in calcite and its comparison to the vitrinite reflectance geothermometer. *Geology* **18**, 1003–1006.

Barnes, C.R., Norford, B.S. & Skevington, D. (1981) The Ordovician System in Canada. *Int. Union. Geol. Sci.,* Publ. 8, 27 pp.

Brookfield, M.E. (1988) A mid-Ordovician temperate carbonate shelf – the Black River and Trenton Limestone groups of southern Ontario, Canada. *Sediment. Geol.* **60**, 137–153.

Budai, J.M. & Wilson, J.L. (1991) Diagenetic history of the Trenton and Black River formations in the Michigan Basin. In: *Early Sedimentary Evolution of the Michigan Basin* (Ed. Catacosinos, P.A. & Daniels, P.A. Jr.) Geol. Soc. Am. Special Paper 256, 73–88.

Burke, W.H., Denison, R.E., Hetherington, E.A., Koepnick, R.B., Nelson, H.F. & Otto, J.B. (1982) Variation of seawater $^{87}Sr/^{86}Sr$ throughout Phanerozoic time. *Geology* **10**, 516–519.

Carter T. (1991) Dolomitization patterns in the Salina A-1 and A-2 carbonate units, Sombra Township, Ontario. *Proc. Ontario Petrol. Inst., 30th Ann. Conf., London, Ontario*, 35 pp.

Carter, T.R. & Easton, R.M. (1990) Extension of Grenville basement beneath southwestern Ontario: lithology and tectonic boundaries. In: *Subsurface Geology of Southwestern Ontario: a Core Workshop* (Ed. Carter, T.R.) Am. Ass. Petrol. Geol. Eastern Section, Core Workshop Volume, London, Ontario, 6–26.

Cercone, K.R. (1984) Thermal history of the Michigan Basin. *Am. Ass. Petrol. Geol., Bull.* **68**, 130–136.

Cercone, K.R. (1988) Evaporative sea-level drawdown in the Silurian Michigan Basin. *Geology* **16**, 387–390.

Cercone, K.R. & Pollack, H.N. (1991) Thermal maturity of Michigan Basin. In: *Early Sedimentary Evolution of the Michigan Basin* (Ed. Catacosinos, P.A. & Daniels, P.A. Jr.) Geol. Soc. Am. Special Paper 256, 1–11.

Choquette, P.W. & James, N.P. (1990) Limestones – the burial diagenetic environment. In: *Diagenesis* (Ed. McIlreath, I.A. & Morrow, D.W.) Geoscience Canada Reprint Series 4, 75–111.

Clifford, M.J. (1973) *Silurian Rock Salt of Ohio*. State of Ohio, Dept. Nat. Res. Div. Geol. Surv. Rep. Invest. 90, 42 pp.

Colquhoun, I. (1991) *Paragenetic History of the Ordovician Trenton Group Carbonates, Southwestern Ontario*. MSc Thesis, Brock University, St Catharines, Ontario, 350 pp.

Coniglio, M. & Williams-Jones, A.E. (1992) Diagenesis of Ordovician carbonates from the Northern Michigan Basin, Manitoulin Island area, Ontario: evidence from petrography, stable isotopes and fluid inclusions. *Sedimentology* **39**, 813–836.

Copper, P. (1978) Paleoenvironments and paleocommunities in the Ordovician–Silurian sequence of Manitoulin Island. In: *Geology of the Manitoulin Area* (Ed. Sanford, J.T. & Mosher, R.E.) Mich. Bas. Geol. Soc. Special Paper 3, 47–62.

Crawford, M.L. (1981) Phase equilibria in aqueous fluid inclusions. In: *Fluid Inclusions: Applications to Petrology* (Ed. Hollister, L.S. & Crawford, M.L.) Mine. Ass. Can., Short Course 75–100.

Davis, D.W., Lowenstein, T.K. & Spencer, R.J. (1990) Melting behaviour in laboratory grown fluid inclusions in the systems $NaCl\text{-}H_2O$, $NACl\text{-}KCl\text{-}H_2O$, $NaCl\text{-}MgCl_2\text{-}H_2O$ and $NaCl\text{-}CaCl_2\text{-}H_2O$. *Geochim. Cosmochim. Acta* **54**, 591–601.

Farquhar, R.M., Haynes, S.J., Mostaghel, M.A., Tworo, A.G., MacQueen, R.W. & Fletcher, I.R. (1987) Lead isotope ratios in Niagara Escarpment rocks and galenas: implications for primary and secondary sulphide deposition. *Can. J. Earth Sci.* **24**, 1625–1633.

Fisher, J.H., Barratt, M.W., Droste, J.B. & Shaver, R.H. (1988) Michigan Basin. In: *Sedimentary Cover – North American Craton: US* (Ed. Sloss, L.L.) Geol. Soc. Am., The Geology of North America D-2, 361–382.

Freeman, E.B., ed. (1979) *Geological Highway Map, Southern Ontario*. Ontario Geol. Surv. Map 2441.

Friedman, G.M. (1989) Case history of deep-burial sulfide mineralization in the northern Appalachian Basin. *Carb. Evap.* **4**, 231–241.

Friedman, I. & O'Neil, J.R. (1977) *Data of Geochemistry: Compilation of Stable Isotope Fractionation Factors of Geochemical Interest*. US Geol. Surv., Paper 440-KK, 1–12.

Fritz, P., Lapcevic, P.A., Miles, M., Frape, S.K., Lawson, D.E. & O'Shea, K.J. (1988) Stable isotopes in sulphate minerals from the Salina Formation in southwestern Ontario. *Can. J. Earth Sci.* **25**, 195–205.

Goldstein, R.H. (1986) Reequilibration of fluid inclusions in low-temperature calcium-carbonate cement. *Geology* **14**, 792–795.

Granath, V.C. (1991) Geochemical constraints on the origin of dolomite in the Ordovician Trenton and Black River Limestones, Albion-Scipio area, Michigan. *Am. Ass. Petrol. Geol. Bull.* **75**, 584–585 (abstract).

GRANATH, V.C. & MCLIMANS, R.K. (1991) Characterization of oil-filled fluid inclusions in the Trenton-Black River Limestones, Albion-Scipio area, Michigan. *Am. Ass. Petrol. Geol. Bull.* **75**, 585 (abstract).

GREGG, J.M. & SIBLEY, D.F. (1984) Epigenetic dolomitization and the origin of xenotopic dolomite texture. *J. Sedim. Petrol.* **54**, 908–931.

GREGORY, R.T. (1991) Oxygen isotope history of seawater revisited: timescales for boundary event changes in the oxygen isotope composition of seawater. In: *Stable Isotopes Geochemistry: A Tribute to Samuel Epstein* (Ed. Taylor, R.P., O'Neil, J.R. & Kaplan, T.N.) Geochem. Soc. Publ. 3, 65–76.

HANOR, J.S. (1987) *Origin and Migration of Subsurface Sedimentary Brines*. Soc. Econ. Paleont. Mineral. Short Course 21, 247 pp.

HARDIE, L.A. (1987) Dolomitization: a critical view of some current views. *J. Sedim. Petrol.* **57**, 166–183.

HOGARTH, C.G. & SIBLEY, D.F. (1985) Thermal history of the Michigan Basin: evidence from conodont coloration index. In: *Ordovician and Silurian Rocks of the Michigan Basin and its Margins* (Ed. Cercone, K.R. & Budai, J.M.) Mich. Bas. Geol. Soc. Special Paper 4, 45–58.

HOLST, T.B. (1982) Regional jointing in the northern Michigan Basin. *Geology* **10**, 273–277.

HURLEY, N.F. & BUDROS, R. (1990) Albion-Scipio and Stoney Point Fields – USA. Michigan Basin. In: *Stratigraphic Traps I. Treatise of Petroleum Geology Atlas of Oil and Gas Fields* (Ed. Beaumont, E.A. & Foster, N.H.) Am. Ass. Petrol. Geol. 1–37.

JOHNSON, M.D. & TELFORD, P.G. (1985) *Paleozoic Geology of the Little Current Area, District of Manitoulin*. Ontario Geol. Surv., Geol. Series – Preliminary Map P.2670.

KEITH, B.D. (Ed.) (1988) *The Trenton Group (Upper Ordovician Series) of Eastern North America*. Am. Ass. Petrol. Geol. Studies in Geology 29, 317 pp.

KENDALL, A.C. (1989) Brine mixing in the Middle Devonian of western Canada and its possible significance to regional dolomitization. *Sediment. Geol.* **64**, 271–285.

LAND, L.S. (1985) The origin of massive dolomite. *J. Geol. Educ.* **33**, 112–125.

LEACH, D.L. & ROWAN, E.L. (1986) Genetic link between Ouachita foldbelt tectonism and the Mississippi Valley-type lead-zinc deposits of the Ozarks. *Geology* **14**, 931–935.

LEE, Y.I. & FRIEDMAN, G.M. (1987) Deep-burial dolomitization in the Ordovician Ellenburger Group carbonates, western Texas and southeastern New Mexico. *J. Sedim. Petrol.* **57**, 544–557.

LILIENTHAL, R.T. (1978) *Stratigraphic Cross-Sections of the Michigan Basin*. State of Michigan, Dept. Nat. Res., Geol. Surv. Div., Rep. Invest. 19, 36 pp.

LOHMANN, K.C. (1988) Geochemical patterns of meteoric diagenetic systems and their application to studies of paleokarst. In: *Paleokarst* (Ed. James, N.P. & Choquette, P.W.) pp. 58–80. Springer-Verlag, New York.

MACHEL, H.G. (1989) Relationships between sulphate reduction and oxidation of organic compounds to carbonate diagenesis, hydrocarbon accumulations, salt domes, and metal sulphide deposits. *Carb. Evap.* **4**, 137–151.

MACHEL, H.G. & MOUNTJOY, E.W. (1986) Chemistry and environments of dolomitization – a reappraisal. *Earth Sci. Rev.* **23**, 175–222.

MACHEL, H.G. & MOUNTJOY, E.W. (1987) General constraints on extensive pervasive dolomitization and their application to the Devonian carbonates of western Canada. *Bull. Can. Petrol. Geol.* **35**, 143–158.

MCNUTT, R.H., FRAPE, S.K. & DOLLAR, P. (1987) A strontium, oxygen and hydrogen isotopic composition of brines, Ontario and Michigan. *Appl. Geochem.* **2**, 495–505.

MIDDLETON, K. (1991) *Fracture-Related Diagenesis of Middle Ordovician Carbonate Reservoirs, Southwestern Ontario*. MSc Thesis, University of Waterloo, Waterloo, Ontario, 141 pp.

MIDDLETON, K., CONIGLIO, M., SHERLOCK, R. & FRAPE, S.K. (1993) Dolomitization of Middle Ordovician carbonate reservoirs, southwestern Ontario. *Bull. Can. Petrol. Geol.* **41**, 150–163.

MOUNTJOY, E.W. & HALIM-DIHARDJA, M.K. (1991) Multiple phase fracture and fault-controlled burial dolomitization, Upper Devonian Wabamun Group. *J Sedim. Petrol.* **61**, 590–612.

NUNN, J.A. (1986) Subsidence and thermal history of the Michigan Basin. In: *Thermal Modeling in Sedimentary Basins* (Ed. Burruss, J.) pp. 417–436. Editions Technip, Paris.

OAKES, C.S., BODNAR, R.J. & SIMONSON, J.M. (1990) The system NaCl-$CaCl_2$-H_2O: I. The ice liquidus at 1 atm total pressure. *Geochim. Cosmochim. Acta* **54**, 603–610.

O'SHEA, K.J., MILES, M.C., FRITZ, P., FRAPE, S.K. & LAWSON, D.E. (1988) Oxygen-18 and carbon-13 in the carbonates of the Salina Formation of southwestern Ontario. *Can J. Earth Sci.* **25**, 182–194.

PHILLIPS, A.R., COLQUHOUN, S.A. & PUTNAM, P.E. (1989) B.A. *Liberty Memorial Core Workshop: Cores of the Trenton–Black River*. Ontario Petrol. Inst., Core Workshop No. 1, London, Ontario, 29 pp.

PREZBINDOWSKI, D.R. & LARESE, R.E. (1987) Experimental stretching of fluid inclusions in calcite – implications for diagenetic studies. *Geology* **15**, 333–336.

PROUTY, C.E. (1988) Trenton exploration and wrenching tectonics – Michigan Basin and environs. In: *The Trenton Group (Upper Ordovician Series) of Eastern North America* (Ed. Keith, B.D.) Am. Ass. Petrol. Geol. Studies in Geology 29, 207–237.

QING, H. & MOUNTJOY, E.W. (1989) Multistage dolomitization in Rainbow buildups, Middle Devonian Keg River Formation, Alberta, Canada. *J. Sedim. Petrol.* **59**, 114–126.

RUSSELL, D.J. & TELFORD, P.G. (1983) Revisions to the stratigraphy of the Upper Ordovician Collingwood beds of Ontario – a potential oil shale. *Can J. Earth Sci.* **20**, 1780–1790.

SAKAI, H. & KROUSE, H.R. (1971) Elimination of memory effect in ^{18}O–^{16}O determinations in sulphates. *Earth Planet. Sci. Lett.* **11**, 369–373.

SANFORD, B.V. (1961) *Subsurface Stratigraphy of Ordovician Rocks in Southwestern Ontario*. Geol. Surv. Can., Paper 60-26, 54 pp.

SANFORD, B.V., THOMPSON, F.J. & MCFALL, G.H. (1985) Plate tectonics – a possible controlling mechanism in the development of hydrocarbon traps in southwestern Ontario. *Bull. Can Petrol. Geol.* **33**, 52–71.

Sears, S.O. & Lucia, F.J. (1980) Dolomitization of northern Michigan Niagara reefs by brine refluxion and freshwater/seawater mixing. In: *Concepts and Models of Dolomitization* (Ed. Zenger, D.H., Dunham, J.B. & Ethington, R.L.) Soc. Econ. Paleont. Mineral. Spec. Publ. 28, 215–235.

Shaw, B. (1975) *Geology of the Albion-Scipio Trend, Southern Michigan*. MS. Thesis, University of Michigan, Ann Arbor, Michigan, 69 pp.

Sonnenfeld, P. & Al-Aasm, I. (1991) The Salina evaporites in the Michigan Basin. In: *Early Sedimentary Evolution of the Michigan Basin* (Ed. Catacosinos, P.A. & Daniels, P.A. Jr.) Geol. Soc. Am. Special Paper 256, 139–153.

Stoakes, F. (1987) Fault controlled dolomitization of the Wabamun Group, Tangent Field, Peace River Arch, Alberta. In: *Devonian Lithofacies and Reservoir Styles in Alberta* (Ed. Krause, F.F. & Burrowes, O.G.) *13th Can. Soc. Petrol. Geol. Core Conf. & Display, & 2nd Int. Symp. on the Devonian System*, 73–85.

Taylor, T.R. & Sibley, D.F. (1986) Petrographic and geochemical characteristics of dolomite types and the origin of ferroan dolomite in the Trenton Formation, Ordovician, Michigan Basin, USA. *Sedimentology* **33**, 61–86.

Trevail, R.A. (1991) Hillman Pool. In: *Oil & Gas Field Manual of the Michigan Basin*, Vol. 2 (Ed. Wollensak, M.S.) Mich. Bas. Geol. Soc. 303–309.

Trevail, R.A., Pakvisis, P.J. & Parker, D.K. (1987) *Oil and Gas Exploration, Drilling and Gas Production Summary, 1983*. Ontario Min. Nat. Res., Oil and Gas Paper 6, 318 pp.

Vugrinovich, R. (1988) Shale compaction in the Michigan Basin: estimates of former depth of burial and implications for paleogeothermal gradients. *Bull. Can. Petrol. Geol.* **36**, 1–8.

Weast, R.C. (1986) *Handbook of Chemistry and Physics*, 66th edn. CRC Press, Boca Raton, Florida.

Wilson, J.L. & Sengupta, A. (1985) The Trenton Formation in the Michigan Basin and environs: pertinent questions about its stratigraphy and diagenesis, In: *Ordovician and Silurian Rocks of the Michigan Basin and its Margin* (Ed. Cercone, K.R. & Budai, J.M.) Mich. Bas. Geol. Soc. Special Paper 4, 1–14.

Yanagisawa, F. & Sakai, H. (1983) Thermal decomposition of barium sulfate–vanadium pentaoxide–silica glass mixtures for preparation of sulfur dioxide in sulfur isotope ratio measurement. *Anal. Chem.* **55**, 985–987.

Spec. Publs Int. Ass. Sediment. (1994) **21**, 255–279

Progressive recrystallization and stabilization of early-stage dolomite: Lower Ordovician Ellenburger Group, west Texas

J.A. KUPECZ* *and* L.S. LAND†

* *ARCO Exploration and Production Technology, 2300 W. Plano Parkway, Plano, Texas 75075, USA; and*
† *Department of Geological Sciences, University of Texas, Austin, Texas 78713-7909, USA*

ABSTRACT

The Lower Ordovician Ellenburger Group of west Texas is composed predominantly of dolomitized mud-supported (bioturbated, mm-laminated, and cryptalgal-laminated) facies. These facies constitute approximately 90% of the Ellenburger section. Dolomites from these facies are very fine to coarse-crystalline, with an increase in non-planar crystal boundaries corresponding to larger crystal sizes. Cathodoluminescence indicates several generations of dolomite. Dark-brown-luminescent dolomite rhombs (E1), are overgrown by non-luminescent dolomite (E2) and dark-brown-luminescent dolomite (E3). These early generations are cemented and replaced by orange-luminescent dolomite (L2). The abundance of E2 and E3 dolomite overgrowths, and the degree of replacement by dolomite-L2, increases with crystal size and the number of non-planar crystal boundaries.

Dolomites from the mud-supported facies are characterized by low concentrations of iron (180–7160 ppm; generally <2000 ppm), strontium (20–210 ppm) and manganese (20–420 ppm), by a wide range in $\delta^{18}O$ values (−2.4 to −8.8‰ PDB), variable stoichiometry (49.53–56.00 mole % $CaCO_3$), and variable $^{87}Sr/^{86}Sr$ ratios (0.70812–0.71238). Although most geochemical parameters are variable, there is a covariance between increasing textural modification, increasing stoichiometry, decreasing $\delta^{18}O$, increasing $^{87}Sr/^{86}Sr$ and decreasing trace element concentrations.

Dolomites of the mud-supported facies probably originated as sabkha-like sequences. However, the covariance between textural changes and geochemistry suggests significant postdepositional modification. Based on petrographic and geochemical data, we conclude that two major diagenetic events have been superimposed following initial dolomite formation: meteoric modification, synchronous with early Middle Ordovician regional exposure and karstification; and late modification via basinal-derived Pennsylvanian pore fluids expelled upon thrusting during the Ouachita Orogeny. These major diagenetic events have resulted in the progressive recrystallization and stabilization of early Ellenburger dolomites. Evidence of extensive modification suggests that it is critical that the textural and geochemical evolution be considered when interpreting the origin of the Ellenburger and other dolomites on the basis of their present geochemical signatures.

INTRODUCTION

Determining whether present-day geochemistry reflects original geochemistry or diagenetic resetting is critical for interpreting the origin of ancient dolomites. To interpret the origin of ancient dolomite, one must ‘see through’ any subsequent diagenesis which has altered the original dolomite (Land, 1985; Hardie, 1987). Some studies of dolomite have suggested that present-day geochemical signatures are the product of recrystallization, not original precipitation (Land *et al.*, 1975; Zenger, 1981, 1983; Gregg & Sibley, 1984; Land, 1985; Banner *et al.*, 1988; Cander *et al.*, 1988; Dorobek & Filby, 1988; Holail *et al.*, 1988; Moore *et al.*, 1988; Zenger & Dunham, 1988; Gregg & Shelton, 1990; Gao & Land, 1991; Kupecz & Land, 1991; Kupecz *et al.*, 1992; Montañez & Read, 1992a).

Early dolomites of the mud-supported (bioturbated, cryptalgal and mm-laminated) facies of

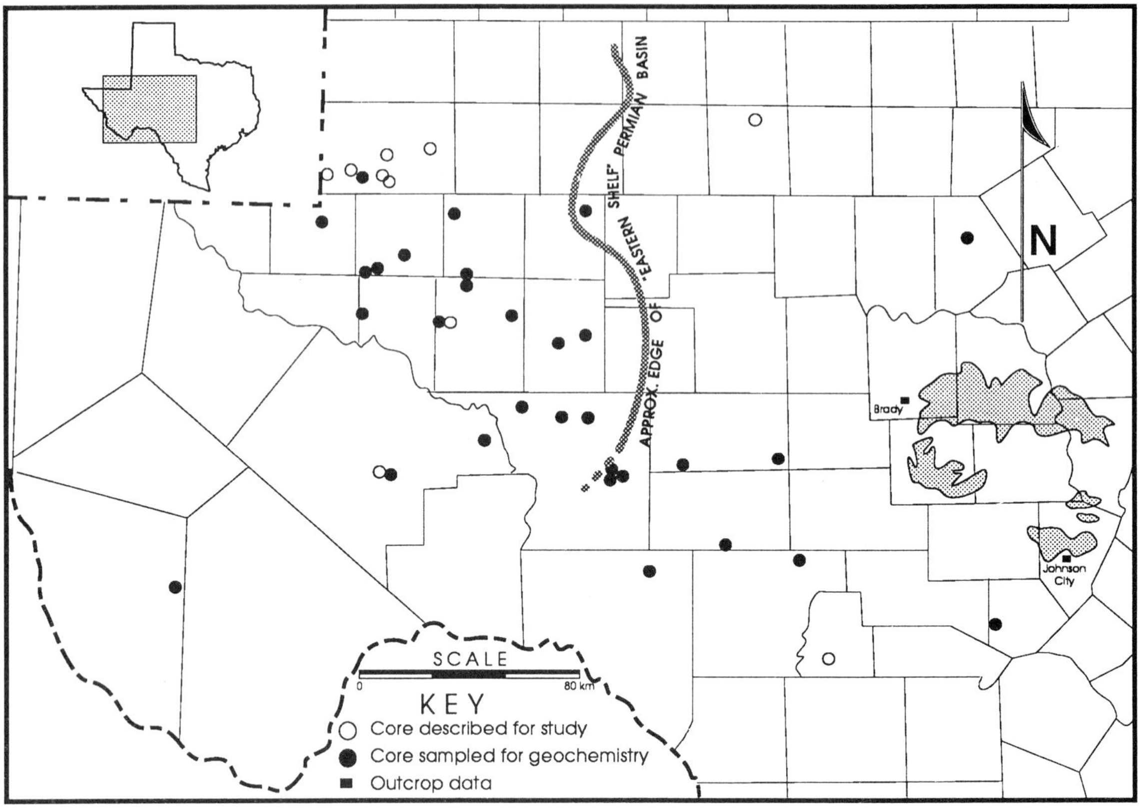

Fig. 1. Location of study area. Circles represent cores described for this study. ● represents cores sampled for geochemical analyses. Stippled area represents the Precambrian outcrop of the Llano Uplift.

the Ellenburger Group (Lower Ordovician, west Texas) provide an excellent setting to examine the recrystallization of dolomite. The thick, regionally extensive Ellenburger Group contains several types of dolomite. The database is regional, extending across central and west Texas, with both outcrop and subsurface cores available for study (Fig. 1). The Ellenburger is presently at depths ranging from outcrop to 4800 m (Kupecz, 1989), providing insights into conditions under which dolomites can be modified.

This study will document the petrography and geochemistry of dolomites from the mud-supported facies of the Ellenburger. These facies comprise the majority (90%) of the stratigraphic section and, consequently, the majority of the dolomitized section. We have integrated standard and cathodoluminescence petrography with major element, trace element and isotopic data $\delta^{13}C$, $\delta^{18}O$, $^{87}Sr/^{86}Sr$). Mean crystal size is compared with the geochemistry. Samples were taken from both karstified and non-karstified intervals to determine whether karst breccias acted as fluid-flow conduits (*cf.* Buchbinder *et al.*, 1984; Stepanek, 1987; Kupecz & Land, 1991).

ANALYTICAL METHODS AND SAMPLE BASE

Subsurface cores extending over an area of west Texas of approximately 500 km (E–W) by 250 km (N–S) were sampled for this study. Of 39 cores examined, 30 were sampled for geochemical analyses (Fig. 1). The total length of examined cores comprises approximately 3600 m of section, with individual cores ranging from 15 to 400 m (Kupecz, 1989).

Dolomite samples (69) from the study area were

analysed for $\delta^{18}O$ and $\delta^{13}C$ by reaction with anhydrous H_3PO_4 and analysis of evolved CO_2. Data are presented relative to the PDB standard. Samples were reacted at 50°C and corrected to fractionation at 25°C by adding +1.22‰ to the machine-corrected $\delta^{18}O$ values (Anderson, 1985). X-ray diffraction (XRD) and inductively coupled argon plasma spectrometry (ICAP) ensured that samples were pure dolomite prior to isotopic analysis.

For Sr isotopic analyses (67), powdered samples were washed in double-distilled water, treated three times with 0.2 N ammonium acetate to remove exchangeable Sr from possible clay contamination, washed three times with ultrapure distilled water, and reacted with 8% distilled acetic acid for 5 min to preferentially dissolve dolomite without incorporating non-carbonate strontium (G.K. Hoops, 1988, personal communication). After separation by standard cation exchange procedures, approximately 200 ng of Sr were analysed for $^{87}Sr/^{86}Sr$ by simultaneous multicollection on a thermal ionization mass spectrometer, which yielded 0.71024 ± 0.00004 (1 σ about the mean) for the NBS-987 standard during the period of analysis. To determine the effects of incorporated Rb on $^{87}Sr/^{86}Sr$, a 'worst-case' analysis was performed on a radiogenic sample (W31A; see Kupecz, 1989), which was dissolved completely in 2 N HCl with no ammonium acetate pretreatment. $^{87}Sr/^{86}Sr_{today}$ is more radiogenic than $^{87}Sr/^{86}Sr_{500ma}$ by 0.0004. As ammonium acetate pretreatment was performed on all samples, original Rb is considered not to have affected the data.

Dolomites (71) were analysed for Ca, Mg, Fe, Mn, Sr, Na, K, Al and Ba concentrations by ICAP spectrometry. Na, K and Al were used as an additional check to indicate the presence of contaminating clays. Samples were prepared by dissolving 10 mg of powdered sample in 10 ml of 5% HCl, followed by filtering through 0.4 μm nuclepore filters. Values are reported using NBS-88 dolomite as a standard.

Approximately 450 thin sections were examined (including 200 from the Bureau of Economic Geology). Of these, 250 polished thin sections were studied for cathodoluminescence on a Technosyn MK-II luminoscope; operating conditions were 17–20 kV beam energy and 400–600 mA beam current. Determination of mean, minimum and maximum crystal size of samples was performed on 26 thin sections, with each sample having 300 measurements. Measurements were based on maximum crystal dimension along a traverse.

STRATIGRAPHY, DIAGENESIS AND REGIONAL TECTONICS OF THE ELLENBURGER

The Lower Ordovician Ellenburger Formation was named by Paige (1911, p. 24; 1912, p. 7). The Ellenburger Formation was elevated to Group status and subdivided by Cloud *et al.* (1943) into the Tanyard (Threadgill and Staendebach Members), Gorman and Honeycut Formations, in ascending order. Detailed studies of the Ellenburger were undertaken by Cloud *et al.* (1943), Cloud & Barnes (1948, 1957) and Barnes *et al.* (1959).

Depositional environments and facies

The Ellenburger was deposited on a broad shallow platform in and/or proximal to an extensive epicontinental sea (Cloud & Barnes, 1957). Approximately 500 m of carbonates (present-day postcompaction measurements) accumulated, primarily as mud-rich deposits. Environments of deposition. interpreted by Loucks and Anderson (1980, 1985), include a complex of alluvial fan, fan delta, supratidal, intertidal and shallow subtidal environments established after transgression on to eroded Precambrian highlands. Rare silicified nodular evaporites in the supratidal facies suggest a sabkha-like setting (Loucks & Anderson, 1980).

The depositional facies of the Ellenburger have been presented by Loucks and Anderson (1980, 1985), Kerans (1988) and Kupecz (1989). Seven facies are recognized in this study: subarkose, mixed siliciclastic–carbonate packstone–grainstone, ooid–peloid grainstone, bioturbated mudstone, mm-laminated mudstone, cryptalgal-laminated mudstone, and gastropod–intraclast packstone–grainstone (Fig. 2). The subarkose and mixed carbonate–siliciclastic facies (ranging in thickness from 12 to 30 m, and 12 to 90 m, respectively) are interpreted to represent deposition in alluvial fan and fan delta complexes developed on, and grading upward from, the Precambrian granitic basement. Bioturbated mudstones (up to 270 m thick), mm-laminated mudstones (up to 90 m thick) and cryptalgal-laminated mudstones (up to 90 m thick) are interpreted to have been deposited in shallow subtidal, subtidal/intertidal and intertidal/supratidal environments, respectively. The bioturbated mudstone, mm-laminated mudstone and cryptalgal-laminated mudstone facies are the focus of this study, and will be referred to as the 'mud-supported

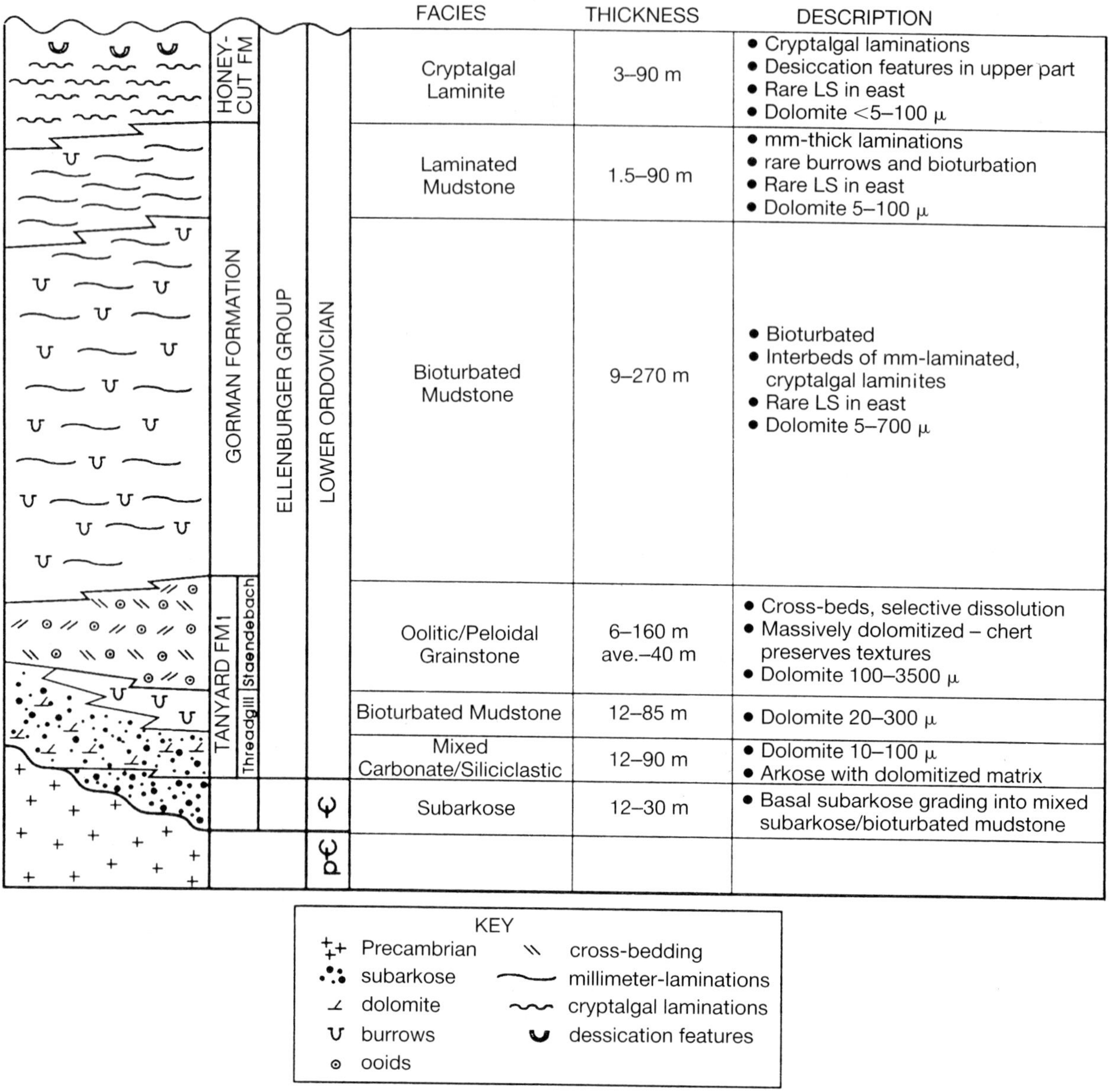

Fig. 2. Generalized stratigraphic column illustrating the relationship between formations of the Ellenburger Group and depositional facies. Thicknesses and summary descriptions of facies are included.

facies'. The ooid–peloid grainstone facies is a thick (up to 160 m) widespread unit deposited in high-energy subtidal shoals and bars. The gastropod–intraclast packstone facies is rare, and is observed only in the eastern part of the study area. It represents more open-marine conditions, nearer the platform edge.

Cyclicity, karstification and associated diagenesis

The majority of subsurface cores have been subjected to such strong diagenetic overprints that it is impossible to recognize depositional fabrics and cyclicity. However, multiple subaerial exposure horizons associated with cyclicity (interpreted here to be third-order) have been recognized locally within the Ellenburger by Loucks and Anderson (1980, 1985). Age equivalents such as the El Paso, Arbuckle and Knox Groups have been studied in detail for their cyclicity (Montañez, 1989; Kerans Lucia, 1989; Wilson *et al.*, 1991; Raymond & Osborne, 1991; Montañez, 1992; Montañez & Read, 1992b; Goldhammer *et al.*, in press). On the basis of relative amounts of Knox Group dolomite, and the relative stratigraphic position of dolomite within individual cycles, Montañez and Read (1992b) suggest that the majority of Knox dolomite is synsedimentary in origin. Because the age-equivalent Ellenburger is not conducive to detailed cycle analysis, it is not possible to determine whether the earliest Ellenburger dolomites were also synsedimentary; however, similarities between the Ellenburger and the Knox Groups in observed depositional fabrics and in paragenesis suggest that this is a reasonable conclusion.

Significant regional karstification within the Lower Ordovician carbonates of west Texas was first documented in the El Paso Group of the Franklin Mountains (Lucia, 1968, 1969, 1987). The interpretation was extended to include the Ellenburger of central and west Texas (Kerans, 1988; Kupecz, 1992), where karstification was interpreted to be related to exposure at a second-order sequence boundary. Cave systems and collapse features infilled and overlain by Middle Ordovician Simpson Group sediments suggest that widespread karst features existed prior to the Middle Ordovician. This karstification event will be referred to as 'pre-Middle Ordovician' karstification.

Correlation of karst breccias across the study area is not possible due to the distance between cores (Fig. 1), and the variable vertical core coverage (Kupecz & Land, 1991; their Fig. 2). However, karstification is widespread. Of 3600 m of core described for this study, approximately 40% are karst breccias, testifying to the importance of karst breccias as exploration targets.

Regional tectonics

The Pennsylvanian to earliest Permian Ouachita Orogeny involved northward thrusting and regional metamorphism of thick basinal sediments that had accumulated near the continental margin (Flawn *et al.*, 1961). Subsequent sediment loading resulted in downwarping and the delineation of central and west Texas into its present-day configuration of basins and platforms (Flawn *et al.*, 1961). Kupecz and Land (1991) concluded that reactive basinal diagenetic fluids were introduced into the Ellenburger system through regional thrusting during the Ouachita Orogeny. These fluids probably caused the dissolution of early-stage dolomites and precipitation of late-stage dolomites in the study area (Kupecz & Land, 1991).

DESCRIPTION AND DISTRIBUTION OF DOLOMITES FROM MUD-SUPPORTED FACIES

Mud-supported facies (Fig. 2) comprise approximately 90% of the Ellenburger section, based on determination of maximum facies thicknesses in cores (Kupecz, 1989). Maximum thicknesses for bioturbated, mm-laminated and cryptalgal-laminated facies are 270 m, 90 m and 90 m, respectively (Fig. 2). These facies are almost exclusively dolomite in the subsurface, except where local dedolomitization and silicification have occurred. Complex limestone–dolomite transitions exist in outcrop surrounding the Llano Uplift.

Multiple generations of dolomite are present within the mud-supported facies. Using the paragenetic sequence nomenclature of Kupecz and Land (1991), these dolomites include early-stage dolomite E1 and late-stage dolomite L2 (late-stage dolomite L1 has not been recognized in samples included in this study). Also observed in samples for this study, but not previously documented, are early-stage dolomites E2 and E3.

Dolomite E1 predates pre-Middle Ordovician karstification, based on the observation that this dolomite can be found in both non-karstified areas and in angular karst breccia clasts (Kupecz & Land, 1991, their Fig. 10). Dolomite E1 (composed of dark-brown-luminescent rhombs) does not extend beyond angular breccia clast boundaries (Fig. 3), suggesting that the E1 dolomite was present before brecciation. Dolomite E1, in turn, is modified by two generations of early-stage dolomite (E2 and E3) that predate dolomite L2 (Figs 4 and 5). Because the E1 dolomite predates karstification, regional distribution of E1 is not controlled by the geometry of the overlying pre-Middle Ordovician unconformity.

Non-luminescent dolomite E2 corrodes and overgrows host E1, and is followed by dark-brown-luminescent overgrowths of dolomite E3 (Fig. 5). Dolomite E2 can be shown to predate dolomite L2,

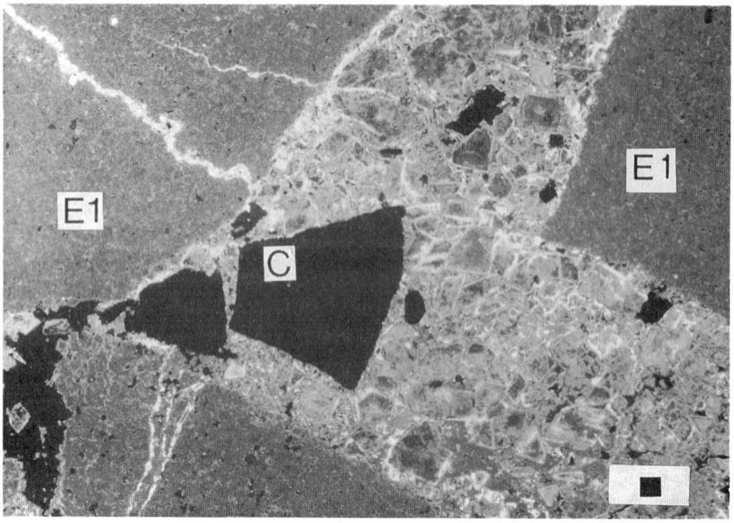

Fig. 3. Cathodoluminescence photomicrograph of karst breccia composed of clasts of chert (C) and dolomite E1 (E1) in dolomitized mud matrix. Phillips Petroleum #1 Wilson, Val Verde Co., Texas, 4837.8 m (15 872 ft). Scale bar = 100 μm.

as E2 lines fractures that are later filled with dolomite L2 cement. However, it is not clear whether dolomite E2 pre- or postdates karstification. Although dolomite E2 is not exclusive to karst breccias, and does not line breccia clasts as does dolomite L2, the thickness of overgrowths (and resulting crystal size) appears to correlate with relative amounts of karst brecciation (Figs 4 and 5).

Dolomite L2 includes orange-luminescent intercrystalline cement, orange-luminescent overgrowths and orange-luminescent replacement dolomite. Replacement dolomite is gradational, from selective irregular replacement of dark-brown-luminescent dolomite E1 rhombs and E2 and E3 overgrowths (Fig. 5), to complete replacement (Fig. 6). This results in a mottled appearance of the host dolomite.

Dolomite L2 postdates pre-Middle Ordovician karstification, as evidenced by L2 intercrystalline dolomite cement grading into L2 clast-lining cement (Kupecz & Land, 1991, their Fig. 8). This suggests that the L2 dolomite precipitated after brecciation and formation of clasts. Replacement of E1, E2 and E3 by the L2 dolomite postdates the L2 intercrystalline cement, and is therefore interpreted to have postdated pre-Middle Ordovician karstification.

Early dolomites E1, E2 and E3 are exclusive to the mud-supported facies. In contrast, dolomite L2 is not exclusive to the mud-supported facies, but also comprises dolomites within the ooid–peloid grainstone, subarkose and mixed carbonate–siliciclastic facies (*cf.* Kupecz & Land, 1991). Within mud-supported facies, orange-luminescent medium- to coarse-crystalline dolomite L2 is observed in core and outcrop in the following manner: in cryptalgal laminae where most laminae are composed of non-luminescent very fine- to fine-crystalline dolomite-E1, but where specific (more permeable?) laminae have been replaced by orange-luminescent medium- to coarse-crystalline L2; where mud-supported facies have been brecciated and/or fractured; and where the mud-supported facies are in contact with porous and permeable facies, such as the ooid–peloid grainstone, subarkose and mixed carbonate–siliciclastic facies. Dolomites in these adjacent permeable facies are also composed of medium- to coarse-crystalline L2 dolomite, and display orange-luminescent replacement fabrics.

Dolomites within the mud-supported facies of the Ellenburger range in crystal size from 5 to 700 μm. Mean crystal sizes range from 13 to 167 μm. On the basis of petrographic characteristics (plane-light and cathodoluminescent), they can be subdivided into two groups: those with mean crystal size from 13 to 59 μm (very fine- to fine-crystalline) and those with mean crystal size from 95 to 167 μm (medium-crystalline). Mean crystal size values of 60–95 μm were not observed.

Very fine- to fine-crystalline dolomite

Very fine- to fine-crystalline dolomites make up the bulk of the Ellenburger mud-supported facies. They are composed predominantly of dolomite E1, having crystals with planar boundaries which exhibit straight extinction. Dolomite E1 is modified by dolo-

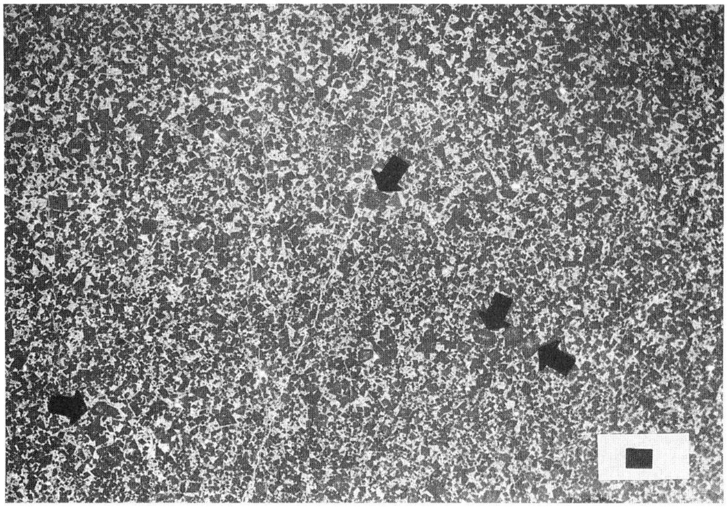

Fig. 4. Cathodoluminescence photomicrograph of non-karstified mm-laminated mudstone facies. Sample contains dark-brown-luminescent rhombs (dolomite E1), with the larger crystals having non-luminescent overgrowths (dolomite E2). Note that larger crystal cores also have a mottled appearance due to partial replacement by orange-luminescent dolomite L2 (arrows). Samples contain orange-luminescent dolomite L2 intercrystalline and fracture-filling cement. Phillips Petroleum #1 Wilson, Val Verde Co., Texas, 4774.7 m (15 665 ft). Scale bar = 100 μm.

mites E2 and L2, resulting in a patchy distribution of larger crystals composed of host E1 with variable amounts of E2 overgrowths, and L2 replacement (Fig. 4). An increase in the abundance of dolomite L2 also corresponds to an increase in abundance of non-planar crystal boundaries.

Crystals of dolomite E1 and L2 contain abundant inclusions, whereas dolomite E2 is inclusion-free. Under cathodoluminescence, host dolomite E1 has mottled dark-brown luminescence. Dolomite E2 overgrowths and cement are non-luminescent; corrosion and infill of dolomite E1 by dolomite E2 results in the mottled texture observed in all samples. Intercrystalline and fracture-filling cement and mottled replacement fabrics of dolomite L2 are orange-luminescent.

Medium- to coarse-crystalline dolomite

Medium- to coarse-crystalline dolomites are composed of interlocking crystals with non-planar crystal faces. Carbonate inclusions are abundant and, under plane light, crystals are generally unzoned; however, some crystals have inclusion-free overgrowths. Dolomites within this subset comprise E1, E2, E3 and L2. Under cathodoluminescence, sample end-members range from dolomite E1 rhombs with E2 and E3 overgrowths (Fig. 5) to orange-luminescent dolomite L2 (Fig. 6). Intermediate to these end-members are samples in which dolomites E1, E2 and E3 are partially replaced by irregular patches of dolomite L2 (Fig. 7). These orange-luminescent patches appear to be more abundant

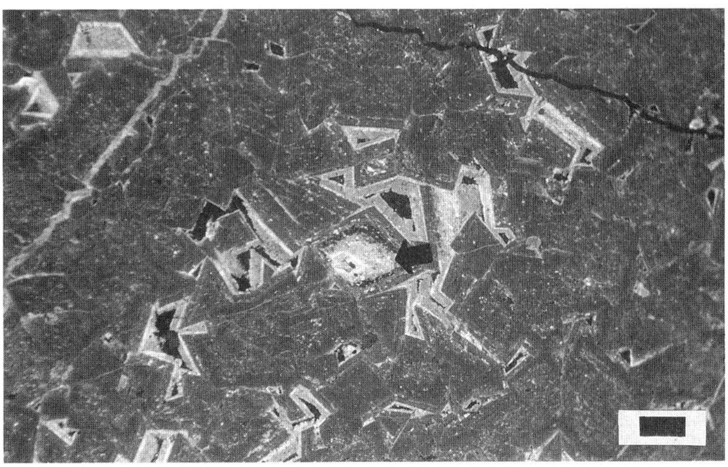

Fig. 5. Cathodoluminescence photomicrograph of a breccia clast of the bioturbated facies, containing dark-brown-luminescent dolomite E1 rhombs with non-luminescent dolomite E2 overgrowths, followed by dark-brown-luminescent dolomite E3 overgrowths. The sample contains varying amounts of intercrystalline dolomite L2 cement. Isolated crystals have cores preferentially replaced by orange-luminescent dolomite L2 (arrow). Phillips Petroleum #1 Wilson, Val Verde Co., Texas, 4720.7 m (15 488 ft). Scale bar = 100 μm.

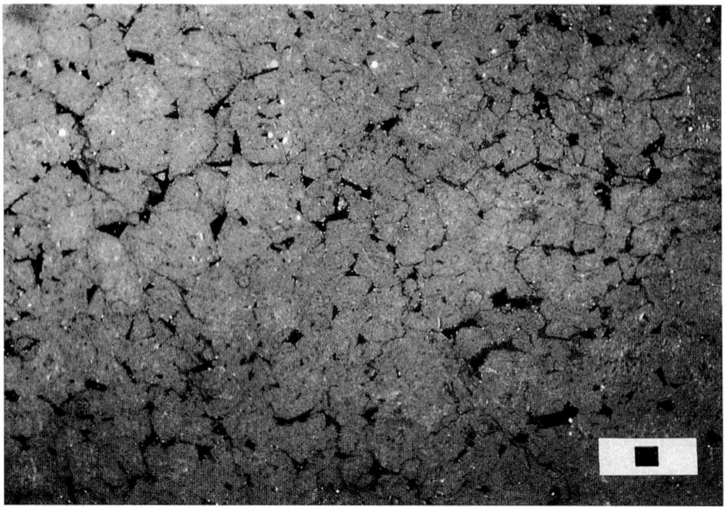

Fig. 6. Cathodoluminescence photomicrograph of a breccia clast from the bioturbated facies. Orange-luminescent dolomite L2 rhombs have completely replaced earlier generation(s) of dolomite. Intercrystalline porosity is infilled with pyrite. Phillips Petroleum #1 Wilson, Val Verde Co., Texas, 4832 m (15 853 ft). Scale bar = 100 μm.

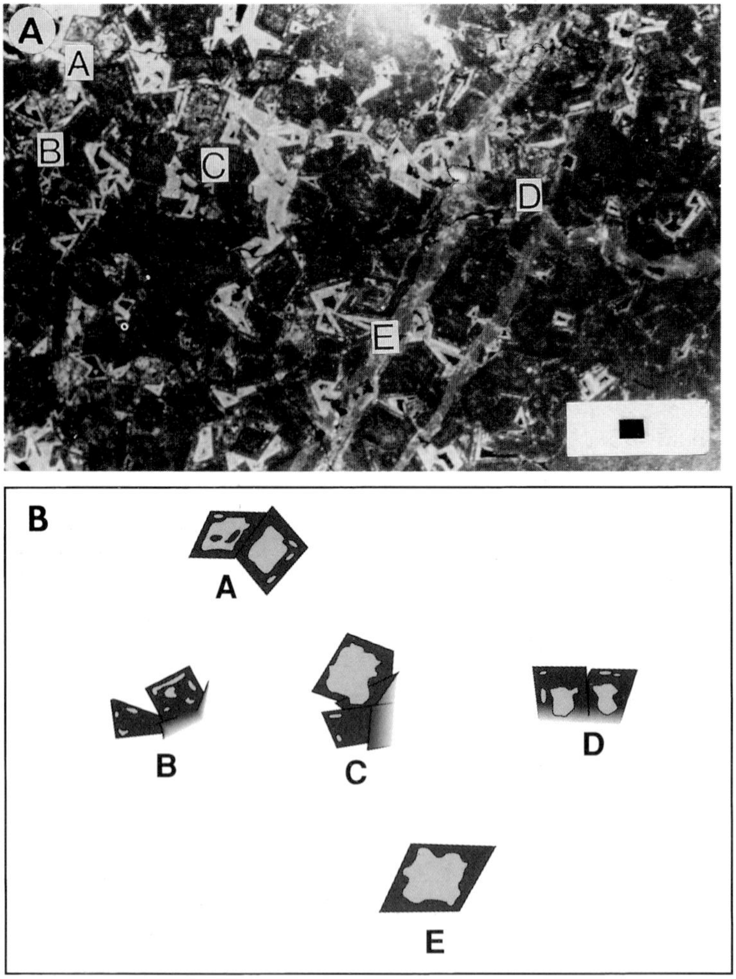

Fig. 7. Cathodoluminescence photomicrograph with corresponding cartoon diagram of selected crystals. (A) Cathodoluminescence photomicrograph of a breccia clast from the bioturbated facies. Rhombs of dolomites E1, E2 and E3 have varying amounts of dolomite L2 intercrystalline and fracture-filling cement. Note that cores of larger rhombs are being preferentially replaced by orange-luminescent dolomite L2 (examples are lettered A–E). Phillips Petroleum #1 Wilson, Val Verde Co., Texas, 4966.4 m (16 294 ft). Scale bar = 100 μm. (B) Cartoon of dolomite crystals (A–E), corresponding to photomicrograph of (A). Note the irregular cross-cutting nature of the contacts.

within crystal cores, although replaced patches within overgrowths also exist (Fig. 7). The abundance of dolomite L2 correlates with crystal size, the coarsest crystals containing the most orange-luminescent replacement L2 (Fig. 6).

GEOCHEMISTRY

Geochemical data for dolomites of the Ellenburger mud-supported facies are given in Table 1. $\delta^{18}O$ (25°C) values range from −2.4 to −8.8‰ (PDB), whereas $\delta^{13}C$ ranges from −0.6 to −3.6‰. Mole % $CaCO_3$ ranges from 49.53 to 56.00. Fe ranges from 180 to 760 ppm; Sr from 20 to 210 ppm and Mn from 20 to 420 ppm. $^{87}Sr/^{86}Sr$ values range from 0.70812 to 0.71238.

Stable isotopes

Dolomites from mud-suported facies display a covariant trend of decreasing $\delta^{18}O$ with decreasing $\delta^{13}C$ (Fig. 8A, B). These data have been subdivided into samples from karst breccias and non-karstified samples; cross-plots illustrate that samples from karst breccias have a much greater range in $\delta^{18}O$ values and average a lighter value of $\delta^{18}O$.

The very fine- to fine-crystalline group has $\delta^{18}O$ values from −2.4 to −8.6‰ (Fig. 8C), and $\delta^{13}C$ values from −0.8 to −2.8‰. Of the medium-crystalline group, $\delta^{18}O$ values range from −2.4 to −8.8‰ (Fig. 8D), and $\delta^{13}C$ values range from −0.6 to −3.1‰. Both crystal size groups illustrate covariance between increasing crystal size and decreasing $\delta^{18}O$. Subdivision of the crystal size data into karst vs. non-karst illustrates that all medium-crystalline samples have been karstified (Fig. 8D). Of the very fine- to fine-crystalline group (Fig. 8C), the non-karstified samples have a tighter range in $\delta^{18}O$ (averaging heavier values) and smaller mean crystal sizes. Karstified samples display a greater range in $\delta^{18}O$ and have larger mean crystal sizes, with the largest crystal fraction having the lightest values. The same covariant trend holds for both karstified and non-karstified samples (Fig. 8C,D).

Dolomite textures also display covariance with $\delta^{18}O$. With depletion in ^{18}O and a corresponding increase in crystal size, both crystal size groups display an increase in non-planar crystal boundaries. Although the very fine- to fine-crystalline samples show corrosion and overgrowths by dolomite E2, they are superimposed by a progressive increase in abundance of orange-luminescent mottled textures, due to replacement by dolomite L2, which corresponds to increasing crystal size and decreasing $\delta^{18}O$.

Major and trace elements

Mole % $CaCO_3$

Mole % $CaCO_3$ vs. $\delta^{18}O$ illustrates weak covariance between increasing stoichiometry and decreasing $\delta^{18}O$ in karstified samples (Fig. 9A). Non-karstified samples have more variability in stoichiometry than do karstified samples, and weaker to no covariance (Fig. 9B). Similar trends hold true for the crystal size subdivisions. Very fine- to fine-crystalline dolomites display high variability and no covariance (Fig. 9C), whereas medium-crystalline (karstified) samples show covariance between increasing stoichiometry and decreasing $\delta^{18}O$ (Fig. 9D).

Cross-plots of mole % $CaCO_3$ vs. crystal size illustrate no covariance. Both size groups approach stoichiometry, but the very fine- to fine-crystalline group has more variability than the medium-crystalline group (Figs 9E,F). Within the very fine- to fine-crystalline group, the least stoichiometric samples (56.0 mole % $CaCO_3$) have the smallest mean crystal size, whereas the most stoichiometric samples (50.17 mole % $CaCO_3$) have the largest crystals (Fig. 9E). Increasing crystal size (and decreasing variability in stoichiometry) also corresponds with increasing numbers of non-planar crystal boundaries and increasing overgrowth and mottled replacement by dolomites E2 and L2.

In contrast to the very fine- to fine-crystalline group, all samples of the medium-crystalline group (Fig. 9F) approach stoichiometry (50.12–52.00 mole % $CaCO_3$), and show no trend with increasing crystal size. Though stoichiometry is consistent, increasing crystal size corresponds to an increase in non-planar crystal boundaries and replacement by dolomite L2.

Iron

The very fine- to fine-crystalline group displays weak covariance between decreasing ppm Fe and increasing crystal size (Fig. 10A), whereas the medium-crystalline group shows no covariance and consistently low values (Fig. 10B). Concentrations from the medium-crystalline group (180–1020 ppm Fe) are lower than those from the very fine- to fine-crystalline group (380–7160 ppm Fe). Within the

Table 1. Data from mud-supported (bioturbated, mm-laminated and cryptalgal-laminated) facies.

Sample	Depth (ft)	Facies	Mean crystal size	σ	Karst?	^{18}O (50°C)	^{18}O (25°C)	^{13}C	^{87}Sr/^{86}Sr	ppm Sr	ppm Mn	ppm Fe	mole % $CaCO_3$
1-27 Univ.	7335	Biot.			Clast	−5.2	−4.0	−2.3	0.708957	90	180	1600	51.62
So. Roy. 16-2	12 146	Biot.			Clast	−6.3	−5.1	−2.4	0.709637				51.09
Mitchell Bro.	15 502	Biot.	131	65	Clast	−3.6	−2.4	−0.6	0.708115	60	30	1020	52.00
Mitchell Bro.	15 673	Biot.			Clast	−6.7	−5.5	−2.2	0.708355	30	50	870	50.37
McElroy	12 505	Biot.			Clast	−5.7	−4.5	−2.4	0.708759	50	80	420	
McElroy	12 247	Biot.	162	82	Clast	−4.4	−3.2	−1.9		40	40	180	51.01
McElroy	12 070	Biot.			Clast	−5.7	−4.5	−1.9	0.708674	50	80	420	
Stevens	2937	Biot.			Clast	−6.5	−5.3	−2.7	0.709097	40	110	1590	52.28
TXL-SS-3E	13 364	Biot.			Clast	−5.3	−4.1	−1.1	0.708582	50	60	780	
TXL-SS-3E	13 365	Biot.	95	36	Clast	−4.9	−3.7	−0.9	0.708777	40	70	520	51.12
TXL-SS-3E	13 403	Biot.	30	11	Clast	−5.2	−4.0	−0.8	0.708914	40	130	1070	50.52
Univ. 1M	10 904	Biot.			Clast	−6.8	−5.6	−1.7	0.709055	30	150	1940	52.20
Wilson	16 294	Biot.	118	62	Clast	−7.3	−6.1	−1.3	0.709275	60	70	500	50.12
Wilson	16 279	Biot.			Clast	−9.9	−8.7	−2.6	0.709025	40	90	570	49.67
Wilson	16 092	Biot.	106	34	Clast	−6.0	−4.8	−2.4	0.709054	120	70	440	50.86
Wilson	15 915	Biot.			Clast	−7.0	−5.8	−3.6	0.709877	120	90	1330	54.30
Wilson	15 853	Biot.	167	66	Clast	−10.0	−8.8	−3.1	0.709663	80	80	640	50.61
Wilson	15 819	Biot.			Clast	−6.7	−5.5	−2.8	0.709306	100	70	630	51.05
Wilson	15 802	Biot.	57	27	Clast	−9.8	−8.6	−2.6	0.709067	50	150	910	50.42
Wilson	15 708	Biot.	59	25	Clast	−7.9	−6.7	−2.5	0.708953	100	150	880	50.39
Wilson	15 610	Biot.	40	14	Clast	−9.4	−8.2	−2.8	0.708948	160	90	670	50.28
Wilson	15 488	Biot.			Clast	−6.4	−5.2	−2.1	0.709348	180	60	620	50.55
Wilson	15 221	Biot.	29	10	Clast	−7.9	−6.7	−2.8	0.712382	160	110	1280	50.17
Wilson	14 983	Biot.	42	17	Clast	−5.8	−4.6	−2.0	0.709082	210	60	560	49.79
1-27 Univ.	7052	Biot.			No	−5.1	−3.9	−2.5	0.708940	120	60	380	50.18
1-27 Univ.	7077	Biot.			No	−4.9	−3.7	−2.5	0.709016	110	70	560	50.75
Below	4188	Biot.			No	−4.8	−3.6	−1.6	0.708969	150	20	660	57.00
Connell	9017	Biot.			No	−4.1	−2.9	−1.8	0.711549	50	80	1160	51.14
COX-1E	8616	Biot.	15	7	No	−3.6	−2.4	−2.0	0.708879	110	80	940	51.79
COX-1E	8628	Biot.	13	5	No	−4.5	−3.3	−2.5	0.708938	200	80	760	56.00
COX-1E	8664	Biot.			No	−4.3	−3.1	−2.2	0.709024	120	30	1380	51.62
Estes	8069	Biot.			No	−5.1	−3.9	−2.4	0.709281	60	80	1010	51.95
Keystone	9000	Biot.			No	−4.6	−3.4	−2.0	0.710073	50	100	1710	50.53
Knippa	7366	Biot.			No	−3.9	−2.7	−0.8	0.709038	80	30	830	51.77
Lockhart	8677	Biot.			No	−4.4	−3.2	−1.2	0.708438	60	120	360	51.17
McElroy	11 768	Biot.	36	16	No	−5.5	−4.3	−1.7	0.709256	40	80	470	50.60

Noble	13 439	Biot.			No	−6.3	−5.1	−1.4	0.708903	20	170	740	50.57
Oswalt	12 047	Biot.			No	−6.0	−4.8	−2.0	0.708768	100	70	600	51.55
Owens	9 655	Biot.	35	16	No	−4.2	−3.0	−1.4	0.709126	90	50	540	51.55
Sawyer	10 372	Biot.			No	−4.9	−3.7	−1.5	0.708384	80	50	510	51.44
Shannon Hsp	7 276	Biot.			No	−4.2	−3.0	−0.8	0.708789	80	50	880	52.13
Univ. 1M	10 930	Biot.	24	10	No	−6.5	−5.3	−1.6	0.709521	40	120	1570	52.19
Wilson	16 266	Biot.	42	25	No	−7.4	−6.2	−2.3		80	90	2210	51.04
Wilson	15 948	Biot.			No	−5.3	−4.1	−3.5	0.709169	200	60	1160	56.00
Wilson	15 559	Biot.			No	−5.8	−4.6	−2.3	0.710237	120	70	1550	51.33
Windham A2	12 742	Biot.	33	15	No	−3.8	−2.6	−1.2	0.708963	80	70	380	50.97
Windham A2	13 033	Biot.	44	17	No	−4.5	−3.3	−1.4	0.711495	80	80	460	51.40
Wilson	15 336	Crypt.	140	56	Clast	−9.2	−8.0	−2.6	0.709292	50	140	990	50.41
Yarb & Allen	10 625	Crypt.			Clast	−4.4	−3.2	−2.0	0.709798	70	50	550	51.28
COX-1C	8 292	Crypt.	41	17	No	−5.2	−4.0	−2.4	0.708942	70	80	1060	52.10
COX-1C	8 637	Crypt.			No	−5.0	−3.8	−1.8	0.709093				
COX-1C	8 637	Crypt.			No	−4.5	−3.3	−1.8	0.708980	100	100	790	53.00
Foster	13 009	Crypt.			No	−4.0	−2.8	−1.6	0.709226	80	60	520	51.52
Mitchell Bro.	15 547	Crypt.			No	−4.5	−3.3	−1.5	0.708954	110	30	1580	53.00
Mitchell Bro.	15 845	Crypt.			No	−4.8	−3.6	−1.6	0.709021	30	50	1740	50.50
Owens	9 502	Crypt.			No	−4.1	−2.9	−1.1	0.709155	50	80	1470	50.57
Owens	9 575	Crypt.	14	6	No	−3.7	−2.5	−1.8	0.708651	70	100	720	
Harris	10 176	mm-lam.			Clast	−4.8	−3.6	−1.9	0.708846	80	220	3440	51.26
Wilson	15 362	mm-lam.			No	−6.2	−5.0	−2.3	0.709300	80	70	540	51.58
1-27 Univ.	7 334	mm-lam.	39	20	Clast	−4.6	−3.4	−2.3	0.708911	70	420	3500	51.96
Wilson	16 388	mm-lam.			Clast	−5.3	−4.1	−1.0	0.709604	80	90	690	49.53
COX-1E	8 630	mm-lam.			No	−5.2	−4.0	−2.8	0.708794	130	220	2060	53.81
COX-1E	8 642	mm-lam.			No	−4.3	−3.1	−1.3	0.709053	170	70	960	51.92
Puckett	13 235	mm-lam.	15	10	No	−4.7	−3.5	−3.1	0.710117	150	120	7160	55.00
Puckett	13 235	mm-lam.			No	−5.0	−3.8	−3.3	0.709427	140	70	3280	
Puckett	13 349	mm-lam.	14	8	No	−4.8	−3.6	−2.5	0.710082	70	20	1020	50.81
Puckett	13 469	mm-lam.			No	−4.8	−3.6	−2.3	0.709003	50	40	550	51.40
Puckett	13 469	mm-lam.			No	−4.9	−3.7	−2.2	0.709105	60	60	860	50.52
Wilson	15 665	mm-lam.			No	−5.9	−4.7	−2.7	0.709024	150	60	500	51.33

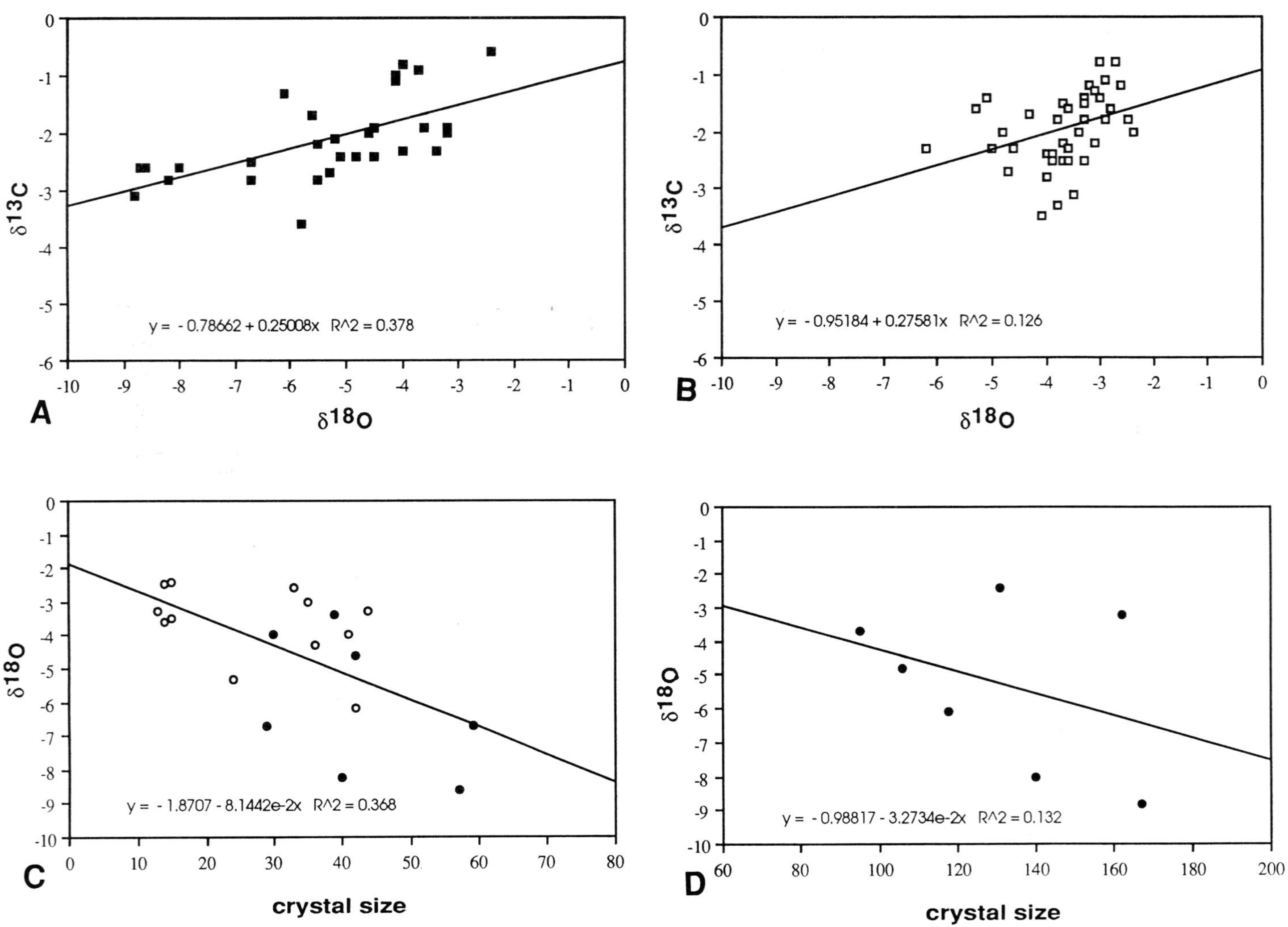

Fig. 8. Cross-plots of $\delta^{18}O$ versus $\delta^{13}C$, crystal size and stoichiometry. In curve fitting with the program Cricket Graph, R^2 represents R^2, and the closer this number is to 1, the more reliable is the curve fit. (A) $\delta^{18}O$ vs. $\delta^{13}C$ from karstified samples. (B) $\delta^{18}O$ vs. δ^{18} C from non-karstified samples. (C) $\delta^{18}O$ vs. crystal size for very fine- to fine-crystalline samples. ○, samples from non-karstified intervals; ●, samples from karst breccias. (D) $\delta^{18}O$ vs. crystal size for medium-crystalline samples. All samples have been brecciated.

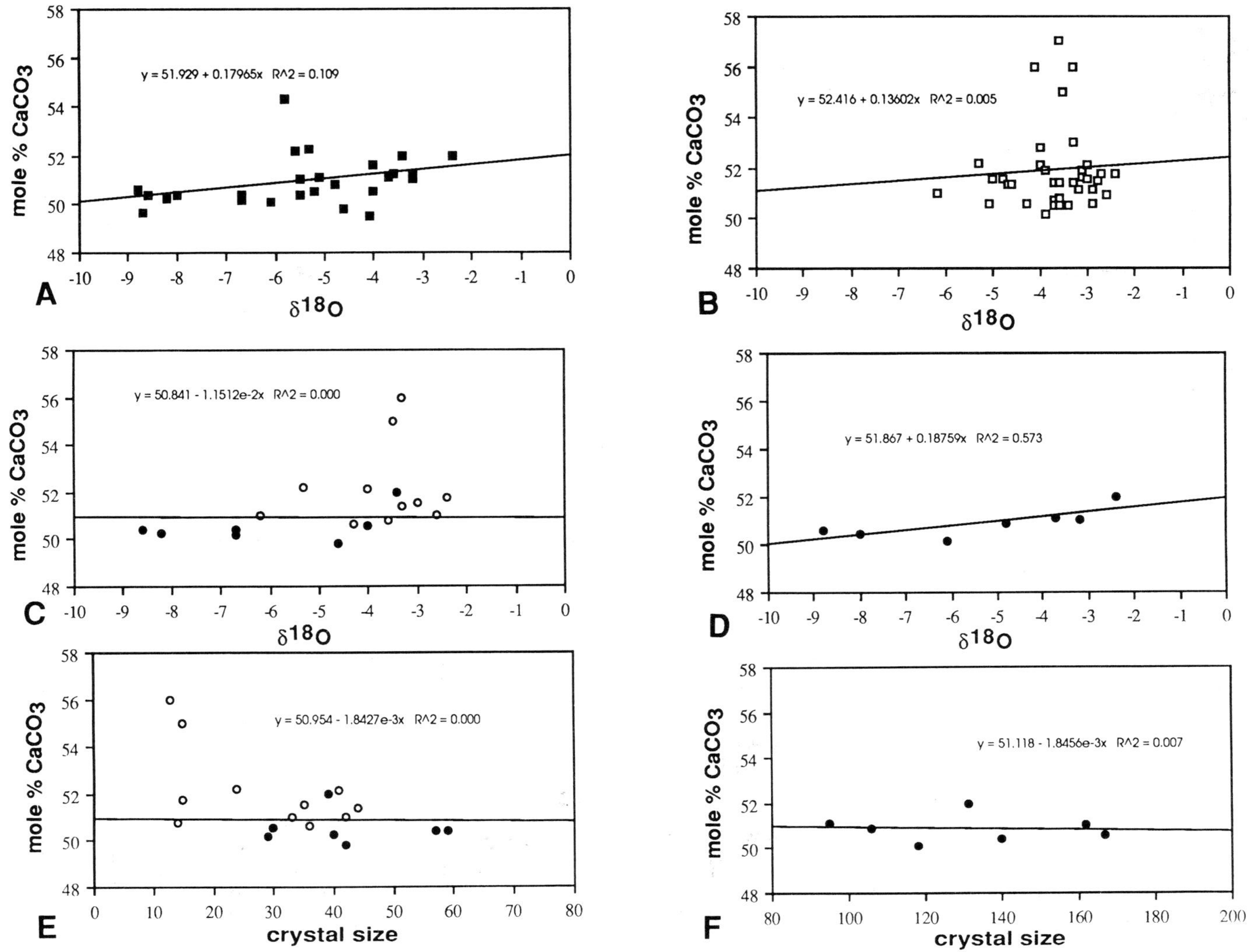

Fig. 9. Mole % $CaCO_3$ versus $\delta^{18}O$ and crystal size. (A) Karstified samples. (B) Non-karstified samples. (C) Very fine- to fine-crystalline samples. ○, samples from non-karstified intervals; ●, samples from karst breccias. (D) Medium-crystalline samples. All samples have been karstified. (E) Mole % $CaCO_3$ vs. crystal size for very fine- to fine-crystalline samples. ○, samples from non-karstified intervals; ●, samples from karst breccias. (F) Mole % $CaCO_3$ vs. crystal size for medium-crystalline samples. All samples have been karstified.

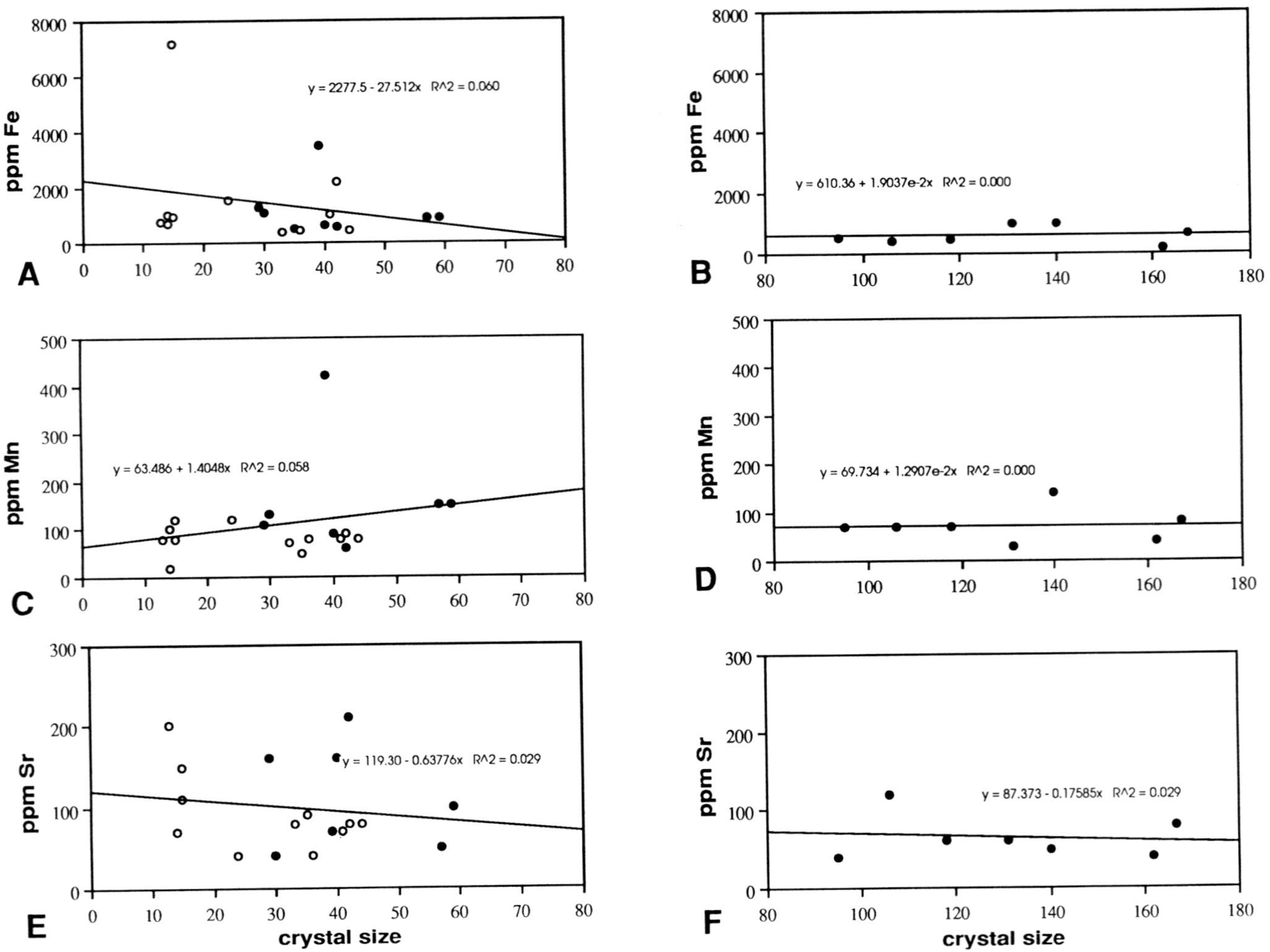

Fig. 10. Trace element concentration vs. crystal size (A) Fe concentration (ppm) vs. crystal size for very fine- to fine-crystalline subset. ○, Samples from non-karstified intervals; ●, samples from karst breccias. (B) Fe concentration (ppm) vs. crystal size for medium-crystalline subset. All samples have been karstified. (C) Manganese concentration (ppm) vs. crystal size for very fine- to fine-crystalline subset. ○, Samples from non-karstified intervals; ●, samples from karst breccias. (D) Mn concentration (ppm) vs. crystal size for medium-crystalline subset. All samples have been karstified. (E) Sr concentration (ppm) vs. crystal size for very fine- to fine-crystalline subset. ○, Samples from non-karstified intervals; ●, samples from karst breccias. (F) Strontium concentration (ppm) vs. crystal size for medium-crystalline subset. All samples have been karstified.

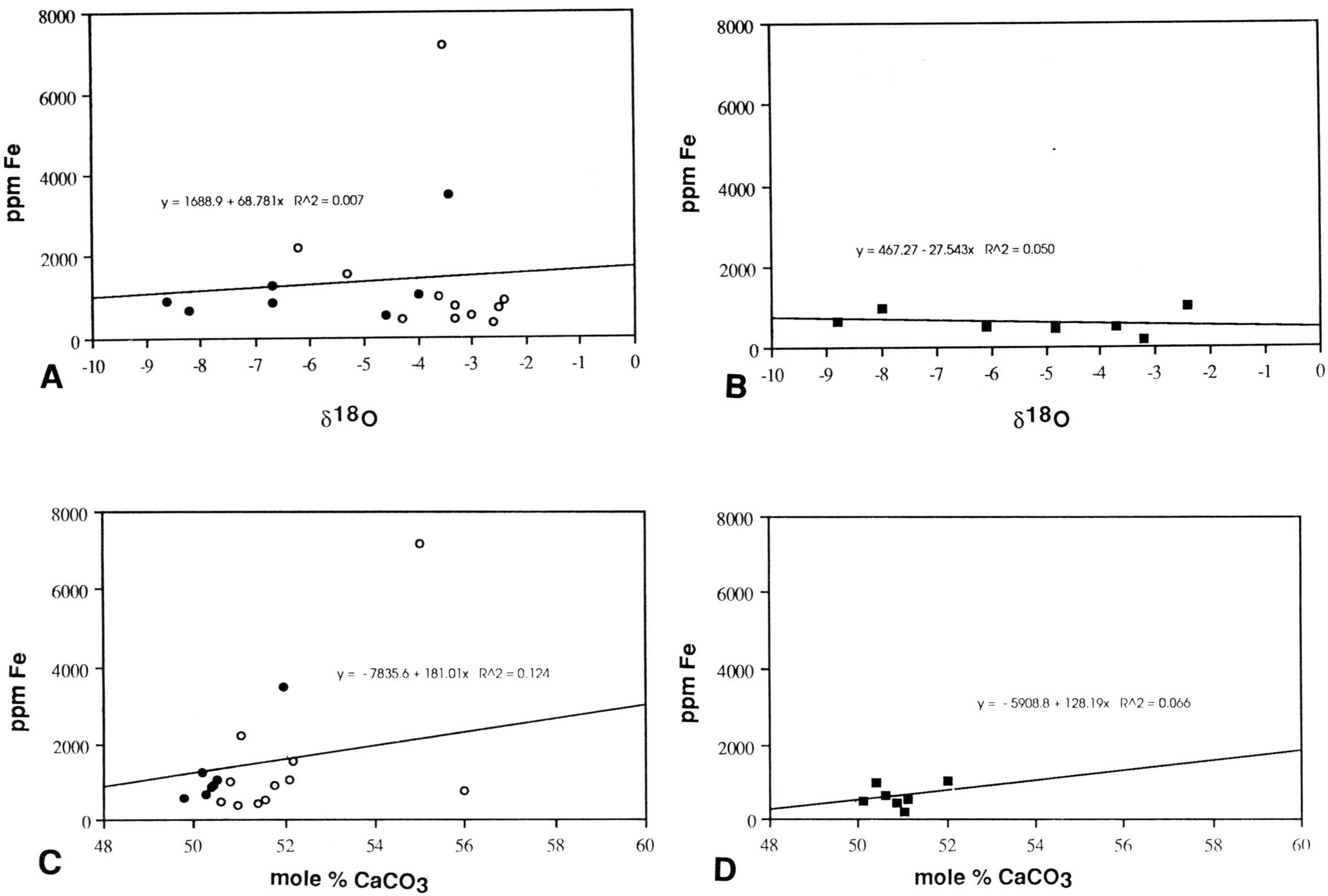

Fig. 11. Iron concentration vs. $\delta^{18}O$ and stoichiometry. (A) ppm Fe vs. $\delta^{18}O$ for very fine- to fine-crystalline subset. ○, Samples that have not been karstified; ●, karstified samples. (B) Fe concentration (ppm) vs. $\delta^{18}O$ for medium-crystalline samples. All samples have been karstified. (C) Fe concentration (ppm) vs. mole % $CaCO_3$ for very fine- to fine-crystalline samples. ○, Data from non-karstified samples; ●, data from karstified samples. (D) Fe concentration (ppm) vs. mole % $CaCO_3$ for medium-crystalline samples. All samples have been karstified.

very fine- to fine-crystalline group, karstified samples have somewhat lower values than non-karstified samples (Fig. 10A).

Iron concentration vs. $\delta^{18}O$ illustrates no covariance between decreasing iron and decreasing $\delta^{18}O$ for both the very fine- to fine-crystalline group (Fig. 11A) and the medium-crystalline group (Fig. 11B). Iron vs. mole % $CaCO_3$ shows covariance between decreasing Fe and increasing stoichiometry for the very fine- to fine-crystalline group (Fig. 11C), and weak covariance for the medium-crystalline group (Fig. 11D).

Manganese

Manganese concentrations range from 20 to 220 ppm, with one value of 420 ppm. Cross-plots of ppm Mn vs. crystal size (Figs 10C,D) display a weakly covariant trend of increasing concentration with increasing size for the very fine- to fine-crystalline group, and no covariance for the medium-crystalline group. Manganese vs. mole % $CaCO_3$ plots show no covariance for the very fine- to fine-crystalline group (Fig. 12A), whereas the medium-crystalline group shows increasing Mn with increasing stoichiometry (Fig. 12B). A weakly covariant trend exists between increasing Mn and decreasing $\delta^{18}O$ for the medium-crystalline group (Fig. 12D), whereas no covariance is observed in the very fine to fine-crystalline group.

Strontium

The very fine- to fine-crystalline group shows weak covariance between decreasing Sr concentration and increasing crystal size (Fig. 10E), whereas the medium-crystalline group shows very weak covariance (Fig. 10F). Cross-plots illustrate very weak to no covariance between decreasing Sr and decreasing $\delta^{18}O$ (Fig. 13A,B). Sr vs. mole % $CaCO_3$ illustrates weak covariance between decreasing concentration and increasing stoichiometry (Fig. 13C) for the very fine- to fine-crystalline group, and no covariance for the medium-crystalline group (Fig. 13D). Strontium concentrations for the medium-crystalline group (to 120 ppm) are lower than for the very fine- to fine-crystalline group (up to 210 ppm), as illustrated in Figures 10E and F. All values are low relative to Cenozoic evaporative dolomites, as will be discussed.

Radiogenic isotopes

Many dolomites from mud-supported facies exhibit $^{87}Sr/^{86}Sr$ values within the range of Early Ordovician seawater (0.7087–0.7091; Burke *et al.*, 1982), whereas four data points are less radiogenic than Early Ordovician seawater (Table 1). Covariant trends between increasing values of $^{87}Sr/^{86}Sr$ and increasing stoichiometry exist for the medium-crystalline group (Fig. 14B); the trend within the very fine- to fine-crystalline group (Fig. 14A) is not covariant. Increasing $^{87}Sr/^{86}Sr$ also covaries with decreasing $\delta^{18}O$ for the medium-crystalline group (Fig. 14D), whereas the trend within the very fine- to fine-crystalline group (Fig. 14C) is very weak to non-covariant. Very fine- to fine-crystalline samples have significantly more variability than do the medium-crystalline samples.

TEXTURAL AND GEOCHEMICAL MODIFICATION OF DOLOMITE: INTERPRETATION AND SIGNIFICANCE

Origin of dolomite in mud-supported facies

Ellenburger dolomite has been diagenetically altered enough to mask most depositional textures. However, age-equivalent Knox Group and Arbuckle Group dolomites are thought to be valid analogues for deposition of the Ellenburger. Studies of cycles within the Knox (Montañez & Read, 1992b) suggest, on the basis of relative amounts of dolomite in individual cycles, the stratigraphic position of these dolomites within the cycles and the reworking of dolomitized clasts at the base of overlying cycles, that the earliest dolomite is syndepositional, precipitated by evaporated seawater. Arbuckle Group dolomites are also interpreted to have formed syndepositionally in a sabkha-like setting (Gao & Land, 1991). As facies are similar in the Ellenburger, Knox and Arbuckle Groups, and silicified evaporite nodules are present in all groups, the earliest dolomites within the Ellenburger may also have been syndepositional. 'Mud-supported' facies comprise cryptalgal-laminated, mm-laminated and bioturbated facies. Of these, the cryptalgal facies is probably the best analogue for sabkha-like dolomitization. The remaining two facies could have been dolomitized in a similar manner, but could also have been dolomitized by unmodified seawater, or mixed seawater–meteoric water. A marine or modified-marine origin for early-stage Ellenburger mud-supported dolomites is consistent with $^{87}Sr/^{86}Sr$ data. Many samples, including the texturally least-altered samples, have values within the range of

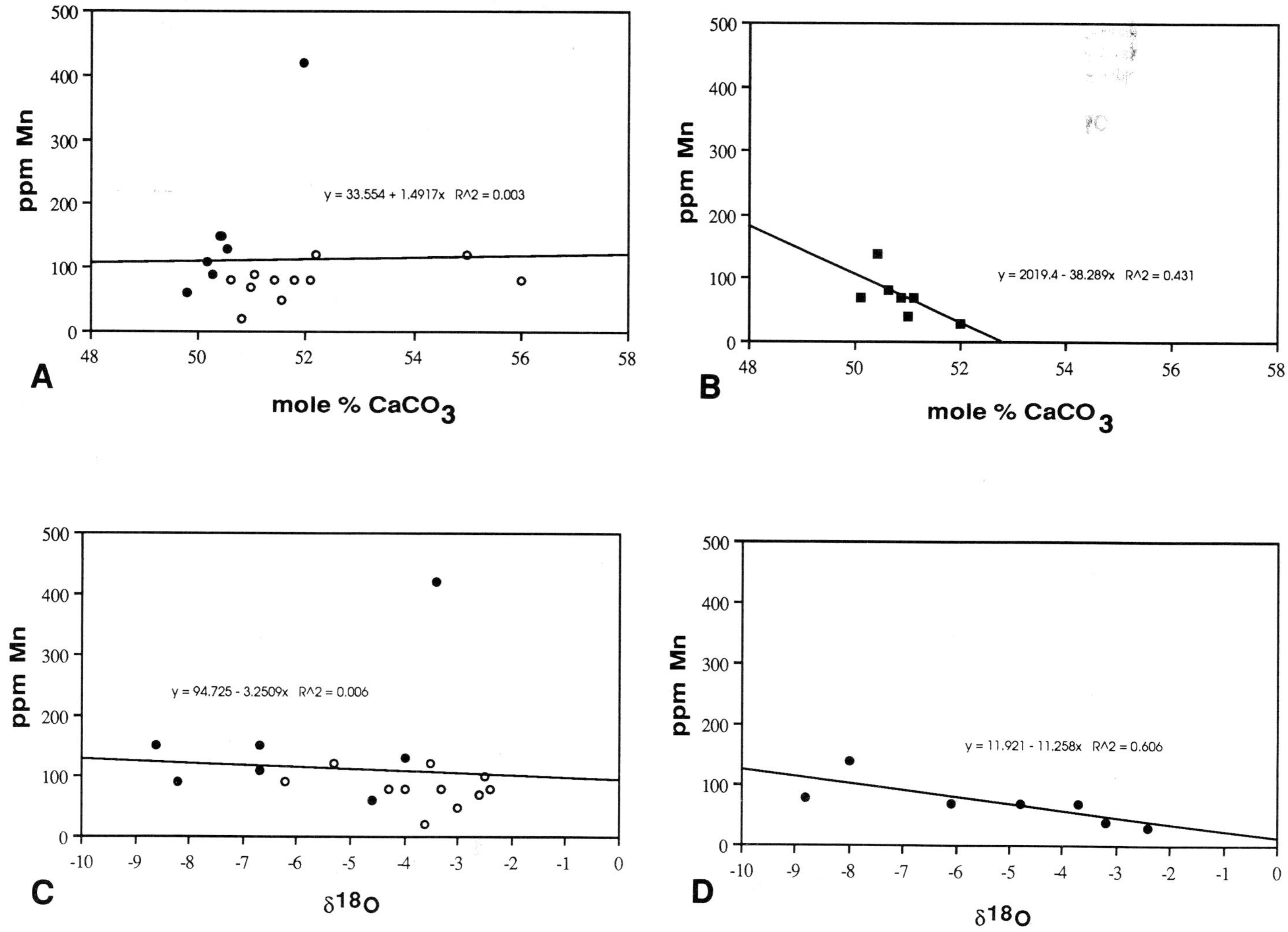

Fig. 12. Manganese concentration vs. stoichiometry and $\delta^{18}O$. (A) Mn (ppm) vs. mole % $CaCO_3$ for very fine- to fine-crystalline subset. ○, Samples that have not been karstified; ●, karstified samples. (B) Mn concentration (ppm) vs. mole % $CaCO_3$ for medium-crystalline samples. All samples have been karstified. (C) Mn concentration (ppm) vs. $\delta^{18}O$ for very fine- to fine-crystalline samples. ○, Samples that have not been karstified; ●, karstified samples. (D) Mn concentration (ppm) vs. $\delta^{18}O$ for medium-crystalline samples. All samples have been karstified.

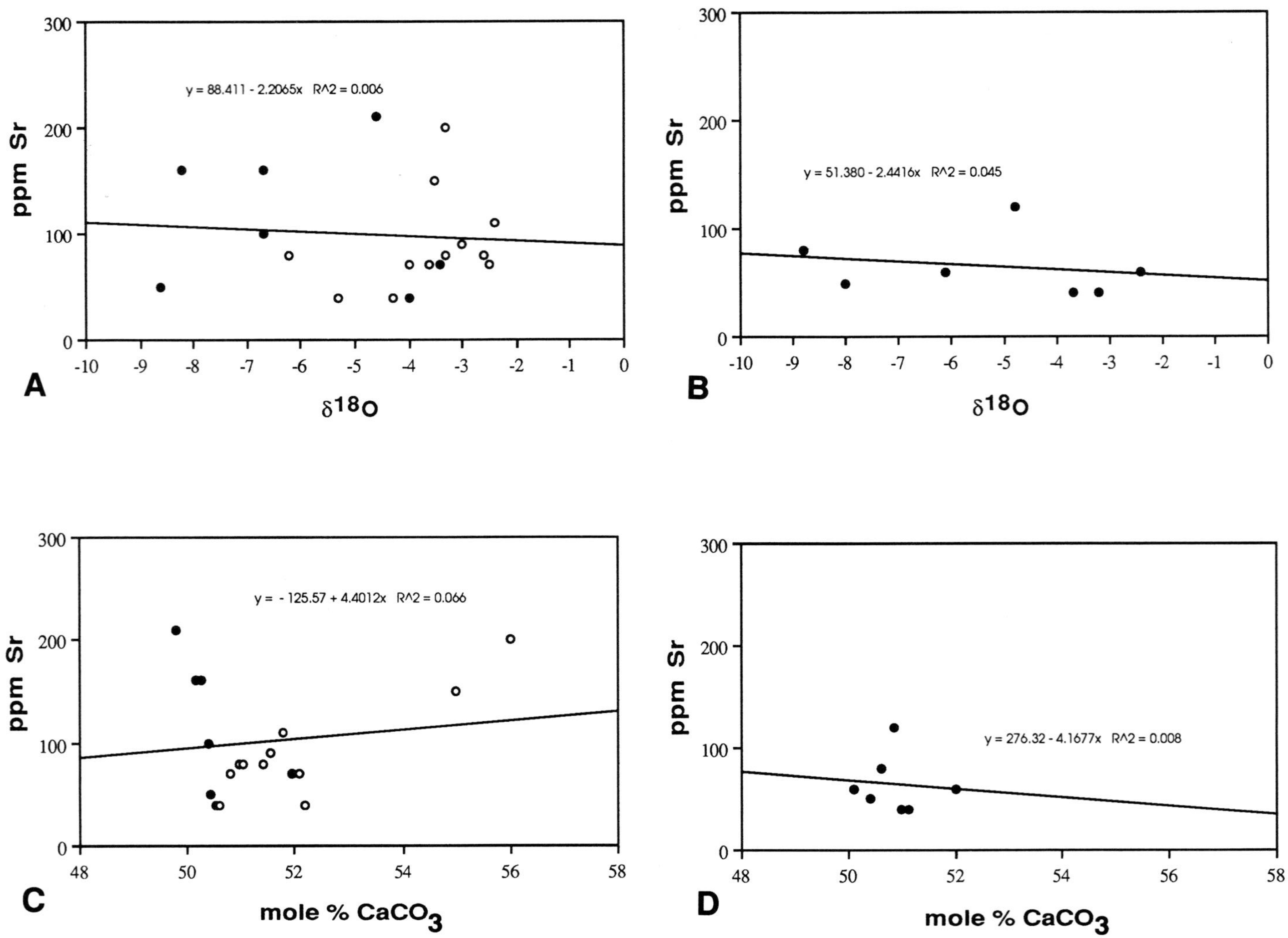

Fig. 13. Strontium concentration vs. $\delta^{18}O$ and stoichiometry. (A) Sr concentration (ppm) vs. $\delta^{18}O$ for very fine- to fine-crystalline samples. ○ Samples that have not been karstified; ●, karstified samples. (B) Sr concentration (ppm) vs. $\delta^{18}O$ for medium-crystalline samples. All samples have been karstified. (C) Sr (ppm) vs. mole % $CaCO_3$ for very fine- to fine-crystalline subset. ○, Samples that have not been karstified; ●, karstified samples. (D) Sr concentration (ppm) vs. mole % $CaCO_3$ for medium-crystalline samples. All samples have been karstified.

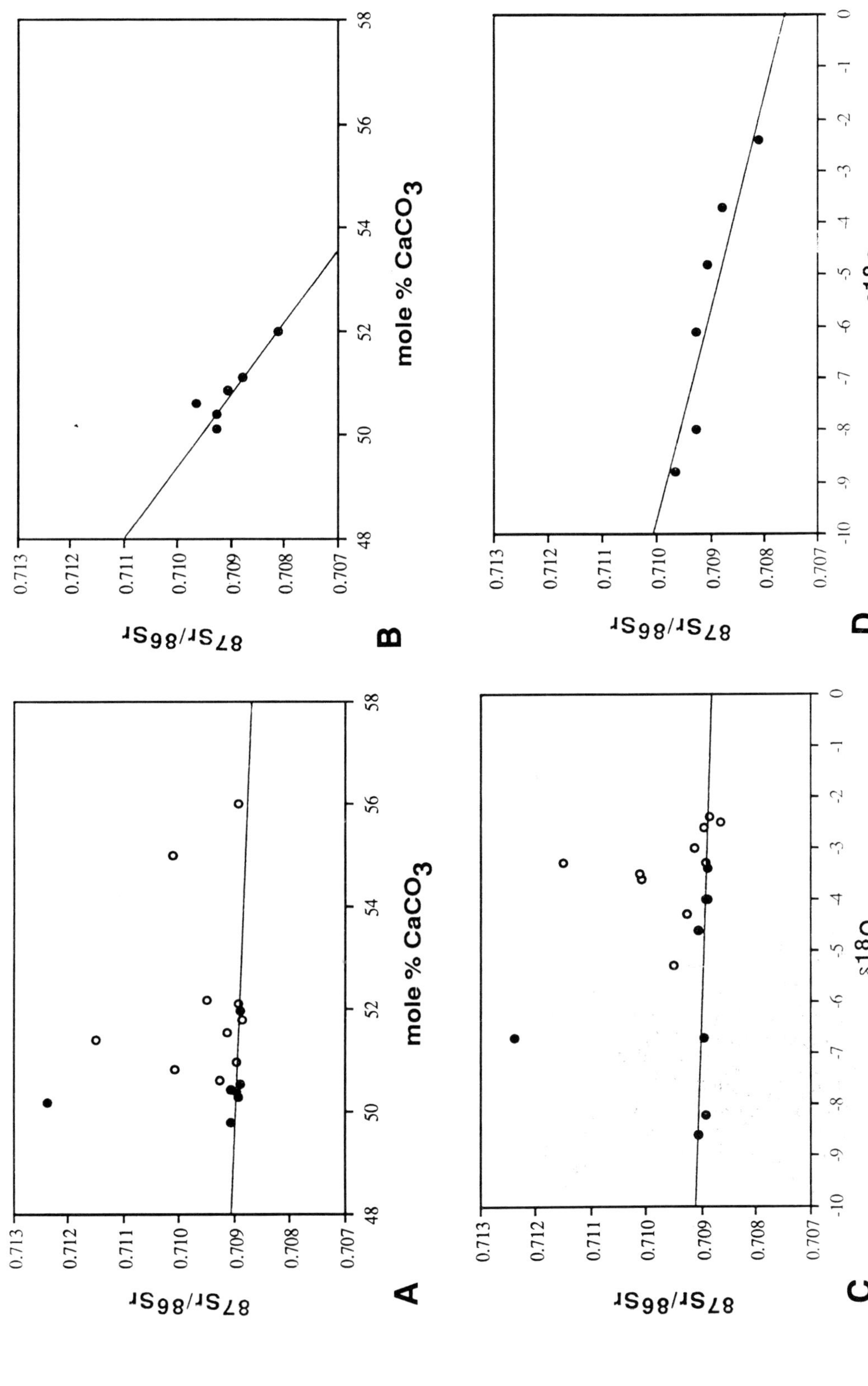

Fig. 14. $^{87}Sr/^{86}Sr$ vs. stoichiometry and $\delta^{18}O$. (A) $^{87}Sr/^{86}Sr$ vs. mole % $CaCO_3$ for very fine- to fine-crystalline subset. ○, Samples that have not been karstified; ●, karstified samples. (B) $^{87}Sr/^{86}Sr$ vs. mole % $CaCO_3$ for medium-crystalline samples. All samples have been karstified. (C) $^{87}Sr/^{86}Sr$ vs. $\delta^{18}O$ for very fine- to fine-crystalline samples. ○, Samples that have not been karstified; ●, karstified samples. (D) $^{87}Sr/^{86}Sr$ vs. $\delta^{18}O$ for medium-crystalline samples. All samples have been karstified.

Early Ordovician seawater (0.7087–0.7091; Burke *et al.*, 1982).

Evidence for modification of mud-supported dolomite

Petrography

The increase in non-planar crystal boundaries with increasing crystal size, and the covariance between increasing crystal size with both depletion in ^{18}O and decreasing trace element concentration, suggest systematic change in the texture and chemistry of these dolomites. Evidence from cathodoluminescence petrography strongly suggests that these dolomites have been modified more than once from their original textures and chemistry. Replacement fabrics are observed in even the finest-crystalline samples. There appear to be two distinct modification events: the first associated with regional exposure and pre-Middle Ordovician karstification, and the second associated with the expulsion of basinal fluids during the Pennsylvanian Ouachita Orogeny. The earlier event manifests itself by corrosion of the dark-brown-luminescent host E1 dolomite, and infilling and overgrowing by non-luminescent dolomite L2. This results in a 'mottled' appearance of the host under cathodoluminescence. Following this is precipitation of dolomite E3 (dark-brown-luminescent overgrowths). Dolomites E2 and E3 are more abundant in samples from karst breccias, suggesting that this modification either postdated, or was penecontemporaneous with, karstification.

Superimposed on these textures is orange-luminescent dolomite L2. It is present as a replacement of host E1, E2 and E3 dolomites (Fig. 7), and grades to complete replacement in some medium-crystalline samples (Fig. 6). Increasing amounts of dolomite L2 correspond to increasing crystal size within each of the crystal size subsets. Within the very fine- to fine-crystalline and non-karstified dolomites, complete replacement does not occur. Complete replacement does occur within some samples of the medium-crystalline karstified subset. The correlation between texture and geochemistry (stoichiometry, $\delta^{18}O$, Sr, Fe, Mn, $^{87}Sr/^{86}Sr$) suggests that the coarser the dolomite, the more extensive its modification.

Stoichiometry

In contrast to Ellenburger dolomites from mud-supported facies, Holocene dolomites are non-stoichiometric and poorly ordered (Behrens & Land, 1972; McKenzie, 1981; Carballo *et al.*, 1987; Mazzullo *et al.*, 1987; Mitchell *et al.*, 1987). The texturally least-altered Ellenburger samples are the least stoichiometric, although very different from presumed precursor phases. The texturally most-altered samples are the most stoichiometric.

$\delta^{18}O$

Present-day geochemical signatures are inconsistent with those expected for syndepositional dolomite precipitated from fluids isotopically similar to modern seawater. $\delta^{18}O$ (25°C) values of Ellenburger mud-supported facies range from −2.4 to −8.8‰ (PDB), significantly depleted relative to reported values for Cenozoic marine dolomites (Behrens & Land, 1972; Supko, 1977; Saller, 1984; Carballo *et al.*, 1987; Mazzullo *et al.*, 1987; Mitchell *et al.*, 1987; Land, 1991), and evaporitive dolomites (McKenzie, 1981; Pierre *et al.*, 1984; Major *et al.*, 1992).

Evaporated seawater can have oxygen isotopic compositions from +0.7‰ enriched (Major *et al.*, 1992) to greater than 11‰ enriched relative to normal surface seawater (Aharon *et al.*, 1977), and elevated temperatures of up to 30–40°C (McKenzie, 1981). Holocene dolomites precipitated from these fluids have mean values ranging from +1.0‰ PBD (Major *et al.*, 1992) to +9.0‰ PDB (Aharon *et al.*, 1977). In contrast, dolomite from unmodified seawater at 25°C should be approximately +1.3‰ PDB (based on the dolomite fractionation equation of Land, 1980). It is possible that Early Ordovician seawater was depleted in ^{18}O relative to modern seawater (Popp *et al.*, 1986; Hudson & Anderson, 1989; Lohmann & Walker, 1989), with seawater depleted by as much as 5‰. Using end-member values of both isotopic compositions of saline brines and temperatures, evaporative dolomite from originally depleted Early Ordovician seawater should have values of approximately −5.7 to +7.1‰ PDB. Given an initially depleted Early Ordovician sea and precipitation from unmodified seawater at 25°C, dolomite values should be approximately −3.5‰ PDB. The range of $\delta^{18}O$ values of mud-supported Ellenburger samples encompasses a significantly lighter range in values, and averages a lighter value, than expected for a pristine precipitate from either unmodified or evaporated Ordovician seawater, which leads us to conclude that dolomites from the mud-supported facies have been significantly modified after initial precipitation.

Trace elements

The strontium concentrations of Ellenburger samples range from 20 to 210 ppm, and are significantly lower than concentrations observed in Cenozoic evaporative dolomites (approximately 400–1000 ppm). Ellenburger dolomites should have had high initial concentrations of Sr if they precipitated during rapid crystallization, such as during evaporative reflux, as kinetic effects have been demonstrated to increase Sr concentrations (Kitano *et al.*, 1971).

It has been postulated that dissolution and reprecipitation should result in progressively depleted Sr concentrations as dolomite becomes more stoichiometric, due to a lower effective distribution coefficient during stabilization reactions (Jacobsen & Usdowski, 1976; Katz & Matthews, 1977; Land, 1980; Bein & Land, 1983). Low Sr concentrations are consistent with the observed petrographic evidence of recrystallization of the Ellenburger. However, Vahrenkamp and Swart (1990) proposed that dolomites may be initially precipitated as stoichiometric, with low Sr concentrations (as low as 50 ppm) the result of a very low distribution coefficient (K_D = 0.0118). If this is correct, some Ellenburger samples may exhibit pristine concentrations. However, covariance between decreasing Sr concentrations and increasing stoichiometry, and the correlations between decreasing Sr concentrations and increasing crystal size, increasing non-planar crystal boundaries and increasing amounts of replacement by dolomites E2 and L2, suggests that diagenetic modification of early non-stoichiometric dolomite is responsible for these trends.

Iron concentrations from Ellenburger samples (180–7160 ppm) are within the range of Cenozoic evaporative dolomites (10–2000 ppm). They exhibit a progressive decrease with increasing crystal size, increasing replacement textures and stoichiometry, suggesting modification of original dolomites by oxidizing and/or low-Fe fluids. This could be the result of meteoric recrystallization, followed by later rock-buffered recrystallization (the trend is muted in medium-crystalline samples), as will be discussed in more detail below.

$^{87}Sr/^{86}Sr$

Many samples have $^{87}Sr/^{86}Sr$ values that fall within the range of Early Ordovician seawater (0.7087–0.7091; Burke *et al.*, 1982). Within the very fine- to fine-crystalline group, values are highly variable. This variability is consistent with modification (corrosion and overgrowths) and recrystallization by two different fluids, first by Sr-poor fluids (with resultant rock-buffering and retention of coeval $^{87}Sr/^{86}Sr$ values), and later by a more radiogenic fluid. The later event is interpreted as the same recrystallization event that more completely affected the medium-crystalline group. Therefore, the variability of $^{87}Sr/^{86}Sr$ values within the very fine- to fine-crystalline group is interpreted to be the result of superposition of multiple diagenetic events, and a lack of complete recrystallization by radiogenic dolomite L2.

As shown most clearly in the medium-crystalline group, $^{87}Sr/^{86}Sr$ values become more radiogenic with increasing stoichiometry and decreasing $\delta^{18}O$. The correspondence of radiogenic values with karstification and with recrystallization by the L2 dolomite suggests late recrystallization by a more radiogenic fluid, which probably involved karst breccias as fluid conduits. Consequently, non-karstified samples are more variable, probably because they were less permeable and less able to be completely recrystallized. This resulted in $^{87}Sr/^{86}Sr$ values between coeval seawater and the most radiogenic end-member.

Diagenetic fluids and relative timing of superimposed recrystallization events

Meteoric fluids

Textural modification (corrosion, mottled replacement and overgrowths) by the E2 and E3 dolomites occurred during or soon after karstification, and before late-stage L2 dolomite. The diagenetic fluid responsible for modification resulted in dolomites (E1, E2, E3) having low trace element concentrations and depleted $\delta^{18}O$ values. The timing (during or soon after exposure) and the setting of the Ellenburger at this time (exposure during second-order sea-level low-stand) are consistent with textural and chemical modification of Ellenburger E1 dolomites by meteoric fluids during early Middle Ordovician exposure. Low Sr concentrations in the water probably resulted in rock-buffering of $^{87}Sr/^{86}Sr$ values.

During the Early Ordovician west Texas is interpreted to have been at a latitude of approximately 30° (Bambach *et al.*, 1980). Meteoric water near this latitude is expected to be approximately 3–4‰ depleted relative to seawater. At surface temperatures of 25°C, $\delta^{18}O$ values for dolomite should be approximately -1.6 to -2.6‰ PDB. Shallow-burial temperatures (40°C; *cf.* Kupecz & Land, 1991) would have precipitated dolomite of approximately -4.4 to -5.4‰ PDB. Note that the range in $\delta^{18}O$ values for early dolomites (E1, E2, E3) that have not been overprinted by dolomite L2 falls within the range of these postulated values, and that this holds for both crystal size subsets. Modification of dolomite E1 by dolomites E2 and E3 is not complete, as shown by cathodoluminescence, which is in agreement with the wide range in $\delta^{18}O$ values observed.

Basinal-derived fluids

Cathodoluminescence evidence for a second recrystallization event superimposed on the interpreted meteoric recrystallization event is striking. Increasing replacement by dolomite L2 (as observed under cathodoluminescence) covaries with increasing crystal size, increasing non-planar crystal boundaries, increasing stoichiometry, decreasing $\delta^{18}O$ and increasing $^{87}Sr/^{86}Sr$. Based on studies of other facies within the Ellenburger that have been replaced by dolomite L2, the timing and distribution of replacement dolomite has been shown to be associated with fluid expulsion during the Pennsylvanian Ouachita Orogeny (Kupecz & Land, 1991). $\delta^{18}O$ values are consistent with hot (60–110°C) reactive basinal fluids, and $^{87}Sr/^{86}Sr$ values are consistent with Pennsylvanian pore fluids progressively modified toward more radiogenic values (Kupecz & Land, 1991). As observed under cathodoluminescence, late replacement of dolomites within the mud-supported facies is incomplete: the degree of replacement corresponds with increasing crystal size, increasing stoichiometry, depletion in ^{18}O and increase in $^{87}Sr/^{86}Sr$.

Significance of fluid conduits in diagenesis of mud-supported facies

Karst breccias appear to have played an important role in the dissemination of diagenetic fluids within the Ellenburger. They controlled modification by both Middle Ordovician meteoric fluids as well as Pennsylvanian basinal fluids. Very fine to fine-crystalline, non-karstified samples, however, did not remain unmodified, but are significantly less altered. For example, dolomite E2 and E3 overgrowths, although present in non-karstified samples, are better developed in breccias; this is also true for replacement by dolomite L2. This could be the result of increased permeability that allowed greater circulation of fluids and more interaction with the E1 host. Both non-karstified and karstified samples show covariance with geochemistry and textures, but samples from karst breccias trend toward complete replacement by dolomite L2 with increasing crystal size.

Karstified and non-karstified samples are texturally and chemically different. Karstified samples are coarser, with more non-planar crystal boundaries, and the coarsest samples are completely replaced by dolomite L2. In karstified samples, trace element concentrations are lower and trends (except $\delta^{18}O$) are more muted. This could be due either to multiple episodes of recrystallization by trace element-poor fluids, and/or to rock-buffering during the latest recrystallization event. $\delta^{18}O$ values are more fluid-buffered and more depleted, with the lightest values corresponding to complete replacement by hot basinal fuids.

CONCLUSIONS

Dolomites from mud-supported (cryptalgal, bioturbated and mm-laminated) facies of the Lower Ordovician Ellenburger Group of west Texas are thought to have formed initially by syndepositional replacement of precursor carbonate by evaporated seawater. This is suggested by facies associations similar to those in age-equivalent strata, the presence of silicified evaporite nodules, and $^{87}Sr/^{86}Sr$ values that are similar to coeval seawater values.

Petrographic and geochemical data suggest that host E1 dolomites have been modified subsequent to precipitation by at least two major diagenetic events. Cathodoluminescence clearly illustrates two separate modification events, the first by dolomites E2 and E3, and the second by dolomite L2. Dolomites display covariant trends between increasing textural modification and increasing stoichiometry, decrease in trace element concentration, and depletion in ^{18}O.

Modification of early Ellenburger dolomites during the first major diagenetic event is suggested by light $\delta^{18}O$ values, low trace element concentrations and rock-buffered Sr isotopes. On the basis of cross-cutting relationships, the timing of modification is demonstrated to be post-karstification. Host E1 dolomite is interpreted to have been recrystallized by meteoric water, which is substantiated by the petrography and geochemistry, as well as the regional setting of the Ellenburger during the early

Middle Ordovician, where the Ellenburger was exposed and karstified during a relative lowering of sea-level.

The second major diagenetic event resulted in lighter $\delta^{18}O$ values, lower and less variable trace element values, and $^{87}Sr/^{86}Sr$ values that become more radiogenic with increasing degree of recrystallization. Recrystallization by hot reactive Pennsylvanian pore fluids, related to fluid expulsion during the Ouachita Orogeny, is consistent with cross-cutting relationships, geochemistry and the regional setting of the Ellenburger. The textures and geochemical trends of dolomite L2 from the mud-supported facies are similar to those of L2 from replaced ooid–peloid grainstone, subarkose and mixed carbonate–siliciclastic grainstone facies (Kupecz & Land, 1991).

The superposition of dolomite recrystallization events has produced geochemically complex rocks. It is critical that the textural and geochemical evolution be considered when interpreting the origin of the Ellenburger, and perhaps many other dolomites, on the basis of their present geochemical signatures.

ACKNOWLEDGEMENTS

We would like to acknowledge the following sources for funding this study: the Geology Foundation of the University of Texas (Owen Coates Fund), Gulf Coast Association of Geological Societies, Mobil Oil Corporation, and Sigma Xi. We would like to thank Bruce Purser, Art Saller, Maurice Tucker and Don Zenger for providing critical reviews of this manuscript. Thanks to Jeffrey Copley for illustrations, to Larry Mack for the Rb analysis and to Isabel Montañez for helpful discussions.

REFERENCES

Aharon, P., Socki, R.A. & Chan, L. (1986) Dolomitization of atolls by sea water convection flow: test of a hypothesis at Niue, south Pacific, *J. Geol.* **95**, 87–203.

Aharon, P., Kolodny, Y. & Sass, E. (1977) Recent hot brine dolomitization in the 'Solar Lake,' Gulf of Elat, isotopic, chemical, and mineralogical study. *J. Geol.* **85**, 27–48.

Anderson, J. H. (1985) *Depositional Facies and Carbonate Diagenesis of the Downslope Reefs in the Nisku Formation (U. Devonian), Central Alberta, Canada*. PhD Dissertation, University of Texas at Austin, 393 pp.

Bambach, R.K., Scotese, C.R. & Zieglar, A.M. (1980) Before Pangea: the geographies of the Paleozoic world. *Am. Sci.* **68**, 26–38.

Banner, J.L., Hanson, G.N. & Meyers, W.J. (1988) Determination of initial Sr isotopic compositions of dolostones from the Burlington-Keokuk Formation (Mississippian): constraints from cathodoluminescence, glauconite paragenesis and analytical methods. *J. Sedim. Petrol.* **58**, 673–687.

Barnes, V.E., Cloud, P.E. Jr, Dixon, L.P., Folk, R.L., Jonas, E.C., Palmer, A.R. & Tynan, E.J. (1959) *Stratigraphy of the Pre-Simpson Paleozoic Subsurface Rocks of Texas and Southeast New Mexico*. University of Texas Austin Bureau of Econ. Geol. Publ. 5924, 2 vols, 836 pp.

Behrens, E.W. & Land, L.S. (1972) Subtidal Holocene dolomite, Baffin Bay, Texas, *J. Sedim. Petrol.* **42**, 155–161.

Bein, A. & Land, L.S. (1983) Carbonate sedimentation and diagenesis associated with Mg-Ca-chloride brines: the Permian San Andres Formation in the Texas panhandle. *J. Sedim. Petrol.* **53**, 243–260.

Buchbinder, L.G., Magaritz, M. & Goldberg, M. (1984) Stable isotope study of karstic-related dolomitization: Jurassic rocks from the coastal plain, Israel. *J. Sedim. Petrol.* **54**, 236–256.

Burke, W.H., Denison, R.E., Hetherington, E.A., Koepnick, R.B., Nelson, H.F. & Otto, J.B. (1982) Variation of seawater $^{87}Sr/^{86}Sr$ throughout Phanerozoic time. *Geology* **10**, 516–519.

Cander, H.S., Kaufman, J., Daniels, L.D. & Meyers, W.J. (1988) Regional Dolomitization of Shelf Carbonates in the Burlington-Keokuk Formation (Mississippian), Illinois and Missouri: Constraints from Cathodoluminescent Zonal Stratigraphy. In: *Sedimentology and Geochemistry of Dolostones* (Ed. Shukla, V. & Baker, P.A.). SEPM Special Publication **43**, 129–144.

Carballo, J.D., Land, L.S. & Miser, D.E. (1987) Holocene dolomitization of supratidal sediments by active tidal pumping, Sugarloaf Key, Florida. *J. Sedim. Petrol.* **57**, 153–165.

Cloud, P.E. Jr. & Barnes, V.E. (1948) *The Ellenburger Group of Central Texas*. University of Texas Bureau of Economic Geology Publ. 4621, 473 pp.

Cloud, P.E. Jr. & Barnes, V.E. (1957) Early Ordovician sea in central Texas. *Geol. Soc. Am. Mem.* **67**, 163–214.

Cloud, P.E. Jr., Barnes, V.E. & Bridge, J. (1943) Stratigraphy of the Ellenburger Group in central Texas – a progress report. In: *Texas Mineral Resources*. University of Texas Bureau of Economic Geology Publication 4301, 390 pp.

Dorobek, S.L. & Filby, R.H. (1988) Origin of dolomites in a downslope biostrome, Jefferson Formation (Devonian), central Idaho: evidence from REE patterns, stable isotopes, and petrography. *Bull. Can. Petrol. Geol.* **36**, 202–215.

Flawn, P.T., Goldstein, A. Jr., King, P.B. & Weaver, C.E. (1961) *The Ouachita System*. University of Texas Bureau of Economic Geology Publ. 6120, 401 pp.

Gao, G. & Land, L.S. (1991) Early Ordovician Cool Creek dolomite, middle Arbuckle Group, Slick Hills, SW Oklahoma: origin and modification. *J. Sedim. Petrol.* **61**, 161–173.

Goldhammer, R.K., Lehmann, P.J. & Dunn, P.A. (in press) Third-order sequences and cycle stacking patterns of Lower Ordovician platform carbonates, El Paso

Group (Franklin Mountains, Texas). In: *Influence of Sea Level on Carbonate Deposition, Diagenesis, and Sequence Stratigraphy* (Ed. Major, R.P. & Halley R.B.) SEPM Special Publication.

Gregg, J.M. & Shelton, K.L. (1990) Dolomitization and dolomite neomorphism in the back reef facies of the Bonneterre and Davis Formations (Cambrian), southeastern Missouri. *J. Sedim. Petrol.* **60**, 549–562.

Gregg, J.M. & Sibley, D.F. (1984) Epigenetic dolomitization and the origin of xenotopic dolomite texture. *J. Sedim. Petrol.* **54**, 908–931.

Hardie, L.A. (1987) Dolomitization: a critical view of some current views. *J. Sedim. Petrol.* **57**, 166–183.

Holail, H., Lohmann, K.C. & Sanderson, I. (1988) Dolomitization and dedolomitization of Upper Cretaceous carbonates: Bahriya oasis, Egypt. In: *Sedimentology and Geochemistry of Dolostones* (Ed. Shukla, V. & Baker, P.A.) SEPM Special Publication 43, 191–207.

Hudson, J.D. & Anderson, T.F. (1989) Ocean temperatures and isotopic compositions through time. *Trans. Roy. Soc. Edin., Earth Sci.* **80**, 183–192.

Jacobson, R.L. & Usdowski, H.E. (1976) Partitioning of strontium between calcite, dolomite, and liquids. *Contrib. Mineral. Petrol.* **59**, 171–185.

Katz, A. & Matthews, R.K. (1977) The dolomitization of $CaCO_3$: an experimental study at 252°C–295°C. *Geochim. Cosmochim. Acta* **41**, 297–308.

Kerans, C. (1989) Karst-controlled reservoir heterogeneity in Ellenburger Group carbonates of west Texas. *Am. Ass. Petrol. Geol. Bull.* **72**, 160–1183.

Kerans, C. & Lucia, F.J. (1989) Recognition of second, third, and fourth/fifth-order scales of cyclicity in the El Paso Group and their relation to genesis and architecture of Ellenburger reservoirs. In: *The Lower Paleozoic of West Texas and Southern New Mexico – Modern Exploration Concepts* (Ed. Cunningham, B.K. & Cromwell, D.W.) Permian Basin SEPM Publ. 89-31, 105–110.

Kitano, Y., Kanamore, N. & Oomori, T. (1971) Measurements of distribution coefficients of strontium and barium between carbonate precipitate and solution: abnormally high value of distribution coefficients at early stages of carbonate formation. *Geochem. J.* **4**, 183–206.

Kupecz, J.A. (1989) *Petrographic and Geochemical Characterization of the Lower Ordovician Ellenburger Group, West Texas*. PhD Dissertation, University of Texas at Austin, 158 pp.

Kupecz, J.A. (1992) Sequence boundary control on hydrocarbon reservoir development, Ellenburger Group, Texas. In: *Paleokarst, Karst-Related Diagenesis, and Reservoir Development: Examples from Ordovician–Devonian Age Strata of West Texas and the Mid-Continent* (Ed. Candelaria, M.P. & Reed, C.L.) Permian Basin Section, SEPM Spec. Publ. 92-33, 55–58.

Kupecz, J.A. & Land, L.S. (1991) Late-stage dolomitization of the Lower Ordovician Ellenburger Group, west Texas. *J. Sedim. Petrol.* **61**, 551–574.

Kupecz, J.A., Montanez, I.P. & Gao, G. (1992) Recrystallization of dolomite with time. In: *Carbonate Microfabrics* (Ed. Rezak, R. & Lavoie D.) Frontiers in Sedimentology. Springer-Verlag, New York.

Land, L.S. (1980) The isotopic and trace element geochemistry of dolomite: the state of the art. In: *Concepts and Models of Dolomitization* (Ed. Zenger, D.H., Dunham, J.B. & Ethington, R.L.) SEPM Special Publ. 28, 87–110.

Land, L.S. (1985) The origin of massive dolomite. *J. Geol. Educ.* **33**, 112–125.

Land, L.S. (1991) Dolomitization of the Hope Gate Formation (North Jamaica) by seawater: reassessment of mixing-zone dolomite. In: *Stable Isotope Geochemistry: A Tribute to Samuel Epstein* (Ed. Taylor, H.P. Jr., O'Neil, J.R. & Kaplan, I.R.) Geochem. Soc. Spec. Publ. **3**, 121–133.

Land, L.S., Salem, M.R.I. & Morrow, D.W. (1975) Paleohydrology of ancient dolomites: geochemical evidence. *Am. Ass. Petrol. Geol. Bull.* **59**, 1602–1625.

Lohmann, K.C. & Walker, J.C.G. (1989) The $\delta^{18}O$ record of Phanerozoic abiotic marine calcite cements. *Geophys. Res. Lett.* **16**, 319–322.

Loucks, R.G. & Anderson, J.H. (1980) Depositional facies and porosity development in Lower Ordovician Ellenburger dolomite, Puckett Field, Pecos County, Texas. In: *Carbonate Reservoir Rocks* (Ed. Halley, R.B. & Loucks, R.G.) SEPM Core Workshop 1, 1–31.

Loucks, R.G. & Anderson, J.H. (1985) Depositional facies, diagenetic terrains, and porosity development in Lower Ordovician Ellenburger dolomite, Puckett Field, west Texas: In: *Carbonate Petroleum Reservoirs* (Ed. Roehl, P.O. & Choquette, P.W.) pp. 21–37. New York, Springer Verlag.

Lucia, F.J. (1986) Sedimentation and paleogeography of the El Paso Group. In: *Delaware Basin Exploration* (Ed. Stuart, W.J.) West Texas Geological Society Guidebook 60-55, 61–75.

Lucia, F.J. (1969) Lower Paleozoic history of the Diablo Platform of west Texas and south-central New Mexico. In: *The Geologic Framework of the Chihuahua Tectonic Belt* (Ed. Seewald, K. & Sundeen, D.) West Texas Geological Society Field Trip Guidebook, 39–56.

Lucia, F.J. (1987) Lower Paleozoic collapse brecciation and dolomitization, Franklin Mountains, Texas. In: *Mega-Collapse Breccia and Associated Late-Stage Dolomitization of Ordovician Carbonates, Franklin Mountains, West Texas* (Ed. Kerans, C.) SEPM Midyear Meeting Guidebook, 16 pp.

McKenzie, J.A. (1981) Holocene dolomitization of calcium carbonate sediments from the coastal sabkhas of Abu Dhabi, UAE: a stable isotope study. *J. Geol.* **89**, 185–198.

Major, R.P., Lloyd, R.M. & Lucia, F.J. (1992) Oxygen isotope composition of Holocene dolomite formed in a humid hypersaline setting. *Geology* **20**, 586–588.

Mazzullo, S.J., Reid, A.M. & Gregg, J.M. (1987) Dolomitization of Holocene Mg-calcite supratidal deposits, Ambergris Cay, Belize. *Geol. Soc. Am. Bull.* **98**, 224–231.

Mitchell, J.T., Land, L.S. & Miser, D.G. (1987) Modern marine dolomite cement in a north Jamaican fringing reef. *Geology* **15**, 557–560.

Montañez, I.P. (1989) *Regional Dolomitization of Early Ordovician, Upper Knox Group, Appalachians*. PhD Dissertation, Virginia Polytechnic Institute, 284 pp.

Montañez, I.P. (1992) Use of sea level history and stratigraphic facies distribution in Lower Ordovician Upper Knox carbonates for prediction of diagenetic facies distribution and reservoir heterogeneity. In: *Paleokarst,*

Karst-Related Diagenesis, and Reservoir Development: Examples from Ordovician–Devonian Age Strata of West Texas and the Mid-Continent (Ed. Candelaria, M.P. & Reed, C.L.) Permian Basin Section, SEPM Spec. Publ. 92-33, 165–181.

MONTAÑEZ, I.P. & READ, J.F. (1992a) Fluid–rock interaction history during stabilization of early dolomites, Upper Knox Group (Lower Ordovician), US Appalachians. *J. Sedim. Petrol.* **62**, 753–778.

MONTAÑEZ, I.P. & READ, J.F. (1992b) Eustatic control on early dolomitization of cyclic peritidal carbonates: evidence from Upper Knox Group, Appalachians. *Geol. Soc. Am. Bull.* **104**, 872–886.

MOORE, C.H., CHOWDHURY, A. & CHAN, L. (1988) Upper Jurassic Smackover platform dolomitization, northwestern Gulf of Mexico: a tale of two waters. In: *Sedimentology and Geochemistry of Dolostones* (Ed. Shukla, V. & Baker P.A.) SEPM Spec. Publ. **43**, 175–189.

PAIGE, S. (1911) *Mineral Resources of the Llano-Burnet Region, Texas*. US Geol. Surv. Bull. 450, 103 pp.

PAIGE, S. (1912) *Description of the Llano and Burnet Quadrangles*. US Geol. Surv. Geologic Atlas, Llano-Burnet Folio (No. 183), 16 pp.

PIERRE, C., ORTLIEB, L. & PERSON, A. (1984) Supratidal evaporitic dolomite at Ojo de Liebre lagoon: mineralogical and isotopic agruments for primary crystallization. *J. Sedim. Petrol.* **54**, 1049–1061.

POPP, B.N., PODOSEK, F.A., BRANNON, J.C., ANDERSON, T.F. & PIER, J. (1986) $^{87}Sr/^{86}Sr$ ratios in Permo-Carboniferous seawater from the analyses of well-preserved brachiopod shells. *Geochim. Cosmochim. Acta* **50**, 1321–1328.

RAYMOND, D.E. & OSBORNE, W.E. (1991) *Stratigraphy and Exploration of the Knox Group in the Appalachian Fold and Thrust Belt and Black Warrior Basin of Alabama*. Arbuckle Group Core Workshop and Field Trip, Oklahoma Geol. Surv. Spec. Publ. 91-3, 163–179.

SALLER, A.H. (1984) Petrologic and geochemical constraints on the origin of subsurface dolomite, Enewetak Atoll: an example of dolomitization by normal seawater. *Geology* **12**, 217–220.

STEPANEK, B.E. (1987) Dolomitization of paleokarst collapse-breccias – the Ordovician El Paso Group, Franklin Mountains, west Texas. In: *Mega-Collapse Breccia and Associated Late Stage Dolomitization of Ordovician Carbonates, Franklin Mountains, west Texas* (Ed. Kerans, C.) SEPM Midyear Meeting Guidebook, 16 pp.

SUPKO, P.R. (1977) Subsurface dolomites, San Salvador, Bahamas. *J. Sedim. Petrol.* **47**, 1063–1077.

VAHRENKAMP, V.C. & SWART, P.K. (1990) New distribution coefficient for the incorporation of strontium into dolomite and its implications for the formation of ancient dolomites. *Geology* **18**, 387–391.

WILSON, J.L., FRITZ, R.D. & MEDLOCK, P.L. (1991) *The Arbuckle Group – Relationship of Core and Outcrop Analyses to Cyclic Stratigraphy and Correlation*. Arbuckle Group Core Workshop and Field Trip, Oklahoma Geol. Surv. Spec. Publ. 91-3, 133–139.

ZENGER, D.H. (1981) *Stratigraphy and Petrology of the Little Falls Dolostone (Upper Cambrian), East-Central New York*. New York State Museum Map and Chart Series 34, 138 pp.

ZENGER, D.H. (1983) Burial dolomitization of the Lost Burro Formation (Devonian), east-central California, and the significance of late-diagenetic dolomitization. *Geology* **11**, 519–522.

ZENGER, D.H. & DUNHAM, J.B. (1988) Dolomitization of Siluro-Devonian limestones in a deep core (5350 m), southeastern New Mexico. In: *Sedimentology and Geochemistry of Dolostones* (Ed. Shukla, V. & Baker, P.A.) SEPM Spec. Publ. 43, 161–173.

Dolomite Reservoirs

Spec. Publs Int. Ass. Sediment. (1994) **21**, 283–308

Nature, origins and evolution of porosity in dolomites

B.H. PURSER*, A. BROWN† *and* D.M. AISSAOUI*

* *Laboratoire de Pétrologie Sédimentaire, UA. CNRS.723, Bâtiment 504, Université de Paris Sud, 91405 Orsay, France; and*
† *ARCO Oil & Gas Co., 2300 W. Plano Parkway, Plano, Texas 75075, USA*

ABSTRACT

Porosity in dolomites is not necessarily related to dolomitization. Many interparticle and dissolutional voids are inherited from precursor limestones. Voids related to dolomitization, such as intercrystalline pores in sucrosic dolomites, may result from the transfer of porosity due to dissolution at one site and incomplete cementation at another. Still other pores clearly postdate dolomitization, such as fractures, dolomite crystal moulds and evaporite dissolution moulds.

During the processes of dolomitization, the rock may gain, lose or conserve its initial porosity, even in the same dolomite body. The actual gain or loss depends upon the scale of observation and the position of investigated samples in the dolomite body. The Plio-Quaternary dolomite lens at Mururoa Atoll demonstrates this principle. In the areas on the downflow side of the hydrological regime, interparticle porosity is preserved and intraparticle porosity is created, whereas in areas in the upflow part porosity is occluded by excess dolomite cementation.

As with all other rock types, the average porosity of dolomite decreases with burial, probably owing to the generation of dolomite cements by pressure-dissolution. However, dolomite loses porosity at a slower rate than limestone and, in many deeply buried or tectonically active areas, dolomite porosity is selectively preserved. Fractures are both beneficial and detrimental for dolomites. They enhance permeability, especially in dolomites with poorly connected pore systems. However, fractures may also provide pathways for fluids, which may precipitate cement selectively occluding nearby porosity. In general, fracturing occurs more readily in dolomites than in limestones.

Because porosity in dolomites has multiple origins, exploration strategy should be adapted to the type of pore system in question. Prediction of dolomite reservoirs whose porosity is inherited from that of the precursor sediment clearly requires an understanding of the sedimentary framework. Intercrystalline pore systems formed during dolomitization require the application of a hydrodynamic dolomitization model for predicting porosity within the dolomite body. Given our limited understanding of ancient hydrodynamics, this could be difficult. Finally, porosities preserved or created under burial or tectonic conditions can best be understood by studying the geodynamic history of the basin.

INTRODUCTION

The chemical and physical factors determining the creation of porosity in dolomites are controversial. This is understandable. In limestones porosity is related to primary sedimentary textures or to subsequent diagenesis, whose effects on pore space are fairly obvious. On the contrary, in many dolomites the entire fabric and associated porosity are seemingly diagenetic in origin. As a result, porosity has often been considered to be genetically related to the dolomitizing process itself, this being particularly true for fabric-destroying dolomites characteristic of many reservoirs. To some workers, dolomitization inherently creates a porous rock. This conclusion overlooks the fact that most ancient dolomites, like most ancient limestones, have low porosities.

The simplicity of the assumption that porosity is the result of dolomitization becomes readily apparent when one examines dolomites, especially those of post-Palaeozoic age, whose pore geometries are highly variable. Although intercrystalline pores are frequent, most dolomites exhibit a vuggy porosity resulting from dissolution, whereas other pure dolomites have pore geometries virtually identical to those of the sediments they have replaced. As most workers realize, pore systems in dolomites are multiple and, as such, have diverse origins whose nature and implications are the subject of this contribution.

There are three possible relationships between dolomitization and porosity evolution. First, dolomitization may essentially be porosity-destructive, i.e. it decreases the porosity of the precursor carbonate. Because it obviously concerns crystal growth within an open system involving the supply of $CaCO_3$ and Mg, porosity destruction seems inevitable. Where dolomitization occurs during burial, mechanical compaction may also reduce porosity. Porosity destruction during dolomitization is amply demonstrated by the existence of many low-porosity dolomites whose precursor primary voids are occluded by dolomite cement. However, highly porous dolomites may also have porosities lower than those of the precursor limestone, although these cases are more difficult to identify.

Secondly, dolomitization may simply 'redistribute' pre-existing pore space but does not necessarily increase or decrease it. At first glance this would appear logical, because dolomitization involves the dissolution of precursor carbonates and the precipitation of dolomite within an open system. A comparison may be made with early vadose diagenesis in limestones, where mineralogical stabilization involves the dissolution of aragonite (increasing porosity) and the contemporaneous precipitation of calcite (decreasing it), as shown by Land *et al.* (1967) and many others. This diagenesis essentially concerns the transfer of carbonate, involving the rearrangement of pre-existing pore space. However, this comparison may be somewhat misleading because calcite and dolomite diagenesis are not identical processes: the diagenesis of limestone, although closely related to water movement, does not require an external source of additional reactant, whereas dolomitization obviously does; limestone diagenesis can be a local (closed) system with respect to calcium and carbonate ions, but dolomitization must be open, at least to the magnesium ion. Thus, a simple rearrangement of pre-existing pore space would require a volume balance between input of $MgCO_3$ and output of $CaCO_3$ – an improbable situation, especially on a local scale. However, as most limestones have some porosity prior to dolomitization, porosity in most dolomites cannot be created entirely by the dolomitization process. Even an approximate conservation of porosity during dolomitization would lead to a strong depositional facies control on the distribution of porosity in a dolomite reservoir.

Thirdly, dolomitization may increase porosity. Although the precipitation of a dolomite cement, by definition, occludes porosity within a pore, the net change in porosity on a larger scale depends upon the source of the material. If carbonate ions are conserved during dolomitization, and if the rock does not compact, porosity will increase due to a change in molar volume. However, because dolomitization is an open diagenetic system, one may envisage with equal ease either excessive dissolution of precursor carbonate during dolomitization or excessive precipitation of dolomite cement. Intercrystalline voids and some types of mouldic porosity formed within carbonate grains and matrix are evidence of porosity creation during dolomitization.

All three relationships seem to exist, although it is not easy to demonstrate which relationship dominates within a given dolomite body. On a pore scale all relationships are observed. Some grains are dolomitized volume for volume, dolomite cements occlude predolomite porosity, and moulds formed during dolomitization enhance porosity, all within the same thin section. In most thin sections one process generally dominates over others. The problem arises when interpreting the porosity change for a large dolomite body. It is likely that some parts of a dolomite body gain porosity while other parts lose it, and therefore it is difficult to measure the total balance.

The unequivocal demonstration that porosity is increased or decreased by dolomitization can be made – at least theoretically – by comparing dolomite porosities with those in equivalent non-dolomitized limestone, e.g. on either side of a dolomitization front. However, this is not a simple matter. The fact that dolomite has a higher (or lower) porosity than its non-dolomitized equivalent is not absolute proof that dolomitization is responsible for this amelioration: lower porosities in the limestone may result from the selective postdolomitization compaction of these latter, a common phenomenon. Inversely, lower porosities in the dolomite relative to the adjacent limestone may result from subsequent evolution of the dolomite, notably the excessive development of dolomite cement.

The seemingly limited progress in the understand-

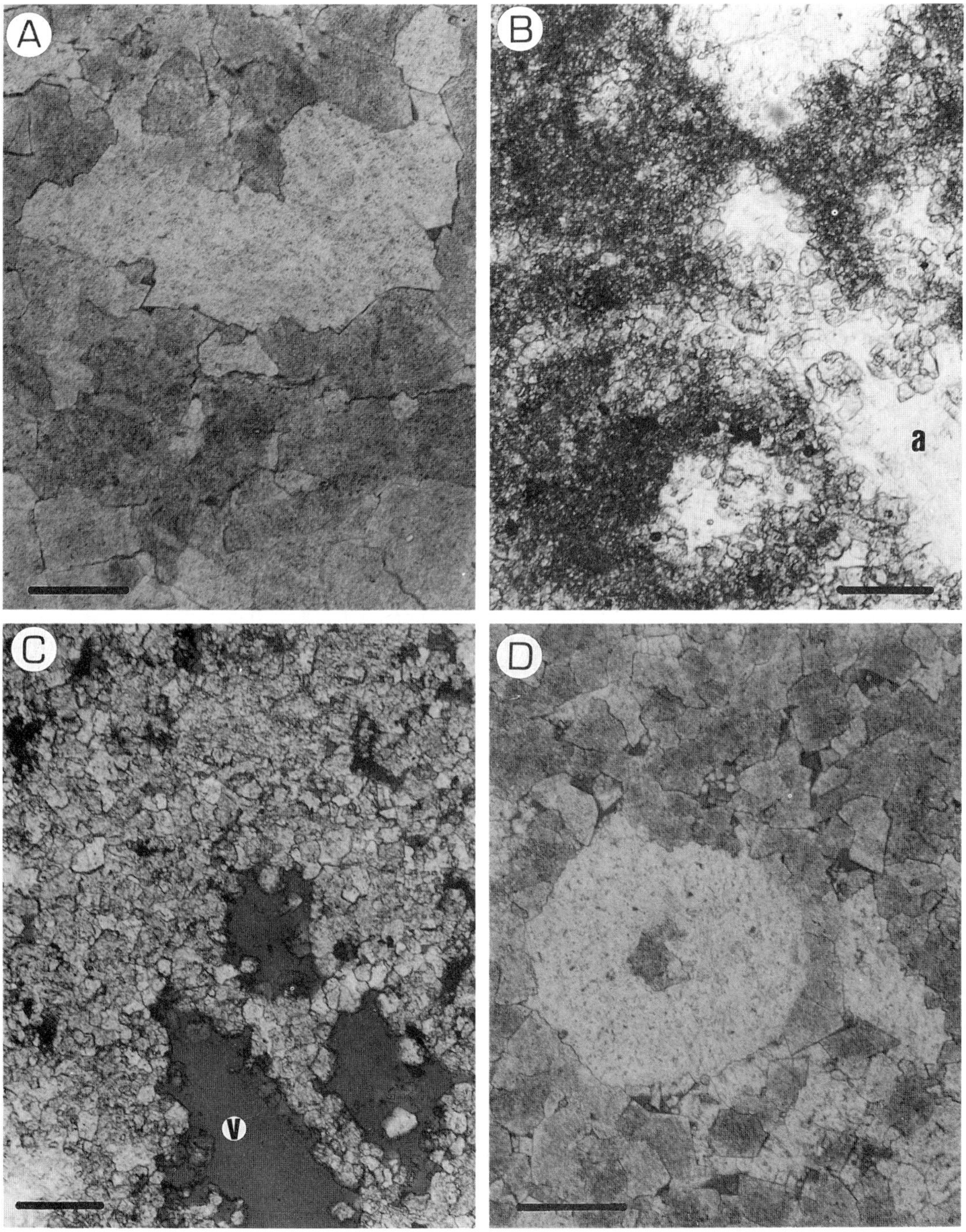

Fig. 1. Varied fabrics in Palaeozoic dolomites of North America. (A) Low-porosity non-planar dolomite with local planar fabric. Ordovician, Scipio-Albion trend, Trenton Formation, Michigan. (Plane-polarized light, scale=0.75 mm). (B) Mimetic non-planar dolomite replacing peloidal–skeletal packstone, and planar dolomite cements occluding most porosity. Anhydrite (a) fills residual voids. Permian, San Andres Formation, Goldsmith Field, Texas. (Plane-polarized light, scale=0.25 mm). (C) Non-planar replacement dolomite with residual fossil moulds (v) enlarged by evaporite replacement followed by dissolution. Permian, San Andres Formation, Wasson Field, Texas. (Plane-polarized light, scale = 0.3 mm). (D) Mixed planar dolomite replacing crinoidal packstone, and planar fabric probably representing cemented pores. Siluro-Devonian, Hunton Formation, Oklahoma. (Plane-polarized light, scale = 0.75 mm).

ing of the relationships between dolomitization and porosity development may be due partly to the fact that many studies are concerned with Palaeozoic dolomites whose fabrics are generally modified by burial diagenesis; logically, to understand porosity evolution, one should look first at the early phases of dolomite formation, rather than the final phase of a long and complex history, much of which probably concerns only porosity destruction (Fig. 1).

The evolution of porosity in dolomites (in common with dolomitization itself) is further complicated by the fact that we are dealing with a complex mineral of variable composition: variable stoichiometry is characteristic not only of recent dolomites but also of many ancient ones. Because dolomites have variable stability due to compositional and structural differences, porosity in different dolomites may evolve in various ways following the initial dolomitization event. Some may dissolve to form porosity, others may recrystallize and lose it, and still others may resist alteration and conserve porosity.

This contribution, taking into account these well known problems, has three principal objectives: to demonstrate the highly variable nature of pore types by briefly examining a number of dolomites, generally of Mesozoic and Tertiary age; to examine the initial growth of dolomite crystals in Plio-Quaternary sediments, notably at Mururoa Atoll (French Polynesia); and finally, to demonstrate the highly variable evolution of porosity in dolomites. We will conclude by suggesting implications regarding the exploration and development of various dolomite reservoirs.

PORE TYPES IN DOLOMITES AND THEIR RELATION TO DOLOMITIZATION

Although generally admitted, pore types in dolomite are not limited to the classic interrhombohedral voids typical of many oil reservoirs. Study of many porous dolomites reveals that the nature and origins of these voids relative to the process of dolomitization is variable, especially if one considers relatively young dolomites. There exist at least five distinct categories of pore space.

Fabric-replacive porosity

In many young dolomites the sedimentary texture and microstructure of grains are perfectly preserved, to such a degree that the dolomitic composition is not immediately obvious under the petrographic microscope. Both at Mururoa Atoll and at Abu Shaar (Gulf of Suez), the perfect preservation of predolomite fabric is due to two factors: minimal bulk dissolution during dolomitization, and high density of crystal nucleation, giving a microcrystalline dolomite. Under scanning electron microscopy, component crystals with planar faces associated with fine intercrystalline voids are clearly visible. In both examples the fabric-preserving dolomite has not been visibly modified by dolomite cement, so that present porosity must be very close to that of the precursor carbonates. Similar inherited fabrics within Late Neogene dolomites of the Bahamas are described by Vahrenkamp and Swart (this volume).

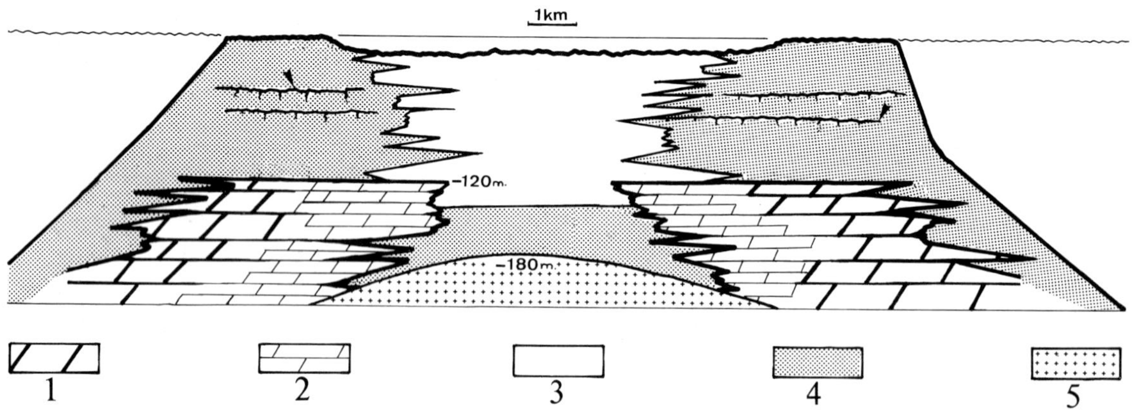

Fig. 2. Simplified section across Mururoa Atoll showing the geometry and major petrographic variations within the dolomites and limestones. 1, Dense cemented dolomite (see Fig. 3C,D); 2, porous mimetic dolomite (see Fig. 3A,B); 3, non-dolomitized lagoonal sands and muds; 4, non-dolomitized reef and reef debris; 5, volcanic basement.

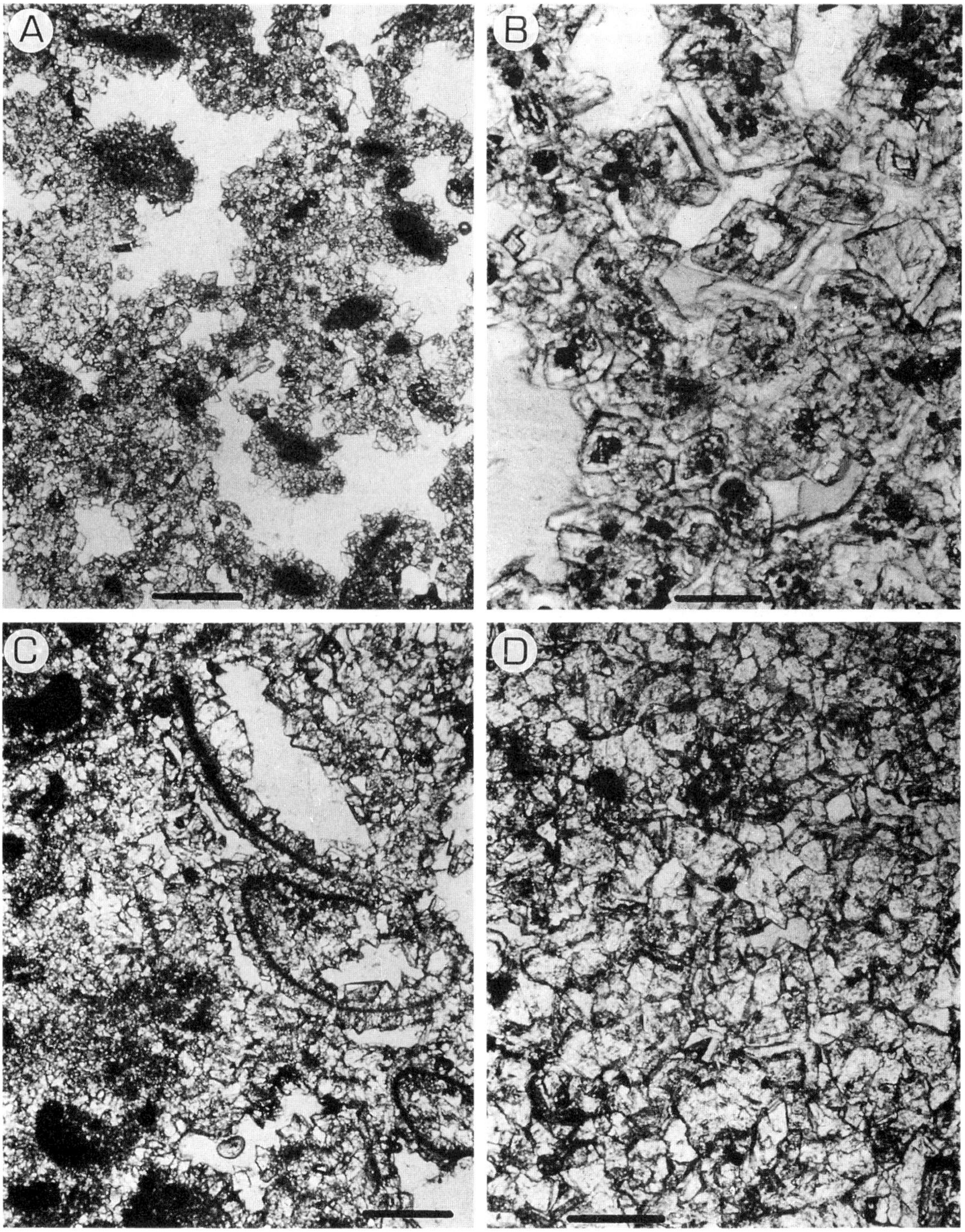

Fig. 3. Divers dolomite fabrics at Mururoa Atoll. (A) Highly porous replacement dolomite whose porosity is essentially mimetic. Scale=0.2 mm. (B) Porous replacement dolomite with clear rims of incipient dolomite cement. Scale=0.1 mm. (C) Dolomite cement in vuggy crystalline dolomite with well developed cement. Scale=0.2 mm. (D) Planar dolomite fabric composed essentially of cement which has occluded most pore spaces. Scale=0.2 mm. Facies C and D are typical of the peripheral (oceanic) parts of the dolomite body.

Mururoa Atoll, French Polynesia

The nature and geometry of the sediments and their diagenetic modifications have been established by Buigues (1982) and by Aissaoui *et al.* (1986a). The dolomite constitutes an annular zone (Fig. 2) fringed by relatively coarse non-dolomitized carbonate sands and silts. The dolomite body itself comprises a series of distinct fabrics and related porosities. The dolomites, whose pore spaces are essentially inherited, are localized mainly within the interior parts of the platform where dolomite has replaced relatively fine-grained porous carbonate. The C and O isotopic compositions of the porosity-preserving dolomites are essentially marine. Averaging 30% porosity, these friable white chalky dolomites pass laterally into progressively denser dolomites towards the periphery of the platform, the decrease being due essentially to the excessive development of dolomite cements (Fig. 3).

At Mururoa, the highly porous dolomite essentially reflects the petrophysical properties of the precursor sediment. However, this depends on scale: originally dense molluscan grains have become chalky as a result of increased porosity. The less-porous peripheral dolomite, in spite of predolomite submarine cementation, has lost at least 10% porosity during subsequent dolomite cementation. Thus, porosity in the dolomites at Mururoa Atoll, although locally inheriting the initial sedimentary fabric, does not correlate simply with the antecedent limestone porosity on a platform-wide scale.

Abu Shaar platform, SW Gulf of Suez

Miocene carbonates have accumulated on a fault block created by extensional rift tectonics (Fig. 4). Dominating the coastal plain bordering the southwest Gulf of Suez (Egypt), the carbonates comprise both platform sands and muds, peripheral reefs exhibiting much dissolution, and spectacular talus deposits, the latter highly affected by Miocene submarine cementation (Aissaoui *et al.*, 1986b). All sedimentary and diagenetic fabrics have been dolomitized. Unlike Mururoa Atoll, the dolomites at Abu Shaar are petrographically homogeneous, virtually all being microcrystalline mimetic replacements of sedimentary and early diagenetic fabrics which are perfectly preserved. The dolomites have variable but generally negative isotopic signatures, suggesting dolomitization from essentially nonmarine fluids (Aissaoui *et al.*, 1986b).

The highly porous dolomites at Abu Shaar, as at Mururoa, occupy most of the inner platform. Porosity may exceed 30%, which is close if not identical to that of the precursor sediments (Fig. 5A). Laterally, porous platform dolomites pass rapidly into less porous platform periphery facies and then into dense (10% porosity) dolomites forming the adjacent slope deposits. The low porosity in these latter reflects the presence of dolomitized fibrous submarine cements whose fabric is perfectly preserved (Fig. 5C) in spite of the mineralogical evolution. Thus, the nature and distribution of porosity in the dolomites of this Miocene platform are a direct expression of the precursor fabric, whose variations are the result of both sedimentary texture and early predolomite diagenesis. This pore system has not been modified significantly by dolomitization. It differs from that at Mururoa Atoll, where the dense dolomite, also occupying the periphery of the platform, is the result of dolomite cementation whose isotopic signatures are essentially marine.

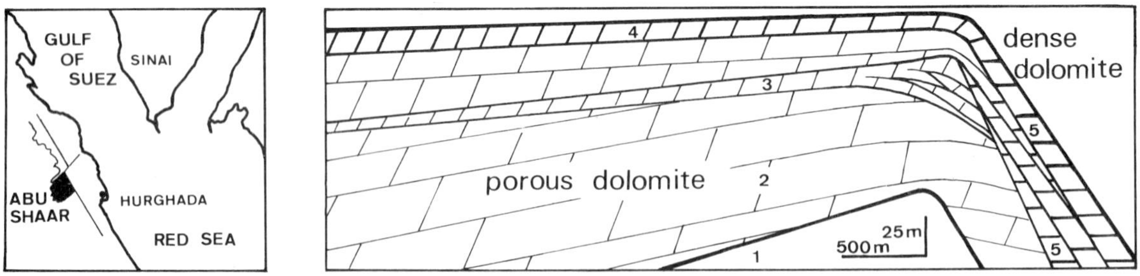

Fig. 4. Schematic profile across the Miocene Abu Shaar platform, southeast Gulf of Suez. This block, which is tilted towards the periphery of the rift, has multiple sedimentary and diagenetic discontinuities. Virtually all carbonates are dolomitized. 1, Palaeozoic basement; 2, porous mimetic dolomite (see Fig. 5A); 3, reef body and talus; 4, dense dololaminite; 5, dense dolomite mimetically replacing marine-cemented slope sands.

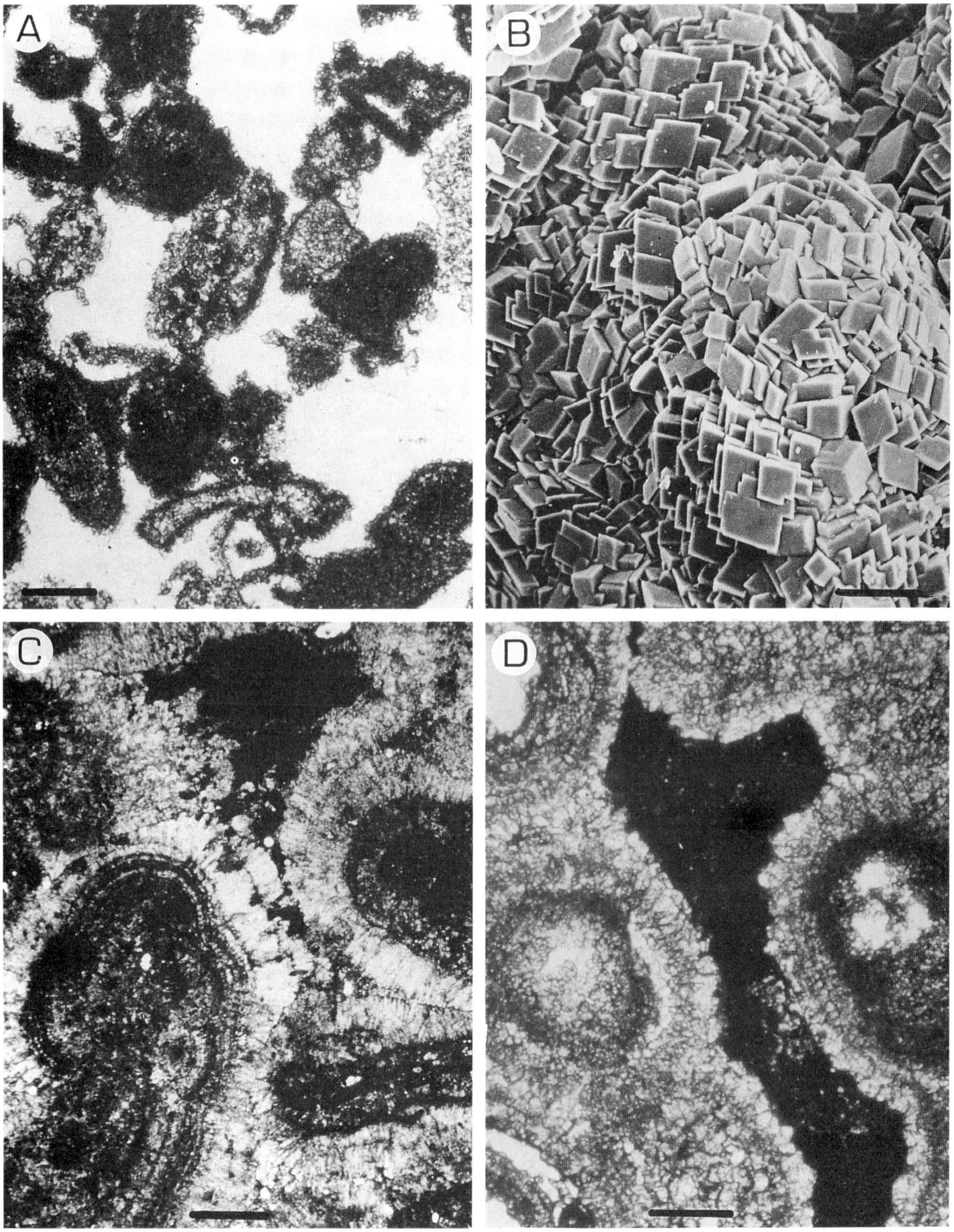

Fig. 5. Fabric-replacive porosity in dolomites. (A) Pure dolomite whose pore system (white) is essentially that of the precursor grainstone. Miocene, Abu Shaar platform, southeast Gulf of Suez. Scale = 1 mm. (B) Scanning electron microphotograph of porous mimetic dolomite showing two pore systems: an inherited interparticle porosity between pellets, and a finer intercrystalline porosity within the dolomitized particles, the latter probably being formed during dolomitization. Pleistocene, Maré Atoll, New Caledonia. Scale = 20 μm. Photo D. Carrière. (C) Mimetic dolomite whose dense fabric replaces predolomite fibrous submarine cement in Miocene slope deposits, Abu Shaar platform. Scale = 1 mm. (D) Fabric-replacive bitumen-filled (black) porosity in dolomitized 'Arab D' reservoir, Dukhan Field, Qatar. Scale = 0.3 mm.

Dissolution, vugs and moulds

Most dolomites, including those forming in sabkhas of the Persian Gulf (B.H.P., personal observations), exhibit a secondary porosity generally due to the dissolution of skeletal components. Vugs and moulds (Choquette & Pray, 1970) are typical of most dolomite reservoirs, including the prolific Upper Jurassic 'Arab' zones of Arabia, the Devonian of Alberta (Mattes & Mountjoy, 1980) and the Silurian dolomites of the Michigan Basin (Huh *et al.*, 1977). In many dolomites, vugs and moulds are the major void types. Their economic importance generally depends on the presence of associated pore systems, including fractures.

There exist three possible relationships between dolomitization and the development of vuggy porosity: obviously, secondary voids may predate dolomitization, they may be formed during dolomitization and thus may be part of that process, or they may postdate dolomite formation. The distinction is generally possible.

Predolomite vugs and moulds

Dissolution voids may be formed during emersion of the buildup prior to dolomitization and thus be one of the many expressions of major sequence boundaries. Emersion may have resulted in the selective dissolution of aragonitic skeletons by meteoric or mixed waters, as in the Miocene carbonates of Abu Shaar platform, where initial porosity is both primary and mouldic. Most corals and molluscs are dissolved, whereas associated coralline algae and foraminifera are not. That this dissolution occurred prior to dolomitization (and not during it) is indicated by the fact that, although all carbonates are now dolomite, dissolution is variable, being best developed on the crest of the platform where secondary voids are filled locally with fine internal sediment. Ancient examples are common, such as Yates Field, Permian, west Texas, (Craig *et al.*, 1986), and Vacuum Field, San Andreas Formation, Permian, southeast New Mexico (Purves, 1986).

Syndolomite vugs and moulds

Where voids are localized only within the dolomitized part of the section there may be a genetic relationship between dissolution and dolomite precipitation. This may be demonstrated by comparing limestones and dolomites across intervening boundaries, where the presence of mouldic porosity only within the dolomites, as shown in crinoidal carbonates by Murray and Lucia (1967), may indicate its syndolomitic origin. However, the contemporaneity of these two processes is difficult to prove convincingly. Two possible examples are given.

At Mururoa, where vuggy and mouldic porosity is ubiquitous (most probably being of predolomite origin), the local paragenetic sequence begins with a fibrous Mg-calcite cement. This phase is followed by dissolution of aragonitic bioclasts and specific laminae within the multiple layers of submarine cement (Aissaoui *et al.*, 1986a). The first traces of dolomite may appear within various microfacies, including these isolated dissolution cavities (see Fig. 13A). It is possible that part of the carbonate liberated during dissolution of the aragonite is incorporated into the adjacent dolomite crystals, but proof of the exact contemporaneity of these two processes (dissolution and dolomitization), admittedly, does not exist. Dissolution and dolomitization may be followed by precipitation of sparitic calcite cement.

A second example concerns Middle Jurassic carbonates of Burgundy (France) where stratiform dolomites locally replace both oolitic barrier sands and inner platform peloids of Bathonian age (Purser, 1975, 1985). The dolomite units (Fig. 6) generally underlie well developed discontinuities locally associated with vadose fabrics. These Bathonian dolomites are invariably associated with leached skeletal debris. Interestingly, dissolution, although not totally lacking within the limestones, is best developed within the dolomites, notably in an abandoned quarry situated within vineyards about 3 km south of Meursault. The interrhombohedral pore spaces and the large (1–5 mm) moulds are partially filled with postdolomitic sparitic cement, probably of telogenetic origin. Again, there is a close temporal association between dolomitization and dissolution.

The Ordovician dolomitization associated with regional fracturing, such as the Scipio-Albion Trend in Michigan, is an example of late dolomitization forming vugs contemporaneous with dolomitization. This is also characteristic of MVT dolomitization.

Postdolomite vugs and moulds

These generally result from the local dissolution of unstable dolomite or evaporite minerals and, as such, have little to do with the process of dolomitization. However, their existence may be the indirect

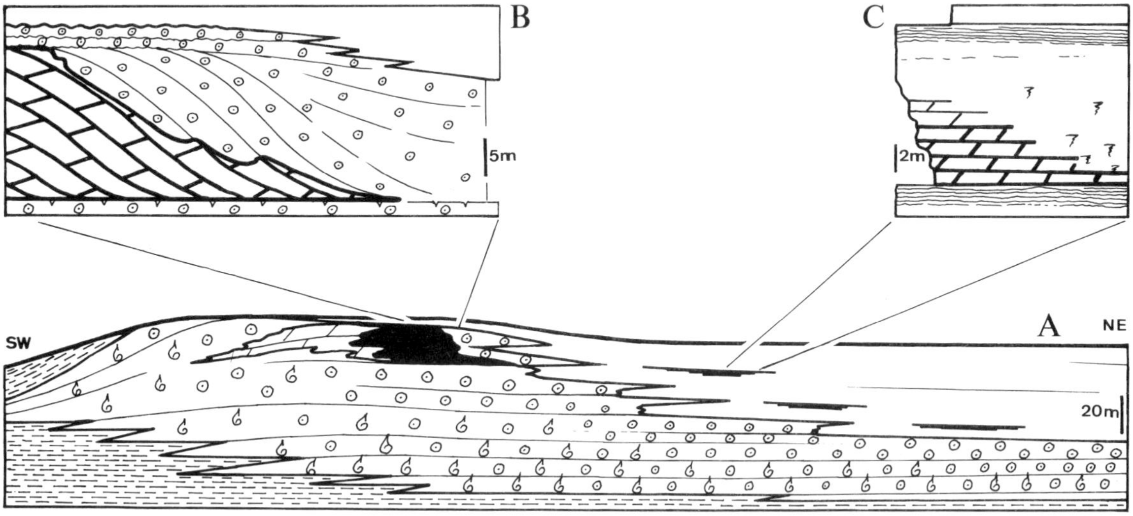

Fig. 6. Schematic section across the southeast periphery of a Middle Jurassic platform in the southeast Paris Basin, showing the relationships between oolitic barrier sands and dolomite geometry. Note that dolomite replaces both the inner parts of the oolite barrier and tidal-flat units within the lagoon. Black indicates ferroan dolomite.

result of dolomite formation. In common with smaller-scale dissolution of individual dolomite rhombs (see below), this postdolomite dissolution may contribute to reservoir potential. Two examples are given.

The Upper Jurassic 'Arab D' reservoirs of Saudi Arabia and Qatar comprise some the largest fields in the Middle East (Powers, 1962). They are multiple reservoirs and include carbonate sands as well as sucrosic and microcrystalline dolomites. These latter, within the Dukhan Field (Qatar), although porous, have relatively low permeabilities due to the finely crystalline nature of the dolomite. Porosity is enhanced by the presence of numerous millimetre-sized vugs and moulds (Fig. 7C) which have been created by the dissolution of a syndolomite mineral – probably anhydrite or halite – by aggressive formation waters. Similar evaporite dissolution is common in evaporite-associated dolomites in the western USA.

A second example concerns small Middle Jurassic reservoirs of the Paris Basin (Purser, 1978, 1985), whose porosity is locally due to the dissolution of unstable dolomite (Fig. 7B). Groups of rhombohedra, localized within a calcite matrix, are dissolved; the resultant plurimillimetre vug may exhibit only vague traces of the initial crystal forms. Because postdolomite vacuoles may be filled with Jurassic internal sediments, their origin closely follows the formation of the precursor dolomite.

In summary, vugs and moulds are typical of dolomite reservoirs and, although contributing little to the permeability of the reservoirs, nevertheless may play an important role in terms of total reserves. Providing that they occur within a permeable matrix, as is often the case, their contribution to recoverable reserves may be considerable.

Intercrystalline porosity

This type of porosity is the most typical of dolomite reservoirs and its origin has often been assumed to result from the dolomitization process. Because of the frequently coarse crystallinity and well developed porosity, it generally coincides with the almost complete destruction of the initial sedimentary fabric.

Porosity and permeability obviously depend on the intensity of crystal nucleation, e.g. the number of crystals per unit volume and their size. Many dolomites which have replaced tidal-flat muds have a high crystal density which, if not destroying porosity, results in low permeability. Porous, coarsely crystalline dolomite very frequently has replaced a carbonate whose initial fabric may occasionally be observed under reflected light. Two examples are given. The first concerns the Upper Jurassic 'Arab

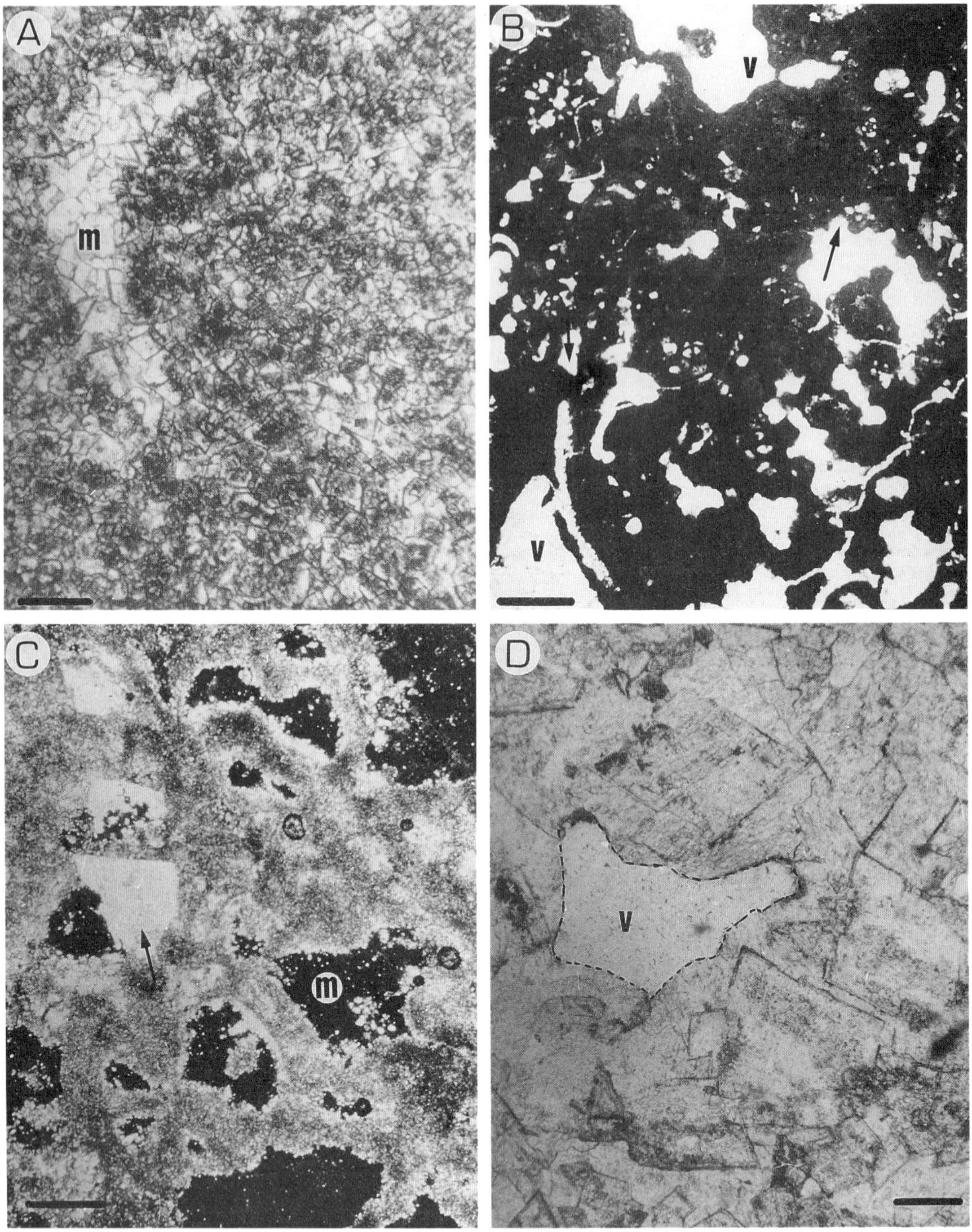

Fig. 7. Postdolomite moulds and vugs. (A) Planar replacement dolomite enclosing a fossil mould (m) partially filled with clear dolospar cement. The mould is possibly syn- or even predolomite in origin. Upper Jurassic, Jura Mountains. Scale=0.25 mm. (B) Peloidal packstone with numerous irregular vugs (v) showing traces (arrows) of dissolved dolomite crystal faces. Middle Jurassic, Coulommes Field, Paris Basin. Scale=0.5 mm. (C) Fine replacement dolomite with numerous moulds (m) resulting from dissolution of dolomite and/or gypsum crystals (arrow). Upper Jurassic, 'Arab D' reservoir, Dukhan Field, Qatar. Scale=0.5 mm. (D) Postdolomite vug (v) whose limits cut planar dolomite crystals. Middle Jurassic, Meursault, Burgundy. Scale=0.3 mm.

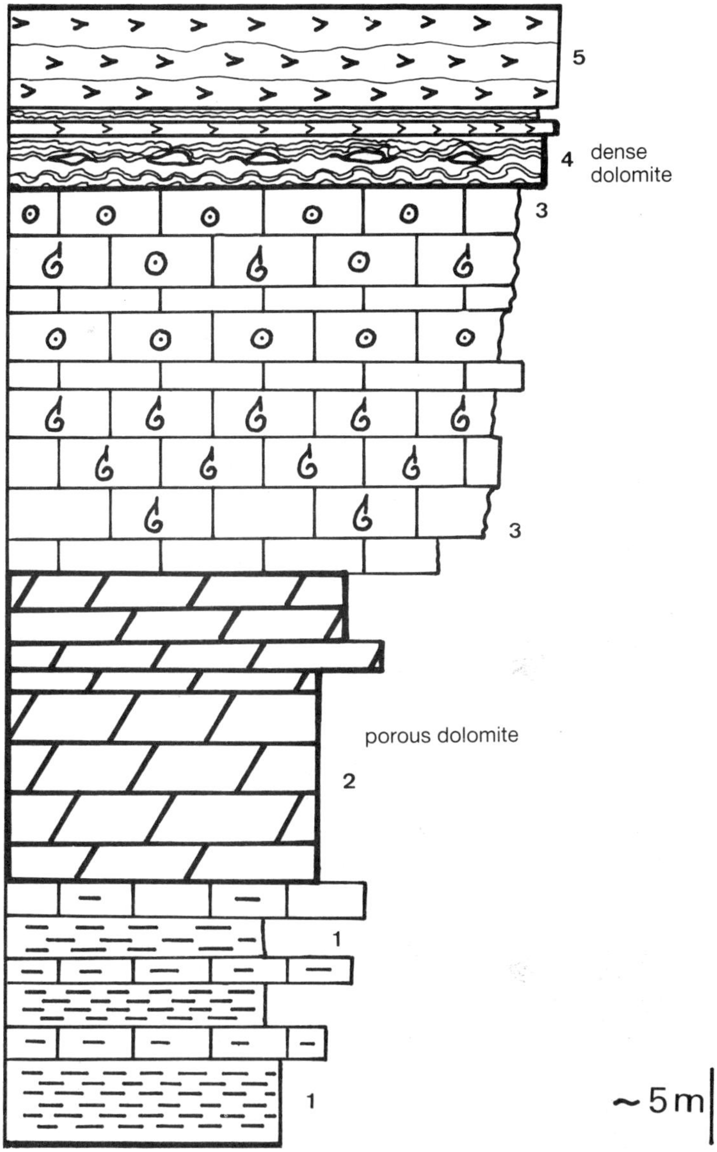

Fig. 8. Schematic log of Upper Jurassic 'Arab D' cycle, typical of Qatar and southeastern Saudi Arabia, showing the stratigraphic distribution of the main dolomite units. 1, Open-marine shale and argillaceous limestone; 2, porous dolomitized open-marine lime mud/ wackestone (see Fig. 9A); 3, porous skeletal and oolitic grainstone; 4, dense dolomitized sabkha-type laminites with anhydrite nodules; 5, massive (Hith) anhydrite.

D' dolomites of Dukhan Field, Qatar. These prolific reservoirs constitute part of a classic shallowing-upwards sequence (of which there are four), measuring about 50 m in thickness (Fig. 8). The lower and middle parts of the sequence are burrowed carbonate muds, which may be intensely dolomitized. The dolomite (Fig. 9A) is very porous, with a well sorted, finely sucrosic fabric. This dolomite is overlain by porous undolomitized carbonate grainstones, the sequence terminating with laminated dolomite mudstone with evaporites and stromatolites. These latter dolomites are finely crystalline but dense. In this example, burrowed mudstone has developed excellent intercrystalline porosity, whereas the laminated upper dolomite body has very low porosity. This contrast is mainly the result of fewer crystals (per unit volume) within the burrowed mudstones, where porosity remains because of incomplete crystal growth. In the dense upper unit, complete growth of the more numerous crystals has occluded porosity, although the ultimate crystal is smaller.

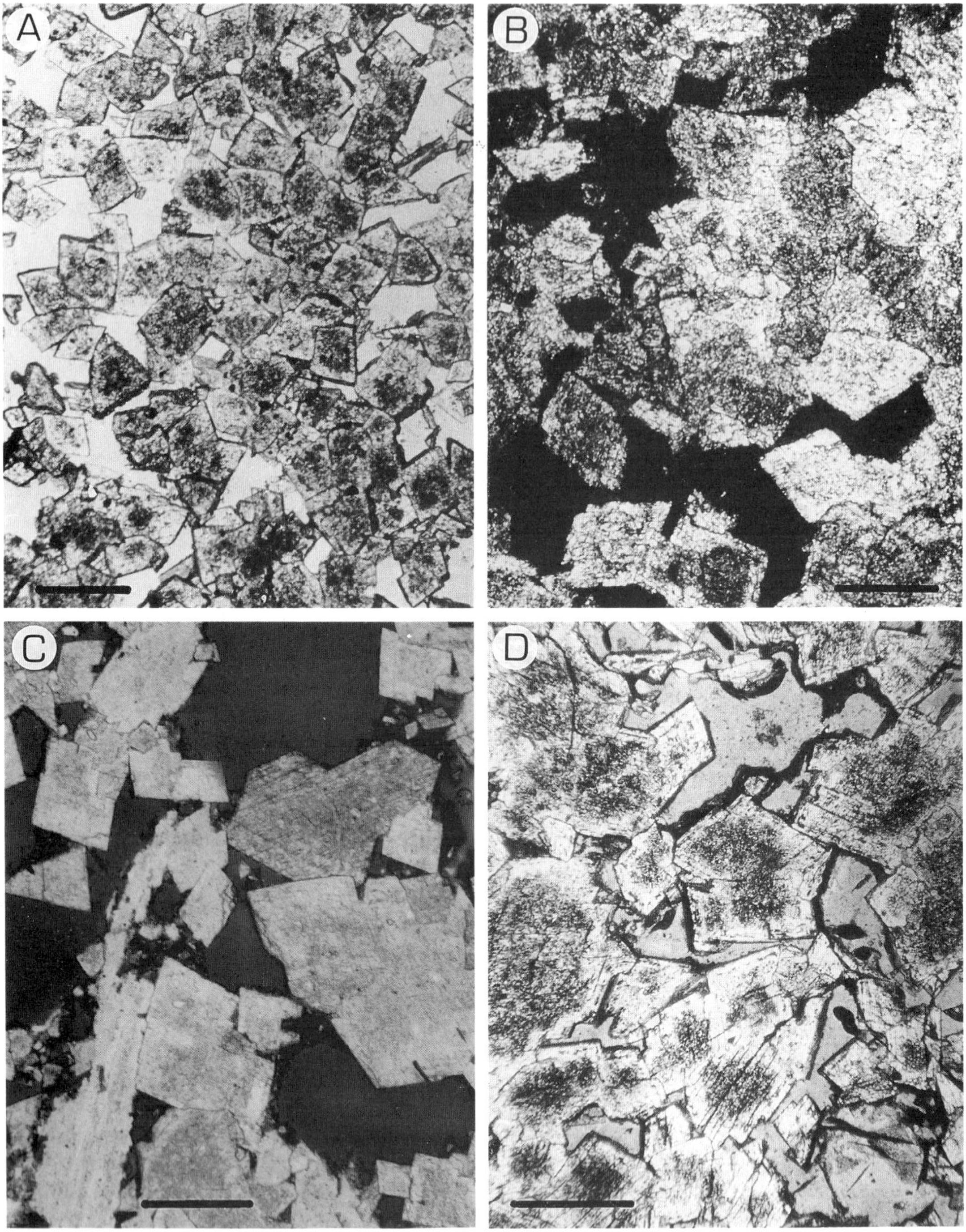

Fig. 9. Intercrystalline porosity. (A) Saccharoidal dolomite whose component crystals exhibit cloudy centres and clear rims, the latter best developed adjacent to pore spaces. 'Arab D' reservoir, Dhukhan Field, Qatar. Scale=0.2 mm. (B) Planar dolomite whose crystal faces show traces of dissolution. Lower Cretaceous Pinda Formation, Angola. Scale= 0.5 mm. (C) Euhedral dolomite with a residual silicified brachiopod (lower left). Part of the intercrystalline porosity is due to dissolution of the brachiopod. Ordovician, Scipio-Albion trend, Trenton Formation, Michigan. Scale=0.3 mm. (D) Planar dolomite crystals whose well developed limpid peripheries have considerably cemented the intercrystalline pores. Cretaceous reservoir, Syria. Scale=0.5 mm.

The second example concerns the Middle Jurassic (Bathonian) dolomites of Burgundy (already discussed). These 10–15 m thick dolomites form a lenticular body localized along the lagoonal side of a periplatform oolite barrier complex (Fig. 6). The dolomitized sands, which include various dissolution moulds, are exceedingly porous and permeable, to such a degree that the dolomite may be sampled with a shovel. The average crystal size in these replaced grainstones is about 0.5 mm, considerably larger than the crystal size of the 'Arab D' dolomitized mudstones discussed above (which have replaced carbonate muds).

Many examples of highly porous sucrosic dolomites could be given. They include part of Mururoa Atoll, the Miocene fields of Iran, parts of the Eocene in Central Tunisia, the Lower Cretaceous Pinda Formation of Angola (Fig. 9B), etc. They also include many Palaeozoic dolomites of North America (Fig. 9C): Yates Field, Permian, west Texas (Craig *et al.*, 1986), Whitney Field, Lower Carboniferous, southwest Wyoming (Harris *et al.*, 1988), Red River Formation in eastern Montana and western North Dakota, etc. However, in a number of these Palaeozoic examples, porosity is relatively low (12–20%) compared with those of younger, Mesozoic and Tertiary dolomites of Europe and the Middle East. Loss of porosity in many Palaeozoic dolomites would appear to be the result of diagenetic evolution involving both crystal growth and recrystallization. The origins and evolution of intercrystalline porosity are discussed further below.

Intracrystalline dolomouldic porosity

Secondary porosity resulting from partial or complete dissolution of dolomite rhombohedra is very common in Mesozoic and younger dolomites, particularly those enriched with iron (Evamy, 1967; Al-Hashimi & Hemingway, 1973). This dissolution, typical of dolomite outcrops, is generally the result of near-surface telogenetic diagenesis. In Quaternary dolomites, dissolution may also affect calcium-rich zones within the crystal (Ward & Halley, 1985). That this dolomite dissolution is not limited to near-surface dolomites is amply demonstrated by its importance in a number of Mesozoic fields, two examples of which are given.

In common with a number of peri-Atlantic Cretaceous carbonates, the Lower Cretaceous Pinda Formation of Angola is extensively dolomitized (B.H.P., personal observations). Reservoirs are characterized by rapid petrophysical changes due in part to the presence of a very heterogeneous dolomite pore system which includes both vuggy, intercrystalline and intracrystalline voids. This is particularly true where ferroan dolomite has selectively replaced large ooids. The relatively sombre inclusion-rich dolomite has been followed by a limpid dolospar which cements the dolomitized ooids. Subsequent dissolution has affected only the initial inclusion-rich dolomite, resulting in a dolomouldic porosity (Fig. 10A). Intracrystalline pores (Fig. 10B) have resulted from the dissolution of dolomite cores, certain crystal lamina or the total dissolution of rhombohedra, giving a dolomouldic porosity comparable to that described by Evamy (1967). It is not possible to quantify this type of dolomouldic porosity within the Pinda Formation, but in certain levels it is the predominant pore type. The permeability of this dolomouldic porosity depends largely on the original sedimentary texture: where ooids were grain-supported the resultant vacuoles, each composed of partially dissolved rhombohedra, are permeable, but where initial textures were matrix-supported the dissolved dolomite forms isolated oomouldic patches floating in an impermeable dolospar cement.

The Middle Jurassic carbonates of the Paris Basin and Burgundy constitute a large platform comprising both peripheral oolite and inner-platform peritidal peloidal sands and muds (Purser, 1985). Palaeotidal flats are underlain locally by stratiform dolomites, which may attain 5 m in thickness (Fig. 6C). This dolomitization is generally incomplete. Dissolution of unstable ferroan dolomite results in rhombohedral voids and vugs, as already discussed. In several fields, including Coulommes, these voids have distinct crystal forms clearly expressing the existence of a dolomite precursor (Fig. 10C). In most cases, however, the millimetre-sized moulds, which cut adjacent grains, are more ovoid in form and their origin is less evident. Careful inspection may reveal flat inframillimetric surfaces or pointed indentations which are probably the remnants of rhombohedral moulds. That this dissolution is a result of Jurassic diagenesis is confirmed not only by its occurrence at depth (−1500 m), but also by the local infilling with Jurassic internal sediment, as already noted. It would seem that both dolomitization and dissolution were both eogenetic processes associated with Jurassic tidal-flat accretion, dissolution reflecting increasing meteoric influences along the landward parts of the tidal-flat system.

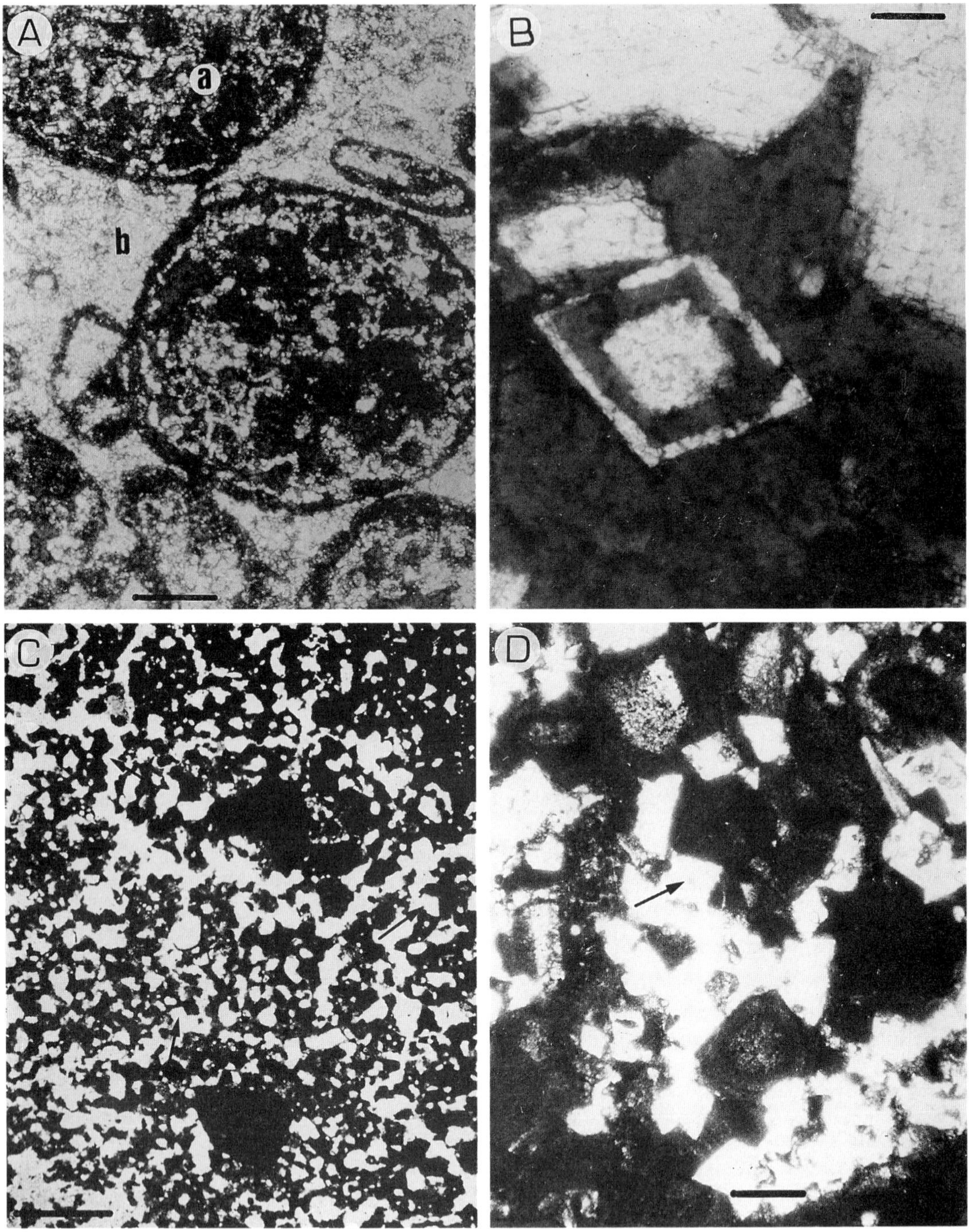

Fig. 10. Intracrystalline porosity resulting from the partial or total dissolution of ferroan dolomite. (A) Totally dolomitized oolitic grainstone comprising two dolomite petrotypes: unstable ferroan dolomite (a), subsequently dissolved, has replaced the ooids, and clear non-ferroan dolomite (b) cements interooid space. Lower Cretaceous Pinda Formation, Angola. Scale=1 mm. (B) Detail of (A) showing the partial dissolution with a ferroan dolomite crystal (dark). Scale=0.1 mm. (C) Dolomouldic voids which locally show crystal forms (arrows); matrix is calcite, white is porosity. Middle Jurassic, Coulommes Field, Paris Basin. Scale=0.5 mm. (D) Rhombohedral pore spaces in oolitic limestone; porosity results from the selective dissolution of unstable ferroan dolomite. Middle Jurassic, Massangis, southeast Paris Basin. Scale=0.3 mm.

Middle Jurassic cross-bedded oolitic and skeletal sands best developed near the periphery of the Burgundy Platform, as already noted, are also dolomitized locally. When this dolomite is non-ferroan, as at Meursault, it resists dissolution and the high porosity is intercrystalline. However, in other parts of the same barrier system the dolomitized grainstone, in common with its tidal-flat equivalents, is highly ferroan and susceptible to dissolution. In large quarries near Massangis, on the southeastern edge of the Paris Basin, some 15 m of Bathonian oolitic grainstone have been partially dolomitized, this being related to a series of local discontinuities within the prograding oolite barrier system (Fig. 6B). Dissolution of gastropods whose moulds are filled with oolite demonstrates that these surfaces were locally emergent. The selective dissolution of ferroan dolomite has produced an extremely porous limestone. Voids are typically rhombohedral (Fig. 10D). Permeability depends on the intensity of dolomitization, which never attained 100%; where this partial dolomitization resulted in rhomb-supported fabric, selective dissolution of the metastable dolomite has resulted in a friable limestone whose permeabilities attain 4 D. Although also present within producing fields, on outcrop at least part of this dissolution may be Quaternary.

The geometry of the porous dolomouldic barrier sands depends on several factors, notably the distribution of ferroan dolomite and the lateral extension of the discontinuities and associated meteoric diagenesis. On a profile across the barrier oolite complex (Fig. 6), the dolomites are seen to have an asymmetric distribution, having maximum thickness on the inner (lagoonal) flank of the barrier. Importantly, only the dolomites on the lagoonal side are ferroan, whose selective dissolution has created the porous body. The open-marine side is composed of non-ferroan dolomite and the rock is dense.

The examples discussed concern partial or total dissolution of metastable dolomite under preburial conditions. However, there exist well documented examples of dolomite dissolution under relatively deep burial conditions (Budai *et al.*, 1984). It is possible that the evolution of organic matter, including oil-source rocks, may be conductive to the dissolution of certain dolomites. In other cases, dolomite may be replaced by anhydrite or other minerals which may dissolve during burial, as in the 'Arab D' zone of Dukhan Field (Fig. 7C). The economic importance of this deep-burial reservoir type remains to be demonstrated.

Fracture and breccia porosity

That dolomite is highly susceptible to fracturation is well known. The preferential fracturation of dolomite relative to limestone, conditioned by the more crystalline, brittle nature of this rock, cannot be treated satisfactorily here. Fractures nevertheless must be included in the list of pore types, not only because they are very important for reservoir development, but also because they are typical of dolomites. In many fields the producibility of dense dolomite (and other rocks) depends on the density of the fracture system (Arguilera, 1980). However, tectonic fractures alone, affecting a rock lacking matrix porosity, are rarely of sustained economic value.

Tectonic fracturation

Fractures have many origins. Those of tectonic origin are predictable and, obviously, are not limited to dolomites. Cataclastic porosity may occur in regions of intense deformation, such as the Overthrust Belt of North America, where coarsely crystalline dolomites may not only be intensely fractured but individual crystals may also be broken (Fig. 11C). Within the Mississippian Madison Group of Wyoming, both at the base of the Darby nappe at outcrop and in producing reservoirs, elevated porosity (10–20%) and permeability are the result of intense dislocation of crystals; within any given thin section many crystals are broken into a cataclastic dolomite fabric (Bureau, 1988). This intense microfracturation, well developed close to individual overthrusts, seems to result from intrastratal adjustment during overthrusting; the dolomite is sufficiently heterogeneous that stress results in the rearrangement within the crystal mass via the breakage and rotation of individual components. Frequently, microfissures affecting individual crystals are healed by a subsequent generation of highly luminescent dolomite. Although some of the subsurface fabrics could be artefacts, the presence of cataclastic fabrics at outcrop clearly precludes this eventuality.

Breccias

Current studies of Miocene sediments of the northwestern Red Sea coast (Plaziat *et al.*, 1990) have shown that stratiform dolomite breccias probably result from Miocene earthquake instability. These highly porous breccias (Fig. 11A,B), locally attaining 10 m in thickness, are enclosed within undis-

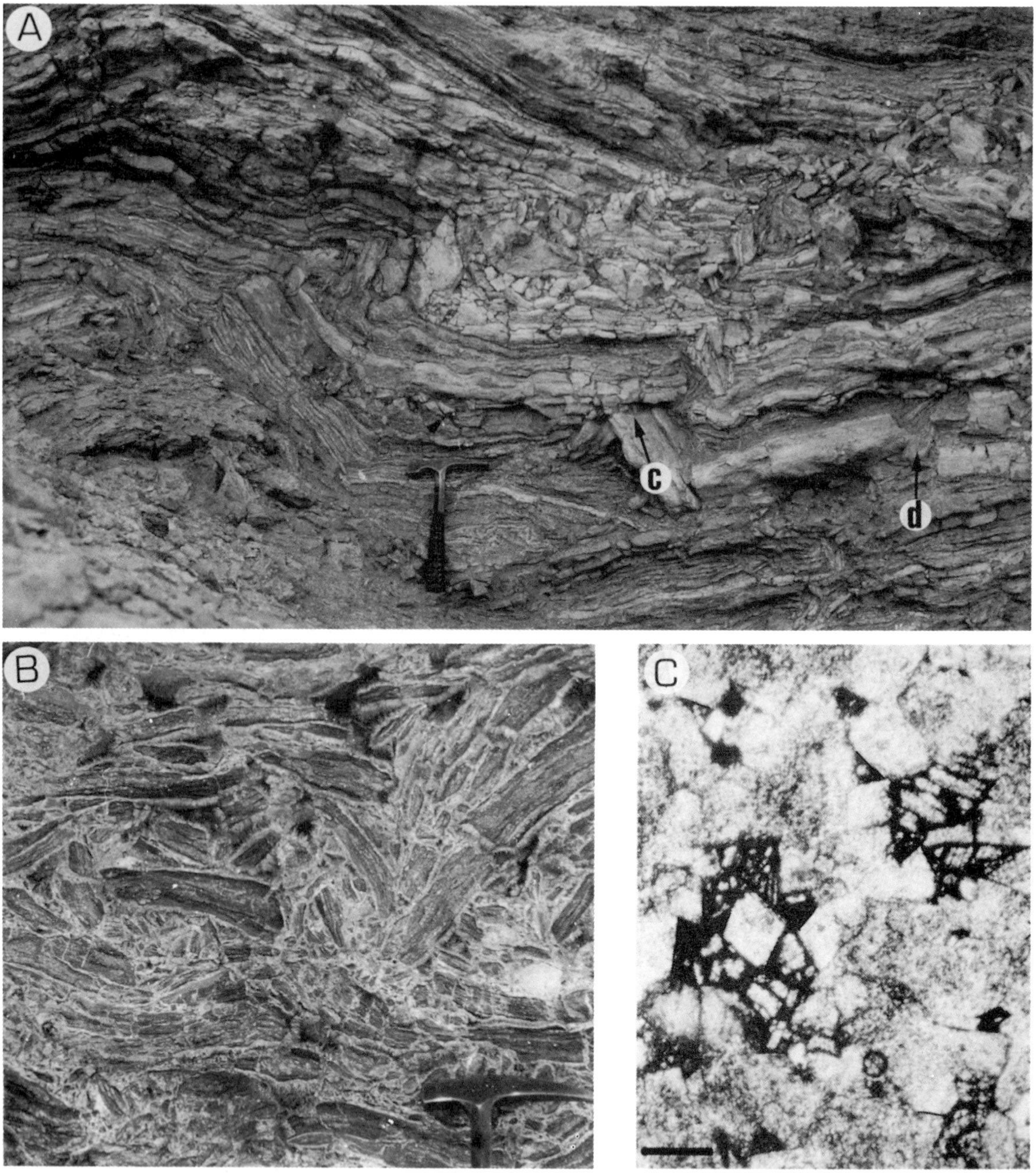

Fig. 11. Fracture porosity. (A) and (B) Field photos of brecciated dololaminites. Note that the deformation typically involves both distensional (d) and compressional (c) displacements, interpreted as the result of palaeoearthquake (seismite) deformation. These multiple units are subhorizontal stratiform phenomena of considerable lateral extension (100 km). Miocene, Wadi Asal, northeastern Red Sea. (Photos J.-C. Plaziat). (C) Pore spaces in cataclastic dolomite fabric resulting from the fracturation of crystals. Base of Darby thrust, Mississippian Madison Group, Wyoming. Scale=0.5 mm; photo S. Bureau.

turbed subhorizontal dolomites and sandstones, confirming their synsedimentary origins. Series of flat clasts exhibit both distensional (opening) and compressional overthrusting and folding. These latter attributes, although occasionally present, are not typical of collapse breccias. Individual breccia zones in Miocene and Pliocene synrift sediments may be followed laterally along the Red Sea coast for more than 100 km, and occur in dolomitized sediments of open-marine and in restricted laminite facies. They appear to record intense episodic deformation of early lithified dolomite crusts, and may be typical of synrift dynamics.

Brecciated dolomite formed by the dissolution of evaporites is a well documented feature of many Palaeozoic strata in North America (Simpson, 1988). Breccias associated with collapse during Mississippi Valley-type dolomitization (Ohle, 1985) also occur in some reservoirs, such as the Ordovician reservoirs of the Scipio-Albion Field in Michigan. These various types of breccia have much in common, although not necessarily being associated with evaporites, nor with dissolution.

Preliminary conclusions concerning porosity in dolomites

We have demonstrated, very briefly, the considerable variety of well known voids in dolomites. High porosity associated with these pore types is most apparent in relatively young, post-Palaeozoic dolomites, probably because subsequent burial diagenesis modifies or destroys many of the earlier near-surface fabrics (see Fig. 1). If one considers the entire range of pore types, it is clear that most have formed through processes that are only indirectly related to, or even totally independent of, the processes of dolomitization.

Intercrystalline and dissolutional pore tyes are by far the most common in hydrocarbon reservoirs. These voids can result from a number of different diagenetic processes occurring either before, during or after the major dolomitizing event. The porosity balance associated with the formation of these dolomite pore types merits additional appraisal.

POROSITY EVOLUTION AND DOLOMITIZATION

This discussion concerns the evolution of intercrystalline and dissolution porosity, typical of most dolomite reservoirs. As already noted, the basic questions concerning the origin of this porosity are:

1 Does dolomitization play a positive role by creating porosity during the transformation of limestone into dolomite?

2 Does it play a passive role, involving only a rearrangement of existing porosity by dissolution of aragonite or calcite and reciprocal volume-for-volume precipitation of dolomite?

3 Does dolomitization play only a negative role, tending to destroy pre-existing porosity?

As will be discussed, dolomitization may play all three roles, sometimes within the same dolomite body. Dolomite replaces calcite or aragonite, but this replacement does not imply, *a priori*, an increase in porosity due to the differences in molar volume of the precursor mineral and dolomite. First, dolomitization does not *necessarily* involve a molecule-for-molecule replacement conserving CO_3. Delivery of external Mg is required for complete dolomitization and a system open to Mg exchange for Ca is not likely to be closed to additional Ca and CO_3. Rarely, conditions approaching closure to CO_3 occur where the total concentration of carbonate species in the porewater is very small, the Mg concentration is high and the water flow through the system is such that the amount of CO_3 delivered is small compared to the amount of Mg. Secondly, the chemical process of replacement, by itself, does not necessarily imply that a rigid framework is present. Thus, the difference in molar volume of CO_3-conservative dolomitization which may, theoretically, result in porosity creation, could also favour compaction.

Theoretically, there are three possible evolutions of porosity, depending on the precise composition of the dolomitizing fluid (Fig. 12).

1 If fluids are such that dissolution of the precursor equals precipitation of the dolomite, then there is no change in porosity, although porosity may be rearranged and permeability increased. This balance is highly improbable because, as already noted, the process is an open system involving an external source, at least for Mg (Murray, 1960; Weyl, 1960). However, a thin-film replacement front during dolomitization provides a volume-conservative boundary condition. Theoretically, no greater volume of material can be precipitated than there is room created by dissolution. In fact, the situation is more complicated because of the presence of predolomite voids into which excessive dolomite may grow. Furthermore, as noted in a preceding paragraph, dissolu-

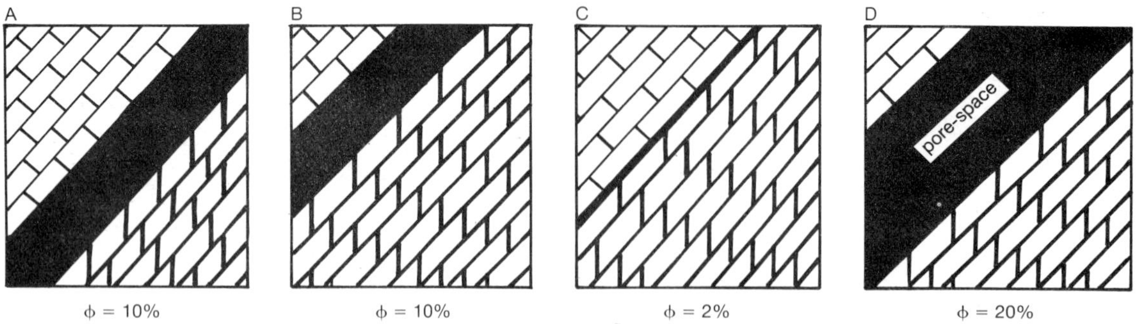

Fig. 12. Simple schematic illustration of three possible evolutions of porosity as a function of relative dissolution and precipitation, depending mainly on the quality of the dolomitizing fluids.

tion of calcite and precipitation of dolomite, in spite of their differences in density, will not necessarily increase porosity because the dissolution/precipitation is not necessarily a mole-for-mole replacement. The presence of volume-for-volume replaced grains may not indicate that the entire rock has been replaced in a volume-for-volume manner, because excess material may have been imported by fluids or porosity may have been created at other sites within the rock.

2 If, on the contrary, rates of dissolution are exceeded by the rates of dolomite precipitation, notably when all material is derived from an external source, clearly any pre-existing porosity will diminish as the dolomite becomes cemented.

3 Conversely, if the volume of precursor carbonate dissolved exceeds the volume of dolomite precipitated, and the rock does not compact enough to account for this volume change, porosity will increase.

Porosity created by dolomitization

If one examines by scanning electron microscopy divers peloids and bioclasts in dolomitized Plio-Quaternary sediments, virtually all are seen to be microporous, *irrespective* of their initial microstructure (Fig. 13). In certain cases, initially dense molluscan skeletons have been transformed into microporous dolomite, the increase in porosity being evident. Nevertheless, the possibility that this porosity was created before dolomitization during the inversion of metastable carbonate into calcite admittedly cannot be denied, although the absence of an intermediate calcite phase in many cases (Fig. 13) militates against this eventuality.

There is also the important problem of scale. Dolomitization of initially dense particles may lead to a microporous texture expressing a local increase in porosity. However, even within the same sample, the growth of identical dolomite crystals into primary microvoids clearly results in a local decrease in porosity. This same problem concerning the redistribution of porosity probably occurs on all scales, and the overall balance is virtually impossible to evaluate.

At Mururoa Atoll, porosity visible under the petrographic microscope is seen to be inherited from the primary sedimentary fabric, as already noted. Thus, at first glance, dolomitization has neither created nor destroyed significant porosity. However, under scanning electron microscopy (Fig. 13) virtually all particles are seen to be porous, indicating a probable increase in porosity on a local scale. The importance of this intragranular porosity may be considerable. Minute dolomite crystals appear initially within secondary voids (Fig. 13A) whose origins could, admittedly, be independent of dolomitization. Whatever their origin, their presence has increased porosity. That at least part of this dissolution is contemporaneous with dolomite growth is indicated by the reciprocity between the two processes: as the aragonite dissolves the dolomite occupies the space created (Fig. 13C). In many voids, the secondary pore space exceeds the volume of adjacent dolomite crystals (Fig. 13A) (i.e. porosity has increased) whereas in other cases voids are progressively filled by rhombohedra, dolomite crystal growth clearly decreasing pore space (Fig. 13C,D). In other cases, a very narrow void separates biogenic aragonite from 'replacement dolomite' (Fig. 13B). In sum, the dissolution/precipitation

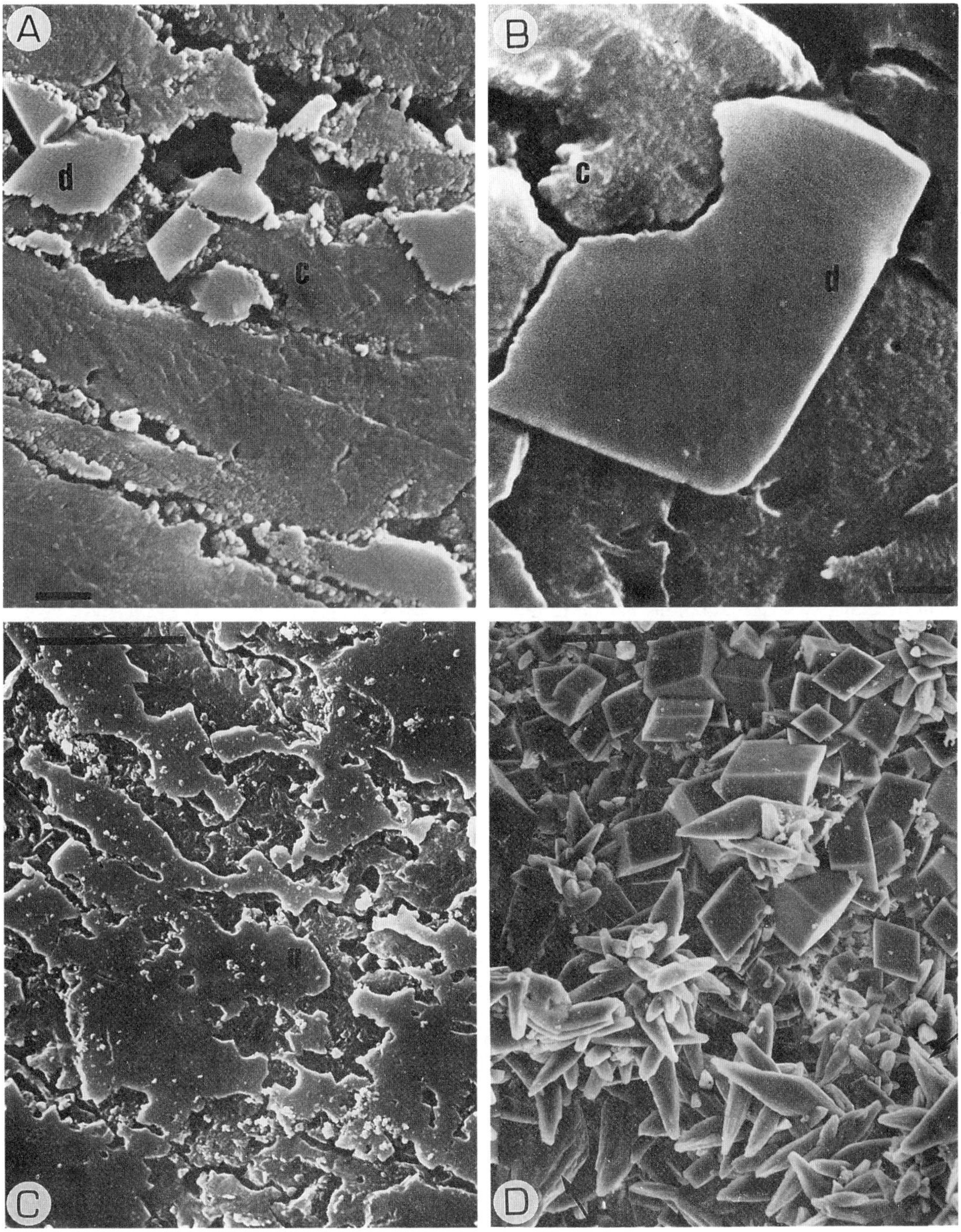

Fig. 13. Porosity evolution in Plio-Quaternary carbonates at Mururoa Atoll. (A) Dolomite crystals (d) form initially within microdissolution voids affecting an aragonitic mollusc (c); porosity is created, although this could have been formed before dolomitization. Scale=2 μm. (B) Dolomite crystal (d) in bioclast (c), a microvoid separating the two phases suggesting syndolomite dissolution. Scale=5 μm. (C) Advanced stage of dolomite replacement (d) progressively occupying (and reducing) pore space. Scale=20 μm. (D) Euhedral dolomite and scalenohedral calcite crystals in primary pore space. The diagenetic calcite shows traces of dissolution (arrows). Scale=2 μm.

fabrics are variable. The very frequent association of microdissolution phenomena with initial stages of dolomite growth suggests that the process of dolomitization may increase porosity, at least temporarily.

The situation concerning permeability is less complex than that of porosity. Dolomitization frequently increases the size of component particles, notably where the process affects micrite. Thus, a porous but relatively impermeable lime mudstone may be transformed into a porous and permeable dolomite, an example being given by Negra *et al.* in this volume.

Some shallow-buried Cenozoic rocks seem to support the concept of porosity creation during dolomitization. Beach (1982) collected porosity data for Plio-Pleistocene limestones in the shallow subsurface of the Bahamas. His data are derived from core-sized samples which do not tabulate cavernous porosity represented by poor core recovery. The data can be tabulated to compare the average matrix porosity of limestones and dolomites having different depositional fabric. These data indicate that the average matrix porosities of dolomites (calculated by A.B.) are consistently higher than the average matrix porosities of limestones, although the average porosity of both limestone and dolomite is less than that of the primary sedimentary porosity:

Texture	*Dolomite*	*Limestone*
Overall (no particular class)	28.4%	19.0%
Packstone	25.9%	16.6%
Wackestone	40.5%	15.1%

Porosity destroyed by dolomitization

Most dolomites, irrespective of age, owe their crystalline fabric to pore-destroying dolomite cement. Pre-existing porosity, whether this be inherited primary or interrhombohedral voids, is diminished by the growth of dolomite crystals. If there is a continued supply of dolomite reactants within the system, coarsely crystalline dolomite cement will eventually occlude porosity. Thus, in a porous dolomite body such as that at Mururoa Atoll, or within the Upper Cretaceous carbonates of Central Tunisia (see contribution by Negra *et al.*, this volume), the permeable zones favouring the passage of dolomitizing fluids have become cemented with dolomite.

Furthermore, if dolomitization is a surface reaction-controlled process of relatively short duration, then porosity destruction will depend upon the number of dolomite crystals nucleated in the sediment. Where there are many nuclei, many smaller dolomite crystals form. If the rate of dolomitization is proportional to the total surface of growing dolomite crystals, porosity reduction will be faster in sediments with many nuclei. Interference in growth between closely packed nuclei will result in overall smaller crystal size and the result will be a tightly packed, low-porosity, low-permeability fabric. Inversely, fewer nucleation sites, which do not interfere with each other, will favour larger crystals. Given a finite time of dolomitization, individual dolomite crystals may not grow sufficiently to fill the primary porosity. The ultimate result is a higher-permeability rock with partially preserved primary pores. However, if there is a continued supply of dolomite reactants to a hydrodynamic system, coarsely crystalline dolomite cement will occlude porosity.

The variable relationships between porosity and dolomitization are illustrated at Mururoa Atoll, where a lenticular body of Plio-Quaternary dolomite has been penetrated by many cored wells which demonstrate major lateral and vertical changes in petrographic fabric and related porosities (Buigues, 1982; Aissaoui *et al.*, 1986a). These variations are summarized in Figures 13 and 14. Because of the young age of the dolomite body, variations in porosity and diagenetic fabric are interpreted to be contemporaneous with dolomitization, at least on the geological scale, and are not due to postdolomite recrystallization.

During the earliest stages of dolomitization, dolomite porosities were inherited from the precursor sediment. The first stages of dolomitization 'replaced' constituent particles and probably improved intraparticle porosity. However, this fabric-preserving phase was followed rapidly by continued crystal growth and porosity occlusion, especially towards the periphery of the platform. Because all pre-existing carbonate was already 'replaced' by dolomite, the growth of dolomite cement depended on the external supply of reactants and thus on water circulation. Furthermore, because the Mg must have an oceanic source, a centripetal flow is envisaged, oceanic waters moving inwards from the

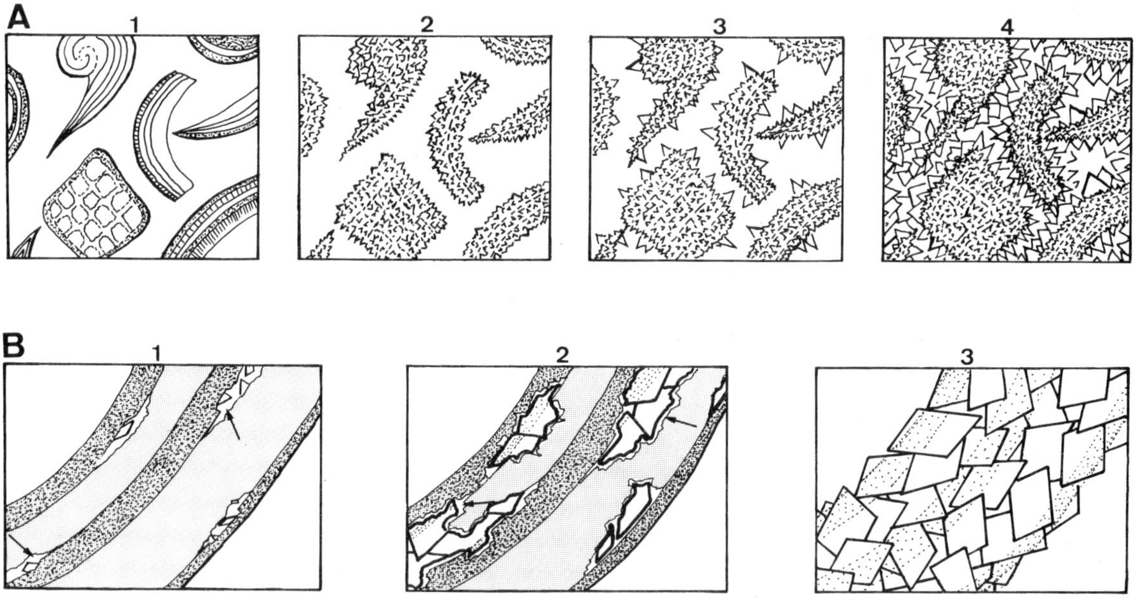

Fig. 14. Schematic evolution of dolomite and related porosity at Mururoa Atoll. (A) As observed under petrographic microscope, 1 and 2 indicating 'replacement' followed by (3 and 4) cementation, the latter towards the periphery of the atoll. (B) As seen under scanning electron microscope, indicating progressive dissolution of an aragonitic mollusc and contemporaneous growth of 'replacement' dolomite.

periphery of the platform driven either by oceanic currents or by thermal convection systems. The upcurrent peripheral parts of the platform have become over-dolomitized, a term initially proposed by Deffeyes (personal communication). Although this zone of intense dolomite cementation will, in time, progress towards the inner parts of the platform, presumably its advance will be slowed and, eventually, stopped as the upcurrent parts of the dolomite body become impermeable.

In summary, the roles of dolomitization *vis-à-vis* porosity evolution are multiple. As demonstrated by the dolomite lens at Mururoa Atoll and elsewhere, dolomite may preserve and even improve initial sedimentary porosity during the early phases of dolomitization. However, crystal growth does not cease when all local source carbonate has been exhausted. Because the system is 'open', the continued supply of reactants via an active hydrological system stimulates further crystal growth and, eventually, the total destruction of dolomite porosity. Thus, there does appear to be a relationship between the dolomitizing model and the distribution of porosity within it.

PRESERVATION AND DESTRUCTION OF POROSITY IN DOLOMITE

Porosity may be destroyed, both in limestones and in dolomites, either by eogenetic cementation (discussed in the preceding section), by pressure dissolution or by a combination of both. Pressure dissolution not only compacts but may also liberate the carbonate necessary to cement the rock. Both in limestones and in dolomite, the presence of porosity, especially in older, deeply buried formations, may be due to its selective preservation (Purser, 1978).

Selective preservation of porosity during burial

In limestones, porosity may be preserved selectively within sediments affected by near-surface (preburial) lithification, which tends to slow or prevent burial compaction. This may also be the case with dolomite. Because eogenetic dolomitization is generally associated with dolomite cementation, most (if not all) dolomites are buried as rock, and thus may resist mechanical compaction.

In spite of limited documentation, there probably exists a relationship between depth and porosity evolution, both in limestones and in dolomites. Studies by Schmoker and Halley (1982) and by Schmoker (1984) of the Tertiary carbonates of southern Florida indicate that porosities decrease with depth. Their comparative study of porosity evolution in limestones and dolomites indicates that porosity is initially lower in dolomites, but the rate of porosity loss with depth is higher in limestones, so that at depths greater than 2000 m the dolomites have a higher average porosity than limestone.

The same selective preservation of porosity in dolomite may also occur in basins affected by a relatively high degree of structural deformation. In the French Jura fold belt limestones are invariably dense, irrespective of their initial sedimentary textures. Porosity destruction appears to be the result of intense pressure-dissolution conditioned by both lithostatic and tectonic pressures; stylolites are both oblique and parallel to the sedimentary stratification. Within these same dense Upper Jurassic limestones, dolomites often retain an important porosity, these porous zones having fewer stylolites than the adjacent limestones (B.H.P., personal observations).

Dolomites, in general, commonly exhibit fewer pressure-dissolution fabrics than limestones, due to several factors including the less soluble nature of near-stoichiometric dolomite relative to calcite in Mg-rich fluids. Porosity, notably in older or deeply buried dolomites, is a question not only of creation but, especially, of selective preservation. This being the case, dolomite is possibly the only potential carbonate reservoir in deeply buried or highly tectonized formations.

Destruction of dolomite porosity during burial or tectonic stress

That dolomite itself is susceptible to diagenetic modification has been demonstrated many times. This diagenesis may involve the rearrangement of dolomite fabrics (recrystallization), as discussed by Land (1985), or it may involve the cementation of dolomite by other minerals, notably calcite or sulphates.

Dolomite diagenesis during burial most frequently involves the addition of a late mesogenetic dolomite cement, generally characterized by relatively light oxygen isotope signatures and by high-temperature fluid inclusions (Machel & Mountjoy, 1986; Moore, 1989). Cathodoluminescence and petrographic studies show that the dolomite cement may fill fractures and matrix porosity within a given dolomite body. This is frequently the case within Mississippian dolomites of the Overthrust Belt in Wyoming (Bureau, 1988). Similarly, the addition of late dolomite cements within the Devonian Miette reefs of Alberta has played a negative role in porosity evolution (Mattes & Mountjoy, 1980).

In many cases dolomite cementation may be related to fracturing and faulting. These features form conduits, favouring the passage of dolomitizing fluids. Thus, Middle Jurassic dolomites outcropping 5 km northeast of the town of Saïda in Algeria are cut by a major fault system associated with an important cementation of the porous dolomite. Not only is the fracture system cemented with dolomite (Fig. 15B), but the adjacent porous dolomite matrix may be similarly transformed into a dense crystalline dolomite due to excessive growth of constituent dolomite rhombohedra.

Many Palaeozoic and younger dolomites exhibit xenomorphic and other fabrics resulting partly from the recrystallization of pre-existing dolomite (Gregg & Sibley, 1984). Although the effects of this late diagenesis on pre-existing porosity are not well documented, it would seem that the dense interlocking mosaics characteristic of many Palaeozoic dolomites (see Fig. 1) are not conducive to porosity preservation.

Non-dolomitic calcite and sulphate cements

These are frequent agents of porosity destruction in dolomite. Calcite may be related to the presence of fractures, from which it may invade the adjacent porous dolomite. Coloration by alizarine of polished surfaces may demonstrate that the patchy distribution of pore-destroying calcite is related geometrically to the fracture system (Fig. 15C). This is frequently the case within Mississippian dolomites of the Wyoming Overthrust Belt, where the final paragenetic phase is a white sparitic calcite which cements both fractures and matrix porosity within the dolomite. Postfracturation calcite and dolomite cements often constitute the final diagenetic phase in certain Cretaceous dolomite reservoirs in Syria (Fig. 15A)

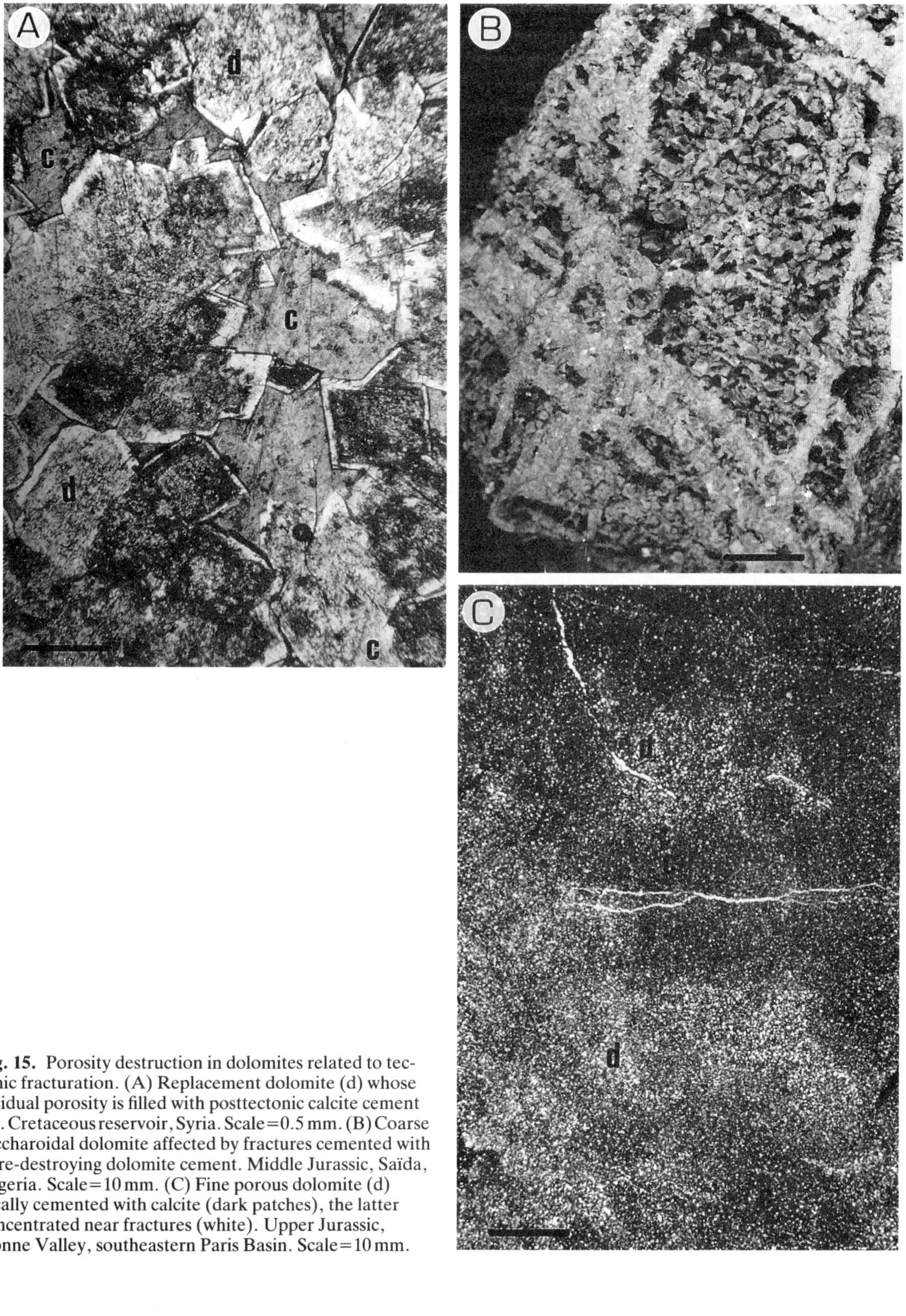

Fig. 15. Porosity destruction in dolomites related to tectonic fracturation. (A) Replacement dolomite (d) whose residual porosity is filled with posttectonic calcite cement (c). Cretaceous reservoir, Syria. Scale=0.5 mm. (B) Coarse saccharoidal dolomite affected by fractures cemented with pore-destroying dolomite cement. Middle Jurassic, Saïda, Algeria. Scale=10 mm. (C) Fine porous dolomite (d) locally cemented with calcite (dark patches), the latter concentrated near fractures (white). Upper Jurassic, Yonne Valley, southeastern Paris Basin. Scale=10 mm.

and in Oxfordian limestones of the southeast Paris Basin (Fig. 15C), suggesting that tectonics play both positive and negative roles in porosity evolution: in creating fractures tectonics obviously enhances permeability, but fracture-associated cements, possibly derived from pressure dissolution of nearby limestones, delivered by fractures, may contribute to the total destruction of the pore system.

DISCUSSION AND CONCLUSIONS

We have briefly reviewed the various pore types that occur in dolomites and have shown that their temporal relationships *vis-à-vis* the dolomitization process are variable. Much (perhaps most) porosity in dolomites is inherited from a precursor sedimentary fabric, and thus has little to do with the process of dolomitization. Some porosity may be created during dolomitization, and certainly much may be destroyed. Porosity may be improved by postdolomite dissolution, whereas porosity in most dolomites may be destroyed by late burial diagenetic processes long after the initial stages of dolomite formation. Obviously the geometry of the porous dolomite body and its location within the basin will depend to a very large degree on its origin. Because this is variable, understanding and prediction will require different tactics according to the type of dolomite reservoir.

Predolomite voids

Inherited from the precursor sediments and their predolomite diagenetic modifications, this type of pore space is generally easy to recognize. Theoretically, prediction involves the understanding of the sedimentary textures and mineralogy; the use of a dolomite model is unnecessary. For some dolomites this may be true. However, for most, an inherited porosity probably will not be true for the entire dolomite body. As demonstrated at Mururoa Atoll, porous dolomites have indeed inherited their petrophysical qualities from parental sediments, but their lateral transition into dense dolomite – and hence the overall geometry of the porous mass – is determined by dolomitization and not by the presence of a less porous sedimentary facies. Nevertheless, even these more complicated relationships, involving both sedimentary fabrics and dolomite cements, are related more or less directly to the overall platform geometry, dolomitizing fluids and their circulation being conditioned by marine morphology.

The prediction of inherited dolomite porosity requires both an understanding of sedimentary facies patterns and a dolomite model involving a flow direction. In some cases porous dolomite may be situated downflow within grainy sediment (and vice versa). Thus, it is essential to know whether water flow is lateral or vertical with respect to the sedimentary body.

Syndolomite voids

These will depend essentially on the dissolution of a precursor aragonite or calcite, and thus on the saturation of the dolomitizing fluids *vis-à-vis* these minerals. Where porous dolomite has replaced a porous limestone, voids may be intercrystalline and therefore petrophysically different from those of the initial limestone. This porosity therefore involves rearrangement and, possibly, an increase in permeability due to increase in crystal size.

Where porosity is considered to be syndolomitic, its prediction logically requires the formulation of a dolomite model. However, porosity created via dolomitization, although strongly suspected, remains rather speculative: our current understanding of its evolution relative to dolomitizing systems, in principle, does not facilitate its prediction. Nevertheless, as already noted, because sustained crystal growth will lead to progressive destruction of porosity, whatever its origins, dense dolomite is most likely to occur where dolomitizing fluids are constantly renewed, i.e. in an upflow direction.

Postdolomite porosity

Postdolomitization porosity may be created chemically by the dissolution of metastable dolomite, relict calcite or evaporite minerals. Porosity may also be created mechanically by brecciation or fracturation of dolomite. Brecciation of dolomite is commonly associated with dissolution of associated evaporites; in other cases brecciation and fracturation are related to regional or local tectonics. The processes of postdolomitization porosity formation are best understood in the context of basin evolution: regional stresses, structural growth and movement of subsurface fluids. Porosity created by postdolomitization processes is difficult to predict from depositional or diagenetic models.

Selectively preserved porosity

Finally, porosity in dolomites, as in any rock, will be determined not only by creational processes but, especially, by selective preservation, which may depend on burial depth and related diagenetic changes in the dolomite fabric, or the position of the dolomite with respect to faults and fractures favouring the delivery of pore-destroying cements.

Conclusions

With few exceptions, no specific pore type can be assigned *a priori* to either a pre-, syn- or postdolomitization event. The distinction between the three dominant types is a conceptual aid to interpreting and predicting on a regional or local scale. The type of dolomite porosity is determined by integrating petrographic, paragenetic and geochemical data with porosity data pertaining to the dolomite body and surrounding carbonates. The conceptual relationship between the dolomite porosity and its setting then allows a limited prediction of porosity distribution.

ACKNOWLEDGEMENTS

The authors wish to thank Don Zenger and Maurice Tucker for their useful suggestions and, especially, Jerry Lucia for his critical remarks, all of which have considerably improved this contribution.

REFERENCES

AISSAOUI, D.M., BUIGUES, D. & PURSER, B.H. (1986a) Model of reef diagenesis: Mururoa Atoll, French Polynesia. In: *Reef Diagenesis* (Ed. Schroeder, J.H. & Purser, B.H.) pp. 27–52. Springer-Verlag, Heidelberg.

AISSAOUI, D.M., CONIGLIO, M., JAMES, N.P. & PURSER, B.H. (1896b) Diagenesis of a Miocene reef-platform: Gebel Abu Shaar, Gulf of Suez, Egypt. In: *Reef Diagenesis* (Ed. Schroeder, J.H. & Purser, B.H.) pp. 112–131. Springer-Verlag, Heidelberg.

AL-HASHIMI, W.S. & HEMINGWAY, J.E. (1973) Recent dedolomitization and origin of the rusty crusts of Northumberland. *J. Sedim. Petrol.* **43**, 82–91.

ARGUILERA, R. (1980) *Naturally Fractured Reservoirs*. Elsevier Publishing Co., Amsterdam.

BEACH, D.K. (1982) *Depositional and Diagenetic History of Pliocene–Pleistocene Carbonates of Northwestern Great Bahama Bank*. Ph D Dissertation, University of Florida, Rosenstiel School of Marine & Atmospheric Science, 600 pp.

BUDAI, J.M., LOHMANN, K.C. & OWEN, R.M. (1984) Burial dedolomite in the Mississippian Madison Limestone, Wyoming and Utah thrust belt. *J. Sedim. Petrol.* **54**, 276–288.

BUIGUES, D. (1982) *Sédimentation et Diagenèse des Formations Carbonatées de l'Atoll de Mururoa, Polynésie Française*. Doctoral Thesis, University of Paris Sud, Orsay.

BUREAU, S. (1988) *Stratigraphie, Sédimentologie, Diagenèse et Paléogéographie du 'Madison Group', Mississippien, dans 'L'Overthrust Belt' de l'Ouest du Wyoming, Etats-Unis*. Doctoral Thesis, University of Paris Sud, Orsay, 301 pp.

CHOQUETTE, P.W. & PRAY, L.C. (1970) Geologic nomenclature and classification of porosity in sedimentary carbonates. *Bull. Am. Assoc. Petrol. Geol.* **54**, 207–250.

CRAIG, D.H., MRUK, D.H., HEYMANS, M.J., CREVELLO, P.D. & LANZ, R.C. (1986) Stratigraphy and reservoir geology of the San Andres Dolomite – Yates Field, West Texas. In: *Hydrocarbon Reservoir Studies, San Andres/Grayburg Formations, Permian Basin* (Ed. Debout, D.G. & HARRIS, P.M.) *PBS–SEPM Publ.* 86–26, 139–143.

EVAMY, B.D. (1967) Dedolomitization and the development of rhombohedral pores in limestones. *J. Sedim. Petrol.* **37**, 1204–1215.

GREGG, J.M. & SIBLEY, D.F. (1984) Epigenetic dolomitization and the origin of xenotopic dolomite textures. *J. Sedim. Petrol.* 54, 908–931.

HARRIS, P.M., FLYNN, P.E. & SIEVERDING, J.L. (1988) Mission Canyon (Mississippian) reservoir study, Whitney Canyon–Carter Creek field, southwestern Wyoming. In: *Giant Oil and Gas Fields: A Core Workshop* (Ed. Lamando, A.J. & Harris, P.M.). SEPM Core Workshop 12, 695–740.

HUH, J.M., BRIGGS, L.I. & GILL, D. (1977) Depositional environments of Pinacle reefs, Niagara and Salina Groups, Northern Shelf, Michigan Basin. In: *Reefs and Evaporites – Concepts and Depositional Models* (Ed. Fisher, J.H.) AAPG Studies in Geology 5, 1–22.

LAND, L.S. (1985) The origin of massive dolomite. *J. Geol. Educ.* **33**, 112–125.

LAND, L.S., MACKENZIE, F.T. & GOULD, S.J. (1967) The Pleistocene history of Bermuda. *Bull. Geol. Soc. Am.* **78**, 993–1006.

MACHEL, H.-G. & MOUNTJOY, E.W. (1986) Chemistry and environments of dolomitization: a reappraisal. *Earth Sci. Rev.* **23**, 175–222.

MATTES, B.W. & MOUNTJOY, E.W. (1980) Burial dolomitization of the Upper Devonian Miette Buildup, Jasper National Park, Alberta. In: *Concepts & Models of Dolomitization*. (Ed. Zenger, D.H., Dunham, J.B. & Ethington, R.L.) Spec. Publ. Soc. Econ. Paleont. Mineral. **28**, 259–299.

MOORE, C.H. (1989) *Carbonate Diagenesis and Porosity. Developments in Sedimentology*. Elsevier, Amsterdam, 338 pp.

MURRAY, R.C. (1960) Origin of porosity in carbonate rocks. *J. Sedim Petrol.* **30**, 59–84.

MURRAY, R.C. & LUCIA, F.J. (1967) Cause and control of dolomite distribution by rock selectivity. *Bull. Geol. Soc. Am.* **78**, 21–35.

Ohle, E.L. (1985) Breccias in Mississippi Valley-Type deposits. *Econ. Geol.* **80**, 1736–1752.

Plaziat, J.C., Purser, B.H. & Philobbos, E.R. (1990) Seismite deformation structures (seismites) in the syn-rift sediments of the NW Red Sea (Egypt). *Bull. Soc. Géol. France* **VI**(3), 419–434.

Powers, R.W. (1962) Arabian Upper Jurassic carbonate reservoir rocks. *Am. Assoc. Petrol. Geol. Mem.* **1**, 122–192.

Purser, B.H. (1975) Tidal sediments and their evolution in the Bathonian carbonates of Burgundy, France. In: *Tidal Deposits* (Ed. Ginsburg, R.N.) pp. 335–343. Springer-Verlag, New York.

Purser, B.H. (1978) Early diagenesis and the preservation of porosity in Jurassic limestones. *J. Petrol. Geol.* **1**, 83–94.

Purser, B.H. (1985) Dedolomite porosity and reservoir properties of Middle Jurassic carbonates in the Paris Basin, France. In: *Carbonate Petroleum Reservoirs* (Ed. Roel, P.O. & Choquette, P.W.) pp. 343–355. Springer-Verlag, New York.

Purves, W.J. (1986) Depositional and diagenetic control on porosity, Upper San Andres Formation–Bridges State leases, Vacuum field, Lea Co., New Mexico. In: *Hydrocarbon Reservoir Studies, San Andres/Grayburg Formations, Permian Basin* (Ed. Debout, D.G. & Harris, P.M.) PBS–SEPM Publ. 86–26, 49–54.

Schmoker, J.W. (1984) Empirical relation between carbonate porosity and thermal maturity: an approach to regional porosity prediction. *Bull. Am. Ass. Petrol. Geol.* **68**, 1697–1703.

Schmoker, J.W. & Halley, R.B. (1982) Carbonate porosity versus depth: a predictable relation for South Florida. *Bull. Am. Ass. Petrol. Geol.* **66**, 2561–2570.

Simpson, F. (1988) Solution-generated collapse (SGC) structures associated with bedded evaporites: significance to base metal and hydrocarbon localization. *Geosci. Can.* **15**(2), 89–93.

Ward, W.C. & Halley, R.B. (1985) Dolomitization in a mixing zone of near-seawater composition, Late Pleistocene, Northeastern Yucatan Peninsula. *J. Sedim. Petrol.* **55**, 407–420.

Weyl, P.K. (1960) Porosity through dolomitization: conservation of mass requirements. *J. Sedim. Petrol.* **30**, 85–90.

Spec. Publs Int. Ass. Sediment. (1994) **21**, 309–323

Permeability and porosity evolution in dolomitized Upper Cretaceous pelagic limestones of Central Tunisia

M.H. NEGRA*, B.H. PURSER† *and* A. M'RABET‡

* *Faculté des Sciences de Tunis, Département de Géologie, Laboratoire de Sédimentologie et Bassins Sédimentaires, 1020-Tunis;*
† *Faculté des Sciences, Université de Paris Sud, Laboratoire de Pétrologie Sédimentaire et Paléontologie, 91405 Orsay, France; and*
‡ *ETAP, 27 bis, avenue Khéreddine Pacha, 1002 Tunis*

ABSTRACT

Campanian–Maastrichtian sedimentation in Tunisia is dominated by well-bedded micritic limestones rich in planktonic foraminifera. However, in Central Tunisia, notably in Wadi Abiod, lensoid bodies of massively bedded carbonates are frequently intercalated within the chalky pelagic lime mudstones. These lenses, whose properties are both sedimentary and diagenetic, correspond mainly to conglomeratic and bioclastic bodies related to gravity flows. They are affected by dolomitization, which modified several original textures. In terms of reservoir properties, this dolomitization can be destructive or constructive. The pelagic lime mudstones have median porosities and permeabilities which are markedly lower than those of the dolomites that have replaced them locally. Although the lower porosities in the non-dolomitized micrites could result from differential compaction, the absence of visible burial compaction fabrics suggests that the more elevated porosities in the dolomites are the consequence of dolomitization. The amelioration of permeability (0.2–300 mD) is probably the direct result of an increase in crystal size, from an average of 2 μm in the micrite to more than 150 μm in the dolomite, and, as such, is the consequence of dolomitization.

INTRODUCTION

Upper Senonian chalky limestones provide potential reservoir rocks in several oilfields, such as the Ekofisk field in the North Sea or the Sidi el Kilani field in central Tunisia. Porosity values are frequently high in chalk (average values reach 35% in the Ekofisk area fields; Scholle, 1977) and in other Senonian micrites (Negra & M'Rabet, 1988; Negra & Purser, 1989). However, permeability values generally are low, generally less than 1 mD. The reservoir potential of these chalky facies is increased by fracturing and elevated pore-fluid pressures (Scholle, 1977). Another process which is important in enhancing the petrophysical properties of micritic facies involves dolomitization and associated dissolution. These diagenetic processes can increase both the porosity and the permeability of chalky pelagic limestones.

This contribution demonstrates that dolomitization can either improve or destroy the effective porosities and permeabilities of Senonian limestones in Central Tunisia. Enhancement of petrophysical properties (Evamy, 1967; Purser, 1985; Negra *et al.*, 1991) is influenced by the initial rock fabric (Purser, 1985; Amthor & Friedman, 1991). In Central Tunisia, planktonic foraminiferal mudstones and wackestones are selectively dolomitized. Dolomitization of deep marine wackestones is not common and requires special conditions. These conditions are related to the primary physical properties of the sediment and their geochemical and hydrodynamic settings. Paradoxically, more permeable grainy sediments, when dolomitized, have reduced reservoir quality.

The Campanian–Maastrichtian Abiod Formation

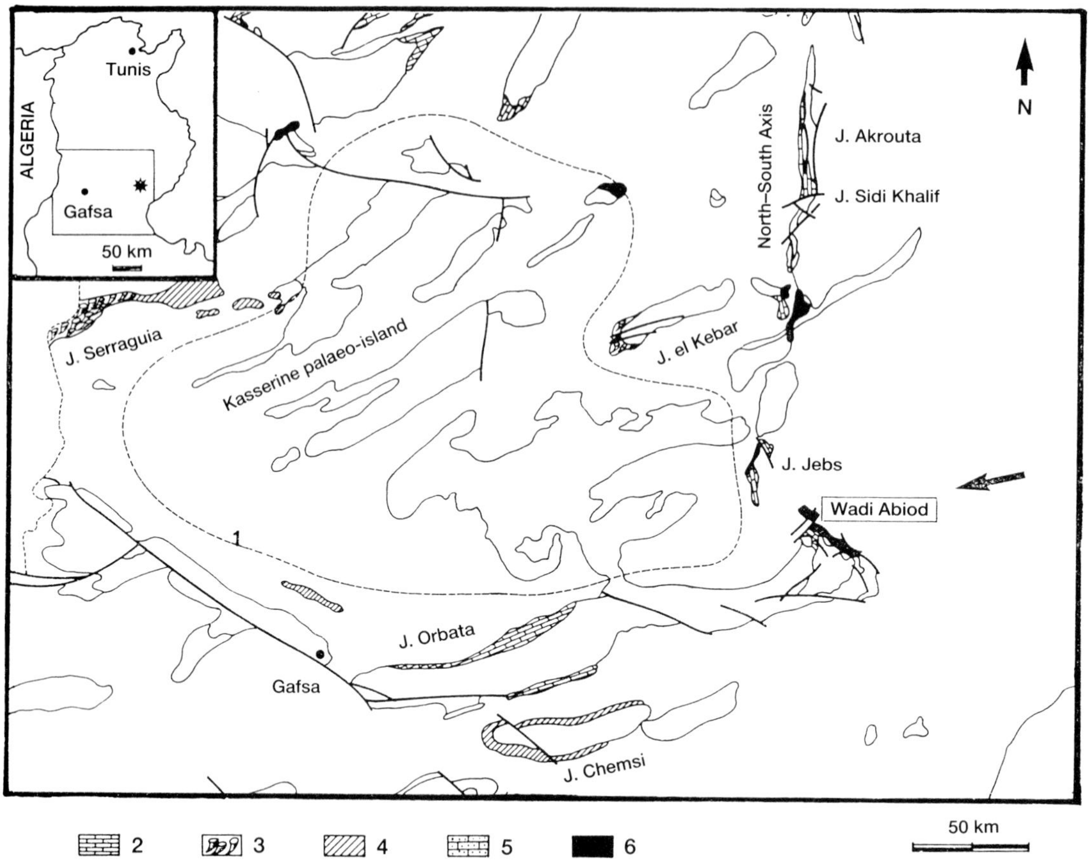

Fig. 1. Geological map showing the Campanian–Maastrichtian facies within and around the area studied. 1, Limit of the Kasserine palaeoisland; 2, well-bedded pelagic facies; 3, rudist facies; 4, lensoid micritic facies; 5, sandy limestone facies; 6, Triassic evaporites.

(Burollet, 1956; Negra, 1989; Negra & M'Rabet, 1991) in Tunisia is composed mainly of well-bedded chalky limestones rich in planktonic foraminifera (*Globotruncana*, *Heterohelix*) and coccoliths. Locally, as in Wadi Abiod (Fig. 1) situated near Maknassy (Khessibi, 1978; Boukadi, 1985), lenticular bodies of massively bedded carbonates are included within the fine chalky limestones.

SEDIMENTOLOGICAL PROPERTIES OF THE ABIOD FORMATION

In Wadi Abiod (Fig. 1), the Abiod Formation comprises two carbonate members separated by a marly unit. The lower member, about 30 m in thickness, is composed mainly of lenticular bodies (Figs 2 and 3) of conglomeratic carbonate whose bases are eroded into the open-marine marls of the underlying Aleg Formation (Burollet, 1956). The conglomeratic carbonates are polygenic debris-flow deposits overlain by turbiditic sequences. These debris-flow deposits consist of subrounded pebbles enveloped by a micritic matrix. The pebbles are millimetric to decimetric in size, and occur in a well sorted packstone rich in peloids, calcispheres and echinoderm debris. The enveloping matrix, in contrast, is a wackestone rich in planktonic foraminifera and coccoliths associated with benthic foraminifera. The overlying turbiditic sequences include bioclastic packstones topped by laminated mudstones and marls (Fig. 2). Sedimentary structures, including flute casts, slump folds associated with the conglomeratic bodies and

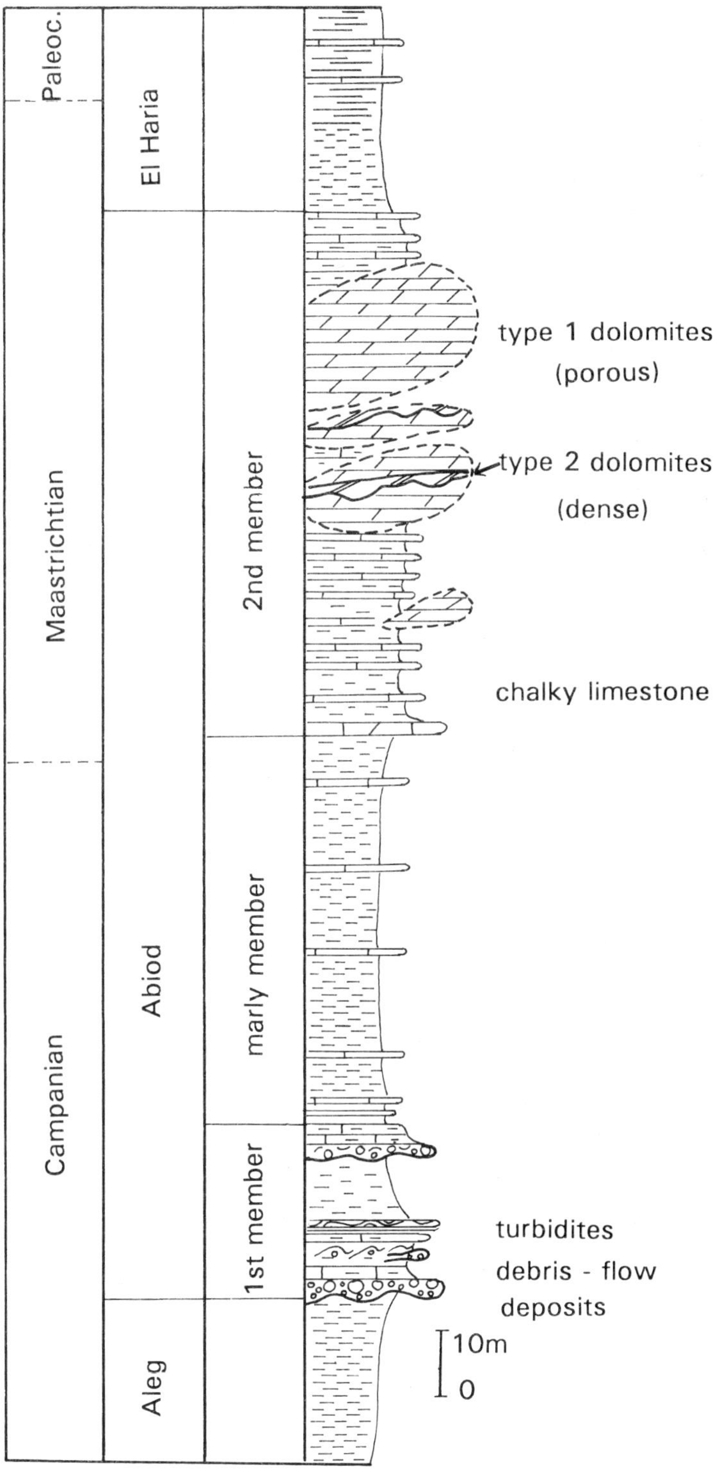

Fig. 2. Section showing the stratigraphy of the Campanian–Maastrichtian dolomites in Wadi Abiod.

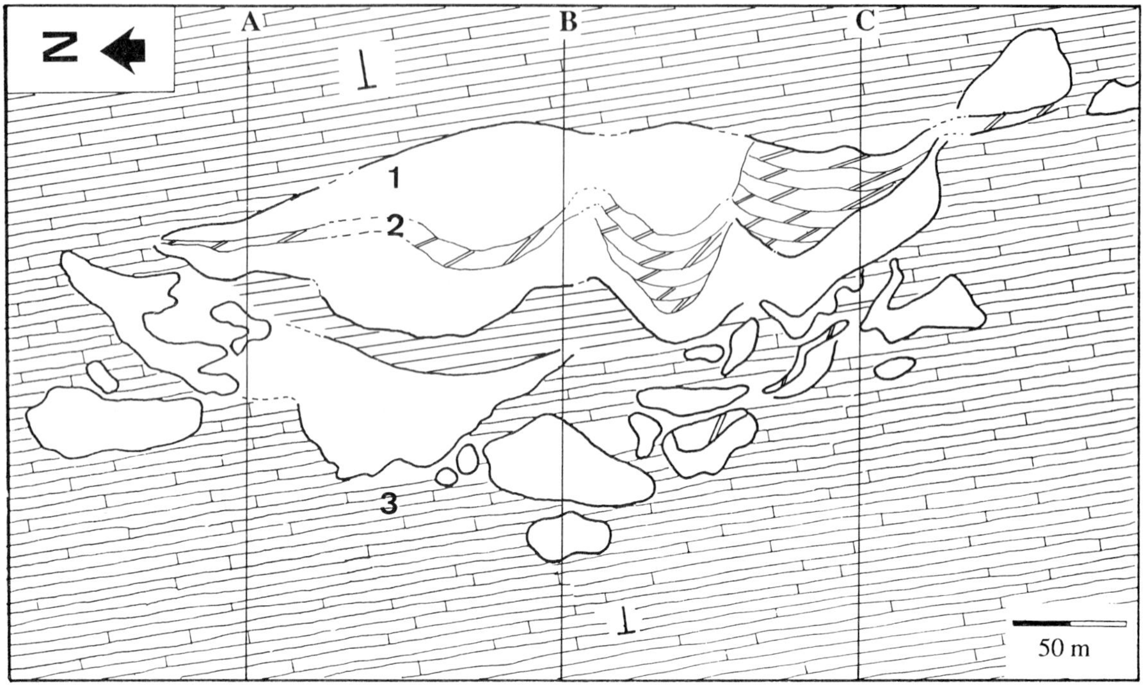

Fig. 3. Lithological map showing the geometry of the dolomitized lenses intercalated within the pelagic chalky limestones. 1, Type 1 dolomite; 2, type 2 dolomite; 3, well-bedded pelagic limestone.

ripple marks associated with the overlying laminated packstone, all indicate flow toward the southeast.

The upper carbonate member, about 80 m in thickness, is composed mainly of centimetric to decimetric marl–limestone alternations. The wackestones are rich in planktonic foraminifera and *Inoceramus* moulds. Brownish dolomitized lenses, the subject of this paper, are included within these marl–chalky limestone alternations.

MORPHOLOGY AND COMPOSITION OF THE DOLOMITE BODIES

The well-bedded light-grey pelagic lime mud/wackestones of the upper member of the Abiod Formation at Wadi Abiod include a number of brownish lenticular masses (Figs 3 and 4) whose boundaries cut across the bedding of the surrounding pelagic limestones. At first glance, the concave-upward aspect of the base of each lens suggests a channel-fill. Closer inspection, however, indicates that these lenticular bodies are more than simple channel-fill: their overall shape results from a combination of lenticular sedimentary accumulation and its subsequent dolomitization, the latter modifying the initial geometry.

Nature and geometry of the sediments

These lenses attain a maximum thickness of about 20 m, extend laterally for several hundreds of metres (Figs 3 and 4), and are situated within fine chalky carbonates rich in planktonic foraminifera. Each lens has a sharp, locally erosional base (Figs 5 and 6A) overlain by 1–2 m of cross-bedded grainstone/packstone composed of benthic foraminifera, molluscan debris, lithoclasts and scattered planktonic forams. This basal cross-bedded facies grades upward into horizontally laminated finer packstone and, untimately, into wackestones with finely broken benthic debris associated with numerous planktonic forams. The top of the unit is generally flat and gradational, contrasting with the slightly concave and generally sharper base. Clearly, these lenticular sedimentary bodies exhibit many aspects typical of submarine channel fillings, probably relating to gravitational slope sedimentation within an open-marine, possibly moderately deep, environment.

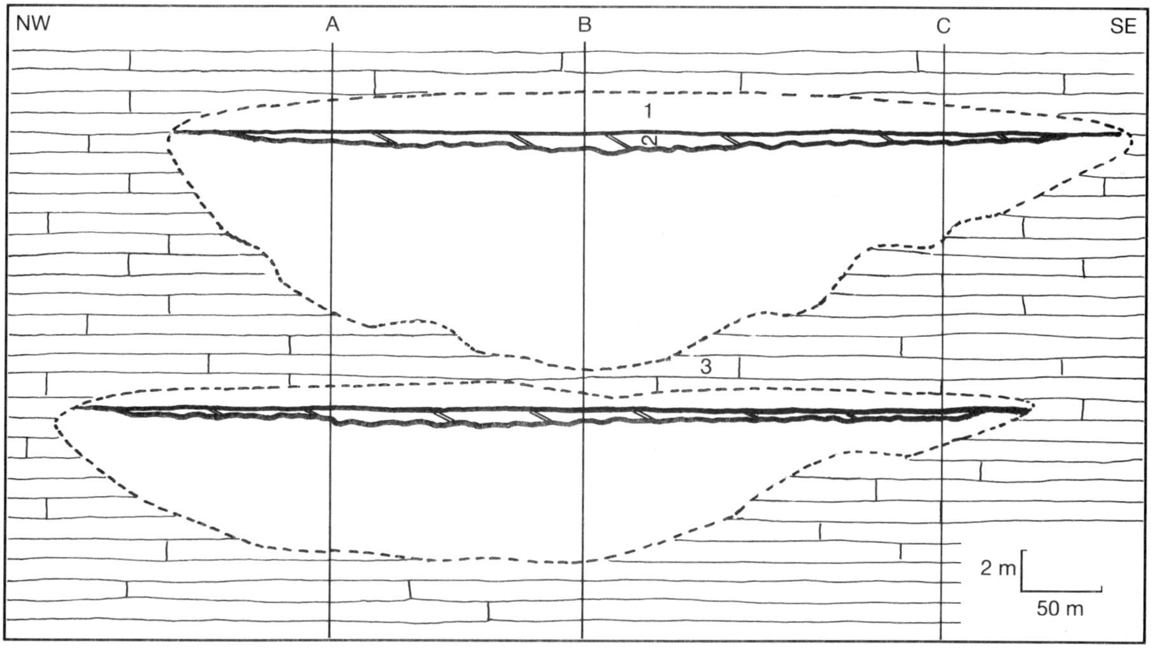

Fig. 4. Lithological section showing the geometry and the dimension of the dolomitized lenses. 1, Type 1 dolomite; 2, type 2 dolomite; 3, well-bedded pelagic limestone.

Cross-bedding directions (Fig. 6A,B) suggest that the sediment has moved in a southeasterly direction. Because the strike of the outcrops is generally oblique to this general transport direction, the 'width' of the channel deposits (Figs 4 and 5) is only an approximation. Each of the two principal channels at Wadi Abiod is overlain by pelagic lime muds which thus seal the lenticular channel sand bodies. These lenticular open (possibly deep)-marine carbonate sands have been dolomitized.

Nature and geometry of the dolomites

In common with the sediment it replaces, the dolomite also has lenticular geometry. However, this brownish dolomite also extends into the pelagic muds adjacent to each channel-fill, so that sedimentary and diagenetic boundaries, although in proximity, do not coincide (Fig. 5B).

The boundary between non-dolomitized pelagic wackestones and the brownish dolomite, although fairly clear, is nevertheless transitional (Fig. 5B) from pure limestone to almost pure dolomite over an interval of less than 1 m. The pelagic limestones are dolomitized completely only in proximity to the channel-fills: at localities where there is no channel-fill, as is the general case, the Abiod Formation is essentially limestone. Clearly, there is a genetic relationship between the distribution of dolomites and the presence of the lime-sand-filled channels.

Dolomitization of the pelagic muds generally occurs both above and below the lenses. Frequently, however, it is generally better developed below each channel, where up to 15 m of pelagic muds are dolomitized. Conversely, only 1–3 m of pelagic muds are dolomitized above the channels. This asymmetry results in a lenticular dolomite body with a markedly concave-upwards base and a relatively flat or slightly convex top (Fig. 4). This preferential dolomitization below the channels appears to be related to the flow of dolomitizing waters along the grainy basal parts of the channel. The more muddy upper parts of the channel tend to retard dolomitization in the overlying pelagic limestones.

Dolomitization is highly variable. The granular cross-bedded basal fillings are almost totally dolomitized, whereas the fine-grained upper and peripheral parts of the channels are only partially dolomitized.

Fig. 5. Field views showing the geometry and the relationship between the different Campanian–Maastrichtian facies in Wadi Abiod. (A) Geometry of a brownish dolomitized lens (grey on the photo) intercalated within the white chalky limestone (1); the dolomite lens showing irregular limits is composed of type 1 porous dolomite (2) and type 2 dense dolomite (arrow). (B) Progressive vertical change from the white chalky limestone (1) to the partly dolomitized limestone (2) and the type 1 porous dolomite (3).

Dolomite petrography, mineralogy and isotope geochemistry

There are two basic and several intermediate dolomite fabrics. The two basic dolomite fabrics are discussed below.

Type 1 dolomite

This dolomite has a polymodal fabric (Sibley & Gregg, 1987), consisting of well-developed crystals of variable size (Fig. 7C,E–H). Each crystal has a cloudy centre and a clear rim, whose thickness is variable. Under scanning electron microscopy, the nuclei of crystals are often partially or totally dissolved, giving well-developed intracrystalline porosity (Fig. 7F). Microprobe analyses (Fig. 8D–F) of non-dissolved nuclei reveal an enrichment in iron relative to the limpid periphery. Oxidation of this iron is the main cause of the typical brownish colour of outcrops. The limpid peripheral parts of each rhombohedron are locally zoned, reflecting minor variations in iron. Microprobe analyses indicate that these clear peripheral zones have higher Mg/Ca ratios than the nuclei.

Type 1 dolomites generally replace pelagic Abiod limestones (Fig. 7A). The transition from relatively pure lime-mud to partially dolomitized carbonate generally coincides with a progressive increase in dolomite crystal size (Fig. 7E,F) from 75 to 150 μm. However, within the totally dolomitized (type 1) sediments, the size of rhombohedra is typically poly-

Fig. 6. Sedimentary structures characterizing the dolomitized bodies of Wadi Abiod. (A) Erosional base (arrows) of the type 2 dense dolomite exhibiting cross-bedding. (B) Cross-bedding in fining-up bioclastic microsequences which constitute the dense type 2 dolomites.

modal. Type 1 dolomite may partially replace the finer, upper parts of channel fillings. It is very porous and may be a potential hydrocarbon reservoir.

Type 2 dolomite

This dolomite is characterized by a xenotopic fabric consisting of equant crystals whose mean dimensions are about 190 μm. As with type 1 dolomites, many crystals exhibit cloudy centres (Fig. 7D). However, type 2 dolomites tend to have a more fully developed clear periphery. The gradational crystal growth and resulting interference give a xenotopic fabric. Thus, the basic difference between the two fabrics is the more complete crystal growth in type 2, leading to a coarser, more homogeneous crystal fabric. This more 'mature' dolomitization has led to the almost total destruction of intercrystalline porosity.

Type 2 dolomites occur in the basal grainy parts (Fig. 7B) of each channel, and grade downwards into porous type 1 dolomites. This transition is generally fairly sharp (Fig. 5B). This diagenetic type 1–type 2 contact coincides with and accentuates the basal limits of the channel. Conversely, the upwards transition from type 2 to type 1 is generally more transitional.

The isotopic composition of both types of dolomite (Fig. 9) is generally very similar. With the exception of three samples, the isotopic ratios for

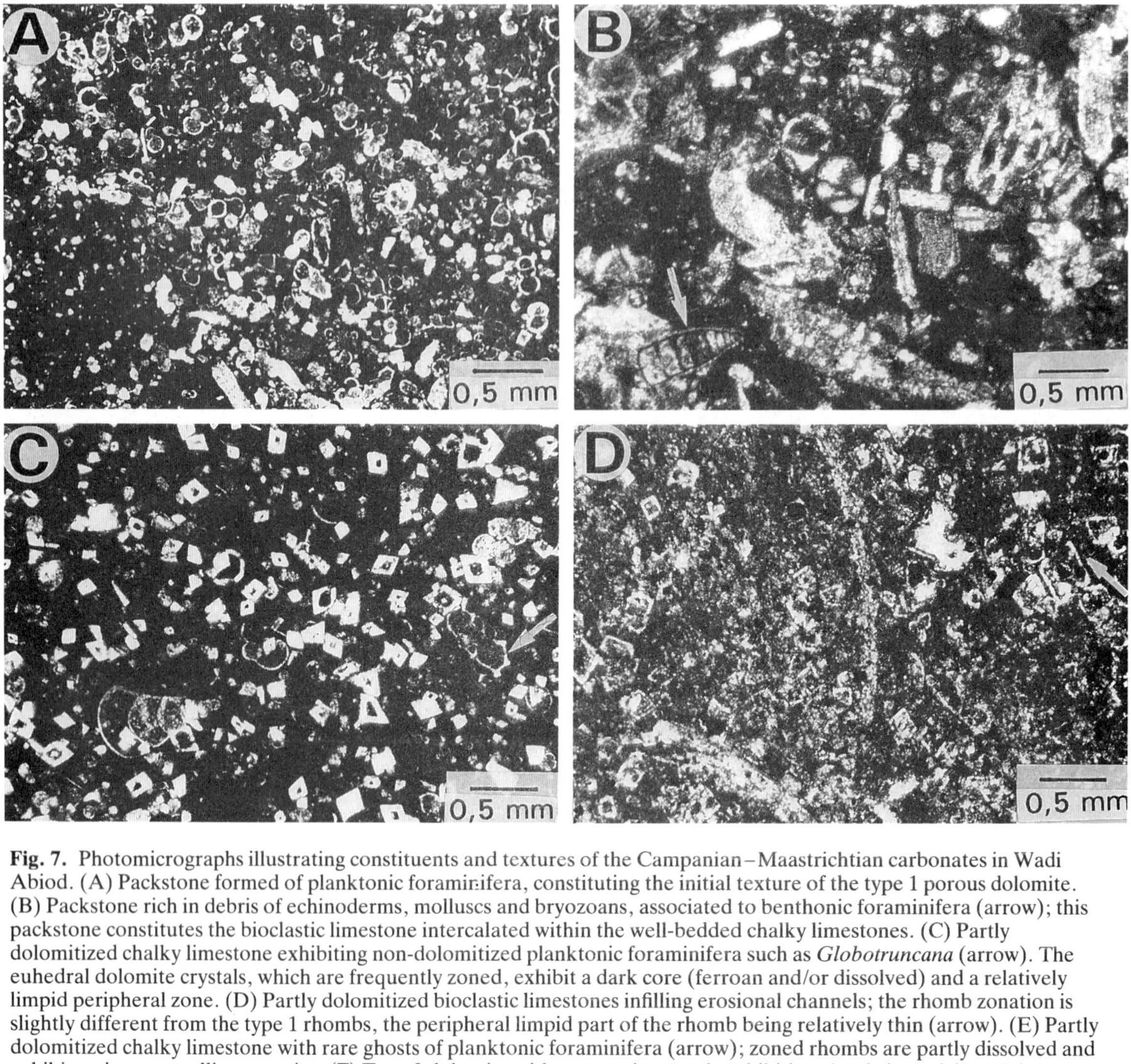

Fig. 7. Photomicrographs illustrating constituents and textures of the Campanian–Maastrichtian carbonates in Wadi Abiod. (A) Packstone formed of planktonic foraminifera, constituting the initial texture of the type 1 porous dolomite. (B) Packstone rich in debris of echinoderms, molluscs and bryozoans, associated to benthonic foraminifera (arrow); this packstone constitutes the bioclastic limestone intercalated within the well-bedded chalky limestones. (C) Partly dolomitized chalky limestone exhibiting non-dolomitized planktonic foraminifera such as *Globotruncana* (arrow). The euhedral dolomite crystals, which are frequently zoned, exhibit a dark core (ferroan and/or dissolved) and a relatively limpid peripheral zone. (D) Partly dolomitized bioclastic limestones infilling erosional channels; the rhomb zonation is slightly different from the type 1 rhombs, the peripheral limpid part of the rhomb being relatively thin (arrow). (E) Partly dolomitized chalky limestone with rare ghosts of planktonic foraminifera (arrow); zoned rhombs are partly dissolved and exhibit an intracrystalline porosity. (F) Type 2 dolomite with xenotopic crystals exhibiting cloudy iron-rich nuclei; pore spaces are absent. (G) Type 1 dolomite showing connected interrhombohedral pore spaces (1). (H) Detailed view of a type 1 dolomite with intercrystal porosity (1).

both O and C are closely grouped: the average $\delta^{18}O$ is -1.5 and the $\delta^{13}C$ is $+1.5$, suggesting dolomitization either from Cretaceous seawater or slight atmospheric modification of these marine waters.

Dissolution of the dolomites (dedolomitization)

Many type 1 dolomite crystals exhibit traces of dissolution, clearly visible under the scanning electron microscope (Fig. 8A–C). As is often the case with dolomites, the central parts of individual crystals are dissolved preferentially with respect to the more stable limpid peripheries (Fig. 8E,F). This local dissolution apparently coincides with the iron-enriched part of the crystal.

Dissolution has not greatly affected the more crystalline type 2 dolomites, despite the presence of dark, iron-enriched nuclei. This preservation of type 2 crystals, in spite of their apparent susceptibility, may best be explained by the dense impermeable

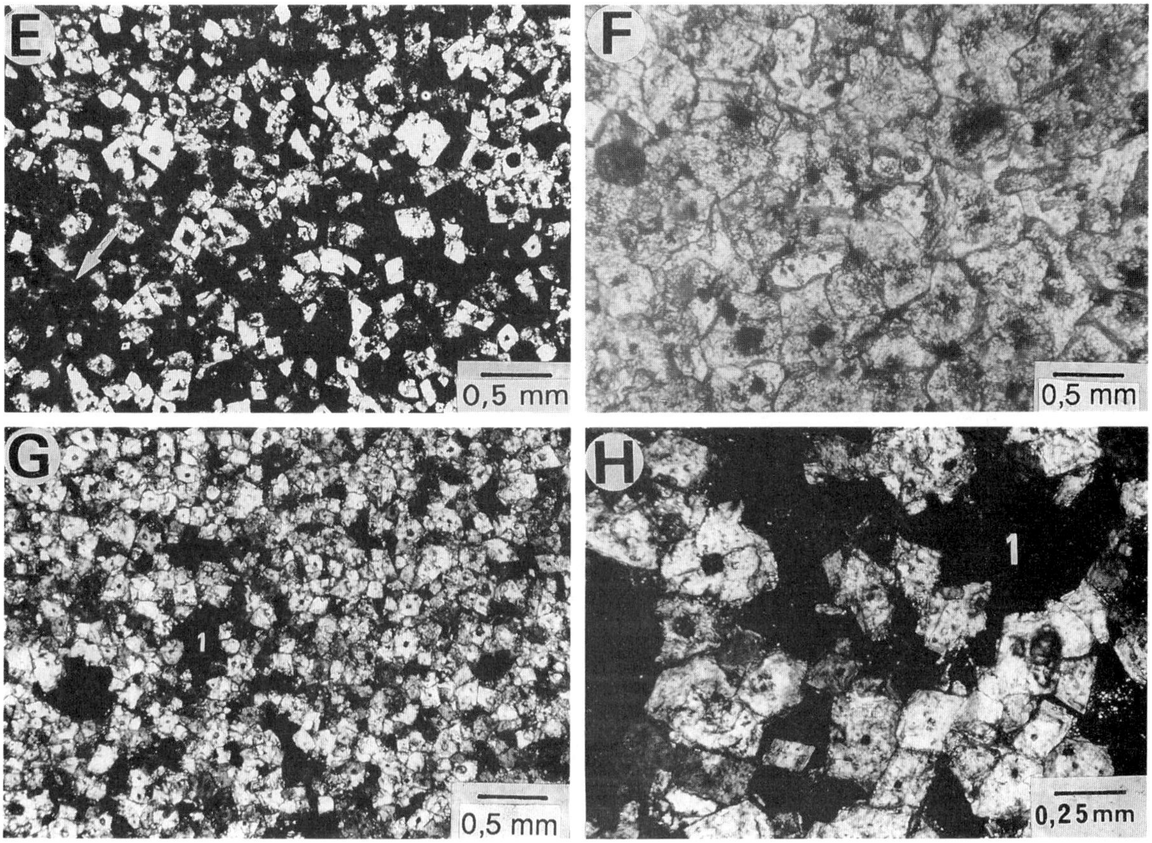

Fig. 7. *Continued.*

nature of the dolomite: dedolomitizing fluids have not had free access to the crystals. Conversely, in type 1 dolomites, partial dissolution of certain rhombohedra coincides generally with the presence of a pre-existing intercrystalline porosity that has facilitated the movement of dedolomitizing fluids. These fluids may also have resulted in the dissolution of coccoliths within adjacent limestones (Fig. 8A–C).

The exact timing of dedolomitization is not established. However, the fact that dissolution affects unweathered samples of type 1 dolomite indicates that it is not a modern surface effect. This is confirmed by the lack of dissolution in outcrop of type 2 dolomite. Furthermore, the presence of minor quantities of probably authigenic clay within this secondary porosity may be a further indication of its fossil origin.

ENVIRONMENTS AND PROCESSES OF DOLOMITIZATION AND DEDOLOMITIZATION

Dolomitization is local and related genetically to the channels. However, because it affects both the channel fillings and the surrounding pelagic muds, dolomitization clearly occurred during burial, and not within an active channel system. The hydrodynamic processes leading to dolomitization are not readily explained. The movement of dolomitizing waters through a buried lenticular sand body whose distal downslope parts were buried in mud and therefore impermeable, poses an obvious problem concerning continuous fluid circulation. In addition, the change from dolomite to dedolomite indicates that fluid composition has evolved.

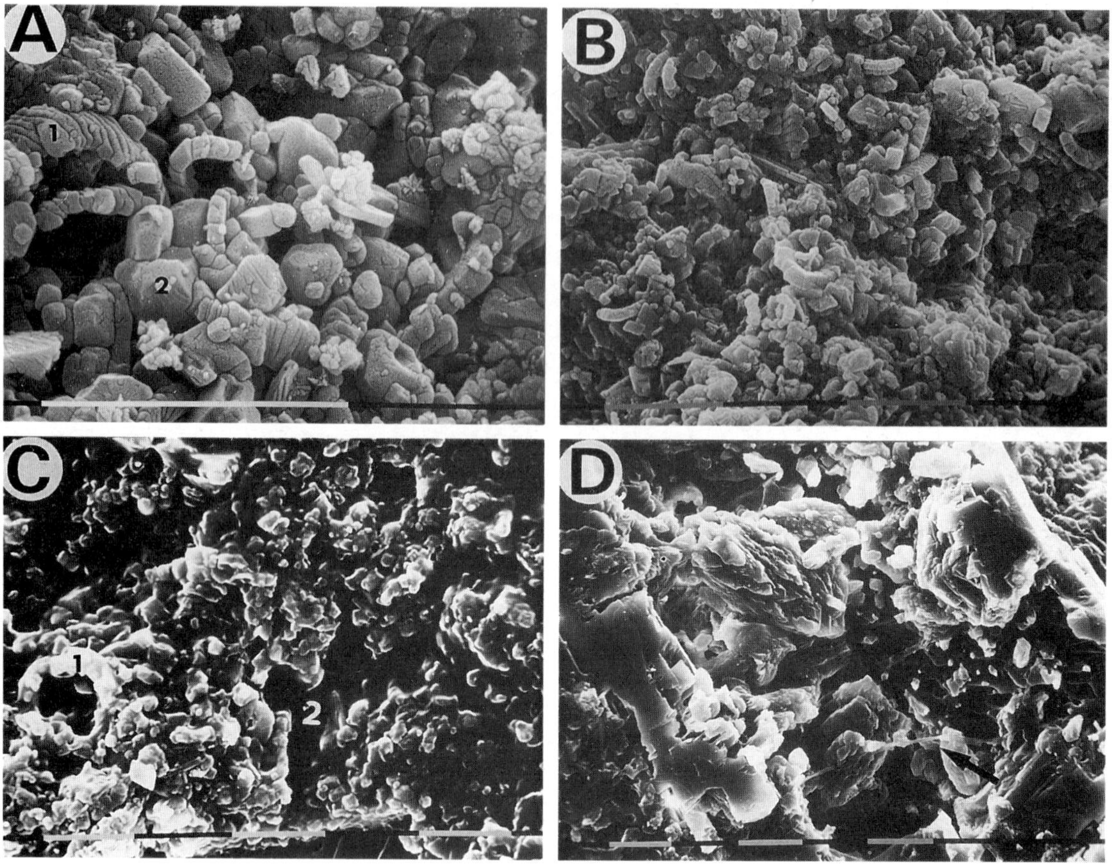

Fig. 8. Scanning electron micrographs illustrating the initial textures of the non-dolomitized chalky limestone and showing the relationship between dolomitization and dissolution within the dolomitized chalky pelagic limestone (type 1 dolomite). (A) Loosely packed to interlocking texture of fragmented coccoliths (1) and anhedral to subhedral micrite grains which are partly surrounded by spar crystals (2); an intergranular porosity is preserved. Scale bar = 10 μm. (B) Loosely packed to interlocking texture formed mainly of relatively well preserved coccoliths and their debris, characterizing chalky pelagic limestones. Scale bar = 10 μm. (C) Partly dolomitized chalky limestones in which coccoliths (1) are leached; intergranular pores are partly connected (2). Scale bar = 10 μm. (D) Detail of a type 1 dolomite crystal; the rhomb core is partly dissolved and partly filled with clay minerals (arrow), scale bar = 10 μm. (E) Detailed view of a partly dissolved type 1 dolomite crystal. Scale bar = 10 μm; the peripheral part (1) of the rhomb is partly dissolved, the core (2) is totally dissolved. (F) Detail of (E); the peripheral part (1) of the rhomb is partly dissolved, the core (2) is totally dissolved. Scale = 10 μm. (G) Type 1 dolomite exhibiting a well-developed intercrystalline porosity (1). Interrhombohedral pore spaces are partly connected. Scale bar = 100 μm. (H) Type 1 dolomite showing a polymodal fabric; size of crystals varies from 50 (1) to 200 μm (2). Scale bar = 100 μm.

At least four possible models exist:

1 Because dolomitization occurs during burial, diagenetic fluids may have passed through the permeable channel conduits during compaction of the pelagic Abiod lime-muds. However, this model, in common with other compaction-type models, does not provide an adequate source of magnesium. Furthermore, the positive $\delta^{18}C$ and slightly negative $\delta^{18}O$, suggesting a moderate freshwater influence, do not strongly support this hypothesis. Also, the change from a dolomitizing to a dedolomitizing system is not easily explained by this system.

2 The region is affected by a series of major faults, notably in the vicinity of Jebel Jebs (Fig. 1). Therefore, a hydrothermal source for the dolomitizing fluids passing through the permeable conduits is theoretically possible. The absence of baroque dolomite crystals, a common hydrothermal dolomite type and the isotope values argues against this hypothesis.

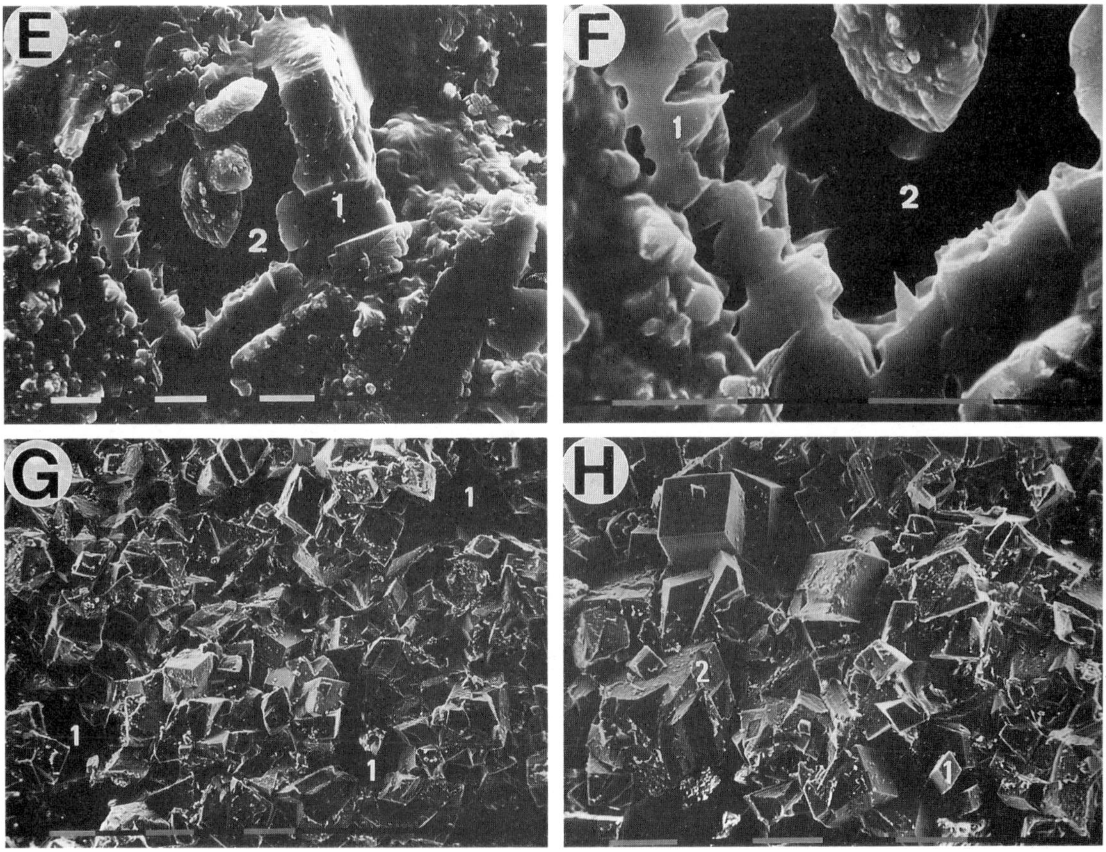

Fig. 8. *Continued.*

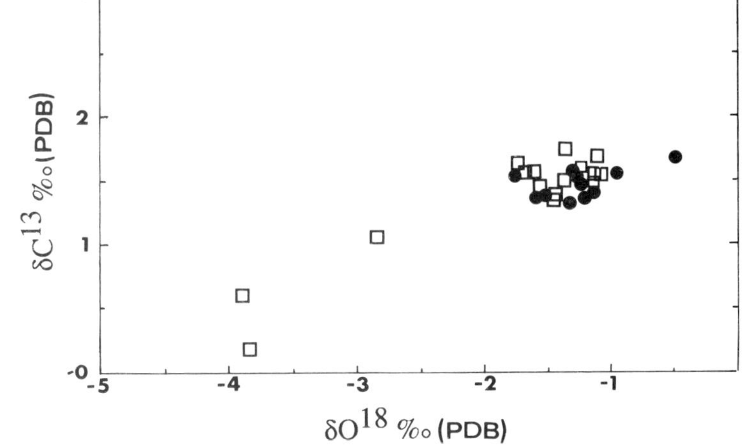

Fig. 9. Cross-plot of carbon versus oxygen stable isotope composition for type 1 (●) and type 2 (□) dolomites in Wadi Abiod.

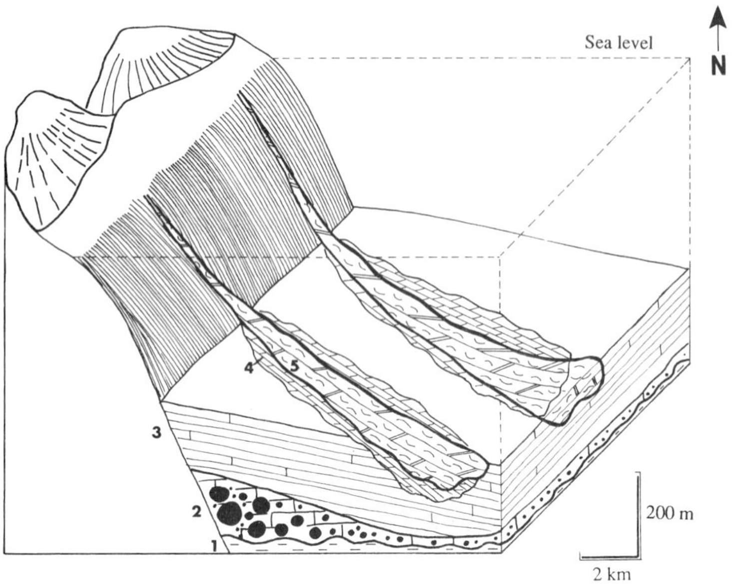

Fig. 10. Block diagram showing the distribution of Campanian–Maastrichtian dolomitized facies in Maknassy area, Central Tunisia. 1, Open-marine marls of Aleg Formation; 2, conglomeratic carbonates; 3, well-bedded pelagic limestones; 4, dolomitized micrites (type 1); 5, dolomitized bioclastic limestones (type 2 dolomites).

3 Resedimentation under the control of relative sea-level fall is common at the study sector, mainly in Jebel Jebs. Upper Palaeocene clays and marls unconformably overlie Upper Maastrichtian marls of the *Gansserina gansseri* biozone. Just below this discontinuity, corresponding to a global sea-level fall (Haq *et al.*, 1988), the microfauna is clearly reworked and associated to phosphatic grains. Upper Maastrichtian foraminifera such as *Gansserina gansseri* are mixed with Lower Maastrichtian foraminifera. The environment of deposition appears to be relatively restricted, as attested by the frequency of the genus *Haplophragmoides*. However, the dolomitized channels are Lower Maastrichtian in age (*Globotruncana tricarinata* biozone; Haq *et al.*, 1988) and the occurrence of dolomite just below the pre-Tertiary unconformity suggests that it may been related to the relative sea-level fall associated with the unconformity.

4 A hydrodynamic flow stimulated by emergence of the 'Kasserine Island' (Fig. 1) situated 10 km to the northwest during much of the Upper Cretaceous, although not satisfying all the constraints, nevertheless offers the best explanation, for the following reasons:

(a) dolomitization is locally more intense towards the northwest, notably in the vicinity of Jebel Jebs, where up to 100 m of carbonate sand and conglomerates are massively dolomitized;

(b) isotopic values are possibly indicating a moderate freshwater influence;

(c) finally, this model best explains the dissolution that subsequently affects the dolomite: with increasing relief on the emerged Kasserine Island, the presumably marine interstitial waters were progressively replaced by dolomitizing diluted seawater and, finally, by meteoric waters, causing partial dissolution of the dolomite.

In sum, the authors envisage a near-surface origin for the dolomitizing and dedolomitizing fluids (Fig. 10) that passed downwards through the moderately buried permeable channel-fills. In so doing they have not only dolomitized these carbonate sand lenses but have also affected the less permeable pelagic muds in proximity to these conduits. The latter are most affected below the base of the channels, due to the passage of dolomitizing fluids through the more permeable channel-base sands.

The presence of cloudy iron-rich nuclei in most dolomite crystals indicates a reducing environment, particularly within the type 1 dolomites that replace the pelagic muds. This ferroan component, although present, is lower within the highly crystalline (cemented) type 2 dolomite that has replaced the channel sands. The increase in crystal size and the occlusion of virtually all intercrystalline voids, resulting in the dense nature of type 2 dolomites, is best explained by a more sustained fluid flow

Table 1. The variation of permeability and porosity with respect to initial textures and dolomitization–dedolomitization processes.

Lithology	Porosity type	Porosity values $\phi_{min}-\phi_{max}$ ($\bar{\phi}$) in %	Permeability values $K_{a\,min}-K_{a\,max}$ ($\bar{K}_a$) in mD
Chalky limestone	Intergranular Intraparticular	9.8–18.7 14	0.1–0.3 0.2
Partially dolomitized chalky limestone	Intergranular Intercrystal	8.6–15.6 12.1	0.1–0.4 0.2
Type 1 dolomite (dolomitized chalk)	Intercrystal Intracrystal	14.5–28.3 19.1	27.4–952.3 302.6
Type 2 dolomite (dolomitized packstone)	Mouldic Intercrystal	2.9–7.3 4.8	<0.1

through these permeable conduits. Subsequent dedolomitizing fluids have circulated through the permeable peripheral type 1 dolomites, leading to their partial dissolution.

PETROPHYSICAL ATTRIBUTES OF THE DOLOMITE

As already noted, dolomite replaces both pelagic lime-muds and grainy channel-fills. Porosity within the non-dolomitized chalky pelagic muds averages 14% (Table 1), with permeabilities below 1 mD. Initial porosity and permeability values for the carbonate sands can only be estimated, as all sands have been replaced by dolomite. Porosity values, more than 14% but less than 30%, and permeability values, more than 27 mD but less than 960 mD, were both clearly higher than those of the surrounding muds. As will be discussed, these initial petrophysical properties of muds and sands have been inversed during dolomitization.

Pores in the dolomite are both inter- and intracrystalline, the former being the most dominant (Fig. 7F–H). Both inter- and intrarhombohedral porosity is concentrated within type 1 dolomite. These dolomites have permeabilities of an average of 300 mD. Intercrystalline porosity is often associated with a heterogeneous fabric (Fig. 7F) whose constituent crystals vary in size from 50 to 200 μm. The incomplete growth of the small crystals has resulted in the preservation of intercrystalline pore spaces.

The porosity (average 20%) within the type 1 dolomites is also somewhat higher than that of the non-dolomitized lime-muds (average 14%). In both the limestone and the dolomite this porosity is somewhat enhanced by the presence of channel-like microvugs (Fig. 8C), created by dissolution both of the dolomite and the calcitic coccoliths.

DOLOMITIZATION AND THE EVOLUTION OF POROSITY AND PERMEABILITY

The non-dolomitized micrites have somewhat lower porosities (14%) than the dolomites that have replaced them (20%). Furthermore, they have distinctly lower permeabilities (0.2 mD) than these dolomites (300 mD). The amelioration of the petrophysical quality in the dolomites may be due to either to the dolomitization process or to the subsequent differential compaction of the limestones. However, the absence of styloliths and other visible compaction fabrics (which can be frequent in other sectors) in the studied micrites suggests that porosity has been little modified or improved by dolomitization. More importantly, the marked difference in permeability is most probably due to the considerable increase in crystal size. The average diameter of the lime-mud components is about 1–10 μm, whereas that of the type 1 dolomite is about 100 μm (50–200 μm). Clearly, this increase in crystal size is the direct result of dolomitization and, as such, the process has considerably improved the reservoir potential of the original micrite.

Porosity and permeability are very low within the dense crystalline type 2 dolomites that have replaced the initially permeable channel sands. Porosity occlusion results from the complete growth of dolomite rhombohedra, giving dense interlocking xenotopic fabrics (Fig. 7D). Clearly, dolomite crystal growth has been more sustained, leading to the complete cementation of type 2 fabric by external cations and anions.

It is suggested that the ultimate transitions from porous (type 1) to dense (type 2) dolomite are related to fluid circulation. Where fluids flowed preferentially through initially permeable channel conduits, dolomitization was more efficient, leading ultimately to the virtual destruction of the dolomite pore system. The less permeable pelagic muds, although completely dolomitized, have preserved an intermediate porous stage; had dolomitization continued, these porous type 1 dolomites would also have been cemented by dolomite. That this was not the case may have been due to the fact that these parental fluids have changed in composition, leading subsequently to the partial dissolution of type 1 dolomites.

CONCLUSIONS

The Campanian–Maastrichtian dolomite bodies at Wadi Abiod, Central Tunisia, although small, are nevertheless unusual. Their localization within an open, possibly relatively deep, marine carbonate poses problems concerning the precise nature of the dolomitizing system. The authors have proposed that the dolomitizing fluids affecting buried sedimentary units were related to an artesian flow from an emerged elevated palaeoisland located to the northwest.

This example therefore demonstrates the important role of primary sedimentary facies in influencing the localization and the geometry of certain dolomite bodies. It also demonstrates that this primary lithological control may indirectly determine the ultimate petrophysical variations within the dolomite. Paradoxically, these diagenetic processes have inverted the petrophysical properties of the initial sediments: the impermeable lime-muds have been converted to permeable dolomite, whereas the more permeable channel sands have been 'over-dolomitized' into a dense crystalline impermeable dolomite.

ACKNOWLEDGEMENTS

The authors express their thanks to Jean-Charles Fontes (Laboratoire d'Hydrologie et de Géochimie Isotopique, Université de Paris Sud, Orsay) for results and discussions concerning the isotope values indicated in this contribution. They also thank M. Tucker, D.H. Zenger and Dr Wendt for a critical review which has considerably improved the quality of this contribution.

REFERENCES

Amthor, J.E. & Friedman, J.M. (1991) Dolomite rock textures and secondary porosity development in Ellenburger Group carbonates (Lower Ordovician), west Texas and southeastern New Mexico. *Sedimentology* **38**, 343–362.

Badiozamani, K. (1973) The dorag dolomitization model – application to the Middle Ordovician of Wisconsin. *J. Sedim. Petrol.* **43**, 965–984.

Boukadi, N. (1985) *Evolution Géométrique et Cinématique de la Zone d'Interférence de l'Axe Nord–Sud et de la Chaine de Gafsa (Maknassy-Mezzouna et Jebel Bouhedma), Tunisie*. Thèse d'Université de Strasbourg, France, 155 pp.

Burollet, P.F. (1956) Contribution à l'étude stratigraphique de la Tunisie centrale. *Ann. Mines Géol.* **18**, 345.

Evamy, B.D. (1967) Dedolomitization and the development of rhombohedral pores in limestones. *J. Sedim. Petrol.* **37**, 1204–1215.

Haq, B.U., Hardenbol, J. & Vail, P.R. (1988) Mesozoic and Cenozoic chronostratigraphy and cycles of sea-level change. In: *Sea-Level Changes: An Integrated Approach* (Ed. Wilgos, C.K. *et al.*) SEPM Spec. Publ. 42, 71–108.

Hardie, L.A. (1987) Perspectives: dolomitization, a critical view of some current views. *J. Sedim. Petrol.* **57**, 166–183.

Khessibi, M. (1978) *Etudes Géologiques du Secteur de Maknassy-Mezzouna et du Jebel Kébar (Tunisie Centrale)*. Thèse d'Université de Lyon, France, 175 pp.

Negra, M.H. (1989) La formation Abiod dans le secteur Faïd-Sidi Khalif, Tunisie centrale. *Actes des II èmes Journées de Géologie Tunisienne Appliquée à la Recherche des Hydrocarbures*, *ETAP*, Tunis, **3**, 331–344.

Negra, M.H. & M'Rabet, A. (1988) Effets de la diagenèse sur la porosité et la perméabilité des calcaires Sénoniens à rudistes des complexes récifaux des Jebel el Kébar et Serraguia, Tunisie centrale. *Actes des 1ères. Journées Tunis. Géol. Appl. et Jubilé de A. Azzouz*. T. **II**. 180–189.

Negra, M.H. & M'Rabet, A. (1991) Découverte de faciès riches en matière organique dans le Sénonien supérieur de Tunisie, indices de conditions anoxiques à proximité de la limite Crétacé-Tertiaire. *C. R. Acad. Sci. Paris*, t. **312**, série II, 1033–1039.

Negra, M.H. & Purser, B.H. (1989) Les monticules Sénoniens à rudistes du Jebel el Kébar, Tunisie centrale. Anatomie, diagenèse et géométrie. *Geol. Medit.* T. **XVI**, no. 2–3, 99–119.

Negra. M.H., M'Rabet, A. & Purser, B.H. (1991) Porosity evolution in dolomitized pelagic Senonian limestones of Central Tunisia. *Dolomieu Conference*, *Ortisei, Italy*, *Abstracts*, pp. 192–193.

PURSER, B.H. (1985) Dedolomite porosity and reservoir properties of middle Jurassic carbonates in the Paris Basin, France. In: *Carbonate Petroleum Reservoirs* (Ed. Roehl, P.O. & Choquette, P.W.) pp. 343–355. Springer Verlag, New York.

SCHOLLE, P.A. (1977) Chalk diagenesis and its relation to petroleum exploration: oil from chalks, a modern miracle? *Am. Ass. Petrol. Geol. Bull.* **61**, 982–1009.

SIBLEY, D.F. & GREGG, J.M. (1987) Classification of dolomite rock textures. *J. Sedim. Petrol.* **57**, 967–975.

Spec. Publs Int. Ass. Sediment. (1994) **21**, 325–341

Porosity evolution through hypersaline reflux dolomitization

F.J. LUCIA *and* R.P. MAJOR
Bureau of Economic Geology, The University of Texas at Austin, Austin, Texas 78713-7508, USA

ABSTRACT

Dolomite of the Plio-Pleistocene foreslope carbonate deposits on the island of Bonaire, Netherlands Antilles, has a mean dolomite porosity of 11%, whereas limestone has a mean porosity of 25%, suggesting that dolomitization does not create but reduces porosity. The shallow burial of these young carbonates precludes burial compaction and cementation as a cause for the low-porosity dolomites. The reduced porosity of dolomite relative to limestone indicates that more dolomite was formed than the original amount of carbonate in the limestone would allow. Therefore, carbonate as well as magnesium must have been imported into the system.

Little difference in porosity is observed between dolomitic limestones and adjacent dolomite bodies, suggesting that sufficient carbonate was imported into the system during the replacement phase to offset the increase in porosity due to changes in molar volumes between calcite and dolomite. Dolomite porosity ranges from 25% to 5%, suggesting a pore-filling phase. The dolomite beds are concentrated at the top of the outcrops, with the lowest porosity values at the top and the highest porosity downdip near the change to limestone. Cathodoluminescent zones, Mg/Ca ratios, dolomite unit cell dimensions and strontium content also show systematic changes from updip to downdip. The presence of sulphate crystals included in the dolomite, the geometry of the dolomite bodies and the stable isotopic data are all consistent with the hypothesis that the dolomite bodies and the observed systematic geochemical and porosity changes are in response to downward-flowing hypersaline fluids.

INTRODUCTION

The origin of dolomite is one of the most fascinating topics for geological research, and theories and models concerning dolomitization abound in the literature. One aspect of dolomite that has intrigued geologists since dolomite was first described is porosity. Elie de Beaumont was intrigued by the presence of vugs in the Triassic dolomites of the Tyrol Alps, and his study of these rocks resulted in the theory of mole-for-mole replacement of limestone by dolomite (Beaumont, 1837). With the advent of the petroleum industry, interest in the origin of porosity in dolomite increased as significant reserves were discovered in dolomite formations, and the myth that dolomitization creates porosity became well established among petroleum geologists as the only explanation for the origin of dolomite porosity. The research results presented in this paper demonstrate that dolomite porosity is simply inherited from the precursor limestone, and that the dolomitization process occludes pore space rather than creating porosity.

To explain the vuggy porosity of about 12% in the Triassic dolomites of the Tyrol Alps, Elie de Beaumont (1837) proposed the theory of mole-for-mole replacement of limestone by dolomite, which results in an increase in porosity of 12.5% due to the greater molar volume of calcite. Landes (1946) reviewed all the arguments against the mole-for-mole replacement theory, including the observation by Twenhofel (1926) and Murray (1930) that dolomite beds can be porous at one location and non-porous at another. This observation should have been sufficient to demonstrate the inadequacy of the mole-for-mole replacement theory but, as commented by Landes (1946), 'In spite of its many flaws this theory still appears in articles and textbooks'.

Any model of the development of porosity in dolomite must account for both porous and non-porous dolomites. Murray (1960) and Weyl (1960) used the mole-for-mole concept in their porosity-through-dolomitization models, but did not account for non-porous dolomite. Weyl (1960) supported the mole-for-mole theory by arguing that, because of the small volume of carbonate in most groundwaters compared to calcium and magnesium, conservation of mass requires that dolomitization must proceed without net import of carbonate. Weyl went on to assume a precursor composed of unconsolidated lime-mud with 40% porosity, and a compaction model that maintained 40% porosity in the lime-mud until there were sufficient dolomite crystals to form a support network. Weyl's (1960) model clearly shows the importance of the nature of the precursor limestone to any dolomitization model.

Murray (1960), referring to mole-for-mole replacement as 'local source' dolomitization, used this theory to explain the origin of sucrosic porosity in a sequence of Mississippian limestones and dolomites. Non-porous dolomites and limestones were ascribed to burial compaction. Murray (1960) presented a model for the development of intercrystalline porosity in dolomites without postdolomitization dissolution of intercrystalline calcite. This model describes the growth of dolomite crystals in a lime-mud matrix by the mole-for-mole exchange of magnesium for calcium, and the dissolution of lime-mud adjacent to the growing dolomite crystals, which provides the excess carbonate needed by the growing dolomite crystal to fill the engulfed matrix pore space. A porosity of 21% for the precursor limestone was assumed to produce a dolomite with 30% porosity, the porosity of the Mississippian dolomites studied by Murray. As with Weyl's (1960) model, Murray's (1960) model clearly shows the importance of the nature of the precursor limestone.

Halley and Schmoker (1983) observed that, in young carbonate rocks of south Florida, dolomite has porosity values equal to or less than limestone. They concluded that mole-for-mole replacement cannot account for this observation and proposed that dolomitization proceeds by incorporating new carbonate in addition to magnesium. In addition, Schmoker *et al.* (1985) demonstrated that the commonly held view that dolomite hydrocarbon reservoirs are more porous than limestone reservoirs is incorrect. In fact, they show that, in the United States, the most likely porosity value of a dolomite reservoir is 7–9%, as compared with 11–13% porosity for limestone reservoirs.

Although solution chemistry predicts that mole-for-mole replacement of calcium carbonate minerals by dolomite produces a loss of mineral volume and an accompanying gain in pore volume, there are no data to support the contention that mole-for-mole replacement is a reasonable model for dolomitization. Any porosity-through-dolomitization study of a rock sequence must account for the physical and chemical characteristics of the precursor limestone, postdolomitization effects such as physical and chemical compaction (pressure dissolution), and both porous and non-porous dolomites. Only the study of south Florida carbonates (Halley & Schmoker, 1983) can make a reasonable case regarding the nature of the precursor limestone. However, this study cannot separate the porosity-destroying effects of pressure dissolution and net import of carbonate.

Both porous and low-porosity dolomite occur on the Island of Bonaire, Netherlands Antilles. The Plio-Pleistocene age and lack of extensive burial, as well as the intimate association of dolomite and limestone, minimize the possibility for postdolomitization compaction effects and provide an excellent opportunity for direct comparisons between dolomites and precursor limestones. In addition, the island setting constrains the source of magnesium to seawater, and the types of water to seawater, evaporated seawater and rainwater.

LOCATION AND GEOLOGICAL SETTING

The island of Bonaire, Netherlands Antilles, is located in the Caribbean Sea 50 miles north of Venezuela (Fig. 1). The island consists of a core of folded upper Cretaceous volcanic flows and tuffs and igneous intrusions (Pipers, 1933). Cretaceous rocks are exposed in low-lying hills in the northern part of the island, and are overlain and flanked by Eocene and Plio-Pleistocene carbonates (Bandoian & Murray, 1974). The southern part of Bonaire contains flat-lying exposures of Pleistocene terrace limestones, partially covered by Holocene carbonates and evaporites deposited in a tidal-flat environment (Lucia, 1968). Holocene dolomite was first documented in these tidal-flat deposits by Deffeyes *et al.* (1964).

Plio-Pleistocene carbonates in the northern part of the island occur as inclined and horizontal beds of

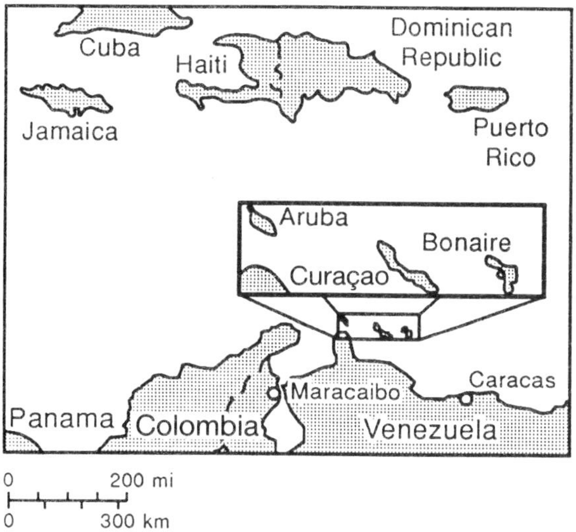

Fig. 1. Map of the southern Caribbean Sea, showing the location of Bonaire, Netherland Antilles.

the Seroe Domi Formation (DeBuisonje, 1974), which are onlapped by younger flat-lying terrace deposits. The Seroe Domi Formation (Fig. 2) was deposited in a foreslope to lagoonal setting encasing the volcanic core, whereas the terrace deposits were deposited in a lagoonal environment. Large corals and grain-dominated sediment composed of fragments of corals, foraminifera, echinoids, molluscs and coralline algae indicate an open-marine depositional environment. Mud-dominated sediments are uncommon. Pebbles and boulders of Cretaceous volcanic material occur in the Seroe Domi foreslope deposits near the contact with the Cretaceous core (Bandoian & Murray, 1974).

Dolomitized Plio-Pleistocene limestones occur in the uppermost exposures of the inclined beds. A karst surface caps the dolomite and the presence of dolomite fragments in the overlying limestone terra rosa indicates that dolomitization preceded karsting. Bandoian and Murray (1974) reported that infill

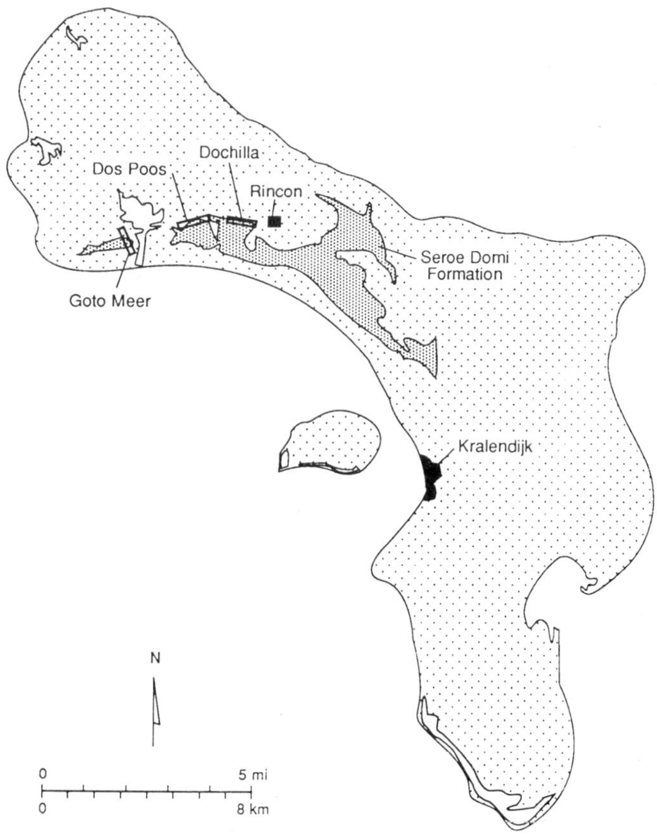

Fig. 2. Map of Bonaire, NA, showing the location of the Seroe Domi Formation and the Dochilla, Dos Poos and Goto Meer sections.

sediment in a younger karst surface has been dolomitized, suggesting a separate dolomitization episode.

METHODS

Samples were slabbed for macroscopic description and thin-sectioned for microscopic description using both light and cathodoluminescence microscopy. The porosity and permeability of Dochilla and Dos Poos samples were measured from plugs using standard laboratory techniques. The porosity of Goto Meer samples was measured by point-counting thin sections. Point-count porosity values were normalized with laboratory porosity values by comparing values measured on similar samples. Strontium, magnesium and calcium compositions of cements and replacement dolomite were measured from polished thin sections by microprobe analysis.

Analysis for sulphate was conducted on samples that were acid-digested and analysed by ion chromatography. Whole-rock samples for stable carbon and oxygen isotope analysis were first analysed for mineralogy by X-ray diffraction. Isotope analysis was made on carbon dioxide extracted by phosphoric acid digestion, following the methods of McCrea (1950). Stable isotopic analysis was by gas ratio mass spectrometry. Samples for strontium, magnesium and calcium analysis were first leached in dilute EDTA (ethylenediamine tetra-acetic acid) to remove calcite (method of Glover, 1961) and verified to be 100% dolomite by X-ray diffraction. Samples were then dissolved in dilute HCl and analysed for cation composition by inductively coupled plasma mass spectrometry. Dolomite unit cell dimensions were measured by X-ray diffraction using the method of Goldsmith *et al.* (1961). Strontium isotope analysis was by solid-source mass spectrometry.

TEXTURAL HISTORY

Limestone

Limestones adjacent to dolomites are mouldic packstones and grainstones composed of coral, foraminifera, echinoids, molluscs and coralline algae fragments (Fig. 3). The average porosity is 25% and average permeability is 1000 mD (Fig. 4). Some moulds are clearly dissolved mollusc fragments, whereas others, by analogy with samples of modern sediment from offshore Bonaire, are dissolved aragonitic coral fragments. Although the limestones are composed of low-Mg calcite, the skeletal fabric of echinoids, coralline algae and foraminifera is well preserved. In addition to mouldic porosity, intergranular, intragranular and intragranular–microporosity pore types are present.

Dissolution of aragonitic coral and mollusc fragments and precipitation of calcite rim and bladed cements in the intergranular, intragranular and

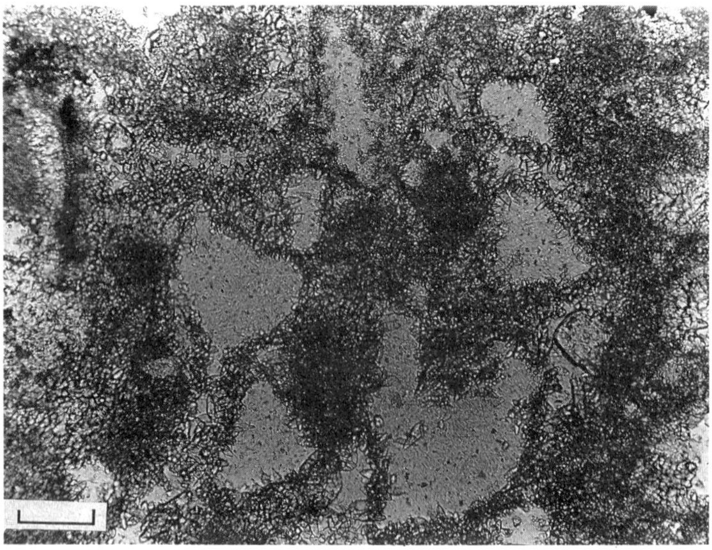

Fig. 3. Photomicrograph of limestone from the inclined beds, showing intergranular and mouldic porosity due to dissolution of probable aragonitic coral fragments and bladed calcite crystals that have grown in the mouldic intergranular pores. Scale bar = 100 μm.

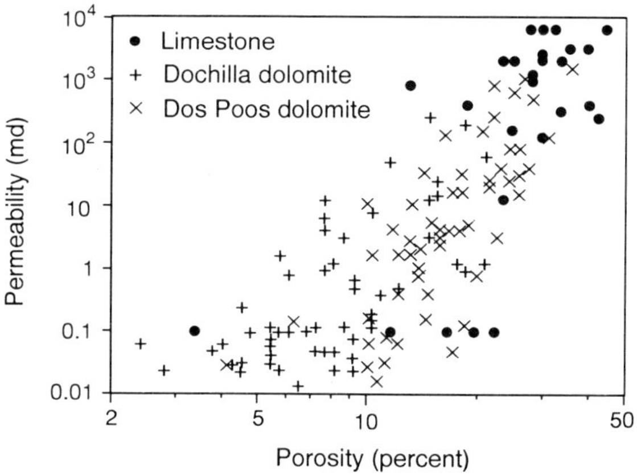

Fig. 4. Cross-plot of porosity and permeability from dolomite samples at Dochilla and Dos Poos and from limestone samples.

mouldic pore space are important predolomitization diagenetic events. Many of the grains have been completely removed, and the grain shape is preserved by intergranular cement. These same moulds, however, have bladed calcite cement within them, suggesting that the grains were being dissolved within the same timespan that the cement was being deposited. This observation is consistent with the concept that the dissolution of unstable aragonite grains is coeval with precipitation of stable low-Mg calcite cement in juxtaposed pore space.

Dolomite

Dolomite is composed of crystals that range in size from a few μm to approximately 200 μm, with a mean size of 75 μm. Pore types present are intercrystalline (Fig. 5), intrafossil and mouldic. Porosity varies from 5% to 25%. Permeability varies from 0.01 to 1000 mD, and is correlated with porosity (Fig. 4). The variation in crystal size is, in part, related to the nature of the material being replaced: coralline algae are composed of microcrystalline dolomite, echinoderms commonly are single large dolomite crystals, and corals are replaced by small dolomite crystals. Voids are filled with large dolomite crystals. Large dolomite crystals commonly include small crystals, most of which appear to be bladed calcite crystals similar to those found in limestone samples. X-ray diffraction data suggest that these calcite crystals are low-Mg calcite. Observations by R.L. Folk (personal communication, 1991) suggest that at least some of the included crystals are anhydrite, based on crystal shape and birefringence. Geochemical analysis confirms the presence of sulphate in the dolomite crystals, as discussed below.

The fabric of the limestone at the time of dolomitization is a key issue for understanding the history of porosity during dolomitization. A key observation is the change that occurs in the transitional area between the limestone and the dolomite. Dolomite crystals preferentially grow in the mud matrix and microporous grains. It is common, however, to see large dolomite crystals extending from

Fig. 5. Photomicrograph of high-porosity dolomite taken in cross-polarized light, showing intercrystalline porosity. Scale bar = 200 μm.

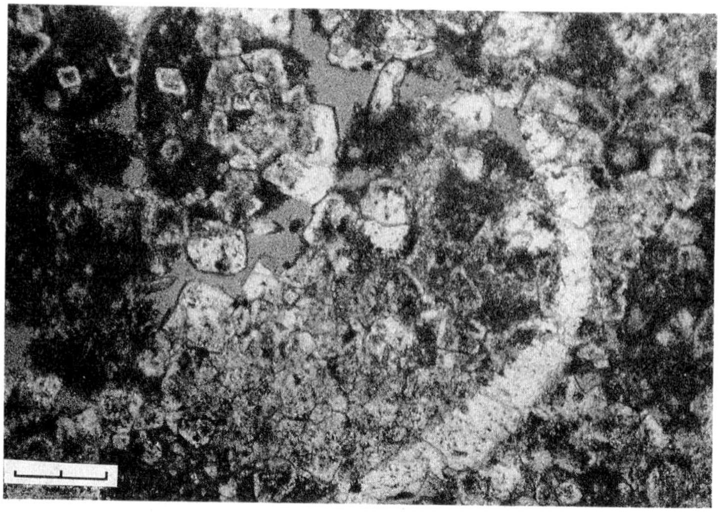

Fig. 6. Photomicrograph of dolomitic limestone, showing dolomite crystals growing in microporous limestone and extending into intergranular and fossil mouldic pore space. Scale bar = 200 μm.

the replacement host into adjacent pore space (Fig. 6). Dolomitization is not uniform, but dolomite crystals tend to be clustered in patches. The last to be dolomitized are coralline algae and echinoderm fragments.

A second key observation is the relict fabrics in the dolomite. At Dochilla, Dos Poos and Goto Meer, the relict fabrics suggest a fabric of the precursor limestone that is similar to that of the adjacent limestones. The amounts of coralline algae, echinoderm fragments and coral fragments are similar between the limestone and the dolomite. The patterns of inclusions suggest a leached fabric similar to the present limestone fabric (Fig. 7). This interpretation is supported by the presence of many bladed calcite crystals included in dolomite crystals, extending in both directions from the 'wall' of a relict mould. In partially dolomitized limestones, dolomite crystals extend into leached voids and include bladed calcite cement. Thus, it appears reasonable that the rock at the time of dolomitization was similar to the present-day limestone, and therefore had porosities similar to the present-day limestone.

Poikilotopic calcite is found in inclined-bed dol-

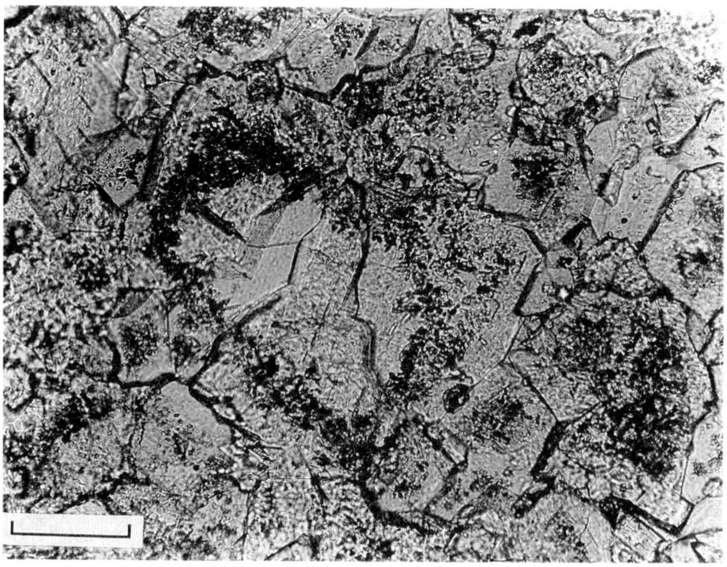

Fig. 7. Photomicrograph of relict structures in non-porous dolomite, showing mouldic pore space (leached coral fragment?) with bladed calcite crystals extending into the grain mould and the mouldic pore space occluded by dolomite crystals. No pore space is visible in this view. Scale bar = 100 μm.

Fig. 8. Photomicrograph of postdolomite sparry calcite and dolomite crystals with included bladed calcite crystals. Scale bar = 100 μm.

omite in the top of the Dochilla and Goto Meer sections. Relict patterns and bladed calcite crystals are included in the associated dolomite crystals (Fig. 8), and the coralline algae and echinoid fragments are dolomite similar to dolomites lower in the section. These textural relationships suggest that the poikilotopic calcite is post dolomitization. Lucia (1972) has shown that postdolomite calcite can be formed by dissolution of sulphate minerals (gypsum and/or anhydrite) and subsequent precipitation of calcite. The presence of sulphate inclusions in the dolomite crystals points to the past presence of sulphate minerals, and it is suggested that the postdolomite poikilotopic calcite is a replacement of gypsum or anhydrite.

DISTRIBUTION OF DOLOMITE AND POROSITY

The distribution of dolomite and porosity was mapped in three localities: Dochilla, Dos Poos and Goto Meer, and samples were collected for porosity measurement, geochemical analysis and thin-section description. The cross-section at Dochilla, approximately 1800 ft long and 160 ft high, shows a 100-ft thick dolomite unit changing to limestone in a downdip direction (Fig. 9). This dolomite unit overlies a limestone bed. The detailed cross-section at Dos Poos is approximately 350 ft long and 100 ft high and shows an 80-ft thick dolomite unit changing downdip to limestone (Fig. 10). Both cross-sections are oblique to true dip.

A porosity map of the Dochilla cross-section based on the results of core-plug measurements (Fig. 9) shows that low porosities are concentrated updip, and high porosity values are found downdip near the limestone–dolomite contact. The Dos Poos cross-section is focused on the limestone–dolomite boundary, and no updip porosity values are available due to outcrop constraints. The porosity map of the Dos Poos cross-section (Fig. 10) clearly shows that porosity values in the dolomite (mean porosity of 17%) are much higher than the updip porosity values in the Dochilla dolomite (mean porosity of 5%). In contrast to the dolomite porosity, the limestones adjacent to the dolomite at Dochilla and Dos Poos have porosity values ranging from 10% to 45%, with a mean value of 25%.

The geology of the area west of the Goto Meer is described by Bandoian and Murray (1974). The vertical sequence consists of approximately 360 ft of inclined foreslope deposits directly overlying volcanic rocks. Southeastern dips decrease upsection, from 30° to 38° in the lower section to 22°–10° in the upper section. Dolomite beds are located in the uppermost inclined beds. The dip of the beds at the crest of Goto Meer hill is nearly horizontal, suggesting that dipping foreslope deposits have been replaced by flat-lying lagoonal sediments. The uppermost unit that contains the dolomite beds is overlain by a karst surface, and blocks of dolomite

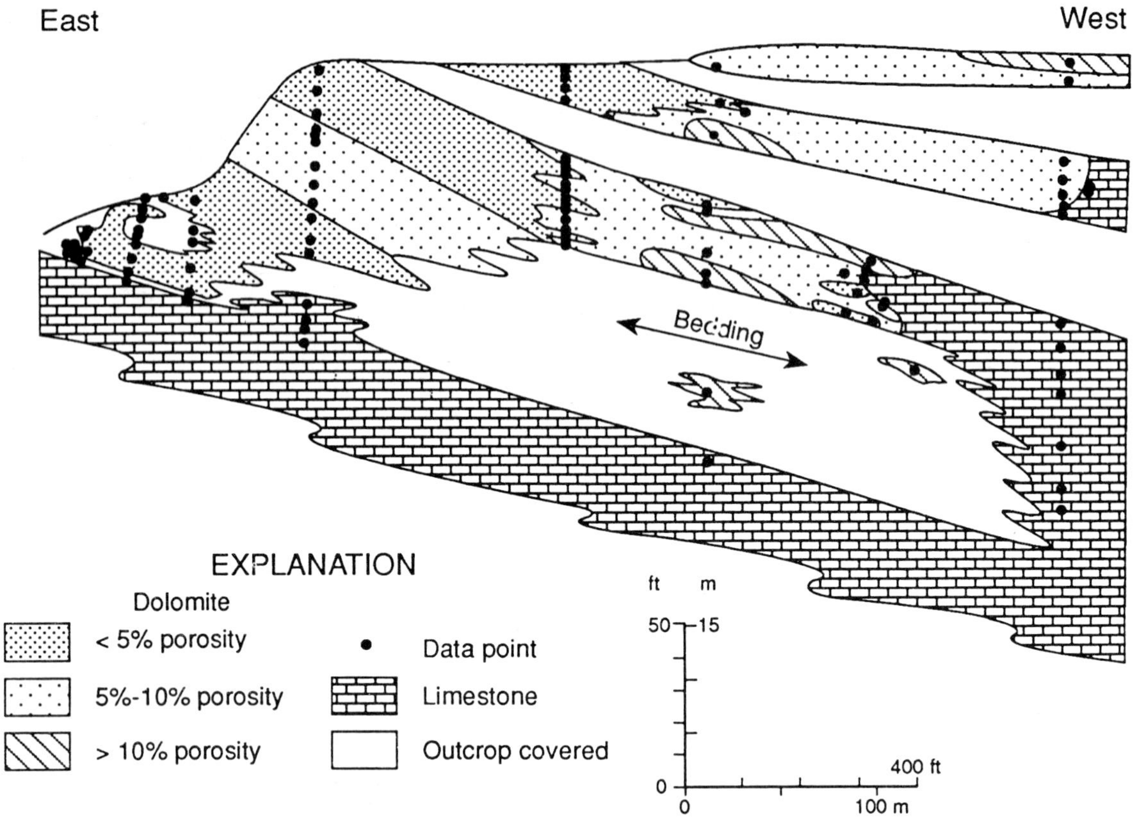

Fig. 9. Cross-section of the Dochilla outcrop, showing lithology and porosity distribution in dolomite.

are present in the overlying red terra-rosa material. The karst surface can be traced downward to an elevation of approximately 300 ft (Bandoian & Murray, 1974) and the underlying dolomite has been locally calcified associated with this karst (Fig. 11). The karst surface is overlain by a nearly horizontal partially dolomitized limestone.

A map of the dolomite in the Goto Meer section shows a massive dolomite unit that fingers out downdip into limestone beds (Fig. 11). Porosity measurements were made on 20 samples from five sample locations (Fig. 11) by point-counting pore space in thin sections. Comparisons with laboratory porosity values measured on similar samples show that values from dolomite samples are similar, whereas values from limestone and partially dolomitized limestone samples are underestimated by approximately 5% porosity due to the presence of microporosity. These results show that the partially dolomitized limestones located at the downdip limit of dolomite have an average normalized porosity value of 20%, the dolomites near the limestone–dolomite contact have an average porosity of 14% (referred to as high-porosity dolomite in Fig. 11), and the updip dolomites have an average porosity of 4% (referred to as low-porosity dolomite in Fig. 11).

GEOCHEMISTRY

Cathodoluminescence of the carbonates at Goto Meer shows systematic changes in a downdip direction. Limestone is non-luminescent, but dolomite has excellent luminescence that can be divided into three luminescence zones. Although there is considerable colour banding in each zone, the inner zone is characterized as bright yellow, the middle zone as dull orange and the outer zone as bright yellow. The bright inner zone is not present in the downdip partially dolomitized limestones, whereas

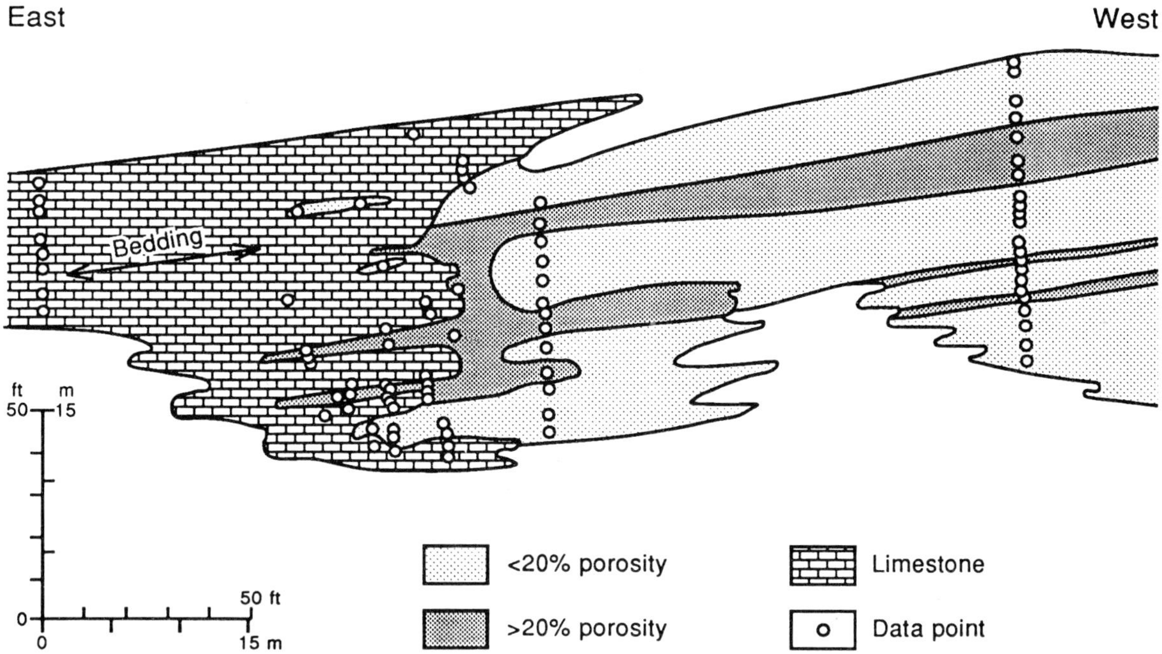

Fig. 10. Cross-section of the Dos Poos outcrop, showing lithology and porosity distribution in dolomite. The dolomite has a mean porosity value of 17%.

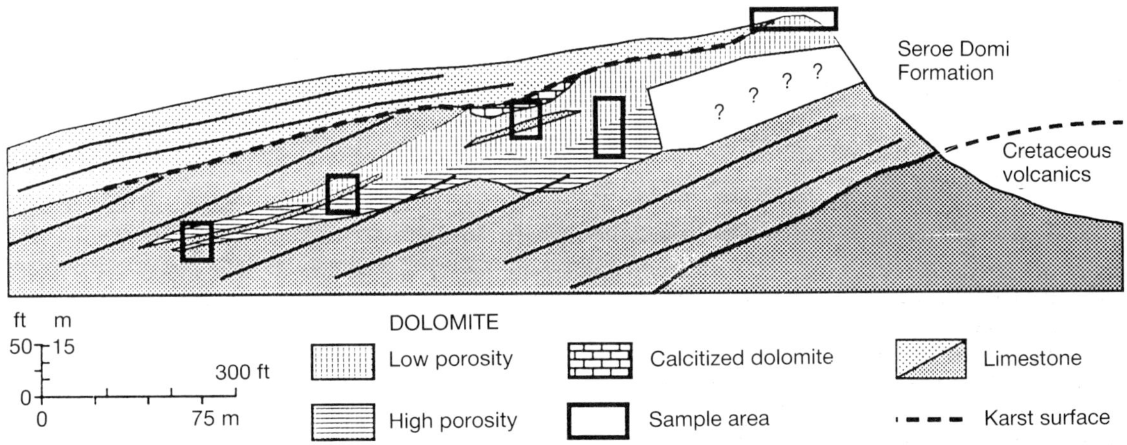

Fig. 11. Cross-section of the Goto Meer outcrop, showing lithology, karst surface, porosity distribution in dolomite and location of sampling areas. The high-porosity area has an average porosity of 14% and the low-porosity area has an average porosity of 4%.

the bright outer zone is not well developed in the updip, low-porosity zone (Fig. 12). This observation suggests that the dolomitizing water did not reach the extremities of the dolomite body during the initial dolomitization stage, and that there was limited pore space available for precipitation of dolomite in the updip dolomite during the last stage of dolomitization.

Dolomite unit cell dimensions, Mg/Ca ratios and strontium compositions have a spatial pattern some-

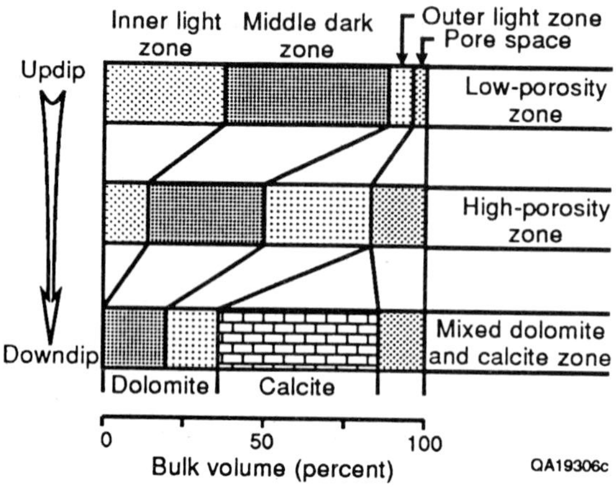

Fig. 12. Diagrammatic representation of the changes in cathodoluminescence zones from updip to downdip at the Goto Meer outcrop.

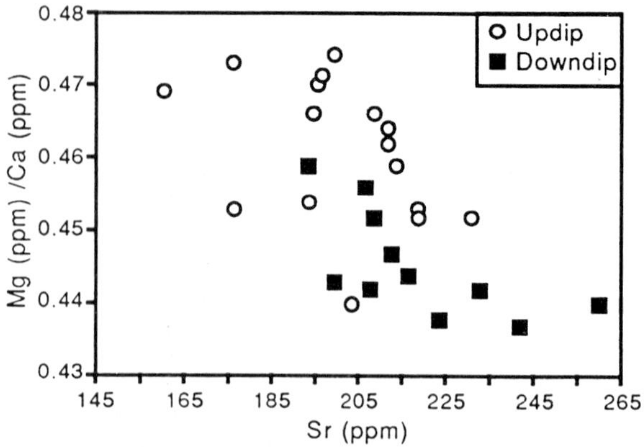

Fig. 14. Cross-plot of Mg/Ca ratios and strontium concentrations in dolomite crystals from Goto Meer samples, showing changes from updip to downdip.

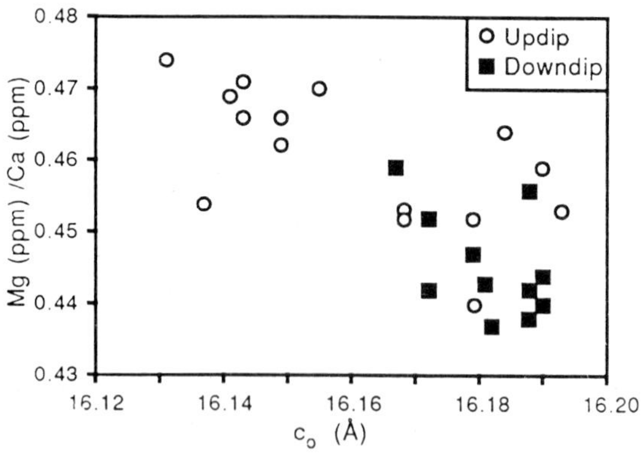

Fig. 13. Cross-plot of Mg/Ca ratios and unit cell dimensions of dolomite crystals from Goto Meer samples, showing changes from updip to downdip.

what similar to porosity. Updip dolomites have more compact unit cells, higher Mg/Ca compositions (are more nearly stoichiometric) and lower strontium concentrations, whereas downdip dolomites, adjacent to the limestone contact, have more expanded unit cell dimensions, lower Mg/Ca ratios (are less stoichiometric) and higher strontium concentrations (Figs 13 and 14).

The observation that dolomite porosity increases from updip to downdip suggests that updip dolomite contains more cement. This idea was tested by collecting magnesium, calcium and strontium concentrations by electron microprobe analysis of both dolomite cement and replacement dolomite. Although there is some overlap in these data, cement generally has a higher Mg/Ca ratio than does replacement dolomite (Fig. 15). Thus, dolomite cement is more nearly stoichiometric than replacement dolomite, and we suggest that the whole-rock relationships (Figs 13 and 14) are principally a consequence of a greater amount of dolomite cement in the updip parts of these outcrops.

A subset of six dolomite samples containing small, highly birefringent inclusions were analysed for sulphate. The results of these analyses indicate bulk-rock sulphate concentrations of between 95 and 180 ppm, with a mean value of 146 ppm, strongly suggesting that the inclusions are calcium sulphate.

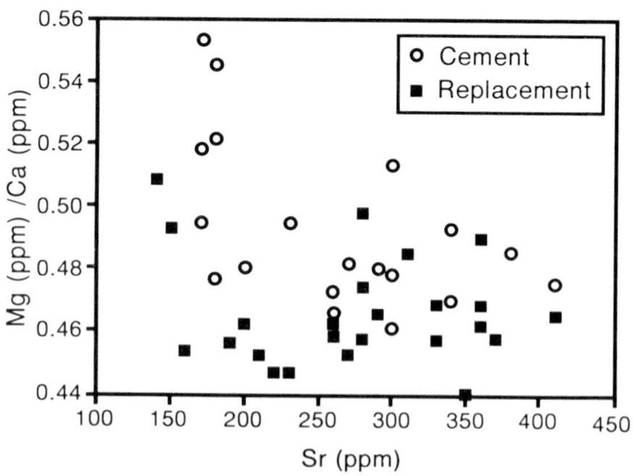

Fig. 15. Cross-plot of Mg/Ca ratios and strontium concentrations in dolomite cements and replacement dolomite from Goto Meer samples.

Dolomite and limestone samples have distinctly different oxygen and carbon isotope compositions. Dolomite compositions cluster in the range $\delta^{18}O$ = 0.5–2.0‰ and $\delta^{13}C$ = 0–1.5‰, and no spatial correlation of these values was observed (all isotopic values reported relative to the PDB standard). Limestones, in contrast, have a wider range of isotopic compositions: $\delta^{18}O$ = −4.5 to −2.0‰ and $\delta^{13}C$ = −8.0 to −2.0‰. Dolomite isolated from dolomitic limestone samples has the same isotopic composition as 100% dolomite samples (Fig. 16).

Two samples of the Seroe Domi dolomite from the top of the Goto Meer section were chosen for strontium isotope analysis. The two samples have $^{87}Sr/^{86}Sr$ compositions of 0.708931 ± 0.000012 and 0.708989 ± 0.000011, values which fall within the range of the Seroe Domi Formation dolomite values reported by Vahrenkamp *et al.* (1991) and by Fouke *et al.* (1992). By comparison with the seawater secular $^{87}Sr/^{86}Sr$ values of Hodell *et al.* (1991), these values correspond to seawater values as old as late Miocene. However, inasmuch as theolitic basalt fragments from the underlying volcanics are common minor constituents of the Seroe Domi dolomite, contamination of the dolomitizing water by non-radiogenic strontium is very possible. Thus, the strontium isotopic composition of the dolomite probably does not represent seawater composition at the time of dolomitization.

The postdolomite age of the sparry calcite is supported by a study of the magnesium content of the calcites using X-ray diffraction. The limestones have $MgCO_3$ values of 1 mole % or less, typical of low-Mg calcite. The calcite included in the dolomite crystals also has a $MgCO_3$ value of approximately 1 mole %. The sparry calcite, however, has $MgCO_3$ values of between 1.6 and 4.4 mole %, indicating a unique origin for these calcites.

DISCUSSION

Porosity evolution during dolomitization

Porosity occlusion by dolomitization occurs on three scales. Individual dolomite crystals replace limestone that contains microporosity (Fig. 6). Therefore, the addition of sufficient magnesium and carbonate from outside the crystal is needed to fill the pore space, as proposed by Murray (1960). Within highly porous limestones there are millimetre-sized patches of low-porosity dolomite. It is reasonable that the

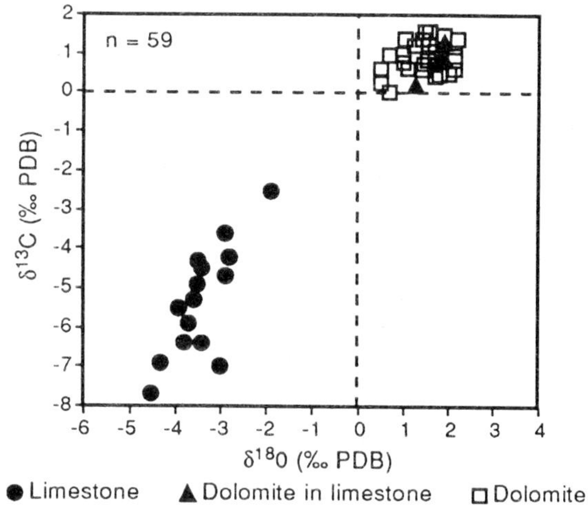

Fig. 16. Cross-plot of carbon and oxygen stable isotopic compositions from calcite and dolomite.

low-porosity dolomite areas replaced areas of high-porosity limestone, and therefore the addition of sufficient magnesium and carbonate from outside the dolomite area is required to fill the pore space. Porosity increases systematically in a downdip direction from low- to high-porosity dolomite on the scale of hundreds of feet. Again, the addition of sufficient magnesium and carbonate from outside the dolomite body is required to fill pre-existing pore space.

The requirement to add carbonate from outside the dolomite on all three scales suggests that the mole-for-mole replacement model does not apply to these rocks. The mean porosity of the precursor limestone is 25%. The resulting dolomite should have a mean porosity of 35% if dolomitization occurred by adding only magnesium to the system (Fig. 17). The resulting dolomite has a mean value of 11%, however, suggesting that carbonate as well as magnesium has been added to the system.

The change in porosity and resulting chemical requirements can be illustrated by a ternary diagram of porosity, limestone and dolomite (Fig. 18). Partially dolomitized limestones have porosity values similar to the porosity values of pure limestone. To maintain a porosity similar to the precursor limestone, sufficient magnesium and carbonate must be added to the system to offset the porosity gain due to the loss in mineral volume resulting from changes in molar volumes. This amount is 12.5% of the mineral volume. Once the limestone is completely converted to dolomite, occlusion of the

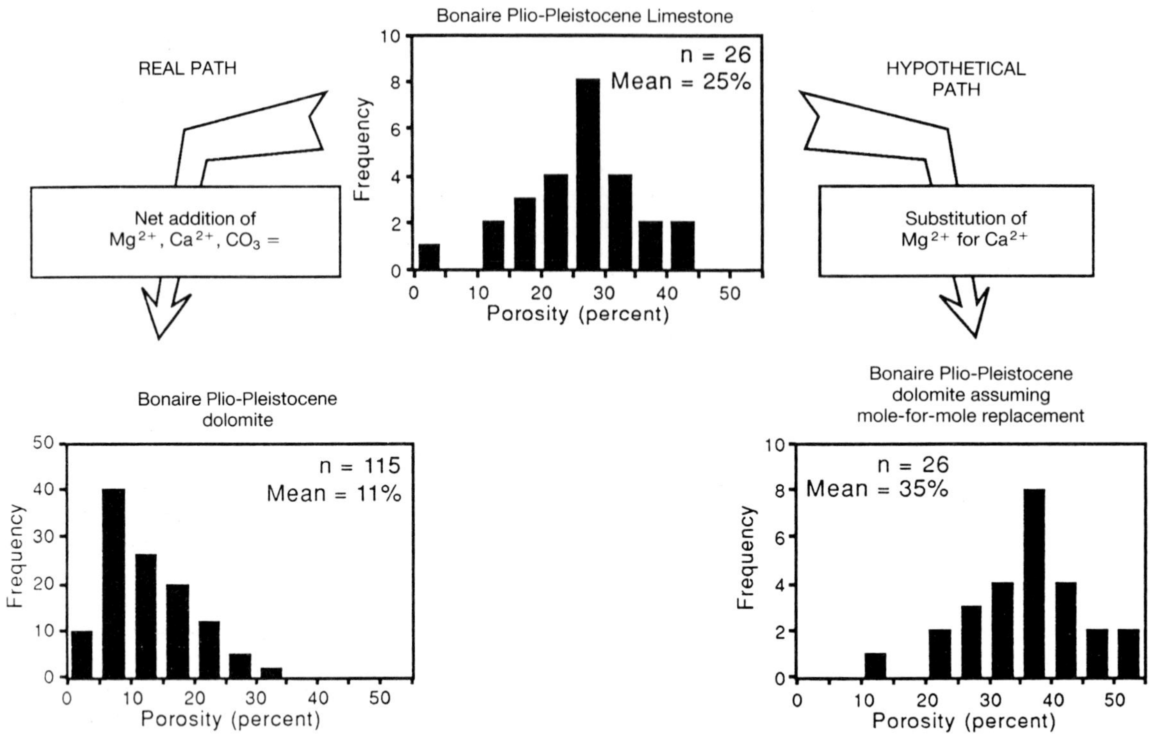

Fig. 17. Porosity frequency plots for dolomite, the precursor limestone, and for a hypothetical dolomite calculated from the limestone frequency plot assuming mole-for-mole replacement.

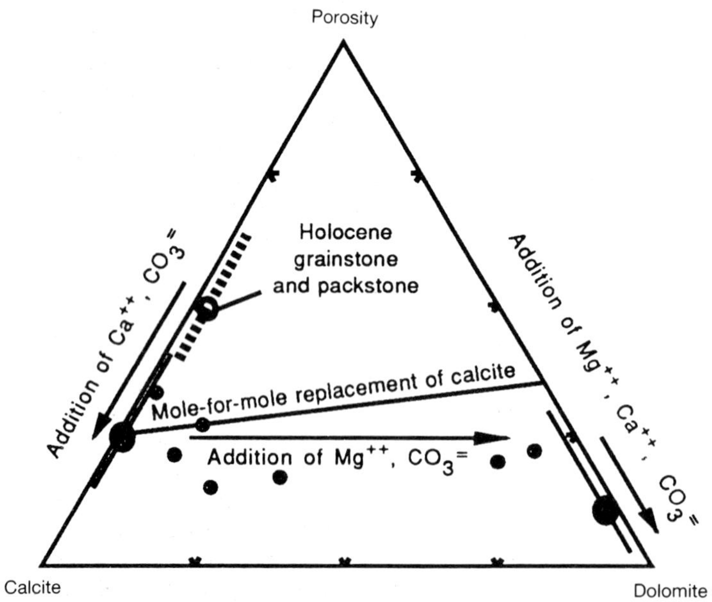

Fig. 18. Ternary diagram showing the evolution of porosity through dolomitization during (1) the replacement phase, during which little change in porosity occurs, and (2) the pore-filling phase, during which considerable porosity can be lost. Holocene grainstone and packstone data are from Enos and Swatsky (1981).

remaining pore space continues by the addition of calcium, magnesium and carbonate. In this case, the resulting mean dolomite porosity of 11% and the initial dolomite porosity of 25% suggests the addition of approximately 14% of the dolomite volume during this phase.

The above conclusions assume that aragonite has been converted to low-Mg calcite before dolomitization. Because aragonite has a density of 2.95 and calcite 2.71, the molar volumes are different and, therefore, the gain in pore volume from mole-for-mole replacement by dolomite is 12.5% if the precursor is calcite and 4.5% if it is aragonite. The petrographic data presented here support the assumption that the limestone was composed of calcite at the time of dolomitization.

Sibley (1980), however, suggested that, although mouldic porosity was formed before dolomitization through the dissolution of aragonite grains in meteoric water, the limestone at the time of dolomitization was composed primarily of aragonite and high-Mg calcite. It seems unreasonable that most of the sediment would retain aragonite and high-Mg calcite after undergoing a meteoric diagenetic event that created extensive mouldic porosity. Steinen and Matthews (1973) concluded from their study of Barbados meteoric diagenesis that any carbonate that was exposed to a freshwater phreatic lens was completely converted to low-Mg calcite, contained significant calcite cement, and developed mouldic and solution-enlarged porosity. Indeed, studies of modern freshwater lenses imply a half-life of 6000–7000 years for aragonite-to-calcite transformation (Vacher *et al.*, 1990), and Budd (1988) concluded from his study of Schooner Cays, Bahamas, that all aragonite would be converted to calcite within the timespan of a Milankovich sea-level cycle. Because high-Mg calcite is less stable than aragonite, it seems reasonable that the transformation rate to low-Mg calcite would be even faster. In the partially dolomitized limestones on Bonaire, low-Mg calcite coralline algae and echinoderm fragments are the last to be replaced, suggesting that these grains were converted to low-Mg calcite before dolomitization. This conclusion was also reached by Sibley (1980).

Hydrology and geochemistry of dolomitizing fluids

The location of dolomite in the uppermost units of the inclined foreslope beds and in the more horizontally bedded lagoonal units found at the crest of Goto Meer hill suggests that the dolomitizing fluids migrated from an overlying formation down into the lagoonal and foreslope beds (Deffeyes *et al.*, 1965). In addition, the presence of thin dolomite beds intercalated within limestone at the downdip dolomite limit is suggestive of gravity flow.

Sibley (1980) concluded that the dolomites on Bonaire did not form from downward flow of water but by lateral flow in a mixing zone between a fresh- or brackish-water lens and seawater. He suggested that, during regression, the freshwater lens descended and dolomitized the section from the top downwards.

The geometry of the dolomite body does not support dolomitization confined to a mixing zone descending through the section during regression. The geometry of a mixing zone can be generalized as saucer-shaped, with a freshwater phreatic zone above and connate seawater below. The thickness of the mixing zone and the freshwater phreatic zone is difficult to estimate, but Matthews and Frohlich (1987), from their studies on Barbados, suggested a mixing-zone thickness of 3 ft and a freshwater phreatic zone of 16 ft. Dolomitization is restricted to the mixing zone. The freshwater phreatic zone remains limestone, according to the mixing-zone model, and can only be dolomitized by raising the mixing zone, presumably during a relative rise in sea level. The dolomite on Bonaire is continuous to the top of the exposures, suggesting that there is no freshwater phreatic-zone limestone. In addition, the dolomite beds conform to the dip of the limestone beds in the foreslope deposits, and do not cut across bedding, as might be expected from a flat-lying to saucer-shaped mixing-zone geometry. For these reasons, it is difficult to accept Sibley's (1980) mixing-zone model of dolomitization during regression.

Geochemical data support the hypothesis that the dolomitizing fluid was hypersaline seawater. Anhydrite inclusions in the dolomite crystals, as confirmed by chemical analysis, are conclusive evidence that hypersaline water was present during dolomitization. Poikilotopic calcite in the uppermost dolomites, which have characteristics suggesting calcitized anhydrite, supports this interpretation. The positive $\delta^{18}O$ values in the dolomite are consistent with a marine origin for the dolomitizing fluid. The stable oxygen isotope values for the dolomite are similar to those found in Holocene dolomite in the hypersaline lake on South Bonaire (the Pekelmeer), suggesting a semiarid, humid climate during the dolomitization period (Major *et al.*, 1992).

Predictive model

Key constraints for a geological model that predicts the distribution of dolomite and dolomite porosity are the chemistry of the dolomitizing fluid, the source of the fluid, the flow paths of the fluid, the nature of the precursor limestone, and the timespan of the dolomitization process. In a hypersaline-reflux dolomitization system, the source of hypersaline water is an evaporitic tidal flat. Fluid flow downward and basinward results from the high topographic position of the tidal flat and the high density of the hypersaline water (Deffeyes *et al.*, 1965; De Groot, 1973). The evaporation of seawater and the resulting precipitation of $CaSO_4$ concentrates magnesium relative to calcium, providing a concentrated source of magnesium required for dolomitization.

The precursor limestone can have a wide range of porosity, mineralogy and texture, and output of the predictive model will depend heavily upon the predolomite history. There are a number of possible predolomite histories. Subtidal sediments underlying evaporitic tidal flats may have porosity values typical of uncemented sediments, as described by Enos and Sawatsky (1981), or they may contain marine cement and have significantly reduced porosity values (Shinn, 1969; James & Ginsburg, 1979). The underlying sediments may have undergone a period of meteoric diagenesis before dolomitization, and have characteristics similar to the Bonaire Plio-Pleistocene carbonates described in this study. The sediments may have undergone extensive karsting and burial diagenesis, significantly reducing porosity before being covered by an evaporitic tidal flat.

The timespan of the hypersaline-reflux dolomitization process will be equivalent to the length of time that evaporitic tidal-flat conditions exist. The maximum timespan for a single Milankovich sea-

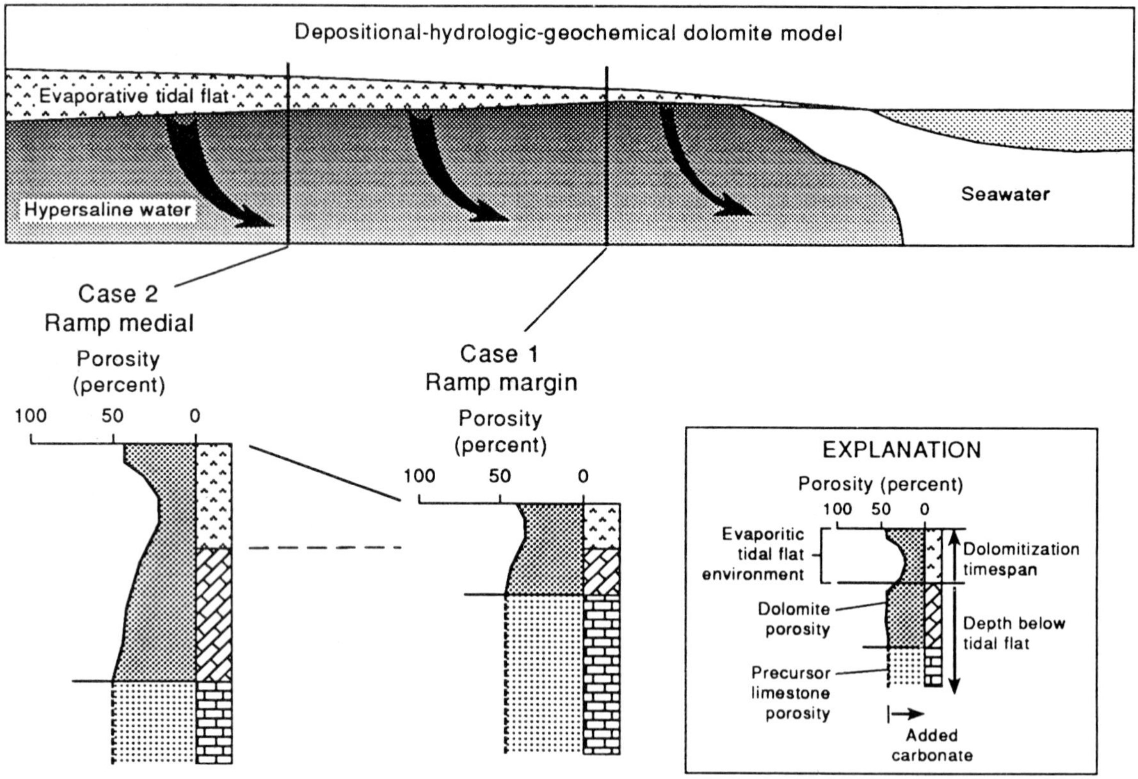

Fig. 19. Predictive dolomite-porosity model of one tidal-flat-capped cycle, showing hydrochemical model and resulting porosity profiles.

level cycle is 100 000 years. During this period the tidal flat will prograde basinward, and the timespan for tidal-flat conditions will be longest landward and shortest basinward. Landward tidal-flat deposition will occupy most of the 100 000 years; basinward, very little.

A significant volume of dolomite can be produced in 100 000 years. The Plio-Pleistocene age of the Bonaire dolomites constrains the dolomitization time to the scale of a million years or less. Bodies of dolomite 100 ft thick were made within this timespan. Deffeyes *et al.* (1965) calculated that sufficient magnesium could be provided to form the Bonaire dolomite in 100 000 years, using the Pekelmeer hydrochemical model and assuming no net import of carbonate. The requirement that a significant amount of carbonate will also have to be imported will have a major impact on the dolomitization rate. The amount of carbonate available for dolomitization is at present not known.

A predictive model for one tidal-flat-capped cycle is illustrated in Figure 19. The timespan for dolomitization is shown diagrammatically by the thickness of the tidal flat, the depth of dolomitization is shown by the distance below the tidal flat, and the porosity of the dolomite is shown schematically in the tidal-flat interval and is plotted against depth below the tidal flat. The estimated porosity for the precursor limestone is shown, and the loss of porosity in the dolomite interval represents the addition of carbonate (porosity occlusion by dolomite). The amount of porosity occlusion is related to the volume of fluid flow, which is related to the timespan of evaporitic tidal-flat conditions. Near the ramp

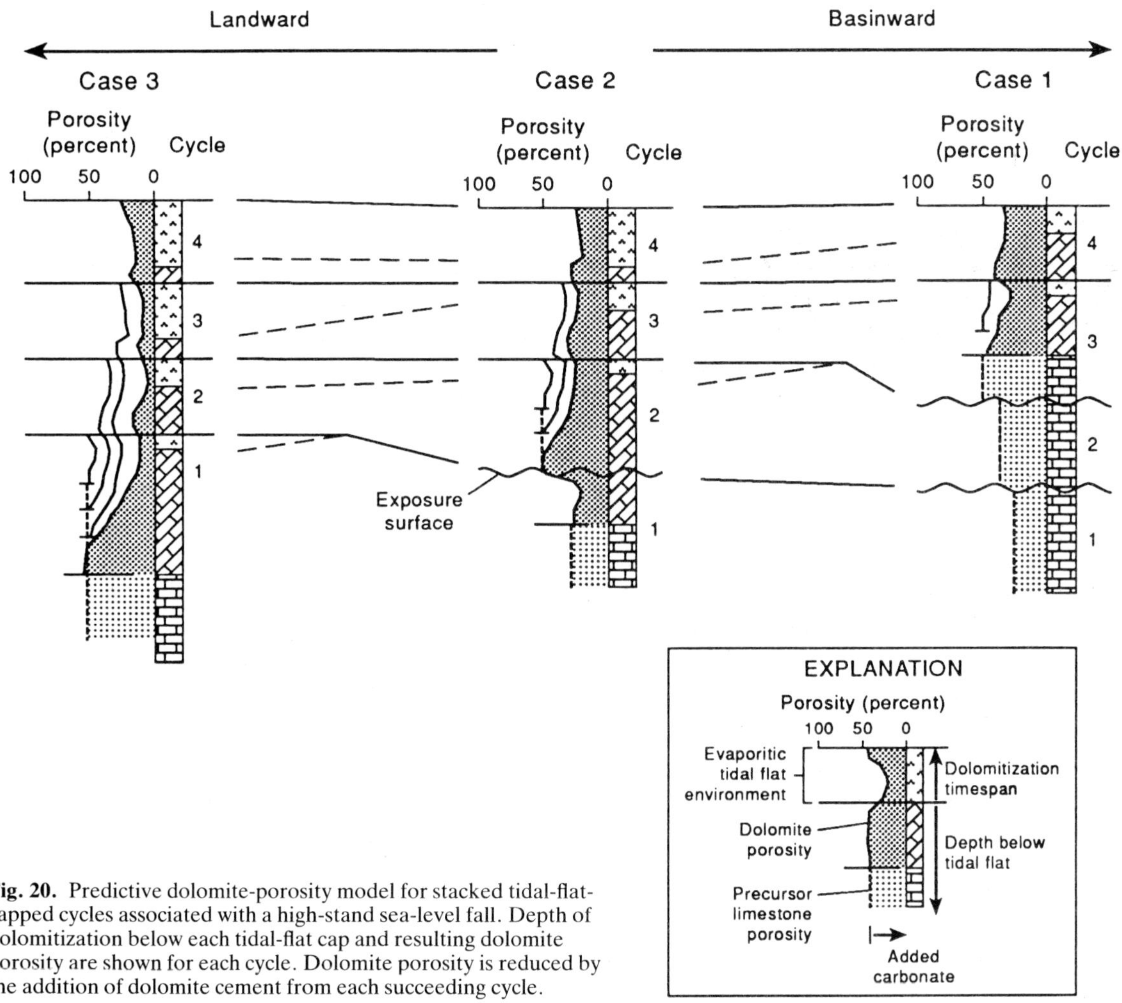

Fig. 20. Predictive dolomite-porosity model for stacked tidal-flat-capped cycles associated with a high-stand sea-level fall. Depth of dolomitization below each tidal-flat cap and resulting dolomite porosity are shown for each cycle. Dolomite porosity is reduced by the addition of dolomite cement from each succeeding cycle.

margin (case 1) the tidal-flat timespan is short, the volume of fluid flow is small, the depth of dolomitization shallow, and little porosity occlusion by dolomitization will occur. In a ramp medial position between basin and land (case 2), the timespan will be longer, the volume of fluid flow larger, the depth of dolomitization deeper, and porosity occlusion will be significant in and immediately underlying the tidal-flat sediments.

Hypersaline-reflux systems are commonly associated with stacked tidal-flat-capped cycles. A predictive model of stacked tidal-flat-capped cycles associated with a high-stand sea-level fall is illustrated in Figure 20. In this model, the volume of porosity occluded during each cycle is subtracted from the previous porosity profile. In addition, the timespan for tidal-flat conditions increases with each subsequent cycle, resulting in more porosity occlusion in the tidal-flat interval. The model (Fig. 20) shows the highest porosity values in the lowest subtidal intervals. The porosity varies with position landward and seaward, as well as with precursor limestone porosity. Case 1 is basinward and contains only two tidal-flat-capped cycles. It has higher porosity than case 2, which is in a more medial position and contains three tidal-flat capped cycles. The effect of an exposure surface and associated reduced porosity in the precursor limestone is illustrated in case 2 (Fig. 20). In this case, instead of increasing with depth, dolomite porosity decreases with depth. Case 3 (Fig. 20) is the most landward example, has four tidal-flat-capped cycles and the lowest dolomite porosity.

These conceptual models illustrate how porous and non-porous dolomites can form without compaction associated with burial diagenesis. In all cases the dolomite contains less porosity than the precursor limestone, and in all cases the most porous dolomite is located in the subtidal intervals. Increased overburden pressure due to burial will undoubtedly cause compactional loss of porosity, and it is likely that limestone will compact more than dolomite, producing less porosity in limestones than in adjacent dolomites, as observed by Schmoker and Halley (1982).

CONCLUSIONS

The mole-for-mole replacement theory of dolomitization does not apply to the Plio-Pleistocene carbonates of Bonaire. Indeed, there is no published geological evidence that the porosity-through-dolomitization theory applies to any sequence of carbonate rocks. Dolomite porosity values will always be equal to or less than the precursor porosity values, suggesting that the characteristics of the precursor limestone are perhaps the single most important factor in the evolution of porosity during dolomitization.

Dolomitization can be divided into two phases: a replacement phase, which exists as long as calcite is present, and a pore-filling phase, which occurs after all calcite has been replaced. Together with the addition of Mg, there is a net addition of carbonate during both phases and a net addition of calcium during the pore-filling phase. Sufficient volumes of dolomite are added during the replacement phase to offset the volume loss produced by the difference in molar volumes between calcite and dolomite. The volume of dolomite added during the pore-filling phase will vary depending upon the chemistry and volumes of dolomitizing water passing through the dolomite. On Bonaire, sufficient volumes of dolomite have been added to reduce the porosity of some dolomites to less than 5%.

Permeability is systematically reduced from 1000 mD to 0.01 mD as the porosity is reduced. The scatter in the data shown in Figure 4 probably results from variations in the size and sorting of dolomite crystals and the volume of mouldic and intraparticle porosity.

The Bonaire dolomite was formed from a hypersaline marine water flowing downwards and outwards from an evaporitic tidal-flat environment and mixing with interstitial normal marine waters. Because a higher volume of dolomitizing fluids flowed through the updip carbonates than through the downdip carbonates, the updip dolomites have more occluded porosity than the downdip dolomites.

The flow direction of dolomitizing fluids can be traced by mapping changes in cathodoluminescent zones, Mg/Ca ratios in dolomite crystals, dolomite unit cell dimensions, and strontium content. Upstream dolomites are more stoichiometric and have lower strontium concentrations than downstream dolomites. Dolomite cement is more stoichiometric and contains less strontium than does replacement dolomite.

ACKNOWLEDGEMENTS

Funding for this research was received from Texas Advanced Research Program, project no. 003658-254. Part of the work was done by the senior author while he was at Shell Development Company, Houston, Texas. We would like to acknowledge the support of Shell Development Company, specifically T.V. Wilson, who provided us with thin sections and stable isotope data from Bonaire, and R. Michael Lloyd who assisted in interpreting the data.

REFERENCES

BANDOIAN, C.A. & MURRAY, R.C. (1974) Pliocene–Pleistocene carbonate rocks of Bonaire, Netherlands Antilles. *Geol. Soc. Am. Bull.* **85**, 1243–1252.

BEAUMONT, E. DE (1837) Application du calcul à l'hypothèse de la formation par épigénie des anhydrites, des gypses et des dolmies. *Soc. Geol. France Bull.* **8**, 174–177.

BUDD, D.A. (1988) Aragonite-to-calcite transformation during fresh-water diagenesis of carbonates: insights from pore-water chemistry. *Geol. Soc. Am. Bull.* **100**, 1260–1270.

DEBUISONJE, P.H. (1974) *Neogene and Quaternary Geology of Aruba, Curacao, and Bonaire*. Martinus Nijhoff, The Hague, 261 pp.

DEFFEYES, K.S., LUCIA, F.J. & WEYL, P.K. (1964) Dolomitization observations on the island on Bonaire, Netherlands Antilles. *Science* **143**, 678–679.

DEFFEYES, K.S., LUCIA, F.J. & WEYL, P.K. (1965) Dolomitization of recent and Plio-Pleistocene sediments by marine evaporite waters on Bonaire, Netherlands Antilles. In: *Dolomitization and Limestone Diagenesis, A Symposium* (Ed. Pray, L.C. & Murray, R.C.) Spec. Publs Soc. econ. Paleont. Miner., Tulsa, 13, 71–88.

DEGROOT, K. (1973) Geochemistry of tidal flat brines at Umm Said, SE Qatar, Persian Gulf. In: *The Persian Gulf, Holocene Carbonate Sedimentation and Diagenesis in a Shallow Epicontinental Sea* (Ed. Purser, B.E.) pp. 377–394. Springer-Verlag, Heidelberg.

ENOS, P. & SAWATSKY, L.H. (1981) Pore networks in Holocene carbonate sediments. *J. Sedim. Petrol.* **51**, 961–985.

FOUKE, B.W., MEYERS, W.J. & HANSON, G.N. (1992) Seawater and basalt-derived $^{87}Sr/^{86}Sr$ in multiple dolomitization events, Seroe Domi Formation, Curacao, Netherlands Antilles (abs). *Geol. Soc. Am. Abs. with Programs* **27**, A38.

GLOVER, E.D. (1961) Method of solution of calcareous materials using the complexing agent EDTA. *J. Sedim. Petrol.* **31**, 622–626.

GOLDSMITH, J.R., GRAF, D.L. & HEARD, H. C. (1961) Lattice constants of the calcium-magnesium carbonates. *Am. Mineral.* **46**, 453–457.

HALLEY, R.B. & SCHMOKER, J.W. (1983) High-porosity Cenozoic carbonate rocks of South Florida: progressive loss of porosity with depth. *Am. Ass. Petrol. Geol. Bull.* **67**, 191–200.

HODELL, D.A., MUELLER, P.A. & GARRIDO, J.R. (1991) Variations in the strontium isotopic composition of seawater during the Neogene. *Geology* **19**, 24–27.

JAMES, N.P. & GINSBURG, R.N. (1979) *The Seaward Margin of Belize Barrier and Atoll Reefs*. Int. Ass. Sedimentol. Spec. Publ. 3, 191 pp.

LANDES, K.K. (1946) Porosity through dolomitization. *Am. Ass. Petrol. Geol. Bull.* **30**, 305–318.

LUCIA, F.J. (1968) Recent sediments and diagenesis of South Bonaire, Netherlands Antilles. *J. Sedim. Petrol.* **38**, 845–858.

LUCIA, F.J. (1972) Recognition of evaporite–carbonate shoreline sedimentation. In: *Recognition of Ancient Sedimentary Environments* (Ed. Rigby, J.K. & Hamblin, W.K.). Spec. Publs Soc. Econ. Paleont. Mineral., Tulsa 16, 160–191.

MCCREA, J.M. (1950) On the isotopic chemistry of carbonates and a paleotemperature scale. *J.Chem. Phys.* **18**, 849–857.

MAJOR, R.P., LLOYD, R.M. & LUCIA, F.J. (1992) Oxygen isotope composition of a Holocene dolomite formed in a humid hypersaline setting. *Geology* **20**, 586–588.

MATTHEWS, R.K. & FROHLICH, C. (1987) Forward modeling of bank-margin carbonate diagenesis. *Geology* **15**, 673–676.

MURRAY, A.N. (1930) Limestone oil reservoirs. *Econ. Geol.* **25**, 167–198.

MURRAY, R.C. (1960) Origin of porosity in carbonate rocks. *J. Sedim. Petrol.* **30**, 59–84.

PIPERS, P.J. (1933) *Geology and Paleontology of Bonaire (NWI)* PhD Dissertation, University of Utrecht, The Netherlands.

SCHMOKER, J.W. & HALLEY, R.B. (1982) Carbonate porosity versus depth: a predictable relation for South Florida. *Am. Ass. Petrol. Geol. Bull.* **66**, 2561–2570.

SCHMOKER, J.W., KRYSTINIK, K.B. & HALLEY, R.B. (1985) Selected characteristics of limestone and dolomite reservoirs in the United States. *Am. Ass. Petrol. Geol. Bull.* **69**, 733–741.

Shinn, E.A. (1969) Submarine lithification of Holocene carbonate sediments in the Persian Gulf. *Sedimentology* **12**, 109–144.

SIBLEY, D.F. (1980) Climatic control of dolomitization, Seroe Domi Formation (Pliocene), Bonaire, NA. In: *Concepts and Models of Dolomitization* (Ed. Zenger, D.H., Dunham, J.B. & Ethington, R.L.) Spec. Publ. Soc. Econ. Paleont. Mineral., Tulsa 28, 247–258.

STEINEN, R.P. & MATTHEWS, R.K. (1973) Phreatic vs vadose diagenesis: stratigraphy and mineralogy of a cored borehole on Barbados, WI. *J. Sedim. Petrol.* **43**, 1012–1020.

TWENHOFEL, W.H. (1926) *Treatise on Sedimentation*. Williams & Wilkins, Baltimore, 641 pp.

VACHER, H.L., BENGTSSON, T.O. & PLUMMER, L.N. (1990) Hydrology of meteoric diagenesis: residence time of meteoric ground water in island fresh-water lenses with application to aragonite–calcite stabilization rate in Bermuda. *Geol. Soc. Am. Bull.* **102**, 223–232.

VAHRENKAMP, V.C., SWART, P.K. & RUIZ, J. (1991) Episodic dolomitization of Late Cenozoic carbonates in the Bahamas: evidence from strontium isotopes. *J. Sedim. Petrol.* **6**, 1002–1014.

WEYL, P.K. (1960) Porosity through dolomitization: conservation-of-mass requirements. *J. Sedim. Petrol.* **30**, 85–90.

Petrology and Geochemistry of Dolomites

Spec. Publs Int. Ass. Sediment. (1994) **21**, 345–360

Synthesis of dolomite and geochemical implications

E. USDOWSKI

Geochemisches Institut der Universität, 34 Göttingen, Goldschmidtstrasse 1, Germany

ABSTRACT

Dolomite and magnesite have been synthesized at 60°C and 90°C by the reaction of $CaCO_3$ with saturatured $CaCl_2-MgCl_2$ solutions for periods up to 7 years. At 60°C the phase boundaries of calcite–dolomite and magnesite–dolomite are at $Ca^{2+}/Mg^{2+} = 0.445$ and 0.107, respectively. At 90°C the corresponding boundaries are at $Ca^{2+}/Mg^{2+} = 3.08$ and 0.304.

These and previous data show that the dolomite field coincides with the Ca^{2+}/Mg^{2+} ratios of the majority of subsurface solutions at about 120°C. It follows that the probability of 'burial' dolomitization by an encounter between migrating pore solutions and carbonate sediments increases with increasing temperature. The Ca^{2+}/Mg^{2+} ratios of the most abundant solutions and the boundary of calcite–dolomite suggest that the most frequent temperatures of 'burial' dolomitization are between 80°C and 90°C.

Mineral assemblages and compositions of solutions from sabkha-type environments reflect specific reaction paths. The evaporation of seawater and other solutions provides low Ca^{2+}/Mg^{2+} ratios and high ionic strengths, so that the resulting brines are located in either the dolomite or the magnesite field. Thus, precursor calcium carbonate may react to dolomite and magnesite or to metastable equivalents. However, as reaction rates are slow, sufficient time has to be provided. From the assemblage dolomite + magnesite the Ca^{2+}/Mg^{2+} ratio of the phase boundary may be obtained. The assemblage Mg–calcite + protodolomite is ambiguous.

INTRODUCTION

The rock-forming mineral dolomite was discovered by the French nobleman, geologist and mineralogist Dieudonné-Sylvain-Guy-Tancrède de Gratet de Dolomieu (1750–1801) who, after the French revolution, called himself simply Déodat de Dolomieu. The botanist and chemist Nicolas-Théodore de Saussure detected that the samples collected by Dolomieu contained a small amount of magnesium, and suggested the name 'dolomie' (see Zenger *et al.*, this volume). Leopold von Buch in Freiberg, Sachsen, introduced the name dolomite (Lacroix, 1921; Mutschlechner, 1953; and others).

The crystal structure of dolomite ($R\bar{3}$) is similar to that of calcite and magnesite ($R\bar{3}c$). However, as the cation layers are alternately occupied by either Ca^{2+} or Mg^{2+}, the symmetry is lower than that of calcite and magnesite. Besides this ordered structure there is the disordered variety 'protodolomite'. In this structure Ca^{2+} and Mg^{2+} are randomly distributed, as in Mg-calcite.

Dolomite occurs in a variety of different geological, petrographic and geochemical settings: carbonatites, metamorphic rocks, hydrothermal deposits, sedimentary rocks and sediments. With respect to the high-temperature and pressure environments, the origin of dolomite is fairly well understood, at least in principle. However, with respect to the sedimentary environment there is the so-called 'dolomite question'. This question, one of the global problems of sedimentary petrology and geochemistry, is at once most appealing, puzzling and frustrating. The essence of the dolomite question is the fact that dolomite has not been synthesized at, or near, earth surface temperatures. Nevertheless, geological evidence suggests that it has originated at such conditions. This is also true for sedimentary magnesite.

The first comprehensive work on sedimentary dolomite is that by Van Tuyl (1916). Many of his conclusions are still valid today, even though modern analytical techniques were not then available, and the occurrence of recent dolomitic and magnesitic sediments was still essentially unknown. Since Van Tuyl (1916), the volume of literature has grown tremendously. The reader is referred to compilations and reviews by Fairbridge (1957), Purser (1973), Zenger *et al.* (1980), Zenger and Mazullo (1982), Morrow (1982a, 1982b), Land (1983), Machel and Mountjoy (1986), Hardie (1987), Füchtbauer (1988), Tucker (1990), and others.

THE GEOCHEMICAL SYSTEM

The dolomite question must be viewed in the light of chemical interactions between carbonate minerals and aqueous solutions. Field observations suggest that there may be various ways to form dolomite. Thus, the geochemical system of dolomite and magnesite formation is bracketed by all dissolved chemical species that occur in the surface waters and the more deeply buried aqueous fluids of the earth. However, a closer examination shows that, on a global scale, the majority of dolomite must have originated in only two basic geological and geochemical environments: within buried sediments and in the surface sediments of coastal plains such as sabkhas and similar superficial environments. The solutions that occur in these surroundings are essentially 'seawater-like', e.g. solutions with predominant Cl^- and SO_4^{2-}. Neglecting potassium, which occurs in relatively low concentrations, and trace elements, these solutions and the pertinent solids are described by subsystem 1: $CaCO_3$–$CaSO_4$–$CaCl_2$–$MgCO_3$–$MgSO_4$–$MgCl_2$–H_2O ± NaCl of the general system Na^+–Ca^{2+}–Mg^{2+}–Cl^-–SO_4^{2-}–CO_3^{2-}–H_2O. The less abundant terrestrial solutions are described essentially by subsystem 2: Na_2CO_3–Na_2SO_4–$CaCO_3$–$CaSO_4$–$MgCO_3$–$MgSO_4$–H_2O of the general system. Brines, carbonates and other minerals from the northern Great Plains of Canada are an example of subsystem 2 (± chloride) (Last, 1989).

THE SYSTEM $CaCO_3$–$CaSO_4$–$CaCl_2$–$MgCO_3$–$MgSO_4$–$MgCl_2$–H_2O

The phase relations of the NaCl-free subsystem 1 are shown in Figure 1. In fact, these are not changed by the presence of NaCl, in principle. The length of any edge of the prism corresponds to 100 mole %, and any point within it or on its surface indicates compositions of solutions in terms of the pertinent ions, which are in equilibrium with one or several solids. At the $CaCO_3$–$MgCO_3$ edge the system is linked to HCO_3^-, H_2CO_3 and CO_2.

Concentrations are highest at the chloride edge and decrease considerably towards the sulphate and carbonate edges. At the $CaCl_2$ point the concentration is 0.93 mole $CaCl_2$/100 g H_2O between 180° and 32°C. The concentration at the $MgCl_2$ point is 0.93 mole $MgCl_2$/100 g H_2O above 116°C. Below this point it decreases to 0.58 mole $MgCl_2$/100 g H_2O at 30°C. At the $MgSO_4$ point concentrations are 0.043, 0.49 and 0.32 mole $MgSO_4$/100 g H_2O at 180°, 67.5° and 30°C, respectively. At the $CaSO_4$ point they are in the orders of 10^{-4} and 10^{-3} mole $CaSO_4$/100 g H_2O between 180° and 30°C. They are somewhat lower than above at the $CaCO_3$–$MgCO_3$ edge.

Planes 8–10–22–16 and 7–9–23–17 represent equilibrium between calcite and dolomite, and between dolomite and magnesite, respectively, according to the overall reactions

$$2CaCO_3 + Mg^{2+} = CaMg(CO_3)_2 + Ca^{2+},$$

$$K_{C-D} = \frac{(Ca^{2+})}{(Mg^{2+})} \quad (1)$$

and

$$CaMg(CO_3)_2 + Mg^{2+} = 2MgCO_3 + Ca^{2+},$$

$$K_{D-M} = \frac{(Ca^{2+})}{(Mg^{2+})} \quad (2)$$

where the parentheses indicate ion activities. The pertinent anions are CO_3^{2-} and HCO_3^-, or mixtures of these with Cl^- and/or SO_4^{2-}. Depending upon the Ca^{2+}/Mg^{2+} ratio of the solution, the following unidirectional reactions may occur:

$$CaCO_3 + Mg^{2+} \rightarrow MgCO_3 + Ca^{2+} \quad (3)$$

and

$$MgCO_3 + Ca^{2+} \rightarrow CaCO_3 + Mg^{2+} \quad (4)$$

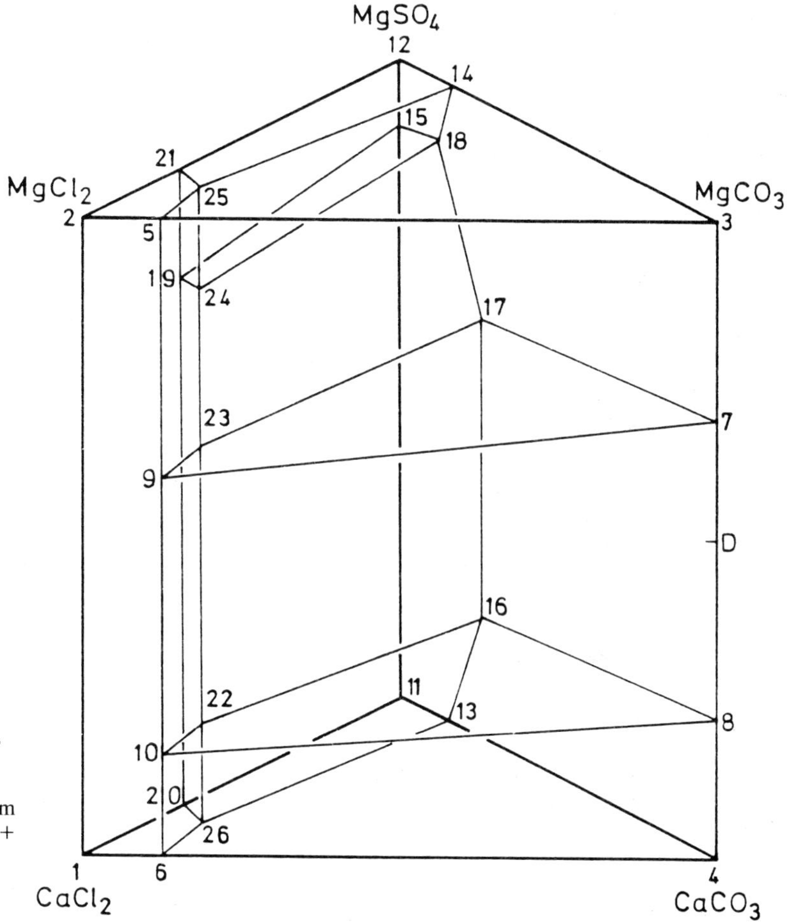

Fig. 1. Phase relations in the system $CaCO_3$–$MgCO_3$–$CaSO_4$–$MgSO_4$–$CaCl_2$–$MgCl_2$–H_2O (from Usdowski, 1967). Dolomite: D. Calcite + dolomite: 8–10–22–16. Magnesite + dolomite: 7–9–23–17. $CaSO_4$ (gypsum or anhydrite): 11. Calcite + dolomite + $CaSO_4$: 16. Magnesite + dolomite + $CaSO_4$: 17. Magnesite + $CaSO_4$ + $MgSO_4$: 18.

It should be noted that reaction 1 proceeds in a complicated way. As $CaCO_3$ is dissolving, Mg-poor solids are precipitated initially, which increases their Mg content via successive steps of redissolution and precipitation until the composition of dolomite is attained (Katz & Matthews, 1977; unpublished results by the present author). Thus, dolomite is formed by a stepwise dissolution and crystallization of metastable Ca–Mg carbonates, the precursor phases being less stable than their successors. Most probably reaction 2 occurs similarly. It should be kept in mind that in none of the reactions do cations diffuse out of the crystal lattice into the solution, or vice versa. The temperatures under consideration are too low for solid-state diffusion.

From Figure 1 it may be deduced that the above reactions occur with any mixtures of carbonate, sulphate and chloride. Theoretically, this is possible but it should be pointed out that the reactions can be realized experimentally only with respect to a specific set of boundary conditions.

SYNTHESIS OF DOLOMITE AND MAGNESITE

Historical experiments

Most probably Von Morlot (1847) was the first to be concerned with the synthesis of dolomite. Calcium carbonate was reacted with epsom salt at 15 atmospheres and 220°C in a sealed glass tube (system $CaCO_3$–$MgCO_3$–$CaSO_4$–H_2O) which was placed inside a sawn-off gun barrel, plugged at the ends. Favre (1849) reported that Marignac reacted calcium carbonate with a solution of magnesium chloride at 200°C (system $CaCO_3-MgCO_3-CaCl_2-MgCl_2-H_2O$). Unfortunately, the concentrations of the $MgSO_4$ and $MgCl_2$ solutions which were used in the above experiments are unknown.

Experiments between 100° and 410°C

Experiments in subsystem 1, e.g. employing Ca–Mg-chloride solutions, in which dolomite was produced from precursor $CaCO_3$, have been made by Rosenberg and Holland (1964) and Rosenberg *et al.* (1967) between 275° and 410°C at $CaCl_2-MgCl_2$ concentrations from 0.16 to 2 M. Similar experiments with saturated Ca–Mg-chloride solutions were conducted by Usdowski (1964a,b; 1967) at 120° and 180°C. Experiments in which dolomitization of $CaCO_3$ was monitored as a function of time have been made by Sureau (1974) (150°C, 0.2 – 0.3 M $CaCl_2-MgCl_2$ solutions), Katz and Matthews (1977) (252–295°C, 2 M $CaCl_2-MgCl_2$), and Gaines (1980) (100°C, 2 M $CaCl_2-MgCl_2$, 0.05 M Li^+ as catalyst, protodolomite seed crystals). Baker and Kastner (1981) have shown an inhibiting effect of sulphate (200°C, ionic strength = 0.7, $CaCl_2 + MgCl_2 \approx 0.14$, $\pm$NaCl, Na_2SO_4).

Experiments at 60° and 90°C

The present experiments employed the same techniques as those utilized at 120° and 180°C (Usdowski, 1964a,b, 1967). Initial solids were calcite powder (Merck p.a.), natural aragonite which was ground in an agate mortar to a grain size of about 20 µm, and fine-grained structurally defective calcite, produced by grinding ordinary calcite for a long time in a 'high-energy' steel mortar until broad X-ray reflections appeared. Pyrex glass tubes of 50 cm^3 volume were filled with about 1 g of a solid and about 30 cm^3 of saturated Ca–Mg-chloride solutions of different Ca^{2+}/Mg^{2+} ratios in individual runs. After sealing, the tubes were installed in a rack inside an oven and were rotated continuously at about 30 rpm by an external electric motor. The temperature was controlled at ±1°C. After defined periods of time a number of tubes were cut open. The solids were separated from the solutions by ultrafiltration and identified by X-ray diffraction (Cu-K_α radiation). Aliquots of the solutions were analysed for Ca^{2+} and Mg^{2+} using EDTA (ethylenediamine tetraacetic acid) methods. About 100 mg of the reaction products were dissolved in dilute hydrochloric acid and analysed as above.

Dolomite, protodolomite and Mg-calcite were obtained according to the forward reaction 1, and magnesite was produced via the unidirectional reaction 3 after periods of about 3 and 7 years (from January/February 1982 to 1985 and 1989). The dolomite showed weak but distinct superstructure reflections; the chemical composition is about 0.5 $Me^{2+} \pm 0.03$.

Results for 90°C are summarized in Tables 1, 2 and 3. The presence of protodolomite and Mg-calcite, after a period of about 3 years, shows that reaction 1 occurs via intermediate phases (see above), independent of the starting material. Dolomite was produced from aragonite in runs 90-A-3 to 90-A-7 (3 years; Table 1). Structurally defective calcite yielded protodolomite and Mg-calcite in the same range of liquid compositions (Table 2), whereas calcite reacted to dolomite

Table 1. Experimental results. Starting material: aragonite. Final solids and solutions at 90°C.

Run	Weeks	Solid	$X_{Mg^{2+}}$	Weeks	Solid	$X_{Mg^{2+}}$	Mol/100 g H_2O
90-A-1	161	M, D	0.932	363	M	0.944	0.720
90-A-2	168	M	0.836	370	M	0.845	0.737
90-A-3	168	D	0.755	370	D, M	0.767	0.755
90-A-4	168	D	0.677	370	D	0.677	0.771
90-A-5	168	D	0.593	370	D	0.583	0.790
90-A-6	168	D	0.484	370	D	0.481	0.810
90-A-7	168	D	0.384	370	D	0.385	0.831
90-A-8	168	PD, MC	0.297	370	C, D	0.296	0.852
90-A-9	168	C	0.186	370	C	0.197	0.875
90-A-10	168	C	0.115	370	C	0.081	0.899

C, calcite; D, dolomite; M, magnesite; PD, protodolomite; MC, Mg-calcite.
$X_{Mg^{2+}} = [Mg^{2+}]/([Mg^{2+}] + [Ca^{2+}])$ of solution.
Mol/100 g H_2O = moles ($CaCl_2 + MgCl_2$)/100 g H_2O.

Table 2. Experimental results. Starting material: structurally defective calcite. Final solids and solutions at 90°C.

Run	Weeks	Solid	$X_{Mg^{2+}}$	Weeks	Solid	$X_{Mg^{2+}}$	Mol/100 g H_2O
90-CB-1	161	M	0.944	363	M	0.944	0.720
90-CB-2	171	D, M	0.849	372	D, M	0.861	0.737
90-CB-3	171	PD, MC	0.766	372	D	0.778	0.753
90-CB-4	171	PD, MC	0.676	372	D	0.692	0.771
90-CB-5	171	PD, MC	0.585	372	D	0.588	0.790
90-CB-6	171	PD, MC	0.482	372	D	0.491	0.810
90-CB-7	171	PD, MC	0.382	372	D	0.407	0.831
90-CB-8	171	C	0.290	372	D	0.289	0.852
90-CB-9	171	C	0.187	372	C	0.201	0.875
90-CB-10	171	C	0.074	372	C	0.097	0.899

C, calcite; D, dolomite; M, magnesite; PD, protodolomite; MC, Mg-calcite.
$X_{Mg^{2+}} = [Mg^{2+}]/([Mg^{2+}] + [Ca^{2+}])$ of solution.
Mol/100 g H_2O = moles ($CaCl_2$ + $MgCl_2$)/100 g H_2O.

Table 3. Experimental results. Starting material: calcite. Final solids and solutions at 90°C.

Run	Weeks	Solid	$X_{Mg^{2+}}$	Weeks	Solid	$X_{Mg^{2+}}$	Mol/100 g H_2O
90-C-1	161	M	0.935	363	M	0.942	0.720
90-C-2	161	M	0.842	368	M	0.845	0.737
90-C-3	161	D, M	0.772	368	D, M	0.775	0.753
90-C-4	161	D, PD	0.671	368	D	0.680	0.771
90-C-5	161	PD, C	0.590	368	D	0.575	0.790
90-C-6	161	C	0.505	368	C, D	0.500	0.810
90-C-7	161	C	0.408	368	C	0.412	0.831
90-C-8	161	C	0.300	368	C	0.307	0.852
90-C-9	161	C	0.220	368	C	0.210	0.875
90-C-10	161	C	0.099	368	C	0.101	0.899

C, calcite; D, dolomite; M, magnesite; PD, protodolomite.
$X_{Mg^{2+}} = [Mg^{2+}]/([Mg^{2+}] + [Ca^{2+}])$ of solution.
Mol/100 g H_2O = moles ($CaCl_2$ + $MgCl_2$)/100 g H_2O.

and/or protodolomite only in three cases (Table 3). This suggests that the capability of the initial solids to yield stable minerals increases in the order calcite, structurally defective calcite, aragonite. Apparently, this does not apply to magnesite, which was obtained in runs A-1, A-2, CB-1, CB-2, C-1 and C-2. However, it should be kept in mind that the above order, as well as metastable Mg-carbonates, may be observed in runs of shorter duration.

After a period of about 7 years, metastable solids were no longer detectable. This suggests that equilibrium was approached. All experiments conducted over 3 and 7 years indicate that the phase boundary between dolomite and magnesite is located between runs 2 and 3. As aragonite yields stable solids with less difficulty than the other initial phases, and as both dolomite and magnesite are present in experiment 90-A-3 (370 weeks; Table 1) the composition of the solution of this run is taken as the equilibrium composition. The phase boundary calcite–dolomite is more difficult to assess. Run 90-C-6 (370 weeks; Table 3) suggests that the boundary is located at $X_{Mg^{2+}} = 0.5$. However, runs 90-A and 90-CB (370 weeks) show that dolomite occurs in a wider range of liquid compositions than in runs 90-C. In run 90-A-8, which yielded both calcite and dolomite, the boundary is at $X_{Mg^{2+}} = 0.296$, whereas it is between $X_{Mg^{2+}} = 0.289$ and 0.201 in the experiments which used structurally defective calcite. Assuming that in run 90-A-8 the reaction is still incomplete, the average value $X_{Mg^{2+}} = 0.245 \pm 0.044$ may be considered to represent the phase boundary.

Table 4 summarizes the results for 60°C from Usdowski (1989), together with other data. After a period of about 3 years aragonite had reacted to an undefinable phase in run 60-A-1, whereas magnesite

Table 4. Experimental results at 50° and 60°C.

Run	°C	Solids Initial	Solids Final	$X_{Mg^{2+}}$	Mol/100 g H_2O	Weeks
60-A-1	60	A	U	0.986	0.636	157
60-A-1	60	A	M	0.945	0.636	363
60-A-2	60	A	D	0.868	0.659	365
60-A-3	60	A	D	0.788	0.681	365
60-A-4	60	A	D, C, A	0.692	0.703	365
60-A-5	60	A	A	0.613	0.729	365
60-A-6	60	A	A	0.522	0.755	365
60-A-7	60	A	A	0.413	0.784	365
60-A-8	60	A	A	0.326	0.815	365
60-A-9	60	A	C, A	0.217	0.849	365
60-A-10	60	A	C	0.105	0.885	365
60-C-1	60	C	C, M	0.998	0.636	157
60-C-1	60	C	C, M	0.985	0.636	363
60-CB-1	60	CB	C, C	0.987	0.636	157
60-CB-1	60	CB	C	0.971	0.636	363
50-A-1	50	A	A	0.975	0.614	157
50-A-1	50	A	A	0.980	0.614	363
50-C-1	50	C	C	0.993	0.614	157
50-C-1	50	C	C	0.993	0.614	363
50-CB-1	50	CB	C, M	0.983	0.614	157
50-CB-1	50	CB	C, M	0.773	0.614	363

A, aragonite; C, calcite; CB, structurally defective calcite; D, dolomite; M, magnesite; U, undefined.
$X_{Mg^{2+}} = [Mg^{2+}]/([Mg^{2+}] + [Ca^{2+}])$ of solution.
Mol/100 g H_2O = moles ($CaCl_2 + MgCl_2$)/100 g H_2O.

was produced in an identical run after a time interval of about 7 years. Aragonite was partially and completely transformed into calcite in runs 60-A-9 and 60-A-10, respectively, whereas it was not transformed in runs 60-A-5 to 60-A-8. In run 60-A-4 a small amount of the initial aragonite was observed. As both stable phases, dolomite and calcite, are present in run 60-A-4, the composition of the solution of this run is taken as the equilibrium composition for the phase boundary between dolomite and calcite. The boundary between dolomite and magnesite may be interpolated from the liquid compositions of runs 60-A-1 and 60-A-2.

At 60°C and periods of about 3 and 7 years, normal calcite with sharp X-ray reflections and magnesite were obtained from structurally defective calcite. At 50°C aragonite and calcite were still unchanged, although small changes are indicated

Table 5. Molar compositions of saturated $CaCl_2$–$MgCl_2$ solutions at the phase boundaries dolomite–calcite and dolomite–magnesite.

°C	10^3/T (°K)	$X_{Mg^{2+}}$ C + D	$X_{Mg^{2+}}$ D + M	$[Ca^{2+}]/[Mg^{2+}]$ C + D	$[Ca^{2+}]/[Mg^{2+}]$ D + M	Reference
60	3.00	0.692	0.903	0.445	0.107	Usdowski (1989)
90	2.75	0.245	0.767	3.08	0.308	This paper
120	2.54	0.129	0.484	6.75	1.07	Usdowski (1967)
180	2.21	0.077	0.318	11.99	2.14	Usdowski (1967)

C, calcite; D, dolomite; M, magnesite.
$X_{Mg^{2+}} = [Mg^{2+}]/([Ca^{2+}] + [Mg^{2+}])$, $[Ca^{2+}]/[Mg^{2+}] = (1/X_{Mg^{2+}} - 1)$.

by the compositions of the solutions. Structurally defective calcite had reacted as above. Experiments with other liquid compositions are still in process.

Table 5 summarizes compositions of solutions for the phase boundaries calcite–dolomite and dolomite–magnesite from 60°C to 180°C.

FACTORS INFLUENCING THE FORMATION OF DOLOMITE AND MAGNESITE

Dolomite and magnesite have never been precipitated spontaneously at room temperature. This fact is based on numerous experimental failures by a number of authors, including the present one. Even 'magic ingredients' and seed crystals do not trigger the crystallization according to the reactions

$$Ca^{2+} + Mg^{2+} + 2CO_3^{2-} = CaMg(CO_3)^2,$$
$$K_D = (Ca^{2+})(Mg^{2+})(CO_3^{2-})^2 \quad (5)$$

and

$$Mg^{2+} + CO_3^{2-} = MgCO_3,$$
$$K_M = (Mg^{2+})(CO_3^{2-}) \quad (6)$$

Instead of magnesite, hydrated and hydroxylated Mg carbonates are formed, and instead of dolomite Ca–Mg carbonates of variable compositions are precipitated spontaneously, depending upon the composition of the solution (e.g. Glover & Sippel, 1967).

Similarly, reactions 1, 2 and 3, which involve a precursor carbonate, cannot be realized at any concentration and at any temperature. At higher temperatures they do occur off the chloride edge of subsystem 1 (Fig. 1) down to concentrations of about 0.1 M. However, as the temperature decreases they can be realized only close to chloride saturation (line 25–26, Fig. 1). This is specially true for 90° and 60°C. The relation between the temperature and a critical concentration at which the reactions are still occurring has not yet been studied. As the temperature decreases the rate of the reaction slows considerably. Figure 2 shows the time that is required to react 1 g of $CaCO_3$ with about 30 cm^3 of saturated Ca–Mg-chloride solution. This is due to the hydration of Ca^{2+} and Mg^{2+}, which occurs via the reaction

$$Me^{2+} + nH_2O = Me(H_2O)_n^{2+},$$

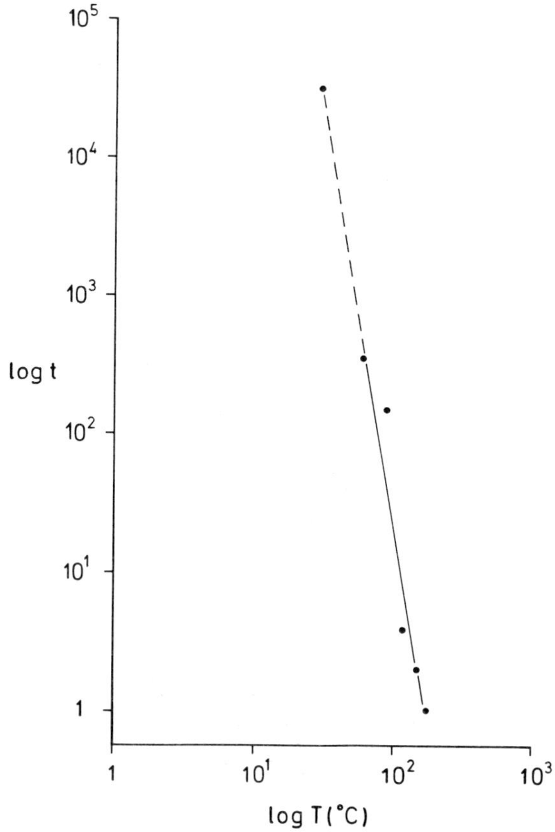

Fig. 2. Relation between temperature and time required to transform $CaCO_3$ into dolomite under a specific set of experimental conditions (see text). The points are related by the expression log t(weeks) = $-5.12 \cdot \log T(°C) + 13.25$.

$$K_{Hy} = \frac{(Me(H_2O)_n^{2+})}{(Mg^{2+})(H_2O)^n} \quad (7)$$

From the composition of the solid hydrates of Ca- and Mg-chloride, it is generally accepted that the smallest hydration number is $n = 6$ below about 100°C. Stability constants for the above reaction are not known. Nevertheless, the electric charge (z) per surface area ($z/4\pi r^2$; r = ionic radius, $z = 2$) provides some information about the strength of the bond between a cation and the water dipole. This quantity is 0.26 (Å^{-2}) for Mg^{2+}, and 0.14 (Å^{-2}) for Ca^{2+}. Thus, Mg^{2+} is hydrated somewhat stronger than Ca^{2+}. For Sr^{2+} and Ba^{2+} the $z/4\pi r^2$ values are 0.1 and 0.08 (Å^{-2}), respectively. Obviously, the above values are small enough for Ca^{2+}, Sr^{2+} and Ba^{2+}, so that equilibria analogous to reactions 5 and 6 can be approached from both sides at low ionic strengths, e.g. from either under- or supersaturation, with respect to calcite, strontianite and witherite. However, it is not approached from supersaturation with respect to magnesite and dolomite at moderate temperatures and any ionic strengths, although it may be attained from undersaturation.

It follows that the formation of dolomite and magnesite from a precursor $CaCO_3$ requires high ionic strengths. Concentrations (ionic strengths) are highest at the chloride edge of subsystem 1 (Fig. 1) and decrease considerably towards the sulphate and carbonate edges. As the concentrations of the dissolved species are decreasing the 'concentration' of the water increases, so that less and less 'dehydrated' and thus thermodynamically 'active' Mg^{2+} is available. At higher temperatures rather 'dilute' solutions provide enough 'active' Mg^{2+}, because the chemical bonds between the cation and its shell of water molecules are relatively weak. However, as the temperature decreases the strength of the bonds increases, so that the concentration must be increased in order to provide the necessary amount of 'active' Mg^{2+}. This requirement is best fulfilled by $MgCl_2$, which has a higher solubility than $MgSO_4$. As the temperature decreases the solubilities of both compounds decrease, so that saturated solutions are increasing their 'water concentration'. Thus, reaction rates slow considerably, due to a rather small amount of 'active' Mg^{2+}. Unfortunately, temperature–time relations are available only for saturated Ca–Mg-chloride solutions. Thus, it is unknown whether, at a given temperature, the time to complete the reaction increases continuously as the ionic strength decreases, or whether the reaction is blocked at a critical ionic strength or hydration state of the cations. Another unknown factor is the relationship between the reaction rate and the nature of the precursor. Obviously, the reaction rate increases somewhat in the order calcite, structurally defective calcite, aragonite.

GEOCHEMICAL IMPLICATIONS

The experiments described suggest that sedimentary dolomite and magnesite may be formed from precursor calcium carbonate in any geochemical environment in which highly concentrated solutions occur. Such conditions may be observed in recent sediments of coastal plains, and the occurrence of subsurface brines suggests that dolomite may be forming today at some depth and at elevated temperatures. It is most likely that such conditions have existed throughout geological time. At low temperatures the formation of dolomite and magnesite from a precursof $CaCO_3$ requires a considerable amount of time and a high ionic strength, which is given by saturated chloride solutions. At elevated temperatures the reactions occur faster and at smaller ionic strengths. Most probably the same applies to the transformation of directly precipitated Ca–Mg carbonates into dolomite, and to the transformation of directly precipitated hydrated and hydroxylated Mg carbonates into magnesite.

Formation of dolomite under subsurface conditions

The evolution of the chemical composition of subsurface solutions is controlled by a number of processes which cause a broad spectrum of concentrations and Ca^{2+}/Mg^{2+} ratios. In order to evaluate the potential of these fluids for dolomitization, their Ca^{2+}/Mg^{2+} ratios have to be compared with the solid–solution equilibria. A comparison comprising solubility data has been made earlier (Usdowski, 1967), whereas the present discussion is based on reactions which have been realized in the temperature range between 60°C and 180°C.

Figure 3 shows the abundance of molar Ca^{2+}/Mg^{2+} ratios of about 500 pore fluids, taken from Usdowski (1967), and as a function of the temperature of the phase boundaries calcite–dolomite and dolomite–magnesite. It should be pointed out that not all fluids presented in Figure 3 have a very high ionic strength, but at any concentration the abundance of the Ca^{2+}/Mg^{2+} ratios follows a rather similar pattern.

At 60°C dolomite is stable with solutions which have Ca^{2+}/Mg^{2+} ratios between 0.11 and 0.45 (broken lines in Fig. 3). However, such solutions are relatively rare, whereas at 120°C the dolomite field lies between Ca^{2+}/Mg^{2+} ratios from 1.07 to 6.75, covering the majority of the solutions. It follows that the probability of dolomitization by a statistical encounter between migrating pore fluids and carbonate sediments increases with increasing temperature, provided that sufficient permeability exists. The maximum of the abundance of the Ca^{2+}/Mg^{2+} ratios (from about 1.5 to 2.3; Fig. 3) suggests that the most frequent temperature of dolomitization lies between about 80°C and 90°C. This agrees well with temperatures obtained from fluid inclusions in dolomite rocks (Liedmann & Koch, 1990).

Figure 3 may also serve to trace the evolution of

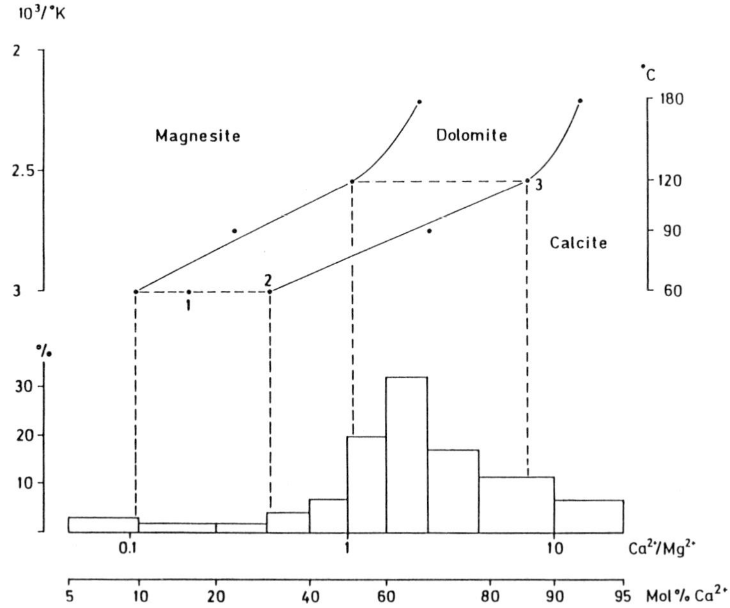

Fig. 3. Phase boundaries and the abundance of molar Ca^{2+}/Mg^{2+} ratios of subsurface solutions. Points 1, 2, 3: see text. Ca^{2+}/Mg^{2+} ratios from Usdowski, 1967.

the Ca^{2+}/Mg^{2+} ratio of brines during the formation of sedimentary basins. It is assumed that a brine has the Ca^{2+}/Mg^{2+} ratio of seawater (point 1) and that the temperature is 60°C. Following the formation of dolomite, the Ca^{2+}/Mg^{2+} ratio increases according to reaction 1, and, if excess calcium carbonate is present, the phase boundary between calcite and dolomite will be attained at point 2. At a later stage, when the solution has been removed, a limestone may be found which is in contact with a dolomite rock. The solution at point 2 is the equilibrium solution with respect to 60°C. However, it becomes reactive again if the temperature increases by further burial. If excess calcium carbonate is present the Ca^{2+}/Mg^{2+} ratio of the solution will change along the line of the calcite–dolomite equilibrium, for example from point 2 to point 3. A homogeneous dolomite may be preserved if the initial excess of calcite was exhausted at this point. Thus, increasing temperature will lead to a continuously progressing dolomitization because the equilibrium between calcite and dolomite is shifted towards higher Ca^{2+}/Mg^{2+} ratios as the temperature increases.

Formation of dolomite under surface conditions

Phase boundaries

Temperatures in sabkhas and similar environments are generally about 30°C and higher. Reactions 1, 2 and 3 are rather sluggish at such temperatures. From Figure 2 a period of about 600 years is extrapolated for 30°C. This estimate is not precise, but it agrees principally with time evaluations for modern dolomitic and magnesitic sediments. It follows that the geochemical environment must remain unchanged for a considerable amount of time, so that the above reactions may proceed.

In order to compare phase boundaries with solutions from sabkhas and similar environments, the equilibria calcite–dolomite and dolomite–magnesite must be known. As these cannot be measured at the temperature under consideration, they must be calculated. This may be done by combining the solubility products for calcite, dolomite and magnesite. Thus equations 5 and 6 yield the equilibrium constant for reaction 2, according to the expression

$$K_{D-M} = \frac{K_D}{K^2_M} \qquad (8)$$

Similarly, the equilibrium constant for reaction 1 can be calculated from

$$K_{C-D} = \frac{K^2_C}{K_D} \qquad (9)$$

where $K_C = (Ca^{2+})(CO_3^{2-})$. Figure 4 shows the range of solubility products for calcite, ordered and disordered dolomite, and magnesite calculated from thermodynamic data (Robie & Waldbaum, 1968;

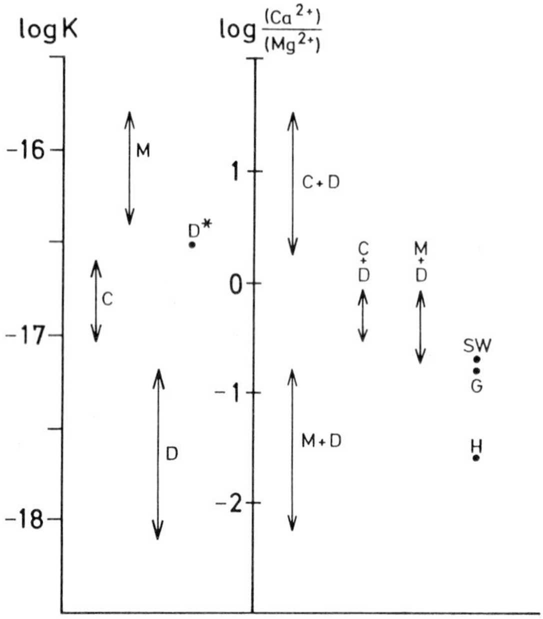

Fig. 4. Range of the phase boundaries calcite–dolomite and dolomite–magnesite derived from solubility constants (25°C). C, calcite; D, ordered dolomite; D*, disordered dolomite; M, magnesite. log K: $\log K_D$, $2 \cdot \log K_C$, or $2 \cdot \log K_M$. $\log (Ca^{2+})/(Mg^{2+})$: $\log K_{C-D}$ or $\log K_{D-M}$. SW, seawater; G, H, gypsum- and halite-saturated seawater.

Helgeson *et al.*, 1978; Robie *et al.*, 1978). This range also covers most of the constants obtained from solubility measurements. It appears that K_{C-D} and K_{D-M} values calculated from equations 8 and 9 depend very much upon the choice of solubility data. The maximum width of the stability field of ordered dolomite is comprised between $\log K_{C-D} = 1.5$ and $\log K_{D-M} = -2.3$, corresponding to Ca^{2+}/Mg^{2+} activity ratios from about 32 to 0.005. The minimum width ranges between $\log K_{C-D} = 0.2$ and $\log K_{D-M} = -0.8$, corresponding to Ca^{2+}/Mg^{2+} activity ratios from about 1.6 to 0.16. If disordered dolomite is considered, the phase boundaries calcite–dolomite and dolomite–magnesite are overlapping.

Figure 4 also shows the Ca^{2+}/Mg^{2+} activity ratios for the evaporation path of seawater (SW–gypsum (G) – halite (H)). Concentrations have been taken from Braitsch (1962). For seawater an activity coefficient ratio $\gamma_{Ca}/\gamma_{Mg} \approx 0.83$ has been used (Garrels & Thompson, 1962; and others). For brines, activity coefficients are most difficult to assess. Nevertheless, values $\gamma_{Ca}/\gamma_{Mg} \approx 0.8$ and ≈ 0.75 may be considered for solutions saturated with respect to gypsum and halite.

With respect to ordered dolomite, point SW (Fig. 4) is in the dolomite field, either distinctly or close to the boundary dolomite–magnesite. The same applies to the gypsum point (G). Point H is either in the dolomite or in the magnesite field. Considering disordered dolomite, point SW is close to the boundary dolomite–magnesite, whereas all other points are in the magnesite field.

It must be admitted that the situation is confusing, and that thermodynamic calculations and solubility measurements are not very helpful. Nevertheless, it can be stated that the evaporation of seawater causes conditions that favour the formation of dolomitic and magnesitic sediments. High ionic strengths of brines facilitate the dehydration of cations, especially Mg^{2+}. Regardless of where the 'true' phase boundaries are located, the low Ca^{2+}/Mg^{2+} ratio of the brines provides sufficient thermodynamic drive $\Delta G = \Delta G° + RT \cdot \ln(Ca^{2+})/(Mg^{2+})$ for the reactions under consideration. Thus, any precursor calcium carbonate, preferentially aragonite, that has been precipitated directly or may be present as remains of organisms, should react to dolomite or magnesite or, in a first step, to metastable phases such as protodolomite and hydromagnesite. Similarly, directly precipitated Ca–Mg carbonates and hydrated or hydroxylated Mg carbonates will be transformed into dolomite and magnesite. However, as reaction rates are slow, sufficient time has to be provided.

Sabkhas and similar environments

Concentrations and Ca^{2+}/Mg^{2+} ratios. Figure 5 shows molar Ca^{2+}/Mg^{2+} ratios for shallow interstitial solutions at Abu Dhabi, Persian Gulf, which exhibit high Cl^-/SO_4^{2-} ratios (Butler, 1969). Some of the solutions plot on the evaporation path of seawater. The majority have attained a stage of high concentration. Brines with Ca^{2+}/Mg^{2+} ratios higher than that of seawater are assumed to be mixtures of terrestrial waters and seawater. Figure 6 shows analytical data from De Groot (1973) for Umm Said, Persian Gulf. The range of molar Ca^{2+}/Mg^{2+} ratios is comparable to that of Figure 5. A number of solutions have attained a high concentration, whereas others have not. Compared to the molar Ca^{2+}/Mg^{2+} ratios of the phase boundaries (calculated from the Ca^{2+}/Mg^{2+} activity ratios shown in Figure 4, using the above activity coefficients)

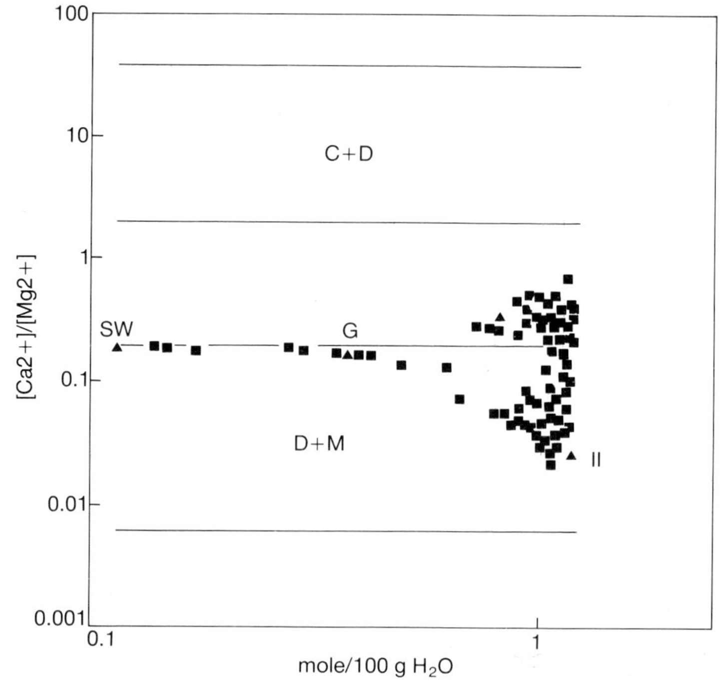

Fig. 5. Molar Ca^{2+}/Mg^{2+} ratios and total concentrations of solutions at Abu Dhabi, Persian Gulf (Butler, 1969). Upper and lower lines: range of calculated calcite–dolomite and dolomite–magnesite phase boundaries (Fig. 4). SW, seawater; G, H, gypsum- and halite-saturated seawater.

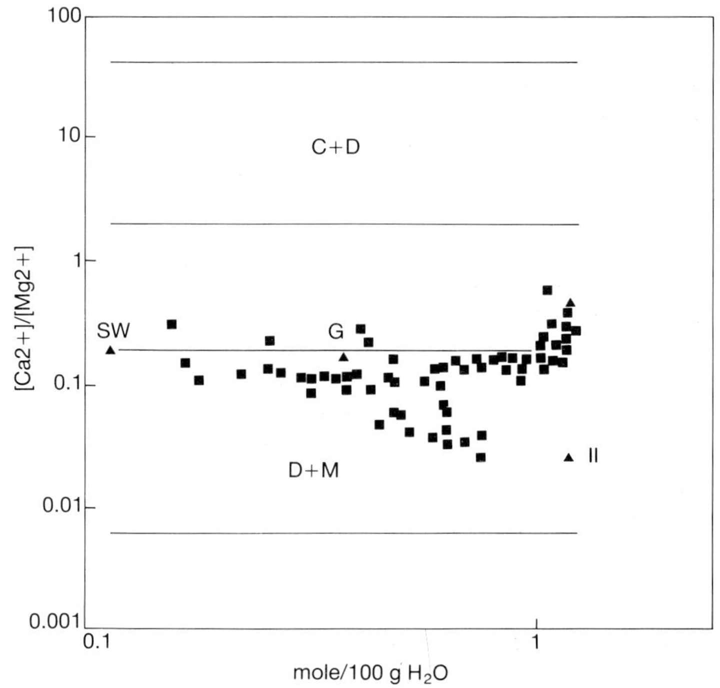

Fig. 6. Molar Ca^{2+}/Mg^{2+} ratios and total concentrations of solutions at Umm Said, Persian Gulf (De Groot, 1973). Upper and lower lines: range of calculated calcite–dolomite and dolomite–magnesite phase boundaries (Fig. 4). SW, seawater; G, H, gypsum- and halite-saturated seawater.

virtually all solutions are located either in the dolomite field or partially in the dolomite and partially in the magnesite field. The interpretation depends upon the choice of equilibrium data. Nevertheless, it appears that most of the brines should be capable of producing dolomite or magnesite (or their metastable equivalents) from a precursor calcium carbonate.

The left-hand side of Figure 7 shows analytical data for the Coorong area (Alderman & Skinner, 1957; Skinner, 1963) and for evaporite playas in the Western Plains District, South Australia (De Dekker & Last, 1989). Total concentrations are rather low. However, the above authors report that concentrations vary considerably in an annual cycle, reaching saturation with respect to halite. The right-hand side of Figure 7 shows mineral assemblages and pertinent Ca^{2+}/Mg^{2+} ratios of solutions given by Alderman (1965) and Von der Borch (1965). Unfortunately, total concentrations are not known, and it is not clear whether the reported Ca^{2+}/Mg^{2+} ratios refer to molar or to weight ratios. Nevertheless, the data by Alderman and Skinner (1957) suggest that weight ratios have been given, so that recalculated molar units are shown in Figure 7. With respect to weight units, the points are shifted towards higher Ca^{2+}/Mg^{2+} ratios by a factor of 1.65.

Both sets of data show that the solids are changing from aragonite or Mg-calcite over dolomite or protodolomite to magnesite or hydromagnesite as the Ca^{2+}/Mg^{2+} ratios of the solutions decrease.

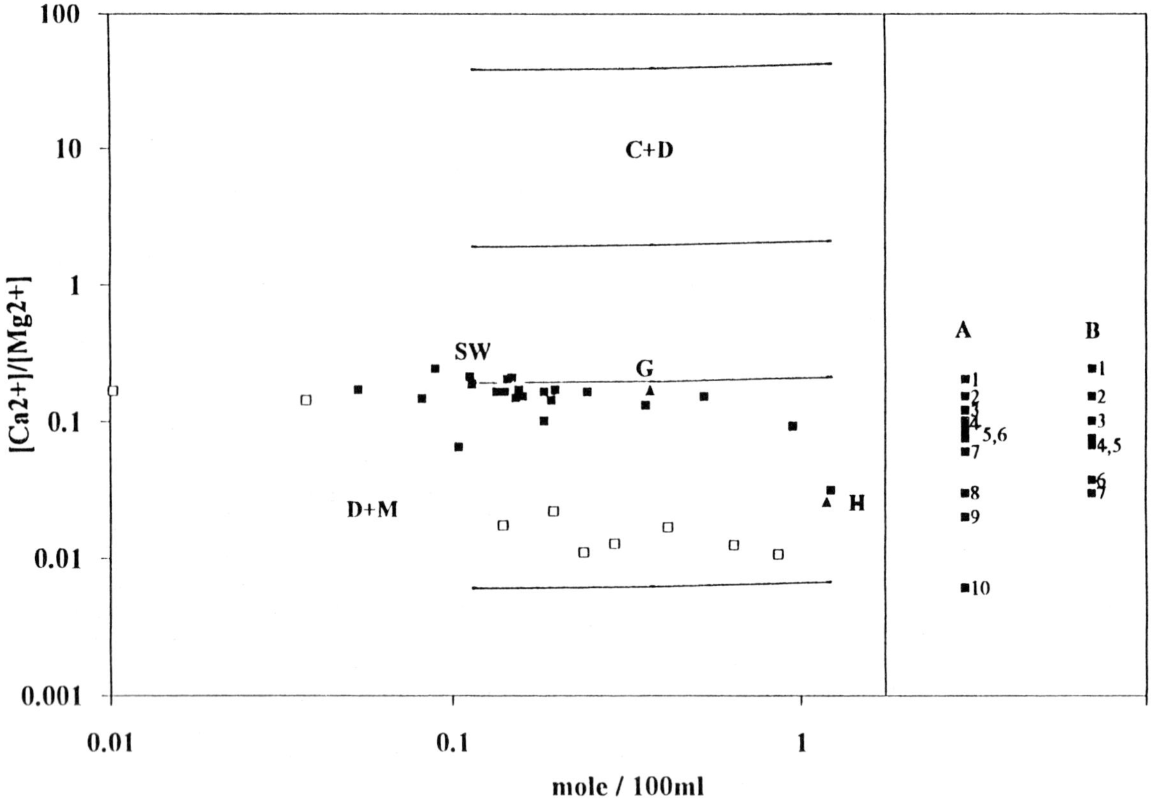

Fig. 7. Left-hand side: compositions of solutions from the Coorong area (Alderman & Skinner, 1957; Skinner, 1963; closed symbols) and from evaporite playas, South Australia (De Dekker & Last, 1989; open symbols). Upper and lower lines: range of calculated calcite–dolomite and dolomite–magnesite phase boundaries (Fig. 4). Mole/100 ml = mole/100 ml solution; SW, seawater; G, H, gypsum- and halite-saturated seawater. Right-hand side: solutions and phase assemblages from the Coorong area. A1–3, aragonite + Mg-calcite; A4, Mg-calcite; A5–6, Mg-calcite + protodolomite; A7, dolomite; A8, dolomite + magnesite; A9, aragonite + magnesite; A10, dolomite + magnesite (Alderman, 1965). B1–2, aragonite + Mg-calcite; B2–3, Mg-calcite + protodolomite; B4–5, dolomite; B6, dolomite + magnesite; B7, aragonite + hydromagnesite (von der Borch, 1965).

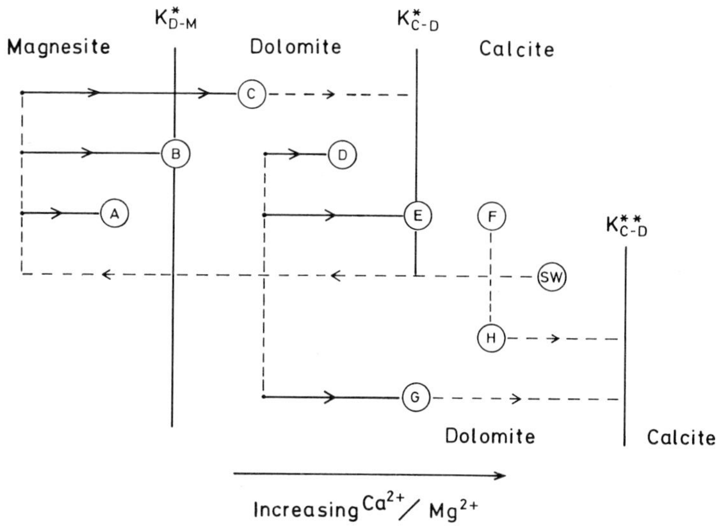

Fig. 8. Reaction paths of brines produced by the evaporation of seawater (SW). A, initial solution in the magnesite field, reaction 3; B, initially reaction 3, at K^*_{D-M} forward reaction 1 and backward reaction 2, Ca^{2+} produced by reaction 1 is consumed by reaction 2; C, as B, but reaction 1 continues with excess $CaCO_3$ and stops in the dolomite field when the $CaCO_3$ is consumed; D, initial solution in the dolomite field, reaction 1, deficient $CaCO_3$; E, reaction 1, excess $CaCO_3$, the boundary calcite–dolomite (K^*_{C-D}) is attained; F, solution in the $CaCO_3$ field, no reaction; G, as E, but the boundary calcite–dolomite is at K^{**}_{C-D} and has not yet been attained; H, solution in the dolomite field, reaction 1 is not yet detectable. See text for details.

This has been interpreted by the above authors as a 'reaction series'. Figure 8 may serve to trace possible reaction paths.

Reaction paths. As seawater evaporates its Ca^{2+}/Mg^{2+} ratio decreases due to the precipitation of calcium carbonate and gypsum. If the resulting brine is located in the magnesite field it will react with a precursor calcium carbonate according to reaction 3. Reaction path 'A' (Fig. 8) represents a situation in which excess calcium carbonate is still present in addition to the newly formed magnesium carbonate. This applies to the phase assemblages aragonite + magnesite and aragonite + hydromagnesite at points A-9 and B-7 (Fig. 7), respectively.

Along path 'B' (Fig. 8) the Ca^{2+}/Mg^{2+} ratio of the solution has increased via reaction 3, so that the molar equilibrium value $Ca^{2+}/Mg^{2+} = K^*_{D-M}$ is attained. At this point dolomite becomes stable and excess calcium carbonate reacts according to the forward reaction 1. Simultaneously, dolomite is forming via the backward reaction 2 from Ca^{2+} produced in reaction 1. Thus, as the reactions proceed Ca^{2+}/Mg^{2+} = constant = K^*_{D-M}. If the amount of calcium carbonate that forms dolomite in reaction 1 is smaller than the amount of magnesite previously produced by reaction 3, the overall reaction will stop. The final products will be dolomite, magnesite and the equilibrium solution exhibiting $Ca^{2+}/Mg^{2+} = K^*_{D-M}$. Obviously, this situation is reflected by the phase assemblage dolomite + magnesite at points A-8 and B-6 (Fig. 7). The erratic point A-10 is disregarded. The average of the Ca^{2+}/Mg^{2+} ratios of points A-8 and B-6 suggests that $K^*_{D-M} = 0.034 \pm 0.004$. A value $Ca^{2+}/Mg^{2+} = 0.037$ is obtained if the phase boundaries dolomite–magnesite from Table 5 are extrapolated to 30°C, using the expression

$$\ln[Ca^{2+}]/[Mg^{2+}] = 9.58 - 3900 \cdot T^{-1}\ (°K),$$

$$r = 0.98.$$

The above values lie well within the range of the boundary dolomite–magnesite, which is calculated from the solubility products.

Along reaction path 'C' (Fig. 8) excess calcium carbonate provides sufficient Ca^{2+} via forward reaction 1, such that at point K^*_{D-M} all magnesite is consumed by backward reaction 2. If there is sufficient calcium carbonate, reaction 1 proceeds and the solution will enter the dolomite field. The reaction stops when the calcium carbonate has been consumed. The products are dolomite and a solution with a Ca^{2+}/Mg^{2+} ratio pertaining to the dolomite field. The same result will be obtained along reaction path 'D', where the starting solution is in the dolomite field and reacts with deficient calcium carbonate. Paths 'C' and 'D' are reflected by the occurrence of exclusively dolomite at points A-7, B-4 and B-5 (Fig. 7).

Along reaction path 'E' (Fig. 8) a solution located in the dolomite field has reacted with excess calcium carbonate according to reaction 1, so that the boundary calcite–dolomite is attained at line K^*_{C-D}. A solution 'F' is located in the calcite field and will not react. It may be considered that path 'E' is represented by the assemblage Mg-calcite + protodolomite at points A-5, A-6 and B-3, and that point 'F' is represented by the occurrence of aragonite and/or Mg-calcite (points A-1, A-2, A-3, A-4, B-1, B-2, Fig. 7).

The assemblage Mg-calcite + protodolomite yields an average ratio $Ca^{2+}/Mg^{2+} = 0.10$ (max. 0.15; min. 0.08). From Table 5 a value $Ca^{2+}/Mg^{2+} = 0.21$ is extrapolated for 30°C using the expression

$$\ln[Ca^{2+}]/[Mg^{2+}] = 11.78 - 4040 \cdot T^{-1}\ (°K),$$
$$r = 0.94.$$

If these data represent equilibrium (or quasi-equilibrium) the seawater point ($Ca^{2+}/Mg^{2+} = 0.191$) is located in the vicinity of the boundary calcite–dolomite. On the other hand, the solubility data suggest that this point is close to the boundary dolomite–magnesite or, more likely, within the dolomite field (Figs 4 and 7). In this case the boundary calcite–dolomite is at K^{**}_{C-D} (Fig. 8) and the assemblage Mg-calcite + protodolomite represents reaction path 'G', along which equilibrium according to reaction 1 has not yet been attained. The assemblage aragonite + Mg-calcite is not stable with the solution, and represents an early stage of reaction 1 with no, or very little, dolomite. Given sufficient time, both solids should react to dolomite along path 'H'. If they are in excess, point K^{**}_{C-D} should be attained. It appears that the situation is ambiguous.

SUMMARY AND DISCUSSION

The present and previous experiments show that, at low temperatures, the formation of dolomite and magnesite requires considerable time and a high ionic strength, which is given by chloride solutions. At elevated temperatures the reactions occur faster and may proceed at lower ionic strengths.

Such conditions are fulfilled by two different geological and geochemical environments: within the subsurface, where brines react with calcium carbonate at depth, and in sediments of coastal plains, where brines produced by the evaporation of seawater are reacting. Unfortunately, there is no well established estimate concerning the predominance of either process throughout geological time.

A comparison of experimentally obtained phase boundaries and Ca^{2+}/Mg^{2+} ratios of subsurface solutions shows that the majority are located in the dolomite field at about 120°C. This means that the probability of 'burial' dolomitization by an encounter between migrating pore fluids and carbonate sediments increases with increasing temperature. Comparing the Ca^{2+}/Mg^{2+} ratios of the most abundant solutions and the boundary calcite–dolomite suggests that the most frequent temperatures of 'burial' dolomitization are between 80° and 90°C. This agrees with fluid-inclusion data. Occasionally, it is considered that the temperatures should be lower. The arguments are based on oxygen isotope data, in general. However, it should be kept in mind that the interpretation of oxygen isotope compositions of dolomites includes a number of assumptions and fundamental uncertainties. Since the critical review by Land (1980), the 'state of the art' has not improved.

For the low-temperature environment the phase boundaries calcite–dolomite and dolomite–magnesite are not known precisely. Thermodynamic calculations and solubility measurements leave rather large uncertainties. Nevertheless, it can be stated that brines resulting from the evaporation of seawater have low Ca^{2+}/Mg^{2+} ratios and high concentrations, and may thus react with a precursor calcium carbonate to form dolomite or magnesite, or their metastable equivalents. However, the geochemical environment must remain unchanged during several hundred years. Mineral assemblages and pertinent Ca^{2+}/Mg^{2+} ratios may serve to trace distinct reaction paths. From the assemblage dolomite + magnesite the Ca^{2+}/Mg^{2+} ratio of the phase boundary may be obtained. The assemblage Mg-calcite + protodolomite is ambiguous.

The state of the art of experimental research is not consistent with the view that dolomite may be formed from 'dilute' solutions, at low temperatures. This standpoint should consider the possibility that 'dilute' solutions in contact with dolomite or magnesite may not represent the original solution from which the solids have formed. The situation becomes worse if 'old and dry' rocks are taken into account. Fluid inclusions may be helpful. In fact, chemical mechanisms that may lead to the formation of dolomite in 'dilute' solutions are unknown, although much research has been done. If such mechanisms are detected, the present statements must be revised.

ACKNOWLEDGEMENTS

The author appreciates critical reviews by Professor Dr J. Hoefs, Geochemisches Institut, Universität Göttingen, Professor Dr E. Sass, The Hebrew University, Jerusalem, and the editors of this volume. The project has been supported by Deutsche Forschungsgemeinschaft.

REFERENCES

Alderman, A.R. (1965) Dolomitic sediments and their environment in the South East of South Australia. *Geochim. Cosmochim. Acta* **29**, 1355–1365.

Alderman, A.R. & Skinner, H.C.W. (1957) Dolomite sedimentation in the South East of South Australia. *Am. J. Sci.* **255**, 561–567.

Baker, P.A. & Kastner, M. (1981) Constraints on the formation of sedimentary dolomite. *Science* **213**, 214–216.

Braitsch, O. (1962) *Entstehung und Stoffbestand der Salzlagerstätten. Mineralogie und Petrographie in Einzeldarstellungen* 3. Springer Verlag, Berlin, 232 pp.

Butler, G.P. (1969) Modern evaporite deposition and geochemistry of coexisting brines, the sabkha, Trucial Coast, Arabian Gulf. *J. Sedim. Petrol.* **39**, 70–89.

De Dekker, P. & Last, W.M. (1989) Modern, non-marine dolomite in evaporitic playas of Western Victoria, Australia. *Sedim. Geol.* **64**, 223–238.

De Groot (1973) Geochemistry of tidal flat brines at Umm Said, SE Qatar, Persian Gulf. In: *The Persian Gulf* (Ed. Purser, B.H.) pp. 377–394. Springer Verlag, Berlin.

Fairbridge, R.W. (1957) The dolomite question. In: *Regional Aspects of Carbonate Deposition* (Ed. Le Blanc R.J. & Breeding, J.G.) Soc. Econ. Paleont. Mineral. Spec. Publ. 5, Tulsa, Oklahoma.

Favre, A. (1849) Note sur l'origine de la dolomie. *C. R. Héb.* **58**, 364–366.

Füchtbauer, H. (Ed.) (1988) *Sedimente und Sedimentgesteine, Sedimentpetrologie* Teil II, 1141 pp., 660 Abb. 113 Tab. Schweizerbart, Stuttgart.

Gaines, A.M. (1980) Dolomitization kinetics: recent experimental studies. In: *Concepts and Models of Dolomitization* (Ed. Zenger, D.H. *et al.*) Soc. Econ. Paleont. Mineral. Spec. Publ. **28**, 81–86.

Garrels, R.M. & Thompson, M.E. (1962) A chemical model for seawater at 25°C and one atmosphere total pressure. *Am. J. Sci.* **260**, 57–66.

Glover, E.D. & Sippel, R.F. (1967) Synthesis of magnesium calcites. *Geochim. Cosmochim. Acta* **31**, 603–613.

Hardie, L.A. (1987) Dolomitization: a critical review of some current views. *J. Sedim. Petrol.* **57**, 166–183.

Helgeson, H.C., Delany, J.M., Nesbitt, H.W. & Bird, D.K. (1978) Summary and critique of the thermodynamic properties of rock-forming minerals. *Am. J. Sci.* **278-A**, 229 pp.

Katz, A. & Matthews, A. (1977) The dolomitization of $CaCO_3$: an experimental study at 252–295°C. *Geochim. Cosmochim. Acta* **41**, 297–312.

Lacroix, A. (1921) *Déodat Dolomieu.* Académie des Sciences. Librairie Académique Perrin et Cie., Paris, 2 vols.

Land, L.S. (1980) *The Isotopic and Trace Element Geochemistry of Dolomite: the State of the Art.* Soc. Econ. Paleont. Mineral. Spec. Publ. 28, 87–110.

Land, L.S. (1983) *Dolomite.* Education note series **24**, 1–20 AAPG, Tulsa, Oklahoma.

Last, W.M. (1989) Continental brines and evaporites of the northern Great Plains of Canada. *Sedim. Geol.* **64**, 207–221.

Liedmann, W. & Koch, R. (1990) Diagenesis and fluid inclusions of Upper Jurassic sponge–algal reefs in SW Germany. *Facies* **23**, 241–268.

Machel, H.G. & Mountjoy, E. (1986) Chemistry and environments of dolomitization – a reappraisal. *Earth Sci. Rev.* **23**, 175–222.

Morrow, D.W. (1981) Diagenesis 1. Dolomite – part 1: The chemistry of dolomitization and dolomite precipitation. *Geosci. Can.* **9**, 5–13.

Morrow, D.W. (1982) Diagenesis 2. Dolomite – part 2. Dolomitization models and ancient dolostones. *Geosci. Can.* **9**, 95–107.

Mutschlechner, G. (1953/54) Déodat de Dolomieu. *Der Bergsteiger*, 21, Jahrg., 6–13.

Purser, B.H. (Ed.) (1973) *The Persian Gulf.* Springer Verlag, Berlin, 471 pp.

Robie, R.A. & Waldbaum, D.R. (1968) *Thermodynamic Properties of Minerals and Related Substances at 298.15°K (25.0°C) and One Atmosphere (1.013 bars) Pressure and at Higher Temperatures.* US Geol. Surv. Bull. 1259, 256 pp.

Robie, R.A., Hemingway, B.S. & Fisher, J.R. (1978) *Thermodynamic Properties of Minerals and Related Substances at 298.15°K and 1 bar Pressure and at Higher Temperatures.* US Geol. Surv. Bull. 1452, 456 pp.

Rosenberg, P.E. & Holland, H.D. (1964) Calcite–dolomite–magnesite stability relations in solutions at elevated temperatures. *Science* **145**, 700–701.

Rosenberg, P.E., Burts, D.M. & Holland, H.D. (1967) Calcite–dolomite–magnesite stability relations: the effect of ionic strength. *Geochim. Cosmochim. Acta* **31**, 391–396.

Skinner, H.C.W. (1963) Precipitation of calcian dolomites and magnesian calcites in the South East of South Australia. *Am. J. Sci.* **261**, 449–472.

Sureau, J.F. (1974) Etude expérimentale de la dolomitization de la calcite. *Bull. Soc. Fr. Minéral. Cristallog.* **97**, 300–312.

Tucker, M.E. (1990) Dolomites and dolomitization models. In: *Carbonate Sedimentology* (Ed. Tucker, M.E. & Wright, V.P.) pp. 365–400. Blackwell Scientific Publications, Oxford.

Usdowski, E. (1964a) Dolomite im System Ca^{2+}–Mg^{2+}–$CO_3{}^{2-}$–$Cl_2{}^{2-}$–H_2O. *Naturwissenschaften* 51, Jahrg. 357.

Usdowski, E (1964b) Die Phasenbeziehungen der Systeme Ca^{2+}–$Mg^{2+}CO_3{}^{2}$–$SO_4{}^{2-}$–$Cl_2{}^{2-}$–H_2O und Na^{2+}–Ca^{2+}–Mg^{2+}–$CO_3{}^{2}$–$SO_4{}^{2}$–$Cl_2{}^{2-}$–H_2O. *Nachr. Akad. Wiss. Göttingen*, II. Math-phys. Kl. Nr. **20**, 263–265.

Usdowski, E. (1967) *Die Genese von Dolomite in Sedimenten.* Springer Verlag. Berlin, 95 pp.

Usdowski, E. (1989) Synthesis of dolomite and magnesite

at 60°C in the system $Ca^{2+}-Mg^{2+}-CO_3{}^{2-}-Cl_2{}^{2}-H_2O$. *Naturwiss*. **76**, 374–375.

Van Tuyl, F.M. (1916) The origin of dolomite. *Iowa Geol. Surv. Ann. Rept*. **25**, 251–422.

Von der Borch, C. (1965) The distribution and preliminary geochemistry of modern carbonate sediments of the Coorong area, South Australia. *Geochim. Cosmochim. Acta* **29**, 781–799.

Von Morlot, A. (1847) Ueber Dolomit und seine künstliche Darstellung aus Kalkstein. *Naturwissenschaftliche Abhandlungen* (Ed. Haidinger, W.) **1**, 306–319.

Zenger, D.H. & Mazullo, S.J. (Eds.) (1982) *Dolomitization*. Benchmark Papers in Geology 65, Hutchinson and Ross Publ. Co.

Zenger, D.H., Dunham, J.B. & Ethington, R.L. (Eds.) (1980) *Concepts and Models of Dolomitization*. Soc. Econ. Paleont. Mineral. Spec. Publ. 28, Tulsa, Oklahoma.

Spec. Publs Int. Ass. Sediment. (1994) **21**, 361–376

Discontinuous solid solution in Ca-rich dolomites: the evidence and implications for the interpretation of dolomite petrographic and geochemical data

A. SEARL

Department of Earth Sciences, University of Birmingham, Edgbaston, Birmingham, B15 5TT, UK

ABSTRACT

The collection of about 2000 electron microprobe spot analyses of mole % $CaCO_3$ in dolomite from several different dolomite occurrences indicates that sets of compositional data are typically polymodal. Similar preferred levels of $CaCO_3$ uptake into dolomite occur as modes in different data sets, and thus may reflect some underlying lattice constraint on dolomite compositions. The rarity of continuous compositional gradients as opposed to sharply bounded growth zones within dolomite crystals also suggests that solid solution within dolomite minerals is discontinuous. This non-ideal solid solution behaviour may be related to the size difference between Ca and Mg ions and the lattice strain associated with substitution of a large Ca ion into a smaller Mg site. It is suggested that there may be an increase in lattice stability associated with evenly spaced as opposed to random substitution of Ca ions into Mg lattice layers. The observed modes in the data sets generally correspond to the preferred levels of Ca uptake predicted by this model. The spread of analyses about predicted modes may be due partly to the inefficiency of this kind of ordering at low temperatures, such that the size of individual ordered domains is less than the width of the probe beam. An underlying bimodal distribution of mole % $CaCO_3$ in dolomite may reflect a significant increase in dolomite stability due to intralayer cation ordering at intermediate levels of Ca uptake into Mg layers. At lower levels of Ca substitution into Mg layers, Ca ions are sufficiently dispersed for there to be little energetic advantage attached to ordering. At higher levels of Ca uptake, the overall distortion associated with the substitution of Ca ions into Mg sites vastly exceeds the energetic effects of ordering. A smaller probe data set for calcite suggests that analogous patterns of cation substitution are also favoured in calcite. Some of the microstructures observed in transmission electron microscopy of carbonates may be due to the presence of domains of differing intralayer cation ordering. Their size might reflect either crystal growth rates or degree of recrystallization during burial. More significantly, if there are gaps within the range of dolomite solid solution, the use of Ca/Mg ratios as an index of crystal growth rate or as a measure of Ca/Mg ratios in the precipitating solution, and the interpretation of compositional growth zonation within dolomite, will have to be re-evaluated.

INTRODUCTION

Dolomite Ca/Mg ratios are frequently analysed during investigations of dolomitization. A number of different significances have been given to dolomite stoichiometry in individual studies (e.g. Mattes & Mountjoy, 1980; Lumsden & Chimahusky, 1980; Machel & Mountjoy, 1986), but there appears to be no simple relationship between Mg/Ca ratios and interpreted environments of dolomitization. Most authors assume that a continuous metastable solid solution series exists between about $Ca_{0.98}Mg_{1.02}(CO_3)_2$ and $Ca_{1.18}Mg_{0.82}(CO_3)_2$, and frequently increasing degrees of Ca/Mg stoichiometry are equated with increasing degrees of cation ordering into distinct Ca and Mg layers. Cathodoluminescence and backscatter scanning electron microscopy, however, indicate that many dolomite crystals show compositional zonation and these compositional zones are typically sharply bounded. The rarity of continuous gradients of compositional variation within dolomite crystals suggests that growing crystals prefer only certain compositions, and intermediate compositions may be less readily precipitated. Compilations of dolomite Ca/Mg compositions determined by X-ray diffraction (XRD) have established that, globally, dolomite compositions form a bimodal distribution with respect to mole % $CaCO_3$ in dolomite (Sperber *et al.*, 1984). This has been interpreted to indicate that most dolomites form initially with relatively Ca-enriched compositions, and then recrystallize to more stoichiometric dolomite (Sperber *et al.*, 1984). This simple model of diagenetic modification does not, however, explain why compilations of compositional data show bimodal rather than the skewed unimodal distributions that would result from a progressive decrease in excess Ca content during diagenesis.

The data presented here are the first compilation of a large number of electron microprobe point analyses of dolomites and ankerites, and suggest that, in addition to an underlying bimodality of dolomite compositions, a number of compositions within the total range of Ca/Mg variation in dolomite occur as modal compositions that reflect preferred levels of Ca uptake. This type of polymodality of dolomite compositions has not previously been noticed, probably because many methods of dolomite analysis, such as XRD, mask fine-scale compositional heterogeneity. These modal compositions occur in dolomites interpreted to have formed in very different environments, and are not therefore likely to be a simple consequence of the conditions of formation. It also seems improbable that the conditions governing dolomite composition should vary in a discontinuous rather than continuous fashion. Hence, the presence of these apparently 'preferred' dolomite compositions, like the absence of continuous compositional gradients within dolomite crystals, suggests that there may be some underlying mineralogical control on the composition of sedimentary dolomites. Possible discontinuities in dolomite solid solution have previously been demonstrated by transmission electron microscopy (TEM) data showing that many dolomites have fabrics comprising fine-scale intergrowths of variably calcian dolomite (Barber & Khan, 1987; Barber & Wenk, 1984).

This article presents a simple model that might explain the apparent discontinuity of solid solution in dolomites through the localized development of cation ordering within Ca and Mg layers. The significance of possible discontinuous solid solution in the interpretation of dolomite stoichiometry and of growth zonation is also discussed.

ANALYTICAL METHODS AND SAMPLES

The dolomite data presented here are taken from three different study areas: a Mesozoic carbonate sequence in southern Turkey, dolostones in a Carboniferous deltaic succession in Scotland, and some dolostones from a Carboniferous carbonate succession in South Wales. The Turkish samples are of non-ferroan dolomite and have isotopic compositions compatible with precipitation from seawater at near-surface temperatures ($\delta^{18}O$ = −1.9 to +0.7‰; Searl, in preparation). The Scottish dolostones have generally undergone a two-stage process of dolomitization in which early mixing-zone dolomites have recrystallized during later hydrothermal activity (Searl, 1990; 1991; Searl & Fallick, 1990). These dolomites are variably ferroan, and the samples include some late-stage ankerite veins. The Welsh samples include early mixing-zone and pedogenic dolomites as well as later hydrothermal ferroan dolomite (Searl, 1988a,b). Each data set includes both replacive dolomite and dolomite cements, but in each case there is no compositional distinction between primary and secondary dolomite of the same age.

In addition to the probe data sets for the three groups of dolomite, two smaller probe data sets for calcite are also presented: one that includes calcites associated with the Turkish and Fife dolomites and a separate set of data for calcites from Lower Jurassic sediments on Skye (Scotland). The first data set includes early meteoric and late-stage burial cements and veins in dolomitized and partially dolomitized limestones from both locations. The second data set includes early meteoric calcite cements in limestones and burial calcite cements in both limestones and sandstones (Searl, 1992 and in preparation). The calcites contain variable amounts of Fe, Mn and Mg, but only negligible amounts of Sr (close to the detection limit). The calcite data are presented because the underlying similarity of the calcite and dolomite lattices (Lippmann, 1973) suggests that anomalous patterns of cation substitution in dolomite may be reflected by analogous lattice constraints on calcite compositions.

Most of the data presented are wavelength-dispersive (WDS) electron probe data collected using a Jeol 733 Superprobe operating at 15 kV and a beam current of 15 nA. The beam was defocused to reduce specimen damage, and calcites and dolomites were analysed under similar conditions. The data have formed part of several published and current studies of dolomitization and carbonate diagenesis, and were not originally intended for a mineralogical study. Consequently, due to the difficulties involved with finding and using carbonate probe standards, the data have been collected using non-carbonate standards. This limits their absolute accuracy, although the data are, themselves, self-consistent. The data for the Welsh dolomites were collected using an energy-dispersive system.

In addition to examining her own sets of carbonate compositional data, the author attempted to conduct a literature survey of published analyses of sedimentary dolomites. This search was conducted by checking all issues of the *Journal of Sedimentary Petrology*, *Sedimentary Geology* and *Sedimentology* from 1980 to 1990 (inclusive) for tabulations and frequency histograms of mole % $CaCO_3$ in dolomite.

ANALYTICAL DATA

Each of the three dolomite data sets shows a clumped distribution of mole % $CaCO_3$ in dolomite (Figs 1, 2 and 3). The combined data set is similarly clumped (Fig. 4), and is composed of two distinct subpopulations, neither of which has the character of a

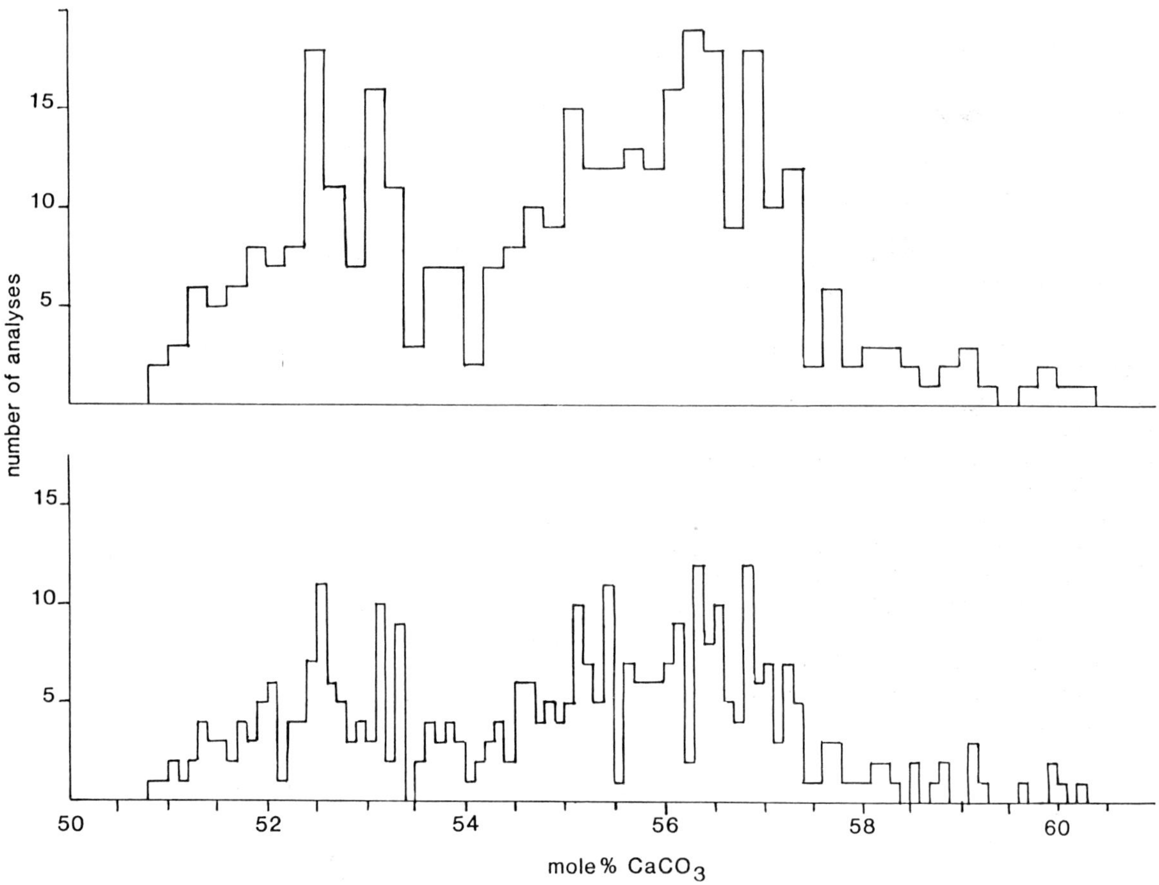

Fig. 1. Frequency histograms of mole % $CaCO_3$ in dolomite: total population of 365 WDS microprobe analyses of 15 samples of non-ferroan dolomite from a Mesozoic sequence in southern Turkey. Top: data plotted with a 0.2 mole % class interval; bottom: the same data plotted with a 0.1 mole % class interval to illustrate the irregular pattern of spikedness.

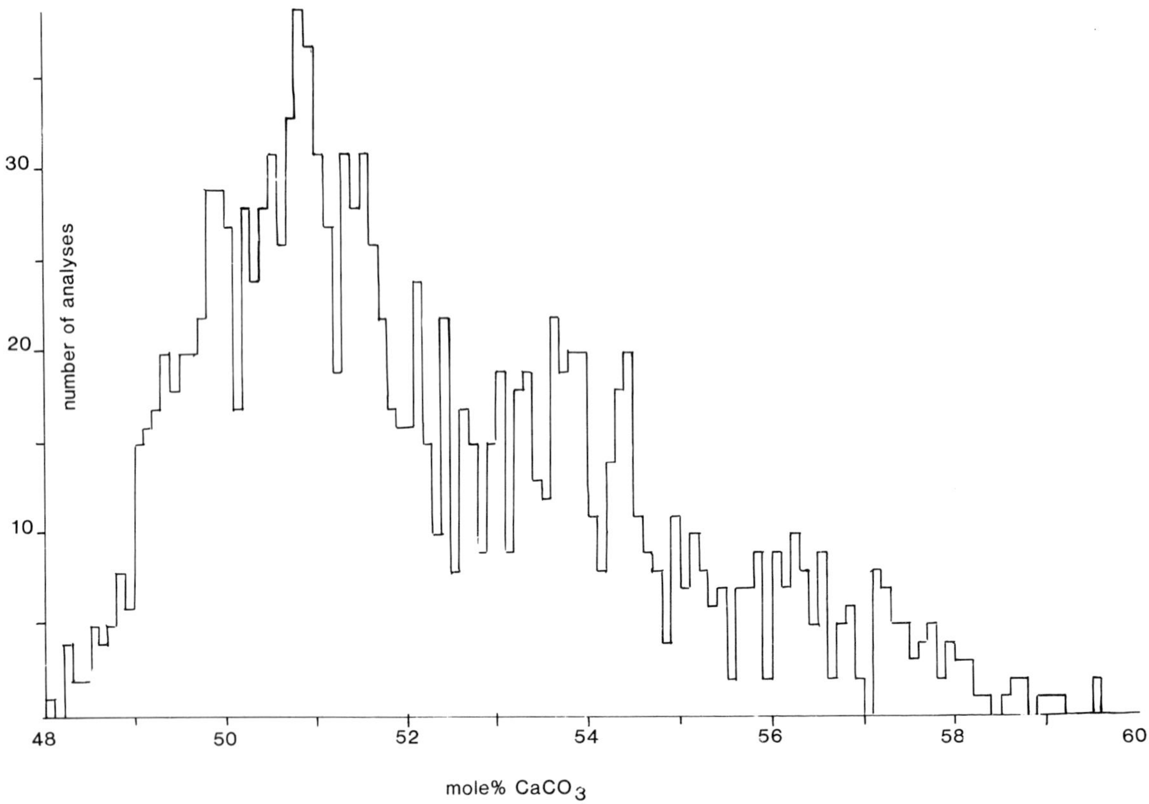

Fig. 2. Frequency histogram of mole % $CaCO_3$ in ferroan dolomite and ankerites: total population of 1459 WDS microprobe analyses of 58 samples of dolostones and carbonate veins from a Carboniferous succession in eastern Scotland.

normal distribution (Fig. 5). The Turkish data set shows the most marked bimodality of the individual data sets, with modes at 52.6 and 56.4 mole % $CaCO_3$ (Fig. 1). Backscatter imaging of the Turkish dolomites indicates that, although crystals are commonly zoned with respect to mole % $CaCO_3$ in dolomite, there are no consistent trends in dolomite composition. This, combined with the limited degree of variation of isotopic composition of these dolomites (above), suggests that the bimodality of dolomite compositions is unlikely to reflect precipitation from two distinct fluids. The Scottish dolomites show a much less marked bimodality and, although dolomitization occurred under a wide range of conditions, dolomite compositions are not a simple function of the environment of formation: variably

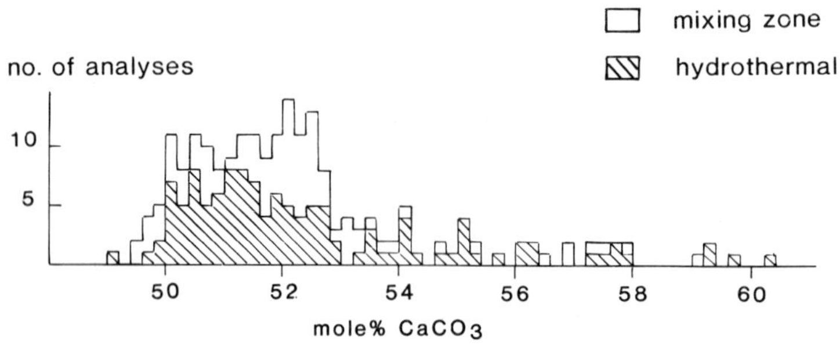

Fig. 3. Frequency histogram of mole % $CaCO_3$ in dolomite: total population of 211 EDS microprobe analyses of 13 samples of dolostones of Carboniferous age from South Wales. The mixing-zone dolomites are non-ferroan and the hydrothermal dolomites are slightly ferroan (<2 mole % $FeCO_3$).

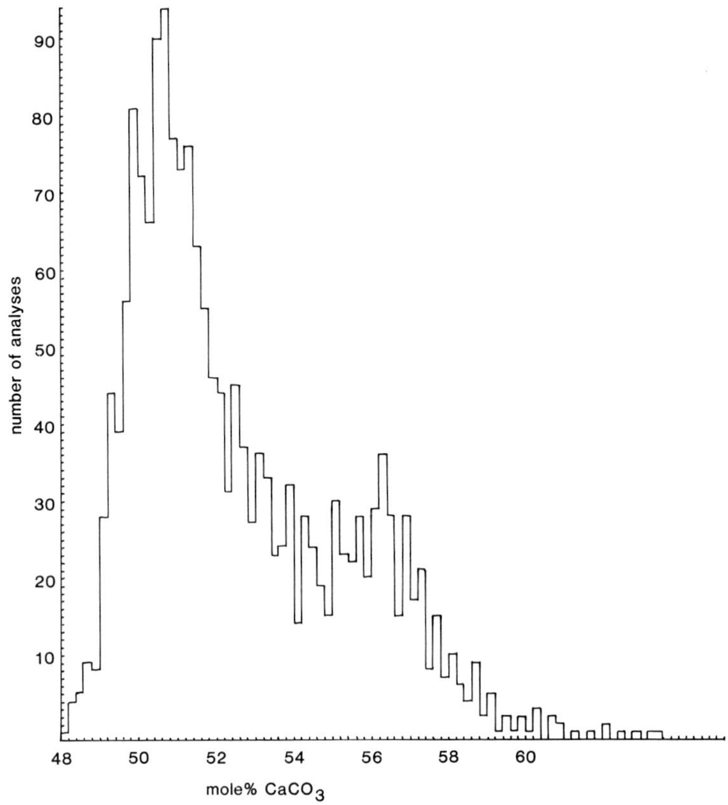

Fig. 4. Frequency histogram of combined WDS data set for mole % $CaCO_3$ in dolomite (i.e. an amalgamation of Figures 1 and 2).

recrystallized mixing-zone dolomites show the same range of Ca compositions as dolomite and ankerite veins (Searl, 1990, 1991; Searl & Fallick, 1990). The Welsh data show an overall difference between early mixing-zone and hydrothermal dolomite compositions, but both data sets are weakly bimodal with modes of similar composition (Fig. 3). The statistical significance of individual spikes in any of the data sets is difficult to establish: it would be possible to 'prove' that each spike was statistically significant or, equally, that each data set was a noisy single population by using (or misusing) an appropriate methodology (Titterington *et al.*, 1985). The use of running averages to smooth the distribution histograms (Marsal, 1987) has been a simple method of investigating the underlying polymodality of each distribution (Fig. 6): smoothing should remove some of the statistical noise and thus reveal the more significant modes. For example, the smoothed 'total' distribution curve shows distinct modes at 50.8 mole % and 56.2 mole % $CaCO_3$ in dolomite, and the presence of minor modes at 50.0, 51.4, 54.0, 54.5, 55.2 and 57.0 mole % $CaCO_3$ in dolomite (Fig. 6).

The calcite data sets show similarly clumped distributions (Fig. 7), with modes at largely similar levels of impurity uptake in both data sets. Both the calcite and dolomite data sets show an irregular pattern of clumping, which suggests that the clumping is not a function of the processing of the probe signal. This was also verified by checking the primary count data for analyses.

LITERATURE DATA

The effectiveness of the literature search was limited, as many authors only illustrate rather than tabulate compositional data and relatively few people have made microprobe spot analyses of dolomite. Most dolomite workers have used 'bulk' analytical techniques such as XRD or ICP (see Fairchild *et al.*, 1988) to determine the composition of several milligrams of powdered sample. These techniques

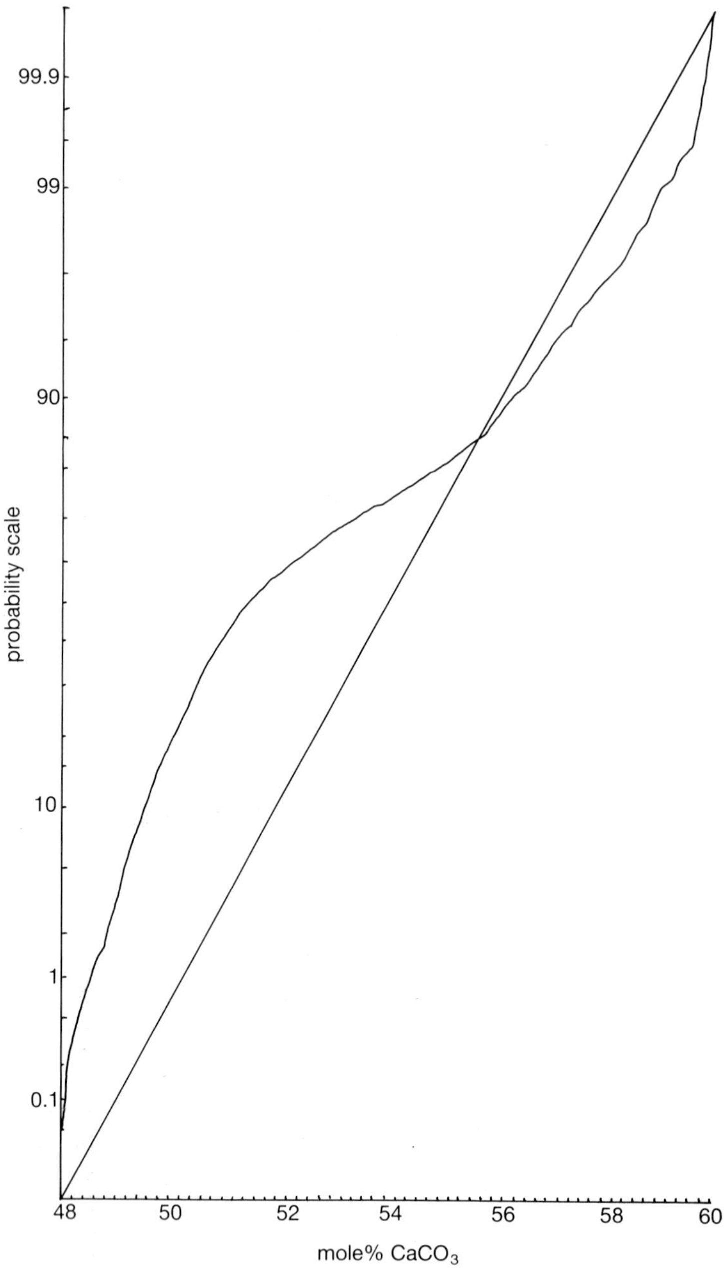

Fig. 5. Arithmetic probability plot of cumulative frequency for the combined dolomite data set illustrated in Figure 4. This divides the total population into two unequal parts, neither of which approximates to a normal distribution (i.e. a straight line). Alternatively, the total population can be interpreted to comprise a large number of overlapping normal distributions with a degree of overlap such that the individual subpopulations (i.e. what would be straight-line segments) are hard to distinguish.

probably mask a large degree of sample heterogeneity (see Searl & Fallick, 1990) and the data are often quoted with a low degree of precision. A further limitation of the literature data is the extremely small size of most data sets. Figure 8 illustrates some published dolomite data: previously published probe data (Schofield & Adams, 1986; Taylor & Sibley, 1986) show modal compositions comparable to those illustrated by the data presented here, and there is a general tendency for all data sets to show clumped rather than smooth distributions with respect to mole % $CaCO_3$ in dolomite.

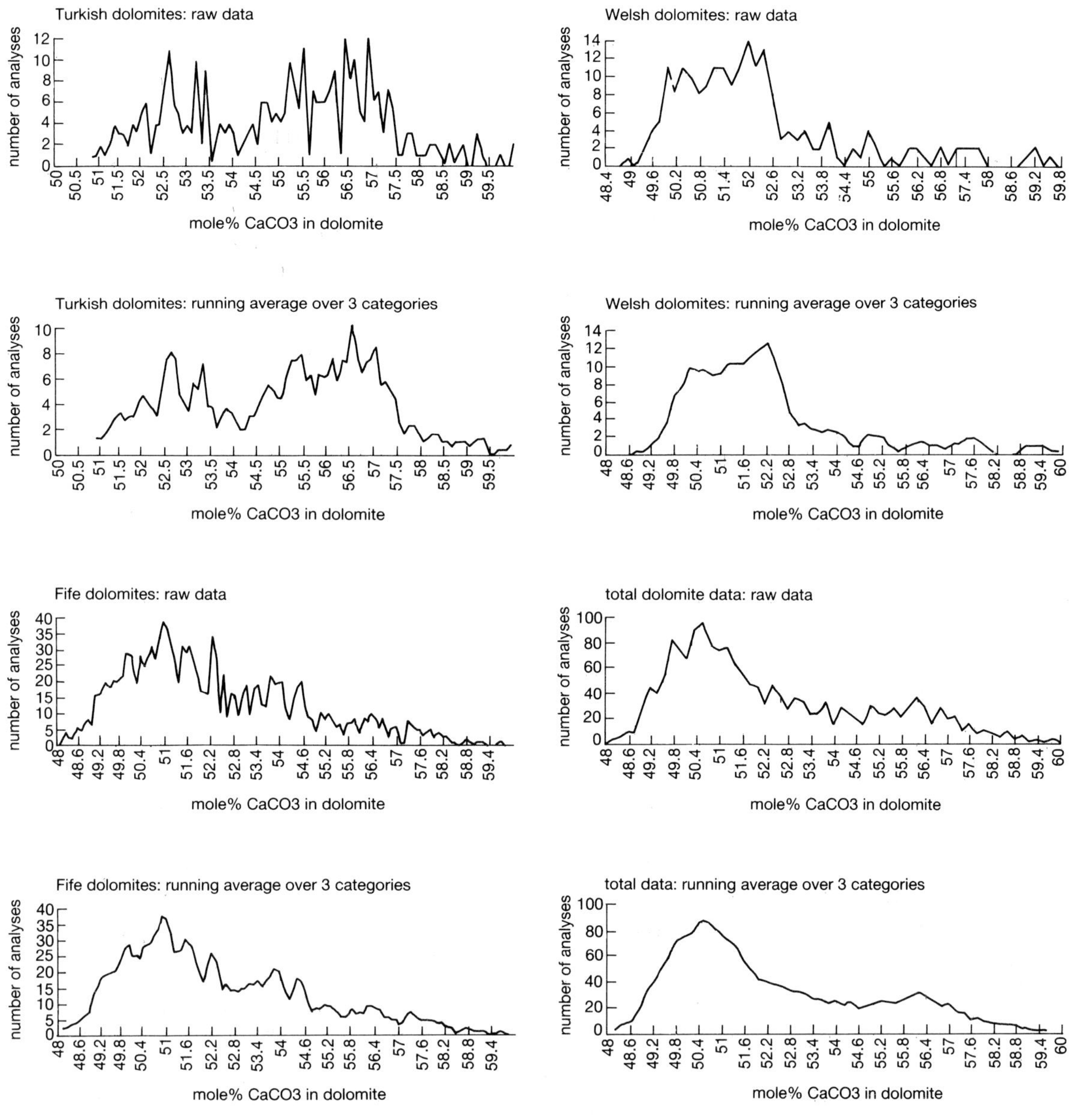

Fig. 6. The distributions of dolomite compositional data illustrated in Figures 1–4 replotted as frequency distribution curves, and also as smoothed curves drawn using running averages over three class intervals. Peaks that disappear in the smoothed curves are assumed to be of less statistical significance than those that are retained in the smoothed distributions.

LIMITATIONS ON DOLOMITE SOLID SOLUTION: A POSSIBLE MODEL

The dolomite lattice is composed of alternating layers of metal cations and carbonate anions, with the Ca and Mg ordered into alternating layers of A (Ca) and B (Mg) cation sites (Lippmann, 1973; Fig. 9). In Ca-rich dolomites the excess Ca substitutes for Mg in Mg layers (or vice versa in Ca-poor dolomites). Given that Ca ions are considerably larger than Mg, the substitution of a Ca ion into an

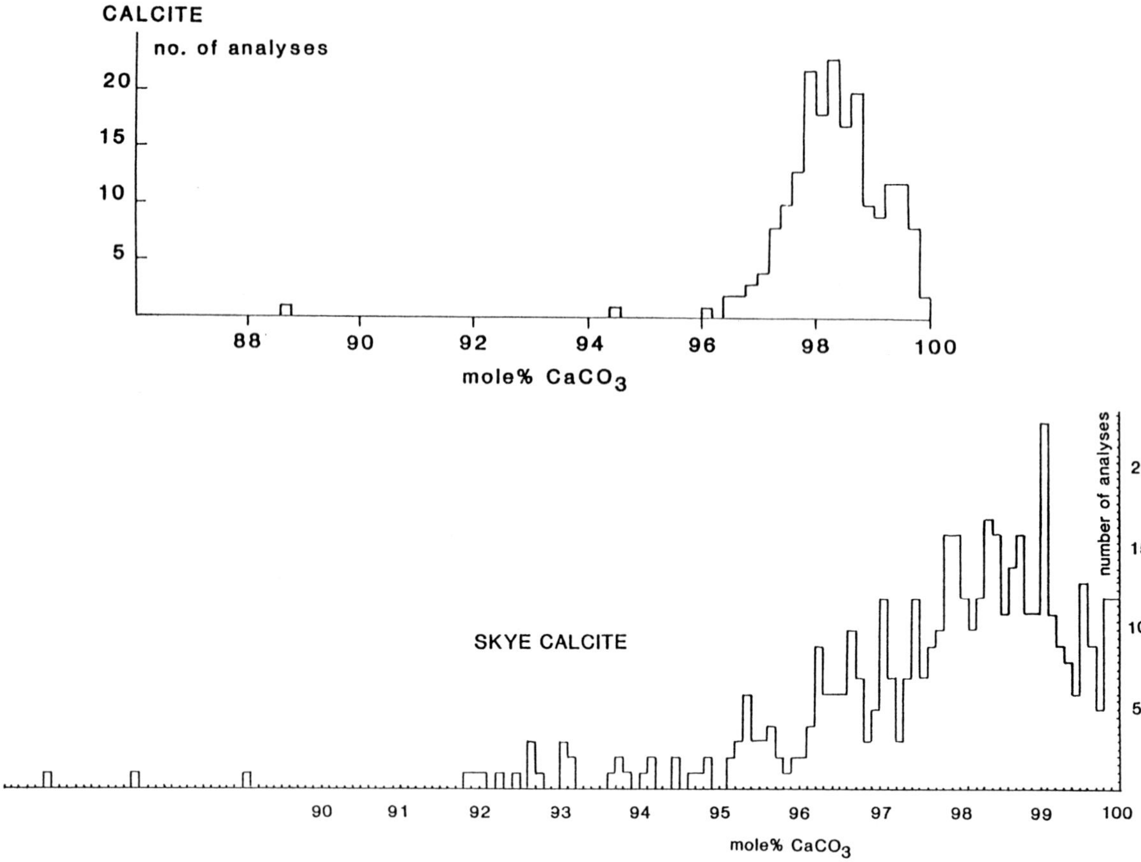

Fig. 7. Frequency histograms of mole % $CaCO_3$ in calcite as determined by WDS microprobe analysis. The other cations present in calcite are Mg, Fe and Mn. Top: calcite veins, early meteoric and burial calcite cements associated with the dolomite samples from southern Turkey and Scotland; bottom: calcite cements and veins from Lower Jurassic rocks on the Island of Skye, Scotland.

Mg site (or vice versa) must create a certain amount of lattice strain. This may be minimized within individual Mg layers if the Ca ions are distributed evenly within the lattice plane, rather than in a clumped or otherwise irregular pattern (Fig. 10). A number of patterns of evenly spaced substitution can be predicted (Fig. 11) and these predictions tested through comparison with the observed data sets. The relatively large size of Ca ions may prevent Ca substitution into two adjacent sites within an Mg layer. This would limit the upper bounds of Ca substitution into dolomite to 62.5 mole % $CaCO_3$ (Fig. 12): close to the maximum observed levels of Ca uptake. In general, however, at such high levels of Ca uptake the intervening Mg sites would be so expanded as to destabilize the lattice, and the ordering of Ca ions within Mg layers would not significantly increase lattice stability. Similarly, at very low levels of excess Ca uptake, Ca ions are likely to be highly dispersed without the specific development of intralayer cation ordering. Hence, if this type of cation ordering does occur, it is most likely to be important at intermediate levels of Ca uptake into Mg layers, and thus 52.6, 53.12, 54.2, 55.0, 56.25 and 57.1 mole % $CaCO_3$ should all occur as modal compositions within compositional data sets (Fig. 11). The size difference between Mg and Fe (or Mn) is considerably less than between Mg and Ca, and thus it is assumed that Mg and Fe are more readily interchangeable within the limits of ferroan dolomite–ankerite solid solution. Microstructural studies of ankerite suggest that crystal field stabilization

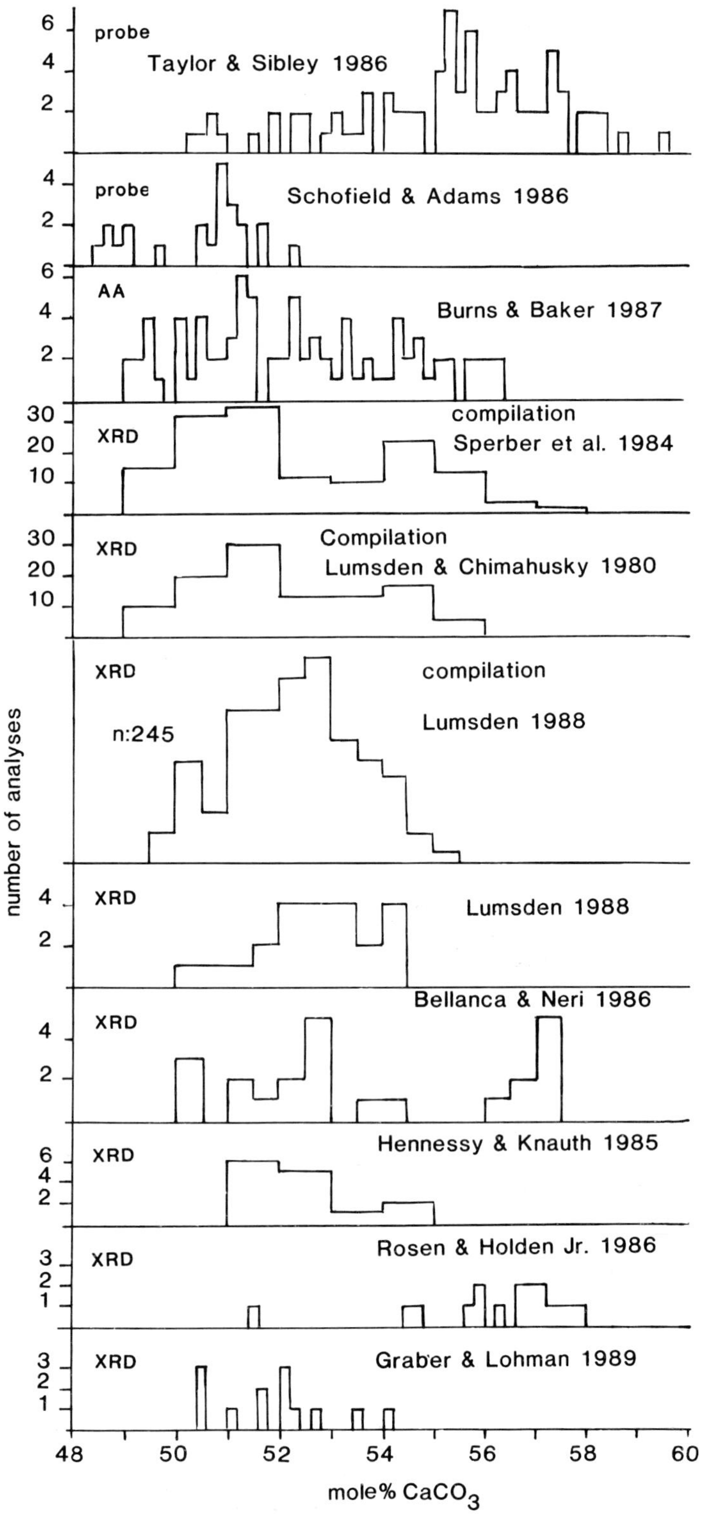

Fig. 8. Compilation of dolomite compositional data from the literature, showing the common tendency of dolomite data sets to show clumped polymodal distributions within respect to mole % $CaCO_3$ in dolomite. The method of analysis is noted for each data set, and this highlights the relatively low precision of XRD, the most widely used technique.

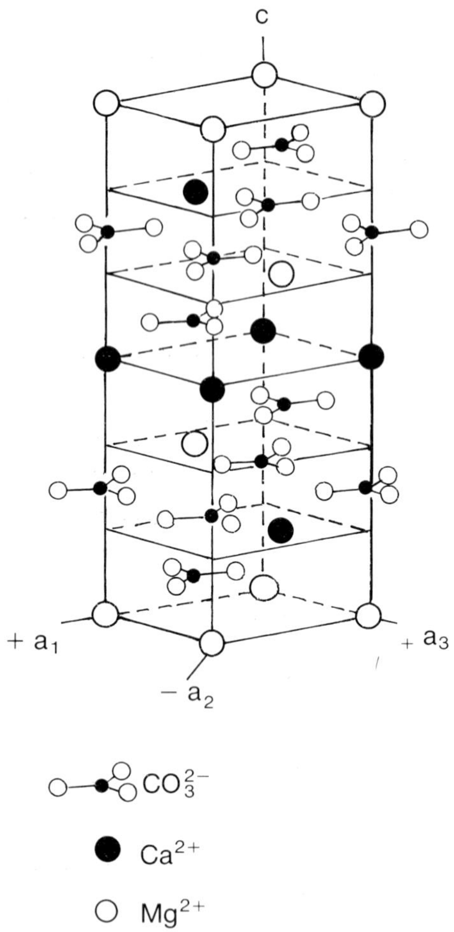

Fig. 9. Lattice structure of dolomite based on Lippman (1973): the cations occur in sheets that alternate with sheets of planar CO_3^{2-} ions. Cation layers are alternately composed of A and B sites. Ca usually occupies the A site and Mg the B site. In ferroan dolomites and ankerites Fe (and Mn) readily substitute for Mg in the B site.

energies do not greatly affect Fe substitution into Mg sites, and there is no evidence to suggest that Fe and Mg are not randomly distributed in the 'B' lattice sites (Reeder & Dollase, 1989).

DISCUSSION

Evaluation of model in light of published mineralogical data

Cations are commonly poorly ordered between Ca and Mg layers within dolomites precipitated at low temperatures (e.g. McKenzie, 1981) and thus it seems likely that any ordering within cation layers will be similarly poorly developed. It does seem probable, however, that ionic substitution into one cation site will affect ionic uptake into adjacent sites, and thus local lattice composition. The dolomite and calcite lattices are not dissimilar, and thus the non-ideal solid solution behaviour of Mg-calcites (Mackenzie *et al.*, 1983) suggests that Ca-rich dolomites may similarly form non-ideal solid solutions. Although there is no published evidence of intralayer ordering in dolomite, TEM data suggest that calcites may minimize lattice strain through maximizing the distances between impurity cations (Reksten, 1990) in a similar but not identical fashion to the model presented here. Some of the unexplained lath and ribbon features in published TEM images of dolomite (e.g. Wenk *et al.*, 1983) may be due to the presence of differently ordered cation domains. Dolomites commonly undergo recrystallization during burial and this may permit the attainment of higher degrees of intralayer cation ordering, with a consequent increase in the size of ordered domains. This is compatible with the observation that lamellar micro-

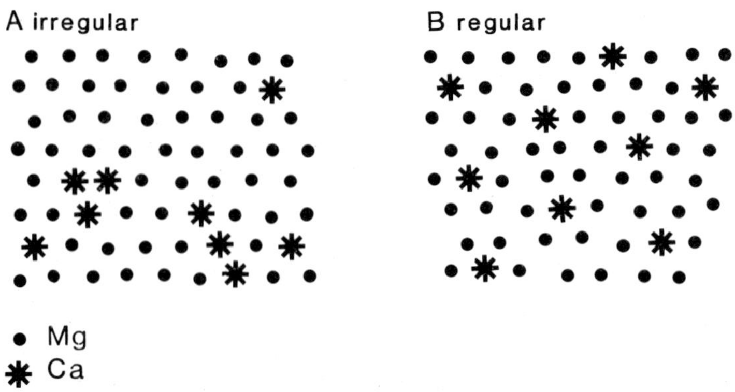

Fig. 10. Plan view of B cation sites in a single *c*-axis perpendicular layer. (A) Random distribution of Mg and a few Ca ions in B sites; (B) Evenly spaced distribution of Ca ions to maximize interatomic distances between these larger cations.

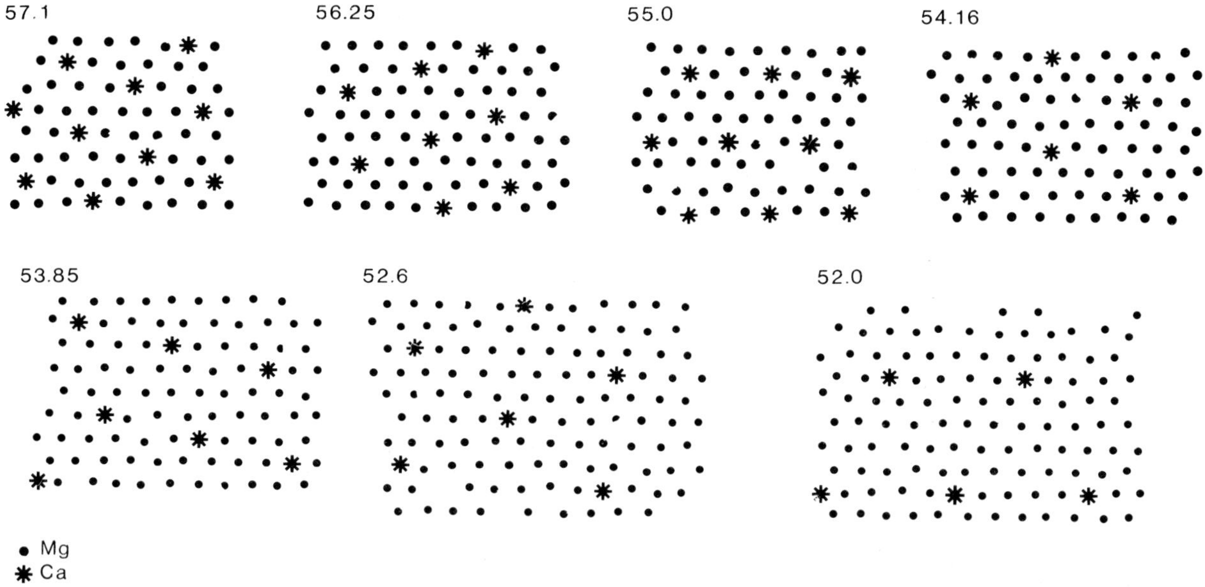

Fig. 11. Suggested patterns of evenly spaced Ca substitution in a layer of B (Mg) cation sites: resultant dolomite compositions are shown as mole % $CaCO_3$ in dolomite.

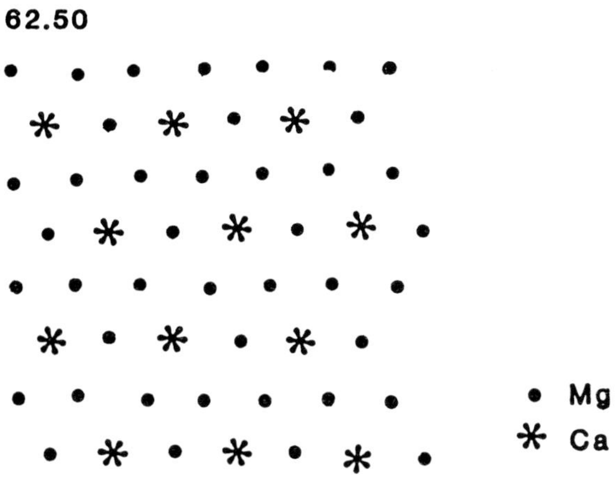

Fig. 12. Maximum possible Ca uptake into a B (Mg) cation layer, based on the assumption that the lattice will be unable to accommodate the strain that would result if two of the large Ca ions were allowed to occupy adjacent B sites. The resultant composition of 62.5 mole % $CaCO_3$ in dolomite corresponds closely to the upper bounds of observed composition (Fig. 4).

structures in dolomite appear to coarsen with age or with heating. However, the various microstructures reported for dolomite are all much narrower than the width of a probe beam. Hence, although it is likely that some intralayer cation ordering may occur in dolomite, both the poor interlayer ordering at low temperatures and the TEM data suggest that the ordered domains are probably considerably smaller than the width of a probe beam. If, however, such ordered domains do exist, then despite the averaging effect of the probe beam, given a sufficiently large number of analyses, the preferred compositions of such domains might stand out as modal compositions.

Comparison of model predictions with data sets

There is some correlation between predicted preferred dolomite compositions and modal compositions within the individual raw data sets, and a slightly better degree of correlation if the predictions are compared to smoothed frequency distribution curves (Fig. 13). Part of the spread of data about predicted modal compositions may reflect the relatively large diameter of the probe beam compared with the size of ordered domains (see above), and some slight offsets between predicted and observed modes may be due to analytical inaccuracy caused by the use of non-carbonate standards. The model, however, does not predict the overall bimodality of dolomite compositions or the greater abundance

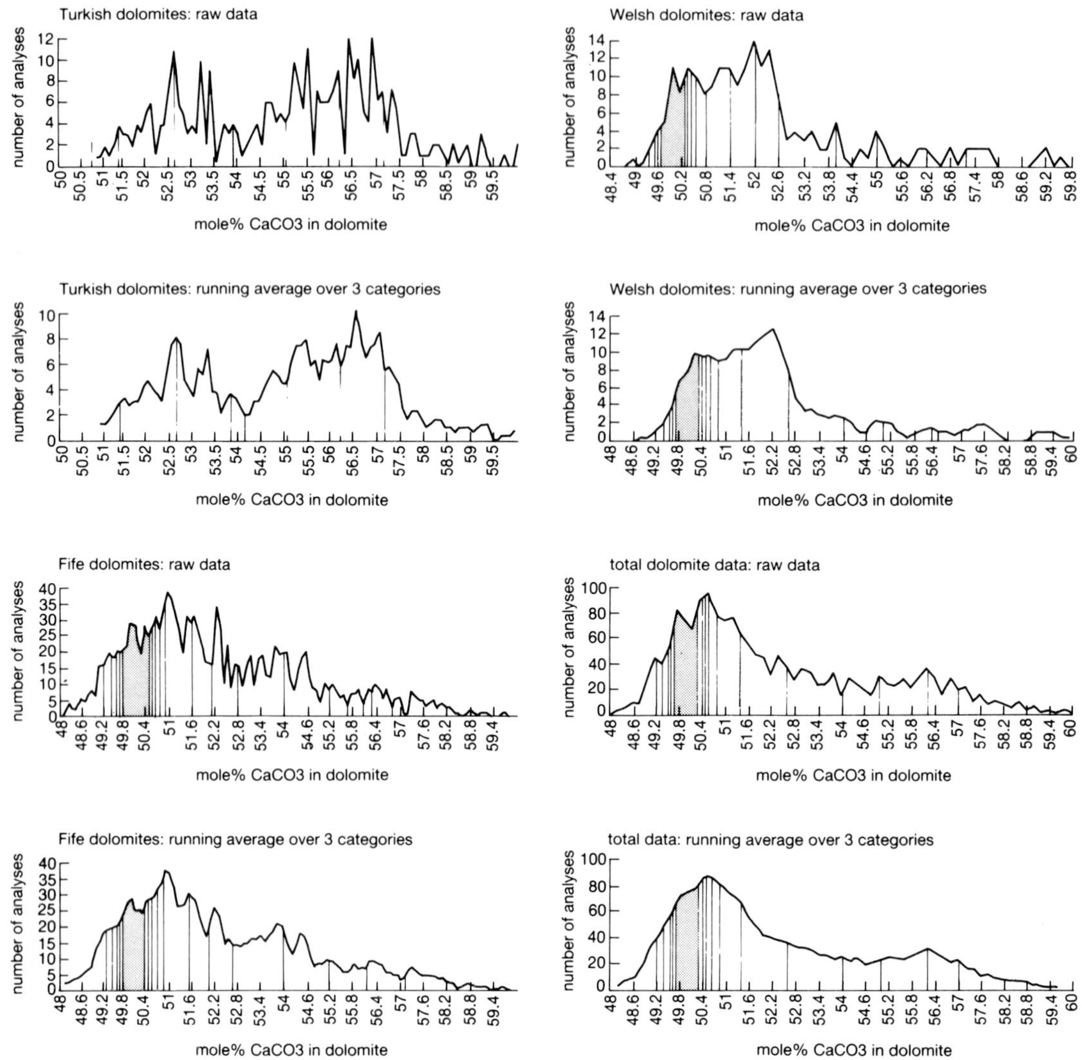

Fig. 13. Comparison of predicted preferred dolomite compositions as determined by drawing patterns of evenly spaced substitution, such as those illustrated in Figure 12, with the observed data sets (Figs 1–4). The density of predicted modal compositions increases as compositions tend towards 50 mole %, such that within the range of composition 50 ± 0.4 mole %, there are too many possible modes to distinguish each individually, and consequently this zone has been highlighted by shading. The level of Ca (or Mg) substitution in this shaded zone is also so low that it is unlikely that significant cation ordering leading to distinct modal compositions would occur. Overall, there is a general coincidence of predicted modes with modal compositions, and some of the offsets between predicted and observed modes may be due to inaccuracies in the data caused by the use of non-carbonate standards. There are, however, several observed modes, most noticeably at around 53.2 mole % $CaCO_3$, that the model does not appear to predict.

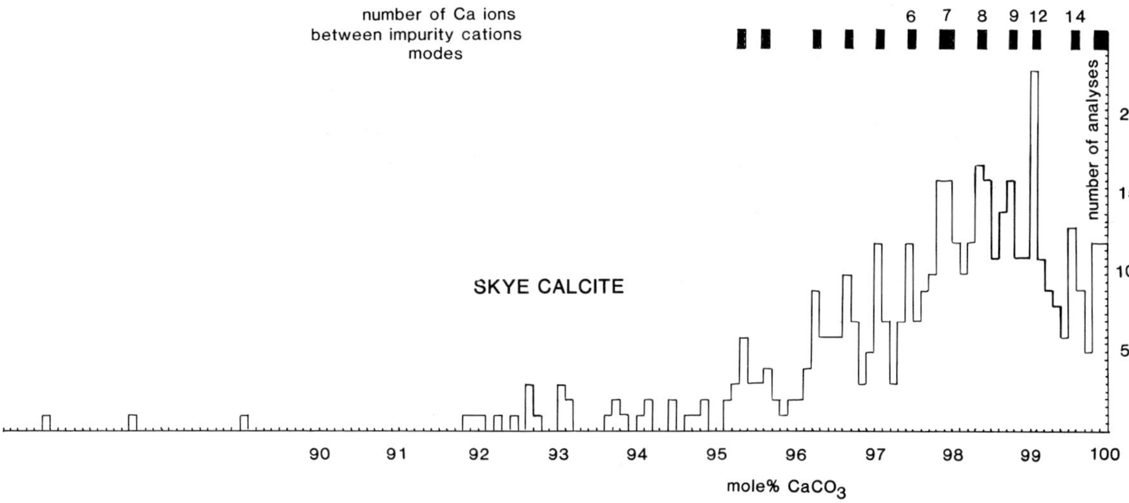

Fig. 14. Interpretation of calcite probe data in terms of evenly spaced substitution of impurity cations (Fe, Mn and Mg) within cation layers (i.e. analogous patterns of substitution to those illustrated in Figure 11). Modal compositions are highlighted along the top of the histogram and, where modal compositions correspond to those predicted by the model, the number of Ca ions between impurity cations is shown: as with the dolomite data, a model of evenly spaced cation substitution fails to predict all the peaks.

of some preferred compositions than others. It is possible that, as suggested above, intralayer cation ordering causes a more significant increase in lattice stability at intermediate levels of excess Ca uptake than at very low or very high levels of Ca uptake. The net effect of such a discontinuous variation in lattice stabilities might favour near-stoichiometric or moderately Ca-enriched compositions, but not slightly Ca-enriched compositions. All the individual data sets contain some modal compositions that are not predicted by the model, and a number of the predicted preferred compositions are not present. This indicates that, even if some of the modes reflect intralayer cation ordering, other factors have an equal or more significant effect on dolomite composition. The relatively good agreement between the model's predictions and the distribution of mole % $CaCO_3$ compositions in the strongly ferroan dolomites and ankerites that comprise the Scottish data set justifies the assumption that Fe and Mg behave as equivalents in the B site.

In contrast to the dolomite data, the calcite data show a more straightforward preference for low levels of trace element uptake. The model compositions correspond to simple patterns of evenly spaced ionic substitution of Fe, Mn and Mg for the larger Ca ions (Fig. 14). The preference for particular spacings of substituted cations may reflect the influence of adjacent cation layers on trace element uptake. The Mg layers in dolomite are isolated from each other by the presence of intervening Ca layers, such that cation substitutions are unaffected by those in adjacent Mg layers. In contrast, the Ca layers in calcite are separated only by intervening layers of carbonate ions, such that ionic substitution in an individual Ca layer is likely to affect that in the layers above and below. In particular, the substitution of an impurity in one layer is likely to preclude substitution in adjacent ionic sites in adjacent layers. The presence of Sr might further complicate patterns of cation ordering in some calcites: unlike Fe, Mn and Mg, Sr ions are considerably larger than Ca ions, and this may give rise to a highly complex mineralogical control on calcite composition.

Ideally, one would test the statistical significance of the apparent correlations between the model and the data sets reported here. However, it is impossible to quantify the degree to which the data set should be skewed towards lower levels of excess Ca substitution, or to estimate reasonable degrees of spread and skewedness about each predicted modal composition.

Discontinuous solid solution in dolomites: relevance to studies of dolomitization

Interpretation of dolomite compositional data

The probe data, in common with previous compilations of whole-rock dolostone compositional data, suggest that, as it is unlikely that solution conditions vary in a similarly discontinuous fashion, there is some underlying mineralogical constraint on the composition of sedimentary dolomites. The possibility of such a mineralogical control on composition has some obvious implications for the interpretation of dolomite compositional data. Caution has already been advised with respect to the interpretation of such data through the recognition that dolomites commonly recrystallize subsequent to their initial precipitation. Their chemical characteristics may therefore be only indirectly related to the original environment of dolomitization (Hardie, 1987). Recrystallization is thought to be driven by the increased stability associated with the attainment of higher degrees of interlayer Ca/Mg ordering and the reduction of excess surface free energy in finely crystalline dolostones. Recrystallized dolomites commonly show a high degree of Ca/Mg stoichiometry (e.g. Sperber *et al.*, 1984), but it should be noted that the extent to which excess Ca can be eliminated is limited by the water/rock ratio during recrystallization. Increased lattice stability associated with the attainment of ordered patterns of excess Ca uptake into Mg layers, and the enlargement of intralayer cation-ordered domains, might be further driving forces in recrystallization that have not been previously discussed. It is possible that lattice constraints on dolomite stability, either those suggested in this article or some other unknown control, may drive recrystallization at low water/rock ratios and promote the breakdown of homogeneous dolomite into an intergrowth of two different, more stable, dolomite compositions.

The Mg/Ca ratio of dolomite is generally interpreted to reflect environmental factors such as fluid temperature and chemistry (e.g. Mattes & Mountjoy 1980; Machel & Mountjoy 1986) and thus, a homogeneous set of Ca/Mg values may be interpreted to reflect dolomitization within a stable diagenetic environment, or complete dolomite recrystallization in a homogeneous fluid regime. Conversely, a polymodal set of Ca/Mg data would generally be interpreted to reflect marked changes in diagenetic environment, variable degrees of recrystallization or recrystallization in a changing fluid regime. If, however, there is not a smooth variation in dolomite stability with increasing degrees of Ca enrichment, then both homogeneous and polymodal distributions could arise under conditions of gradually evolving porewater compositions. It has already been demonstrated that surface geometry during crystal growth has a significant effect on the chemistry of carbonates precipitated at diagenetic temperatures (e.g. Dickson, 1991; Paquette & Reeder, 1990). These studies have highlighted the importance of surface site geometry as determined by differing layer-spreading mechanisms (Reeder, 1991). The model of discontinuous solid solution in dolomites presented here suggests that the composition of the preceding lattice layer might be an important component of surface geometry. Thus the composition of an existing lattice layer may partly determine the composition of a succeeding lattice layer, such that crystal chemistry may be only indirectly related to fluid environment. A similar previous lattice layer control on succeeding lattice layer chemistries has been proposed for oscillatory zoned plagioclase feldspars that appear to grow out of equilibrium with bulk melt compositions (Allegre *et al.*, 1981).

Previous studies of surface effects on carbonate composition during crystal growth (Reeder & Prosky, 1986; Reeder & Grams, 1987; Dickson, 1991; Hendry & Marshall, 1991; Paquette & Reeder, 1990) have suggested that these are sufficiently important to limit the use of carbonate trace element and isotopic data to determine the nature of precipitating fluids. The possible lattice controls on Ca/Mg uptake on growing surfaces of dolomite suggest that the value of dolomite Ca/Mg data may be similarly limited. However, if discontinuous solid solution in dolomite is due to the development of intralayer cation ordering, then Mg/Ca ratios may still be of some use in studies of dolomitization: the degree to which cation ordering within Mg layers is attained may reflect how quickly the crystal has grown, and may also reflect the extent of subsequent recrystallization. It might be possible to use the spread of analytical data about predicted compositions as a measure of the degree of intralayer cation ordering, and the interpretation of TEM data in terms of crystal growth and recrystallization could be greatly extended.

Interpretation of patterns of compositional zonation in dolomite

The presence of compositional zonation within dolomite is readily determined using imaging techniques such as backscatter electron microscopy (e.g. Searl, 1989) or cathodoluminescence (e.g. Kaufman *et al.*, 1991). It is generally assumed that homogeneous crystals are likely to have grown under uniform conditions, and concentrically banded crystals reflect episodes of growth under markedly changed conditions. If, however, dolomites exhibit discontinuous solid solution, then crystals are likely to grow slightly out of equilibrium with bulk fluid conditions, and the composition of new lattice layers may be largely determined by the preceding layer (above). Thus a relatively homogeneous crystal need not reflect growth within a stable diagenetic environment, and sharp charges of composition between adjacent concentric growth zones may be the result of a gradually evolving fluid environment. This implies that the interpretation of concentric growth banding in terms of discontinuous change in the precipitational environment may only be possible if there is other evidence of a hiatus in growth history, such as the presence of a layer of fluid or silt inclusions or an etched surface between adjacent growth bands. Possible lattice controls on dolomite composition may also be relevant to the understanding of the origin of fine-scale oscillatory zonation in dolomite, as suggested for plagioclase by Allegre *et al.* (1981). However, given that oscillatory zonation is generally believed to arise from rapid crystal growth, it is unlikely that subtle mineralogical controls on dolomite surface chemistry are a significant factor in its origin.

CONCLUSIONS

Sets of dolomite compositional data typically show polymodal distributions with respect to mole % $CaCO_3$ in dolomite. Modal compositions are often similar in unrelated data sets, suggesting that there is an underlying mineralogical constraint on dolomite compositions. The implied discontinuous variation of dolomite stability with increasing levels of Ca substitution may be due to cation ordering within Mg layers. Patterns of substitution in which Ca ions are evenly distributed within Mg layers are likely to be more stable than those in which lattice strain is unevenly distributed within Mg layers. The existence of discontinuities of solid solution in Ca-rich dolomites has important consequences for the interpretation of dolomite compositional data and of patterns of compositional zonation within dolomite. Current models of dolomitization are largely based on the interpretation of dolostone geochemical and petrographic data, with little consideration of subtle mineralogical effects on such data types.

ACKNOWLEDGEMENTS

I thank the Geology Department at the University of St Andrews for many hours of use of their electron microprobe. I articularly thank Donald Herd and Ed Stephens for their help with analyses. I thank L. Dever, B. Purser, M. Tucker and D. Zenger for reviewing the manuscript.

REFERENCES

Allegre, C.J., Provost, A. & Jaupart, C. (1981) Oscillatory zoning: a pathological case of crystal growth. *Nature* **294**, 223–228.

Barber, D.J. & Khan, M.R. (1987) Composition-induced microstructures in rhombohedral carbonates. *Mineral. Mag.* **51**, 71–87.

Barber, D.J. & Wenk, H.R. (1984) Microstructures in carbonates from the Aloe and Fen carbonatites. *Contrib. Mineral. Petrol.* **88**, 233–245.

Dickson, J.A.D. (1991) Disequilibrium carbon and oxygen isotope variations in natural calcite. *Nature* **353**, 842–844.

Fairchild, I.J., Hendry, G., Quest, M. & Tucker, M.E. (1988) Chemical analysis of sedimentary rocks. In: *Techniques in Sedimentology* (Ed. Tucker, M.) pp. 274–354. Blackwell Scientific, Oxford.

Hardie, L.A. (1987) Dolomitization: a critical view of some current views. *J. Sedim. Petrol.* **57**, 166–83.

Hendry, J.P. & Marshall, J.D. (1991) Disequilibrium trace element partitioning in Jurassic sparry calcite cements: implications for crystal growth mechanisms during diagenesis. *J. Geol. Soc. London* **148**, 835–848.

Kaufman, J., Hanson, G.N. & Meyers, W.J. (1991) Dolomitisation of the Devonian Swan Hills Formation, Rosevear Field, Alberta, Canada. *Sedimentology* **38**, 41–66.

Lippmann, F. (1973) *Sedimentary Carbonate Minerals.* Springer-Verlag, Berlin, 228 pp.

Lumsden, D.N. & Chimahusky, J.S. (1980) Relationship between dolomite nonstoichiometry and dolomite facies parameters. In: *Concepts and Models of Dolomitization* (Ed. Zenger, D.H., Dunham, J.B. & Ethington R.L.) Spec. Publs. Soc. Econ. Paleont. Mineral., Tulsa 28, 123–137.

Machel, H.G. & Mountjoy, E.W. (1986) Chemistry and environments of dolomitization – a reappraisal. *Earth Sci. Rev.* **23**, 175–222.

McKenzie, J.A. (1981) Holocene dolomitization of calcium carbonate sediments from the coastal sabkhas of Abu Dhabi, UAE: a stable isotope study. *J. Geol.* **89**, 185–198.

Mackenzie, F.T., Bischoff, W.D., Bishop, F.C., Loijens, M., Schoonmaker, J. & Wollast, R. (1983) Magnesian calcites: low-temperature occurrence, solubility and solid solution behaviour. In: *Carbonates: Mineralogy and*

Chemistry (Ed. Reeder, R.J.) Reviews in Mineralogy 11, 94–144. Mineralogical Society of America, Washington.

Marsal, D. (1987) *Statistics for Geoscientists*. Pergamon Press, Oxford, 176 pp.

Mattes, B.W. & Mountjoy, E.W. (1980) Burial dolomitization of the Upper Devonian Miette buildup, Jasper National Park, Alberta. In: *Concepts and Models of Dolomitization* (Ed. Zenger, D.H., Dunham, J.B. & Ethington, R.L.) Spec. Publs. Soc. Econ. Paleont. Mineral., Tulsa 28, 259–297.

Paquette, J. & Reeder, R.J. (1990) New type of compositional zonation in calcite: Insights into crystal growth mechanisms. *Geology* **18**, 1244–1247.

Reeder, R.J. (1991) Surfaces make a difference. *Nature* **353**, 797–798.

Reeder, R.J. & Dollase, W.A. (1989) Structural variation in the dolomite–ankerite solid solution series: an X-ray, mossbauer and TEM study. *Am. Mineral.* **74**, 1159–1167.

Reeder, R.J. & Grams, J.C. (1987) Sector zoning in calcite cement crystals: implications for trace element distribution coefficients in carbonates. *Geochim. Cosmochim. Acta* **51**, 187–194.

Reeder, R.J. & Prosky, J.L. (1986) Compositional sector zoning in dolomite. *J. Sedim. Petrol.* **56**, 237–247.

Reksten, K. (1990) Superstructures in calcite. *Am. Mineral.* **75**, 807–812.

Schofield, K. & Adams, A.E. (1986) Burial dolomitisation of the Woo Dale Limestones Formation (Lower Carboniferous), Derbyshire, England. *Sedimentology* **33**, 207–220.

Searl, A. (1988a) Pedogenic dolomites from the Oolite Group (Lower Carboniferous) South Wales. *Geol. J.* **23**, 157–169.

Searl, A. (1988b) Mixing-zone dolomites in the Gully Oolite, Lower Carboniferous, South Wales. *J. Geol. Soc. London* **145**, 891–899.

Searl, A. (1989) Saddle dolomite – a new view of its nature and origin. *Mineral. Mag.* **53**, 547–555.

Searl, A. (1990) Dolomitization of the Ardross Limestones (Dinantian), East Fife, Scotland. *Sedim. Geol.* **69**, 77–94.

Searl, A. (1991) Early Dinantian dolomites from East Fife: hydrothermal overprinting of early diagenetic fabrics? *J. Geol. Soc. London* **148**, 737–747.

Searl, A. (1992) Sedimentology and early diagenesis of the Broadford Beds (Lower Jurassic), Skye, north-west Scotland. *Geol. J.* **27**, 243–270.

Searl, A. & Fallick, A.E. (1990) Geochemistry of some Dinantian dolomites from East Fife: hydrothermal overprinting of early mixing zone stable isotopic and Fe/Mn compositions. *J. Geol. Soc. London* **147**, 623–663.

Sperber, C.M., Wilkinson, B.H. & Peacor, D.R. (1984) Rock composition, dolomite stoichiometry and rock/water reactions in dolomitic carbonate rocks. *J Geol.* **92**, 609–622.

Taylor, T.R. & Sibley, D.F. (1986) Petrographic and geochemical characteristics of dolomite types and the origin of ferroan dolomite in the Trenton Formation, Ordovician, Michigan Basin, USA. *Sedimentology* **33**, 61–86.

Titterington, D.M., Smith, A.F.M. & Makov, V.E. (1985) *Statistical Analysis of Finite Mixture Distributions*. John Wiley & Sons, New York, 143 pp.

Wenk, H.R., Barber, D.J. & Reeder, R.J. (1983) Microstructures in carbonates. In: *Carbonates: Mineralogy and Chemistry* (Ed. Reeder, R.J.) Reviews in Mineralogy 11, 301–368. Mineralogical Society of America, Washington.

Spec. Publs Int. Ass. Sediment. (1994) **21**, 377–386

Rates of dolomitization: the influence of dissolved sulphate

D.W. MORROW *and* H.J. ABERCROMBIE

Institute of Sedimentary and Petroleum Geology, 3303–33rd St NW, Calgary, Alberta T2L 2A7, Canada

ABSTRACT

Dissolved sulphate at a concentration of 0.005 M retards, but does not prevent, the dolomitization of calcite at 230°C and equilibrium vapour pressure. Reaction simulations using geochemical modelling programs indicate that calcite is somewhat greater than an order of magnitude more undersaturated in sulphate-free solutions at 230°C than in sulphate-bearing solutions. This divergence in the level of calcite saturation is not apparent at low temperatures. One explanation for the differences in the observed rates of dolomitization is that dissolution of calcite is more rapid in a sulphate-free environment at high temperatures because of its greater degree of undersaturation.

INTRODUCTION

Dolomite, a common carbonate mineral, forms dolostone reservoir facies for many gas and oil reservoirs in carbonate strata of sedimentary origin (Roehl & Choquette, 1985) and is the host rock for large Mississippi Valley-type (MVT) lead-zinc deposits (Anderson & Macqueen, 1982). Some coarsely crystalline white dolomites, which are associated with producing gas fields in sedimentary rocks, are known to have been precipitated at temperatures greater than 200°C (Aulstead *et al.*, 1988).

The inability of experimenters to obtain unequivocal precipitates of dolomite at temperatures less than 100°C (but see Usdowski, this volume) has often been cited as an important hindrance to our understanding of dolomitization at these temperatures (Lippman, 1973; Land, 1980). Equally important, however, is our understanding of dolomitization at the elevated temperatures typical of deep-burial diagenetic environments. Reliable results for the synthesis of ordered dolomite have been obtained for reactions between silicate minerals and dolomite at temperatures ranging from 400°C to 525°C (Gordon & Greenwood, 1970; Skippen, 1971). Helgeson *et al.* (1978) and Berman *et al.* (1985) have used these high-temperature data to derive values for the enthalpy, entropy and Gibbs free energy of formation of ordered dolomite. These values can be used to predict the thermodynamic properties of dolomite through a wide range of temperatures and pressures using computer-based programs such as SUPCRT (Kirkham *et al.*, 1978) and are used as part of the mineral database in programs that simulate aqueous chemistry at elevated temperatures, such as PHREEQE (Parkhurst *et al.*, 1980) and SOLMINEQ.88 (Kharaka *et al.*, 1988). However, some experimental studies of dolomitization at temperatures in the range of deep-burial diagenesis—100–300°C—have shown that the dolomitization reaction may be significantly influenced by kinetic factors (e.g. Gaines, 1980; Rosenberg & Holland, 1964; Rosenberg *et al.*, 1967; Baker & Kastner, 1981; Katz & Matthews, 1977).

The purpose of this study is to investigate the influence of dissolved sulphate on the dolomitization reaction at temperatures above 200°C. Baker and Kastner (1981), on the basis of a series of experiments at 200°C, reported that the presence of even small amounts of dissolved sulphate în Ca–Mg–Na chloride solutions greatly inhibited, and even prevented, the reaction of calcite (and aragonite) to dolomite. This was attributed to a kinetic inhibition effect caused by the presence of the dissolved sulphate anion. They suggested that a key factor in dolomitization lies in the presence or absence of the sulphate anion in natural solutions that participate in dolomitization, regardless of temperature. Subsequent work by Morrow and Ricketts (1988) verified this effect to the extent that the dolomitization reaction proceeds at a much slower rate in the presence of dissolved sulphate at 225°C. The experiments reported here involve larger solution-to-solid mass ratios which permit more exact characterization of the thermodynamic state of the reaction system than was possible with previous experiments. Measurements of pH, total dissolved carbonate species and of cation concentrations performed after quenching of reactions to room temperature were used as input into programs such as PHREEQE or SOLMINEQ.88 to estimate the thermodynamic state of the solution under experimental conditions. Estimates of the saturation states of calcite and dolomite and other minerals in the reaction system provide critical information for the assessment of the nature of the inhibiting effect of dissolved sulphate.

This inhibition effect could have two distinctly different causes. Reaction rates in these experiments may be governed solely by differences in the saturation states of calcite and dolomite in reaction solutions of different compositions. In other words, the observed inhibitory effect might be due purely to differences in the thermodynamic states of these reaction systems. In this case, the rates of precipitation or dissolution may be governed by the rate of diffusion of components to or from the mineral surface (Morse, 1983). Alternatively, the inhibition of the dolomitization reaction might be the consequence of processes affecting crystal surfaces during their precipitation or dissolution. A well known example of this type of kinetic inhibitor is the pronounced retardation of calcite dissolution by phosphate ions adsorbed on calcite, irrespective of the degree of calcite undersaturation (Berner & Morse, 1974).

METHODS

Eight experimental runs were performed in four sets of two runs each at vapour saturation pressure for reaction times ranging from 21 h to 69 h (Tables 1 and 2, Figs 1 and 2). Each set of runs involved one run without dissolved sulphate (e.g. run 1, Fig. 1) and one with dissolved sulphate (e.g. run 2, Fig. 1). Two sets of runs, including runs 1, 2, 3 and 4, were short-term experiments with durations of less than 30 h (Fig. 1), whereas the other two run sets, including runs 5, 6, 7 and 8, were longer-term experiments with durations of more than 60 h (Fig. 2).

Table 1. Reaction products.

Run no.	Time (h)	Dolomite (%)	Calcite (%)	Anhydrite (%)
1	21.0	082	018	n.a.*
2	21.0	001	007	092
3	23.5	099	001	n.a.
4	23.5	002	008	090
5	64.0	094	006	n.a.
6	64.0	006	011	083
7	69.0	100	trace	n.a.
8	69.0	005	002	093

* Not applicable to runs 1, 3, 5 and 7 without sulphate. Other runs 2, 4, 6 and 8 contained sulphate and precipitated anhydrite.

Table 2. Solution analyses at end of runs.

Run no.*	Time (h)	Temp. (°C) ±0.1	pH ±0.02	Ca^{2+} (mM/l) ±1.8	Mg^{2+} (mM/l) ±0.4	CO_2 (mM/l) ±1.1
1	21.0	34.0	4.92	123.5	64.4	10.2
2	21.0	31.4	4.92	121.2	62.4	11.0
3	23.5	23.8	5.03	130.5	63.7	14.4
4	23.5	24.0	5.09	120.1	58.8	16.6
5	64.0	24.1	5.11	134.9	67.4	9.1
6	64.0	24.1	5.05	132.2	65.4	12.1
7	69.0	25.4	4.97	122.2	64.1	8.5
8	69.0	29.0	4.92	121.8	62.3	11.4

* Runs 1, 3, 5 and 7 are sulphate-free. Runs 2, 4, 6 and 8 are sulphate-bearing.

Stainless steel reaction vessels (300 ml volume) manufactured by the Parr Instrument Company were used in all runs. Almost all runs were performed in reaction vessels fitted with blank heads and Teflon liners. In addition, loosely fitting Teflon lids were made to cap the liners in order to isolate the solution from the vessel walls. Heating and temperature control was provided by an insulated bench-type gravity convection laboratory oven manufactured by Blue M that is rated for temperatures up to 288°C. All runs were performed at 230°C.

In these experiments, 125 ml of $CaCl_2$–$MgCl_2$

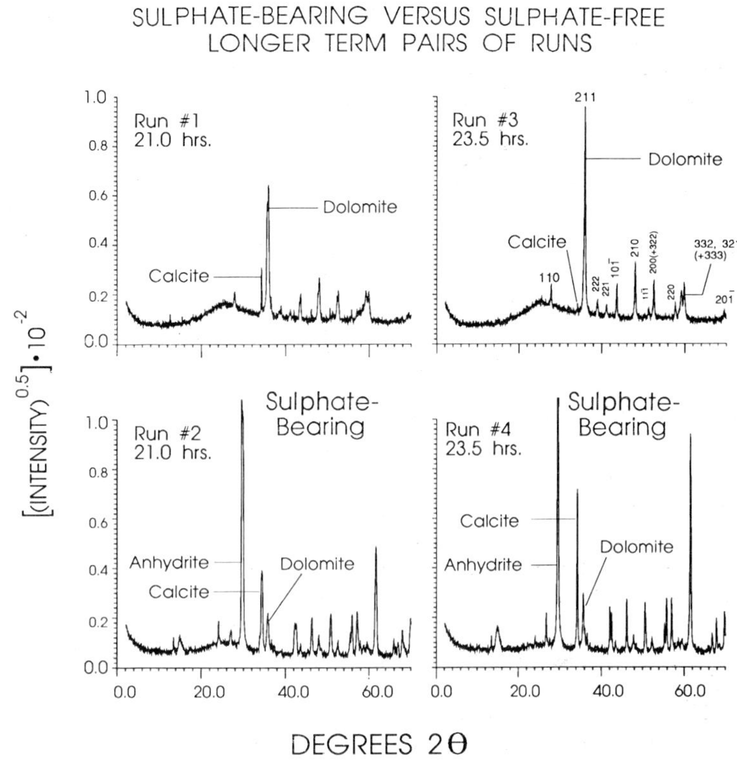

Fig. 1. X-ray diffractograms of the reaction products of the short-term pairs of runs (1, 2, 3 and 4; Table 1). Calcite in the sulphate-free runs (1 and 3) is largely dolomitized, but only a small proportion of the reactant calcite has been dolomitized in the sulphate-bearing runs (2 and 4). Anhydrite is an abundant precipitate in the sulphate-bearing runs and about 30 mg of anhydrite is estimated to have precipitated, based on the decrease in the concentration of sulphate in the reaction solutions of these runs (Table 3).

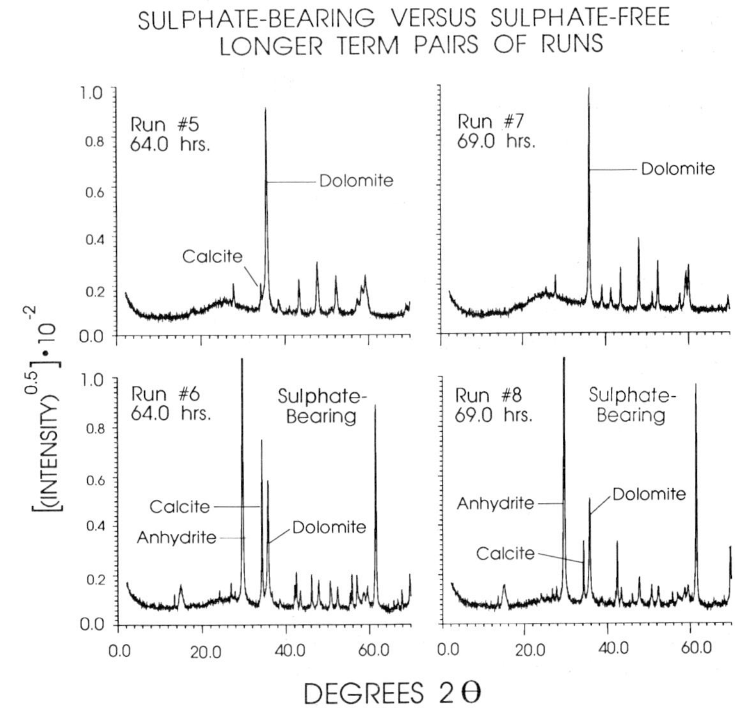

Fig. 2. X-ray diffractograms of the reaction products of the longer-term pairs of runs (5, 6, 7 and 8; Table 1). Calcite in the sulphate-free runs (5 and 7) is largely to completely dolomitized, but only part of the reactant calcite has been dolomitized in the sulphate-bearing runs (6 and 8). About 35 mg of anhydrite is estimated to have precipitated, based upon the decrease in the concentration of sulphate in the reaction solutions of these runs (Table 3). Note that the proportion of dolomite relative to calcite in runs 6 and 8 is much greater than that in the short-term runs 2 and 4 (Fig. 1; Table 1).

solution were added to reaction vessels containing 50 mg of calcite reactant (reagent-grade calcite powder, Fisher Scientific Ltd). This solution was prepared gravimetrically using reagent-grade $CaCl_2$ ($CaCl_2 \cdot 2H_2O$) and $MgCl_2$ ($MgCl_2 \cdot 6H_2O$), so that the solution composition was 0.125 M $CaCl_2$ and 0.0625 M $MgCl_2$ for a molar Ca/Mg ratio of 2/1. This solution was used for all sulphate-free experimental runs. Reagent-grade anhydrous Na_2SO_4 was added to this solution, so that the calculated concentration of Na_2SO_4 in the initial solution for sulphate-bearing runs was 0.005 M. Initial solutions for both sulphate-free and sulphate-bearing runs were analysed to contain 123.1 mM/l Ca^{2+} and 61.5 mM/l Mg^{2+}, and 124.0 mM/l Ca^{2+} and 62.1 mM/l Mg^{2+}, respectively. The initial sulphate-bearing solution was analysed to contain 6.34 mM/l SO_4^{2-}. These analyses are close to the concentrations expected from gravimetric calculations.

Carbon dioxide gas at atmospheric pressure was bubbled through the initial solutions for 1 h before the beginning of all runs and before the addition of calcite, in order to suppress the formation of brucite. Reactions were quenched by refrigeration at the ends of runs. The time required for reduction of the external temperature of the reaction vessels from 230°C down to room temperature (24°C) or slightly less was 2.5 h, but external temperatures fell below 100°C in less than 1 h. Measurements of temperature, pH and total CO_2 content of the reaction solutions were recorded immediately after reactions were quenched. Solutions were then filtered and centrifuged to recover solids.

Temperature and pH were measured with an Orion Research ROSS combination 8104 pH electrode attached to an Orion Research ORION 960 Autochemistry System. The total CO_2 equivalent of all carbonate species in solution was measured with an Orion Research CO_2 gas-sensing electrode attached to an Orion Research Microprocessor Ionalyzer 901.

Filtered reaction solids were washed with double-distilled water and pipetted on to glass slides for X-ray analysis. Solids were analysed by powder X-ray diffraction on a Phillips X-ray diffractometer with iron-filtered Co Kα radiation and are reported in Table 1. Semiquantitative estimates of mineral abundances as percentages of total reaction products were estimated from the ratios of major X-ray peak heights. Mineral products were also examined by scanning electron microscopy on a Cambridge Stereoscan 150 MK2, and some crystals were analysed for their CaO, MgO and SO_2 contents with an energy-dispersive X-ray spectrometer (KEVEX).

All final reaction solutions were analysed for their Ca^{2+} and Mg^{2+} contents by atomic absorption using a Perkin Elmer #603 atomic absorption spectrophotometer. Sulphate-bearing solutions were analysed for their SO_4^{2-} contents by ion chromatography with a 2120I Dionex ion chromatograph using an AS 4A column with a sodium carbonate–sodium bicarbonate eluent (Table 2).

RESULTS

Reactant calcite in sulphate-free runs tended to be almost completely dolomitized, and dolomite formed between 82% and 100% of the reaction products (Table 1). In all runs containing 0.005 M dissolved sulphate, dolomitization of the reactant calcite was only partial and dolomite constituted only 12.5–71% of carbonate phases in the reaction products (Figs 1 and 2; Table 1).

Anhydrite was the dominant reaction product in all sulphate-bearing runs (Tables 1 and 2). This is consistent with the experimental results reported by Morrow and Ricketts (1988) for reactions in $MgCl_2$–$CaCl_2$–$NaCl_2$ solutions. Dolomitization appears to be further advanced in the longer-term sulfate-bearing runs. The reaction products of runs 6 and 8 (Fig. 2) contain a much greater proportion of dolomite than do their shorter reaction time period counterparts, runs 2 and 4 (Fig. 1). The sulphate contents of the quenched reaction solutions of runs 2 and 4 are also higher than are the sulphate contents of runs 6 and 8 (Table 3).

Dolomite formed in the sulphate-free runs tends to occur as intimate intergrowths of euhedral 2–20 μm-sized crystals (Fig. 3A,B). High-magnification views of these crystals showed that crystal surfaces are not perfectly planar but have a submicrometre stepped or terrace-like surface morphology (Fig. 3B) that is typical of the growth

Table 3. Sulphate in final reaction solutions.

Run no.	2	4	6	8
SO_4*	2.80	3.26	2.72	2.60

* Concentration in mM/l (±0.0013 mM/l).

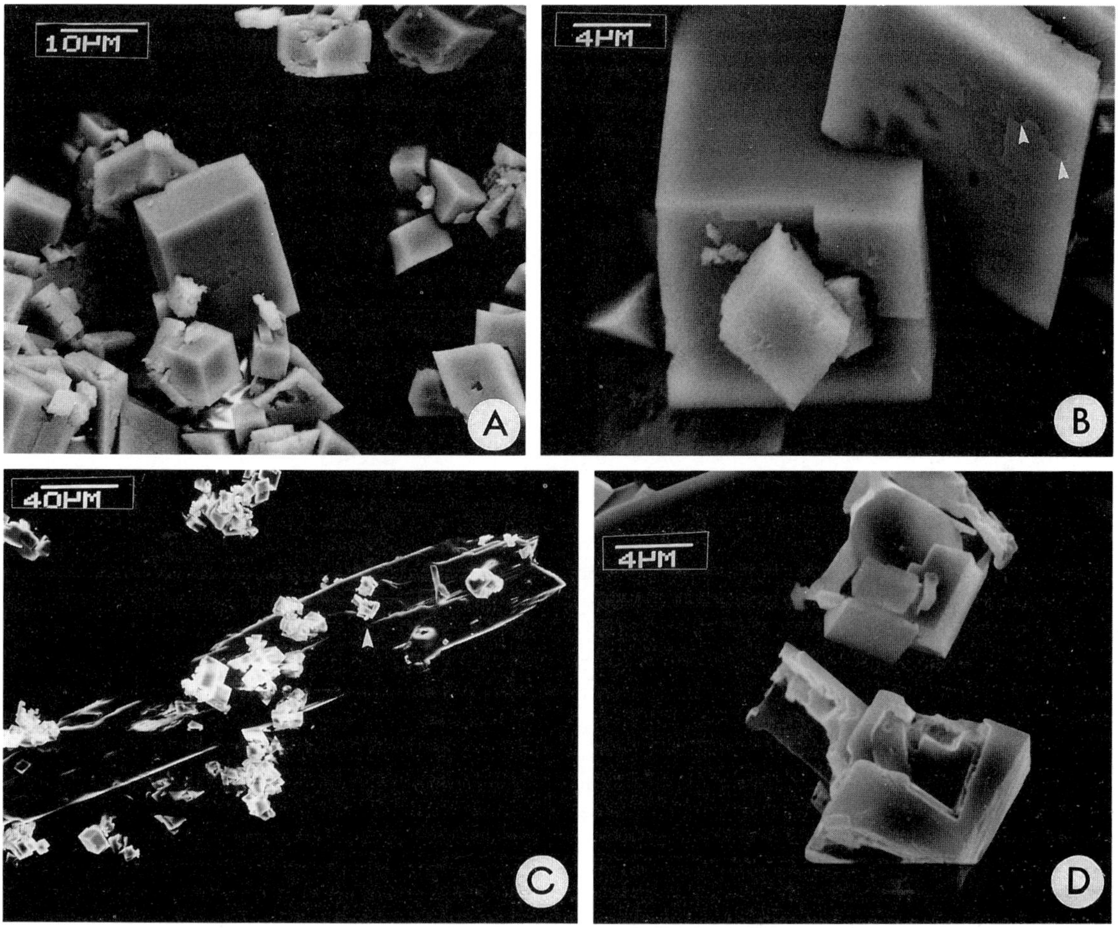

Fig. 3. Scanning electron photomicrographs of reaction products. (A) Intergrown aggregates of euhedral dolomite crystals formed from sulphate-free solution (run 1). Small pits occupy the centres of some of these otherwise perfectly euhedral crystals. (B) A closer view showing the inward-facing terraces or steps occurring on crystal faces (white arrows). The central hollows of some crystal faces may indicate that crystal growth occurred by the accretion of successive layers of terraces that grew incompletely from the edges towards the centres. (C) Aggregates of dolomite and calcite crystals among large euhedral lathes of anhydrite. (D) A closer view of partly leached calcite crystals shown by the white arrow in (C).

pattern that occurs on the surfaces of many different types of crystals (Atkins, 1986). The central parts of crystal faces on many otherwise well-formed crystals contain holes of indeterminate depth.

Dolomite formed in the sulphate-bearing runs (Fig. 3C,D) is similar in crystal size and general appearance to dolomite in the sulphate-free runs. Some crystals exhibit an almost skeletal outline, with hollow interiors (Fig. 3D).

The measured pH of final reaction solutions were all close to 5.0 (±0.15). The final calcium and magnesium concentrations departed from the calculated initial solution concentrations of 125.0 mM/l Ca^{2+} and 62.5 mM/l Mg^{2+} by about ±5.0 mM/l and ±3.3 mM/l, respectively. The calcium concentrations of the sulphate-bearing run final solutions (2, 4, 6, 8) are all lower than in their sulphate-free counterparts.

REACTION MODELS

The aqueous chemistry reaction modelling program PHREEQE (Parkhurst *et al.*, 1980) was used to simulate the reaction path followed by the experiments reported here. The temperature dependence of the equilibrium solubility products of calcite, aragonite, anhydrite and dolomite in the mineral database of PHREEQE were modified to correspond to more accurate values from SOLMINEQ.88 (Kharaka *et al.*, 1988). Many minerals in the PHREEQE mineral database, such as dolomite, use the approximation provided by the Van't Hoff expression to extrapolate their solubility products throughout the temperature domain. The inaccuracy of this approximation increases progressively with temperature divergence from 25°C (i.e. from Standard Temperature). Another shortcoming of PHREEQE and its successor program PHRQPITZ (Plummer *et al.*, 1988) is that they do not incorporate algorithms to estimate the effect of pressure on the dissociation constants of aqueous complexes. However, the vapour pressure of these reactions calculated from the SUPCRT program (Kirkham *et al.*, 1978) is only 27 bars, which has an insignificant effect on mineral solubility products or on the dissociation constants of aqueous complexes (Kharaka *et al.*, 1988).

Two sets of four simulations each were made to model sulphate-free versus sulphate-bearing runs (Fig. 4). In both the sulphate-free and the sulphate-bearing simulation sets the initial chloride solutions were equilibrated under a $p\text{CO}_2$ ranging from 0.3 to 0.9 bars (i.e. 30–90 kPa), in increments of 0.2 bars at 23°C, in order to bracket the probable range of $p\text{CO}_2$ values for the initial solutions that had had CO_2 gas bubbled through them. Figure 4 records changes in the levels of saturation of calcite and dolomite during the progress of this simulated reaction sequence. The families of saturation curves shown for calcite and dolomite in the sulphate-bearing and sulphate-free simulations in Figure 4 reflect the different initial $p\text{CO}_2$ values that were applied to each simulation.

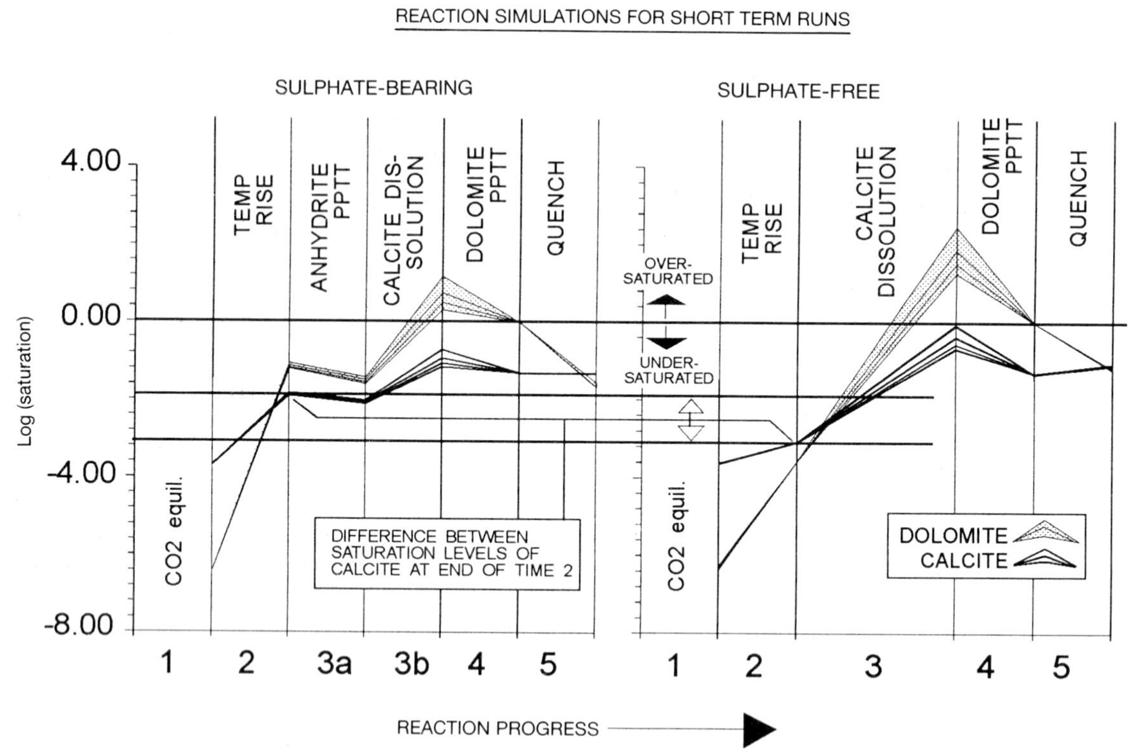

Fig. 4. Simulations of reaction pathways for calcite dolomitization in the sulphate-bearing and sulphate-free solutions of this study. Note the greater undersaturation of the sulphate-free solution immediately after the reaction system is raised to 230°C but before any reactions have occurred. Based on observed data, 40 mg of calcite are dissolved in the sulphate-free simulation, but only 20 mg in the sulphate-bearing simulation. The numbered reaction steps reflect changes in the reaction system. See text for discussion.

In the sulphate-free simulations, the first reaction step is the equilibration of CO_2 with a solution composed of 125 mM/l Ca^{2+}, 62.5 mM/l Mg^{2+} and 50 mM/l Cl^- at 23°C. Step 2 records the effect of the rise in temperature of the reaction system from 23° to 230°C. Step 3 involves the dissolution of 0.004 M (i.e. ≈40 mg) of $CaCO_3$, causing calcite to reach saturation and dolomite to be greatly oversaturated. Step 4 involves the precipitation of dolomite until the system equilibrates with dolomite. Finally, the reaction system is quenched from 230°C to 23°C (Fig. 4, step 5).

The sulphate-bearing simulations are similar to the sulphate-free simulations except for the addition of 5.0 mM/l of Na_2SO_4 to the initial solution and the addition of a reaction step involving the precipitation of 1.7 mM/l of anhydrite ($CaSO_4$) before calcite dissolution at 230°C (Fig. 4, step 3a). This amount of anhydrite is equal to that required to reduce the dissolved sulphate content of the reaction solution to 3.3 mM/l, an amount similar to that observed in the short-term runs. After anhydrite precipitation, the sulphate-bearing simulations follow a path similar to that followed by the sulphate-free simulations, except that only 0.002 M of calcite are dissolved in the calcite dissolution step (Fig. 4, step 3b). These amounts of anhydrite precipitation and calcite dissolution are similar to those that occur in the short-term runs, and represent only a partial approach to equilibrium.

Calculated anhydrite saturation at 23°C ranges from −0.6275 to −0.6278; in other words, the solutions are slightly undersaturated with respect to anhydrite at room temperature. At 230°C, however, the calculated saturation of anhydrite rises to values of 1.9284–2.0154, and the solutions are considerably oversaturated.

In the sulphate-free simulations the amounts of dolomite precipitated after calcite dissolution range from 23 to 33 mg. In the sulphate-bearing simulations the amount of precipitated dolomite ranges from 5 to 13 mg. These amounts correspond with what is actually observed in the short-term runs (Fig. 1).

DISCUSSION

The application of the PHREEQE program for the simulation of reactions involving the chloride solutions of these experiments is justifiable on the basis that all the solutions modelled have ionic strengths less than 0.6 or, in other words, slightly less than that of seawater ($I_{seawater} = 0.7$). The Debye–Hückel equation employed by PHREEQE provides a reasonable approximation for the calculation of ion activity coefficients and, hence, mineral saturations in solutions at ionic strengths below that of seawater. PHRQPITZ (Plummer *et al.*, 1988), a program developed from PHREEQE, calculates activity coefficients using the method of Pitzer (1979) and is significantly more accurate for the estimation of ion activity coefficients in brine solutions. However, the application of this program to reaction systems at temperatures above 100°C is undocumented for solutions such as those of this study. Trial simulations following the reaction sequences plotted in Figure 4 using PHRQPITZ were less consistent with observed data. In particular, the reaction quench step caused the calculated solution pH to decrease, which is contrary to the expectation of a higher pH at lower temperature (Sillén, 1964), and at variance with the observed pH values after quenching of the reactions.

The step involving the dissolution of calcite in these simulations is an obvious oversimplification. Dissolution of calcite is probably accompanied by dolomite precipitation as soon as dolomite becomes oversaturated, and calcite dissolution and dolomite precipitation occurring simultaneously as soon as the dissolution of calcite (about 0.0015 M) has caused the solution to become oversaturated with respect to dolomite. There is no evidence to indicate that the reactant calcite has completely dissolved before dolomite precipitation. Instead, dolomitization has proceeded in the presence of calcite (Figs 1 and 2). This is in agreement with calculations that indicate that, at the temperature of these reactions (230°C), dissolution of the entire 50 mg of calcite reactant would cause the solution to be oversaturated with respect to calcite. Dissolution of 40 mg of calcite, as shown in the sulphate-free simulations of Figure 4, is sufficient for the reaction solution to be either saturated, or very nearly saturated, with calcite.

Also, it is possible that a small amount of anhydrite precipitated at high temperature dissolved during the reaction quench. Substantial anhydrite dissolution is unlikely because no features indicative of dissolution were observed on anhydrite crystal surfaces (Fig. 3C) and because, even at temperatures above 150°C, the anhydrite dissolution reaction takes more than 24 h to reach equilibrium (Dickson *et al.*, 1963). At lower temperatures, anhydrite reacts slowly with solutions. At 70°C, anhydrite requires 3 months to equilibrate with solutions (Zen, 1965). The much shorter time involved in the quenching of these reactions to room temperature indicates that only a minor amount of anhydrite dissolved during quenching. A technique involving a more rapid reaction quenching, such as the use of an internal cooling coil, might provide an even more unequivocal answer to the question of anhydrite dissolution during the reaction quench.

At the elevated temperature of the experiments almost complete precipitation of anhydrite to saturation would be expected, particularly for the longer-term runs of more than 60 h. At 230°C, the calculated concentration of dissolved sulphate in equilibrium with anhydrite is about 0.1 mM/l, which is much lower than the observed concentrations of sulphate in the quenched and filtered solutions. However, longer-term runs do have lower sulphate concentrations (Table 3), which is consistent with a slow approach to anhydrite saturation.

One explanation for the apparent slow rate of anhydrite precipitation in these experiments might involve the large difference between the concentrations of calcium (125 mM/l) and sulphate (5 mM/l) in the experimental solutions. This relatively low concentration of dissolved sulphate may in itself be a significant kinetic impediment to anhydrite precipitation. All previous determinations of anhydrite solubility (e.g. Dickson *et al.*, 1963) involved solutions in which calcium and sulphate were present in stoichiometric amounts for anhydrite precipitation.

The hollows observed in the centres of dolomite crystals precipitated in these experiments (Fig. 3) may reflect a diffusion control of ions. The microenvironment near the centre of a face of a growing crystal may be relatively depleted of ions during precipitation, whereas the reaction zone adjacent to edges and corners will be less depleted. Once precipitation of ions in the centres of crystal faces begins to lag, a hollow or low area should form.

Divergences between observed and calculated pH values and element concentrations (Table 2 versus Table 4) are not great. Some initial solutions may have become more concentrated because of the removal of water as H_2O-saturated CO_2 vapour during the CO_2 bubbling procedure. Alternatively, the absorption of water by $CaCl_2$ and $MgCl_2$ during weighing may have caused the original solutions to be slightly more dilute than that calculated under the assumption that pure compounds were weighed. The observed pH values (Table 2) are slightly lower than their calculated counterparts (Table 4). This may partly be due to the fact that the observed pH values of the final solutions were recorded at temperatures higher than 23°C (Table 2).

Table 4. Calculated calcite saturations, element concentrations and pH*.

Model type	Calcite saturation at 230°C	Final Ca^{2+} (mM/l)	Final Mg^{2+} (mM/l)	Final pH	Final CO_2 (mM/l)
Sulphate-free	$8.2 * 10^{-4}$ (±0.00001)	130.1–130.5	62.4–62.8	5.38–5.12	10.7–30.6
Sulphate-bearing	$1.3 * 10^{-2}$ (±0.001)	127.4–127.7	63.2–63.6	5.25–5.01	10.4–30.1

* These calculations are for reaction solutions equilibrated with a pco_2 ranging from 0.3 to 0.9 bars (i.e. 30–90 kPa).

The degree of calcite undersaturation at 23°C is virtually the same in both the sulphate-bearing and sulphate-free simulations, but at 230°C calcite in the sulphate-free runs is more than ten times more undersaturated than in sulphate-bearing runs before any calcite dissolution occurs (Fig. 4). It is well known that the rate of dissolution or precipitation of a phase is a function of its degree of under- or oversaturation (Atkins, 1986). In particular, over a wide range of solution compositions the dissolution behaviour of calcite fits the equation

$$R_d = k_d(1-\Omega)^n \qquad (1)$$

where R_d is the rate of calcite dissolution, Ω is the saturation index or degree of calcite undersaturation, n is the empirical reaction order and k_d is the dissolution rate constant. For example, Sjöberg (1978) and Morse (1983) found that dissolution of calcite in seawater-like solutions at earth surface temperatures conforms to equation 1 with an empirical reaction order n of about 2.5. The exact value of n is dependent on the reaction mechanism and is assumed to be largely independent of temperature, whereas the value of k_d is temperature-dependent (Atkins, 1986).

Calcite saturation indices calculated for sulphate-free and sulphate-bearing simulations at 230°C before the onset of calcite dissolution (Table 4) may be substituted into equation 1, and ratios may be calculated for the rates of calcite dissolution in sulphate-bearing relative to sulphate-free solutions. Table 5 shows that the dissolution rate of calcite in the 0.005 M Na_2SO_4 solutions of this study is about 10% less than in sulphate-free solutions at 230°C. This

Table 5. Relative rates of calculated calcite dissolution at 230°C based on equation 1.

Na_2SO_4 (M/l)	Calcite saturation	Dissolution ratios*
0.0051	0.0471	0.90
0.0102	0.0923	0.80
0.0153	0.1387	0.70
0.0205	0.1849	0.61
0.0256	0.2312	0.53
0.0307	0.2780	0.45

* Ratios of rates of calcite dissolution in sulphate-bearing solutions versus sulphate-free solutions calculated for SOLMINEQ.88 based on equation 1. Solution compositions are similar to that used in Figure 3 for the 0.5 bar p_{CO_2} simulation, except for variable Na_2SO_4 concentrations.

divergence in dissolution rates increases to about 50% for sulphate-bearing $CaCl_2$–$MgCl_2$ solutions with a sulphate content similar to modern-day seawater (0.028 M). SOLMINEQ.88 was used for these calculations because of its more comprehensive database of aqueous species and algorithms for the extrapolation of solubilities to elevated temperatures and pressures, and because of its calculation of activity coefficients by the more accurate modified Debye–Hückel 'B-dot' method of Helgeson (1969).

It should be clearly appreciated that this divergence between the levels of calcite saturation in sulphate-free and sulphate-bearing solutions at 230°C does not exist at room temperature (Fig. 4). At a p_{CO_2} of 0.5 bars and 23°C, calcite saturation in sulphate-free solutions is $2.3 * 10^{-4}$, as calculated by both SOLMINEQ.88 and PHREEQE, which is within the narrow range of $2.0 * 10^{-4}$ to $3.2 * 10^{-4}$ for calcite saturations at this temperature in the sulphate-bearing solutions of Table 5. Consequently, differences in calcite dissolution rates driven by differences between the degree of calcite undersaturation in sulphate-bearing and sulphate-free solutions apply only to higher-temperature geological settings such as those of deep burial, and cannot be extrapolated to lower-temperature earth surface diagenetic environments.

It is possible that a surface inhibition type of kinetic control may influence calcite dissolution at low temperatures (Sjöberg, 1978), but this effect should be greatly diminished at high temperature because of the greater kinetic energy of ionic motion at higher temperatures. In this regard, dissolution rates of calcite at high temperature in solutions with variable phosphate content might indicate whether the adsorption of sulfate on calcite crystal surfaces is likely to be an important factor in governing the rate of calcite dissolution at high temperature.

Whatever the exact cause for the striking differences in the rates of dolomitization in these experiments and the experiments of Baker and Kastner (1981), it is evident that dolomitization is not halted in sulphate-bearing runs. At high temperatures, sulphate in calcium-bearing solutions will ultimately be controlled by equilibrium with anhydrite, and will be present only in trace amounts of much less than 1 mM/l. At such low concentrations, the influence of dissolved sulfate on the level of calcite saturation will be slight. Consequently, it is unlikely that dissolved sulfate will influence the dolomitization potential of hot subsurface solutions to any significant degree.

CONCLUSIONS

The salient conclusions of this study are:

1 The existence of a sulphate inhibitory effect on dolomitization at high temperature has again been documented. These experiments show that dolomitization is slower in the presence of up to 0.005 M sulphate, but is not halted, as inferred by Baker and Kastner (1981) from similar experiments.

2 Geochemical modelling calculations indicate that, in our experiments at high temperature, the extent of calcite undersaturation in sulphate-free solutions is much greater than in sulphate-bearing solutions. The fact that calcite in sulphate-bearing solutions is more saturated than in sulphate-free solutions may indicate that the rate of calcite dissolution is the dominant rate control on dolomitization in these experiments.

At earth surface temperatures, unlike at higher temperatures, there is virtually no difference in the level of calcite saturation in sulphate-free versus sulphate-bearing solutions. Consequently, differences in calcite saturation due solely to variations in sulphate concentrations cannot be a factor in governing differences in dolomitization rates at low temperature.

3 It is unlikely that the dramatic differences in the rates of dolomitization observed in these experiments play any significant role in deep-burial dolomitization. This is because of the short-term nature of this phenomenon in comparison to geological time.

ACKNOWLEDGEMENTS

We wish to acknowledge the assistance of Mr Brad Gorham and Ms Jenny Wong of the Institute of Sedimentary and Petroleum Geology with respect to analyses by atomic absorption, X-ray diffraction and the scanning electron microscope. We gratefully acknowledge the assistance of Ms Pat Longhurst of Shell Canada Ltd in obtaining aqueous sulphate analyses. Helpful suggestions were offered by Lloyd Snowdon (ISPG) and by an anonymous reviewer. Last, but not least, we extend our thanks to the editors, Bruce Purser (University of Paris), Don Zenger (Pomona College) and Maurice Tucker (University of Durham). This work was funded as part of Geological Survey of Canada Project 880002, and is contribution number 28592 of the Geological Survey of Canada.

REFERENCES

Anderson, G.M. & Macqueen, R.W. (1982) Ore deposit models – 6. Mississippi Valley Type lead-zinc deposits. *Geosci. Can.* **6**, 108–117.

Atkins, P.W. (1986) *Physical Chemistry*, 3rd edn. W.H. Freeman and Company, New York, 857 pp.

Aulstead, K.L., Spencer, R.J. & Krouse, H.R. (1988) Fluid inclusion and isotopic evidence on dolomitization, Devonian of Western Canada. *Geochim. Cosmochim. Acta* **52**, 1027–1035.

Baker, P.A. & Kastner, M. (1981) Constraints of the formation of sedimentary dolomite. *Science* **213**, 214–216.

Berman, R.G., Brown, T.H. & Greenwood, H.J. (1985) *An Internally Consistent Thermodynamic Data Base for Minerals in the System Na_2O–K_2O–CaO–MgO–FeO–Fe_2O_3–Al_2O_3–SiO_2–TiO_2–H_2O–CO_2.* Atomic Energy of Canada Ltd, 62 pp.

Berner, R.A. & Morse, J.W. (1974) Dissolution kinetics of calcium carbonate in seawater. IV: theory of calcite dissolution. *Am. J. Sci.* **274**, 108–134.

Dickson, F.W., Blount, C.W. & Tunell, C. (1963) Use of hydrothermal equipment to determine the solubility of anhydrite in water from 100°C to 275°C and from 1 bar to 1000 bars pressure. *Am. J. Sci.* **261**, 61–78.

Gaines, A.M. (1980) Dolomitization kinetics: recent experimental studies. In: *Concepts and Models of Dolomitization* (Ed. Zenger, D.H., Dunham, J.B. & Ethington, R.L.). Spec. Publs Soc. Econ. Paleont. Mineral., Tulsa 28, 81–86.

Gordon, T.M. & Greenwood, H.J. (1970) The reaction: dolomite + quartz + water = talc + calcite + carbon dioxide. *Am. J. Sci.* **268**, 225–242.

Helgeson, H.C. (1969) Thermodynamics of hydrothermal systems at elevated temperatures. *Am. J. Sci.* **267**, 729–804.

Helgeson, H.C., Delaney, J.M., Nesbitt, H.W. & Bird, D.K. (1978) Summary and critique of the thermodynamic properties of rock-forming minerals. *Am J. Sci.* **278-A**, 229 pp.

Katz, A. & Matthews, A. (1977) The dolomitization of $CaCo_3$: an experimental study at 252–295°C. *Geochim. Cosmochim. Acta* **41**, 297–308.

Kharaka, Y.K., Gunter, W.D. & Aggarwal, P.K. (1988) *SOLMINEQ.88: A Computer Program for the Geochemical Modeling of Water–Rock Interactions.* US Geological Survey, Water-Resources Investigations Report 88-4227, 420 pp.

Kirkham, D.H., Walther, J., Delaney, J. & Flowers, G. (1978) *SUPCRT: program #7402*, Computer Program Library, Dept. of Geology and Geophysics, University of California, Berkeley, California.

Land, L.S. (1980) The isotopic and trace element geochemistry of dolomite: the state of the art. In: *Concepts and Models of Dolomitization* (Ed. Zenger, D.H., Dunham, J.B. & Ethington, R.L.). Spec. Publs Soc. Econ. Paleont. Mineral., Tulsa 28, 87–110.

Lippman, F. (1973) *Sedimentary Carbonate Minerals.* Springer-Verlag, New York, 228 pp.

Morrow, D.W. & Ricketts, B.D. (1988) Experimental investigation of sulfate inhibition of dolomite and its mineral analogues. In: *Sedimentology and Geochemistry of Dolostones* (Ed. Shukla, V. & Baker, P.A.) Spec. Publs Soc. Econ. Paleont. Mineral., Tulsa 43, 25–38.

Morse, J.W. (1983) The kinetics of calcium carbonate dissolution and precipitation. In: *Carbonates: Mineralogy and Chemistry: Reviews in Mineralogy* (Ed. Reeder, R.J.) Mineral. Soc. Am. 11, 227–264.

Parkhurst, D.L., Thorstenson, D.C. & Plummer, L.N. (1980) *PHREEQE—A Computer Program for Geochemical Calculations.* US Geological Survey, Water-Resources Investigations 80–96, 193 pp.

Pitzer, K.S. (1979) Theory: ion interaction approach. In: *Activity Coefficients in Electrolyte Solutions* (Ed. Pytkowicz, R.M.) pp 157–208. CRC Press, Boca Raton.

Plummer, L.N., Parkhurst, D.L., Fleming, G.W. & Dunkle, S.A. (1988) *A Computer Program Incorporating Pitzer's Equations for Calculation of Geochemical Reactions in Brines.* US Geological Survey, Water-Reasources Investigations Report 88-4153, 309 pp.

Roehl, P.O. & Choquette, P.W. (1985) *Carbonate Petroleum Reservoirs.* Springer-Verlag, New York, 622 pp.

Rosenberg, P.E. & Holland, H.D. (1964) Calcite–dolomite–magnesite stability relations in solutions at elevated temperatures. *Science* **145**, 700–701.

Rosenberg, P.E., Burt, D.M. & Holland, H.D. (1967) Calcite–dolomite–magnesite stability relations in solutions at elevated temperatures. *Geochim. Cosmochim. Acta* **31**, 391–396.

Sillén, L.G. (1964) *Stability constants of metal ion complexes. Sec. I: Inorganic Ligands.* Chem. Soc. London Spec. Publ. 17.

Sjöberg, E.L. (1978) *Kinetics and Mechanism of Calcite Dissolution in Aqueous Solutions at Low Temperatures.* Acta Universitatis Stockholmiensis, Stockholm Contributions in Geology, XXXII:1, 92 pp.

Skippen, G.B. (1971) Experimental data for reactions in siliceous marbles. *J. Geol.* **79**, 457–481.

Zen, E'an (1965) Solubility measurements in the system $CaSO_4$–NaCl–H_2O at 35°, 50° and 70°C and one atmosphere pressure. *J. Petrol.* **6**, 124–164.

Spec. Publs Int. Ass. Sediment. (1994) **21**, 387–405

Pervasive dolomitization of a subtidal carbonate ramp, Silurian and Devonian, Illinois Basin, USA

J.M. KRUGER* *and* J.A. SIMO

Department of Geology and Geophysics, University of Wisconsin-Madison, 1215 West Dayton Street, Madison, Wisconsin 53706, USA

ABSTRACT

Silurian and Devonian rocks of the Tippecanoe II and Kaskaskia I sequences in the Illinois Basin provide an example of extensively dolomitized subtidal carbonate rocks. Dolomite in predominantly mid-ramp, fine-grained Silurian and Devonian rocks occurs as two types: a pervasive and extensive massive grey dolomite that cross-cuts facies and depositional sequences and extends across an interregional unconformity, and a less abundant brown type that occurs in tabular bodies that cross-cut bedding. Stratigraphic and petrographic data indicate that both brown and grey dolomite range in age from Silurian to Devonian. Brown and grey dolomites have similar $\delta^{18}O$ and $\delta^{13}C$ values and overlapping major, minor and trace element concentrations. The similar or overlapping chemical signatures suggest that all dolomites precipitated from similar or related fluids. Oxygen and strontium isotopic values indicate that contemporaneous Silurian and Devonian seawater or modified seawater was the dolomitizing fluid, and that dolomitization was a continuous, or quasicontinuous, process. Seawater was probably pumped through Silurian and Devonian sediments during rises and falls in relative sea level.

INTRODUCTION

Extensive dolomites in Silurian and Devonian open-marine subtidal ramp-facies rocks in the Illinois basin provide a classic example of thick, widespread, enigmatic dolomite bodies. Most subtidal carbonate ramps, such as the Silurian and Devonian of the Illinois Basin, are unlikely to undergo extensive early dolomitization due to refluxing hypersaline fluids or from schizohaline or mixed meteoric–marine waters (Burchette & Wright, 1992). Thus, dolomite in subtidal rocks is often interpreted as recording later dolomitization by highly modified seawater or other 'special' fluids (Hardie, 1987). For example, dolomite in subtidal rocks of the Silurian Interlake Formation, the Triassic Muschelkalk in Spain and the Miocene Hawthorne dolosilt have been interpreted as reflecting dolomitization in mixing zones or by later hypersaline fluids (Scott, 1988; Shukla, 1988; Calvet *et al.*, 1990). However, petrographic, stratigraphic and geochemical relations of Silurian and Devonian dolomite in the Illinois Basin suggest a different interpretation.

Subtidal rocks from carbonate ramps are bathed by seawater during and after deposition. Published mass-balance calculations suggest that seawater or modified seawater is the only fluid capable of supplying enough Mg^{2+} for massive dolomitization in near-surface settings (e.g. Land, 1985; Vahrenkamp, 1988). Other studies have supported the concept of seawater dolomitization with Sr isotopic data (Saller, 1984; Aharon *et al.*, 1987; Vahrenkamp *et al.*, 1991).

Dolomite in Silurian and Devonian rocks of the Illinois Basin occurs as locally restricted tabular-shaped bodies that parallel a major interregional unconformity on arches surrounding the basin, and as abundant massive dolomite on the arches but also extending into the basin. In this paper we discuss petrographic, stratigraphic and geochemical

* Present address: 1560 Lila Ave., Baton Rouge, Louisiana 70820, USA.

relations that suggest that both dolomite types record dolomitization by seawater or modified seawater in a marine setting. We hypothesize that seawater pumping due to flow in updip submarine meteoric lenses and continuous wave action during rise and fall of relative sea level pumped seawater through the sediments.

REGIONAL SETTING

The Illinois Basin is located near the centre of the North American craton (Leighton *et al.*, 1990), and is outlined by the Ozark Dome, the Sangamon arch, the Kankakee arch and the Pascola uplift (Fig. 1). Basin subsidence began in the late Cambrian (Willman *et al.*, 1975; Droste & Shaver, 1983), and continued throughout much of the Palaeozoic, during which time the basin opened to the south. This southern opening prevented restriction of the basin, and massive evaporite deposits are notably absent. Uplift of the Pascola arch between Permian and Cretaceous times closed the basin's southern margin (Atherton, 1971). On the northern and eastern basin margins, the Sangamon and Kankakee arches, respectively, formed gentle long-lived structures that were uplifted during the latest Ordovician and Early Devonian. Rocks deposited during major transgressions in the early Silurian and Middle Devonian onlap the arches (Whiting & Stevenson, 1965; Atherton, 1971).

STRATIGRAPHIC SETTING

Palaeozoic rocks within the Illinois Basin consist of seven interregional unconformity-bounded sequences (Sloss, 1963, 1984, 1988; Vail *et al.*, 1977; Collinson *et al.*, 1988). This study concerns massive dolomites in predominantly open marine Silurian and Devonian rocks of the Tippecanoe II and the overlying Kaskaskia I sequences (Figs 2 and 3). The Tippecanoe II and lower Kaskaskia I form a wedge that is more than 550 m thick in the basin centre, but locally less than 10 m thick on surrounding arches. The wedge-shaped geometry is due largely to erosional truncation of the Tippecanoe II beneath the pre-Kaskaskia unconformity to the north. The wedge is bounded above and below by thick shales that isolated Silurian and Devonian strata as a hydrologic and diagenetic unit (Bethke, 1986; Stueber *et al.*, 1987; Bethke *et al.*, 1991).

Fig. 1. Thickness and distribution of Tippecanoe II rocks in the Illinois Basin. Isopachs in metres (modified after Collinson *et al.*, 1988). Sangamon and Kankakee arches formed gentle positive areas that were episodically uplifted during the Palaeozoic, whereas the Pascola uplift rose between Permian and Cretaceous times. Main study area shown in west-central Illinois. ●, Location of subsurface Silurian samples (Skiles Immer #1, Madison Co., Illinois). *, Location of subsurface Devonian dolomite sample (National Associated Petroleum, Marion Co. Coal Co. #1, Marion Co., Illinois).

In the Illinois Basin, the Tippecanoe II consists of uppermost Ordovician to Lower Devonian limestone, dolomite and chert (Figs 2 and 3), and lies unconformably on Ordovician shale (Fig. 3) (Rogers, 1972; Willman *et al.*, 1975; Collinson *et al.*, 1988). The Tippecanoe II consists primarily of open-marine subtidal rocks deposited on a broad carbonate ramp (Whitaker, 1988). Uppermost Ordovician and Lower Silurian rocks at the base of the Tippecanoe II (the Edgewood, Kankakee and Sexton Creek Formations) form a depositional sequence

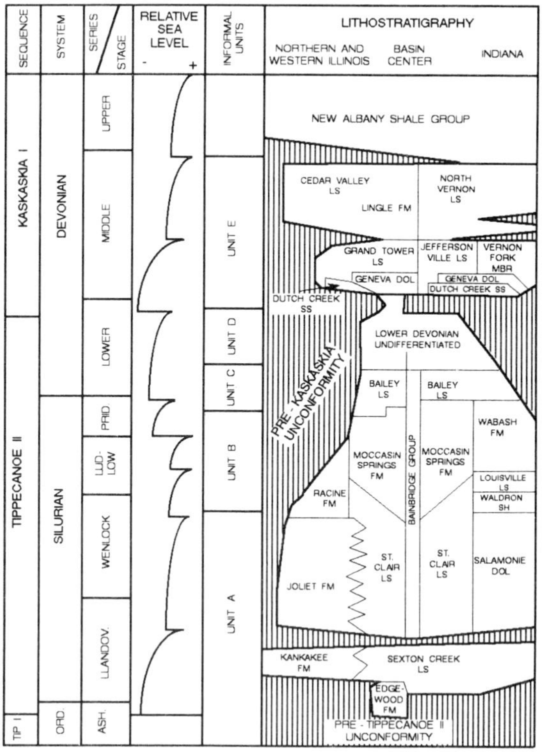

Fig. 2. Stratigraphy of the Tippecanoe II and lower Kaskaskia I cratonic sequences in the Illinois Basin. The post-Kankakee hiatus probably does not record subaerial exposure in the basin. Informal units A–E are for reference to Figure 3. Lower Devonian rocks in basin centre consist of skeletal wackestones and massive chert. Stratigraphy modified after Shaver (1985) and Mikulic (1990). Relative sea-level curve generalized and modified after Ross and Ross (1988) and McKerrow (1979).

with basal retrograding oolitic and pelmatozoan grainstones. The grainstones are overlain by progradational open-marine subtidal wackestones and packstones (Kruger, 1991).

Middle and Upper Silurian rocks in the basin and on surrounding platforms contain scattered pinnacle reefs. Interreef rocks on the Kankakee arch record minor episodes of hypersaline conditions (Droste & Shaver, 1980, 1982), but interreef strata in the basin are dominated by open-marine subtidal mixed skeletal wackestones and packstones (St Clair Limestone, Moccasin Springs Formation, Bailey Limestone and equivalents; Figs 2 and 3). Uppermost Tippecanoe II rocks occur only in the basin centre and consist of Lower Devonian limestone and diagenetic chert. Subaerial unconformities on arches bordering the basin (Kahle, 1988), repetitive sedimentation patterns (Droste & Shaver, 1980, 1982) and episodic reef growth (Droste & Shaver, 1980, 1982; Whitaker, 1988) suggest that Middle and Upper Silurian rocks may contain up to four depositional sequences or packages (Fig. 2) that reflect changes in relative sea level (McKerrow, 1979; Ross & Ross, 1988).

The pre-Kaskaskia sequence-bounding unconformity cuts downsection through the Tippecanoe II to the north (Figs 2, 3 and 4) (Sloss, 1963, 1984; Atherton, 1971; Whitaker, 1988). Although it forms a single omission surface, the unconformity is probably a composite unconformity resulting from several progressive sea-level drops and associated small-scale transgressions and regressions (Fig. 2) (Esteban, 1991). Scattered karst features indicate subaerial exposure of the surface (e.g. Bretz, 1939; Summerson & Swann, 1970; Kruger, 1992a); but palaeosols and karst topography are rare, probably due to an arid climate (Scotese *et al.*, 1985; Witzke & Heckel, 1988) and/or pre-Kaskaskia erosion. In the study area the unconformity is typically a planar surface, but sand-filled pipes locally extend more than 10 m below the unconformity (Kruger, 1992a).

Middle Devonian limestone and dolomite of the Kaskaskia I Sequence onlap the pre-Kaskaskia unconformity, and consist of two depositional sequences separated by exposure surfaces (Fig. 2). The lower sequence onlaps the Sangamon and Kankakee arches and contains shallow-marine locally sandy limestone at the base overlain by regressive tidal and supratidal rocks (Meents & Swann, 1965; North, 1969; Droste & Shaver, 1975). The upper sequence extends across the entire basin and forms a lithologically similar transgressive–regressive cycle (North, 1969; Droste & Shaver, 1975). Upper Devonian black shales of the New Albany Group overlie Middle Devonian rocks throughout the basin and on the arches (Willman *et al.*, 1975) (Figs 2 and 3).

The open-marine subtidal mud-dominated Tippecanoe II rocks in the Illinois Basin lack evaporites (Willman *et al.*, 1975). Consequently, massive dolomite in Silurian and Devonian rocks is difficult to reconcile with the concept of extensive penecontemporaneous dolomitization from hypersaline fluids. More importantly, these open-marine subtidal rocks contain several depositional packages that reflect changes in relative sea level during the Silurian and Devonian.

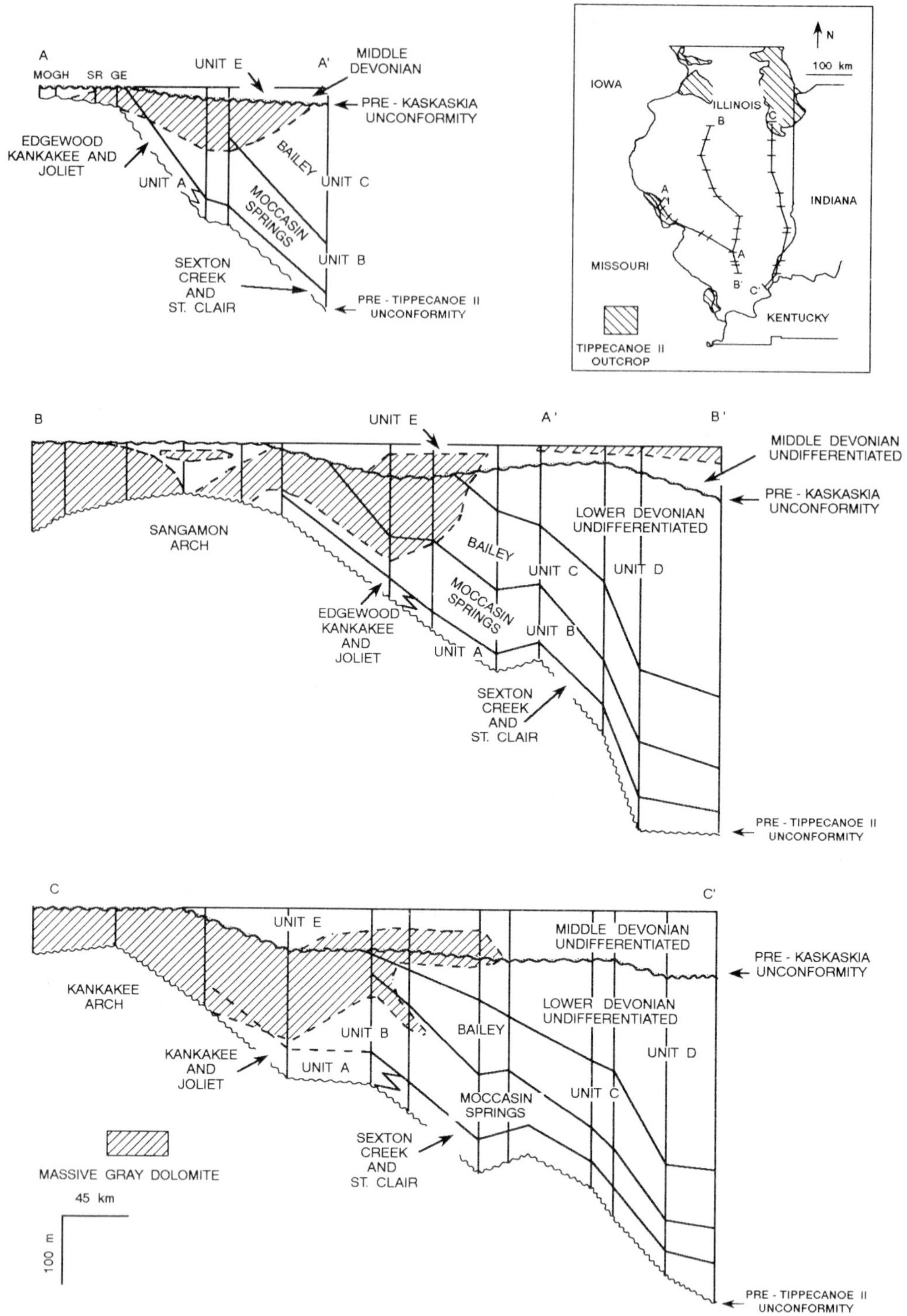

Fig. 3. Generalized distribution of massive grey dolomite in the Tippecanoe II and lower part of the overlying Kaskaskia I sequence in the Illinois Basin. The dolomite is most abundant on arches bordering the basin and cuts across depositional sequences within the Tippecanoe II. Datum is the base of the New Albany Shale Group. Shale above datum is not shown. Dolomite distribution along northern part of line B–B′ is schematic. Informal units A–E are for reference to Figure 2. Compiled from a combination of regional surface and subsurface studies of Palaeozoic rocks (Whiting & Stevenson, 1965; Rogers, 1972; Kolata, 1990), studies of Silurian strata in Illinois and Indiana based on cores and cuttings (Droste & Shaver, 1987), and outcrop and subsurface studies of separate Silurian and Devonian units (Meents & Swann, 1965; North, 1969). Sections MO, GH, SR and GE (line A–A′) from Figure 4.

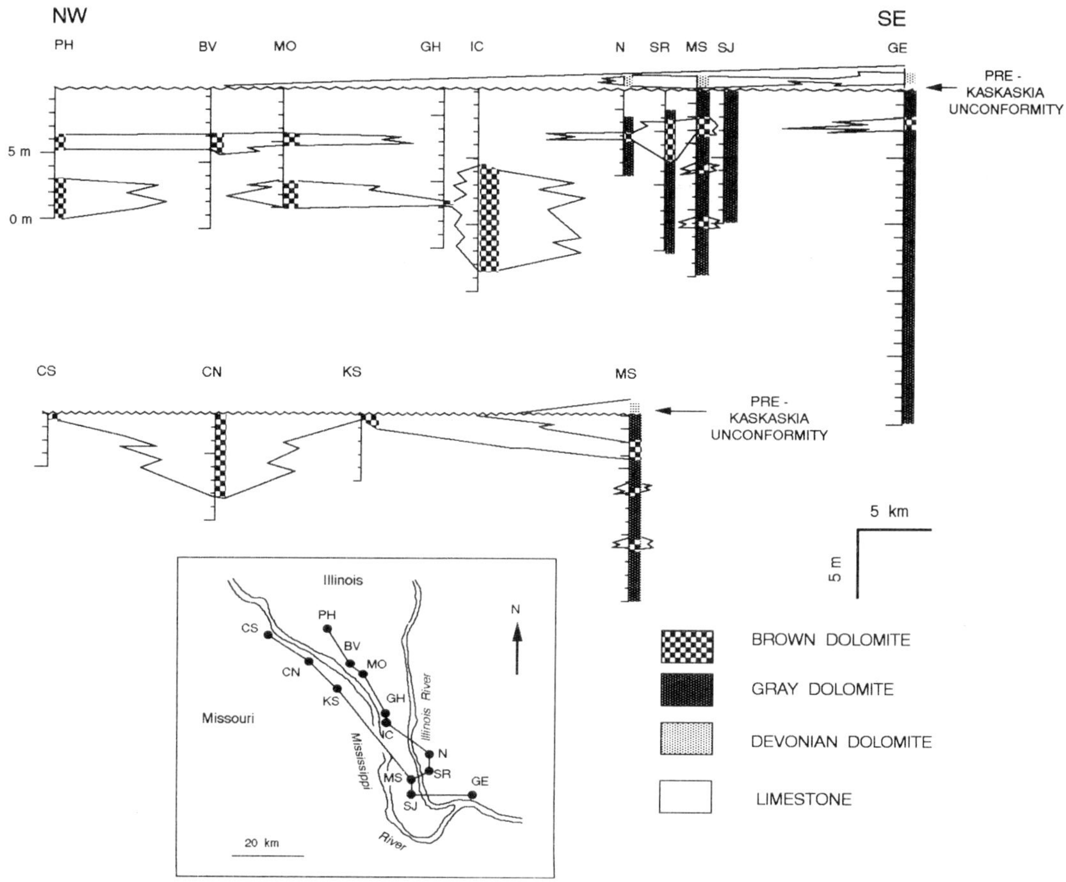

Fig. 4. Distribution of brown and grey dolomite in the study area along northwest basin margin (Fig. 1). The massive grey dolomite forms a northward-tapering wedge. Brown dolomite forms layers or tabular-shaped bodies that generally parallel the unconformity and can be physically traced into the grey dolomite. Datum is pre-Kaskaskia unconformity. The thin wedge of rocks above the unconformity are Middle Devonian in age, and are overlain by the Late Devonian New Albany Shale. PH, Pleasant Hill, IL; BV, Belleview, IL; MO, Mozier, IL; GH, Gresham Hollow, IL; IC, Indian Creek, IL; N, Nutwood, IL; SR, Rosedale, IL; MS, Monterey School, IL; SJ, St Joseph Church, IL; GE, Grafton, IL; CS, Clinton Springs, MO; CN, Clarksville, MO; KS, Kissenger Hill, Missouri.

DOLOMITE STRATIGRAPHY, PETROGRAPHY AND TIMING

Dolomite occurs as two distinct types in the basinward-thickening wedge of Silurian and Middle Devonian rocks: pervasive extensive grey dolomite, and brown dolomite that is restricted to the northern part of the basin and occurs in tabular-shaped bodies that cross-cut bedding (Figs 4 and 5). Although the dolomite types are distinct petrographically and in outcrop, they are geochemically similar and cross-cut facies, stratigraphic units, depositional sequences and an interregional unconformity (Fig. 3).

Grey dolomite is massive and replaces Silurian and Devonian rocks (Fig. 3). The grey dolomite is fabric-destructive, ferroan and consists primarily of 5–40 μm zoned and non-zoned planar-s and anhedral dolomite (Fig. 6). The dolomite is non-mimetic and fabric-selective for muddy textures. Some of the largest dolomite crystals are zoned in plane light, and crystal size is typically greatest in fossil moulds or where grey dolomite replaces fossil fragments (Fig. 6a). Under cathodoluminescence the dolomite is typically dull brown, but some samples show dull-brown and non-luminescent banding (Fig. 6b). Many rhombs have a 2–5 μm-thick bright-luminescent outer zone. The dolomite is volume-

Fig. 5. Line drawing of brown dolomite layer cutting through undolomitized limestone in outcrop. Brown dolomite layer is hachured, karst-filling sandstone shown in dark stipple, karst breccias at top of dolomite layer (to right of tree in centre) shown in medium stipple. Pod of undolomitized limestone at base of dolomite layer (to right of tree in centre) and at far right illustrates that dolomite layers are locally discontinuous. Pre-Kaskaskia unconformity lies at top of outcrop. Jacob staff (1.5 m) to left of centre tree for scale. From Belleview, Illinois (Fig. 4). Modified after Kruger (1992a).

conservative and dolomitization apparently did not significantly increase porosity (Fig. 6c) (*cf.* Lucia & Major, this volume).

Grey dolomite is the dominant type, and its geometry and distribution are shown in Figures 3 and 4. In the outcrop area, grey dolomite is thickest in the south and cuts downsection to the north. On the Sangamon arch, grey dolomite is bounded below by the pre-Tippecanoe II unconformity, and on the western and central parts of the arch grey dolomite pinches out toward the arch crest. In the eastern part of the basin, the dolomite extends north and east on to platforms bordering the basin. The basinward end of the dolomite body is blunt in the western and central parts of the basin, but digitate in the east. In the eastern part of the basin, Middle Devonian rocks above the pre-Kaskaskia unconformity also contain massive grey dolomite (Fig. 3). Although some grey dolomite is Middle Devonian or younger in age, geochemical data discussed below strongly suggest that the grey dolomite records continuous or quasicontinuous dolomitization that began in the Silurian and continued through the Middle Devonian.

Brown dolomite occurs locally on the northwest basin flank, and consists of 40–100 μm planar-e and planar-s dolomite rhombs that are typically zoned in plane light (Fig. 7). In cathodoluminescence, cores have mottled orange and dark luminescence, and the outer parts of rhombs have very thin alternating non- and orange-luminescing bands (Fig. 7b). Dolomitization ranges from partial to complete, is fabric-destructive and, in partially dolomitized rocks, is strongly selective for muddy textures.

Brown dolomite is much less abundant than grey dolomite, and occurs in 1–6 m-thick layers or tabular-shaped bodies that grossly parallel the pre-Kaskaskia unconformity (Figs 4 and 5) (Kruger, 1992a). These layers occur most commonly 5–12 m below the unconformity, and locally cross-cut bedding (Fig. 5). Although layers can be correlated across the area and into the grey dolomite (Fig. 4), locally they have abrupt lateral terminations (Fig. 5). The upper and lower contacts of dolomite layers range from bedding planes to gradational across 2 cm-thick zones. Similar 7 m-thick 2 km-wide dolomite bodies fill palaeovalleys on the pre-Tippecanoe II unconformity, and locally produce hydrocarbons (Seyler & Cluff, 1990).

The ages of dolomite in the study area (Fig. 1) are constrained to Silurian and Devonian by cross-cutting relationships of dolomite and karsting beneath the pre-Kaskaskia unconformity. Karst features in Tippecanoe II rocks include solution-enlarged fractures, solution breccias and caves (Fig. 5) (Kruger, 1992a). Some karst is demonstrably pre-Kaskaskia in age, but other minor karst cuts Middle Devonian rocks and so is Middle Devonian or younger.

Two lines of evidence indicate that some dolomite preceded karsting. First, rhombs of brown dolomite are included as detrital debris in karst-filling sandstone (Fig. 8), indicating that brown dolomite preceded karsting and is most probably Silurian in age. Secondly, coarse aggrading neomorphic fabrics and associated dedolomitized brown dolomite surround some karst fractures, and probably resulted from fluid channelling through caves and solution-enlarged fractures (Fig. 9a,b) (Kruger, 1992a). Similar fabrics have been interpreted as resulting from subaerial diagenesis (Chafetz, 1972; Al-Hashimi & Hemingway, 1973; Jones *et al.*, 1989), and are probably related to pre-Kaskaskia subaerial exposure. Within the neomorphic calcite, dedolomite occurs as cloudy 40–100 μm ghosts that show zoning of the parent dolomite rhombs (Fig. 9a,b). The zoned dedolomite rhombs within karst-related neomorphic calcite crystals, and zoned rhombs included as detrital debris within karst-filling sandstones, indicate that growth of the brown dolomite was complete before karsting.

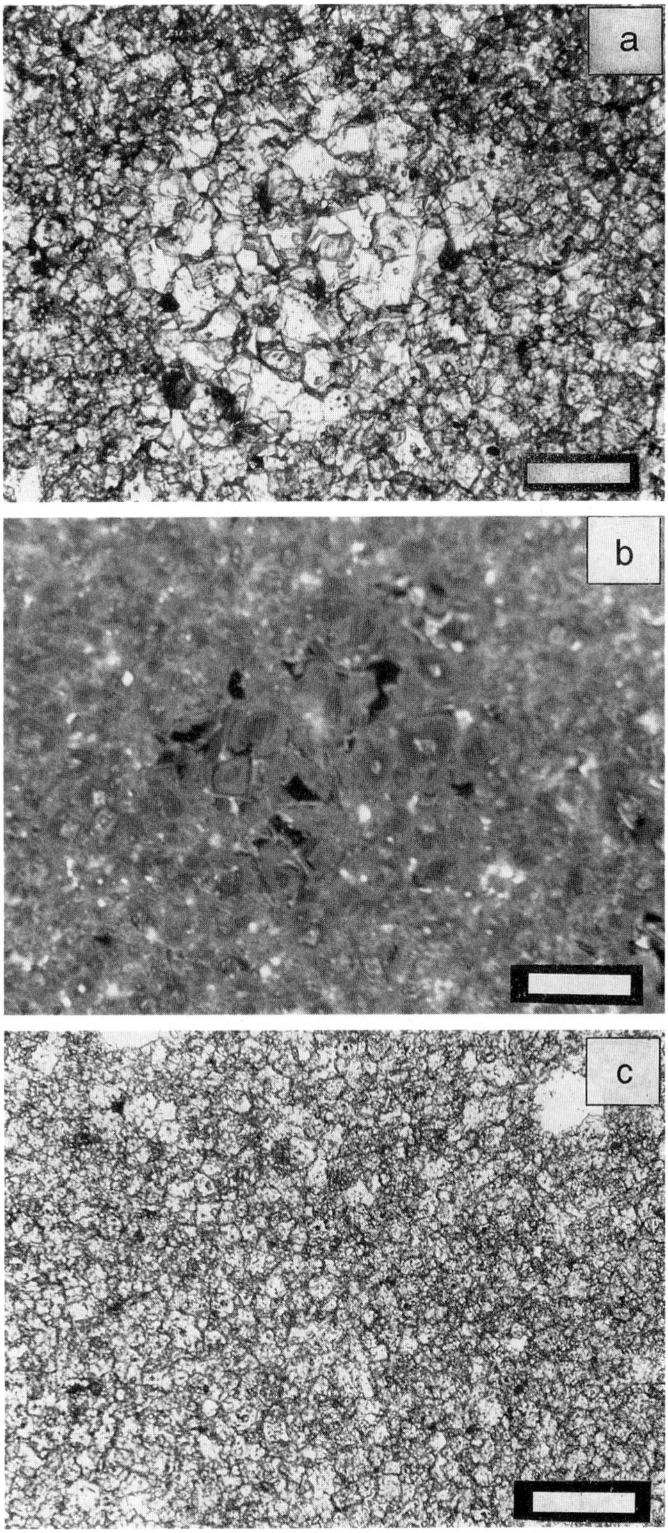

Fig. 6. Fabrics in massive grey dolomite showing complete replacement of precursor muddy packstones. (a) Massive grey dolomite in plane light. Crystal size is greatest where it replaces fossil fragments. Sample from Nutwood, Illinois (Fig. 4). Scale bar = 125 µm. (b) Same view as (a), in cathodoluminescence. Largest rhombs have fine-scale dull- and non-luminescent banding. Scale bar = 125 µm. (c) Finely crystalline planar-s and non-planar massive grey dolomite in plane light. From Rosedale, Illinois (Fig. 4). Scale bar = 125 µm.

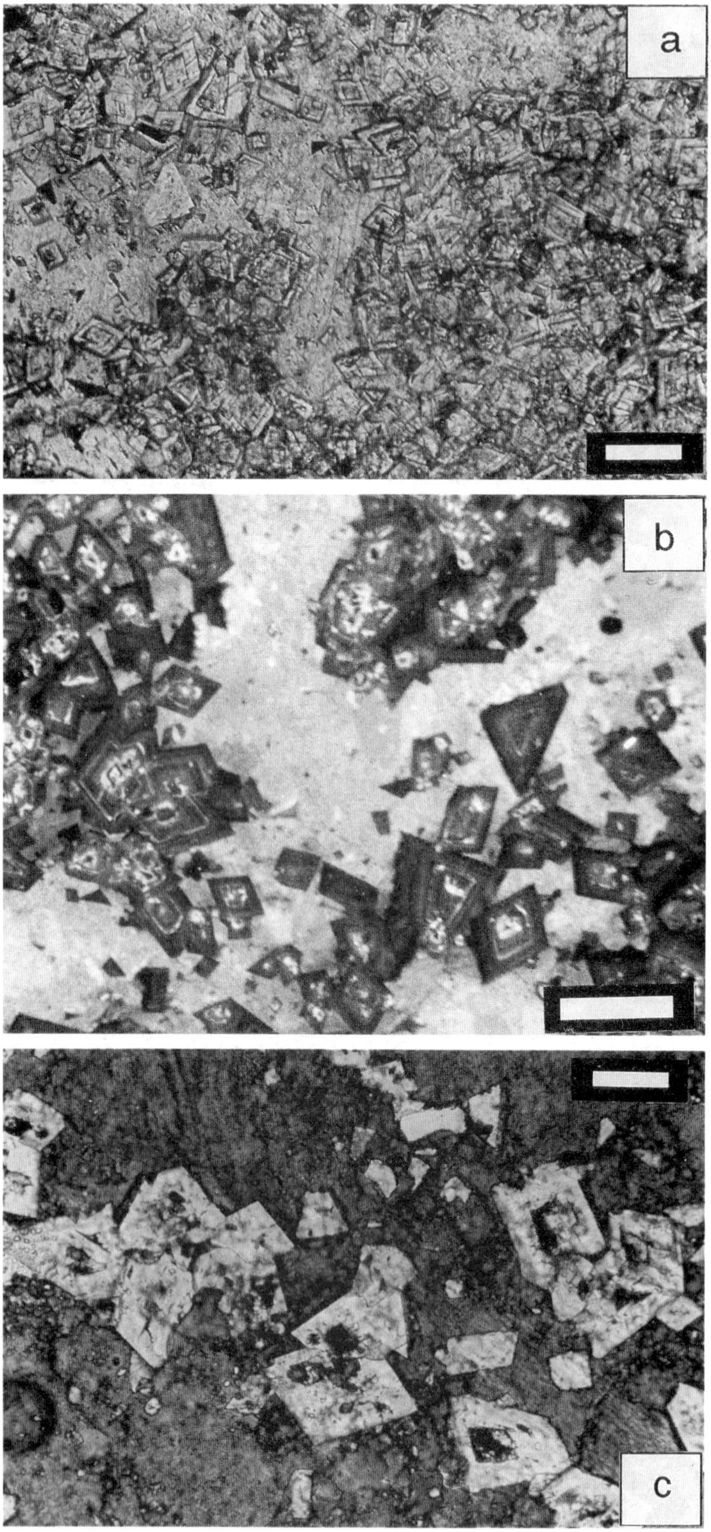

Fig. 7. Fabrics in brown dolomite. (a) Zoned planar-e and planar-s brown dolomite surrounded by late calcite cement. Plane light. Scale bar = 100 μm. From Belleview, Illinois (Fig. 4). (b) Enlarged view of (a) in cathodoluminescence. Brown dolomite has irregular brightly luminescing cores and outer zones with orange and non-luminescent layers. Scale bar = 125 μm. (c) Corroded zoned rhombs of brown dolomite overlain by late calcite cement. Scale bar = 62 μm. From Belleview, Illinois (Fig. 4).

Fig. 8. Rhombs of brown dolomite included as detrital grains in karst-filling sandstone. Crystal size and zoning of rhombs are similar to brown dolomite in host rock, suggesting that dolomite growth was complete before karsting. From Belleview, Illinois (Fig. 4). Scale bar = 62 μm.

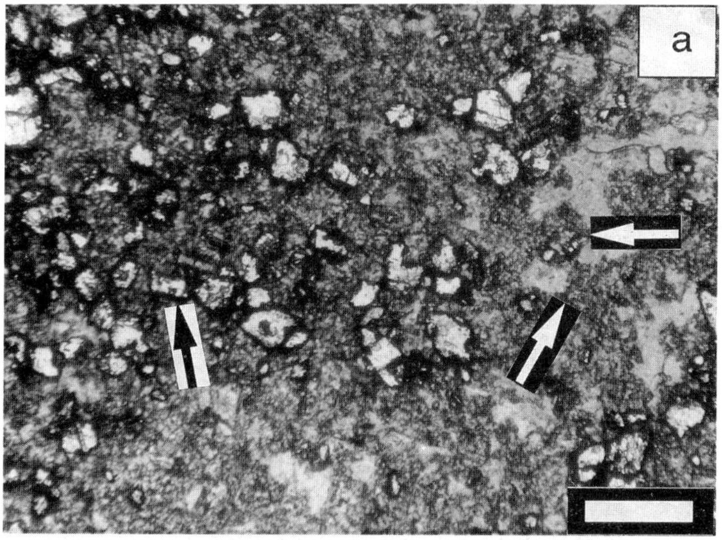

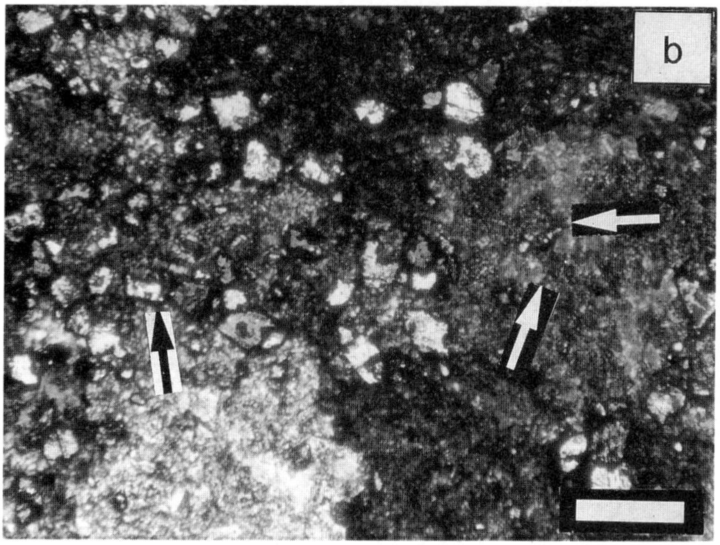

Fig. 9. Dedolomite and aggrading neomorphic fabrics associated with pre-Kaskaskia karsting. The size and remnant zoning in dedolomitized rhombs are similar to brown dolomite in adjacent limestone, suggesting that dolomite growth was complete before karsting. (a) Plane light photomicrograph of cloudy dedolomite (white arrows) and associated aggrading neomorphism adjacent to karst-related solution-enlarged fracture. Some cloudy dedolomite shows relict zoning, indicating that dolomite growth was complete before karsting. Superimposed dolomite (black arrow) indicates later postkarst dolomitization. From Belleview, Illinois (Fig. 4). Scale bar = 125 μm. (b) Same view as (a) in crossed polars showing aggrading neomorphic texture. Scale bar = 125 μm.

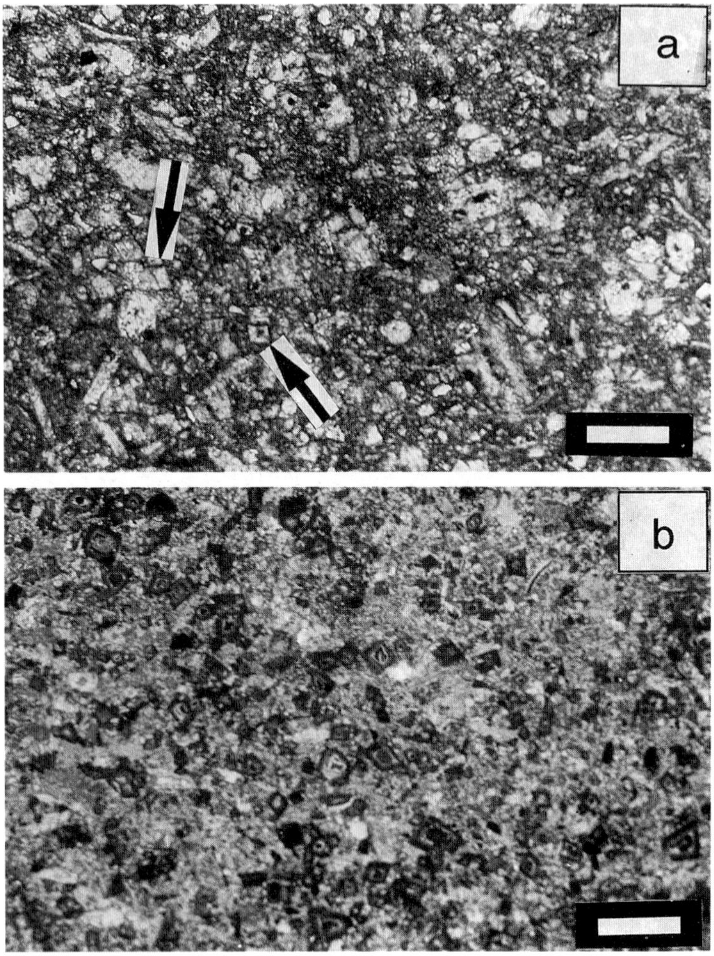

Fig. 10. Dolomite fabrics in Middle Devonian rocks overlying pre-Kaskaskia unconformity. (a) Zoned and unzoned dolomite in partially dolomitized Middle Devonian rocks, similar to brown and grey Silurian dolomite. Plane light. From Monterey School, Illinois (Fig. 4). Scale bar = 250 μm. (b) Devonian dolomite in cathodoluminescence. Largest dolomite rhombs have dark cores, thin brightly luminescing central layers, and dark to banded outer zones. From Monterey School, Illinois (Fig. 4). Scale bar = 250 μm.

Three lines of evidence indicate a post-Middle Devonian age for some brown and grey dolomite. First, subsurface Middle Devonian rocks are dolomitized in parts of the basin (the Geneva Dolomite Member) (Figs 2 and 3). Secondly, scattered finely crystalline dolomite with dull-brown luminescence is superimposed on karst-related aggrading neomorphic fabrics and associated dedolomite (Fig. 9). The superimposed dolomite probably postdates neomorphism, and is Middle Devonian or younger in age. Thirdly, zoned and unzoned 10–40 μm euhedral and anhedral dolomites that are similar to both brown and grey dolomite occur in Middle Devonian rocks in the outcrop area (Fig. 10a,b). In cathodoluminescence, large zoned rhombs of Devonian dolomite have dark cores, irregular brightly luminescing rings, and finely zoned dark and non-luminescent outer zones (Fig. 10b). Smaller crystals have dull luminescence, similar to dolomite superimposed on the aggrading neomorphic calcite and associated dedolomite.

The widespread geographical distribution of grey dolomite, and its distribution that cross-cuts lithofacies, depositional sequences and an interregional unconformity, could be consistent with burial dolomitization or dolomitization by migration of brines during proposed basin dewatering (Bethke, 1986; Bethke *et al.*, 1991). However, the petrographic evidence that points to several generations of dolomite, and dolomite geochemistry (discussed below) is inconsistent with burial dolomitization, but consistent with dolomitization by seawater or modified seawater.

DOLOMITE GEOCHEMISTRY

Carbon and oxygen stable isotope ratios, major, minor and trace element data and Sr isotope ratios indicate similar or overlapping geochemistries for Silurian and Devonian dolomites, and strongly suggest that seawater or slightly modified seawater was the dolomitizing fluid. A total of 31 brown, grey and Devonian dolomite samples from outcrop and the subsurface were analysed for carbon and oxygen isotopic compositions. Major, minor and trace elemental compositions of a total of 27 Silurian and Devonian dolomites from outcrop and the subsurface (Fig. 1) were analysed by inductively coupled plasma spectrometry (ICP). Samples were analysed in a 5% HNO_3 matrix at 25°C using an ARL 3520B ICP sequential analyser. Analyses of seven duplicates indicate errors (1σ) of 0.8%, 0.6%, 0.5%, 1.1% and 0.9% for concentrations of Ca, Mg, Fe, Sr and Mn, respectively.

Dolomite samples analysed for carbon and oxygen isotopic ratios and by ICP were purified using a method modified from E.J. Oswald (personal communication, 1991). Samples were ground to 75 μm and X-rayed. Samples with calcite were reacted with 2% acetic acid for 5 min at about 22°C, filtered with distilled water, dried and X-rayed. The process was repeated until no calcite (104) peak was visible. Leached samples have a less than 0.002° shift in 2Θ towards smaller d_{104}, suggesting minimal preferential leaching. In addition, a sample of pure dolomite that was leached and analysed for carbon and oxygen isotopes showed no significant shift in $\delta^{13}C$ or $\delta^{18}O$, indicating that the leaching technique had little or no effect on isotopic values. Dolomite samples were analysed at EXXON Production Research Company and at the University of Michigan. Two powders from a hand sample that were leached separately and analysed at different laboratories varied by ±0.3‰ for $\delta^{18}O$, and ±0.1‰ for $\delta^{13}C$. Calcite samples were analysed at the University of Wisconsin-Madison stable isotope laboratory and at the University of Michigan. Analyses of laboratory standards and duplicate samples indicate an error of less than 0.1‰ for both $\delta^{13}C$ and $\delta^{18}O$. Values of $\delta^{13}C$ and $\delta^{18}O$ reported here are relative to PDB unless otherwise indicated, and values for dolomite are uncorrected for dolomite–acid fractionation.

Outcrop samples were collected from the sections shown in Figures 1 and 4. Silurian and Devonian dolomites from the subsurface were sampled from cores (Skiles Petroleum Immer #1, Madison Co., IL, and National Associated Petroleum Marion Co. Coal Co. #1, Marion Co., IL; Fig. 1) on repository at the Illinois State Geological Survey, Urbana, Illinois.

Carbon and oxygen isotope ratios

Table 1 and Figure 11 show $\delta^{13}C$ and $\delta^{18}O$ values for grey, brown and Devonian dolomites from outcrop and cores. $\delta^{13}C$ of grey dolomites averages −1.1 ± 1.4‰, and brown dolomites average −0.2 ± 1.1‰. $\delta^{13}C$ values range from −3.7 to 1.5‰ for all dolomites, and bracket values of the parent limestone ($\delta^{13}C$ = 0.3‰ for microspar matrix; Fig. 11). The range in $\delta^{13}C$ values may be an artefact of bulk analysis of finely zoned dolomite, perhaps related to observed cathodoluminescent zoning in brown and grey dolomite (Figs 5b and 7b). The grey dolomite with the lowest $\delta^{13}C$ value (−3.7‰) is from a sample

Table 1. $\delta^{18}O$ and $\delta^{13}C$ values (PDB) for separate dolomite diagenetic phases in Tippecanoe II and Kaskaskia rocks from the study area (Figs 1 and 4).

Section	Sample	Age	Type	$\delta^{18}O$ (PDB)	$\delta^{13}C$ (PDB)
BV	16	Sil.	Brown	−1.6	1.4
BV	18	Sil.	Brown	−1.3	0.4
CS	5	Sil.	Brown	−1.9	1.0
IC	3	Sil.	Brown	−4.6	−0.5
MO	14	Sil.	Brown	−1.0	−1.5
MO	23	Sil.	Brown	−0.3	−1.4
PH	3	Sil.	Brown	−1.3	0.1
PH	9	Sil.	Brown	−0.5	−1.5
PH	13	Sil.	Brown	−1.2	0.2
GE	11	Sil.	Grey	−1.1	0.6
GE	25	Sil.	Grey	−1.6	0.2
GE	28	Sil.	Grey	−1.4	0.5
GE	28	Sil.	Grey	−1.6	0.3
GE	34	Sil.	Grey	−0.7	0.2
MS	47	Sil.	Grey	−1.6	−3.2
MS	50	Sil.	Grey	−0.8	−1.5
MS	54	Sil.	Grey	−1.3	−2.4
MS	56	Sil.	Grey	−1.4	−0.7
N	23	Sil.	Grey	−1.6	−3.6
N	52	Sil.	Grey	−1.4	−1.9
SI	1968	Sil.	Subsurface	−1.6	−0.9
SI	1990	Sil.	Subsurface	−1.8	0.4
SR	1	Sil.	Grey	−2.1	−0.6
SR	2	Sil.	Grey	−2.9	−0.6
SR	5	Sil.	Grey	−1.6	−0.7
GE	9	Dev.	Outcrop	−0.7	−0.4
MO	6	Dev.	Outcrop	0.0	1.6
N	11	Dev.	Outcrop	−0.6	0.3
NA	3278	Dev.	Subsurface	−3.1	1.5

NA, National Associated Petroleum; SI, Skiles Immer #1.

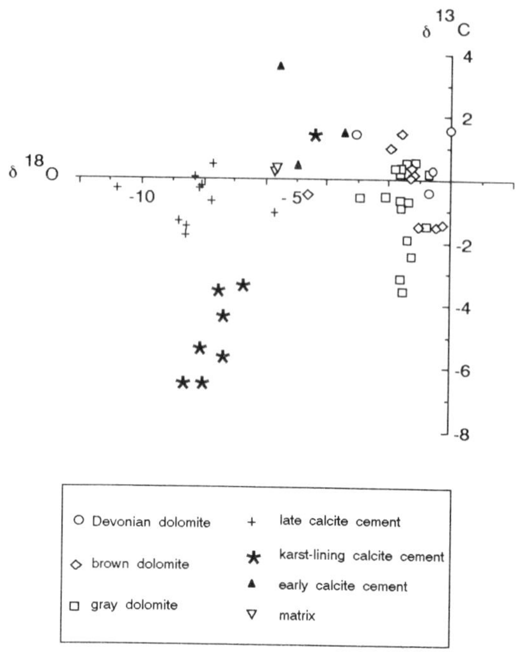

Fig. 11. Stable carbon and oxygen isotopic compositions of grey, brown and Devonian dolomite from outcrop, and different calcite diagenetic phases. All dolomites have similar $\delta^{18}O$ values, suggesting similar dolomitizing fluids, and are too enriched in ^{18}O to record burial dolomitization. Dolomites also do not show the positive correlation between $\delta^{13}C$ and $\delta^{18}O$ expected for mixing-zone dolomite. The range in $\delta^{13}C$ values may be in part an artefact of bulk analyses of finely zoned dolomite, reflect postkarst dolomitization, or indicate modification of prekarst dolomite during subaerial exposure.

containing centimetre-scale karst breccias that may predate or postdate dolomitization.

Grey, brown and Devonian dolomites cluster between −3 and 0‰ for $\delta^{18}O$. Grey dolomite $\delta^{18}O$ values average -1.5 ± 0.5‰, and brown dolomites average -1.5 ± 1.2‰. The relatively small range of $\delta^{18}O$ values for all dolomites suggest dolomitization from a large reservoir of diagenetic fluids with a relatively uniform $\delta^{18}O$ value. One brown dolomite sample is depleted in ^{18}O relative to other dolomites (Fig. 11), and may represent locally restricted conditions, an increased component of parent-rock oxygen relative to other dolomites, increased temperature during dolomitization, or ^{18}O-depleted dolomitizing fluids.

Assuming precipitation at 25°C and no later equilibration, dolomite precipitated in water with $\delta^{18}O$ of about −2‰ SMOW (+1‰ SMOW at 40°C; calculated using the equation of Matthews & Katz, 1977). Early calcite cements and microspar matrix in Silurian rocks from the study area have $\delta^{18}O$ values that average −5.1‰ (Fig. 11). The early calcite phases (microspar matrix and early calcite cements) reflect precipitation from fluids with $\delta^{18}O$ values of about −3‰ SMOW at 25°C (0‰ SMOW at 40°C; calculated using equations of Friedman & O'Neil, 1977). The calculated $\delta^{18}O$ values of the early calcite-precipitating fluids are within 1‰ of those calculated above for dolomitizing fluids. Both Silurian dolomite and early calcite diagenetic phases indicate that diagenetic fluids were too depleted in ^{18}O to reflect extensively evaporated seawater. However, if seawater has not varied by more than 3‰ (Sheppard, 1986), $\delta^{18}O$ values could represent slightly concentrated seawater, unmodified seawater, or seawater diluted slightly with meteoric water.

The relation of $\delta^{18}O$ and $\delta^{13}C$ values of dolomite and calcite cements places additional constraints on the conditions of dolomitization. Karst-lining calcite cements that postdate brown dolomite have a positive correlation between $\delta^{18}O$ and $\delta^{13}C$ (Fig. 11). Similar trends in $\delta^{18}O$ and $\delta^{13}C$ have been interpreted as resulting from the mixing of meteoric water containing ^{13}C-depleted CO_2 from soil zones, with conate fluids in phreatic zones (Allan & Matthews, 1982) (Fig. 11). The $\delta^{18}O$ and $\delta^{13}C$ values of brown and grey dolomite do not show trends suggesting 'mixing-zone' dolomitization, but are consistent with petrographic data that suggest early dolomitization.

Postkarst calcite cements are slightly ^{18}O-depleted relative to dolomite and other calcite cements, suggesting increasing temperature during post-Middle Devonian burial, waters with different iosotopic ratios, or water–rock interaction (Fig. 11). However, the $\delta^{18}O$ values of Silurian and Devonian dolomites suggest dolomitization at low temperatures rather than in deep-burial settings, as interpreted for several other pervasive dolomites (e.g. Gregg, 1985; Zenger & Dunham, 1988; Coniglio *et al.*, this volume; Kupecz & Land, this volume; Miller & Folk, this volume; Mountjoy & Amthor, this volume). Low-temperature or near-surface dolomitization is consistent with regional stratigraphic relationships that indicate that preserved Tippecanoe II rocks north of central Illinois were buried by less than 500 m of Silurian and Lower Devonian strata before pre-Kaskaskia exposure (Whitaker, 1988).

Table 2. Major, minor and trace element compositions of Silurian and Devonian dolomite in Tippecanoe II and Kaskaskia I rocks in the study area and from the subsurface (Figs 1 and 4).

Sample	Dolomite type	$CaCO_3$ (mole %)	$MgCO_3$ (mole %)	$FeCO_3$ (wt%)	Mn (ppm)	Sr (ppm)
GE 09	Dev. outcrop	53.41	44.65	2.18	569	98
MO 6	Dev. outcrop	54.25	44.39	1.49	579	117
MS 09	Dev. outcrop	53.04	42.71	4.69	976	97
N 11	Dev. outcrop	53.54	43.93	2.65	1201	97
NA 3278	Dev. subsurface	50.84	48.70	0.46	408	42
BV 16	Sil. brown	53.92	44.89	1.12	1225	98
BV 18	Sil. brown	53.93	45.08	1.04	580	81
CS 05	Sil. brown	54.04	43.94	2.19	680	154
IC 03	Sil. brown	50.74	47.55	1.94	451	55
MO 14	Sil. brown	53.99	44.95	1.17	398	93
MO 23	Sil. brown	53.89	44.99	1.22	422	80
PH 03	Sil. brown	54.45	44.30	1.35	620	83
PH 13	Sil. brown	54.34	44.42	1.35	582	78
GE 11	Sil. grey	50.20	49.14	0.67	429	44
GE 25	Sil. grey	49.98	49.43	0.61	417	45
GE 28	Sil. grey	50.22	49.28	0.51	352	49
GE 34	Sil. grey	50.04	49.41	0.56	419	46
MS 47	Sil. grey	52.80	46.64	0.60	221	136
MS 50	Sil. grey	49.90	48.67	1.61	480	45
MS 54	Sil. grey	50.23	48.84	1.03	361	54
N 23	Sil. grey	52.49	46.74	0.84	328	139
N 52	Sil. grey	53.22	46.46	0.32	191	169
SR 01	Sil. grey	49.73	49.34	1.06	241	48
SR 02	Sil. grey	50.00	49.07	1.06	262	48
SR 05	Sil. grey	49.76	49.56	0.78	224	51
SI 1968	Sil. subsurface	53.38	45.91	0.78	251	154
SI 1990	Sil. subsurface	53.72	45.82	0.48	178	121

Major, minor, and trace elements

Table 2 summarizes Ca, Mg, Fe, Mn and Sr concentrations for leached brown and grey dolomite from the outcrop area and the subsurface (Fig. 1). Silurian grey dolomite from the subsurface, some grey dolomite from outcrop, and brown and Devonian dolomite from outcrop, are calcium-rich (mole % $CaCO_3$/mole % $MgCO_3$ between 1.15 and 1.25; Fig. 12). Some grey dolomite from outcrop, petrographically similar to other grey dolomite, plots as a group of nearly stoichiometric dolomite (mole % $CaCO_3$/mole % $MgCO_3$ between 1.00 and 1.15; Fig. 12). The Fe and Mn concentrations of calcian and stoichiometric dolomites overlap, but stoichiometric dolomites have less Sr than calcian dolomite (Table 2). Published Sr concentrations for other Illinois Basin subsurface Silurian limestone and grey dolomite range from 51 to 167 ppm (Stueber *et al.*, 1987), close to the values presented in Table 2. None of the elements show systematic variations with stratigraphic position, geographic

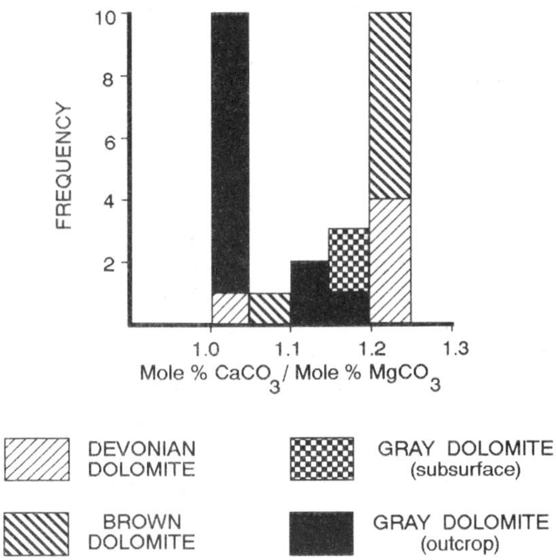

Fig. 12. Stoichiometry of dolomite types expressed as mole % $CaCO_3$/mole % $MgCO_3$ based on ICP analyses. Similar Ca-rich compositions of grey dolomite from subsurface, Devonian, brown and some grey dolomite from outcrop suggest precipitation from similar fluids.

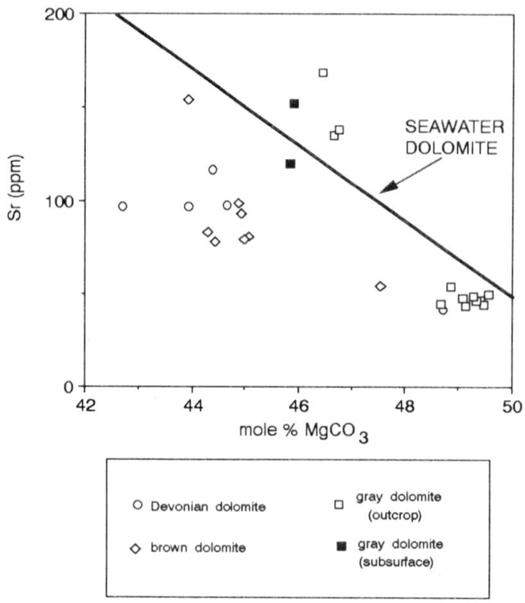

Fig. 13. Strontium vs. $MgCO_3$ for different dolomite fabrics shows weak negative correlation. Grey dolomite from outcrop and subsurface falls close to the proposed line for seawater dolomite (Vahrenkamp & Swart, 1990). Some brown dolomite and dolomite in Devonian rocks also lies near the proposed seawater dolomite line.

location or number of leachings required to purify the dolomite. Neither calcian nor stoichiometric dolomite shows strong correlations between the major elements and Fe or Mn. Concentrations of Mn, Fe and Sr overlap slightly for each dolomite type, suggesting that all dolomite types may form end-members of a single population.

As a group, dolomites show a negative correlation between Sr and Mg, although the correlation is strongest for the grey dolomite (Fig. 13). Similar trends have been noted for other ancient dolomites (Fig. 13). (Vahrenkamp & Swart, 1990). In addition, Silurian and Devonian dolomites, especially Silurian grey dolomite from outcrop and the subsurface, plot near a proposed line for seawater dolomite (Fig. 13) (Vahrenkamp & Swart, 1990).

Strontium isotopic ratios

Published $^{87}Sr/^{86}Sr$ ratios for grey dolomite from subsurface Silurian reefs in the basin range from 0.70810 to 0.70983 ($1\sigma = \pm 0.00006$; $n = 6$) (Stueber *et al.*, 1987). The $^{87}Sr/^{86}Sr$ ratios of undolomitized reef limestones are similar to dolomites, and vary between 0.70826 and 0.70866 ($n = 5$) (Stueber *et al.*, 1987). With the exception of slightly argillaceous dolomites that have elevated $^{87}Sr/^{86}Sr$ ratios (0.70983 and 0.70957) and low Sr concentrations (51 and 66 ppm), dolomite and limestone show general agreement with $^{87}Sr/^{86}Sr$ variations in Silurian seawater (Stueber *et al.*, 1987) that increased from about 0.7080 in the Early Silurian to 0.7089 in the Late Silurian (Burke *et al.*, 1982). Lower Devonian limestone and Middle Devonian dolomite also have $^{87}Sr/^{86}Sr$ ratios consistent with values for Devonian seawater (Stueber *et al.*, 1987).

Other studies (Saller, 1984; Aharon *et al.*, 1987; Banner *et al.*, 1988; Vahrenkamp, 1988) indicate that dolomite assumes the $^{87}Sr/^{86}Sr$ ratio of the dolomitizing fluids, apparently with little or no Sr contribution from the parent rock (Aharon *et al.*, 1987). The general concordance between $^{87}Sr/^{86}Sr$ ratios of subsurface limestone and dolomite with $^{87}Sr/^{86}Sr$ trends for Silurian seawater suggests that Silurian seawater was the major component of the dolomitizing fluid for Silurian grey dolomite. Other potential dolomitizing fluids, such as Devonian seawater, would be unlikely to preserve the positive and concordant Silurian seawater $^{87}Sr/^{86}Sr$ variations observed for Silurian grey dolomite and limestone (Stueber *et al.*, 1987). Conceivably, Silurian grey dolomite could have maintained its original Sr isotopic ratios if dolomitizing fluids, such as seawater diluted with meteoric water, lacked significant Sr. However, such Sr-depleted fluids would be unlikely to carry sufficient Mg to dolomitize the large volumes of grey dolomite.

The relatively low $^{87}Sr/^{86}Sr$ ratios of the dolomites suggest that basinal brines were not the dominant dolomitizing fluids. Modern shale-influenced brines and saline water in Silurian and Devonian formations have elevated $^{87}Sr/^{86}Sr$ ratios (0.70913–0.71084) relative to Silurian seawater, and to Silurian and Devonian limestone and dolomite (Stueber *et al.*, 1987). If these or similar brines were the dolomitizing fluids, dolomites would be expected to have elevated $^{87}Sr/^{86}Sr$ ratios (e.g. Banner *et al.*, 1988).

Summary

Silurian brown and grey dolomite and dolomite in Devonian rocks have similar $\delta^{18}O$ values (Fig. 11) that suggest dolomitization by seawater or modified seawater in a very shallow burial setting. Two lines of evidence suggest that the $\delta^{18}O$ values of Silurian dolomites reflect the $\delta^{18}O$ values of dolomitizing fluids, rather than complete geochemical resetting. First, the size and zoning of brown dolomite rhombs included as karst-filling debris and affected by karst-related neomorphism and dedolomitization are similar to brown dolomite in the host rock (Figs 8 and 9), indicating that dolomite growth was complete before karsting. If growth was complete before karsting, extensive post-Middle Devonian geochemical resetting was not accompanied by petrographic changes and seems unlikely. Secondly, the small crystal size of grey dolomite (5–40 μm) is inconsistent with the increasing crystal sizes commonly associated with dolomitization and/or neomorphism in burial settings (e.g. Gregg & Sibley, 1984; Zenger & Dunham, 1988; Kupecz & Land, this volume).

As a group dolomites do not show covariation of $\delta^{13}C$ and $\delta^{18}O$, as expected for mixing-zone dolomitization. However, the relatively depleted $\delta^{13}C$ value of one grey dolomite sample containing small karst breccias may reflect postkarst dolomitization of ^{13}C-depleted limestone, and supplement petrographic and stratigraphic data suggesting that grey dolomite records multiple episodes of dolomitization. Alternately, the sample may indicate the influence of ^{13}C-depleted CO_2 during minor stabilization or neomorphism of dolomite associated with pre-Kaskaskia exposure.

Brown and Devonian dolomite from outcrop, Silurian subsurface grey dolomite, and some grey dolomite from outcrop have similar Ca-rich compositions (Fig. 12), and all fabrics have similar or overlapping $\delta^{18}O$ values and Fe, Mn and Sr compositions (Figs 12 and 13). The similar and overlapping chemistries suggest the dolomites formed from geochemically similar fluids. Finally, subsurface grey Silurian dolomite and limestone have $^{87}Sr/^{86}Sr$ ratios and trends that reflect $^{87}Sr/^{86}Sr$ changes in Silurian seawater (Stueber *et al.*, 1987), suggesting that Silurian seawater was the source of the dolomitizing fluid for Silurian grey dolomite. The concordance of the $^{87}Sr/^{86}Sr$ seawater curves with $^{87}Sr/^{86}Sr$ trends in Silurian limestone and dolomite also suggests that dolomitization was a relatively early diagenetic event. Finally, some modern Illinois Basin formation waters in Silurian rocks have Br/Cl ratios consistent with modified Ca-enriched Mg-depleted seawater (Walter *et al.*, 1990; Stueber & Walter, 1991). The Ca enrichment and Mg depletion of formation waters are consistent with chemical modifications expected from dolomitization (Walter *et al.*, 1990; Stueber & Walter, 1991).

DISCUSSION

Combined stratigraphic, petrographic and geochemical data are required to interpret the origin of dolomites in Silurian and Devonian rocks from the Illinois Basin. Petrographic and stratigraphic data indicate that some dolomitization (grey and brown dolomite) preceded pre-Kaskaskia Sequence (late Early Devonian) karsting, but other grey and brown dolomites are Middle Devonian or younger in age. Thus, dolomite in Silurian and Devonian rocks cannot result from a single dolomitization event. Combined stable carbon and oxygen isotopic data, and major, minor and trace element data, indicate that dolomite in Silurian and Devonian rocks have similar or overlapping geochemical signatures (Figs 11, 12 and 13). This probably reflects dolomitization from similar or genetically related fluids. Values of $\delta^{18}O$ for Silurian and Devonian dolomites generally cluster between −3 and 0‰. These $\delta^{18}O$ values suggest dolomitization at relatively low temperatures from dolomitizing fluids with uniform $\delta^{18}O$ values and sufficient Mg^{2+} to dolomitize large bodies of rock.

Conceivably, several generations of geochemically similar dolomite could reflect homogenization of pre-existing dolomite during burial, or more than one episode of dolomitization from fluids with similar chemistries during burial or from migration of basin-derived brines. However, several lines of evidence argue against burial neomorphism or burial or brine-induced dolomitization. First, regional stratigraphic relations indicate that Silurian rocks were not deeply buried before pre-Kaskaskia exposure (Whitaker, 1988). Secondly, grey dolomite is fine-crystalline, and petrographic data suggest that brown dolomite growth was complete before karsting. Thus, any extensive geochemical resetting would not have been accompanied by petrographic changes and seems unlikely. Thirdly, $\delta^{18}O$ values of dolomite, prekarst calcites and karst-lining calcite cements are too enriched in ^{18}O to record precipitation or neomorphism at elevated temperatures associated with burial. Finally, basin-derived brines would probably be enriched in ^{87}Sr, similar to modern brines in the Illinois Basin (Stueber *et al.*, 1987).

However, dolomites in Silurian rocks generally show concordance with the systematic increase in $^{87}Sr/^{86}Sr$ during the Silurian (Burke *et al.*, 1982), rather than uniform enrichment in ^{87}Sr. Therefore, stratigraphic, petrographic and geochemical evidence suggests that burial dolomitization or dolomitization from basin-derived brines is unlikely.

Several generations of genetically linked dolomites could represent continuous or quasicontinuous Silurian through Middle Devonian dolomitization from fluids with similar or overlapping $\delta^{18}O$ values and minor and trace element concentrations. Similar $\delta^{18}O$ values and concordant $^{87}Sr/^{86}Sr$ variations between grey dolomite and Silurian seawater suggest that Silurian seawater was the dolomitizing fluid for the Silurian grey dolomite. Based on the overlapping geochemistries of Devonian dolomite and Silurian brown dolomite to the grey dolomite, we believe that dolomite in Devonian rocks also records early dolomitization by seawater or modified seawater.

If seawater was the dominant dolomitizing fluid, some mechanism must have provided a hydrological pump, because the gentle topographic gradients of the Silurian and Devonian carbonate ramps were unlikely to have the same vigorous pumping systems present in modern reefs and atolls (e.g. Saller, 1984; Buddemeier & Oberdorfer, 1986). We hypothesize that the relative sea-level changes that controlled the dominantly subtidal sedimentation in the basin (McKerrow, 1979; Ross & Ross, 1988; Kruger, 1991) provided the necessary pump. During relative sea-level low-stands, fair-weather wave base would have extended part-way into the basin, and much of the basin may have been below storm wave base. Thus, much of the basin, and especially arches surrounding the basin, could have been subject to weak pumping by waves.

As relative sea level rose and fell during the Silurian and Devonian, seawater could also have been flushed through sediments by alternately establishing and replacing submarine meteoric lenses on the arches, similar to those in the modern Arabian Gulf (Chafetz *et al.*, 1988). Water flow within submarine meteoric lenses may also have provided a pump for seawater downdip from lenses, similar to models proposed for the Bahama platform (Vahrenkamp *et al.*, 1991). The area affected by this mechanism would have moved up- and downdip in response to the relative rise and fall in sea level. This process would have been most effective on arches bordering the basin, especially during the Middle and Late Silurian, when much of the basin was covered by shallow epeiric seas (Rogers, 1972; Droste & Shaver, 1983). However, seawater pumping was probably not driven solely by tidal or wave action, because lower Tippecanoe II rocks were deposited in mid- and outer-ramp settings (Rogers, 1972; Whitaker, 1988; Kruger, 1992b). The pump may have been augmented by additional factors, such as extensive bioturbation, gentle submarine currents, compaction of basinal carbonates, convection caused by subtle differences in geothermal gradients, or by other unidentified factors.

It should be noted that the hypothesized pump need not have pushed seawater deep into the sediment column. If dolomitization was a relatively early diagenetic event, water would need only circulate through the upper part of the slowly accumulating sediment pile, perhaps only to depths of several tens of metres. The pumping mechanism need not have been vigorous, but could have operated slowly, repetitively or continuously. The micrite-rich sediments may have been particularly susceptible to dolomitization, since dolomites in Silurian and Devonian rocks, and in general (Bathurst, 1975), are fabric-selective for muddy textures.

Dolomitization by flushing of relatively contemporaneous seawater through the rocks would generate dolomites with Sr isotopic signatures and $\delta^{18}O$ compositions close to those for contemporaneous seawater. Continued operation of the pumping system during the Devonian could have continued to dolomitize Tippecanoe II rocks that escaped dolomitization during the Silurian.

CONCLUSIONS

Dolomite in predominantly mud-rich subtidal Silurian and Devonian rocks in the Illinois Basin occurs above and below a major interregional unconformity, and cross-cuts bedding, facies and depositional sequences. Petrographic data suggest that dolomitization began before subaerial exposure of the pre-Kaskaskia sequence-bounding unconformity, and continued until after deposition of Middle Devonian carbonates that onlap the unconformity. Combined carbon, oxygen and strontium isotopic data, and major, minor and trace element concentrations, suggest that seawater or modified seawater was the dominant dolomitizing fluid.

We hypothesize that whereas relative sea level

rose and fell during the Silurian and Devonian, pumping due to flow in updip meteoric lenses and wave action, augmented by other factors, pushed seawater through the slowly accumulating sediment pile. If these gentle mechanisms pumped seawater through Silurian and Devonian sediments, they may also have operated on similar subtidal carbonate ramps elsewhere, and may hold a key to interpreting the origins of extensive dolomite in other ramp settings.

Finally, the evidence for seawater dolomitization of Silurian and Devonian rocks suggests that subaerial exposure of the interregional pre-Kaskaskia sequence-bounding unconformity played only a minor role in the dolomitization history of underlying rocks. In part this is surprising, because rocks beneath unconformities are likely candidates for mixing-zone dolomitization (Morrow, 1982). It seems possible, then, that thick dolomitized packages of subtidal rocks beneath other interregional unconformities may reflect similar sea-level pumping mechanisms, rather than dolomitization in mixing zones or due to fluid flow along karsted unconformities (Bethke *et al.*, 1991).

ACKNOWLEDGEMENTS

Funding was provided by Sigma Xi, the Geological Society of America, ARCO Oil and Gas Company, EXXON Production Research Company, the University of Wisconsin-Madison Graduate School, and the University of Wisconsin-Madison Department of Geology and Geophysics. Acknowledgment is also made to the Donors of The Petroleum Research Fund, administered by the American Chemical Society, for partial support of this research, and the Gas Research Institute. We are grateful to EXXON Production Research Company, and particularly Patrick Lehmann and Ralph Hockett, for stable carbon and oxygen isotopic analyses of dolomite and calcite performed at EPR and at the University of Michigan. We thank Professor John Valley for discussion and for allowing the use of the stable isotope laboratory at the Department of Geology and Geophysics, University of Wisconsin-Madison. Kevin Baker assisted with isotopic analyses at Madison. Dr J. Burton, Department of Anthropology, University of Wiscönsin, made ICP analyses possible. The Illinois State Geological Survey made cores available for examination and sampling. Thanks also to Dr Greg Nadon, who read an early draft of this paper. Finally, we thank D. Zenger, M. Tucker, B. Purser and R. Ginsburg for helpful reviews.

REFERENCES

Aharon, P., Socki, R.A. & Chan, L. (1987) Dolomitization of atolls by sea water convection flow: test of a hypothesis at Niue, south Pacific. *J. Geol.* **95**, 187–203.

Al-Hashimi, W.S. & Hemingway, J.E. (1973) Recent dedolomitization and the origin of rusty crusts of Northumberland. *J. Sedim. Petrol.* **43**, 82–91.

Allan, J.R. & Matthews R.K. (1982) Isotopic signatures associated with early meteoric diagenesis. *Sedimentology* **29**, 797–817.

Atherton, E. (1971) Tectonic development of the Eastern Interior Region of the United States. *Ill. State Geol. Surv. Ill. Petrol.* **96**, 29–43.

Banner, J.L., Hanson, G.N. & Meyers, W.J. (1988) Water–rock interaction history of regionally extensive dolomites of the Burlington-Keokuck Formation (Mississippian): isotopic evidence. In: *Sedimentology and Geochemistry of Dolostones* (Ed. Shukla, V. & Baker, P.A.) Spec. Publs Soc. Econ. Paleont. Mineral., Tulsa 43, 161–174.

Bathurst, R.G.C. (1975) *Carbonate Sediments and Their Diagenesis*, 2nd edn. Elsevier, New York, 658 pp.

Bethke, C.M. (1986) Hydrologic constraints on the genesis of the Upper Mississippi Valley Mineral District from Illinois Basin brines. *Econ. Geol.* **81**, 233–249.

Bethke, C.M., Reed, J.D. & Oltz, D.F. (1991) Long-range petroleum migration in the Illinois Basin. *Bull. Am. Ass. Petrol. Geol.* **75**, 925–945.

Bretz, J.H. (1939) Geology of the Chicago region – general. *Bull. Ill. State Geol. Surv.* **65**(1), 118 pp.

Buddemeier, R.W. & Oberdorfer, J.A. (1986) Internal hydrology and geochemistry of coral reefs and atoll islands, In: *Reef Diagenesis* (Ed. Schroeder, J.H. & Purser, B.H.), pp. 91–111. Springer-Verlag, New York.

Burchette, T.P. & Wright, V.P. (1992) Carbonate ramp depositional systems. *Sedim. Geol.* **79**, 3–57.

Burke, W.H., Denison, R.E., Hetherington, E.A., Koepnick, R.B., Nelson, H.F. & Otto, J.B. (1982) Variation of seawater $^{87}Sr/^{86}Sr$ throughout Phanerozoic time. *Geology* **10**, 516–519.

Calvet, F., Tucker, M.E. & Henton, J.M. (1990) Middle Triassic carbonate ramp systems in the Catalan Basin, northeast Spain: facies, systems tracts, sequences and controls. In: *Carbonate Platforms* (Ed. Tucker, M.E., Wilson, J.L., Crevello, P.D., Sarg, J.R. & Read, J.F.) Spec. Publ. Int. Assoc. Sedimentol 9, 79–108.

Chafetz, H.S. (1972) Surface diagenesis of limestone. *J. Sedim. Petrol.* **42**, 325–329.

Chafetz, H.S., McIntosh, A.G. & Rush, P.F. (1988) Freshwater phreatic diagenesis in the marine realm of Recent Arabian Gulf carbonates. *J. Sedim. Petrol.* **58**, 433–440.

Collinson, C., Sargent, M.L. & Jennings, J.R. (1988) Illinois Basin Region. In: *The Geology of North America, Sedimentary Cover – North American Craton; US*. Vol. D-2 (Ed. Sloss, L.L.) pp. 383–426. Geological Society of America 506 pp.

Droste, J.B. & Shaver, R.H. (1975) The Jeffersonville Limestone (Middle Devonian) of Indiana – stratigraphy, sedimentation, and relation to Silurian reef-bearing rocks. *Bull. Am. Ass. Petrol. Geol.* **59**, 1217–1221.

Droste, J.B. & Shaver, R.H. (1980) Recognition of buried Silurian reefs in southwestern Indiana – application to the Terre Haute Bank. *J. Geol.* **88**, 567–587.

Droste, J.B. & Shaver, R.H. (1982) *The Salina Group (Middle and Upper Silurian) of Indiana*. Spec. Rept. Ind. Geol. Surv. 24, 41 pp.

Droste, J.B. & Shaver, R.H. (1983) *Atlas of Middle Paleozoic Paleogeography of the Southeastern Great Lakes Area*. Spec. Rept. Ind. Geol. Surv. 32, 32 pp.

Droste, J.B. & Shaver, R.H. (1987) *Upper Silurian and Lower Devonian Stratigraphy of the Central Illinois Basin*. Spec. Rept. Ind. Geol. Surv. 39, 29 pp.

Esteban, M. (1991) Palaeokarst: practical applications. In: *Palaeokarsts and Palaeokarstic Reservoirs* (Ed. Wright, V.P., Esteban, M. & Smart, P.L.) Occ. Publ. Postgraduate Res. Inst. Sedimentol 2, 89–119.

Friedman I. & O'Neil, J.R. (1977) Compilation of stable isotope fractionation factors of geochemical interest. In: *Data of Geochemistry* (Ed. Fleischer, M.) US Geol. Surv. Prof. Paper, 440-KK, 1–12.

Gregg, J.M. (1985) Regional epigenetic dolomitization in the Bonneterre Dolomite (Cambrian), southeastern Missouri. *Geology* **13**, 503–506.

Gregg, J.M. & Sibley, D.F. (1984) Epigenetic dolomitization and the origin of xenotopic dolomite textures. *J. Sedim. Petrol.* **54**, 907–931.

Hardie, L.A. (1987) Dolomitization: a critical view of some current views. *J. Sedim. Petrol.* **57**, 166–183.

Jones, B., Pleydell, S.M. & Ng, K.-C. (1989) Formation of poikilotopic calcite–dolomite fabrics in the Oligocene–Miocene Bluff Formation of Grand Cayman, British West Indies. *Bull. Can. Petrol. Geol.* **37**, 255–265.

Kahle, C.F. (1988) Surface and subsurface paleokarst, Silurian Lockport, and Peebles Dolomites, western Ohio. In: *Paleokarst* (Ed. James, N.P. & Choquette, P.W.) pp. 229–256. Springer-Verlag, New York.

Kolata, D.R. (1990) Overview of sequences. In: *Interior Cratonic Basins* (Ed. Leighton, M.W., Kolata, D.R., Oltz, D.R. & Eidel, J.J.) Mem. Am. Ass. Petrol. Geol., Tulsa 51, 59–74.

Kruger, J.M. (1991) Systems tract framework of cratonic sequences: early Silurian deposition, northwestern Illinois Basin. *Geol. Soc. Am. Abstracts with Programs* **23**, 347–348.

Kruger, J.M. (1992a) *Diagenesis and Porosity Development Beneath the pre-Kaskaskia (Early Devonian) Interregional Unconformity, Northwest Illinois Basin*. Field Trip Guidebook, Permian Basin Section, Soc. Econ. Paleont. Mineral., 182–191.

Kruger, J.M. (1992b) Transgressive systems tract of the Tippecanoe II (Silurian) cratonic subsequence in the Illinois Basin. *Geol. Soc. Am. Abstracts with Programs* **24**, 27.

Land, L.S. (1985) The origin of massive dolomite. *J. Geol. Educ.* **33**, 112–125.

Leighton, M.W., Kolata, D.R., Oltz, D.R. & Eidel, J.J. (1990) *Interior Cratonic Basins*. Mem. Am. Ass. Petrol. Geol., Tulsa 51, 819 pp.

McKerrow, W.S. (1979) Ordovician and Silurian changes in sea level. *J. Geol. Soc. London* **136**, 137–145.

Matthews, A. & Katz, A. (1977) Oxygen isotope fractionation during dolomitization of calcium carbonate. *Geochim. Cosmochim. Acta* **41**, 1431–1438.

Meents, W.F. & Swann, D.H. (1965) *Grand Tower Limestone (Devonian) of Southern Illinois*. Circ. Ill. State Geol. Surv. 389, 34 pp.

Mikulic, D.G. (1990) Tippecanoe II subsequence: Silurian System through Lower Devonian Series. In: *Interior Cratonic Basins* (Ed. Leighton, M.W., Kolata, D.R. Oltz, D.R. & Eidel, J.J.) Mem. Am. Ass. Petrol. Geol. Tulsa, 51, 101–108.

Morrow, D.W. (1982) Diagenesis 2. Dolomite – part 2. Dolomitization models and ancient dolostones. *Geosci. Can.* **9**, 95–107.

North, W.G. (1969) *The Middle Devonian Strata of Southern Illinois*. Ill. State Geol. Surv. Circ. 441, 45 pp.

Rogers, J.E., Jr. (1972) *Silurian and Lower Devonian Stratigraphy and Paleobasin Development: Illinois Basin – Central United States*. PhD Dissertation, University of Illinois, Urbana, Illinois, 144 pp.

Ross, C.A. & Ross, J.R.P. (1988) Late Paleozoic transgressive–regressive deposition. In: *Sea Level Changes: an Integrated Approach* (Ed. Wilgus, C.K., Hastings, B.S., Kendall, C.G. St. C., Posamentier, H.W. Ross, C.A. & Van Wagoner, J.C.) Spec. Publs Soc. Econ. Paleont. Mineral., Tulsa 42, 227–248.

Saller, A.H. (1984) Petrologic and geochemical constraints on the origin of subsurface dolomite, Enewetak Atoll: an example of dolomitization by normal seawater. *Geology* **12**, 217–220.

Scotese, C.R., Van der Voo, R. & Barrett, S.F. (1985) Silurian and Devonian base maps. *Phil. Trans. Roy. Soc. London* **B309**, 55–77.

Scott, T.M. (1988) The lithostratigraphy of the Hawthorn Group (Miocene) of Florida. *Florida Geol. Surv. Bull.* **59**, 148 pp.

Seyler, B. & Cluff, R.M. (1990) Petroleum traps in the Illinois Basin. In: *Interior Cratonic Basins* (Ed. Leighton, M.W., Kolata, D.R., Oltz, D.R. & Eidel, J.J.) Mem. Am. Ass. Petrol. Geol., Tulsa 51, 361–402.

Shaver, R.H. (1985) Midwestern Basin and Arches Region. Correlation of Stratigraphic Units of North America, Am. Ass. Petrol. Geol., Tulsa, Oklahoma.

Sheppard, S.M.F. (1986) Characterization and isotopic variations in natural waters. In: *Stable Isotopes in High Temperature Geological Processes* (Ed. Valley, J.W., Taylor, H.P. & O'Neil, J.R.) Rev. Mineral. 16, 165–183.

Shukla, V. (1988) Sedimentology and geochemistry of a regional dolostone: correlation of trace elements with dolomite fabrics. In: *Sedimentology and Geochemistry of Dolostones* (Ed. Shukla, V. & Baker, P.A.) Spec. Publs Soc. Econ. Paleont. Mineral., Tulsa 43, 145–157.

Sloss, L.L. (1963) Sequences in the cratonic interior of North America. *Bull. Geol. Soc. Am.* **74**, 93–114.

Sloss, L.L. (1984) Comparative anatomy of cratonic unconformities. In: *Interregional Unconformities and Hydrocarbon Accumulation* (Ed. Schlee, J.S.) Mem. Am. Ass. Petrol. Geol., Tulsa 36, 1–6.

Sloss, L.L. (1988) Tectonic evolution of the craton in Phanerozoic time. In: *The Geology of North America, Sedimentary Cover – North American Craton; US*. Vol. D-2, (Ed. Sloss, L.L.), pp. 25–51. Geological Society of America, 506 pp.

STUEBER, A.M. & WALTER, L.M. (1991) Origin and chemical evolution of formation waters from Silurian–Devonian strata in the Illinois basin, USA. *Geochim. Cosmochim. Acta* **55**, 309–325.

STUEBER, A.M., PUSHKAR, P. & HETHERINGTON, E.A. (1987) A strontium isotopic study of formation waters from the Illinois basin, USA. *Appl. Geochem.* **2**, 477–494.

SUMMERSON, C.H. & SWANN, D.H. (1970) Patterns of Devonian sand on the North American craton and their interpretation. *Bull. Geol. Soc. Am.* **81**, 469–490.

VAHRENKAMP, V.C. (1988) *Constraints on the Formation of Platform Dolomites: a Geochemical Study of Late Tertiary Dolomite from Little Bahama Bank, Bahamas*. PhD Dissertation, University of Miami, Coral Gables, Florida, 434 pp.

VAHRENKAMP, V.C. & SWART, P.K. (1990) New distribution coefficient for the incorporation of strontium into dolomite and its implications for the formation of ancient dolomites. *Geology* **18**, 387–391.

VAHRENKAMP, V.C., SWART, P.K. & RUIZ, J. (1991) Episodic dolomitization of Late Cenozoic carbonates in the Bahamas: evidence from strontium isotopes. *J. Sedim. Petrol.* **61**, 1002–1014.

VAIL, P.R., MITCHUM, R.M. & THOMPSON, S. (1977) Seismic stratigraphy and global changes in sea level, part 4: global cycles of relative changes of sea level. In: *Seismic Stratigraphy – Applications to Hydrocarbon Exploration* (Ed. Payton, C.) Mem. Am. Ass. Petrol. Geol., Tulsa 26, 83–97.

WALTER, L.M., STUEBER, A.M. & HUSTON, T.J. (1990) Br–Cl–Na systematics in Illinois basin fluids: constraints on fluid origin and evolution. *Geology* **18**, 315–318.

WHITAKER, S.T. (1988) *Silurian Pinnacle Reef Distribution in Illinois: Model for Hydrocarbon Exploration*. Ill. State Geol. Surv. Illinois Petrol. 130, 32 pp.

WHITING, L.L. & STEVENSON, D.L. (1965) *Sangamon Arch*. Circ. Ill. State Geol. Surv. 383, 20 pp.

WILLMAN, H.B., ATHERTON, E., BUSCHBACH, T.C., COLLINSON, C., FRYE, J.C., HOPKINS, M.E., LINEBACK, J.A. & SIMON, J.A. (1975) *Handbook of Illinois Stratigraphy*. Bull. Ill. State Geol. Surv. 95, 170 pp.

WITZKE, B.J. & HECKEL, P.H. (1988) Paleoclimatic indicators and inferred Devonian paleolatitudes of Euramerica. In: *Devonian of the World* (Ed. McMillan, N.J., Embry, A.F. & Glass, D.J.) Mem. Can. Soc. Petrol. Geol., Calgary 14, 49–63.

ZENGER, D.H. & DUNHAM, J.B. (1988) Dolomitization of Siluro-Devonian limestones in a deep core (5350 m), southeastern New Mexico. In: *Sedimentology and Geochemistry of Dolostones* (Ed. Shukla, V. & Baker, P.A.) Spec. Publs Soc. Econ. Paleont. Mineral., Tulsa 43, 161–174.

Dolomitization and Organic Matter

Spec. Publs Int. Ass. Sediment. (1994) **21**, 409–427

Organic matter distribution, water circulation and dolomitization beneath the Abu Dhabi Sabkha (United Arab Emirates)

F. BALTZER,* F. KENIG,†§ R. BOICHARD,‡
J.-C. PLAZIAT* and B.H. PURSER*

* *URA 723 CNRS, Laboratoire de Pétrologie Sédimentaire et Paléontologie, Bâtiment 504, Université de Paris Sud, 91405 Orsay, France;*
† *IFP; B.P.311, 92506 Rueil Malmaison cédex, France; and*
‡ *TOTAL CFP, CST, Route de Versailles, 78470 St Rémy lès Chevreuse, France*

ABSTRACT

At the northeastern extremity of the Abu Dhabi sabkha, less than 1 m below the surface, a lithified lenticular dolomitic body immediately overlies Holocene regressive mangrove deposits. The persistent association of this dolomite lens with the palaeosol appears to have a genetic significance.

Field measurements of salinities, pH and Eh of interstitial waters clearly indicate lateral interstitial input of water from the lagoon. This movement is favoured by the porosity created by the dense network of buried mangrove roots and by the diagenetic evolution of the sediment. There is no indication of a control of the chemistry of the interstitial waters by flood recharge through the sabkha surface. pH and Eh profiles also indicate the influence of organic matter on the circulating fluids. In the lower part of the section, decay of the organic matter in the mangrove palaeosol results in an anoxic environment with slightly acidic conditions, probably responsible for the dissolution of aragonite observed in the palaeosol and overlying white mud. The evaporative processes, especially active in the upper microbial mat and sabkha deposits, favour reoxidation by diffusion of ions, resulting in slightly basic, oxic conditions. Dolomite occurs in the transitional zone between these anoxic and oxic environments.

The carbon isotopic signature of the dolomite-rich layer is less (down to 0‰) than that of aragonite in the same location (3‰). As proposed by McKenzie (1981), a contribution of carbon from the decaying organic matter to the carbonate pool is indicated.

INTRODUCTION

Many sedimentological studies have been carried out on the Abu Dhabi lagoon–sabkha sedimentary system (Kinsman, 1964, 1966; Kendall & Skipwith, 1968; Kinsman & Park, 1976; Park, 1977; Patterson & Kinsman, 1977, 1981, 1982 etc.). Until recently, few papers were concerned with organic geochemistry in this environment, with the exception of microbial mats (Cardoso *et al.*, 1978). Other organic matter producers, *Avicennia* mangrove and *Halodule* seagrass, were studied only as elements of the sedimentary environment by Kendall and Skipwith (1969a,b), Plaziat *et al.* (1987) and Evans *et al.* (1964, 1969, 1973).

Recent studies of dolomite in Abu Dhabi sabkhas followed the discovery of this mineral in carbonate sediments of Qatar (Wells, 1962) and Abu Dhabi (Curtis *et al.*, 1963). On a global scale present sites of dolomitization are numerous, although dolomite is usually formed in limited amounts. Peritidal dolomite is typically associated with sabkha conditions (McKenzie, 1976), although it also occurs in tidal flats under more humid climates, in the Bahamas and in Florida Bay (Shinn *et al.*, 1965; Shinn 1968). In sabkhas, evaporation provides both the concentrated solutions involved in dolomite formation and the driving force needed to displace significant

§ Present address: Organic Geochemistry Unit, Delft Technical University, de Vries van Heystplantsoen 2, 2628RZ Delft, the Netherlands.

amounts of reactants through sediments (McKenzie, 1976; McKenzie *et al.*, 1980). A series of hydrocirculation models were proposed for 'evaporitic dolomitization', including 'seepage refluxion' (Adams & Rhodes, 1960), concentration by capillary rise from the water-table (Shinn *et al.*, 1965) and 'evaporative pumping' (Hsü & Siegenthaler, 1969; Hsü & Schneider, 1973). This study demonstrates lateral interstitial circulation of marine porewaters over a distance of more than 1.5 km from lagoon to dry land, beneath the sabkha near Ras Ghanada (Abu Dhabi). It is based on stratigraphic, mineralogical (carbonates) and hydrogeochemical data (salinity, pH, Eh), confirming that multiple hydrological processes are active on different scales. Dolomites in Abu Dhabi are often found in association with organic matter, notably from mangrove swamps and microbial mats. This association will be examined in terms of facies distribution, hydrology and geochemistry.

THE RAS GHANADA SABKHA AND LAGOON COMPLEX

The Ras Ghanada sabkha and lagoon complex, located 60 km northeast of Abu Dhabi city, is one of the more mature sediment-filled systems along the Abu Dhabi coast (Fig. 1). The map of Ras Ghanada region (Fig. 2) is based on Spot satellite images reprocessed by the Institut Français du Pétrole and calibrated by field observations (Kenig, 1991). The Ras Ghanada complex includes a nearly closed lagoon connected to the Arabo-Persian Gulf by narrow channels, and a large sabkha constituting the supratidal part of the lagoon. Fieldwork was centred on a 3.5 km transect across the sabkha and the tidal part of the lagoon, and included topographic levelling, digging of pits and coring.

Micromorphology of the sabkha

The general impression that Abu Dhabi sabkhas are very flat plains is correct only when one considers vertical topography on a metric scale. In contrast, the micromorphology defined on a decimetric scale displays a wide variety of microrelief resulting from ancient channels; local depressions are bordered by subtle levees and beach ridges. This microrelief determines the duration of tidal flooding and consequently the salinities of interstitial waters and the ecological zonation. The micromorphology, not readily appreciated in the field, has been defined by careful topographic measurements along transects. The hydrological, geochemical and sedimentological parameters are interpreted within this topographical framework.

Morphosedimentary units and vegetal zonation

Vegetal zonation and modern sediments are intimately related, especially in these arid tidal-flat environments. The morphosedimentary units (Figs 2 and 3) successively comprise a subaquatic lagoonal depression, with sea meadows and local mangrove fringes, a littoral ridge capped with microbial mats,

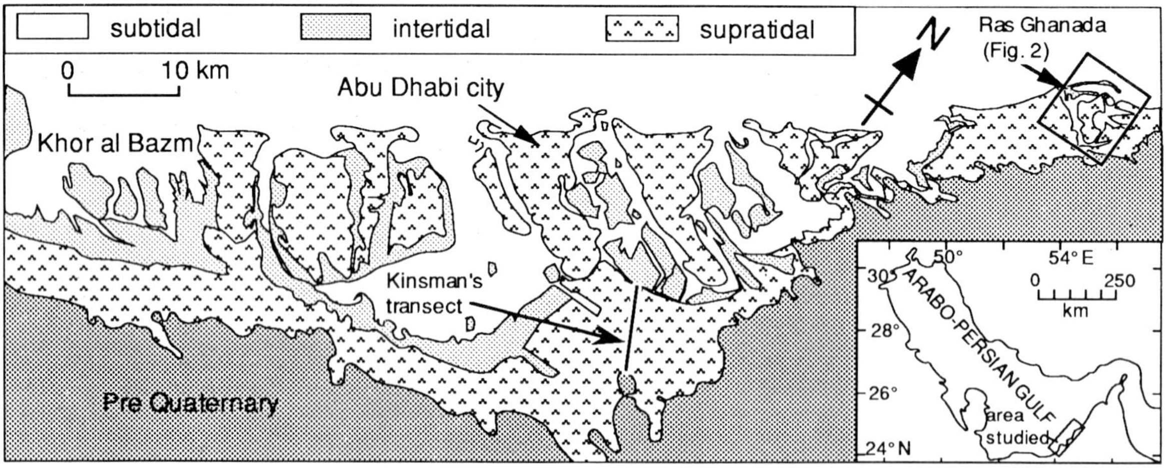

Fig. 1. Location map, sabkhas and lagoons in Abu Dhabi.

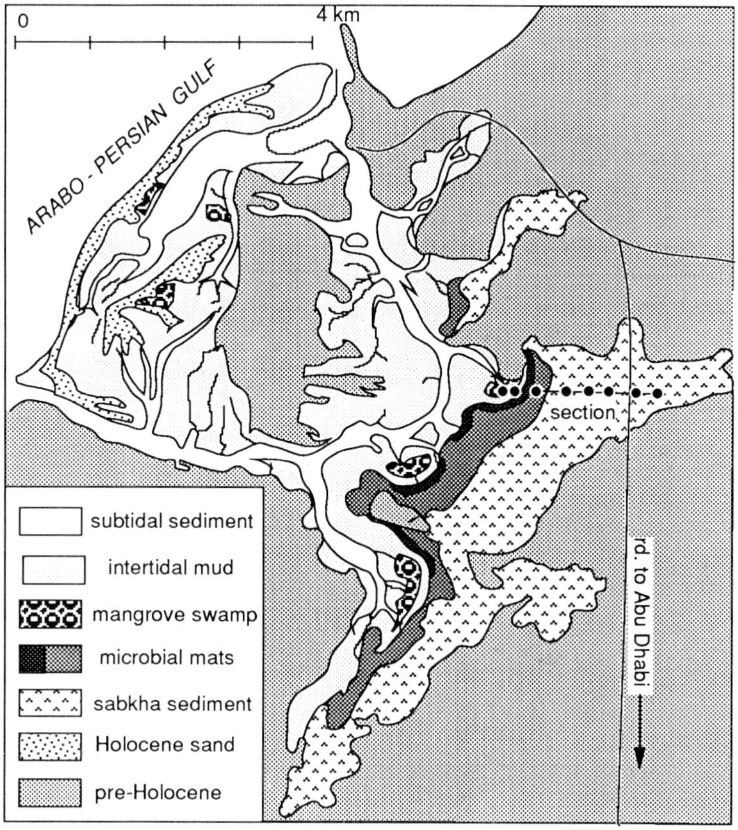

Fig. 2. Map of principal sedimentary facies at Ras Ghanada.

and a wide sabkha which grades into an aeolian dune field on the continental side.

The subtidal lagoonal sediments are composed mostly of muddy calcareous sands, with many forams (*Peneroplis*) and molluscan debris (Lucinids, *Cerithium* and *Potamides*). Living seagrass stands (*Halodule uninervis*) occur immediately below the lower limit of the intertidal zone.

Sediments at the lower part of the intertidal zone are fine, white muds in which aragonitic needles, 1 μm in diameter and 3–5 μm in length, are predominant. At favourable sites the median part of the white mud zone is colonized locally by scattered *Avicennia* mangroves, which may attain 3 m in height. Towards the upper part of the intertidal zone these white muds are covered by a microbial mat, which is blackish and smooth near the shore but becomes brown and highly desiccated inland. The development of this microbial mat coincides precisely with the top of the topographic ridge, which is covered only during high spring-tide. Beyond the top of this ridge, the slope towards the continent marks the beginning of the dry, shallow depression, approximately 1.2 km wide, whose bottom is situated 1.25 m lower than the top of the ridge, i.e. about the mean low-tide level. The bottom of this vague depression comprises a sandy sabkha covered by a thin, broken, brown microbial mat and, locally, a thin saline crust. Beyond this depression begins the flat sabkha, which extends to a dune field.

The importance of bioturbation

The white muds are intensely bioturbated by the endofauna (molluscs, fish, annelids), and especially by crabs producing a dense network of open burrows which are recharged by flood-tides (Fig. 4). 'Crab flats' were mapped by Kinsman (1964) and the sedimentary importance of bioturbation is evident. Mangrove and seagrass also cause bioturbation, their decayed roots forming channels and cavities in the sediments.

Fig. 3. Field views of the peritidal zone at Ras Ghanada. (a) Intertidal zone bordering the lagoon, showing scattered mangroves (background), white muds and microbial mats (foreground); view from the top of the littoral ridge where microbial mats are desiccated. (b) Scattered mangroves in the intertidal zone showing the intense burrowing by crabs and abundant *Avicennia* roots indicated by rows of subaerial, vertical pneumatophores.

METHODS

Field analyses

Field analyses are based on sediments observed in pits and 1.6 m long undisturbed cores, obtained with an Eijkelkamp gouge auger. Sediments were immediately submitted to physical measurements. Salinity was measured with an Atago hand refractometer (317 SC 28) on drops of interstitial water extracted by pressure. This simple equipment provided the numerous salinity measurements with sufficient precision, considering the extreme range of variations in this environment. To correct for the effects of temperature changes, the refractometer was reset prior to each series of measurements (each core). Oxidation–reduction potential (Eh) and pH were measured with a Knick Portamess pH meter, by inserting the probes directly into a section of the core immediately after collection.

Samples of interstitial waters were pressure-extracted in the field (modified NL Baroïd series 300 extractor), using nitrogen instead of CO_2 as the pressure gas in order to limit CO_2 equilibrium changes. Extraction of water for analysis was made from sections of core collected horizontally in the walls of pits dug in the sabkha. An aliquot of the collected water was submitted to alkalinity measurements by HCl titrimetry a few hours after the extraction of interstitial waters. The remaining solutions were refrigerated until chemical analysis could be performed at the Centre de Sédimentologie et Géochimie de la Surface in Strasbourg.

Laboratory analyses

Sediment samples for geochemical analysis were kept on ice in the field, and subsequently deep-frozen the same day. The samples remained frozen until lyophilization at the Laboratoire de Géochimie Organique de l'Institut Français du Pétrole in Rueil Malmaison, France. Total organic carbon (TOC) was measured at the Institut Français du Pétrole by the Rock–Eval process, and confirmed by LECO

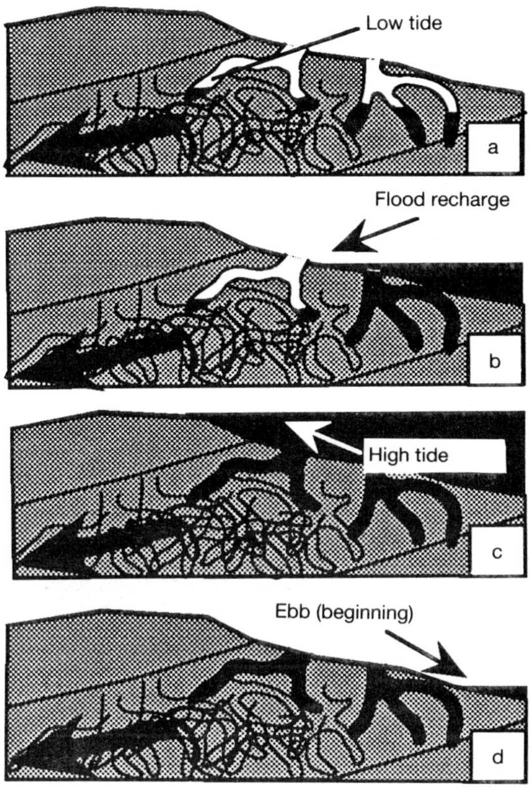

Fig. 4. Accelerating effect of bioturbation on aquifer recharge: crab and fish burrows and hollow root debris, partly empty at low tide (a), are rapidly filled by flood tide (b, c) forming a perched reservoir ensuring water percolation during falling tide (d).

(Kenig, 1991). Mineralogical analyses by X-ray diffractometry were performed by the Institut Français du Pétrole and Orsay University (Purser *et al.*, 1987; Aissaoui *et al.*, 1988).

RESULTS

Coastal stratigraphy

The base of the profile studied (Fig. 5) is composed of light-grey subtidal mud with seagrass root debris (*Thalassia*, *Halodule*) situated between −3 and −1.5 m (below the highest point on this section). This mud is overlain by a 1–1.5 m layer of light-grey muddy sand with *Peneroplis* and potamids, the lowest 20 cm of which contain little organic matter and are partially lithified. The middle and upper parts of these sands contain abundant fibres of mangrove roots.

The overlying layer contains abundant mangrove material forming a dark brown palaeosol (Fig. 6a, a–d) in which lignite debris and mangrove stumps of the species *Avicennia marina* were collected. Field determination of organic debris has been confirmed by microscopic examinations by J.C. Koeniger (Laboratoire de Paléobotanique, Université Pierre et Marie Curie, Paris). Mangrove wood debris collected at 850 m (site V_5) and 250 m from the sea (site V_7) have given radio carbon ages of 2843 ± 166 and 1249 ± 62 BP, respectively. This buried mangrove

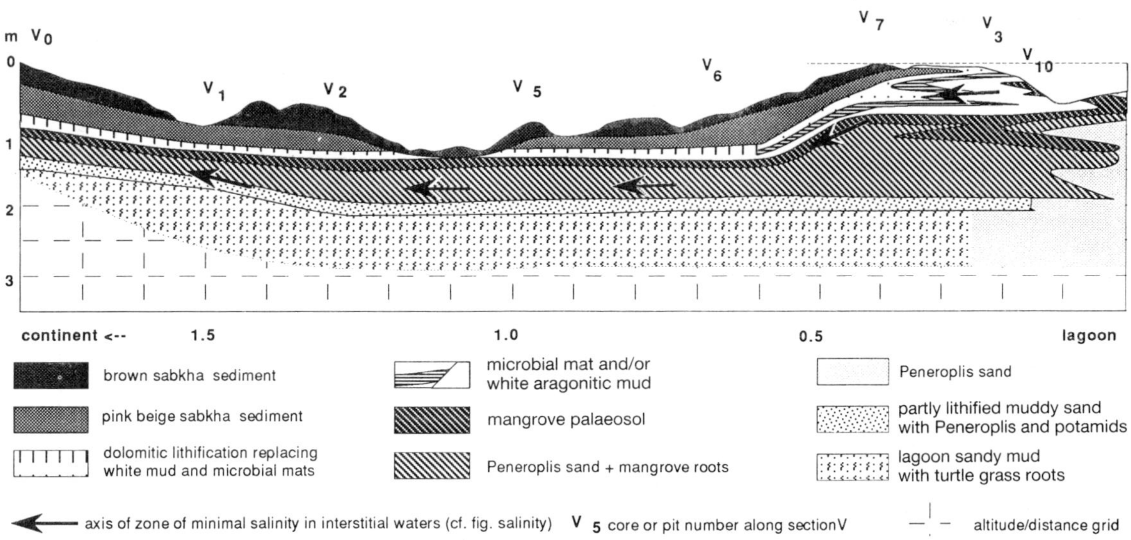

Fig. 5. Stratigraphy of Ras Ghanada sabkha.

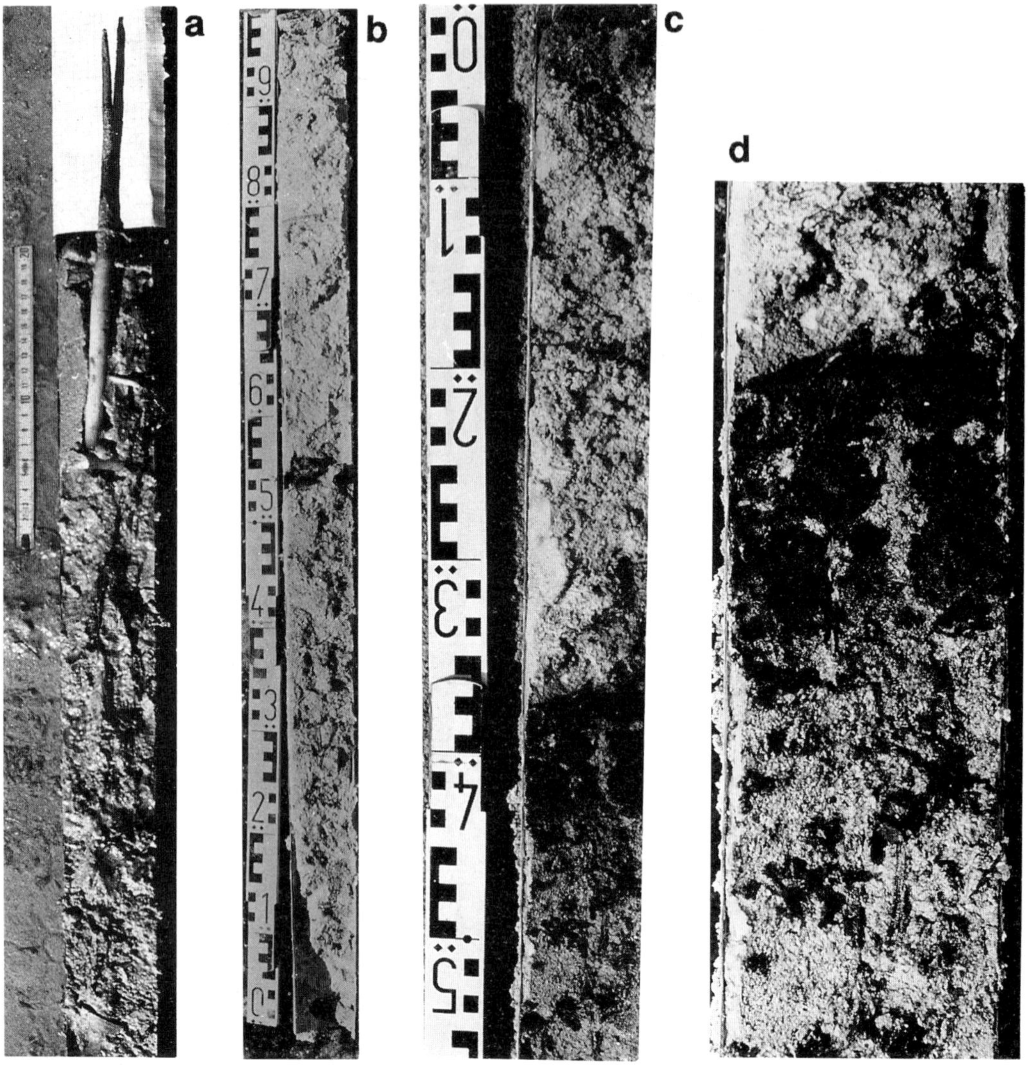

Fig. 6. (A) Sediments from Ras Ghanada. a, core of modern sediment from the mangrove swamp (site V_{10}), showing one subaerial live pneumatophore and the corresponding root system, mixed with black, dead roots; b, core of Recent sediment collected at the outer limit of the microbial-mat zone (site V_3), showing white mud overlying the muddy sand with mangrove debris (black dots); c, core of Recent sediment from the transition zone between dry sabkha and microbial mat (site V_7), showing slightly dolomitized white mud overlying mangrove palaeosol; d, detail of palaeosol in c.

soil rises laterally up to the zone of living mangroves. A detailed examination of this cross-section (Fig. 5) indicates that the position of the mangrove swamps has oscillated seaward and landward during the late Holocene.

The buried mangrove soil is overlain by sediments which change across the sabkha. Near the sea the overlying bed is a white mud associated with microbial layers which, like the mangrove soil, experienced several transgressive and regressive displacements

Fig. 6. (B) A, Holocene mangrove palaeosol from site V_0 (dark layers) overlain by white laminated dolomitic mud, ensuring transition with algal mat; b, pit showing Holocene section at site V_1, with dark brown sabkha mud (from 4 to 5 on scale), pink-beige sabkha mud (from 3 to 4) and white lithified dolomitic layer (up to 3). Metric scale standing on top of harder part of this lithified dolomitic crust above which water has gushed; c, contination of Holocene section in pit at site V_1, lower part, showing the dark compacted mangrove palaeosol underlying the dolomitic crust, and overlying white muddy sand with mangrove root remnants; d, dry sabkha surface being covered by spring-tide arriving from the upper right corner, water limit under lens cap.

of about 100 m. Towards the sabkha these sediments pass laterally into three layers, white, pink-beige and brown, whose characteristic colours express diagenetic processes. These layers locally contain remnants of shells and microbial mats (Fig. 6b, a) reminiscent of the white muds and associated mats. The basal white unevenly lithified crust directly overlies the mangrove palaeosol (Fig. 6b, b and c).

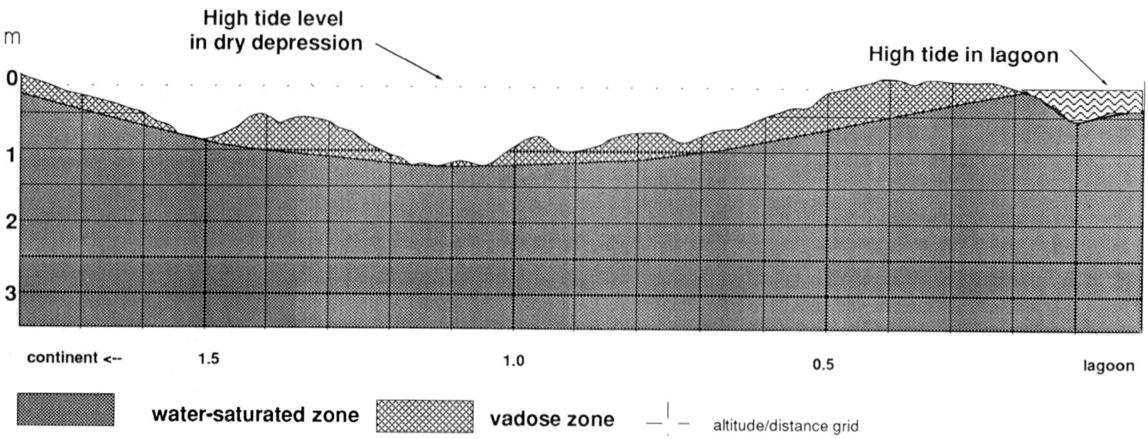

Fig. 7. Piezometric surface in Ras Ghanada sabkha aquifer.

Above this crust occurs a pink-beige sabkha mud (Fig. 6b, b) in which no primary bedding is preserved, except for local remnants of microbial mat. This pink-beige layer and the white lithified carbonate layer have dolomite contents locally attaining 100% of the total carbonate (see Fig. 8).

The upper layer, 20–30 cm thick, is a brown, clotted sabkha mud (Fig. 6b, b and d) containing scattered plurimillimetric gypsum crystals which locally obliterate one or both of the two underlying layers of this transect. It is noteworthy that the contact between this brown horizon and the underlying pink-beige sabkha layer is approximately parallel to the top of the sabkha water-table (Figs 5 and 7).

Composition of sediments

Mineralogy of modern sediments

Modern sediments exposed on the bottom of Abu Dhabi lagoons are formed predominantly of authigenic or bioclastic calcium carbonate (aragonite, high and low Mg-calcite). Other minerals include some dolomite and detrital silicates, which may be abundant in a few restricted locations. X-ray diffraction analyses of sediments from Ras Ghanada and other sites show that aragonite is predominant (Figs 8 and 9), the aragonite/calcite ratio reaching 3/1. The abundance of aragonite is due in part to the skeletal contribution of molluscs, especially in sedi-

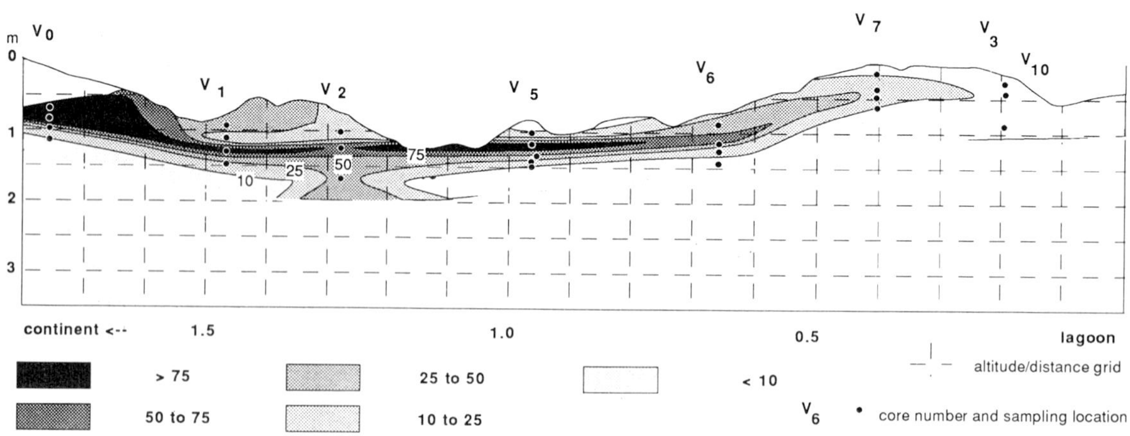

Fig. 8. Distribution of dolomite (% of total carbonates) in Ras Ghanada sediments.

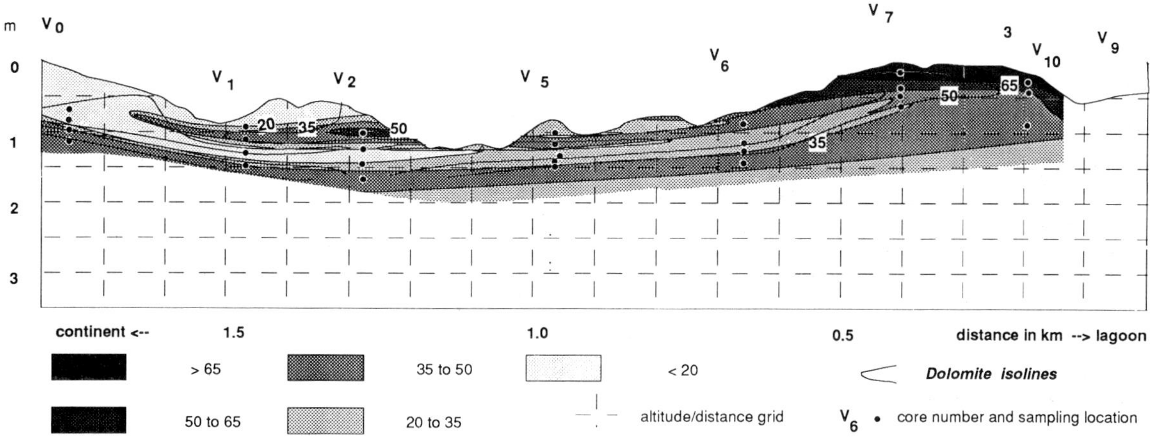

Fig. 9. Distribution of aragonite (% of total carbonates) in Ras Ghanada sediments.

ments in the subtidal zone. However, the white muds of the middle and upper parts of the intertidal zone probably reflect precipitation of aragonite needles within the middle and lower intertidal zones of the lagoon. Chemical precipitation of aragonite in free waters has been documented by Wells and Illing (1964) and Loreau (1982). This process probably plays a critical role in the formation of a carbonate levee at the borders of channels and lagoons of Abu Dhabi, especially in Ras Ghanada.

Diagenetic changes across the sabkha

Within the supratidal zone, aragonite percentages decrease and locally reach zero within the sabkha. This evolution may reflect one or more of the following processes: selective dissolution of aragonite; neoformation of calcite; and/or precipitation of dolomite. All samples contain at least traces ($\leqslant 5\%$) of dolomite, which becomes increasingly abundant beneath the sabkha, where it may constitute 100% of the carbonate fraction. Dolomitization (Fig. 8) coincides with local lithification of sediments (formation of a crust) and with a pink-beige colour. It attains a maximum within the lithified crust, but is also important between the upper limit of the mangrove debris and the pink-beige unit. Dolomite formation is associated with a gradual dissolution of molluscan shells. Scanning electron microscopic examination of the mud shows that the formation of dolomite takes place at the expense of aragonite and calcite (Fig. 10a,b).

Dolomite crystals, generally less than 1 μm in size, develop on the surface of sedimentary particles and grow in the primary pore spaces; in this sense, the dolomite is a cement (Fig. 10a). The subsequent growth of individual dolomite rhombohedrons envelops adjacent aragonite nanoparticles, which are thus included within the dolomite. These may subsequently be dissolved, giving intracrystalline voids. Thus, the aragonitic muds are 'replaced' progressively by Ca-rich dolomite and the process essentially concerns penecontemporaneous dissolution of one mineral phase and the growth of another. The dolomite is not stoichiometric. Ca and Mg percentages were measured on three samples from the lithified level of Ras Ghanada using a diffractometric technique (Mélières, 1973). They are, respectively: 56.0 and 44.0 (± 0.2); 57.0 and 43.0 (± 0.2); and 54.0 and 46.0 (± 0.2).

There are surprisingly few evaporite minerals in the Ras Ghanada profile. Gypsum never exceeds 15% in near-surface sediments, except in the layer overlying the lithified dolomitic carbonate, at site V_1, and anhydrite occurs only as a trace mineral. This situation, suggesting that the brines are not saturated with respect to gypsum or anhydrite, is obviously unexpected considering the aridity. This implies the reflux of a significant amount of brine towards the sea.

Organic carbon in sabkha sediments

Total organic carbon (TOC) percentages greater than 1% (Fig. 11) were measured only in the mangrove palaeosol and microbial mats. In the former it

Fig. 10. Dolomitization in Ras Ghanada sediments. (a) Incipient dolomitization: dolomite crystals in contact with dissolving aragonitic needles. Scale bar = 2 μm. (b) Final stage of dolomitization: dolomite crystals completely replace aragonitic needles. Scale bar = 2 μm.

attains 8.3%, the average TOC of 13 samples of the mangrove palaeosol being 3.6%. Beneath the mangrove palaeosol proper the distribution of organic matter is heterogeneous, due to an uneven distribution of occasional roots, and TOC seldom exceeds 1% (average 0.7% for seven samples). TOC values in the microbial mat overlying white muds are somewhat lower, ranging from 0.2 to 2% and averaging 1.1% (four samples). Very low values, never exceeding 0.3% TOC, were measured in the lithified dolomitized layer.

Phreatic brines

Piezometric surface

The height of the piezometric surface was measured over 2 km and showed that the surface of the water-table is depressed and tends to follow closely the topographic surface, except for local irregularities (Fig. 7). Landward, at site V_0, a few tens of centimetres under the sabkha surface, the piezometric level is lower than mean high-tide level. However, towards the sea it is close to mean sea level. Be-

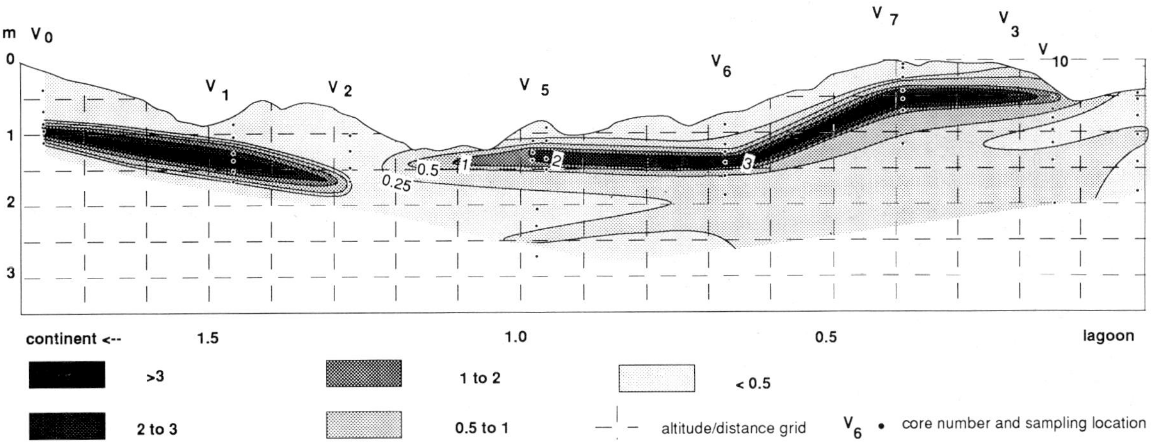

Fig. 11. Distribution of total organic carbon (% of dry weight) in Ras Ghanada sediments.

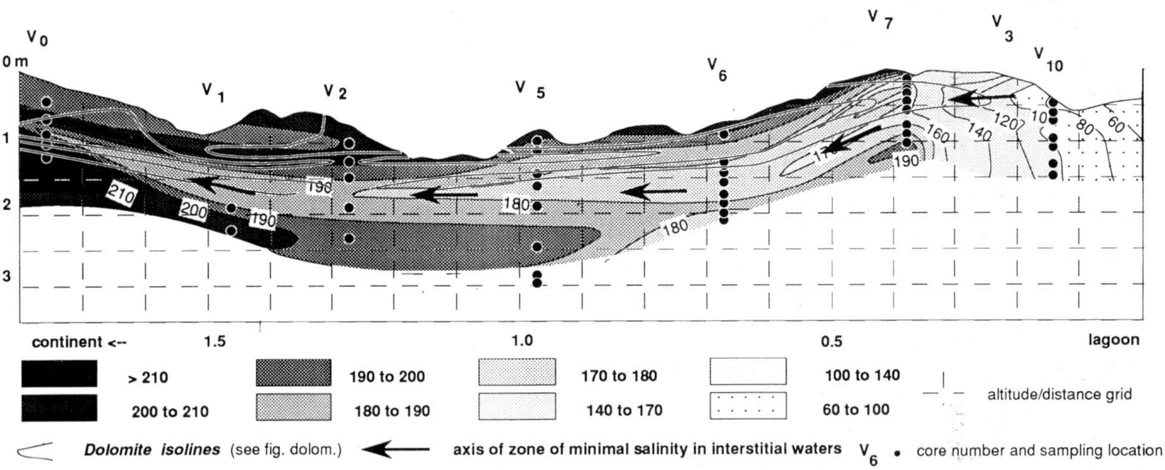

Fig. 12. Distribution of salinity in Ras Ghanada sediments (‰).

tween these two points, the piezometric level gradually descends towards the bottom of the depressed part of the sabkha and locally moistens the surficial mud in the deepest parts of this depression.

At the bottom of the depression (sites V_1, V_2 and V_5), pits dug through the dolomitized layer revealed the artesian behaviour of the water-table, under pressure below the dolomite crust. When this aquiclude was perforated the water rapidly rose to an equilibrium level, several tens of centimetres above the crust.

The slope of the piezometric surface, dipping from the lagoon towards the dry depression, clearly shows that the interstitial water flows in that direction. The recharge of this aquifer occurs during flood-tides, and thus is produced by lagoonal waters whose penetration is probably favoured by bioturbation in the intertidal zone.

Salinity

The distribution of salinities in the aquifer, between the lagoon and the depression, exhibits a marked, regular increase (Fig. 12) due to the concentration of a saline aquifer whose water-table is close to the surface and thus submitted to strong evaporative conditions. Again, this implies that the aquifer flows landwards, in the direction where salinity is higher;

in agreement with Schoeller's (1966) model, the seawater entering littoral sediments replaces water lost through evaporation, and becomes progressively concentrated.

In addition, a vertical salinity gradient increases towards the upper part of the aquifer. The combination of both the horizontal and the vertical gradients results in the tongue-shaped isohaline curves (Fig. 12) that reflect the existence of a distinctive, less saline layer of water, centred 0.5–1 m below the sabkha surface and whose salinity slowly increases from the sea (60–80 g/kg) towards the continent (>210 g/kg). At every site, an increase of roughly 40 g/kg was measured between the centre of the tongue and the surface. This low-salinity lens (Fig. 12) and the mangrove palaeosol and sands rich in mangrove roots (Fig. 5) coincide, suggesting that the higher permeability of the mangrove palaeosol favours water recharge relative to less permeable surrounding carbonate sediments (see below).

Beneath the inland, non-depressed part of the sabkha, at depths from 1 to 3 m, salinities display two opposing trends. They attain the highest value (>210 g/l) at site V_0 and gradually decrease from there towards the sea. This results in a tongue of highly concentrated brine situated below the tongue of relatively less saline water. Conversely, 700 m inland from site V_0, at site V_{20}, salinities as low as 120 g/kg were recorded, indicating dilution by continental or rain waters.

Chemistry of interstitial waters

Two trends are indicated by the distribution of oxidation–reduction potential (Eh) isolines measured on Ras Ghanada profile (Fig. 13). As expected, reducing conditions follow the distribution of organic matter (TOC), i.e. of microbial mat and mangrove deposits. However, they are limited to the outer (lagoonal) half of the transect, between V_5 and the lagoon. No reducing Eh values and only a relative lowering of Eh were measured on the continental half of the transect, although sediments with 4% TOC are present. These less reducing conditions result from the near total consumption of the organic matter available for reducing micro-organisms. The relatively high remaining TOC consists mostly of resistant aliphatic materials unsuitable for bacterial consumption (Kenig *et al.*, 1989; Kenig, 1991).

The pH distribution is more complex. Interstitial waters in sediments containing scattered mangrove roots and in the mangrove palaeosol contain a continuous band of acid pH values, usually close to pH 6 and never exceeding pH 6.75 (Fig. 14). These acid pHs promote the dissolution of carbonates and produce a high alkalinity in interstitial waters, a point that will be discussed later. In contrast, in the dolomitized sediments pH values form a band close to or exceeding 7. Considering the hydrological path stimulated by evaporation, waters in the aquifer pass from slightly acidic in the mangrove-rich deposits to slightly basic in the dolomitized horizon.

The lowest value of alkalinity, 1 mmol/l, was measured at site V_{20}, about 3 km from the sea, in an area where the salinities of interstitial waters are relatively low, probably as a result of the influx of continental water. Closer to the sea, the highest alkalinity value, as expected, was measured at site V_0 on waters extracted from the mangrove palaeosol (3.32 mmol/l). In the underlying muddy sand with

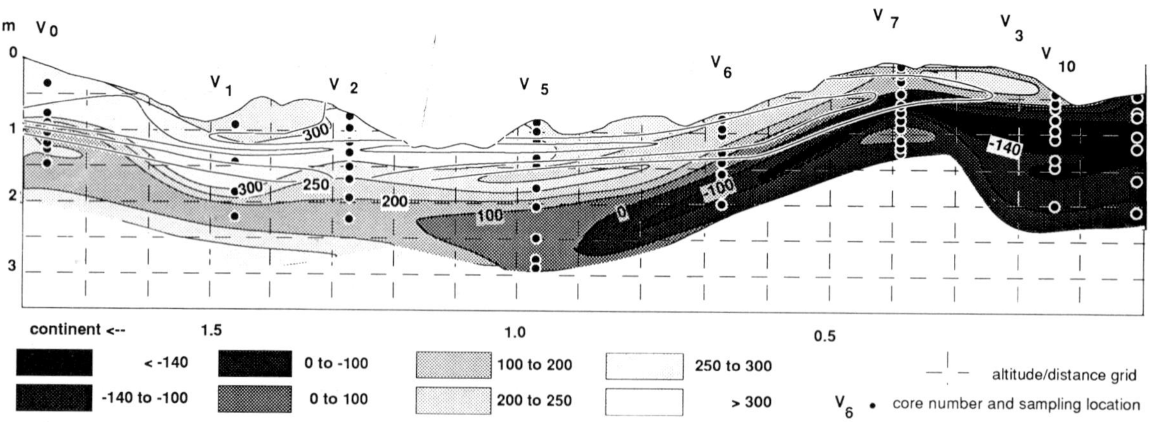

Fig. 13. Eh distribution in Ras Ghanada sediments (Eh in mV).

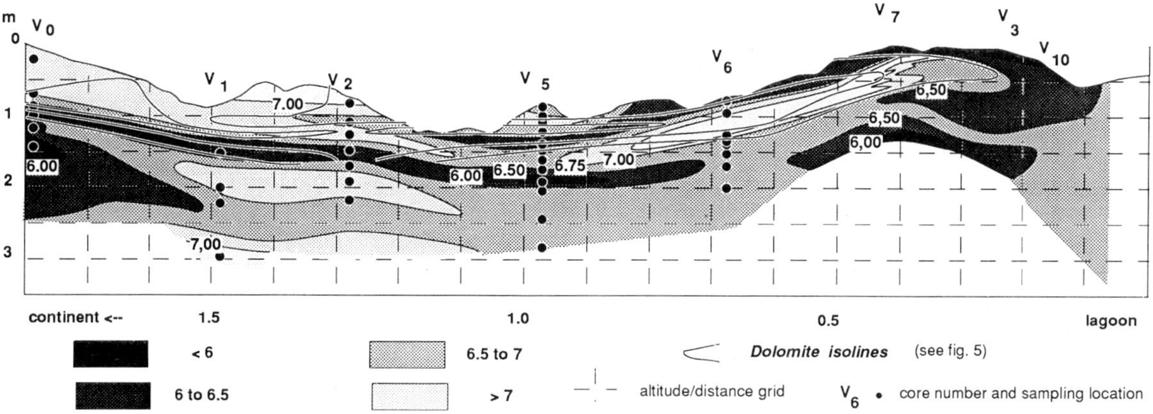

Fig. 14. pH distribution in Ras Ghanada sediments.

mangrove root debris, alkalinity was lower (2.92 mmol/l). Above the mangrove palaeosol, alkalinity decreases to 2.95 mmol/l in the microbial mat and to 1.55 mmol/l in the lithified dolomitized layer. This situation was also observed at sites V_5 and V_7. Therefore, the alkalinity of interstitial waters in the mangrove palaeosol and the microbial mat largely exceeds the value normally measured for seawater (2 mmol/l) but clearly decreases below that of seawater in the lithified dolomitized layer (Kenig, 1991).

Major elements were measured (Kenig, 1991) on samples of interstitial waters collected at three levels of the Ras Ghanada section, from base to top: muddy sand with mangrove root debris; dark-brown mangrove palaeosol; and the lithified dolomitized layer or its unlithified equivalent. The evolution of ionic ratios in these interstitial waters reflects the diagenetic processes taking place within these sediments. These variations are comparable to those described by Butler (1973). Mg^{2+}/Ca^{2+} ratios change drastically from 4.5 in the free water in the lagoon to more than 11 (35 in Butler's transect) in the interstitial water of the levee at site V_7 (Fig. 15). The same ratio subsequently decreases to 4 in sabkha sites (from V_5). The Mg^{2+}/Ca^{2+} ratio increase, from the lagoon to site V_7, can be interpreted as a relative decrease of calcium concentration, probably due to aragonite precipitation in the intertidal zone, where gypsum has not been observed.

Beyond site V_7, the Mg^{2+}/Ca^{2+} ratio returns to values similar to those of the lagoon. This decrease

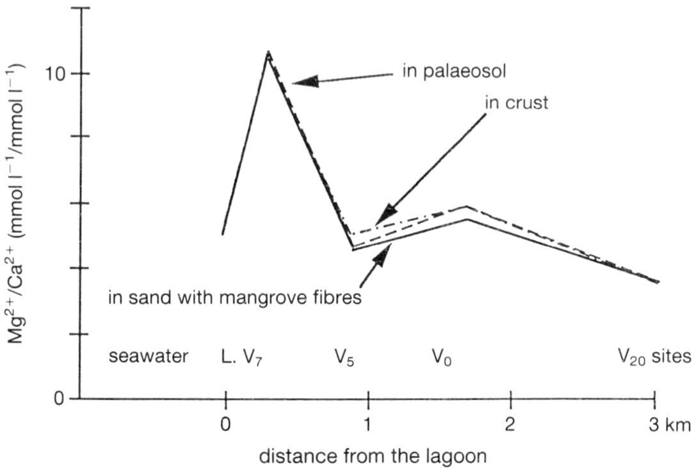

Fig. 15. Variation of the Mg^{2+}/Ca^{2+} molar ratios across the Ras Ghanada sabkha.

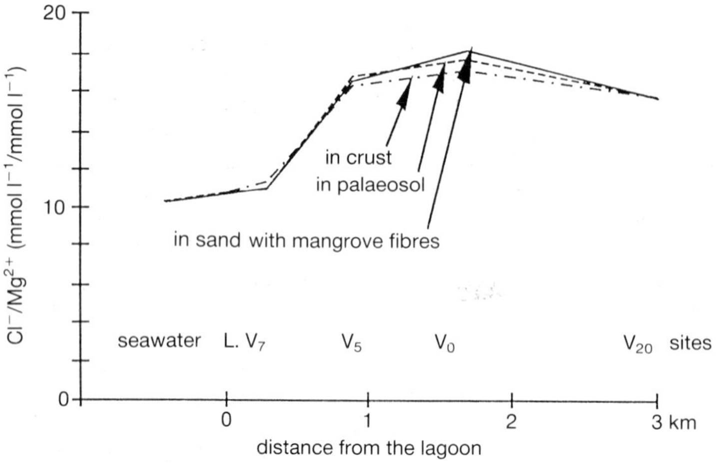

Fig. 16. Variation of the Cl^-/Mg^{2+} molar ratios across the Ras Ghanada sabkha.

in Mg^{2+}/Ca^{2+} ratio is coincident with the area of dolomitized sediments between sites V_7 and V_{20}.

The Cl^-/Mg^{2+} molar ratio increases rapidly from the levee (site V_7) to the centre of the sabkha depression (site V_5) and this is readily interpreted in terms of magnesium depletion due to dolomite formation (Fig. 16). A detailed discussion of the geochemistry of interstitial waters is given elsewhere (Kenig, 1991).

Isotope geochemistry

Isotopic analyses of carbon and oxygen have been carried out on 10 sediment samples from Ras Ghanada (Kenig, 1991) from which interstitial water was extracted for chemical analysis, plus one surface sample from the modern intertidal zone (analyses by J.Ch Fontes' Laboratoire d'Hydrologie de Géochimie Isotopique de l'Université Paris Sud, Orsay, France). A $\delta^{13}C/\delta^{18}O$ plot is given in Figure 17, where the oblique line separates an upper, calcitic field from a lower, predominantly dolomitic field, and shows that dolomite is slightly more depleted in ^{13}C and slightly more enriched in ^{18}O relative to the coexisting calcium carbonate, the latter indicating participation of evaporated seawater. This is valid even when the difference in isotopic fractionation of ^{18}O between $CaCO_3$ and $(Ca,Mg)CO_3$ is consider-

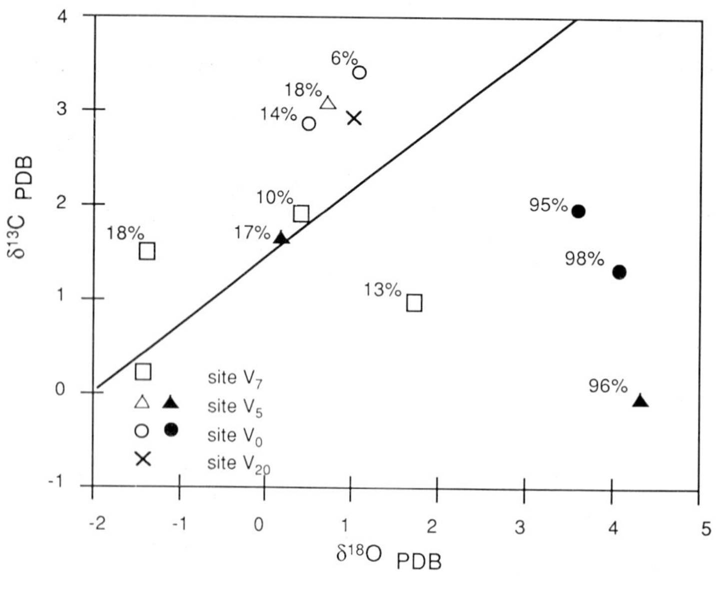

Fig. 17. $\delta^{13}C/\delta^{18}O$ plot for 10 Ras Ghanada samples from which interstitial water was extracted: percentages express dolomite content in the carbonate fraction; in black, samples from the lithified dolomitized horizon.

ed, and confirms results obtained by McKenzie (1976) on sediments from other sites in Abu Dhabi.

As suggested by McKenzie (1976), the more negative $\delta^{13}C$ figures for dolomite-rich samples probably are an effect of CO_3^{2-} ions produced by the bacterial degradation of organic matter which is incorporated in the lattices of authigenic dolomite crystals. The $\delta^{13}C$ measured in Ras Ghanada are somewhat lower than those indicated by McKenzie (1981), probably because of the proximity of the palaeosol to the lithified dolomitic layer. Mangrove organic matter exhibits lower $\delta^{13}C_{org}$ (−25.3‰) than microbial mats ($\delta^{13}C_{org}$, −12‰), which dominated in the area studied by McKenzie (1976), resulting in our $\delta^{13}C_{dol}$ lower values. Carbon derived from organic matter therefore participates in the composition of the CO_3^{2-} ions of dolomite. This point is consistent with the hypothesis that dolomite precipitates from water which has circulated through the adjacent underlying mangrove palaeosol and the associated microbial mat.

DISCUSSION

Hydrocirculation patterns are obviously an important factor in developing a model for the near-surface formation of dolomite in Abu Dhabi. The introduction of seawater is influenced by tidal ranges: 2 m in the gulf and 1.25 m in the lagoons adjacent to sabkhas, beneath which dolomite precipitation occurs.

In Al Salmiyah, west of Abu Dhabi city, and in Ras Ghanada, at approximately mean high tide, lagoonal waters are level with the tops of the levees bordering lagoons and channels. Under such conditions the levee tends to prevent seawater from flowing farther landward and entering the depressed part of the sabkha. This depression remains dry, although its bottom occurs 1.25 m lower than the top of the ridge, i.e. about low-tide level. Only spring-tide waters rise above the levee, and enter the depression as a centimetric sheet of water flowing on the landward flank of the levee. This surface flow, which was observed during several spring-tides, should have a bimonthly frequency: it lasted only a few tens of minutes and wetted the microbial mats of the sabkha without effectively recharging the underlying aquifer.

Exceptional tides, especially when the *Shamal*, a strong (5 − 6 on the Beaufort scale) persistent wind, is active, occasionally invade the depression, but the distribution patterns of salinities, pH and Eh remain unaffected under the lithified dolomitized layer. In contrast, long episodes of drought influence salinities down to these levels. Under such conditions, flood recharge of the aquifer by waters flowing above the sabkha apparently has only limited influence (Fig. 18).

Flood recharge in the intertidal part along the lagoonal shore is favoured by bioturbation. The infilling of the burrows by seawater is nearly instantaneous (Fig. 4b), constituting a water input at a relatively high topographic level, thus building an efficient head for the aquifer. The instantaneously recharged aquifer will probably continue to flow until the next flood-tide. This type of perched high-tide reservoir has also been observed in mangrove swamps from New Caledonia, Iran and Egypt (Baltzer, 1982; Baltzer *et al.*, 1987). In conjunction

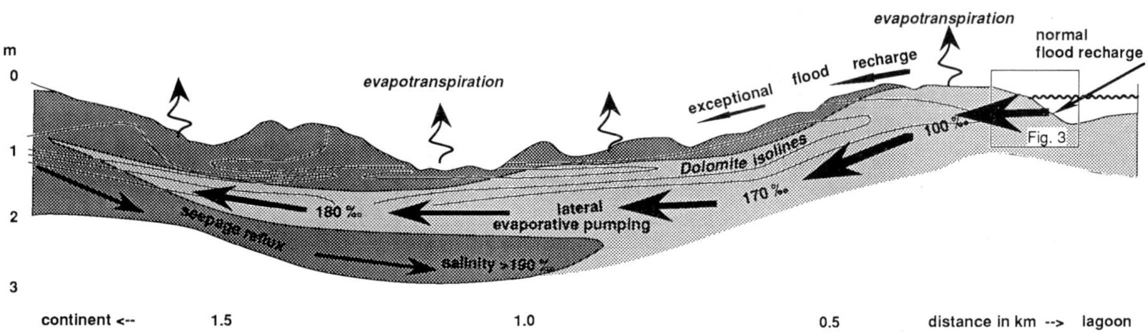

Fig. 18. Schematic hydrocirculation model favouring dolomitization in Ras Ghanada sabkha: seawater percolates from perched bioturbated reservoir (see Fig. 4) downwards and landwards to replace water evaporated from the surface of the sabkha (evaporative pumping). Concentrated brine flows seawards, under the landward entering flow (seepage reflux). Dolomite is forming immediately above the landward flow. Flood recharge occasionally occurs when seawater from the lagoon overflows the algal mat barrier towards the dry depression (Fig. 7).

with evaporative pumping, it probably has a major role in the constant introduction of the lagoonal waters into the sabkha sediments.

The distribution of salinities in interstitial waters is interpreted in terms of hydrocirculation from the lagoon towards the sabkha. The 'axis' of the lens of minimal salinity (nearly marine toward the lagoonal shore: Fig. 12; Fig. 18, black arrows) is considered as the conduit along which circulation in the aquifer is most rapid. This axis is coincident with the sands containing mangrove root debris and the mangrove palaeosol (Fig. 5). A similar hydrocirculation pattern is suggested by tritium measurements made by McKenzie (1976) in 'Kinsman's area', on a much longer transect lacking mangrove palaeosol. The tritium isolines redrawn on McKenzie's data, expressed in TU (tritium units), form a tongue (Fig. 19) similar in shape to that of the relatively low salinities demonstrated at Ras Ghanada. The highest figures of tritium have to be interpreted in terms of very recent ages. In the Kinsman transect, the tritium maximum is coincident with the area where, according to our model, the penetration of new seawater is expected. This supports the idea of a lateral displacement of water similar to that of Ras Ghanada, even though other geochemical data rather point to a downward vertical recharge (McKenzie, 1981). Therefore, we have to consider that the pattern of hydrological circulation typical for Ras Ghanada could have been effective, at least temporarily, in different environments of the Abu Dhabi sabkha, even in the absence of a mangrove palaeosol, and possibly in alternation with the more classical downward recharge.

The increase in permeability resulting from the presence of an ancient mangrove soil, although not indispensable for that type of hydrocirculation, is frequent and does enhance water movement. It was also observed in New Caledonia swamps (Baltzer, 1982) where it favours hydrocirculation in the sediments of the older parts of the mangrove swamps. Increase in permeability is related to both the presence of organic matter and the dissolution of aragonite. Mangrove roots and litter produce an organic-rich layer which decays under the action of micro-organisms. This organic matter is preserved in the form of small tubes and plant fibres, which increase permeability. Moreover, the byproducts of organic decay create and maintain a slightly acid environment in which fine carbonate particles dissolve. This dissolution seems to be confirmed by the quartz concentration, which is significantly higher in the fossil mangrove soils. The abundant (up to more than 40%) quartz fraction also could be due to aeolian transport but the dissolution of shells in sediments rich in organic matter is observable (chalky aspect and incompletely dissolved molluscs observed with scanning electron microscopy), and the calcium content of the interstitial waters of the mangrove palaeosol are higher than that of neighbouring

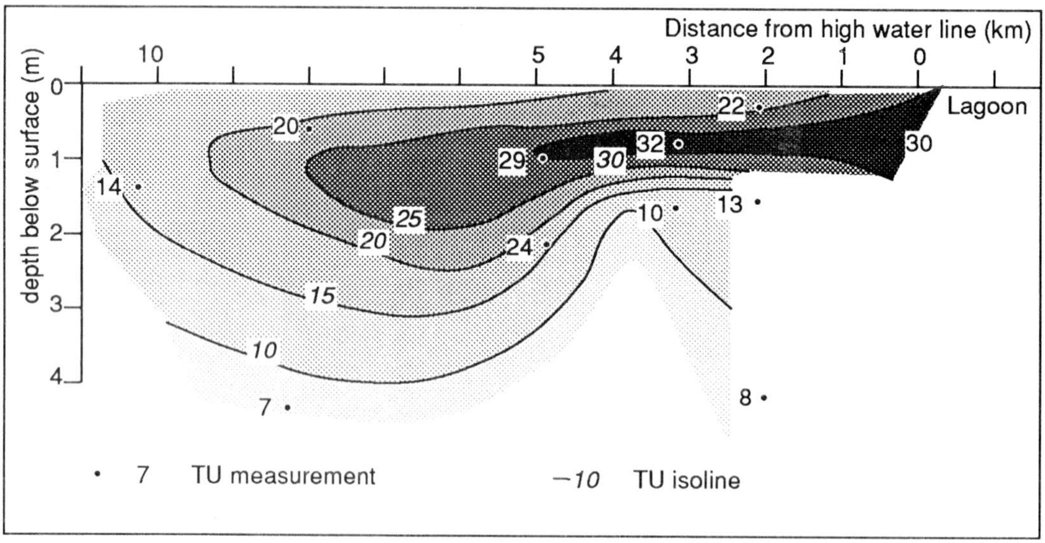

Fig. 19. Tritium content (in TU) of sabkha waters in Kinsman's area along a traverse from the lagoon to Miocene cliffs. (Data from McKenzie, 1976: scale modified, isolines added.)

sediments, whereas the salinity remains similar. These points support the interpretation of dissolution of carbonates (Kenig, 1991, p. 134) and thus, the concentration of insoluble quartz residues.

Water gushes from the aquifer whenever the lithified dolomitized bed is perforated, demonstrating its tendency to rise from mangrove-bearing deposits towards the surface. This rise is normally slow across the relatively impermeable aquiclude, and results in increased salinity towards the sabkha surface. Occasional local dilution occurs near the surface, due to either exceptional tides, storm surges or rainstorms, the latter possibly explaining the relative dilution observed close to the surface at sites V_0, V_1 and, partially, V_2 (Fig. 12), but without significant influence on the distribution of salinity in the deeper interstitial waters.

A zone of higher salinity below the lens of minimal salinity results in a reverse salinity gradient (Fig. 12). This suggests either seepage reflux towards the sea, or a high-salinity lens corresponding to an older, drier period. The first hypothesis is most probable because it explains the paucity of intraformational evaporites. This indication of a reverse percolation (seepage reflux) has been demonstrated by De Groot (1973) on Umm Said sabkha in Qatar. It also compares with advections of supersaline water observed by Burns and Swart (1992) in muds in Florida Bay, USA, where the supersaline water originates in the central lagoon of an emergent mudbank.

CONCLUSIONS

The formation of lithified dolomitized layers in Abu Dhabi is the result of multiple ecological, geochemical and hydrological circumstances, combined with the relative stability of sea level during the last 4000 years. Mesotides are also a major control favouring the development of muddy-shore ecosystems, such as mangroves and microbial mats, which enrich the sediment with organic matter and enhance water percolation (mangroves, crabs).

The vegetal zonation and abundance of carbonates stimulated by the arid climate of Abu Dhabi has led to a regressive sequence whose level is nearly constant due to the stability of the Holocene sea level. In this sequence, calcareous muddy sands, enriched in organic matter by mangroves and intensely bioturbated by crabs and other burrowers, are overlain by a layer of carbonate mud. Seawater fills burrows and other bioturbational cavities at high tide, forming a reservoir situated near high-tide level which enables the downward percolation of water until the subsequent high tide. Seawater, drawn towards the sabkha by evaporative pumping, percolates laterally within the landward-dipping, more permeable mangrove-enriched bed. The chemical evolution of water during this process favours the precipitation of aragonite followed by dissolution of most of this aragonitic material, and then the precipitation of dolomite. The carbon isotopic composition of the Ras Ghanada dolomite ($\delta^{13}C$) confirms the data obtained by McKenzie (1976, 1981), indicating that the low ^{13}C content in the Abu Dhabi dolomites can be explained by the incorporation of some carbon liberated by the bacterial decomposition of organic matter.

The progressive 'replacement' of predominantly aragonitic sediments by metastable dolomite involves both dissolution and crystal growth. Dissolution occurs in evaporated waters (brines) having salinities higher than 100 g/kg, mainly because of the acidic pH conditions resulting from the bacterial decay of associated organic material (dead mangrove roots). The growth of dolomite crystals, essentially in the form of a nanocement fabric, incorporates many inclusions. Since these processes occur on a very small scale, the result favours a good preservation of the original sedimentary fabric.

The time required for the initial formation of the dolomite at Ras Ghanada, and possibly over much of the Abu Dhabi sabkha, is probably less than 1000 years. The mangrove soil situated about 250 m from the present living mangrove zone has an age of 1249 $\pm$ 62 years. Even if we assume that dolomitization postdates mangrove growth and burial, the beginning of dolomite formation is almost certainly less than 1000 years old. This is further supported by the presence of dolomite within surface-living microbial mats over many parts of the Abu Dhabi sabkha.

Dolomitization within these magnesium-rich and organic carbon-rich environments is stimulated by the active circulation of near-surface waters. Since the sabkha progrades, its peripheral sedimentary and diagenetic environments will gradually evolve as these zones become progressively more continental. It is not unlikely that the metastable dolomite itself will suffer from the effects of meteoric or continental waters, resulting in its partial or complete dissolution or even recrystallization to a more ordered dolomite. Future research in Abu Dhabi and other sabkhas should test this hypothesis.

ACKNOWLEDGEMENTS

This study is part of a project concerning the sedimentation and diagenesis of organic matter in the hypersaline environment of Abu Dhabi. It has been carried out in cooperation with the Institut Français du Pétrole, TOTAL CFP and the Université de Paris Sud, Orsay. The authors express their thanks to the Abu Dhabi government for granting research permission, to the ADNOC Company and to Société TOTAL ABK for technical and administrative assistance during field work. Very useful comments and suggestions by referees D.H. Zenger, J. McKenzie and M. Tucker are gratefully acknowledged.

REFERENCES

ADAMS, J.E. & RHODES, M.L. (1960) Dolomitization by seepage refluxion. *Bull. Am. Ass. Petrol. Geol.* **44**, 1913–1920.

AISSAOUI, D.M., BALTZER, F., CUIF, J.P., PLAZIAT, J.C. & PURSER, B.H. (1988) *Sédimentation et Évolution de la Matière Organique dans les Milieux Carbonatés Récents des Lagunes d'Abu Dhabi*. Rapport Interne Lab. Pétro. Sédim. et Pal., Université Paris-Sud, 78 pp.

BALTZER, F. (1982) La transition eau douce–eau salée dans les mangroves. Conséquences sédimentologiques et géochimiques. Symposium 'Transition eaux douces–eaux salées', Ass. Sédim. Français, Paris, 20–21 janvier 1981. *Mém. Soc. Géol. Fr. NS* **144**, 27–42.

BALTZER, F., AUBRY, CH. & PLAZIAT, J.C. (1987) Les marais maritimes à sebkhas et mangroves du Sud de l'Egypte (Mer Rouge). *1er Congrès Français de Sédimentologie, Résumés*, ASF, Paris, Nov. 1987, p. 38, 39.

BURNS, S.J. & SWART, P.K. (1992) Diagenetic processes in Holocene carbonate sediments: Florida Bay mud banks and islands. *Sedimentology* **39**, 285–304.

BUTLER, G.P. (1973) Strontium geochemistry of modern and ancient calcium sulphate minerals. In: *The Persian Gulf* (Ed. Purser, B.H.) pp. 423–452. Springer-Verlag, Heidelberg.

CARDOSO, J.N., WATTS, C.D., MAXWELL, J.R., GOODFELLOW, R., EGLINGTON, G. & GOLUBIC, S. (1978) A biogeochemical study of the Abu Dhabi algal mats: a simplified ecosystem. *Chem. Geol.* **23**, 273–291.

CURTIS, R., EVANS, G., KINSMAN, D.J.J. & SHEARMAN, D.J. (1963) Association of dolomite and anhydrite in the Recent sediments of the Persian Gulf. *Nature* **197**, 679–680.

DE GROOT, K. (1973) Geochemistry of tidal-flat brines at Umm Said, SE Qatar, Persian Gulf. In: *The Persian Gulf* (Ed. Purser, B.H.) pp. 377–394. Springer-Verlag, Berlin.

EVANS, G., KINSMAN, D.J.J. & SHEARMAN D.J. (1964) A reconnaissance survey of the environment of recent carbonate sedimentation along the Trucial coast, Persian Gulf. In: *Deltaic and Shallow Marine Deposits* (Ed. Van Straaten, L.M.J.U.) pp. 129–135. Developments in Sedimentology I, Elsevier, Amsterdam.

EVANS, G., SCHMIDT, V., BUSH, P. & NELSON, H. (1969) Stratigraphy and geologic history of the sabkha, Abu Dhabi, Persian Gulf. *Sedimentology* **12**, 145–159.

EVANS, G., MURRAY, J.W., BIGGS, H.E.G., BATE, R. & BUSH, P.R. (1973) The oceanography, ecology, sedimentology and geomorphology of parts of the Trucial coast Barrier Island complex. In: *The Persian Gulf* (Ed. Purser, B.H.) pp. 233–277. Springer-Verlag, Heidelberg.

HSÜ, K.J. & SCHNEIDER (1973) Progress report on dolomitization: hydrology of Abu Dhabi sabkhas, Arabian Gulf. In: *The Persian Gulf* (Ed. Purser, B.H.) Springer-Verlag, Berlin. pp. 409–422.

HSÜ, K.J. & SIEGENTHALER, C. (1969) Preliminary experiments on hydrodynamic movement induced by evaporation and their bearing on the dolomite problem. *Sedimentology* **12**, 11–25.

KENDALL, C.G. ST. C & SKIPWITH, BT. P.A. D'E. (1968) Recent algal stromatolites of a Persian Gulf lagoon. *J. Sedim. Petrol.* **38**, 1040–1058.

KENDALL C.G. ST. C. & SKIPWITH, BT. P.A. D'E. (1969a) Geomorphology of a recent shallow-water carbonate province: Khor al Bazam, Trucial Coast, South West Persian Gulf. *Geol. Soc. Am. Bull.* **80**, 865–892.

KENDALL, C.G. ST. C & SKIPWITH, BT. P.A. D'E. (1969b) Holocene shallow-water carbonate and evaporitic sediments of Khor al Bazam, Abu Dhabi, Persian Gulf. *Bull. Am. Ass. Petrol. Geol.* **53**, 841–869.

KENIG, F. (1991) *Sédimentation, Distribution et Diagenèse de la Matière Organique dans un Environnement Carbonaté Hypersalin: le Système Lagune–Sabkha d'Abu Dhabi*. Thèse Inst. Français du Pétrole et Université d'Orléans, 311 pp.

KENIG, F., HUC, A.Y., PURSER, B.H. & OUDIN, J.L. (1989) Sedimentation, distribution and diagenesis of organic matter in a recent carbonate environment, Abu Dhabi, UAE. In: *Advances in Organic Geochemistry* (Ed. Behar, F. & Durand, B.) Org. Geochem. 16, 4–6, 735–747.

KINSMAN, D.J.J. (1964) The recent carbonate sediments near Halat el Bahrani, Trucial coast, Persian Gulf. In: *Deltaic and Shallow Marine Deposits* (Ed. Van Straaten, L.M.J.U.) pp. 185–192. Developments in Sedimentology I, Elsevier, Amsterdam.

KINSMAN, D.J.J. (1966) Gypsum and anhydrite of recent age, Trucial coast, Persian Gulf. In: *Second Symp. on Salt*, N. Ohio Geol. Soc., Cleveland, Ohio, **1**, 303–326.

KINSMAN, D.J.J. & PARK, R.K. (1976) Algal belt and coastal sabkha evolution, Trucial coast, Persian gulf. In: *Stromatolites* (Ed. Walter M.H.) pp. 421–433. Developments in Sedimentology, 20, Elsevier Amsterdam.

LOREAU, J.P. (1982) *Sédiments Aragonitiques, Leur Genèse*. Mém. Museum Nat. Hist. Nat. Série C, Paris XLVII, 311 pp.

MCKENZIE, J.A. (1976) *Isotope Study of the Hydrology and the Coexisting Carbonate Phases from Site of Recent Dolomitization, the Coastal Sabkha of Abu Dhabi, Persian Gulf*. PhD Dissertation Eidegenoessischen Technischen Hochschule Zurich, 128 pp.

MCKENZIE, J.A. (1981) Holocene dolomitization of calcium carbonate sediments from the coastal sabkhas of Abu Dhabi UAE: a stable isotope study. *J. Geol.* 89, 185–198.

McKenzie, J.A., Hsü, K.J. & Schneider, J.F. (1980) Movement of surface waters under the sabkha, Abu Dhabi, UAE, and its relation to evaporative dolomite genesis. In: *Concepts and Models of Dolomitization* (Ed. Zenger, D.H., Dunham, R.J. & Ethington, R.L.) Soc. Econ. Paleont. Mineral., Spec. Publ. 28, 11–30.

Mélières, F. (1973) Porte-échantillon tournant pour analyse par diffractométrie X. *Bull. Soc. Franç. Miné. Crist.* **96**, 75–79.

Park, R.K. (1977) The preservation potential of some recent stromatolites. *Sedimentology* **24**, 485–506.

Patterson, R.J. & Kinsman, D.J.J. (1977) Marine and continental ground water sources in a Persian Gulf sabkha. In: *Reefs and Related Carbonates—Ecology and Sedimentology* (Ed. Frost, J.G., Weiss, M.P. & Saunders, J.G.) Am. Ass. Petrol. Geol. Stud. Geol. **4**, 381–390.

Patterson, R.J. & Kinsman, D.J.J. (1981) Hydrologic framework of a sabkha along Arabian Gulf. *Bull. Am. Ass. Petrol. Geol.* **65**, 1457–1475.

Patterson, R.J. & Kinsman, D.J.J. (1982) Formation of diagenetic dolomites in coastal sabkha along Arabian (Persian) Gulf. *Bull. Am. Ass. Petrol. Geol.* **66**, 28–43.

Plaziat, J.C., Baltzer, F., Prieur, A. & Purser, B.H. (1987) Identification de Paléomangroves à Avicennia dans l'Holocène du Golfe Arabo-Persique. *1er Congrès Français de Sédimentologie, Résumés,* ASF, Paris, Nov. 1987, p. 282.

Purser, B.H., Huc, A.Y. & Kenig, F. (1987) *The Nature and Distribution of Organic Material Within the Recent Carbonate Sediments of the Abu Dhabi Lagoons and Sabkhas: a Preliminary Study*. Rapport Inst. Français du Pétrole, réf. 34940, 33 pp.

Schoeller, M. (1966) La concentration chimique des eaux souterraines dans les deltas. *Dacca Symp. on Scientific Problems of the Humid Tropical Zone Deltas,* 1964, Unesco, Paris, 169–173.

Shinn, E.A. (1968) Selective dolomitization of Recent sedimentary structures. *J. Sedim. Petrol.* 38, 612–616.

Shinn, E.A., Ginsburg, R.N. & Lloyd, R.M. (1965) Recent supratidal dolomite from Andros Island, Bahamas. In: *Dolomitization and Limestone Diagenesis* (Ed. Pray, L.C. & Murray, R.C.). Soc. Econ. Paleont. Mineral. Spec. Publ. 13, 112–123.

Wells, A.J. (1962) Recent dolomite in the Persian Gulf. *Nature* **194**, 274–275.

Wells, A.J. & Illing, L.V. (1964) Present-day precipitation of calcium carbonate in the Persian Gulf. In: *Deltaic and Shallow Marine Deposits* (Ed. Van Straaten, L.M.J.U.) pp. 429–435. Developments in Sedimentology I, Elsevier, Amsterdam.

Spec. Publs Int. Ass. Sediment. (1994) **21**, 429–445

Burial dolomitization of organic-rich and organic-poor carbonates, Jurassic of Central Tunisia

M. SOUSSI* *and* A. M'RABET†

Université de Tunis II, Faculté des Sciences, Département de Géologie, Laboratoire de Géologie du Pétrole et Bassins Sédimentaires, Campus Universitaire 1060, Tunis, Tunisia

ABSTRACT

Sedimentological, isotopic, geochemical and fluid-inclusion studies of Jurassic organic-rich and organic-poor carbonates from the Nara Formation in Central Tunisia have led to the definition of three groups of dolomites. The Lower Nara dolomites (Liassic) correspond to peritidal and bioclastic deposits; they are depleted in heavy oxygen ($\delta^{18}O$ of −7‰ PDB) and impoverished in strontium (30 ppm) and contain fluid inclusions whose temperature of homogenization has reached 120°C. The Middle Nara dolomites (Toarcian) are organic-rich, ferroan and relatively enriched in strontium (110 ppm); they have light oxygen and carbon isotopic signatures ($\delta^{18}O$ of −3 to −12‰ PDB and $\delta^{13}C$ of −2 to +1.5‰ PDB). $\delta^{13}C$ values which negatively correlate with TOC indicate that part of the carbon was derived from organic matter, either during early bacterial reduction of sulphates or later during deep thermal decarboxylation. The Upper Nara dolomites (Malm) correspond to hemipelagic carbonates and are impoverished in ^{18}O ($\delta^{18}O$ of −5‰ PDB) and Sr (60 ppm). These Nara dolomites formed in the subsurface, during shallow to deep burial and under increasing temperatures. Dolomitizing fluids seem to have been multiple in origin and include interstitial (connate) seawater, waters expelled from neighbouring shales and updip-flowing basinal marine water. Dolomitization and associated recrystallization started during the Lower Cretaceous, and were completed by the end of the Upper Cretaceous. The massive dolomites have been affected by post-Cretaceous fractures, the latter having been cemented by hydrothermal saddle dolomite. This study confirms that aggradational burial dolomitization by warm and saline waters can be an important process in dolomite genesis, and that organic matter may be an active controlling factor of dolomitization in anoxic environments.

INTRODUCTION

Several studies have demonstrated that dolomitization of carbonate rocks may occur in various environments, either during near-surface early diagenesis or during later burial (Zenger & Dunham, 1980; M'Rabet, 1981). However, the controlling factors of dolomitization and their roles are variable and not always well established. Among these factors, the origin of dolomitizing fluids is still controversial. During the 1970s, the mixing-zone model (Hanshaw *et al.*, 1971; Badiozamani, 1973) was very popular and applied in many cases, sometimes without supporting data. A critical review of the mixing-zone model (Hardie, 1987) has minimized its importance as a major mechanism. As alternatives, seawater models are being developed (Land, 1985, 1991; Tucker & Wright, 1990; and others) in spite of unresolved questions (e.g. kinetics, dehydration of Mg). Furthermore, the effect of temperature, especially during burial dolomitization of ancient carbonates, is being reconsidered (Mattes & Mountjoy, 1980; Hardie, 1987), especially as a result of the

* Present address: Université de Sfax, Faculté des Sciences, Département de Géologie, 3038 Sfax, Tunisia.
† Present address: ETAP, 27 bis Avenue Khereddine Pacha, 1002 Tunis, Tunisia.

laboratory synthesis of hydrothermal dolomites (Katz & Matthews, 1977; Kastner, 1984). On the other hand, the influence of organic matter during dolomitization, except in a few case studies (Kelts & McKenzie, 1984; Slaughter & Hill, 1991), is still poorly documented.

This multidisciplinary study of Jurassic organic-rich and organic-poor dolomites in Central Tunisia confirms the importance of burial dolomitization induced by both seawater-derived fluids and increasing temperatures in producing massive dolomite bodies and, in addition, demonstrates the role of organic matter during dolomitization of organic-rich carbonates.

METHODS

Petrographic and cathodoluminescence study of thin sections has been supplemented by scanning electron microscope observation of fresh fracture surfaces. The mineralogy of the dolomites has been determined by X-ray diffraction, whereas major and trace elements (Ca, Mg, Fe, Mn, Sr and Na) have been quantified by atomic absorption spectrophotometry. In addition, quantitative spot analysis of these elements has been performed by electron microprobe (Link system). Carbon and oxygen isotope ratios were measured by mass spectrometry from the carbon dioxides liberated by prolonged attack (8 days) using 100% phosphoric acid (Fontes *et al.*, 1970). ^{18}O and ^{13}C contents are expressed with respect to the PDB standard. $\delta^{18}O$ ratios were corrected according to Sharma and Clayton (1965). The temperature of homogenization and salinity of fluid inclusions have been measured by microthermometry. Finally, total organic carbon (TOC) contents and T_{max} of the organic-rich dolomites have been determined using Carmograph and Rock Eval, respectively.

GEOLOGICAL SETTING

The Jurassic rocks of Central Tunisia crop out along the so-called north–south axis (Fig. 1). They comprise the Nara Formation (Burollet, 1956), which is tectonically underlain by Triassic evaporites (Rheouis Formation) and conformably overlain by Early Cretaceous deep-marine to prodeltaic shales (Sidi Khalif Formation; M'Rabet, 1984).

The Nara Formation is subdivided into three Members (Fig. 2). The Lower Nara, up to 200 m thick and assigned to Upper Triassic–Lower Liassic (up to Carixian), consists mainly of dolomitized tidalites; only the uppermost part is composed of dolomitized bioclastic carbonates (Soussi, 1990). The Middle Nara, up to 50 m thick, includes three lithologic units: Lower-Middle Toarcian black shales unconformably overlain by Bajocian ironstones, and Lower-Middle Bathonian hemipelagic marls and mud-wackestones. The Upper Nara, up to 300 m thick and dated as Upper Callovian to Lower Tithonian in age (Soussi *et al.*, 1991), is composed of massively dolomitized pelagic carbonates.

The depositional environments of these Jurassic rocks and their palaeogeographic framework may be summarized as follows. The Liassic Lower Nara facies were deposited on a very shallow-marine restricted carbonate platform environment with numerous tidal flats. Prior to the Toarcian, the Liassic platform was fractured, dislocated and drowned, and then outer shelf hemipelagic to pelagic carbonate sedimentation ensued. Organic-rich carbonates were deposited in small anoxic tilt-block basins during the Toarcian Oceanic Anoxic Event (Soussi *et al.*, 1990; Soussi, 1990).

PROPERTIES AND ENVIRONMENTS OF FORMATION OF DOLOMITES

According to their field, petrographic, mineralogical, geochemical and isotopic properties, the Jurassic dolomites are divided into three major groups (Fig. 2): Lower Nara dolomites, Middle Nara organic-rich dolomites and Upper Nara dolomites. In addition, Lower and Upper Nara dolomites are affected by several generations of fractures; some of them are filled by dolomite cement.

Lower Nara dolomites

Occurrence

The major part of the Lower Nara consists of centimetric to decimetric dolomitized peritidal beds. The latter are organized into repetitive inter-/supratidal cycles exhibiting microbial lamination, mud cracks and desiccation breccias, birdseyes, dissolved evaporite micronodules and early vadose internal sediments (Fig. 3A,B,C). The upper part of the Lower Nara is composed of metre-thick bedded and fossiliferous (bivalves, gastropods and

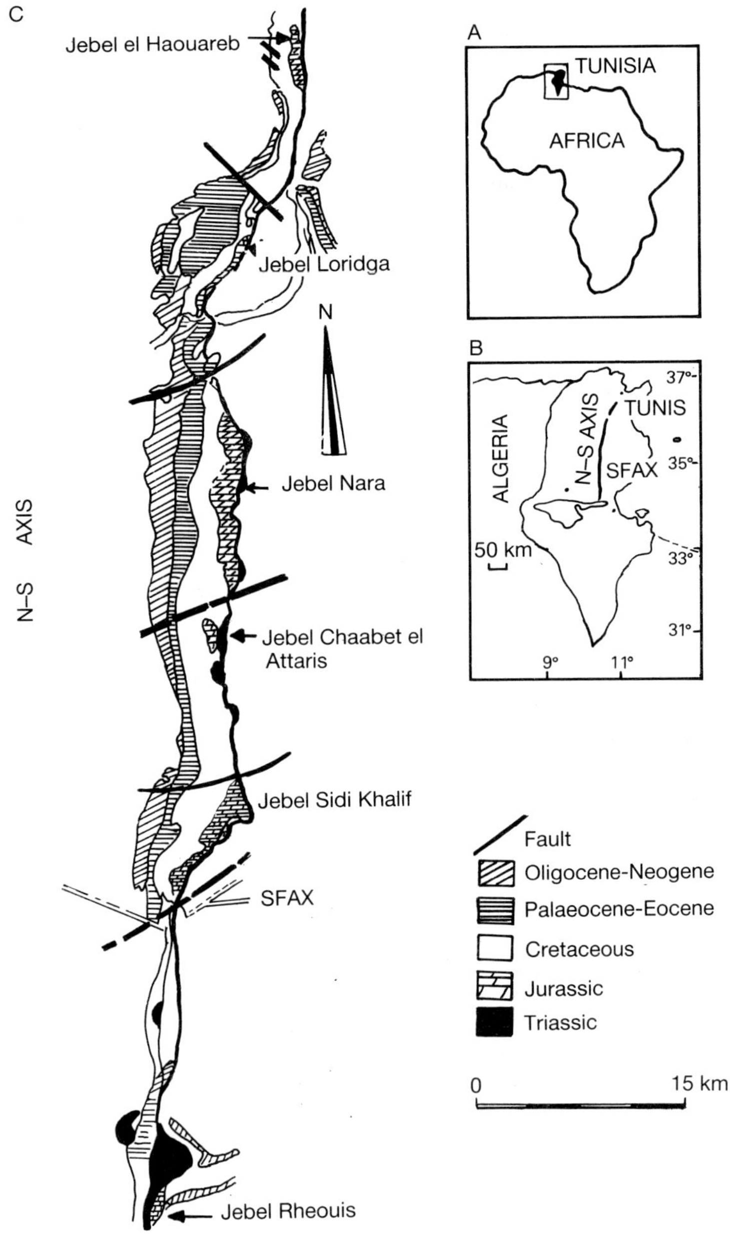

Fig. 1. Location maps of the area studied.

ammonites) dolomites, indicating a subtidal depositional environment.

Petrography

These Liassic dolomites are fine- to coarse-grained. Petrographic study of more than 500 thin sections has led to the distinction of four types of dolomite crystal (petrotypes) according to their size, shape and crystal boundaries.

1 Petrotype 1. This consists of very fine dolomite crystals, less than 20 μm in size (dolomicrite to dolomicrosparite). This petrotype resulted from mimetic replacement and preserved the original micritic

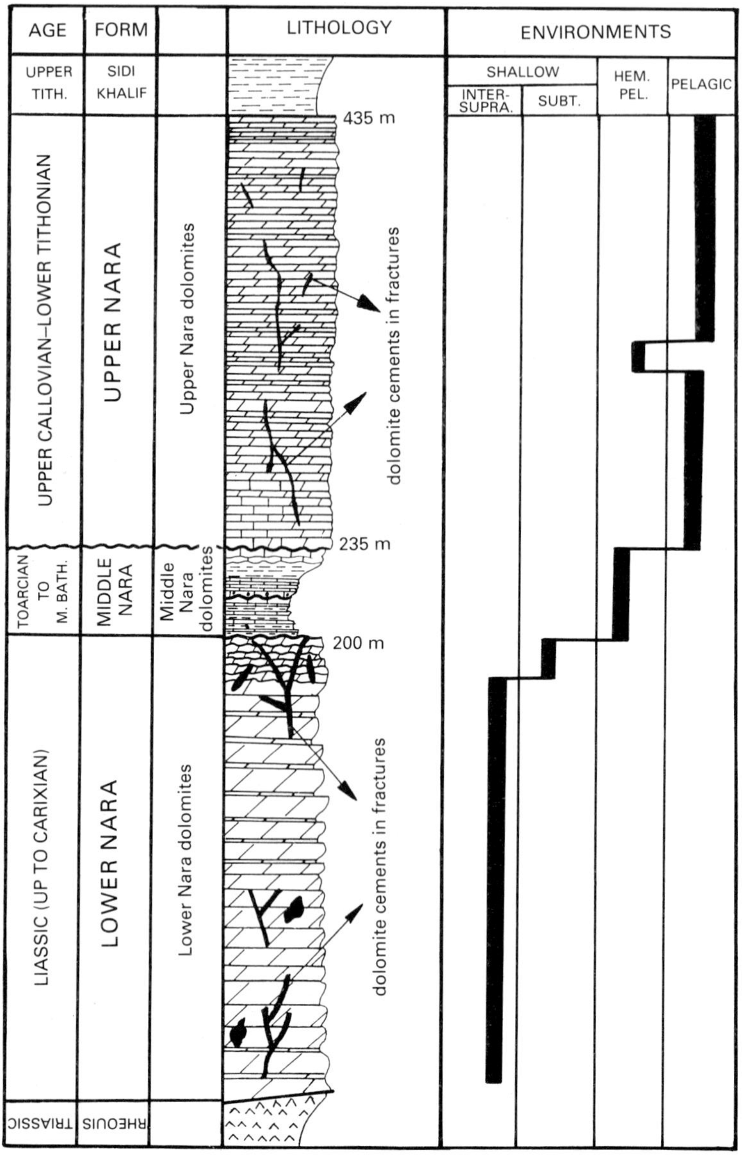

Fig. 2. Stratigraphic succession of the Jurassic Nara rocks in Central Tunisia, showing the lithology, the depositional environments and the distribution of the studied dolomites: the Liassic Lower Nara dolomites, the Toarcian Middle Nara dolomites, the Malm Upper Nara dolomites and the dolomite cements in fractures affecting both Lower and Middle Nara dolomites.

texture as well as the geopetal structure of birdseye fillings (Fig. 3A,B,C).

2 Petrotype 2. This consists of subhedral to anhedral dolosparitic rhombs, 70–150 μm in size, and corresponds to the idiotopic texture of Gregg and Sibley (1984). The dolomite crystals have a cloudy centre surrounded by limpid rims. This petrotype probably formed by overgrowth of petrotype 1, with partial destruction of the original texture.

3 Petrotype 3. This is represented by coarse anhedral dolomite crystals up to 900 μm in size, with compromise boundaries and undulatory extinction. This petrotype corresponds to the xenotopic A texture (Gregg & Sibley, 1984) and includes solid and fluid inclusions (Fig. 3D). It could be formed by non-mimetic replacement, by neomorphism (of petrotypes 1 and 2) and/or under high temperature conditions.

4 Petrotype 4. This is a dolosparitic cement partly to completely filling birdseyes (Fig. 3B), mud cracks, dissolved evaporite micronodules (Fig. 3C) and breccia porosity. This petrotype con-

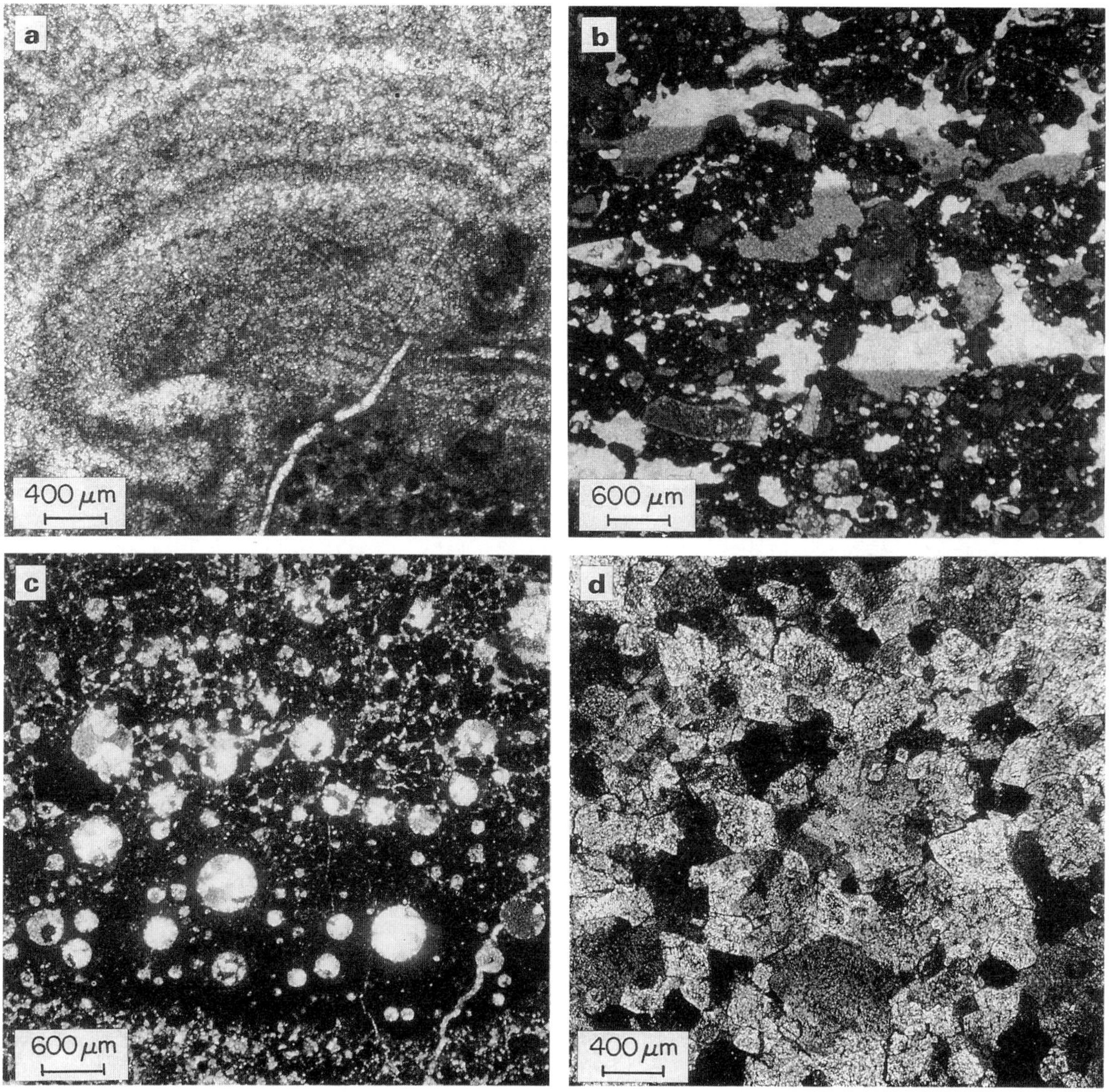

Fig. 3. Petrography of Lower Nara dolomites. (a) Petrotype 1: dolomicrite with preserved stromatolitic texture. (b) Birdseyes filled with dolomicritic internal sediment (petrotype 1) and dolomite cement (petrotype 4, light-coloured). (c) Dissolved evaporitic micronodules cemented by dolosparite (petrotype 4). (d) Mosaic of coarse dolomite crystals (petrotype 3).

sists of white and limpid euhedral crystals, 200–900 μm in size, which are generally zoned under cathodoluminescence.

The distribution of these four petrotypes is not uniform. Within the peritidal deposits of the Lower Nara, the brecciated horizons are composed mainly of petrotypes 1 and 4. The microbial laminated dolomites include petrotypes 1, 2 and 3, whereas the birdseye vugs and mud cracks are filled by petrotypes 1, 2 and 4. The bioclastic dolomites of the upper part of the Lower Nara are mainly, if not exclusively, composed of petrotype 3. This distribution seems to be controlled, at least in part, by original sedimentary textures and associated porosity. However, the coexistence of petrotypes 1, 2 and 3 in a given sample may be interpreted as the paragenetic sequence comprising nucleation, overgrowth and, possibly, neomorphism.

Table 1. Mineralogical, geochemical and isotopic properties of the Liassic Lower Nara dolomites.

Sample no	Mineralogy	$CaCO_3$ (mol %)	$MgCO_3$ (mol %)	Fe (ppm)	Mn (ppm)	Na (ppm)	Sr (ppm)	$\delta^{13}C$ (‰ PDB)	$\delta^{18}O$ (‰ PDB)
CA10	Dolomite	50.0	45.0	552	48	56	20	+1.7	−6.1
CA20a	Calcian dolomite	56.0	43.0	458	52	283	22	+1.5	−6.8
CA21a	Calcian dolomite	56.0	43.0	821	38	370	26	+1.6	−6.7
CA30	Dolomite	51.0	41.0	751	78	112	26	+1.7	−7.1
CA39	Calcian dolomite	54.0	45.0	851	100	146	48	+2.0	−6.8
CA45b	Calcian dolomite	58.0	41.0	1754	244	192	50	+1.0	−6.8
CA50d	Calcian dolomite	58.0	41.0	2017	252	61	28	+0.7	−6.8
CA403	Calcian dolomite	54.0	46.0	nm	50	74	nm	+1.3	−5.2
KH7	Calcian dolomite	54.0	45.0	1149	152	181	nm	nm	nm
KH10	Calcian dolomite	53.0	46.0	853	70	78	nm	nm	nm
KH13	Calcian dolomite	56.0	44.0	255	120	2025	nm	nm	nm
KH22	Calcian dolomite	55.0	45.0	1265	116	71	nm	nm	nm
KH24	Calcian dolomite	59.0	39.0	5510	142	59	nm	nm	nm
KH25	Calcian dolomite	60.0	40.0	1010	64	95	nm	nm	nm
KH47	Calcian dolomite	55.0	42.0	8290	382	61	nm	nm	nm
SK1	Calcian dolomite	53.0	47.0	836	78	363	nm	nm	nm
SK5	Calcian dolomite	53.0	47.0	810	96	404	nm	nm	nm
SK25	Calcian dolomite	54.0	46.0	797	94	121	22	+1.4	−7.0
SK39	Dolomite	51.0	44.0	896	72	326	34	+2.0	−6.8

nm, Not measured.

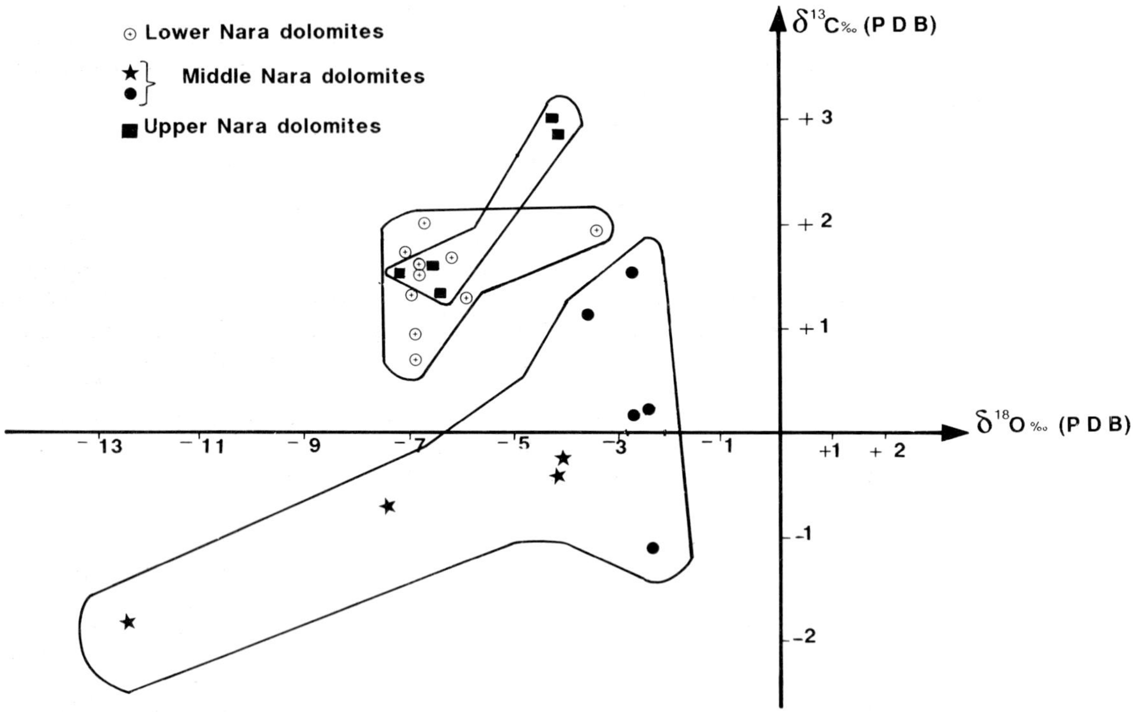

Fig. 4. ^{18}O and ^{13}C contents of the Liassic Lower Nara dolomites, the Toarcian Middle Nara dolomites and the Malm Upper Nara dolomites.

Mineralogy, geochemistry and isotopes

Lower Nara dolomites are calcian and non-ferroan (Table 1). $CaCO_3$ varies from 53 to 59 mole % and $MgCO_3$ from 39 to 47 mole %. Total iron content corresponds to less than 0.5 mole % $FeCO_3$. Strontium contents vary from 20 to 50 ppm, averaging 30 ppm. Mn contents are less than 270 ppm, and Na contents vary from 60 to 2025 ppm. These dolomites have homogeneous oxygen and carbon isotopic signatures. $\delta^{18}O$ varies from −6.1 to −7.1‰ (PDB), and $\delta^{13}C$ varies from +0.7 to +2‰ (PDB) with an average of +1.4‰ (Table 1; Fig. 4). In addition, fluid inclusions in petrotype 3 dolomite crystals are two-phase and derived from homogeneous saline waters (19–25 NaCl equivalents), with temperatures of homogenization varying from 113 to 128°C.

Milieu of dolomitization

According to their sedimentary structures (microbial lamination, mud cracks, birdseyes and dissolved evaporitic micronodules) and early diagenetic fabrics (vadose internal sediments), the peritidal Lower Nara dolomites could have been formed, at least in part, in an inter-/supratidal environment comparable to that described from the Arabo-Persian Gulf (McKenzie *et al.*, 1980). However, this interpretation cannot be applied to the upper bioclastic dolomites of the Lower Nara and, furthermore, is not supported by geochemical data. These dolomites have a low average Sr content (30 ppm) and light oxygen isotopic composition ($\delta^{18}O$ of −6.1 to −7.1‰ PDB). In particular, the impoverishment in ^{18}O may be due to the influence of meteoric waters, to a temperature effect or to a combination of both (Fontes *et al.*, 1970; Friedman & O'Neil, 1977; Land 1980). On the other hand, the fluid inclusions indicate that dolomitization of the Lower Nara occurred under high-temperature (nearly 120°C) conditions associated with saline waters. Thus, the oxygen isotopic signature almost certainly reflects a temperature effect rather than a freshwater influence. This means that Lower Nara dolomites have formed through burial diagenesis. In this case, the non-ferroan character of the dolomites is not related to near-surface oxidizing conditions, but rather to the absence of iron in the dolomitizing fluids. In addition, the wide range of Na content has no special significance since sodium cannot be regarded as a reliable salinity indicator of dolomitizing fluids (Mattes & Mountjoy, 1980; M'Rabet, 1981).

In summary, the overall data support burial dolomitization induced by hydrothermal saline waters. This may not exclude an early inter-/supratidal genesis of some of the Lower Nara metastable dolomites, which could have been neomorphosed in the subsurface under increasingly high-temperature conditions, as suggested by the paragenetic sequence.

Middle Nara organic-rich dolomites

Occurrence

These occur within the Toarcian part of the Middle Nara as decimetric laminated beds alternating with wackestones and shales. These hemipelagic facies contain ammonites, belemnites, fish skeletons and rare benthic fossils. They are locally rich in organic matter (up to 4.1% TOC) and pyrite (Soussi *et al.*, 1990).

Petrography

The Middle Nara dolomites have fine textures including two dolomite petrotypes, similar to petrotypes 1 and 2 in the Lower Nara dolomites. Petrotype 1 consists of anhedral dolomicritic to fine dolomicrosparitic crystals. Petrotype 2 corresponds to anhedral to subhedral rhombs having cloudy centres and limpid rims.

Mineralogy, geochemistry and isotopes

These dolomites are calcian and slightly ferroan (Table 2). $CaCO_3$ varies from 56 to 58 mole %, $MgCO_3$ from 40 to 43 mole % and $FeCO_3$ from 0.9 to 2.2 mole %. Strontium contents vary from 62 to 136 ppm, averaging 110 ppm. Sodium values have a wide range (103–919 ppm) and are not used in the following interpretation.

The oxygen and carbon isotopic compositions vary according to locality and the amount of organic matter. In localities where the Toarcian facies are organic-poor (<0.2% TOC; e.g. Jebel Sidi Khalif, Fig. 1), dolomites are relatively impoverished in heavy oxygen, with $\delta^{18}O$ values ranging from −3.6 to −2.8‰ (PDB), averaging −2.8‰; they are relatively enriched in ^{13}C, with $\delta^{13}C$ values varying from −1.1 to +1.5‰ (PDB), averaging +0.8‰. In localities where the Toarcian carbonates are organic-rich (TOC content ranging from 0.97 to 1.9%; e.g. Jebel Chaabet El Attaris, Figs 4 and 5), dolomites are impoverished both in heavy oxygen and carbon.

Table 2. Mineralogical, geochemical and isotopic properties of the Toarcian Middle Nara dolomites.

Sample no	Mineralogy	$CaCO_3$ (mol %)	$MgCO_3$ (mol %)	$FeCO_3$ (ppm)	Mn (ppm)	Na (ppm)	Sr (ppm)	$\delta^{13}C$ (‰ PDB)	$\delta^{18}O$ (‰ PDB)
CA51	Calcian & ferroan dolomite	58.0	41.0	1.9	880	148	120	−0.2	−4.1
CA52a	Calcian dolomite	59.0	40.0	0.9	nm	nm	nm	nm	nm
CA52b	Calcian & ferroan dolomite	55.0	44.0	1.2	nm	nm	nm	nm	nm
CA52c	Calcian & ferroan dolomite	58.0	41.0	1.9	nm	nm	nm	nm	nm
CA52d	Calcian & ferroan dolomite	56.0	42.0	2.1	nm	nm	nm	nm	nm
CA52e	Calcian & ferroan dolomite	58.0	40.0	2.2	nm	nm	nm	nm	nm
CA53	Calcian & ferroan dolomite	59.0	40.0	1.6	926	102	62	−0.5	−4.2
CA57	Calcian dolomite	nm	nm	nm	nm	183	78	−1.8	−12.3
CA58	Calcian & ferroan dolomite	61.0	36.0	1.2	nm	175	118	−0.7	−7.2
SK295	Calcian dolomite	nm	nm	nm	798	872	136	+1.1	−3.6
SK296	Calcian dolomite	60.0	39.0	nm	688	887	118	+1.5	−2.7
SK298	Calcian dolomite	59.0	40.0	nm	2724	879	74	+0.2	−2.8
SK300	Calcian dolomite	59.0	41.0	nm	3000	919	nm	−1.1	−2.4
SK302	Calcian dolomite	59.0	41.0	nm	2946	352	nm	+0.2	−2.5

nm, Not measured.

$\delta^{18}O$ values vary from −12 to −4.1‰ (PDB), averaging −6.9‰. $\delta^{13}C$ values are negative, and vary from −1.8 to −0.2‰ (PDB), averaging −0.7‰. It is worth noting that $\delta^{13}C$ values show a negative correlation with TOC (Fig. 6).

Milieu of dolomitization

The fact that the Toarcian Middle Nara dolomites are included in deep-marine facies and do not exhibit any near-surface or emergence features suggests subsurface dolomitization, during shallow or deep burial. This interpretation is supported by geochemical and isotopic data. The ferrous iron content (up to 2.2 mole % $FeCO_3$) and the presence of pyrite suggest reducing conditions. The relatively high strontium content, averaging 110 ppm, suggests a relatively closed system with interstitial saline waters comparable to seawater (Land, 1980; M'Rabet, 1981). In addition, $\delta^{18}O$ values, particularly those characterizing the organic-rich dolomites (with an average of −6.9‰ PDB) suggest a temperature effect. In this respect, Rock Eval pyrolysis of the associated organic matter indicates that the latter has reached the oil window (T_{max} ranging from 437 to 440°C; Soussi *et al.*, 1990; Soussi, 1990) at a depth of nearly 2000 m. Within the organic-poor dolomites, $\delta^{18}O$ values are less negative and average −2.8‰ (PDB). If these values are derived from marine dolomitizing fluids, they do not reflect a clear and pronounced temperature effect (Land, 1980). This suggests that dolomitization started early, within the first metres of burial.

On the other hand, the carbon isotopic ratios (especially those characterizing the organic-rich dolomites, with a $\delta^{13}C$ average of −0.8‰ PDB) and their negative correlation with TOC indicate that at least part of the carbon incorporated into the dolomites was probably derived from organic matter (the associated organic-rich limestones are not impoverished in ^{13}C; Fig. 6).

The derived organic carbon may be liberated either during shallow burial and bacterial reduction of sulphates and associated alkalinity increase (Irwin

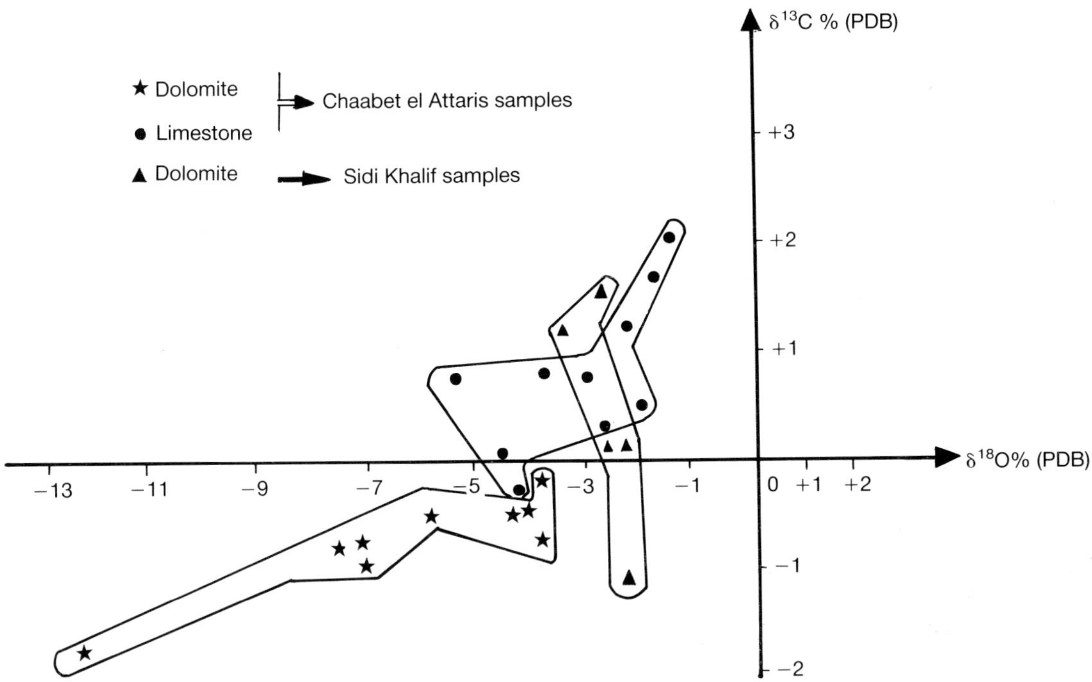

Fig. 5. ^{18}O and ^{13}C contents of the Toarcian Middle Nara dolomites and associated limestones.

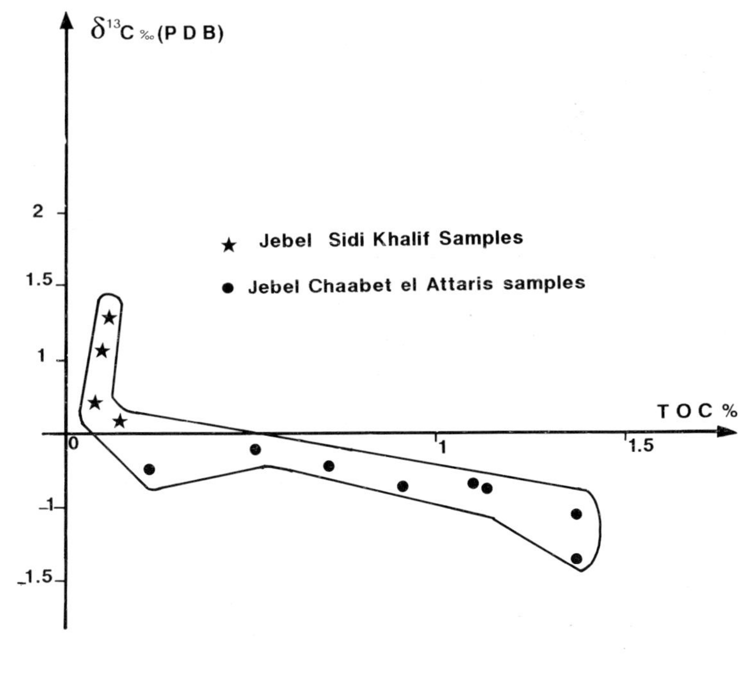

Fig. 6. Negative correlation between total organic carbon (TOC) and ^{13}C content, indicating that at least part of the carbon incorporated in the Toarcian Middle Nara dolomites was derived from the associated organic matter.

et al., 1977; Baker & Burns, 1985; Slaughter & Hill, 1991) represented in a simplified form by

$$2CH_2O + SO_4 \xrightarrow[\text{catalysis}]{\text{bacterial}} 2CO_2 \text{ (light } ^{12}C) + 2H_2O + S$$

or during deep burial and associated thermal decarboxylation (Irwin *et al.*, 1977) represented by

$$RCO_2H \xrightarrow[\text{temperature}]{\text{high}} RH + CO_2 \text{ (light } ^{12}C)$$

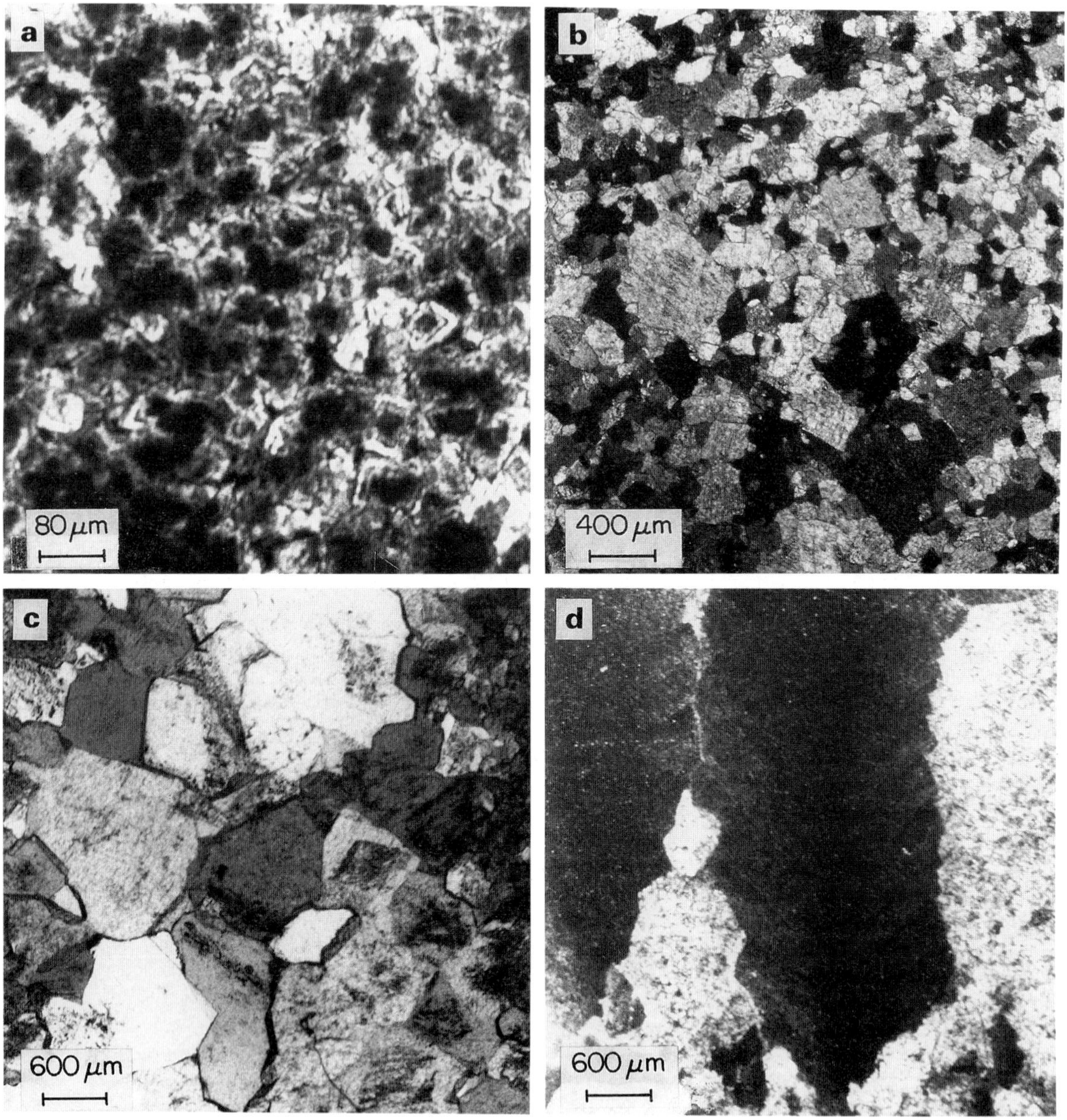

Fig. 7. Petrography of Upper Nara dolomites and dolomite cements in fractures. (a) Dolomite crystals (petrotype 2) showing cloudy centres and zoned limpid rims resulting from overgrowth; Upper Nara dolomites. (b) Mosaic of dolosparitic crystals (petrotype 3). This petrotype 3 probably resulted from neomorphism; Upper Nara dolomites. (c) Dolomite cement in a fracture. (d) Detail of saddle dolomite crystals showing rhombohedral terminations and undulatory extinction.

The derived organic carbon liberated as a CO_2 phase may, at least in part, have favoured the dissolution of the host limestone before reprecipitation of the dolomites. In addition, the organic matter may have acted as local support for fixed or adsorbed Mg (Kelts & McKenzie, 1984).

In summary, the Toarcian Middle Nara dolomites formed under shallow to deep burial conditions associated with increasing temperatures and interstitial waters of near-marine composition.

Upper Nara dolomites

Occurrence

The Upper Nara carbonates consist of decimetric to metric dolomitic beds associated very locally with original limestones (mud-wackestones) and containing pelagic fossils (ammonites, belemnites, crinoids, protoglobigerines and radiolaria) indicating a deep-marine depositional environment; they are conformably overlain by the deep-marine Sidi Khalif shales (M'Rabet, 1984).

Petrography

The Upper Nara dolomites are generally medium- to coarse-grained dolosparites comprising three petrotypes. Petrotype 1, relatively rare, is represented by fine dolosparitic crystals less than 75 μm in size. Petrotype 2 consists of coarser subhedral crystals (100 μm) having cloudy centres surrounded by limpid overgrowth rims (Fig. 7A) and corresponding to the idiotopic E and S textures (Gregg & Sibley, 1984). Petrotype 3 is represented by coarse dolosparitic crystals more than 200 μm in size (xenotopic A texture). The crystals include numerous solid impurities and locally have undulatory extinction (Fig. 7B). They may cut stylolites and are clearly post-compactional. They result either from non-mimetic replacement and/or from neomorphism of previous metastable dolomitic crystals.

Mineralogy, geochemistry and isotopes

The Upper Nara dolomites are calcian and non-ferroan. $CaCO_3$ varies from 52 to 60 mole % and $MgCO_3$ from 39 to 47 mole %. Strontium contents range from 26 to 130 ppm, averaging 60 ppm, and sodium and manganese concentrations are less than 200 ppm and 2000 ppm, respectively (Table 3). These dolomites have $\delta^{13}C$ values ranging from +1.4 to +3.4‰ (PDB), averaging +2.0‰ (PDB). Their $\delta^{18}O$ values vary from −7.1 to −2.4‰ (PDB) and average −5.2‰ (PDB) (Fig. 4).

Milieu of dolomitization

The Upper Nara carbonates have been deposited generally in a deep-marine environment. They are conformably overlain by the pelagic Sidi Khalif shales and do not exhibit any feature suggesting early emergence. These sedimentological attributes strongly suggest burial dolomitization. The abundant petrotype 3 (which is locally postcompactional) and related xenotopic texture resulted either from neomorphism and/or from non-mimetic replacement under high-temperature conditions (Gregg & Sibley, 1984). In this respect, the light oxygen isotopic ratios (with a $\delta^{18}O$ average of −5.2‰ PDB) support a relatively high-temperature effect rather than a meteoric influence. It is to be recalled that

Table 3. Mineralogical, geochemical and isotopic properties of the Malm Upper Nara dolomites.

Sample no	Mineralogy	$CaCO_3$ (mol %)	$MgCO_3$ (mol %)	Fe (ppm)	Mn (ppm)	Na (ppm)	Sr (ppm)	$\delta^{13}C$ (‰ PDB)	$\delta^{18}O$ (‰ PDB)
CA83	Calcian dolomite	nm	nm	4010	508	92	26	+2.9	−4.1
CA86a	Calcian dolomite	59.0	39.0	nm	1936	88	50	+1.4	−2.4
CA87a	Calcian dolomite	58.0	41.0	2670	456	54	28	+1.6	−7.1
CA87j	Calcian dolomite	52.0	47.0	278	114	154	130	+3.0	−4.4
CA87h	Calcian dolomite	59.0	39.0	4065	660	179	nm	nm	nm
CA88	Calcian dolomite	59.0	40.0	3126	716	196	68	+1.3	−6.4
CA92	Calcian dolomite	60.0	39.0	3032	646	72	44	+1.6	−6.6
KH82	Calcian dolomite	57.0	40.0	9310	nm	83	nm	nm	nm
KH89	Calcian dolomite	57.0	40.0	6220	nm	81	nm	nm	nm
KH106	Calcian dolomite	58.0	39.0	7382	nm	59	nm	nm	nm
OG203	Calcian dolomite	54.0	45.0	2479	300	220	44	+1.7	−5.7

nm, Not measured.

this oxygen isotopic signature is comparable to that characterizing Lower and Middle Nara dolomites, for which a temperature effect has been demonstrated by the study of fluid inclusions in the Lower Nara and the pyrolysis of the associated organic matter in the Middle Nara. Concerning the salinity of dolomitizing fluids, the Sr contents range from 26 to 130 ppm and do not reflect a typical marine origin. In fact, these properties could be related to neomorphism (M'Rabet, 1981).

In summary, the Upper Nara dolomites have been formed during burial by warm fluids derived from interstitial formation waters, in addition to probable Mg-rich waters expelled by compaction from adjacent Middle Nara and Sidi Khalif shales.

Dolomite cements in fractures

Generations of fractures and associated fillings

The Jurassic rocks cropping out along the north–south axis are affected by a complex fault system and associated fractures oriented east–west, north–south and northwest–southeast. The fractures are more abundant within Lower and Upper Nara dolomites. According to their filling relationships, three generations of fractures are recognized. Generation 1 fractures may be completely cemented by white dolomite; they are locally associated with dissolution cavities also cemented by white dolomite; these are believed to be relatively early in origin. Generation 2 fractures are partly or completely filled with galena and baryte. They cut, displace and thus postdate the white dolomite-cemented fractures. Generation 3 fractures are still open, and clearly late in origin.

Petrography

Dolomite cement is composed of limpid crystals, up to 3 mm in size, elongated along their *c* axes and often having rhombic terminations (Fig. 7C,D). They are rich in fluid inclusions and particularly characterized by curved faces and undulatory extinction resembling saddle or baroque dolomite (Radke & Mathis, 1980). Locally, this saddle dolomite is zoned, the peripheral zones being more luminescent.

Mineralogy, geochemistry and isotopes

Dolomite cements are calcian and non-ferroan. $CaCO_3$ varies from 54 to 63 mole % and $MgCO_3$ from 36 to 46 mole %. Their strontium contents, based on seven analyses, range from 32 to 106 ppm, averaging 51 ppm. Manganese content is less than 950 ppm and sodium is erratic (Table 4). $\delta^{18}O$ values are negative, averaging −6.5‰ (PDB) and $\delta^{13}C$ values range from +1.0 to +2.1‰ (PDB) (Fig. 8). In addition, fluid inclusions give a temperature of homogenization of nearly 130°C, and suggest crystallization from saline waters (20 NaCl equivalents).

Conditions of dolomite cementation

It is clear that the saddle dolomite cementing generation 1 fractures postdates dolomitization of Lower

Table 4. Mineralogical, geochemical and isotopic properties of dolomites cements in fractures.

Sample no	Mineralogy	$CaCO_3$ (mol %)	$MgCO_3$ (mol %)	Fe (ppm)	Mn (ppm)	Na (ppm)	Sr (ppm)	$\delta^{13}C$ (‰ PDB)	$\delta^{18}O$ (‰ PDB)
CA59	Calcian dolomite	58.0	42.0	0	6	900	nm	+3.8	−6.0
CA401	Calcian dolomite	57.0	43.0	976	100	95	nm	+1.8	−6.9
CA402	Calcian dolomite	57.0	42.0	616	50	86	nm	+1.6	−6.8
CA403	Calcian dolomite	52.0	47.0	599	40	1965	nm	+1.6	−6.8
CA404	Calcian dolomite	nm	nm	nm	nm	nm	nm	+1.7	−6.2
CA405	Calcian dolomite	54.0	45.0	896	110	nm	nm	+1.6	−7.0
CADC1	Calcian dolomite	63.0	36.0	1963	460	nm	56	+1.1	−7.0
CADC2	Calcian dolomite	54.0	45.0	1572	300	163	32	+1.0	−7.7
CADC3	Calcian dolomite	54.0	45.0	1963	350	nm	106	+2.1	−7.1
OG203	Calcian dolomite	54.0	42.0	1612	950	69	26	+2.1	−7.1
OG204	Calcian dolomite	57.0	42.0	2794	850	130	38	+1.7	−6.0
OG207	Calcian dolomite	57.0	42.0	1234	800	104	32	+2.0	−7.3
DCOG1	Calcian dolomite	56.0	42.0	7130	800	nm	nm	nm	nm

nm, Not measured.

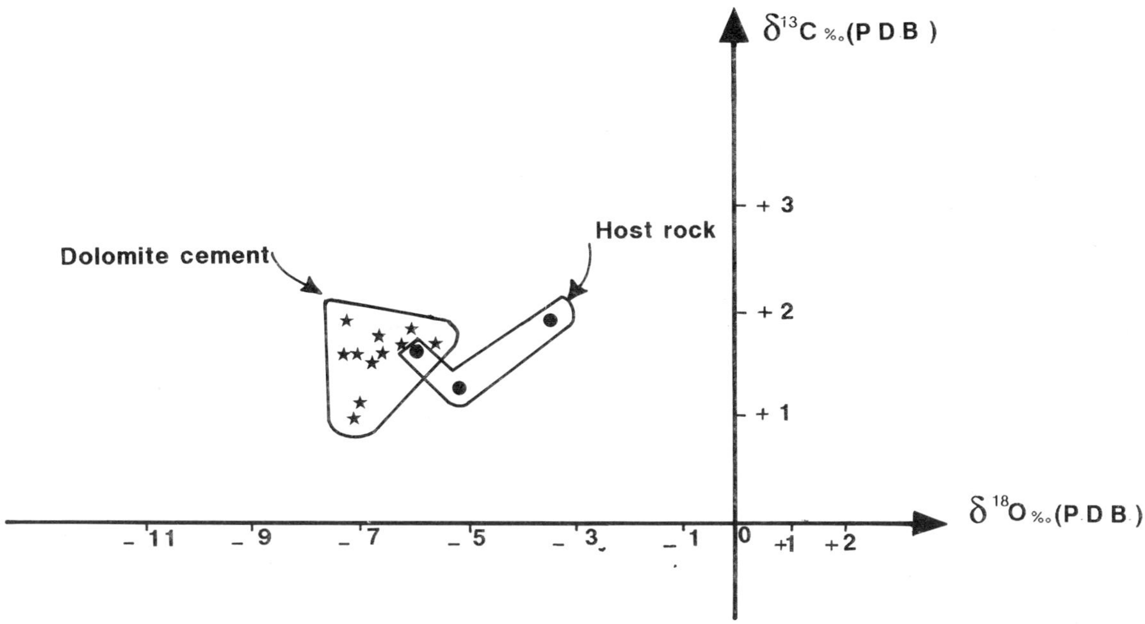

Fig. 8. ^{18}O and ^{13}C contents of the saddle dolomite cementing generation 1 fractures affecting the Lower and Upper Nara dolomites (host rock). The oxygen and carbon isotopic compositions of host rock are given for comparison.

and Upper Nara carbonates. Fluid-inclusion data indicate that the saddle dolomite has precipitated from warm and rather saturated saline waters within the fracture system. This interpretation is also supported by petrographic and geochemical data. Saddle dolomite with curved faces is considered to be a high-temperature fabric, often associated with ore deposits (Radke & Mathis, 1980). The observed luminescence zonation within this saddle dolomite suggests multiphases of cement from fluids having changing chemical composition. In addition, the light oxygen isotopic ratios, with a $\delta^{18}O$ average of −6.5‰ (PDB), reflect a high-temperature effect. A comparison of these oxygen isotopic signatures with those of the Lower and Upper Nara host dolomitic rocks (Tables 1, 3 and 4; Fig. 8) suggests that the fluids involved had similar characteristics.

DISCUSSION

Dolomite properties

The different dolomite petrotypes defined especially within the Lower and Upper Nara confirm that the petrographic expression of dolomitization is quite variable. Petrotype 1, which occurs within the microbial or clayey laminated dolomites of the Lower Nara, resulted from mimetic replacement with numerous sites of nucleation. It seems to be controlled mainly by the presumably poor porosity and permeability of the original fine-grained textures (Sibley, 1991). Chemistry of the dolomitizing solutions does not seem to be a first-order factor determining petrography. Petrotype 2 may reflect an advanced stage of dolomitization and more aggressive dolomitizing fluids. This advanced stage implies dissolution of the original calcium carbonates and/or metastable dolomitic nuclei followed by overgrowth of limpid rims. Petrotype 3, represented by coarse crystals with undulatory extinction and containing numerous dolomitic inclusions, may result either from non-mimetic replacement (including precursor dissolution and competition between repeated dolomite nuclei and overgrowths in porous rocks) or by recrystallization of former metastable dolomitic crystals under high-temperature conditions. The low strontium content of petrotype 3 observed in Upper Nara dolomites favours dolomite neomorphism. The decrease in strontium content and increase in dolomite crystal size due to neomorphism have been observed elsewhere, in the Ordovician–Silurian of

Central Nevada (Dunham & Olson, 1980) and the Lower Cretaceous of Central Tunisia (M'Rabet, 1981).

The geochemical properties and, especially, the oxygen isotopic signatures suggest temperature effects during subsurface dolomitization of Lower, Middle and Upper Nara carbonates. This interpretation is substantiated by fluid-inclusion analysis and Rock Eval pyrolysis data. On the other hand, the carbon isotopic signature, together with the negative correlation between $\delta^{13}C$ and TOC within the Toarcian organic-rich dolomites, demonstrates that dolomitization has been influenced by organic matter; the latter, acting as a donor of organic carbon, may be considered as one of the factors controlling dolomitization in anoxic environments.

Environments and timing of dolomitization

The Lower, Middle and Upper Nara dolomites were formed essentially in the subsurface under increasing burial temperature. However, part of the Liassic Lower Nara peritidal carbonates may have been dolomitized early in an inter-/supratidal to sabkha-like setting; the early-formed dolomites then have been affected by neomorphism during burial. This mesogenetic and multiphase dolomitization of Nara carbonates was followed by tectonic fracturing and subsequent dolomite cementation by hydrothermal solutions.

The timing of this burial dolomitization may be evaluated as follows. Fluid-inclusion study reveals that the temperature of crystallization of Lower Nara dolomites reached 120°C. If a Jurassic surface temperature of 30°C and a regional geothermal gradient of 40°C/1000 m (Ben Dhia, 1987) are assumed, a minimum depth of 2250 m is required in order to attain a burial temperature of 120°C (3000 m considering a normal geothermal gradient of 30°C/1000 m). Such a depth corresponds to the total decompacted thicknesses of the Middle and Upper Jurassic rocks and the overlying Lower and Upper Cretaceous mixtures of shales, sands and carbonates (assuming decompaction factors of 20% and 30% for carbonate-sands and shales, respectively).

Thus, the burial dolomitization of the Liassic Lower Nara facies may have begun by the end of the Lower Cretaceous if not during the Upper Cretaceous. As an alternative, if burial dolomitization was initiated at lower temperatures (say 80°C), as suggested by the experimental data of high-temperature dolomitization (Usdowski, 1968) and thermal cracking of the Toarcian organic matter, a minimum depth of 1300–1700 m is needed. In the latter case, dolomitization could have begun no earlier than the Lower Cretaceous.

In summary, burial dolomitization of the Jurassic Nara carbonates in the north–south axis area began locally as early as the Lower Cretaceous, and was completed by the end of the Upper Cretaceous. Later fracturing of Nara dolomites and subsequent cementation by saddle dolomite appear to be post-Cretaceous in age.

Origin of dolomitizing fluids

The origin of dolomitizing solutions, and especially of magnesium, is still debated in the literature. Since dolomitization of Jurassic carbonates was essentially by burial, it is worth pointing out that experimental data have demonstrated that the Mg/Ca ratio of the waters needed to dolomitize calcium carbonates decreases, becoming Ca-rich at some temperature below 100°C (Baron, 1960; Rosenburg *et al.*, 1967; Usdowski, 1968; Hardie, 1987). Consequently, any warm subsurface water (saline or more dilute) becomes a potential dolomitizing fluid. In the case of the Jurassic Nara carbonates, the origin of dolomitizing fluids seems to be multiple. The Middle Nara organic-rich limestones were converted to dolomites by interstitial marine waters sufficiently rich in Mg. The Lower and Upper Nara carbonates were dolomitized by the contribution of three possible types of water. Formation waters of marine origin contained within the pore system may have constituted the first available solutions for dolomitization. However, since their volume was limited, these warm *in situ* waters were not able to produce massive dolomites. Another potential contributor may have been Mg-rich waters expelled from adjacent shales during burial compaction (i.e. Neocomian Sidi Khalif shales and Middle Jurassic shales). Migration of the expelled solutions may have been vertical (*per ascensum* or *per descensum*; Magara, 1974) or, more likely, lateral. Such lateral migration was probably facilitated by the architecture of tilted blocks and associated boundary faults in Central Tunisia during the Jurassic and Cretaceous periods (Fig. 9). Evidently, the importance of regional mass transfer remains questionable; however, its ability to convert substantial quantities of Nara carbonate to dolomite was probably not negligible. In addition, basinal marine water could be a third origin of dolomitizing fluids. Garven (1985) noted that it

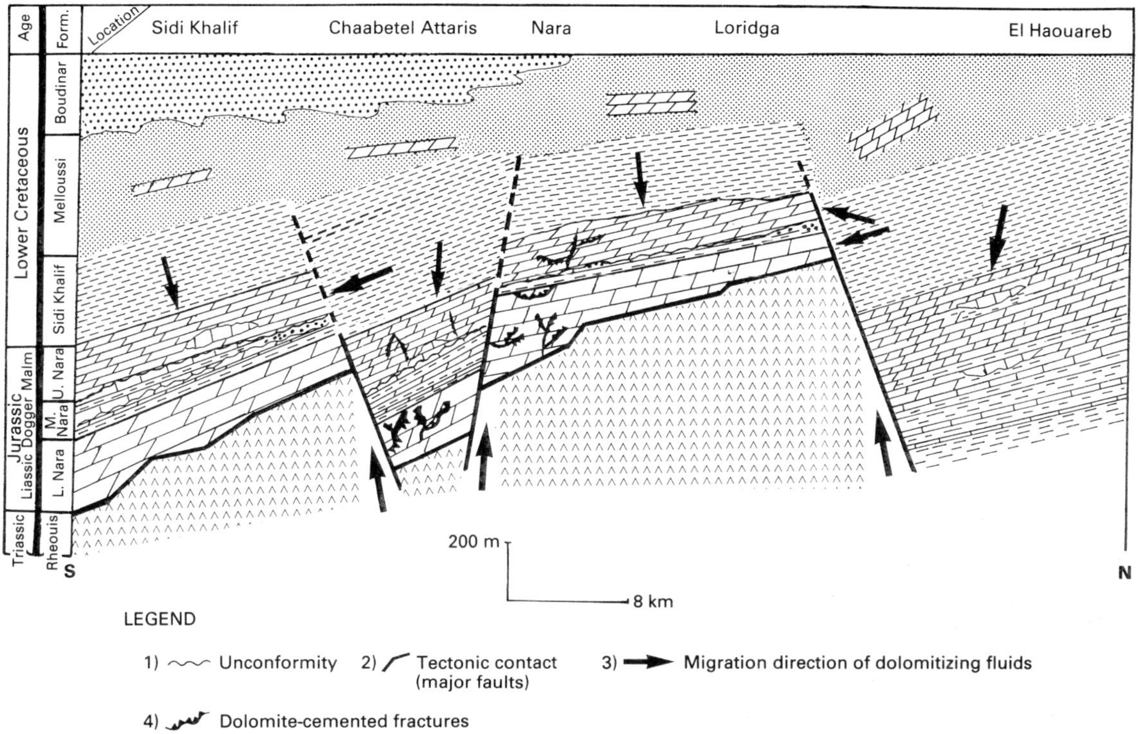

Fig. 9. Interpretative model of dolomitization of the Jurassic Nara carbonates in Central Tunisia. This model shows that the sedimentary and diagenetic frame of the Jurassic carbonates corresponds to tilted blocks separated by major faults and associated fracture system, and that the dolomitizing solutions were derived by vertical and lateral migration (arrows) from waters expelled from adjacent shales and basinal updip-flowing waters.

should be possible for basinal seawater to be driven updip for long distances towards the basin edge, and especially towards thermally anomalous topographic highs. This might have been the case in Central Tunisia, along the north–south axis which, during the Jurassic and Cretaceous, acted as a dislocated shelf characterized by low subsidence rates and a relatively high geothermal gradient (Ben Dhia, 1987). Such updip-flowing basinal waters were able to maintain the dolomitizing pump and, becoming warmer, might have played a major role in the massive dolomitization of Nara carbonates and the neomorphism of earlier shallow-burial metastable dolomites.

Finally, the saddle dolomite cement in fractures has most certainly been precipitated from hydrothermal and relatively saline solutions circulating along fault and fracture systems.

CONCLUSIONS

The Jurassic organic-rich and organic-poor dolomites in Central Tunisia resulted essentially from burial dolomitization. The latter was initiated during early shallow burial, especially for the Toarcian organic-rich carbonates, and continued at depth under higher temperatures (up to 120°C). This burial dolomitization and associated neomorphism were completed by the end of the Upper Cretaceous.

Dolomitizing solutions seem to have been warm, saline and multiple in origin; they include interstitial (connate) seawaters, waters expelled from adjacent shales and updip-flowing deep basinal marine waters. The latter may ensure a sufficient hydrological mass transfer for massive dolomitization.

The massive dolomites were affected by post-Cretaceous fractures which were cemented by hydrothermal saddle dolomite.

Dolomitization of the Toarcian organic-rich carbonates was influenced by organic matter. The latter has acted as a donor of organic carbon, either during shallow burial and bacterial reduction of sulphates or during deep burial and associated thermal decarboxylation. This demonstrates that organic matter may be considered as one of the factors controlling dolomitization in anoxic environments.

ACKNOWLEDGEMENTS

We would like to thank CONOCO Tunisia Ltd for financial support of this study. We extend our thanks to N. Ouazaa-Laaridhi (University of Tunis) for scientific help in the fluid-inclusion study, B.H. Purser and J. Ch. Fontes (University of Paris Sud, Orsay) for access to microprobe and isotope facilities, and ETAP Tunisia and INRST for organic geochemistry analysis. We are particularly grateful to the editors, B.H. Purser, M.E. Tucker and D.H. Zenger, who critically read the manuscript and suggested considerable improvements. Helpful comments from M. Kastner are also acknowledged.

REFERENCES

Badiozamani K. (1973) The Dorag dolomitization model – application to the Middle Ordovician of Wisconsin. *J. Sedim. Petrol.* **43**, 965–984.

Baker, P.A. & Burns, S. (1985) Occurrence and formation of dolomite in organic-rich continental margins. *Bull. Am. Ass. Petrol. Geol.* **69**, 1917–1930.

Baron, G. (1960) Sur la synthèse de la dolomite. Application au phénomène de dolomitisation. *Rev. Inst. Franc. Petrole Ann. Combust. Liquides* **15**, 3–68.

Ben Dhia, H. (1987) The geothermal gradient map of Central Tunisia: comparison with structural, gravimetric and petroleum data. *Tectonophysics* **142**, 99–109.

Burollet, P.F. (1956) Contribution à l'étude stratigraphique de la Tunisie centrale. *Ann. Mine. Geol. Tunis* **18**, 345 pp.

Dunham, J.B. & Olson, E.R. (1980) Shallow subsurface dolomitization of subtidally deposited carbonate sediments in the Hanson Creek Formation (Ordovician–Silurian) of Central Nevada. In: *Concepts and Models of Dolomitization* (Ed. Zenger, D.H., Dunham, J.B. & Ethington, R.L.) Spec. Publ. Soc. Econ. Paleont. Mineral., Tulsa **28**, 139–161.

Fontes, J.C., Fritz, P & Letolle R. (1970) Composition isotopique, minéralogique et genèse des dolomies du Bassin de Paris. *Geochim. Cosmochim. Acta* **34**, 279–294.

Friedman, I. & O'Neil, J.R. (1977) Compilation of stable isotope fractionation factors of geochemical interest. In: *Data of Geochemistry*, 6th edn. US Geol. Surv. Prof. Paper. 440 KK, 12 pp.

Garven, G. (1985) The role of regional fluid flow in the genesis of the Pine Point deposit, western Canada sedimentary basin. *Econ. Geol.* **80**, 307–324.

Gregg, J.M. & Sibley, D.F. (1984) Epigenetic dolomitization and the origin of xenotopic dolomite texture. *J. Sedim. Petrol.* **54**, 908–931.

Hanshaw, B.B., Back, W. & Deike, R. (1971) A geochemical hypothesis for dolomitization by groundwater. *Econ. Geol.* **66**, 710–724.

Hardie, L.A. (1987) Dolomitization: a critical view of some current views. *J. Sedim. Petrol.* **57**, 166–183.

Irwin, H., Curtis, C. & Coleman, M. (1977) Isotopic evidence for source of diagenetic carbonates formed during burial of organic-rich sediments. *Nature* **269**, 209–213.

Kastner, M.J (1984) Control of dolomite formation. *Nature* **311**, 410–411.

Katz, A. & Matthews, A. (1977) The dolomitization of $CaCO_3$: an experimental study at 252–295°C. *Geochim. Cosmochim. Acta* **41**, 297–308.

Kelts, K. & McKenzie J. (1984) A comparison of anoxic dolomite from deep sea sediments; Quaternary Gulf of California and Messinian Tripoli Formation of Sicily. In: *Dolomites of Monterey Formation and Other Organic-Rich Units* (Ed. Garrison, R.E., Kastner M. & Zenger, D.H.) Pacific Section, Soc. Econ. Paleont. Mineral., Tulsa **41**, 19–28.

Land, L.S. (1980) The isotopic and trace elements geochemistry of dolomite: the state of the art. In: *Concepts and Models of Dolomitization* (Ed. Zenger, D.H., Dunham, J.B. and Ethington, R.L.) Spec. Publ. Soc. Econ. Paleont. Mineral., Tulsa **28**, 19–28.

Land, L.S. (1985) The origin of massive dolomite. *J. Geol. Educ.* **33**, 112–125.

Land, L.S. (1991) Dolomitization models: seawater and mixing zones. In: *Dolomieu Conference on Carbonate Platforms and Dolomitization, Ortisei* (Ed. Bosellini, A., Brandner, R., Flugel, E., Purser, B., Schlager, W., Tucker, M. & Zenger, D.) 144 Abstracts.

McKenzie, J.A., Hsü, K.J. & Schneider, J.F. (1980) Movement of subsurface water under the sabkha, Abu Dhabi, UAE and its relation to evaporative dolomite genesis. In: *Concepts and Models of Dolomitization* (Ed. Zenger, D.H., Dunham, J.B. & Ethington, R.L.) Spec. Publ. Soc. Econ. Paleont. Mineral., Tulsa, **28**, 11–30.

Magara, K. (1974) Aquathermal fluid migration. *Bull. Am. Ass. Petrol. Geol.* **58**, 2513–2516.

Mattes, B.W. & Mountjoy, E.W. (1980) Burial dolomitization of the Upper Devonian Miette Buildup, Jasper National Park, Alberta. In: *Concepts and Models of Dolomitization* (Ed. Zenger, D.H., Dunham, J.B. & Ethington, R.L.) Spec. Publ. Soc. Econ. Paleont. Mineral Tulsa **28**, 259–297.

M'Rabet, A. (1981) Differentiation of environments of dolomite formation, Lower Cretaceous of Central Tunisia. *Sedimentology* **28**, 331–352.

M'Rabet, A. (1984) Neocomian deltaic complex in Central Tunisia: a particular example of ancient sedimentation and basin evolution. *Sedim. Geol.* **40**, 191–209.

Radke, B.M. & Mathis, R.L. (1980) On the formation and occurrence of saddle dolomite. *J. Sedim. Petrol.* **50**, 1149–1168.

Rosenburg, P.E., Burt, D.M. & Holland, H.D. (1967)

Calcite–dolomite–magnesite stability relation in solutions: the effect of ion strength. *Geochim. Cosmochim. Acta* **31**, 391–396.

SHARMA, T. & CLAYTON, R.N. (1965) Measurement of $^{18}O/^{16}O$ ratios of total oxygen of carbonates. *Geochim. Cosmochim. Acta* **29**, 1347–1353.

SIBLEY, D.F. (1991) Dolomite mineralogy and texture in time and space. In: *Dolomieu Conference on Carbonate Platforms and Dolomitization, Ortisei* (Ed. Bosellini, A., Brandner, R., Flugel, E., Purser, B., Schlager, W., Tucker, M.E. & Zenger, D.) 245 Abstracts.

SLAUGHTER, M. & HILL, R.J. (1991) The influence of organic matter in organogenic dolomitization. *J. Sedim. Petrol.* **61**(2), 296–303.

SOUSSI, M. (1990) *Les Faciès Argilo-carbonatés Jurassiques en Tunisie Centrale: Stratigraphie, Sédimentologie, Diagenèse (Dolomitisation) et Intérêt Pétrolier*. Unpubl. Thèse Spécialité, Université Tunis, 281 pp.

SOUSSI, M., BEN ISMAIL, M.H. & M'RABET, A. (1990) Les 'Black Shales' toarcians de Tunisie centrale: témoins d'évènement anoxique sur la marge sud téthysienne. *C.R. Hebd. Séanc. Acad. Sci., Paris* **310**(II), 591–596.

SOUSSI, M., ENAY, R., MANGOLD, C., M'RABET, A., RAKUS, M. & RABHI, M. (1991) Datations par ammonites des séries et discontinuités du Jurassique de l'Axe Nord–Sud (Tunisie centrale). *C.R. Hebd. Séanc. Acad. Sci., Paris*, **312**(II), 501–507.

TUCKER, M. & WRIGHT, P. (1990) *Carbonate Sedimentology*, Blackwell Scientific Publications, Oxford, 482 pp.

USDOWSKI, H.E. (1968) *Die Genese von Dolomit in Sedimenten*. Mineral Petrogr. Einzeldarstell 4, Springer-Verlag, 95 pp.

ZENGER, D.H. & DUNHAM, J.B. (1980) Concepts and models of dolomitization: an introduction. In: *Concepts and Models of Dolomitization* (Ed. Zenger, D.H., Dunham, J.B. & Ethington, R.). Spec. Publ. Soc. Econ. Paleont. Mineral., Tulsa **28**, 1–9.

Index

Page numbers in *italic* refer to figures and tables.

Abu Dhabi sabkha 4, 10, 29, 409–25
coastal stratigraphy 413–16
Dolomia Principale compared 63–8
dolomites 65–7, 409–10
hydrocirculation patterns 423
lagoonal water levels 423
chemistry of interstitial waters 420–2
isotope geochemistry 422–3
piezometric surface 418–19
salinity 419–20
phreatic brines 418–23
Ras Ghanada sabkha/lagoon complex 410–11
bioturbation 411, *413*, 423
micromorphology 410
morphosedimentary units 410–11
vegetal zonation 410–11
salinity 424–5
sedimentological studies 409
diagenetic changes across sabkha 417
field analyses 412
laboratory analyses 412–13
mineralogy of modern sediments 416–17
organic carbon in sediments 417–18
Shamal (wind) effect 423
subtidal calcian dolomites 68
Abu Shaar (Gulf of Suez) 286
Angola, Lower Cretaceous Pinda Formation 295
ankerite 368–70
Appenines 80
Arabian Gulf 63
aragonite 6, 15, 33, 43, 300
precipitates 33
aragonitic skeleton dissolution 290
Arctic Ocean 86
Aurunci Mountains 80–1
Avicennia mangrove 409, 413

Bahama Banks 112–30
blue holes *112*, 113, *114*, 120
methods 113–5
North Andros island *112*, 113
saline groundwaters 122–3
circulation 115–18
dolomitization 118–22
rate estimation 123–5
shallow dolomites from South Andros 125–8
study area 112–13
wall-rock samples 129
Bahamas 63
fabrics within Late Neogene dolomites 286
see also Bahama Banks; Little Bahama Bank

basement faulting 217
Bathonian dolomites 290, 295
beachrock 129
Belize dolomites 63
bioturbation 411, *413*
black shale 10
Bonaire, Netherlands Antilles 326–41
carbonate cathodoluminescence 332–3
dolomite 329–32
distribution 331–2
porosity 331–2
dolomitizing fluids, hydrology/geochemistry 337–8
geochemistry 332–5
geological setting 326–8
limestone 328–9
location 326
methods 328
porosity evolution during dolomitization 335–7
predictive model, dolomite/porosity 338–40
breccias 81, 297–8
Brenta dolomites 57–8, 76–9
brine reflux 51
Burgundy, Middle Jurassic carbonates 290, 295
burial dolomitization 11–12
models 30–1
see also Michigan basin

Calcare di Zu 78
calcite 6, 300, 350
dissolution 378, 383, 384–5
laser bombardment 16
non-dolomitic 304–6
undersaturation at 23°C 384
calcium-rich dolomites 56, 361–75
analytical data 363–5, *366*, *367*
analytical methods/samples 362–3
limitations on solid solution 367–70
literature data 265–6, *369*
model evaluation in light of mineralogical data 370–1
model predictions, comparison with data sets 371–3
Campanian–Maastrichtian Abiod Formation 309–22
dedolomitization (dissolution) 316–22
porosity/permeability evolution 321–2
dolomites 312–17
geometry 313–14
isotope geometry 314–16
mineralogy 314–16
nature 313–14
petrography 314–16
petrophysical attributes 321

Campanian–Maastrichtian Abiod Formation (*Contd*)
sediments, nature/geometry 312–13
sedimentological properties 310–12
stratigraphy (Wadi Abiod) *311*
carbonate
authigenic 46
platforms 55–6, 91
sediments 130
Carpathians 82
Cenozoic rocks, shallow-buried 302
Central Tunisia *see* Campanian–Maastrichtian Abiod Formation
Cerithium 411
climatic effects 8
conduit systems 217
Cook Island, South Pacific 9
Coorong lagoon 5, 44
coral dissolution 290

deep-sea sediments 129
Delaware Basin, Permian 12
Dinarids 82
dissolution–reprecipitation reactions 56
dissolution voids 290
Dolomia Principale 29, 57–72, 75–80
Abu Dhabi Sabkha compared 63–8
burial end-member 71
calcian dolomite 69–70, 71
dislocation-rich dolomite 69
dolomitization mechanisms 68–9
hydrothermal event 71–2
inability to re-equilibrate dolomites 71
microstructures 68–72
persistence of facies 83
stable isotope analysis 60–1
textural analysis 58–60
transmission and analytical electron microscopy: experimental procedures 61–3
Dolomieu, Déodat de (1750–1801) 21–7
dolomite 4–9, 133, 377
abundance–global eustasy, genetic link 92
basin morphology 7–8
calcium–magnesium ratios 32, 374
calcium-rich *see* calcium-rich dolomites
cement 6–7, 11, 15
cementation 304
composition 4, 56
compositional data interpretation 374
compositional zonation in, pattern interpretation 374–5
crystals 5, 44, 345
diagenesis during burial 304
discontinuous solid solutions in 374–5
distribution 7–9
global factors 8–9
early 20–9–10
possible neomorphism 209–10
evolution 10–11
fabrics 5–6, 362
recrystallization 304
fenestral fine-crystalline 80, 82
ferroan 295, *296*, 297
formation 50–1
organic matter for 10, 33
fracturation 297
tectonic 297
geochemistry 32–3, 346–7
instant 4–5
late-stage 217
lattice 367–70
limpid 6
luminescent 12
metastable 11
mineralogy 6–8
modern 4, 5
modification 10–11
peritidal 4
petrology 32–3
porosity *see* porosity
postlithification 30–1
properties 11
question 4
recrystallization 56
replacement 6
reservoirs 31–2
sedimentary 346
silicate minerals reaction 377
solid solution limitations 367–70
stability 16
synthesis 347–51
factors influencing 351–2
under subsurface conditions 352–3
under surface conditions 353–8
phase boundaries 353–4
reaction paths 357–8
sabkhas (and similar environments) 354–7
dolomitization 4–5, 6–9, 92, 377–403
alkalinity increase 129
burial *see* burial dolomitization
dissolved sulphate effect 377–85
methods 378–80
reaction models 381–3
from mixed water 9–10
from normal seawater 9
hydrodynamic drive (pump) 9
mixed water model 153
organic matter and 10, 33
porosity evolution 299–303
creation 300–2
destruction 302–3
reflux models 29
sea-level changes *13*
sulphate effect 129
vs calcite, temporal variations in relative abundance 92
Dukhan Field, Qatar 291, 293

Early Ordovician 9
East Salina, West Caicos Island 29
Ellenburger group, West Texas 31
eustatic variation 84
evaporative pumping 51
evaporite dissolution 51–2
evaporite models 29

Florida–Bahamas dolomitization model 86
fractionation factors 145

Gansserina gansseri 320
geological time 9
Globotrucana tricarinata biozone 320
gravity-driven flow 217
Greenwood Formation *see* Michigan Basin
groundwater–seawater mixing 51
Guadalupian Yates Formation 91–105
  dolomite distribution 95–7
  dolomite geochemistry 100–3
    microprobe analyses 100–1
    stable isotopes 101–3
  dolomite petrography 97–8
  dolomitization
    early, relative sea-level changes 104–5
    timing 98–100
  geological setting 92–3
  Upper Yates depositional facies/cycles 93–5
gypsum 38, 40, 51

Halodule seagrass 409, 411, 413
Haplophragmoides 320
Hauptdolomit Formation 75
Hellenids 82
Holocene dolomitization 56
Holocene sediments 29
  dolomitizing fluids in 50–1
Hungary 82
hydrodynamic models 9

Illinois Basin, Silurian/Devonian rocks 387–403
  dolomite geochemistry 397–401
    carbon isotope ratios 397
    major elements 399–400
    minor elements 399–400
    oxygen isotope ratios 397–8
    strontium isotope ratios 400–1
    trace elements 399–400
  petrography 391–6
  regional setting 388
  stratigraphy 391–6
    setting 388–9, *390*
  timing 391–6
Infralias 80
Infraliassic breccias 81
iron 146, 295, *296*, 297

Jurassic, Central Tunisia, organic-rich/organic-poor carbonates 429–44
  dolomite cements in fractures 440–1
    conditions of dolomite cementation 440–1
    fractures generations/associated fillings 440
    geochemistry 440
    isotopes 440
    mineralogy 440
    petrography 440
  dolomite properties 441–2
  dolomitization environments 442
  dolomitization timing 442
  dolomitizing fluids 442–3
  geological setting 430–1, *431*
  Lower Nara dolomites 430–5
    dolomitization milieu 435
    geochemistry 434–5
    isotopes 434–5
    mineralogy *434*–5
    occurrence 430–1
    petrography 431–3
  methods 430
  Middle Nara organic-rich dolomites *432*, *434*, 435–9
    dolomitization milieu 436–9
    geochemistry 435–6, *437*
    isotopes 435–6
    mineralogy 435–6
    occurrence 435
  Upper Nara dolomites *432*, *434*, 439–40
    dolomitization milieu 439–40
    geochemistry 439
    isotopes 439
    mineralogy 439
    occurrence 439
    petrography 439

karstification 130
Kuwait 5

Late Neogene 5
Lattari mountains 78–80
Laurasia supercontinent 84
law of persistence of facies 75
lead-zinc deposits 377
Little Bahama Bank 30, 133–51
  ancient dolomites 148–9
  carbon isotopes 139, 146
  cathodoluminescence 137
  chemical maturity 143
  CMS dolomites 137
  crystalline mimetic dolomites 137
  crystalline non-mimetic dolomites 137
  depositional environment 139–40
  depositional texture 135–6
  diagenetic texture 137
  dolomite formation environment 146–8
  dolomitizing waters 146
  future diagenesis 148, *149*
  geological setting 134
  iron 146
  major elements 137
  manganese 146
  methods 134–5
  oxygen isotopes 143–6
    equilibrium precipitation 144–5
    kinetic controls 145
    phosphoric acid fractionation 145
  permeability 137
  petrographic maturity 140–3
  porosity 137
  stable oxygen 139

Little Bahama Bank (*Contd*)
  sucrosic dolomites 137
  trace elements 137–9, *141*
Lower Viséan Dolomites (Campine Basin, Belgium) 30, 155–65
  geochemistry 161–5
    stable isotopes 161–2
    trace elements 161
  methods 156–7
  palaeogeographical setting 157–9
  paragenetic sequence 159–61
  petrography 159–61
  sedimentological setting 157–9

magnesite
  geochemical system 346–7
  synthesis 347–51
    factors influencing 351–2
magnesium 51, 129
  recycling at mid-ocean ridges 86
magnesium/calcium ratio 32, 374
mangrove swamps 423, 424
Megalodontids 80
Mesozoic dolomites 6
  porosity 4
Michigan Basin, Greenwood Formation 30, 169–85, 231–51
  geochemistry 176–83
    calcite cements *177*
    dolomite types *177*
    fluid inclusions 176–9
    isotopic elements *178–9*
    Sr concentration–Sr isotope variations 181–2
    Sr isotopes 180–1
    Sr–O isotope variations 182–3
    stable isotopes 179–80
    trace elements *178–9*
  lithofacies associations *173*
  paragenetic sequence 174–6
    early diagenesis 174
    late diagenesis 175–6
    middle diagenesis 174–5
  rock types *173*
  sedimentology 170–1
  sources of dolomitizing brine 183–4
  stratigraphy 70–1, *172*
Michigan Basin, Ordovician carbonates 231–51
  cap dolomite 246
  fluid inclusions 242
  fracture-related dolomite/dolomitization 246–50
    mechanism 246–9
    timing 249–50
  geologcial setting 233–5
  hydrothermal diagenesis 250–1
  Manitoulin Island *232*, 237, 239
    microthermometry 244, *245*
    petrography 242–3
  microthermometry 244
  petrography 242–3
  postdolomitization diagenetic history 239
    fracturing 239
    late calcite 239
    pressure–dissolution 239
  predolomitization diagenetic history 234
  replacive dolomite 235
  saddle dolomite cement 238–9
  Southwestern Ontario 31, 235–7
  stable isotope data 239–42
    anhydrite 242
    calcite 240, *241*
    dolomite 240–2
microbial mats 129, 149
microfracturation 297
mid-ocean ridges 86
Mississippi Valley-type mineralization 11
mixing-zone models 30, 429
modern lakes, dolomite precipitation in 5
mole-for-mole replacement 325, 326
mollusc dissolution 290
moulds 290–1
  postdolomite 290–1, *292*
  predolomite 290
  syndolomite 290
Mururoa Atoll 286, *287*, 288
  dolomite evolution, related porosity *303*
  mouldic porosity 290
  porosity evolution 300–2
  vuggy porosity 290

Neogene synrift formation 8
Nevada 12
New Caledonia swamps 424
New Mexico, Silurian–Devonian carbonates 12
Norian Dolomia Principale *see* Dolomia Principale
Norian–Lower Liassic 79

organic matter 10, 33
over-dolomitized platform 303

Pacific atolls 9, 63
  see also Mururoa Atoll
Palaeozoic rocks 3, 6, 11
Pangea fragmentation 83–4
parasequences 14
Paris Basin, Middle Jurassic reservoirs 291
Peneroplis 411
Permian shelf deposit (Yates Formation, West Texas) 29
phosphorites 10
Pleistocene 5
pond geochemistry 47
pore fluid migration 217
porewater geochemistry 47
porosity 3–4, 14–16, 283–341
  Abu Shaar platform 288, *289*
  Bahamas 137
  breccia, fracture and 297
  destruction during burial/tectonic stress 304–6
  development in dolomite 326
  dolomitization effect 284
    creation 300–1
    destruction 302–3
  evolution 284

fabric-replacive 286
highly porous sucrosic 295
intercrystalline 291–5
intracrystalline dolomouldic 295–7
Mururoa Atoll 286, *287*, 288
postdolomite 306
secondary 290
selectively preserved 307
selective preservation during burial 303–4
vuggy 290–1
Portovenere Area, La Spezia 30–1, 187–201
bedding controls 194–5
breccia-controlled dolomitization 195–6
fracture-controlled dolomitization 191–4
postlithification dolomites 189–200
petrography 190–1
precursor rock 188–9
Sr analysis 197–200
stable carbon/oxygen analysis 196–7
stylolite-controlled dolomitization 195
Potamides 411
preburial 303
predolomite voids 306
Proterozoic age 11
protodolomites 32, 140
pyrite 43

Qatar 4

Red Sea 8
reflux (Adams & rhodes) 83
reflux dolomitization models 29
reflux mechanism 51
regional fracturing 290
Rhaetian–Liassic succession 79
Rhaetian period 84, 85–6
Rhaetian (Triassic) successions 29

sabkha models 29, 85
saline groundwater
aqueous geochemistry 122–3
dolomitization rate estimation 123–5
saline lagoons 129
Scipio-Albion Trend, Michigan 290
sea-level
global fluctuations 8
lowering 12–14
stillstands 12
seawater 9, 16, 51, 111–12, 402
dolomitization model 30
models 429
modifications to 129
seepage reflux 50
sequence stratigraphy 12–14
Sicily 81
storm recharge 51
stylolites 195, 239
submarine cement 6
subtidal open-marine carbonate sediments 32
sulphate 10
sulphate cements 304–6
supratidal flats, dolomite crusts 63
syndolomite voids 306

tectonic fracturation *305*
temperature effect 429–30
Tertiary dolomites 6
porosity 4
Tethys 84–6, *87*
Thalassia 413
tidal flats 10
tidal pumping 51
Triasina hantkeni 80
Triassic–Jurassic boundary 84
Turks and Caicos Islands 37–52
dolomite formation 50–1
geochemistry of pond/porewaters 47
gypsum 38, 40, 51
hydrology of East Salina 47–50
salina sediments 38–47
mineralogy 40–4
stable isotopes 44–7
statigraphy 38–9, 51–2

Umm Said sabkha, Qatar 5
Upper Jurassic 'Arab D' reservoirs 291, 291–3
Upper Senonian chalky limestone 309–10
Upper Triassic carbonates 86–7

voids, predolomite/syndolomite 306
vugs 290–1, 325
postdolomite 290–1, 292
predolomite 290
syndolomite 290

Western Canada Sedimentary Basin 203–25
burial dolomites *213*
deeper-burial (late) dolomites 217–25
late-stage calcites 222–3
other examples 222
possible origins 223–4
Presqu'ile barrier 220–1, 223–4
Rimbey-Meadowbrook reef trend 221
massive replacement dolomites 210–17
geochemical characteristics *211*
petrographic characteristics *211*
porosity development 215–16
Rimbey-Meadowbrook reef trend 210–13
widespread, origin of 216–17
sea-floor/very shallow-burial dolomitization 208–10
Pine Point fine-crystalline dolomites 208–9
Wabamun Group 209
Winnipegosis buildups 208
world ocean/atmospheric evolution 8–9
Wyoming Overthrust belt 12

Zechstein, Western Europe 12